WIRELESS

COMMUNICATION

UPENA DALAL

Electronics Engineering Department

Sardar Vallabhbhai National Institute of Technology

Surat

Oxford University Press is a department of the University of Oxford.
It furthers the University's objective of excellence in research, scholarship,
and education by publishing worldwide. Oxford is a registered trademark of
Oxford University Press in the UK and in certain other countries.

Published in India by
Oxford University Press
YMCA Library Building, 1 Jai Singh Road, New Delhi 110001, India

First Edition published in 2009
Seventh impression 2014

ISBN-13: 978-0-19-806066-6
ISBN-10: 0-19-806066-1

Typeset in Times Roman
by Tej Composers, New Delhi
Printed in India by Yash Printographics, Noida 201 301

To

my daughters, Parima and Jahnavee,

and

my mother, Nirmala Kania

Preface

Communication systems have played an important role in our lives for more than a century. The earlier systems, like telegraphy and telephony, were all analog and use electrical wires. They are hence examples of the wired communication systems. With further development, the focus shifted towards digitization of the communication systems. A major breakthrough in the field came with the advent of the wireless technology due to the invention of radio in the late nineteenth century. Since then, wireless technology has evolved at a fast pace and has revolutionized the field of communication systems.

Initially, wireless communication was used for military purposes. Only later it was commercialized resulting in rapid advancement in the physical layer, which has led to the development of the second-generation technology (2G) and, subsequently, the development towards 3G. The earlier analog systems of communication, which laid the foundation of the modern communication systems, comprise the first-generation technology. The second-generation technology has been in use mainly for voice data transmission and slow transmission. And now with the introduction of GPRS, EDGE, UMTS, etc., the stage is set for 3G, and R&D is already aiming for the fourth-generation (4G) technology. New systems are integrated with the Internet applications and data services, telephony, multimedia communication, and many more features. The modern communication systems aim to become all-wireless in future and the users of such systems would have a single and unique identification number, UTN (universal telecommunication number), to ensure maximum network security.

Owing to its dynamic nature and wide application, most universities offer a core course on wireless technology at the undergraduate level in India and abroad. It is also a prerequisite for specialized courses at MTech and higher levels of study.

About the Book

This book is primarily designed to serve as a textbook for students of electronics and communications engineering as well as computer engineering and would be found suitable for courses on mobile communication, wireless communication, and mobile networks.

Beginning with an overview of wireless systems, fundamental concepts of wireless communication, DSP, digital communication, and information theory, the book provides an extensive coverage of wireless channel and modulation schemes with multiple access scenario and advanced wireless systems.

Each block of wireless link has been dealt with in sufficient detail, including source coding and channel coding methods, radio wave propagation over wireless channels, and single-carrier and multicarrier modulation techniques. Recent technologies (such as CDMA, turbo codes, spread spectrum communication, multiplexing, mobile communication, OFDM, MIMO, and software radio), latest coding techniques (such as wavelet and DCT), and most recent standards and systems (such as DAB, DVB, WiMAX, and UWB), which lead towards 4G, have been adequately covered. Thus, the text gives a holistic view of the subject correlating the interdisciplinary areas like wireless networks, digital signal processing, cellular theory, etc.

All the chapters start with simpler topics and gradually build up advanced concepts through detailed explanations and illustrations. The book has over 400 self-explanatory figures, case studies, and solved examples, which will help students revise the concepts through visualization and practice. It also contains model question papers for self-analysis.

Owing to its comprehensive coverage of basic concepts and latest trends in wireless communication, the book will be equally useful for the postgraduate students and practising engineers in the field.

Contents and Coverage

Chapter 1 is the introductory chapter and describes the present scenario and trends in wireless communication.

Chapter 2 starts with the requirements of wireless communications, such as modulation of a carrier, signal, and channel bandwidth as also the fundamental concepts of DSP, such as sampling, time and frequency domain relationship, audio/image/video signal processing, channel impulse response, convolution and correlation between two signals, power and energy spectrum density, and spectrum components. The chapter then covers the fundamentals of digital communication, which are the additional key points to fulfil detailed aspects of the fundamental link model.

Chapter 3 mainly deals with the concept of source coding/waveform coding. Most of the real-time signals are analog in nature. Beginning from the digitization of analog signal through quantization, further processing applied to the digitized signal to compress its database may be using lossy or lossless methods. Time and frequency domain data compression techniques, special voice coders for low bit rate signals, etc. are described in the chapter along with the concepts of information theory.

Chapter 4 delves into error-handling tasks over the noisy channel. Error-detection schemes allowing retransmissions are mostly used in case of wired Internet but sometimes are also used in wireless networks. Due to unreliable wireless channel, error-correcting codes are certainly required. Widely used error-detecting and error-correcting schemes along with their error-correction capabilities are described in the chapter. For the receiving end, various decoding schemes are described.

Chapter 5 describes the radio propagation over wireless channel. Starting from free space propagation model, different types of long-distance radio propagation are described in the chapter along with path loss model. Different fading effects due to delay spread and Doppler spread are very common in multipath environment. Shadowing effect and outage condition cause the pauses during conversation. All these effects are dealt with in this chapter.

Chapter 6 mainly describes channel models, particularly Rayleigh model, Rician model, and Nakagami model. The diversity techniques, equalization methods, and channel estimation to mitigate the channel effects are also explained in the chapter. The chapter is very important as because of these techniques, phase ambiguity due to multipath, frequency dependent effects, fading effects, etc. can be reduced considerably at the receiver side and bit errors reduce.

Chapter 7 explains all the basic single-carrier digital modulation schemes along with their mathematical representation and block diagrams. Eye diagram, constellation diagram, and other important diagrams are described briefly for analysis of modulation schemes. Conventional methods and the modified versions of the conventional modulation schemes (DPSK, OKQPSK, MSK, GMSK, M-FSK, etc.) are described.

Chapters 8 and *9* describe the wideband/broadband modulation techniques—spread spectrum modulation (SSM) and OFDM, respectively. Due to their importance in the latest scenario, both the schemes, which are especially suitable for the 3G and 4G systems, are described in detail in separate chapters. Both the techniques are described in all respects, considering all the related terms and associated mathematics.

Chapter 10 is related to multiple access techniques. From this chapter onwards, the conceptual development is discussed to have multi-user system environment. The various ways by which the multiple users are allowed to access the available wireless channel on the sharing basis with fixed allocations are described here. The random access schemes and reservation-based access schemes are also dealt with.

Chapter 11 is related to infrastructure development for cell-based wireless communication in the multiple user environments. The cell theory is necessary for deciding the size of the cell, location of the transmitter in a cell, and splitting of the cell to cover more population density. Frequency reuse and cochannel interference are the key aspects of the cell theory. Without cellular infrastructure, implementation of the mobile networks is very difficult. At the end of the chapter, traffic engineering is provided in brief.

Chapters 12, *13*, and *14* deal with the fundamentals of all the existing wireless digital systems developed on the basis of certain standards and protocols. The systems are covered in brief, highlighting all the key points. The detailed explanation of all the systems is beyond the scope of this book. Chapter 12 describes the digital broadcast systems and their standards like DAB, DVB, etc. Chapter 13 describes the infrastructure-based/cell-based networks, which are established permanently and support mobility, such as GSM, CDMA, UMTS, and WLL. Chapter 14 describes special categories of the wireless systems. These are ad hoc networks like Bluetooth and ad hoc and cell-based Wi-Fi and WiMAX networks mainly designed for data access or transfer. Zigbee is the special protocol for the wireless sensor network. The UWB is the system for ultra-high speed indoor communication.

Chapter 15 explains how MIMO is different from smart antennas and also explains the aspects of MIMO system like spatial diversity, spatial multiplexing, multipath exploitation, etc., including space time processing and coding.

Chapter 16 gives the details on simulation methodology and software radio. The simulation is an art to develop the system equivalent environment on computer by using mathematical relationships or some logic undergoing systematic steps. Using the software and hardware combination, the new trend is developed to implement the radio, which is known as software-defined radio (SDR). Based on SDR, the universal radio sets can also be designed. These details are covered in this chapter.

Appendices A to D deal with Linear Systems Theory, Algebra for the Linear System, Probability Theory, and DSP Fundamentals Applied to OFDM Processing, respectively. *Appendix E* provides four model question papers and *Appendix F* provides their answers. *Appendix G* gives answers to end-chapter Multiple Choice Questions.

Acknowledgements

It is a great pleasure and honour for me to be associated with Oxford University Press. I express my sincere gratitude and thanks to the entire editorial team and production department at Oxford University Press, New Delhi, for publishing the book maintaining a high degree of precision and accuracy. I want to thank God for providing such good opportunities in my life. I thank the senior teachers of my department, Prof. Mrs Nila Desai, Prof. B.R. Taunk, retired Prof. K.U. Joshi, my PhD supervisor Dr Y.P. Kosta (Professor, EC Dept., CIT, Changa), and my colleague Mrs Jigisha Patel for their encouragement and support. I would also like to thank my husband, Devang, for his continuous motivation and support.

Every effort has been made to produce an error-free text; however, I would be grateful if readers can point out any unintended error or discrepancy. They may also feel free to write to me at upena_dalal@yahoo.com or to Oxford University Press, New Delhi, to give any suggestions or feedback to enhance the quality of the book.

Upena Dalal

Contents

List of Symbols

$\alpha(t)$ = Rayleigh amplitudes of the multipath signals
δ = handoff margin
ε = permittivity of free space (capacitance per unit length measured in farads/meter)
ε_0 = dielectric constant of vacuum
ε_r = relative dielectric constant of the earth
$\phi(t)$ = different phases due to different delays of multipath signals
Φ = phase of noise vector
λ = wavelength (of carrier normally)
λ = call arrival rate
μ = permeability of free space (inductance per unit length measured in henries/meter)
μ = mean number of call arrivals
η = aperture efficiency
η_{fdma} = spectral efficiency of FDMA scheme
$\eta_{fdma\text{-}tdma}$ = spectral efficiency of FDMA-TDMA scheme
η_{tdma} = spectral efficiency of TDMA scheme
σ = standard deviation
σ_g = conductivity of ground
σ_x = standard deviation from the mean value
θ = phase variable
θ = directions variable
θ_{err} = phase error
θ_{max} = maximum allowable phase margin
τ = time variable for autocorrelation function
τ_{max} = maximum delay spread on multipath channel
τ_1 = delay between received multipath signals
ω = angular frequency of a signal
ω_c = angular carrier frequency
a's = filter coefficients of an FIR filter
a_n = value of nth chip in PN code
A = traffic in erlangs
A_0 = peak amplitude
A_e = effective receiving aperture area of antenna
A_m = amplitude of the signalling element
A_r = actual receiving aperture area of antenna
A_t = largest physical dimension of the transmitter antenna
B = spread bandwidth
B = busy period in ISMA scheme
B = grade of service
B_c = coherence bandwidth
B_d = Doppler spread
b_g = number of bits in each guard interval
B_j = interference signal bandwidth
B_m = message signal bandwidth
b_{OH} = overhead bits in TDMA frame
b_p = number of bits in each slot preamble
b_r = number of overhead bits per reference burst
b_T = total bits in TDMA frame

B_s = average bits per symbol
B_s = PN signal bandwidth
c = average number of calls in period of observation
c_n = amplitude of nth reflected component
$c(t)$ = PN sequence (subscript t or r represents transmitter PN sequence and received PN sequence)
C = channel capacity in bits/s
C = total effective number of duplex channels available in cell area
$C_{auto}(k)$ = autocorrelation function for speech for kth sample
d = distance between isotropic source and receiver antenna
d = distance between two antenna elements
d_1 = initial period of the busy period in ISMA
d_2 = processing delay within busy period after packet duration in ISMA
d_f = far-field distance
d_{min} = minimum weight
d_0 = close-in distance
$d(U,V)$ = hamming distance between the codes U and V
D = frequency reuse distance
$e(n)$ = estimation or equalization error signal in discrete form
E = electric field strength w.r.t. transmitter
E_b = bit energy
E_{ms} = mean square error
E_o = field strength while propagation in free space
E_s = signal energy dissipated in time T
f = frequency variable
f_1 = spacing between consecutive hopping frequencies in FHSS
f_1, f_2, etc. = subcarrier frequencies in OFDM
$f_1,\ldots,f_N$ = frequency groups within a cluster f_1 for cell 1 and so on
f_c = carrier frequency
f_d = Doppler spread
f_{max} = highest frequency content of analog in
f_o = centre frequency of the PN signal spectrum
f_s = sampling frequency or Nyquist frequency
$f(t)$ = basic pulse function
$f(x,y)$ = image function (2D)
Δf = carrier spacing between orthogonal subcarriers in OFDM
F = flux density
F = fade margin
$F(x)$ = companding function to compress x integer value
G = offered traffic
G_r = gain of receiving antenna
G_t = gain of transmitting antenna
$G(\theta)$ = gain of antenna at angle θ
$G_s(f)$ = PSD function
H = average call holding time
h_R = receive antenna height
h_T = transmit antenna height
$h(n)$ = channel impulse response (discrete form)
$h(t)$ = channel impulse response
$h(x)$ = polynomial for LFSR design
$h_e(n)$ = estimated channel impulse response in discrete form
$H(\omega)$ = channel impulse response in frequency domain
H = complex $N \times M$ matrix representing the MIMO channel impulse response, subscripts LOS, NLOS represents corresponding components
H or $H(x)$ = entropy (in outcome x)
$i(t)$ = current as a function of time t (variable current)
$i(t)$ = interference signal
$I(t)$ or I = in-phase component of the received modulated (complex) signal
I = mean value of exponentially distributed Poisson arrivals
I_i = interference power of ith cell
I_{int} = total number of interfering cochannel cells
I_o = interference density (per bit)
$I(x_i \text{ or } x_j)$ = information content in outcome x_i or x_j
J = total interference power in CDMA

k = input bits applied to channel coder
k = ratio of message bit duration to chip duration (hops per message bit)
k = number of channels within a group
k = occupied trunks
K = frequency multiplier in FHSS
K = Rician factor
K = number of users in DS-CDMA
K = variable for smart antenna elements in general
l = number of mobile terminals
L = dimension of vector quantization, number of samples in a block or vector
L = maximum resolvable paths for CDMA
L = total number of duplex channels available to the operator
L = number of multipath components received at each antenna element from l mobiles
m = shape factor or gamma distribution (Nakagami model)
$m(t)$ = transmitted message signal
$m'(t)$ = received message signal
M = number of quantization levels, number of signalling elements
M = number of hopping frequencies (in FHSS), number of hopping time slots (in THSS)
M = number of times cluster is repeated
M = number of receiving antennas for MIMO
$M(f)$ = spectrum of message signal
n = number of bits per quantization level (i.e. number of bits per sample)
n = number of bits/symbol
n = output bits from channel coder
n = number of users/slots per channel in TDMA
n = path loss exponent
N = total noise power
N = maximum number of significant reflected components to model the channel
N = number of users in CDMA
N = one period comprising number of samples, IFFT bin size in OFDM
N = number of trunks
N = cluster size in cell theory
N = number of transmitting antennas for MIMO
N_c = number of chips in a full period of PN code
N_c = number of subcarriers in OFDM bandwidth
N_c = total number of carriers in FDM scheme
N_t = number of data slots per TDMA frame
N_o = noise power spectral density
N_r = number of reference bursts per TDMA frame
N_s = total number of slots in TDMA frame
N_u = number of users supported in FDM
N_{us} = number of users per sector
p = probability of occurrence of information in binary symmetric source
p = number of shift registers in an ML sequence generator
p = persistency
p_p = prime number to generate twin prime sequences
P = power (subscript differentiating 's' signal or 'n' noise power)
P_B = bit error probability
P_o = power radiated by test antenna
P_r = received power at the receiver front end
Pr_{handoff} = power level at which handoff is made
$Pr_{\text{min usable}}$ = minimum usable power for acceptable voice quality
P_t = (isotropic source) transmitted power
PG = processing gain
$P(\theta)$ = power radiated in a direction with angle θ
$P(t)$ = power as a function of time t (variable power)
$p(x)$ = probability function
$P(x_i \text{ or } x_j)$ = probability of outcome x_i or x_j
$P(Y/S^{(m)})$ = probability of maximum likelihood with *mth* sequence when Y sequence is received

$\overline{P}$ = local mean power

$\overline{\overline{P}}$ = area mean power

q = integer deciding the q-r sequences and Hall sequences

q_e = quantization error

$Q(t)$ or Q = quadrature component of the received modulated (complex) signal

r = integer deciding the Hall sequences

Δr = resolution, step size in quantizer

$r(n)$ = received signal (discrete form)

$r(t)$ = received time varying signal

R = resistance

R = data rate or information rate in bits/s

R = cell radius

R_b = bit rate in bits/s

R_c = ground reflection coefficient

R_d = data redundancy

$R_{\max}$ = maximum cross correlation

R_T, R_R = correlation matrices of transmitter and receiver, respectively

$R(D)$ = rate of vector quantization in bits per sample (for given distortion D)

$R_c(\tau)$ = autocorrelation function for PN codes

s = number of sectors in a cell

s_1, s_2, etc. = symbols, split from OFDM symbol (frame) to assign the carrier

$s(t)$ = transmitted time varying signal

$s(t)$ = spread baseband at the transmitter

$s'(t)$ = spread baseband at the receiver

S = average signal power

$S(f)$ = frequency spectrum (signal in frequency domain)

$s(n)$ = transmitted signal (discrete form)

$s(k)$ = kth speech sample

S_k = kth frequency sample in the speech spectra

$S^{(m)}$ = mth possible sequences

$s_k(t)$ = kth subcarrier in multicarrier system

$s_q(t)$ = quantized speech signal

t = time variable

t_c = error-correcting capability of the code

t_{chip} = chip duration

t_d = error-detecting capability of the code

t_{di} = delay over the ith channel path in case of multipath reception for SSM system

t_m = message bit duration for spread spectrum system

T or T_o = period of a periodic signal (time for one cycle)

T = traffic observation duration

T_b = One bit interval (pulse duration)

T_{code} = period of PN code

T_f = frame duration containing all time slots

T_g = guard interval between consecutive OFDM symbols

T_m = maximum excess delay over channel

T_{mc} = (multicarrier system) symbol period

T_o = coherent time

T_{rms} = RMS delay spread

T_s = OFDM symbol period

T_{sc} = (single carrier system) symbol period

T^k = shift operator

ΔT_s = sampling interval

v = velocity of mobile

v = velocity of light/velocity of electromagnetic waves

v_f = voice activity factor

V = voltage (subscript differentiate with 's' signal or 'n' noise voltage)

$V(t)$ = voltage as a function of time t (with appropriate identity subscript)

w = variable for channel taps or weights

$w(n)$ = white noise (discrete form)

$w(t)$ = white Gaussian noise varying with time

W = bandwidth

W_{channel} = available bandwidth

W_{guard} = guard band

W_{signal} = per user bandwidth

$W(U)$ = Hamming weight for code word U or V

x = actual number of call arrivals

Y = received sequence

z = number of standard deviations

z = threshold level (or noise threshold level)

chapter 1

Present Scenario in Wireless Communication Systems

Key Topics

- Different types of communication systems (in general)
- Wired vs wireless communication
- Different types of wireless systems
- Present scenario and requirements
- Evolution of wireless systems
- 1G, 2G, 3G, and 4G wireless systems
- Licensed and unlicensed band communication

Chapter Outline

The book mainly covers wireless digital communication. Though it is assumed that the reader of the book is familiar with the basic theory of communication, many required concepts are revised as a ready reference, starting from this chapter. The students of this generation must be familiar with the wireless communication systems, both conventional and latest. This chapter discusses wireless systems. It explores the need for and scope of the best developments in wireless communications in India and other countries, which is possible only if the standards used today for wireless systems are known. Evolution of a system is linked with the previous systems and the new system is designed by looking into the problems of the previous systems and eliminating them. Hence, the development scenario of 1G to 4G is also necessary to be known. Once this background is created and students start studying these from the root level of the wireless link, considering each and every stage of the wireless link, every part of the theory and its application to the system can be correlated and best solutions can be found out for the 'anywhere, anytime' communication scenario.

1.1 INTRODUCTION

There are dual possibilities in all of the following while dealing with communications with the two hardware ends, a *transmitter* and a *receiver*.

- The input (or baseband) signals may be analog or digital.
- The channels may be wired (guided) or wireless (unguided).
- The transmissions may be analog or digital.
- The number of bits sent at a time may be serial (one bit at a time) or parallel (more bits at a time, i.e. symbols).

- The communication may be baseband or passband (general word for broadband/ wideband).
- The mode of communication may be synchronous or asynchronous.
- The information may be real-time or non-real-time (stored data).
- The direction of transmission may be unidirectional or bidirectional.

Out of the two possibilities, only one can exist at a time. To have the combination of both possibilities, either conversion or convergence in the system is required. Because of the two possibilities in the input signals and the two possibilities in the transmissions, according to the theory of binary, four combinations of communication systems are possible, as described in Section 1.1.1. The analysis of the systems may be done by using a qualitative approach first and then a quantitative approach and analysing first the ideal system and then the actual system, with noise.

For studying wireless digital communication, some basic knowledge of other fields of electronics is also required. They are mentioned in Fig. 1.1.

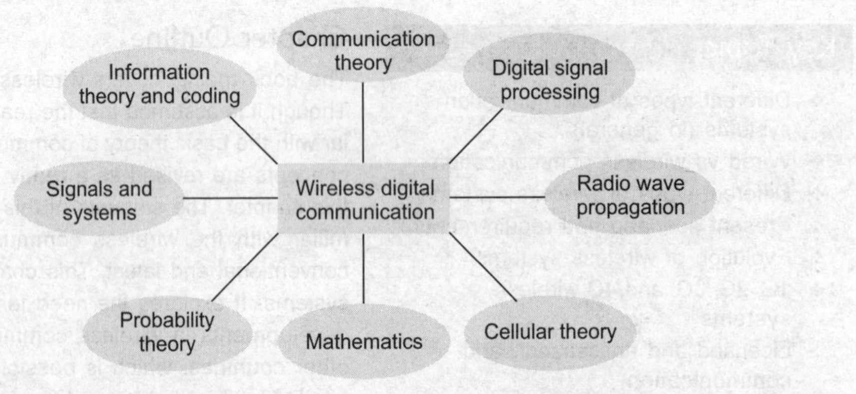

Fig. 1.1 Required knowledge of other fields for understanding wireless digital communication

1.1.1 Different Types of Communication Systems

The communication systems can be of four different types in general as mentioned previously (they may be wired or wireless) and are explained below. Digital input analog transmission type of system is mainly categorized as wireless digital system (except wired line communication through MOdulator+DEModulator (MODEM)). Pulse code modulation (PCM) scheme, which exists for analog-to-digital conversion (ADC), is considered in the source coding stage of wireless communication link, though it is the method for analog input-digital transmission. Here different types of systems and the corresponding modulation schemes are described for a proper visualization.

Analog Input-Analog Transmission

These types of systems were designed for the very first time. Wireless communication commercially started with AM (amplitude modulation) radio broadcasting in the 550 to 1600 kHz range. Thereafter, FM (frequency modulation) transmissions also started commercially in the band from 88 to 108 MHz. In both these systems, the input was in the analog form of audio signal. These broadcast systems still exist. When the analog television standards were framed, AM was selected for video information and FM for audio information, for combined audio and video transmission. These standards are still followed to maintain the compatibility with the older televisions and follow the VHF and UHF ranges. In cable TV also, analog transmission method is used. In local loops of wired telephone lines, analog baseband signal is transmitted without modification in the signal.

The recent age is revolutionary and it seems that in the near future, the analog input-analog transmission systems will be obsolete. Transient period of revolution has already started with digital broadcast systems employing A-D-A conversion stages with the standards, digital audio broadcasting (DAB) and digital video broadcasting (DVB). High definition radio (HD Radio) and digital radio mondiale (DRM) systems are also coming up. All these systems follow orthogonal frequency division multiplexing (OFDM) modulation scheme, which is suitable for long-distance communication and hence for broadcasting.

Analog Input-Digital Transmission

Digital transmission in its baseband form is suitable for transmission only on the wired lines. To achieve this, analog-to-digital conversion is required and the PCM scheme can achieve this. Over the telephone trunk lines or over the integrated services digital network (ISDN) or broadband ISDN (B-ISDN) B channels, PCM signals of 64 kbps bit rate are transmitted. Another method for analog input digital transmission is delta modulation (DM), but because of its practical limitations related to slope overload and sampling rate, it is not standardized in commercial systems. PCM signals can also be converted into frames for transmissions over wired links of computer networks. Differential pulse code modulation (DPCM) and adaptive DPCM (ADPCM) are the modified and efficient versions of PCM.

Thus, PCM is the important scheme in the present scenario. It forms the basis for the source coding stage of the wireless link for digital communication by providing A to D conversion of the real-time input signals, like voice, image, and video. It is described in detail in Chapter 3.

Digital Input-Digital Transmission

When it is necessary to send the digital information in its baseband form, the binary form of transmission may not be always suitable, as it may not be compatible with the transmission channel or because it adds the DC level in the final transmission, which causes more energy in the signal. We need to convert the form of transmission by changing the bit representation format/voltage levels for shaping of signal power and also incorporating the synchronization points in the signal. In short, we can shape the signal for the desired spectrum characteristics for digital baseband communication. *Non-return-to-zero* (NRZ), *return-to-zero* (RZ), Manchester, differential Manchester, bipolar, etc., are the methods that have a final digital form of transmission. These methods are normally suitable for the wired line or mostly in computer networks; however, they are sometimes incorporated in wireless links. These methods are also called *digital signalling*; they are a suitable form for ISDN lines. It is also called *line coding* and can be applied to digital baseband.

Line coding can be applied to digital baseband in wireless communication before the modulation stage. It is described in Chapter 2.

Digital Input-Analog Transmission

This type of transmission is mainly used in the systems that use the MODEMs either over wired line or wireless links. Here modulation scheme converts the input digital signal into an analog form for the transmission. Final wireless communication is always possible in the analog form only. If the wireless transmission is required and the frequency after modulation does not fall into the RF range, it is necessary to use an RF upconversion stage. If the wired communication is used, only data modem is required without upconversion stage. Amplitude shift keying (ASK), frequency shift keying (FSK), M-ary phase shift keying (M-PSK), M-ary quadrature amplitude modulation (M-QAM), minimum shift keying (MSK), spread-spectrum modulation (SSM), orthogonal frequency division multiplexing (OFDM), etc., fall into this category. The details of each modulation scheme are covered in Chapters 7, 8, and 9.

1.1.2 Wired vs Wireless Communication

The existing systems are not all wireless; a few are wired. Fundamentals of both types of media are described here, so that the questions like 'how both the communications differ' and 'what kind of conversions are required for the converged system' may be answered.

The electrical signals on an open wired line, such as a twisted pair, travel at a velocity of light, which is determined by the expression

$$v = \frac{1}{\sqrt{\varepsilon\mu}} \tag{1.1}$$

where ε and μ are the permittivity of free space (capacitance per unit length measured in farads/metre) and the permeability of free space (inductance per unit length measured in henries/metre), respectively. In free space $v = 3 \times 10^8$ m/sec, given that $\varepsilon = 9.854 \times 10^{-12}$ F/m and $\mu = 4\pi \times 10^{-7}$ H/m. The signal travels as an electromagnetic (EM) wave just outside the wires (radiation). It differs from a free space EM wave (such as the one launched by a TV, radio, or mobile antenna, which spreads out in all directions) only in that it is bound to and guided by the wires of the transmission line.

When do two connecting wires become a transmission line? It is when the capacitance and inductance of the wires act as 'distributed' instead of 'lumped'. This begins to happen when the wire approaches dimensions of a wavelength (wavelength λ and frequency f are related by $\lambda = v/f$). At sufficiently high frequencies, when the length of connecting wires between any two devices, such as two computers, is in the order of a wavelength or larger, the voltages and currents between these two devices act as waves that can travel back and forth on the wires. Hence, a signal sent out by one device, propagates as a wave towards the receiving device and the wave is reflected unless the receiving device is properly terminated or matched. Of course, if we have a mismatch, the reflected wave can interfere with the incident wave, making communication unreliable or even impossible. Proper termination of a wired link is important when networking computers, printers, and other peripherals, which must be properly matched to avoid reflections.

Three different wired media are mainly popular:

1. Twisted pair wirelines, unshielded twisted pair (UTP) and shielded twisted pair (STP), for conventional landline telephone system, 10BaseT Ethernet cabling, etc.
2. Coaxial cable for closed circuit TV (CCTV) and cable TV Network, Ethernet 10Base2, 10Base5 cabling, etc.
3. Optical fibres for long distance communications, B-ISDN, fibre distributed data interface (FDDI), local area network (LAN), synchronous optical network (SONET), etc.

The first two wired media provide a reliable, guided link that conducts an electric signal associated with the transmission of information from one fixed terminal to another. Wires act as filters (due to lumped resistance and capacitance) that limit the maximum transmitted data rate of the channel because of band limiting frequency response characteristics. Twisted pair wireline can support typically 250 kbps bit rate while a coaxial cable may support typically 300 Mbps. The signal passing through a wire also radiates EM waves outside of the wire to some extent that can cause interference to nearby radio signals or to other wired transmissions as a noise. These characteristics may differ from one to another wired medium. Laying additional cables in general can double the bandwidth of wired medium and in no any other way.

Optical fibre is a dielectric guided medium which passes the information through itself in the form of a light wave. The carrier frequency range is of the order of 10^{14} Hz. Ideally optical fibres have infinite bandwidth but in practice, due to limitations of sources and detectors and dispersion effect, the bit rate up to Tbps (terabits per second) is achieved over high-grade optical fibres. It exhibits pulse spreading effect due to the dispersion and hence bit errors may occur. Dielectric medium supports more than one frequency to pass through it and such is the case in optical fibres in form of *wavelength division multiplexing* (WDM). Wireless medium (which is also dielectric in nature) supports more than one frequency at a time. All the links undergo the effect of white noise.

Compared to wired media, the wireless medium is unreliable, though ideally infinite, it has a low bandwidth, effectively due to delay spread and ISI effects described in Chapter 5. However, it supports mobility due to its tetherless nature. Different signals through wired media are physically conducted through different wires, but all wireless transmissions share the same medium, air, in form of unguided electromagnetic wave released through an antenna of supporting bandwidth. Thus, it is the frequency of operation and the legality of access to the band that differentiates the variety of wireless services. Wireless networks operate around 1GHz (cellular), 2.4 GHz [personal communication systems (PCS) and wireless LANs], 5 GHz (wireless LANs), 28–60 GHz [local multipoint distribution service (LMDS) and point-to-point base station connections], and 300 GHz satellite ranges and IR frequencies for optical line-of-sight communication/laser communication. These bands are either licensed, like cellular and PCS bands, or unlicensed, like the ISM (**I**ndustrial, **S**cientific and **M**edical) bands or U-NII bands. As the frequency of operation and data rates increase, the hardware implementation cost also increases and the ability of a radio signal to penetrate walls decreases. The electronic cost has become less significant with time. For frequencies up to few GHz, the signal penetrates through the walls, allowing indoor applications with minimal wireless infrastructure inside a building. At higher frequencies, a signal generated outdoors does not penetrate into buildings and a signal generated indoors stays confined to a room. This phenomenon imposes restrictions on the selection of a suitable band for wireless application.

Capacity improvement is a continuous issue for the scientists to combat for wireless systems. Wired media provide an easy means to increase capacity; we can use more wires, where and when required, if it is affordable. With the wireless medium, we are restricted to a limited available band for operation, and we cannot obtain new bands or easily duplicate the medium to accommodate more number of users in a system. As a result, researchers have developed a number of techniques to increase the capacity of that wireless system to support more users with a fixed bandwidth. A method for wireless cellular systems, like frequency reuse, is comparable to laying new wires in wired systems. If the two cells of same frequency are at a sufficient distance, the same frequency can be reused for interferenceless communication in both the cells. The theory is described in Chapter 11. Even one may reduce the size of the cells to overcome the demand of the population. In a wireless system, reducing the size of the cells by half allows twice as many users as in one cell. Reduction of the size of the cell increases the cost and complexity of the infrastructure that interconnects the cells. Capacity issues are highlighted in Chapter 11 for the various technologies implemented over cellular infrastructure. Multiple access schemes also help to accommodate more users.

There appears to be a tremendous potential for improving capacity by using smart antenna systems. Single input-multi output (SIMO), multi input-single output (MISO), and multi input-multi output (MIMO) systems are described in Chapter 15. Capacity increment by 300 to 400 per cent is possible in cellular environments with such techniques; compared to single input single output (single antennas at transmitter and also at receiver) system. Even OFDM can support multiple users with multi carrier communication in the cellular environment.

1.1.3 Types of Wireless Systems

Basically there are two types of wireless communication systems:

1. Wireless broadcast systems, in which the user is always at the receiver end.
2. Wireless networks, where multiple users can exchange their information, being a transmitter or a receiver independently.

Hence, accordingly, the wireless link requirements will be different. The modulation schemes are also selected according to the suitability of the system.

Wireless Broadcast Systems

These kinds of systems do not require the cellular structure or device identification numbers except some special systems. The transmissions are with a single transmitter and of sufficiently high power amplification. Within the predefined range anybody can receive the transmissions with the help of user receiver set. They are mainly frequency tuning based communications. The examples of such systems are AM/FM radio, television, direct-to-home (DTH), DAB, DVB systems.

Wireless Networks

These types of systems are mainly based on cellular infrastructure or ad hoc connections, e.g. mobile telephone network and universal mobile telecommunication system (UMTS), wireless LAN, and mobile Internet, based on personal domain/cell support. For cell-based systems, at least one transmitter per cell is required. They are low power transmitters as compared to the broadcast systems. The transmitters (or transceivers) of different cells may be interlinked to form a path between the destination and source devices. They are frequency tuning plus identification number or address-based communications. Ad hoc networks do not require cellular infrastructure.

With this fundamental understanding, one can start the study of wireless digital communication to have the proper visualization of the system. We shall start our study with a comprehensive presentation on existing scenario of the wireless systems because the readers may be familiar with these applications as they might be the users.

1.2 EXISTING TECHNOLOGIES

There is an increasing demand for broadband/wideband wireless communication systems due to requirement of high-speed communications (mobile Internet, wireless video transmissions, etc.). At the same time, the telecommunications industry faces the problem of providing telephone services to rural areas, where the customer base is small, but the cost of installing a wired phone network is very high. One method of reducing the high infrastructure cost due to a wired system is to use a fixed wireless radio network. The problem with this is that to enable the rural and urban areas to communicate, large cell sizes are required to obtain sufficient coverage. It results in problems caused by the large signal path loss and long delay times in multipath signal propagation due to long distances. If we design more number of cells for the rural area, it would be an inefficient, costly affair due to low population density. Hence, a modulation technique should be introduced in the systems, which covers longer distance eliminating the problems of wireless channel.

However, researchers in wireless digital communication still face problems, such as multipath delay compensation, speed of communication or high bit rate communication, and efficient use of available spectrum and spectrum efficiency improvement for accommodating more users and applications.

Currently, *global system for mobile telecommunications* (GSM) technology is being applied to wireless telephone systems even in rural areas. However, GSM uses frequency division multiple access/time division multiple access (FDMA/TDMA) with frequency reuse, which has limited frequency channels to communicate and with a high symbol rate, leading to problems with multipath causing intersymbol interference (ISI). Hence, there is a need for a scheme that can give no ISI effects at high-speed communications. Even afterwards, the enhanced data rate for GSM evolution (EDGE) technology is also introduced for higher bit rate. Many service providers compete with each other providing maximum possible coverage for mobile telephony. They also try to introduce the latest services to the subscribers to acquire the market. General packet radio service (GPRS) is the protocol by which packet radio is made possible and hence data services are added to the GSM system. It is designed to have wireless Web access through mobile telephony service providers.

All basic methods are tried by the scientists and engineers. Several techniques are under consideration for the fourth generation of digital phone systems—land mobile communication as well as wireless ATM, with the aim of improving cell capacity, multipath immunity, security, and flexibility. These include wideband code division multiple access (WCDMA) and the latest development is the emergence of a new multicarrier modulation (MCM)/multiple access technique, namely orthogonal frequency division multiplexing or multiple access (OFDM or OFDMA). Both these techniques could be applied to provide a fixed wireless system for rural areas. However, each technique has different properties, making it more suited for specific applications. The combinations of both these schemes are also considered to overcome the limitations of both the systems.

With CDMA systems, all users transmit in the same frequency band using specialized separate orthogonal codes as a basis of channelization (described in Chapter 10). The transmitted information is spread over the spectrum by multiplying it with a wide bandwidth pseudo-random sequence. Both the base station and the mobile station know these random codes, which are used to modulate the data sent, allowing it to de-scramble the received signal.

The OFDM is for multiple-user access and allows many users to transmit in an allocated band simultaneously, by subdividing the available bandwidth into many narrow bandwidth carriers (described in Chapter 9). Information is allocated to several carriers in which to transmit their data, so that the bits on each subcarrier are much longer, drastically reducing ISI. Thus, it provides the concept of multicarrier modulation (multiple carriers for one digital baseband signal) rather than the conventional single-carrier modulation. The transmission is generated in such a way that the carriers used are orthogonal to one another, non-interfering with each other and thus allowing them to be packed together much closer than standard frequency division multiplexing (FDM). This leads to OFDM providing a high spectral efficiency. Presently, OFDM is used as a physical layer standard in IEEE 802.11a/g and 802.16x protocols, HIPERLAN protocols (Japan), DAB, and DVB. HDradio is a new concept developed based on OFDM to have CD-like quality of audio reception.

It is the requirement of time that all the existing systems should be converged and total wireless scenario is expected to cover the entire world. It will bring in a new era in communication. Using these technologies, wireless connections with mobility can be maintained directly to the people and information can be made available whenever they want it. Newly designed mobile devices can support many technologies in one device along with conventional mobile telephony. Essential wireless technologies that are coexisting are ultra-wideband (UWB), Wi-Fi, Bluetooth, and various 3G technologies like WCDMA and wireless access protocol (WAP). These technologies are working synergistically to meet unique users' needs. It is likely that no single broadband wireless technology will achieve dominance over another.

It is necessary to know about the suitability of a particular technology in a particular application. The UWB is most suitable for very small networks supporting a very high bit rate. Wireless personal area networks (WPANs) are very small networks within a confined space, such as an office workspace or a room within the home and using Bluetooth standard. The UWB technology offers to WPAN users a much faster service, short-distance connection and are currently under development. WLANs have a broader range than WPANs, typically confined within office buildings, stores, homes, etc. Intel has developed Intel Centrino Mobile Technology for Wi-Fi, which is gaining popularity. Wireless metropolitan area networks (WMANs) cover a much greater distance than WLANs, connecting buildings to one another over a broader geographic area. The emerging WiMAX technology (802.16d today and 802.16e in the near future) will further enable mobility and reduce dependence on wired connections. Wireless wide area networks (WWANs) are the broadest range wireless networks and they are most widely deployed today in the cellular voice infrastructure although they have the ability to transmit data. Next generation cellular services based on various 3G technologies will significantly improve WWAN communications.

It is now a challenge to cover the global wireless communication and for that universal mobile telecommunication system (UMTS) project is undertaken by engineers for ten years. It is the standard for universal telecommunications standardized by IMT 2000. Using WCDMA, the standards are developed for the system even for indoor and outdoor communication. Table 1.1 summarizes some of the present wireless digital communication-based systems that are already in practice. Table 1.2 gives a comparison chart for existing and upcoming technologies for wireless networking.

Table 1.1 Summary of existing wireless digital communication-based applications

Application	Existing Standard/Technology Used
Mobile telephony (digital cellular telephony)	GSM, CDMA (IS-95 to CDMA 2000), WCDMA-UMTS
Wireless LAN/MAN/WAN	IEEE 802.11(Wi-Fi), 802.16(WiMAX), etc.
Personal area communication	Bluetooth
Digital audio broadcast, HD radio, DRM	DAB
Digital video broadcast, DTH through Satellite	DVB
Mobile satellite communication, Global communication	Iridium, UMTS, GPS
Mobile Internet access	GPRS, Mobile IPv6, WAP
Wireless local loops	DECT, CorDECT, CDMA, GSM
Mobile adhoc networks	All WLAN/WMAN standards and Bluetooth, sensor N/w

Table 1.2 Comparison of most recent wireless networking technologies

	EDGE	CDMA 2000/1 x EVDO	Blue-tooth	Wi-Fi	Wi-Fi	Wi-Fi	WiMAX	WiMAX	WCDMA/ UMTS	UWB
Standard	2.5G	3G	802.15.1	802.11a	802.11b	802.11g	802.16d	802.16e	3G	802.15.3a
usage	WWAN	WWAN	WPAN	WLAN	WLAN	WLAN	WMAN Fixed	WMAN Portable	WWAN	WPAN
Throughput	Up to 384 kbps	Up to 2.4 Mbps (typical 300–600 Kbps)	Up to 720 kpbs	Up to 54 Mbps	Up to 11 Mbps	Up to 54 Mbps	Up to 75 Mbps (20MHz BW)	Up to 30 Mbps (10MHz BW)	Up to 2 Mbps (Up to 10 Mbps with HSDPA techno-logy)	110 – 480 Mbps
Range	Typical 1–5 miles	Typical 1–5 miles	Up to 30 feet	Up to 300 feet	Up to 300 feet	Up to 300 feet	Typical 4–6 miles	Typical 1–3 miles	Typical 1–5 miles	Up to 30 feet
Frequency	1900 MHz	400, 800, 900, 1700, 1800, 1900, 2100 MHz	2.4 GHz	5 GHz	2.4 GHz	2.4 GHz	2–11 GHz	2–6 GHz	1800, 1900, 2100 MHz	7.5 GHz

1.3 EVOLUTION IN WIRELESS SYSTEMS: TOWARDS WIRELESS EVERYWHERE

The communication link requires a transmitter, a channel, and a receiver to transfer the information—maybe in unidirectional or bidirectional manner. Here the real-time signals and data must be modified in accordance with the channel characteristics and also in the suitable detectable format so that they can be communicated reliably through the media. For the transmission, wired or wireless media can be chosen. As the wireless systems are more portable and easy to carry, it is becoming more and more popular. In the present scenario, we have combination systems that may have wired infrastructure with an extensive wireless support, e.g., mobile telephony integrated with the landline public switched telephone network (PSTN) and also with the Internet. In future, it is expected that every system will be totally mobile. In Fig. 1.2, it is shown that the highest mobility is achieved with UMTS with a little

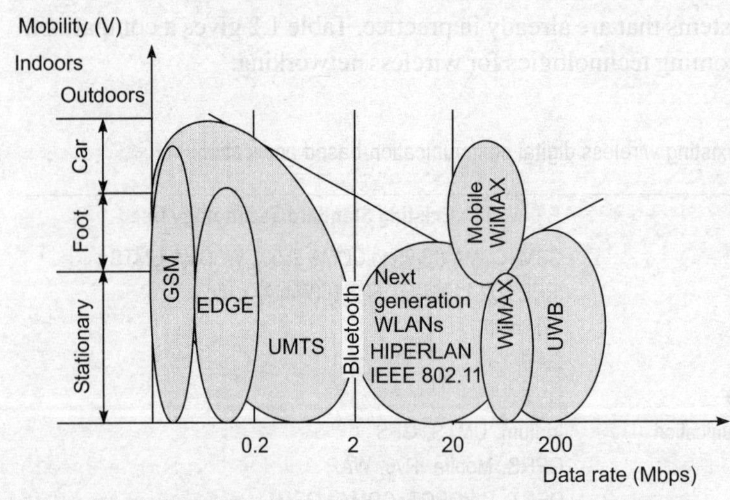

Fig. 1.2 Mobility and bit rate both are increasing with the generations

compromise on the bit rate. However, in IEEE 802.16e mobile wireless broadband access system, the solutions are found to have vehicular mobility with higher bit rate.

At the same time, digital systems are more advantageous in terms of performance compared to analog systems. It is easy to use processor or computer support, have digital signal processing, handle the bit errors, and store the data. It is the age when the concept of 'digital everywhere' is adopted by the masses. The people have experienced the excellent performance of digital systems and they demand such a quality even in wireless applications. In short, wireless digital communication based systems are in great demand.

Basically, by *wireless digital communication,* we mean that the focus is on the main link (transmitter+channel+receiver) and its fundamentals for communication, including various blocks of processing the information signal, which are introduced in Chapter 2 onwards. Here it is necessary to know the various methods of modifying the data or real signals, modulation schemes, channel characteristics, receiving methods, etc. Cellular theory forms the systematic platform to have the infrastructure to develop the wireless digital communication based multiple users' links without interference. (i.e., a system development can be possible). Due to cell concept, users can be identified uniquely even in the mobility mode. In *mobile communications*, the main focus is on the cell-based wireless multiuser telecommunication systems, for which the standards and protocols are developed to get them communicated. Here the user is assumed to be either in steady or in mobility mode. The aim of wireless communication is to optimize the physical link while that of mobile communication is to optimize the whole system, including higher layers.

Actually, wireless communications were started specially for military purpose. Gradually, the development is observed in computers, *digital signal processing* (DSP), VLSI etc., which made possible the development of portable, sophisticated wireless units, such as decent mobile phones, Centrino technology based laptops, and palmtops. Digital signal processing has become the unavoidable part of the existing wireless systems. wireless communication systems of present age are mostly based on DSP processors, VLSI/ASIC/FPGA chips, microstrip RF circuits, and PC interface. In Fig. 1.3, it is mentioned that the faster DSP processors (compared in terms of MIPS) are incorporated with the systems to make the higher bit rate support. Sophisticated software products are developed and hence the amazing applications have been possible in multimedia. People enjoy the services like local/STD/ISD mobile calls, SMS/MMS, chat, wireless FAX and e-mail, and mobile Internet access.

Wireless systems are now popular worldwide to help people and machines to communicate with each other irrespective of their location. Cellular system is the most common platform even for the future systems. Using that, in near future, we shall have the numerous options to set up an unwired connection over radio interface. One of the slogans of the wireless communication systems (4G) is 'always best connected' meaning that wireless equipment should connect to network or system that at the moment is 'best'.

It is necessary to have the frequency planning for various wireless systems to coexist. Wireless channel is an unguided dielectric media and hence the frequency ranges it can support are ideally infinite. Still due to many reasons, full available spectrum cannot be

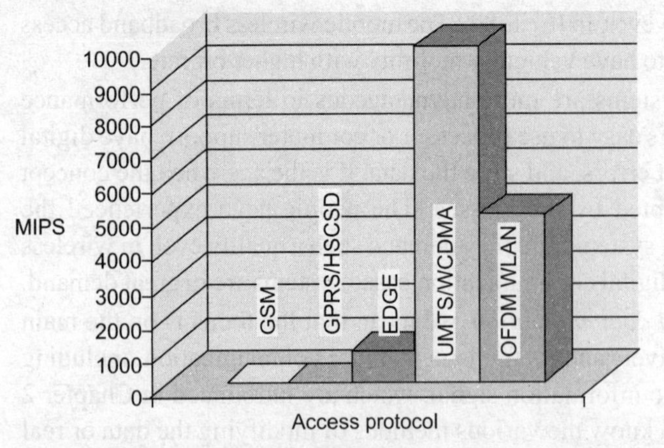

Fig. 1.3 Processing power requirement for wireless protocols and standards according to complexity of hardware MIPS=million instructions per second (It is the measure to compute the speed of DSP processor.)

utilized. The RF and the above range utilized for wireless communication are systematically shared; different ranges are used for different applications. Various connection ranges from the satellites provide a low bit rate but global coverage and cellular system based mobile satellite communication provides connections with a high bit rate. Contradictorily, local area network and personal area network provide maximum range of few to hundred metres. If the systems have to coexist, they would obtain a crowded frequency spectrum, since there are many factors that want their share of the limited frequency resource. To use the signal strategies that are spectrally efficient is thus of utmost importance. The current trend to achieve high spectral efficiency is by using the adaptivity on all four dimensions: time, frequency, power, and phase.

The requirements of wireless communication in short are as follows:

- High speed/high bit rate
- High spectral efficiency
- Zero ISI/ICI
- Converged
- Anywhere, anytime
- Global coverage
- Multimedia supported
- Wireless
- Digital communication systems.

Latest techniques like WCDMA, OFDM, hybrid OFDM, and MIMO will fulfill most of the requirements mentioned above. Also a new approach, software defined radio, is coming up with the same set of hardware but different programs used to have the different channel coding or modulation schemes.

1.3.1 1G, 2G, 3G, AND 4G WIRELESS SYSTEMS

There is no specific measure to calculate the years of generation. Rather than that, the generations are measured on the basis of the considerable innovations in the standards and applications. Analog systems are considered as the start-up and hence they are known as first-generation systems. The systems of other generations are illustrated in Fig. 1.4.

Again it is very difficult to distinguish the systems on the basis of generations. For simplicity, complete analog systems mainly dealing with audio (except television with analog video) are classified as first-generation systems, also analog mobile phone systems (AMPS). Partially analog and digital systems are classified as 2G systems, where audio and images were communicated. Bit rate was very low around 10

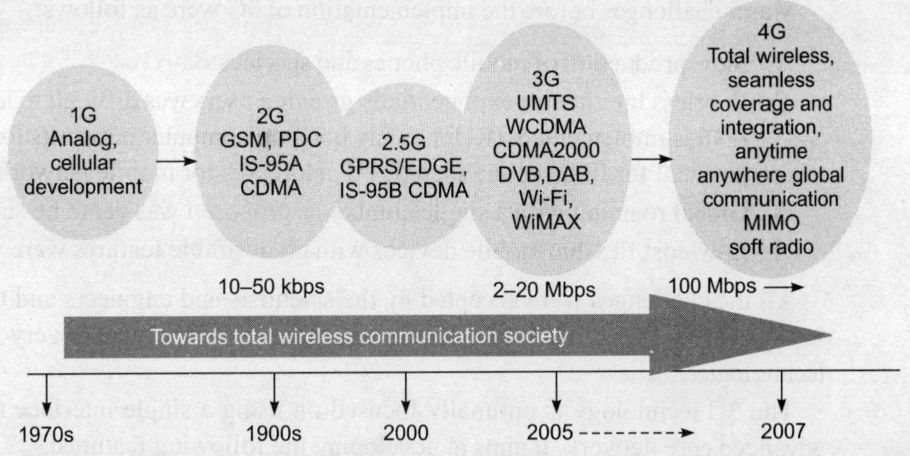

Fig. 1.4 Generations in wireless digital communication: moving towards high bit rate and mobile IP

kbps. Fully digital systems with audio, image, and video are classified as 3G with tremendous rise in the bit rate, of the order of 2 to 20 Mbps, in Wi-Fi and WiMAX even up to 54 Mbps. In 4G high-speed, fully digital, anywhere, anytime, and converged, wireless communication is expected with total multimedia. The expected bit rate may reach up to 100 Mbps or more in wireless manner. With evolution in WiMAX standards and UWB, development in the 4G systems has been started.

Why does a wireless channel face the problem of high bit rate? The channel faces the problem of delay spread due to multipath fading, meaning that the channels are time dispersive; it is discussed in detail in chapter 5. Spreading results in merging of two consecutive pulses. If the bit rate is too high, the bit duration is less and hence due to merging of two consecutive pulses, it is very difficult to identify two separate pulses. This limits the bit rate of the system. Higher order M-PSK and diversity mitigation techniques like MIMO or multicarrier technique like OFDM can eliminate the problem of higher bit rate.

The 2G technology for mobile communication originated during 1990s. Before that the conventional telephony was based on wired line. A few military wireless applications, AM, FM, television, radar, and satellite communication systems were the only implemented and known systems to the people. The revolution started with two new systems—the Internet based on wired lines and cellular-based GSM based on wireless channel mainly for voice communication. In the year 2000, GSM had data-transmission enhancement called GPRS, which could use any number of time slots among the total eight slots for sending the data. The technology exists with a data rate of 14.4 kbps to 64 kbps. People found another high-speed data enhancement in GSM, called EDGE, in which modulation scheme is changed from *Gaussian minimum shift keying* (GMSK) to 8-PSK and the transmission data rate can be up to 500 kbps. The GSM system initially was focused on voice services with circuit switching, whereas current 2.5G technology is focused on circuit-switched voice service and packed switched data services.

Major challenges before the implementation of 3G were as follows:

1. Slow production of mobile phones and services.
2. Wireless Internet for exponentially growing users was difficult to implement until IPv6 is implemented. (Refer to any book on computer networks for IPv6. It is the protocol for IP layer and includes IP addresses for mobile networks also.)
3. Global roaming with a single number as proposed was yet to be standardized.
4. Low-cost flexible mobile devices with all desirable features were yet to evolve.

All the challenges were accepted by the scientists and engineers and they developed 3G systems successfully solving major problems and now we are very near to the 4G technologies.

The 3G technology is optimally focused on using a single interface number and an advanced core network. It aims at developing the following features:

1. Anywhere and any time mobile communication with low-cost and flexible hand-held devices.
2. Wireless data access, particularly with wireless Internet connection. This was motivated by exponential growth of the Internet access.
3. High data rate of 2 Mbps or more compared to the previous 2G systems offering 10 to 50 kbps.
4. High-speed multimedia or broadband services causing shift from voice-oriented services to the Internet access (both data and voice) video, graphics, and other multimedia services.
5. Global roaming support and global communication.
6. Use of spectrum around 2 GHz and higher, whereas spectrum allocation for 2G was 800/900 MHz.

The 2G technology offered a quite satisfactory voice communication, but with growing data traffic, the 3G technology has mainly targeted data services, particularly the Internet traffic. The main service component of the 3G technology is quality and reliable data traffic. The journey from 2G to 3G started with intermediate halt on 2.5G providing reliable services with minimal investment. The UMTS is the typical 3G system that uses WCDMA technology as mentioned previously. It has the following aims:

- Data services up to 2 Mbps in rural or urban environment
- Voice over packet switched IP based network
- Good spectral efficiency and low delay
- Complete mobility to the user
- Typical applications:
  - Speech—Teleconferencing and voice mail
  - Message—SMS, e-mail, etc.
  - Switched data—Low-speed LAN, Internet, etc.

Table 1.3 Important UMTS applications and their requirements

Applications or Services	Data Rate Required	Quality of Service Required	Time Critical Data
Messaging (e-mail, etc.)	Low (1–10 kbps)	High	No
Voice	Low (4–20 kpbs)	Low (BER < 1e-3)	Yes
Web browsing	As high as possible (>10–100 kbps)	High (BER < 1e-9)	Depends on the material; generally not time critical.
Video conferencing	High (100 kbps –2 Mbps)	Medium	Yes
Video surveillance	Medium (50–300 kbps)	Medium	No
High quality audio	High (100–300 kbps)	Medium	Yes
Database access	High (>30 kbps)	Very High	No

- Medium multimedia—E-commerce, LAN, and Internet public messaging
- High multimedia—Video clips, on-line shopping, and fast LAN and Internet
- High interactive multimedia—Video telephony and video conferencing

Some important UMTS applications and their requirements are listed in Table 1.3.

1.3.2 Beyond 3G

During the past 20 years, wireless networks have evolved from analog, single-medium (voice), and low data rate (few kbps) system to digital, multimedia, and high data rate (10 to 100 Mbps) system of today.

The International Telecommunication Union (ITU) in July 2003 had made the following requirements for 4G system:

1. At a standstill condition, the transmission data rate should be 1 Gbps.
2. At a moving condition, the transmission data rate should be 100 Mbps.

With these high-speed data systems, many advanced applications for the users can be realized like video streaming. A potential 4G system could be used in the family of OFDM, because OFDM can have a transmission data rate of 54 to 70 Mbps, which is much higher than what the CDMA system can provide. Comprehensive, broadband, integrated mobile communication will step forward into all-mobile 4G service and communication. The 4G technology is developed to provide high-speed transmission, next generation Internet support (IPv6, VOIP and Mobile IP), high capacity, seamless integrated services and coverage, utilization of higher frequency, low mobile cost, efficient spectrum use, quality of service and end-to-end IP system. In short, the 4G requirements are as follows:

- High-speed data communication
- Best quality voice
- Multimedia on mobile
- LAN and intranet/Internet on mobile

1.4 LICENSED AND UNLICENSED BANDS FOR PRESENT-DAY WIRELESS SYSTEMS

The wireless channel is shared by a number of users and frequency ranges are provided systematically to the users or to the services or applications for reliable communication (refer to Chapters 10 and 11). Suppose if a few frequencies are allocated to some cellular mobile operators like Airtel, Hutch, or Idea, they have to pay heavy charges for the allocated ranges. Even satellite channels also are paid channels because of this. Mobile operators cannot have huge private infrastructures, like satellites, and moreover they have to follow government rules. Hence, they have to get the licensed bands for communication. The GSM, CDMA based mobile telecommunications are made over licensed bands.

Presently, few technologies, limited to the user's area without the need for a huge or global infrastructure are developed. Some applications of these technologies are PAN, based on Bluetooth, UWB, and WLAN, based on WiFi, which are the small-area communication systems. The frequency range of operation is 2.4 GHz–5.6 GHz. Actually these bands are the international bands for the scientists and medical officers. Because the systems are not concerned with other such systems at far distances, independent communication is possible. For example, in Bluetooth application, one device with Bluetooth support will search other active Bluetooth devices within a 10 metres area. The list will be displayed on the screen and the device will be selected from the list for the communication. Beyond this range, if any other *Bluetooth device is active*, it will not be concerned with the devices of the previous 10 metre area. These communications are called *unlicensed band communications*. Since they are based on spread spectrum or OFDM technology, secure communication is possible. In spread spectrum techniques, orthogonal codes are present while in OFDM, orthogonal carriers are present.

Summary

- Study of wireless digital communications requires basic knowledge of many other fields.
- Final RF transmission form is always analog but baseband signal inputted to modulation stage decides whether the wireless communication link is analog or digital.
- Wireless digital communication systems are demanded everywhere because of advantages of digital communication with mobility.
- With different combinations of coding and modulation schemes, different response of the wireless systems can be observed. Hence, selection of an optimum set-up of the protocols and standards is a matter of compromise.
- Wireless systems can be categorized on the basis of generations and the generations are formed on the basis of major changes in the communication systems.
- Many different systems and standards are existing presently, which will converge in future and we shall have 'all in one' type mobile devices.
- The expected trends are 'anywhere, anytime' mobile communication.
- Presently, the 3G systems are coming up and also the UMTS system is under development. Development of 4G system has also started.
- Unlicensed (ISM) band communications are allowed only for personal area communica-

tion systems like Bluetooth and operate at 2.4 and 5.6 GHz. For infrastructure-based mobile networks, licensed bands are utilized, in which frequencies are planned out for coexistence of the systems.

- CDMA, OFDM, and their combination along with MIMO will solve most of the problems of the next generation. These will be the technologies of 4G.

Review Questions

1. How are the communication systems classified in general?
2. How are the wireless systems classified? Find out the major changes in the classified wireless systems.
3. What are the systems at present in which partly wired links and partly wireless communication is incorporated? Can you find out the types of cables used in different wired systems?
4. Prepare the list of all the existing communication systems used in everyday life. Out of these, find out which are wired and which are wireless and then prepare a list of the existing wireless systems and the associated standards along with their modulation schemes, bit rate, frequency range of communication, special features, etc.
5. Represent the electromagnetic wave equation with its amplitude, frequency, and phase assuming that the wave is travelling in any one direction.
6. When will a signal be scalar or vector? How can scalars and vectors be represented in mathematical form?
7. Compare AM, FM, and PM techniques of modulation. What are the drawbacks of these techniques that are eliminated in digital modulation techniques?
8. Why is line coding more important for wired line communication?
9. List out the requirements of 4G and from the analysis of the existing standards, find out the points at which we are lacking.

   or

   Which are the areas that should be concentrated upon by the scientists and engineers to have the reliable 'anywhere anytime' communication scenario?
10. Develop the wireless digital communication transmitter and receiver requirements in the form of blocks and link them to form basic link diagram.
11. List out the basic requirements of the UMTS system.
12. Compare wired and wireless communication and find out why a higher bit rate is the problem of wireless link and not of the wired link. When does a wired link have the problem of higher bit rate?
13. How can we increase the users' accommodation capacity on the wired and wireless links?
14. How does licensed and unlicensed band communication differ?
15. Discuss the major changes that took place in the communication systems from first to fourth generation in general. Also discuss separately the changes in the 1G to 4G wireless systems.
16. How can you say that wireless digital communication exhibits interdisciplinary approach?

Multiple Choice Questions

1. Which of the following is the communication system mainly suitable for wireless digital communication?
   (a) Analog input-analog transmission
   (b) Anlog data-digital transmission
   (c) Digital data-digital transmission
   (d) Digital data-analog transmission
2. Which of the following is the scheme for creating digital database of real signals?
   (a) Pulse code modulation
   (b) Manchester coding
   (c) Binary conversion
   (d) Pulse amplitude modulation
3. Which of the following systems is a 3G system?
   (a) Analog cellular system
   (b) EDGE
   (c) FM
   (d) UMTS
4. The capacity of the wireline system can be increased
   (a) by TDMA
   (b) by random access
   (c) by increasing the number of wires
   (d) by all of the above methods
5. The protocol for Wi-Fi system is
   (a) IEEE 802. 16d
   (b) IEEE 802.15.3
   (c) IEEE 802.11a
   (d) IEEE 802.15.1
6. Which of the following is a system in which long-haul communication is involved?
   (a) Mobile satellite communication system
   (b) GSM system
   (c) WiMAX system
   (d) Bluetooth system
7. The systems that utilize ISM band for communication are
   (a) GPRS and EDGE
   (b) Bluetooth and Wi-Fi
   (c) GPRS and Bluetooth
   (d) Bluetooth and WiMAX

chapter 2

Fundamentals of Wireless Digital Communication

Key Topics

- Fundamental terms of communication
- Modulation and carrier
- Bandwidth and SNR
- Types of signals
- Signal space
- Fundamentals of DSP, such as A/D conversion, correlation, aliasing, antialiasing, convolution, frequency resolution, and time and frequency domain transformations
- Fundamentals of digital transmission, such as Nyquist criteria for zero ISI, pulse shaping, and windowing
- Introduction to complete link diagram for wireless digital communication

Chapter Outline

This chapter begins with a brief revision of basics of communication and moves on to highlight the fundamentals of digital signal processing (DSP) and digital communication and introduce a complete wireless link. The chapter is divided into four major parts. Firstly, efforts are made in such a way that students can revise and correlate the fundamentals of the wireless system design. The terms with which the reader must be acquainted are information, transmitter, channel, noise, receiver, modulation, carrier, bandwidth, and signal-to-noise ratio. Secondly, we must understand the pulse and square wave in all respects as we deal with digital signals. It is necessary to highlight the fundamentals of DSP because most of the blocks of wireless communication system are based on DSP. None of today's systems is possible without DSP processors; spectral consideration, correlation between two signals, estimation of channel, etc., all are based on DSP concepts. Thirdly, as we deal with digital communication, the fundamentals of digital communication are also required, which include the Nyquist's criteria, pulse shaping, energy of the digital signal, etc. This provides the solution for the spectrally efficient transmission that helps to accommodate more number of users. Lastly, complete wireless digital communication link diagram with significant stages is given. The subsequent chapters are developed on the basis of this block diagram. It is difficult to cover all the fundamentals in detail. Readers may use appropriate books to solve some typical problems/questions.

2.1 FUNDAMENTAL TERMS OF COMMUNICATION

This part of the chapter will provide a brief revision to the fundamental terms and is described in general, which also applies to wireless digital communications.

2.1.1 Information

The communication systems exist to convey a message. This message comes from the information source, which originates it. The information may be analog or digital and hence the communication system can be classified as analog or digital system. For the multiple-user system of communication, the set or total number of messages consists of individual messages, which may be distinguished from one another. The sine wave is the fundamental analog information signal. A general sine wave can be represented by three parameters, peak amplitude (A_0), frequency (f), and phase (θ), in the form

$$s(t) = A_0 \sin(\omega t + \theta)$$

where $\omega = 2\pi f$, the angular frequency. The analog information may be voice or video (real-time signals).

The information is to be transmitted from and received at a far distance where it is interpreted. A converted form is required to process and transmit the information. Digital information may be converted into words, groups of words, code symbols, or any other prearranged units. Without applying any interpretation, these units are called *data*, which may be a raw bit stream. Because it is received and interpreted at the other end, it becomes *information,* which is conveyed. Information signal may be represented with its amplitude, frequency, or phase, varying with respect to time. For the digital systems, the data or information transmission rate is measured in bits per second. Along with the useful data, if additional bits are added (for special purpose), the information-transmission efficiency reduces. It must be realized that no real information is conveyed by a redundant message; but redundancy is not wasteful under all conditions, especially when error handling is concerned. In short, a set of information or data with respect to time is the time domain *input signal* for the system, whose frequency contents can be observed in frequency domain by observing the spectrum.

Transmitter and receiver are the systems connected through a channel. These systems process the input signal further.

2.1.2 Transmitter

Unless the message arriving from the information source is electrical in nature, it will be unsuitable for immediate transmission. A lot of work must be done to make a message suitable for transmission, for example conversion of another form of signal into an electrical signal, restricting the range of frequencies, compressing the amplitude ranges, etc. should be done before modulation. In local loop wire telephony, no processing may be required as the mouthpiece of the handset gives an analog electrical signal directly and it is baseband communication, but in a long-distance communication, a transmitter is required to process, possibly encode, and to modulate the incoming information so as to make it suitable for transmission over the desired channel and subsequent reception. Eventually, in a transmitter, the information modulates the carrier, i.e. it is superimposed on a comparatively high-frequency sine wave. The actual method of modulation varies from one system to another. The radio frequency (RF) upconversion may be followed by a modulator stage, especially for wireless link and then the power amplifier stage completes the transmitter part. The signal becomes ready for transmission through an antenna.

2.1.3 Channel

It should be noted that the term *channel* is often used to refer to the frequency range allocated to a particular service for transmission, such as a television channel (the allowable carrier bandwidth with modulation), but in general a channel is a medium through which the signal propagates towards the receiver. Noise and interference are the most serious problems associated with the channel.

It is inevitable that the signal will deteriorate during the process of transmission, propagation, and reception as a result of some distortion in the system, or because of the introduction of noise, which is unwanted energy (usually of random nature) present in the transmission system due to a variety of causes. Since noise will be received together with the signal as shown in Fig. 2.1, it places a limitation on the transmission system as a whole. When noise is severe, it may mask a given signal so much that the signal becomes unintelligible and therefore useless. Due to noise, the bit errors may be introduced, which should be taken care of by the error handling stage.

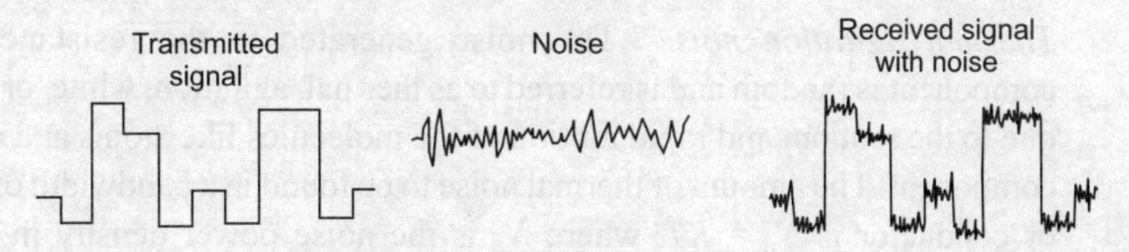

Fig. 2.1 Noise added to the signal (Noise up to certain level is tolerable.)

Noise may interfere with a signal at any point in the communication system, but it will have its greatest effect when the signal is weak. Hence, the noise in the channel or at the input to the receiver is the most noticeable. Reverse is the case when the signal is strong, the noise effects are less. This introduces the parameter signal-to-noise ratio (SNR); better the SNR, stronger is the signal in presence of noise.

Different Types of Noise in Wireless Systems

External noise This is caused by the surroundings of the transmitter and receiver.

Atmospheric noise This noise is the result of spurious radio waves that include voltages in the antenna as the antenna picks up them. The majority of these radio waves come from natural sources of disturbance. The atmospheric noise is generally called *static*. Static is caused by the natural disturbances occurring in the atmosphere. It originates in the form of amplitude modulated impulses, and because such processes are random in nature, it is spread over most of the RF spectrum normally used for broadcasting. It is additive in nature. As it has an infinite spectrum, just like white light, it is also called *white noise* (Chapter 5). It cannot be eliminated but its effect can be reduced by various methods. Atmospheric noise becomes less severe at frequencies above 30 MHz.

Extra-terrestrial noise It is of two types: solar and cosmic noise.

- Solar noise Under normal conditions, there is a constant radiation from the sun, simply because it is a large body at a very high temperature (greater than 6000°C). It, therefore, radiates over a very broad frequency spectrum, which includes the frequencies we use for communication.
- Cosmic noise Since distant stars are also suns with high temperatures, they radiate RF noise in the same manner as our Sun. Though they are far from the earth, they are many in number, which in combination become significant.The galaxies are also responsible for cosmic noise.

Industrial noise In urban, suburban, and other industrial areas, where machines are used significantly, the noise affecting the signal may be in the range of 1 to 600 MHz. This type of noise is caused by automobiles and aircraft ignition, electric motors and switching equipment, leakage from high voltage lines, a multitude of other heavy electric machines, etc.

Internal noise This type of noise in caused by the components and connections used in the hardware of transmitters and receivers.

Thermal agitation noise The noise generated in the resistance or the resistive component is random and is referred to as thermal, agitation, white, or Johnson noise. It is due to the random and rapid motion of the molecules like atoms and electrons inside the component. The amount of thermal noise to be found in a bandwidth of 1 Hz in any device or conductor is $N_o = KT$, where N_o is the noise power density in watts/Hz, K is the Boltzmann's constant (1.3803×10^{-23} J/K), and T is the temperature in kelvins (absolute temperature).

Shot noise The most important of all the other sources of noise is the *shot effect*. It leads to shot noise in all amplifying devices and virtually all active devices like transistors. It is caused by random variations in the arrival of the electrons (or holes) at the output electrode of an amplifying device and appears as a randomly varying noise current superimposed on the output. When amplified, it is supposed to sound as though a shower of lead shot were falling on a metal sheet. Hence the name shot noise.

Transit time noise If the time taken by an electron to travel from the emitter to the collector of a transistor becomes significant to the period of the signal being amplified, it means that at frequencies in the upper VHF range and beyond, the so-called transit time effect takes place, and the noise input admittance of the transistor increases. The minute currents induced in the input of the device by random fluctuations in the output current become of great importance at such frequencies and create random noise and hence frequency distortion.

Miscellaneous noise is due to flicker and resistance also. The flicker noise is most serious at low frequencies.

Apart from this, when signals at different frequencies share the same transmission medium, the result may be an intermodulation noise. For example, the mixing of signals at frequencies f_1 and f_2 might produce energy at the frequency $f_1 + f_2$. This derived signal can interfere with an intended signal at the frequency $f_1 + f_2$. Crosstalk has been experienced by anyone who, while using the telephone, has been able to hear another conversation. This is an unwanted coupling between signal paths.

2.1.4 Receiver

There are a great variety of receivers in communication systems, since the exact form of a particular receiver is influenced by a great many requirements, among the more important requirements are the modulation scheme used, the operating frequency, and its range and the type of output device required, which in turn depends on the destination of the intelligence received. Most of the receivers are of the superheterodyne type. Receivers run the whole range of complexity from a very simple crystal receiver, with headphones, to a far more complex rake receiver. The latest rake receivers used in CDMA system are more complex due to digital signal processing, the same is the case with an OFDM receiver, but both give good performance against channel effects.

As stated initially, the purpose of the receiver and the form of its output influence its construction. The output of a receiver may be fed to a loudspeaker, video display unit, various radar displays, television picture tube, pen recorder, or computer. In each instance different arrangements must be made, each affecting the receiver design. Note that the transmitter and receiver must be in agreement with the modulation and coding methods used (and also timing and synchronization).

2.1.5 Modulation

Modulation is the process by which a signal is transformed into waveforms that are compatible with the characteristics of the channel, in terms of bandwidth support also. Modulation may be of two types, analog and digital. Especially, digital modulation is the process by which (digital) symbols are transformed into the required form. In the case of *digital baseband modulation*, these modulated waveforms usually take the form of shaped pulses [ideally a sinc function shape (sin x/x) in frequency domain]. But in the case of *digital bandpass modulation*, the shaped pulses modulate a sinusoid called a carrier wave, or simply a *carrier*, and for radio transmission the carrier is converted into an electromagnetic (EM) field through an antenna for propagation to the desired destination.

Anyway, the final form of a modulated RF signal will be analog only while transmitted over wireless channels, even though the information signal maybe digital or digital modulation techniques are used.

2.1.6 Carrier

The transmission of EM fields through space is accomplished with the use of antennas. The size of the antenna depends upon the wavelength λ and the application. For cellular telephones, antennas are typically small in size. Wavelengh and frequency are related as $c = f\lambda$, where c is the speed of EM wave in free space. Thus, indirectly antenna dimensions will decide the frequency it can transmit. For sending a baseband signal of a very low frequency would require a very large sized antenna. To transmit a 3 kHz signal through space, without carrier wave modulation, an antenna that spans 15 miles would be required. If the baseband information is first modulated on a high frequency carrier (e.g., 900 MHz) the equivalent antenna diameter would be about 8 cm. For all portable application, RF conversion is necessary.

To transmit a signal over the air, there are three main steps:

1. A carrier is generated at the transmitter (single tone).
2. The carrier is modulated with the information to be transmitted.
3. At the receiver, the signal modifications or changes are detected, corrected and the carrier is demodulated to retrieve the information.

Another advantage of modulation with a carrier is in the multi-user environment. If more than one signal utilizes a single channel, modulation with the different carriers may be used to separate the different signals (FDM). The reception will be based on the tuning of carriers. Systematic frequency band allocation may be possible due to dedicated allocation of carrier. Some modulations can be used to minimize the effects of interference. The class of such modulation schemes, known as SSM and OFDM, requires a transmission bandwidth much larger than the minimum bandwidth that would be required by the message. Bandwidth concepts are highlighted in the following section.

2.2 BANDWIDTH CONCEPT

Information may be analog or digital. Any signal can be represented in time domain (amplitude vs time plots) and frequency domain (amplitude vs frequency plot, also called spectrum). The meaning of digital signal is any signal with discrete values at discrete time. It is produced by sampling of continuous envelope of information and will carry discrete well defined amplitude levels. Binary coded data is one typical case of digital system that takes only two values of amplitude levels—logic 1 and logic 0. It will carry the amplitudes decided for logic 0 and 1. Digital information transfer is measured in bits per second or in symbols per second units. When analog or digital time domain signal is converted into frequency domain signal the significant frequency components decide the bandwidth. *It is observed that analog signal consumes less bandwidth compared to its digital counterpart.*

Bit Rate and Symbol Rate

To understand and compare different modulation scheme efficiencies, it is important to first understand the difference between bit rate and symbol rate. The signal bandwidth needed for the digital communications depends on the symbol rate, not on the bit rate (refer to digital modulation schemes in Chapter 7). *Bit rate* is the frequency of a system bit stream.

$$\text{Symbol rate} = \frac{\text{Bit rate}}{\text{The number of bits transmitted with each symbol}}$$

Symbol rate is also called Baud rate and is represented in symbols/second. Each symbol represents M finite states. Each symbol represents k bits of information, where

$$k = \log_2 M \tag{2.1}$$

Example 2.1 What may be the symbol rates for QPSK, 8PSK, and 16PSK schemes if the bit rate is 256 Mbps? (Refer to Chapter 7 for the schemes.)

Solution

All these schemes modulate symbols rather than bits. For QPSK scheme, 2 bits/symbol are taken in for modulation. For 8PSK, 3 bits/symbol are taken in and for 16PSK, 4 bits/symbol are taken in. Hence

For QPSK, Symbol rate = 256 Mbps/2 = 128 Mbps

For 8PSK, Symbol rate = 256 Mbps/3 = 85.33 Mbps

For 16PSK, Symbol rate = 256 Mbps/4 = 64 Mbps

A signal may have one or more frequency content. These frequencies can be represented in the frequency domain and significant frequencies will decide the bandwidth of that signal.

There is no universally satisfying definition for bandwidth. It gives important information about the signal in the frequency domain. The term *bandwidth* is used in three ways:

1. To characterize a signal (the input signal as well as the baseband or broadband to be transmitted) being the signal or transmission bandwidth correspondingly.
2. A channel is allocated to the user to allow the transmission of maximum frequency content. In this case, the bandwidth is called channel bandwidth. It decides the capacity of transmission.
3. To design a wireless system hardware, including transmitter and receiver. The frequency response of the hardware stages must be such that the total system bandwidth must support the channel bandwidth (hardware frequency response must be set accordingly).

Signal bandwidth and channel bandwidth are of main importance, which will be clear in next topics of discussion. Always, the channel bandwidth must support the (line coded) signal bandwidth in case of baseband communication and channel bandwidth must support modulation/transmission bandwidth in case of broadband communication with modulation techniques used. Transmission bandwidth can be reshaped after modulation by applying windowing or pulse shaping as described afterwards. This way, the bandwidth after the modulation stage and the actual transmission bandwidth may differ.

2.2.1 Bandwidth of Signal and System

A system can be as simple as a low-pass filter or an amplifier, or as complicated as an entire satellite communication link. Bandwidth, when referring to a system or a device, usually means that ability to pass, amplify, or somehow process a band of frequencies. Whereas bandwidth of significant energy for a signal can be subjective—for example, the speech signal bandwidth of maximum energy could be specified as the range between 100 and 6000 Hz, or the bandwidth of significant energy for telephone quality speech as between 100 and 3000 Hz instead of what we mentioned earlier. Bandwidth for a system is usually defined between the –3 dB points assuming 0 dB point as maximum gain plotted against range of frequencies.

Technically speaking, we do not need to differentiate between analog and digital signals. After all, we need only to look at the spectral content of each signal, the extent of which determines bandwidths. Typical analog signals, because of their smooth variations, usually have a finite bandwidth, whereas digital signals, because of their discrete nature, usually have unlimited bandwidth. But also for the digital signals, for practical limitations, it is useful to specify the finite bandwidth. For finding the useful bandwidth for digital signals, it is necessary to know the range of frequencies that contains significant energy of the signal.

We can now make a simple but powerful observation. When the available bandwidth of a transmission system is equal to or larger than the bandwidth of a signal that is to be transmitted over the system, and also the actual transmitted signal frequency contents at an instant of time is less than the maximum frequency content, that is

$$W_{\text{signal}} \leq W_{\text{channel}} \tag{2.2}$$

then the entire information content of the signal can be recovered at the receiving end. Conversely, when the system bandwidth is less than the signal bandwidth, some degradation of the signals always results because of the loss of frequency components due to lack of its capacity to transfer those frequencies.

2.2.2 Bandwidth of Digital Signals

Let us consider digital signals and the bandwidth requirements for pulse transmission. We have to distinguish between the case of an exact reproduction at the receiving end of a transmitted square pulse (which represents a binary digit 1) and a distorted reproduction. An exact reproduction would require a transmission channel with infinite bandwidth, as a square pulse has infinite bandwidth. But if we only need to detect that a pulse has been sent, we can get by with a finite channel bandwidth. For example, if we were to calculate the effect of an ideal low-pass filter on a square pulse, we would find the output to be a distorted pulse that resembles the original pulse better and better with increasing bandwidth W of the filter. Channel acts as a low-pass filter. For wired as well as wireless links, the reasons are different (think). Hence, higher harmonic losses are certain. In addition to this, the attenuation also occurs. Bandwidth of a binary digital signal will always be half of that of its bit rate, as the consecutive 1 and 0 bits will establish the worst-case condition for transitions, which will decide the highest frequency content, making one cycle of frequency and bandwidths being represented in terms of frequencies only. With fixed channel bandwidth W_{channel}, if bit rate varies, different situations are generated as shown in Fig. 2.2 because a change in bit rate will vary the signal bandwidth. Bit rate and signal bandwidth are related mathematically as

$$W_{\text{signal}} = 1/2T_b \tag{2.3}$$

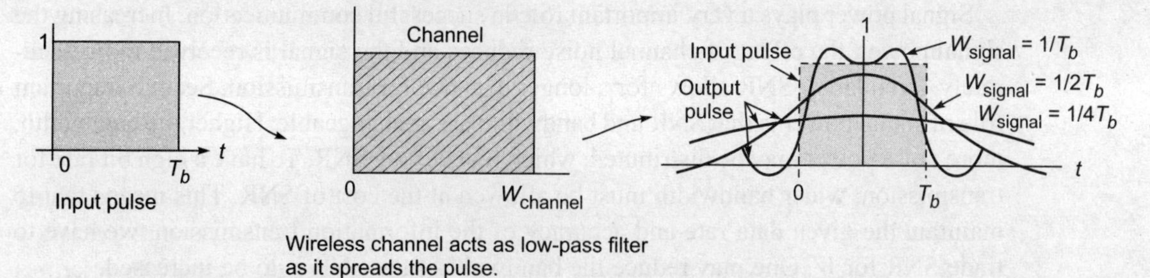

Fig. 2.2 Ideal pulse transmitted through channel gives response according to the bit rate over a fixed channel bandwidth

where T_b is the 1 bit interval (time duration). This bandwidth for many purposes yields a resolution with an acceptable error rate.

Since a wireless transmission channel with multipath effects has bandpass characteristics similar to that of a low-pass filter, a pulse propagating over the channel will be affected similarly due to delay spread. Narrower the bit interval, more errors are likely to occur. Thus, as long as the shape of the received signal can still be identified as the presence or absence of a pulse, the sent message can be identified at the receiving end. This is the heart of the digital transmission as compared to analog transmission in which the signal becomes irreversibly distorted due to addition of noise. In digital transmission, even though the individual pulses become badly distorted during propagation, as long as the distorted pulse that is received can be identified with the presence or absence of a pulse, the original message is preserved.

2.2.3 Signal-to-Noise Ratio and Channel Bandwidth

The amount of information that a channel can carry reliably depends on the bandwidth of the channel and the magnitude of the noise present in the channel. The amount of noise present in any channel limits the number of distinct amplitude levels that a signal propagating may have. For example, if a varying analog signal has a maximum level of 10 V and the noise level is 5 V, the signal may have only two levels. On the other hand, if noise level is only 1 mV, the same signal can be divided into approximately 10 V/1 mV=10^4 levels. Figure 2.1 shows pictorially how noise that has been added during transmission can degrade the signal and hence its resolution at the receiving ends.

The *signal-to-noise ratio* (SNR) is the standard measure of noise level in a system. It is the ratio of received power P_s to noise power P_n and since power is proportional to voltage squared, we can express SNR as

$$\text{SNR} = \frac{P_s}{P_n} = \left(\frac{V_s}{V_n}\right)^2 \tag{2.4}$$

where V_s is the received signal voltage and V_n is the noise voltage (noise voltages, because of their multitude of random amplitudes, are typically given as rms voltages). The SNR is usually expressed in decibels (dB).

$$\text{SNR}_{\text{dB}} = 10 \log_{10} \text{SNR} = 20 \log_{10} \left(\frac{V_s}{V_n}\right) \tag{2.5}$$

Signal power plays a very important role in successful communication. Increasing the signal power, the effect of channel noise reduces and the signal is received more accurately. Even larger SNR allows for a longer distance for transmission. Second important role of signal power is that SNR and bandwidth are exchangeable. Higher the bandwidth, more noise power may be distributed, which reduces the SNR. To have a high bit rate for transmission, wider bandwidth must be allowed at the cost of SNR. This means that to maintain the given data rate and accuracy of the information transmission, we have to trade SNR for W. One may reduce the bandwidth if the SNR is to be increased.

It can be shown that the relationship between the bandwidth expansion factor and SNR is exponential. Suppose SNR bandwidth trade-off is considered.

SNR1 is there with a particular rate of transmission with bandwidth $W1$

SNR2 is another value with different rate and bandwidth $W2$

Then, for the same channel capacity (will be defined very soon), it can be derived that

$$\text{SNR2} \approx \text{SNR1}^{W1/W2} \tag{2.6a}$$

Thus, if $W2 = 2W1$, then

$$\text{SNR2} \approx \text{SNR1}^{1/2} \tag{2.6b}$$

That is, SNR2 is the square root of SNR1.

Example 2.2 Compare the SNR requirements for the BPSK and QPSK systems that have bit rate of 1Mbps.

Solution

Considering first nulls:

For BPSK, transmission bandwidth, $W1$ = Baud rate = 1 Mbps (because it takes in 1 bit per symbol)

For QPSK, transmission bandwidth, $W2$ = Baud rate = 0.5 Mbps $\Rightarrow W1/W2 = 2$

Now, $\quad \text{SNR2} \approx \text{SNR1}^{W1/W2}$

$\Rightarrow \quad \text{SNR2} \approx \text{SNR1}^{2}$

Thus, theoretically, QPSK requires higher value of SNR compared to BPSK scheme for the same bit rate to be transmitted as in BPSK.

Note for BPSK We know that for square signal, the spectrum contains odd harmonics of the fundamental, which here equals $1/2T_b$. Thus, strictly speaking, the signal's bandwidth is infinite. In practical terms, we use the 90% power bandwidth to assess the effective range of frequencies consumed by the signal. The first and third harmonics contain that fraction of the total power, meaning that the effective bandwidth of our baseband signal is $3/2T_b$, or expressing this quantity in terms of the data rate, $3R/2$. Thus, a digital communications signal requires more bandwidth than the data rate: a 1 Mbps baseband system requires a bandwidth of at least 1.5 MHz. However, bandwidth depends upon adopted line coding scheme also. Same fundamentals can be applied to QPSK and higher PSK systems.

Compared to square pulses, Nyquist pulses consume half the bandwidth.

There is the bandwidth and the SNR relationship, given by Shannon's equation. In 1948, Dr. Claude Shannon of Bell Telephone Laboratories published a groundbreaking work entitled The Mathematical Theory of Communication, in which he gave an idea about the development of communication systems that transmit data effectively with limits on the exchange of SNR-bandwidth, the limitations imposed on communication by the transmission with zero errors. We can consider the channel as a pipe through which we send the information. Shannon worked on the channel capacity and found the equation for the bandlimited signal to be transmitted over additive white Gaussian noise (AWGN) channel as follows:

$$C = W_{\text{channel}} \log_2(1 + \text{SNR}) \tag{2.7a}$$

Nyquist has given another formulation:

$$C = 2\,W_{\text{channel}} \log_2 M \tag{2.7b}$$

Channel capacity is the maximum amount of data that can be pumped through the channel in a fixed period of time and can be measured in terms of bits/sec, W_{channel} is the channel bandwidth in hertz and SNR [CNR (carrier-to-noise ratio)] is the power ratio, where S is the signal power and N is the noise power in watts. Equation (2.7) gives the maximum possible information transmission when one bit/symbol is transmitted, $M = 2^k$ are the signalling levels. If more bits per symbols are being transmitted, then the maximum rate of transmission of information in symbols per second is C_s and for k bits/symbols, we can say $C = kC_s$. In other way, combining Nyquist and Shannon relationships, where C is the maximum bit rate capacity and k is the number of bits per signalling element (symbol), $k = C/\text{symbol rate}$, where the symbol rate is two times the bandwidth of the signal according to Nyquist relation. There may be 2^k different possible bit combinations to send in form of symbols.

There is a parameter related to SNR more convenient for determining digital data rates and error rates. The parameter is the ratio of the signal energy per bit to the noise power density (noise power per hertz), E_b/N_o. Consider a signal that contains binary digital data transmitted at a certain bit rate R. Recalling that 1 watt = 1 J/s, the energy per bit in a signal is given by $E_b = ST_b$, where S is the signal power and T_b is the time required to send one bit. The data rate is just $R = 1/T_b$. For thermal noise, $E_b/N_o = S/KTR$.

Example 2.3 A standard 4 kHz telephone channel has signal-to-noise ratio of 25 dB at the input to the receiver. Calculate its information-carrying capacity. Also find the capacity of the channel if its bandwidth is doubled, while the transmitted signal power remains constant.

Solution

$$\text{SNR} = \text{antilog}\,(25/10) = 316$$

Capacity of the channel in the first case,

$$C = 4000 \log_2(1 + 316) = 33.233 \text{ kbps}$$

If the SNR is 316, it means that when signal power is 316 mW, the noise power is 1 mW. Now, the bandwidth is doubled with no change in signal power, but the noise power is doubled due to the bandwidth being increased by two times.

Capacity in the second case,

$$C = 8000 \log_2(1 + 316/2) = 58.503 \text{ kbps}$$

Thus, the capacity of the channel has increased.

2.3 TYPES OF SIGNALS

Appropriate signal processing can be applied if and only if we know the type of signal. By knowing the form of signal, we can treat it in time or frequency domain and can identify the changes applied.

2.3.1 Analog and Digital Signals

(These signals are classified in terms of nature of amplitude possibilities and specified in time domain.)

Though we have used the words 'analog' and 'digital' many times previously, let us define the analog and digital signals again for starting the explanation in terms of DSP. A signal whose amplitude takes all the values over the measuring interval/time and is continuous in time is called the *analog signal*. Here the signal can take infinite number of values and precision is dependent upon the resolution of the system. If the signal amplitude takes finite number of values and not all, it is called the digital signal. Binary is the special case of digital signal, which takes only two values for logic 0 and 1.

2.3.2 Continuous Time and Discrete Time Signals

(These signals are classified on the basis of time and represented in time domain.)

Continuous signal is specified for every value of time, whatever precise time can be resolved, while discrete time signal is specified with the gap of measuring instants.

Mathematical representation:

$s(t) = A_0 \sin \omega t$ (continuous time signal)

$s(n) = A_0 \sin \omega n/N$, where N is a period of n samples (discrete signal)

A discrete time signal is represented at discrete instants of time with its natural value or quantized value. The time variable is not continuous and hence discrete time signal can be represented as a sequence of numbers.

From the above two types of signals, four different signal categories can be formed:

- Continuous time analog signal (real-time signals)
- Continuous time digital signal (square wave representing the binary signal)
- Discrete time analog signal (with natural value of samples)
- Discrete time digital signal (with quantized value of samples)

Discrete time signal is represented as a sequence $s(n)$, where n can take in a set of values in the integer range $-\infty$ to $+\infty$. In most cases, a discrete time signal $s(n)$ is obtained by sampling a continuous time signal $s(t)$ at periodic interval ΔT_s, so we can write $s(n) = s(t)|_{t=n\Delta T_s}$.

A *discrete time system* is one that accepts a set of sequences $s_i[n]$ (i stands for ith sequence) and produces a set of sequences $r_j[n]$ as output.

2.3.3 Periodic and Aperiodic Signals

A signal is said to be periodic for some positive constant T_0 (or N for discrete signal), i.e. a fixed period, if it satisfies the following conditions:

$$s(t) = s(t + T_0) \text{ for all } t \quad \text{(continuous time)} \tag{2.8a}$$

$$s(n) = s(n + N) \text{ for all } n \quad \text{(discrete time)} \tag{2.8b}$$

The smallest value of T_0 that satisfies this condition is called a period in terms of time unit. It is obvious that $s(t)$ will remain same when it is shifted in time by one period. A periodic signal must start at $-\infty$ and continue for ever. Also the periodic signal can be generated by repeating the signal $s(t)$ with period T_0 for infinite number of times. Here, from which instant to which instant of time the period is measured, is immaterial; the shape of $s(t)$ during that period must repeat itself infinite number of times. The signal that occurs for finite duration of time is called *aperiodic signal*. It does not have repetition of the shape of $s(t)$ for infinite number of times. It is a time-limited non-repetitive signal.

Most signals of practical interest can be decomposed into a sum of sinusoidal signal components. For the class of periodic signals, such decomposition is called a *Fourier series*. For the class of finite energy signal (aperiodic), it is called *Fourier transform.* The signal in time domain and the corresponding frequency domain equivalents are given in Table 2.1.

Table 2.1 Time and frequency domain equivalents

Time Domain Signal Type	Frequency Domain Equivalent
Continuous time periodic signal	Discrete spectrum aperiodic
Continuous time aperiodic	Continuous spectrum aperiodic
Discrete time periodic	Discrete spectrum periodic
Discrete time aperiodic	Continuous spectrum periodic

2.3.4 Deterministic and Probabilistic Signals

A signal can be classified as deterministic if there in no uncertainty with respect to its value at any instant of time. Probabilistic signals, also known as random or non-deterministic signals, cannot be predicted, i.e., some degree of uncertainty is there. Deterministic signals can be represented with the mathematical expression, which will be unique. Random signals are generated from random or stochastic processes.

Random function of time is often referred to as stochastic signals. First of all, we must know the stochastic signal. Stochastic signal may be continuous or discrete in time and may have continuous-valued or discrete-valued amplitudes. Stochastic processes are classes of signals whose fluctuations in time are partially or completely random, such as speech, music, image, time-varying channel response, noise, and video. Stochastic signal is completely described in terms of probability model and theory, but can also be characterized with relatively simple statistics, such as the mean or statistical averages, the correlation and the power spectrum. They must deal with the ensemble averages, variance, probability density function (PDF), cumulative distribution function (CDF), etc.

Any book on Statistical Signal Modelling can be used to explore the topic further. More description will be found in Section 5.9.2 in Chapter 5.

2.3.5 Energy and Power Signals

Normally power is related to voltage or current as follows. Here continuous time analog signal is considered, that is why we have to deal with integrations in the subsequent formulas.

$$P(t) = \frac{V^2(t)}{R} \quad \text{or} \quad P(t) = i^2(t) \times R \tag{2.9}$$

where R is the resistance across which power is measured.

When the communication system is to deal with power is often normalized, assuming $R = 1\Omega$, though the resistance may have another value in the actual circuit. Actual value of power is obtained by denormalizing the normalized power value. Conventionally, whether the signal is voltage or current waveform, the normalization convention for power allows us to express the instantaneous power as

$$P(t) = s^2(t) \tag{2.10}$$

Energy dissipated in the time interval ($-T/2$ to $T/2$) of a signal with instantaneous power is measured by the following expression:

$$E_s = \int_{-T/2}^{T/2} s^2(t)dt \tag{2.11}$$

The average power dissipated by the signal during the same interval is

$$P_{\text{av}} = \frac{E_s}{T} = \frac{1}{T} \int_{-T/2}^{T/2} s^2(t)dt \tag{2.12}$$

The performance of the communication link depends on the received signal energy. Higher the energy, more accurately the signal can be detected. At the same time, power is the rate at which the energy is delivered. This is necessary because voltages, currents, or electromagnetic field intensities are related to power and they need to be designed as per power requirements. The signal $s(t)$ can be converted into discrete form by sampling and samples can be written as $s(n)$, where n is the index value. All the above formulas can be rewritten by replacing integration by summation and $s(t)$ by $s(n)$. Similarly, the changes can be applied to energy and power signals also.

With these fundamentals, energy and power signals can be differentiated as follows.

While analysing the signals, it is often desirable to deal with the waveform energy E_s. We can classify $s(t)$ as an energy signal, if and only if it has non-zero but finite energy for all time, i.e. when $T \rightarrow \infty$

$$E_s = \int_{-\infty}^{\infty} s^2(t)dt \qquad \text{(continuous time signal)} \tag{2.13a}$$

$$E_s = \sum_{-\infty}^{\infty} |s(n)|^2 \qquad \text{(discrete time signal)} \tag{2.13b}$$

In the real world, the transmitted signals have finite energy ($0 < E_s < \infty$). Finite energy signal has zero average power. However, in order to describe periodic signals, which by definition exist for all time and thus have infinite energy, these are called the power signals. Even random signals having infinite energy are power signals. If E_s is infinite, the average power P_s may be either finite or infinite. A signal is defined as power signal only if it has finite but non-zero power for all time t.

$$P_s = \lim_{T\to\infty} \frac{1}{T} \int_{-T/2}^{T/2} s^2(t)dt \tag{2.14}$$

For signal $s(n) = Ae^{j\omega n}$, the average power is A^2.

The energy and power signals classification is mutually exclusive.

- Energy signal has finite energy but zero average power (e.g., deterministic and aperiodic signals) and can be generated in a lab.
- Power signal has finite average power but infinite energy (e.g., periodic and probabilistic signals). Obviously, it is impossible to generate a true power signal in practice because such a signal has infinite duration and infinite energy.

However, a few signals are neither energy signals nor power signals; the ramp signal is such an example.

From the theory of linear system, Parseval's theorem tells that the Fourier transform preserves energy and power. However, the energy (or power) in the complex envelope is not equal to the corresponding energy (or power) in the corresponding band-pass signal.

2.4 SIGNAL SPACE

The theory of signals and systems will start with the theory of signal space. Here it is highlighted to establish the requirement of the knowledge of signal space to visualize the *signal* to be treated for digital signal processing. According to Fourier series, it is clear that any signal is made up of number of harmonics. These harmonics can be represented in the frequency domain and also can be defined as individual functions, which are called *basis functions*. These functions are necessary to represent the signal into multi-dimensional signal spaces as the signal is a vector quantity. Basis functions are the minimum number of functions that are necessary to represent any signal. Basis functions of a signal are also called basis set. Orthogonal basis set do not interfere with each other. If the energy of the basis functions are normalized as mentioned above to represent the power, the basis set is called the orthonormal basis set.

2.4.1 Fourier Base Signals/Basis Functions

To visualize signals as vectors, we start with the familiar example of a Fourier series. (The notations used in this section are arbitrary and limited to this topic only and hence are not included in the list of symbols.) For reasons that will be clear later, here we shall deal with only time limited (aperiodic) signal with period of length T. This means that we consider a well-behaved (e.g., integrable) real arbitrary signal $x(t)$ inside the time interval $0 \leq t \leq T$

and set $x(t) = 0$ outside. Inside the interval, the signal can be written as a Fourier series as follows:

$$x(t) = \frac{a_0}{2} + \sum_{k=1}^{\infty} a_k \cos\left(2\pi \frac{k}{T} t\right) - \sum_{k=1}^{\infty} b_k \sin\left(2\pi \frac{k}{T} t\right) \tag{2.15}$$

The Fourier coefficients a_k and b_k are given by

$$a_k = \frac{2}{T} \int_0^T \cos\left(2\pi \frac{k}{T} t\right) x(t) dt \tag{2.16}$$

and

$$b_k = -\frac{2}{T} \int_0^T \sin\left(2\pi \frac{k}{T} t\right) x(t) dt \tag{2.17}$$

These coefficients are the amplitudes of the cosine and (negative) sine waves at the respective frequencies $f_k = k/T$. The cosine and (negative) sine waves are interpreted as a *base* for the (well-behaved) signals inside the time interval of length T. Every such signal can be expanded into that base according to Eq. (2.15) inside that interval. The underlying mathematical structure of the Fourier series is similar to the expansion of an N-dimensional vector $\mathbf{x}$ into a base $\{V_i\}_{i=1}^N$ according to

$$\mathbf{x} = \sum_{i=1}^{N} \alpha_i v_i \tag{2.18}$$

The base $\{V_i\}_{i=1}^N$ is called *orthonormal* if two different vectors are orthogonal (perpendicular) to each other and if they are normalized to length one, that is

$$\mathbf{v}_i \cdot \mathbf{v}_k = \delta_{ik} \tag{2.19}$$

where δ_{ik} is the Kronecker delta ($\delta_{ik} = 1$ for $i = k$ and $\delta_{ik} = 0$ otherwise) and the dot denotes the usual scalar product. We have the relation

$$\mathbf{x} \cdot \mathbf{y} = \sum_{i=1}^{N} x_i y_i = \mathbf{x}^T \mathbf{y} \tag{2.20}$$

for real N-dimensional vectors. In that case, the coefficients α_i are given by

$$\alpha_i = \mathbf{v}_i \cdot \mathbf{x} \tag{2.21}$$

For an orthonormal base, the coefficients α_i can thus be interpreted as the projections of the vector $\mathbf{x}$ onto the base vectors, as depicted in Fig. 2.3 for $N = 2$. Thus, α_i can be interpreted as the *amplitude* of $\mathbf{x}$ in the direction of $\mathbf{v}_i$.

The Fourier expansion illustrated in Eq. (2.15) is of the same type as the expansion illustrated in Eq. (2.18), except that the sum is infinite. It can be proved that the base of signals $v_i(t)$ fulfills the orthonormality condition

$$\int_{-\infty}^{\infty} v_i(t) v_k(t) dt = \delta_{ik} \tag{2.22}$$

The description just means that the Fourier base forms a set of orthogonal signals. With this interpretation, Eq. (2.22) says that the base signals for different frequencies are orthogonal and, for the same frequency $f_k = k/T$, the sine and cosine waves are orthogonal. We note that the orthonormality condition and the formula for α_i are very similar to the case of finite dimensional vectors. One just has to replace sums by integrals. A similar geometrical interpretation is also possible; one has to regard signals as vectors, that is, identify $v_i(t)$ with $\mathbf{v}_i$ and $x(t)$ with $\mathbf{x}$. The interpretation of α_i as a projection on v_i is obvious. For only two dimensions, we have $x(t) = \alpha_1 v_1(t) + \alpha_2 v_2(t)$, and the signals can be adequately illustrated as in Fig. 2.3. In this special case, where $v_1(t)$ is a cosine signal and $v_2(t)$ is a (negative) sine signal, the figure depicts nothing else but the familiar phasor diagram. However, this is just a special case of a general concept that applies to many other scenarios in communications.

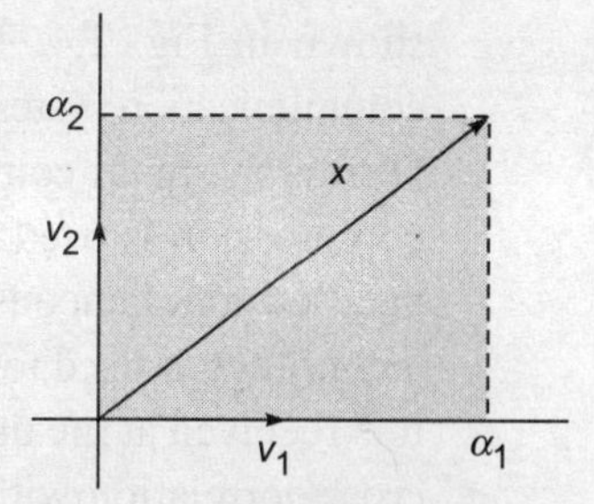

Fig. 2.3 A signal vector representation in two dimensions

In case of reception of transmitted signal $s(t)$, the coefficient α_i is the complex amplitude (i.e. amplitude and phase) of the wave at frequency f_k. It can be interpreted as the component of the signal vector $s(t)$ in the direction of the base signal vector $v_k(t)$, that is, we interpret frequency components as vector components or vector *coordinates*.

2.5 FUNDAMENTALS OF DSP ASSOCIATED WITH WIRELESS COMMUNICATION SYSTEMS

2.5.1 Recent Scenario with DSP

In recent years, a tremendous growth is observed in the intranet, Internet, and wireless systems. Digital signal processing (DSP) played an important role in making multimedia applications possible across the Internet and over mobile telephone systems. The DSP technology is nowadays commonplace in devices such as mobile phones, multimedia computers, video recorders, CD players, hard disc drive controllers and modems, and will soon replace analog circuitry in TV sets and telephones. A few DSP applied devices are

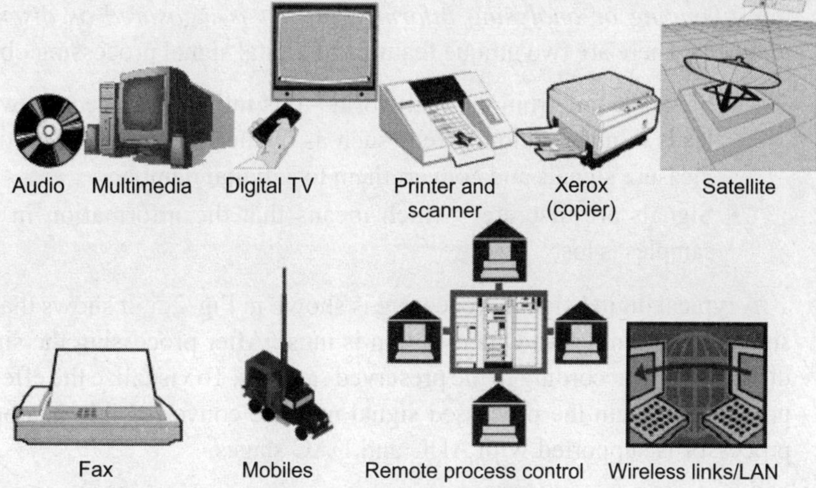

Fig. 2.4 Devices with applied DSP including wireless communication capability

shown in Fig. 2.4. The shown devices are directly or indirectly related to wireless communication systems, may be including Bluetooth features. An important application of DSP is in signal compression and decompression, correlation and convolution. Signal compression is used in digital cellular phones to allow a greater number of calls to be handled simultaneously within each local 'cell'. It is also called source coding. The DSP technology is used to perform complex error detection and correction on the raw data as it is received at the other end. The DSP fundamentals are required to detect the auto and cross correlation with the signals. Sending the signal over the channel is a convolution process. Speech and image processing for desired effects is possible only with DSP techniques. The algorithms required for DSP are sometimes performed using specialized computers, which make use of specialized microprocessors called digital signal processors. They process signals in real time. They are optimized for DSP computations.

Most of the DSP-based systems exhibit some common features:

- They use a number of algorithms [e.g., multiplying and adding signals, maybe for FFT/IFFT(explained afterwards)] and processing techniques like filtering.
- They deal with signals that come from the real world.
- They require a response in a certain time as fast as possible.
- Programmed DSP processor chip is there in each device. General-purpose DSP processors are sufficient for audio applications, while for image and video, high performance and faster DSP processors are required.

After establishing the importance of DSP processors in real world, let us see what kind of processing it does. Actually signal processing may be applied to analog or digital signal but here we concentrate only on digital signal processing.

2.5.2 Digital Signal Processing

Again, *signal* means a variable parameter by which information is conveyed through an electronic circuit, *digital* means discretized signal and *processing* means to perform operations on digital signal, which leads us to a simple definition of digital signal processing, *changing or analysing information that is measured as discrete sequences of numbers.* There are two unique features of digital signal processing observed in general.

- Signals come from the real world—this intimate connection with the real world leads to many unique needs such as the need to react in real time and a need to measure signals and convert them to a digital number.
- Signals are discrete—which means that the information in between discrete samples is lost.

A typical digital signal processing is shown in Fig. 2.5. It shows that, before applying signal processing, A to D conversion is must. After processing the signal may take the different form according to the preserved samples. To visualize the effect of digital signal processing, again the processed signal must be converted into analog form. Any DSP processor is supported with ADC and DAC stages.

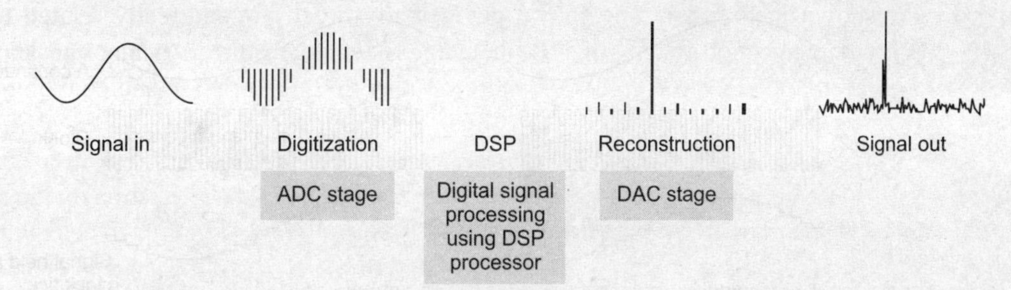

Fig. 2.5 A typical example of digital signal processing on arbitrary analog signal that gives the different required output

The DSP has four major subfields: speech/audio signal processing, digital image processing, video processing, and communication-related processing. The goal of DSP is usually to measure, convert, or filter these continuous real-world analog signals, that is why the first step is usually to convert the signal from an analog to a digital form, using an analog-to-digital converter. After this stage, the signal may be taken in form of discrete samples or discrete samples may be converted into binary database. Both forms of signal are useful for DSP. Sampling is the process in which the input analog signal is multiplied with the chopping/sampling signal that is generally an impulse train. Impulses in practice are of finite duration narrow pulses. Resulting signal will be a chopped analog that can be converted into binary codes.

In DSP, engineers usually study digital signals in one of the following domains: time domain, spatial domain, frequency domain, Z-domain, wavelet domain, etc. They choose the domain in which to process a signal by making an educated guess (or by trying different possibilities) as to which domain best represents the essential characteristics of the signal. A sequence of samples from a measuring device produces a time or spatial domain representation, whereas a discrete Fourier transform produces the frequency domain information that is the frequency spectrum.

Rather than mathematical aspects, the conceptual application-oriented representation is given here. For DSP in detail with mathematics, any good book can be referred. The intention of the following topics is not to introduce fundamentals of DSP but to have comprehensive interpretations of DSP for the applications (especially in communication and multimedia processing) and to have perfect visualization.

2.5.3 Converting Analog Signals into Digital Form

Most DSP-based applications deal with real-time analog signals, which should be converted into digital form. The analog signal, a continuous variable defined with infinite precision, is converted into a discrete sequence of measured values that are represented digitally, but with loss of precision. The converted signal is shown in Fig. 2.6.

Completion of sampling process gives discrete time signal with natural or quantized amplitudes. Information is lost in converting from analog to digital, due to the following factors:

- Inaccuracies in the measurement of input signal
- Uncertainty in timing

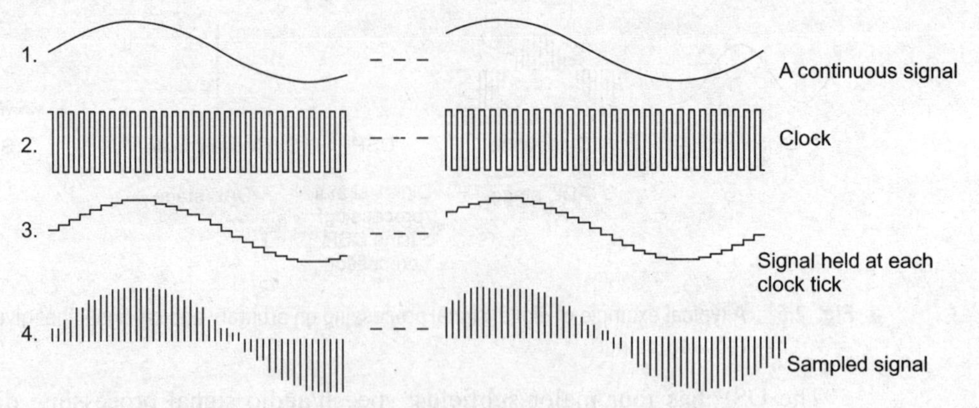

Fig. 2.6 Sampling procedure for analog to digital conversion

- Discretization in the time and levels
- Limits on the duration of the measurements

Discretization in levels results in quantization errors. *Quantization* is the intermediate process while performing analog-to-digital conversion to assign the samples the discrete predefined values near the actual one. Quantization is described in detail in Chapter 3 along with PCM and source coding. Here the discussion about quantization is omitted.

Time constraints are present in A to D conversion. The continuous analog signal has to be held before it can be sampled; otherwise, the signal would be changing during the measurement. Only after it has been held, can the signal be measured, and the measurement converted to a digital value. We do not know what we measure or what we do not. In the process of measuring the signal, some information may be lost. Sometimes, we may have some *a priori* knowledge of the signal, or be able to make some assumptions that will let us reconstruct the loss. Figure 2.7 gives the constraints for measuring the analog signal and converting it into sampled signal with timing constraints, which results into sampling error.

The sampling results in a discrete set of digital numbers that represent measurements of the signal, usually taken at equal intervals of time. Note that the sampling takes place after the hold. This means that we can sometimes use a slower ADC than might seem required at first sight. The hold circuit must act fast enough that the signal is not changing during the time the circuit is acquiring the signal value but the ADC has all the time that the signal is held to make its conversion.

2.5.4 Aliasing

What happens to the input analog voice signal that is undersampled? Undersampling can be understood by the Nyquist criterion given below. Nyquist criteria or sampling theorem says that

$$f_s \geq 2W_{\text{signal}} \tag{2.23}$$

It means that the sampling frequency must be equal to or greater than twice the input analog signal bandwidth. The sampling frequency less than twice the bandwidth of the input analog signal creates an overlap between the frequency spectrum of the samples

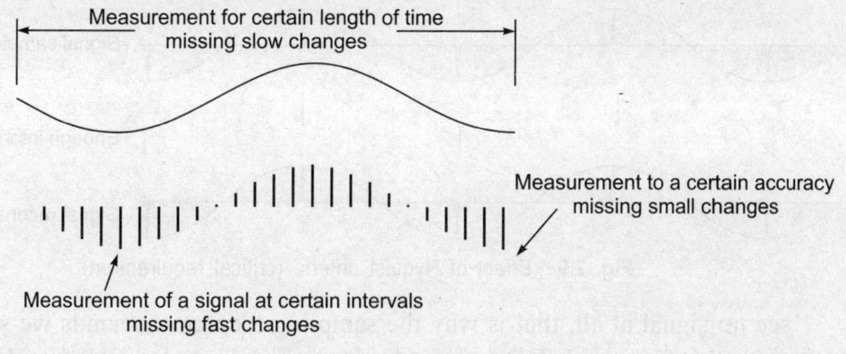

Fig. 2.7 Reasons for errors during sampling (Clock errors may have some timing uncertainty.)

and the associated input analog signal bandwidth. This is nothing but the effect of under sampling. The low-pass output filter, used to reconstruct the original input signal, is not smart enough to detect this overlap. So it creates a new signal that did not originate from the source. This creation of a false signal while under sampling is called *aliasing*.

Why such criteria? We only sample the signal at intervals. We do not know what happened between the samples. Sampling rate must be decided on the basis of how much minor details of the time waveform is to be maintained. This situation is illustrated in Fig. 2.8.

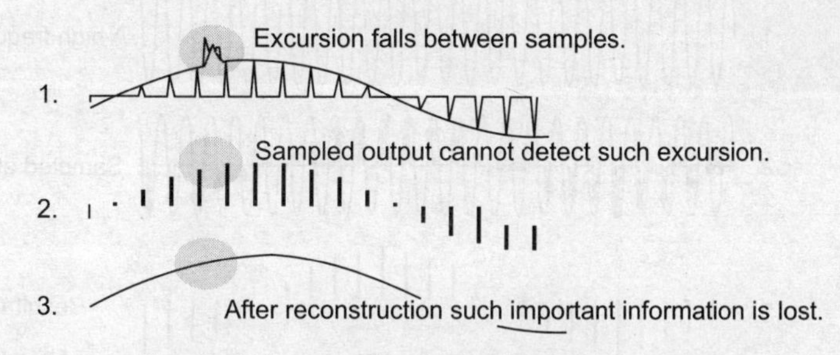

Fig. 2.8 Importance of sampling rate selection

A typical example as shown in Fig. 2.8 is to consider a 'fast change/spike' that happened to fall between adjacent samples. Since we do not measure it, we have no way of knowing if the spike was there at all. Again, we could not track the rapid inter sample variations. Hence, we must sample fast enough to see most of the rapid changes in the signal, so the sampling rate is highest frequency dependent. This is what described in Nyquist criteria/sampling theorem. Sometimes, we may have some a priori knowledge of the signal, or be able to make some assumptions about how the signal behaves in between the samples.

Proof along with mathematics is not given here for Nyquist criteria. It can be referred from any book on signals and systems or basic communication. In Fig. 2.9, a signal is sampled twice every cycle. It says that this way the signal becomes recoverable.

If we draw a smooth connecting line between the samples, the resulting curve looks like the original signal. But if the samples happened to fall at the zero crossings, we would

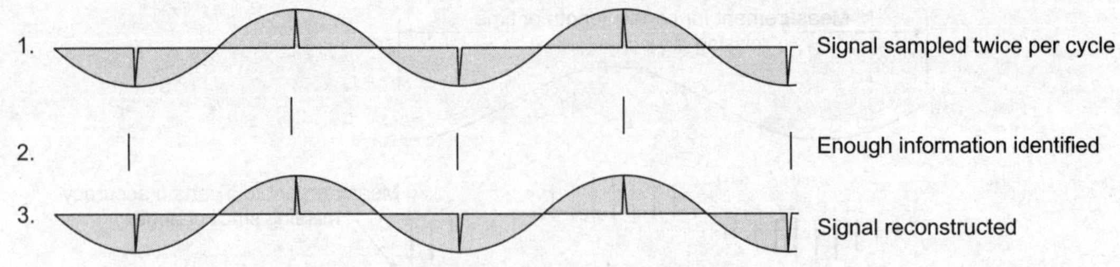

Fig. 2.9 Effect of Nyquist criteria (critical requirement)

see no signal at all, that is why the sampling theorem demands we sample faster than twice that frequency. This avoids aliasing, which is explained below. Highest and lowest frequency of the signal decides the bandwidth. If signal bandwidth starts from DC then the maximum frequency allowed for a given sample rate is called the *Nyquist frequency*.

In Fig. 2.10, the high-frequency signal is sampled just under twice every cycle. The result is that each sample is taken at a slightly later part of the cycle. If we draw a smooth connecting line between the samples, the resulting curve looks like a lower frequency. This is called 'aliasing' because one frequency looks like another. This effect is explained here in time domain. Note that the problem of aliasing is that we cannot tell which frequency is a high frequency because a high-frequency signal may look like a low one after reconstruction.

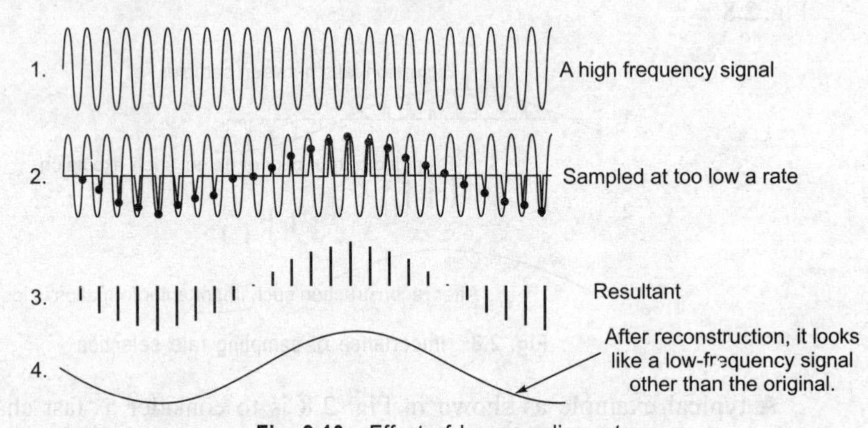

Fig. 2.10 Effect of low sampling rate

Aliasing effect can better be observed in frequency domain as overlapping of the consecutive spectral components. This will be clear when the frequency domain transforms, i.e. discrete time Fourier transform (DTFT), will be studied. For continuity of the topic, the frequency domain representation of the sampled signal and aliasing effect are shown in Fig. 2.11. While converting the sampled signal into frequency domain, the periodic energy bundles are created, as shown in the figure. The gap between the bundles depends upon sampling rate.

If the frequency domain response is observed, the multiple harmonic blocks are generated, centred at the multiples of sampling frequency f_s due to DTFT. Due to sampling rate of a time domain signal, lower than Nyquist rate, the aliasing is observed in

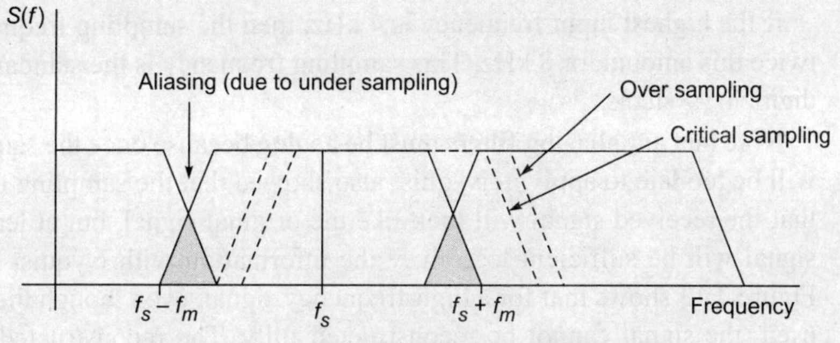

Fig. 2.11 Aliasing effect in the frequency domain

frequency domain, which causes the overlapping of two consecutive blocks. If critical sampling with the Nyquist rate, the two consecutive spectra critically touch each other. With oversampling, a gap is observed between two consecutive blocks. Aliasing shows the interference between corner frequencies, which result in loss. Critical or oversampling reduces the losses due to gaps between two consecutive spectral blocks. It creates a new signal that does not resemble the transmitted signal, but it is very similar.

2.5.5 Antialiasing

Nyquist showed that for the exact reconstruction of the signal, we must sample it at a rate of at least twice the bandwidth. To avoid aliasing, we simply filter out all the high frequency components before sampling, i.e. the antialiasing filter is a low-pass filter. To prevent aliasing, a bandlimiting filter is used to limit the frequency spectrum of the input analog voice signal before it is sampled. Most of the energy of spoken language is somewhere in between 200 or 300 Hz and about 2700 or 2800 Hz. Roughly a 3000 Hz bandwidth for standard speech and standard voice communication is established. A bandwidth of 4000 hertz is made from an equipment point of view. This band limiting filter also makes the signal spectrally efficient, as only the useful frequency components are considered removing the unwanted ones. Figure 2.12 shows the effect of antialiasing filter, which is a low-pass filter and removes the glitches in the signal that may not be carrying useful information but unnecessarily consume the spectrum.

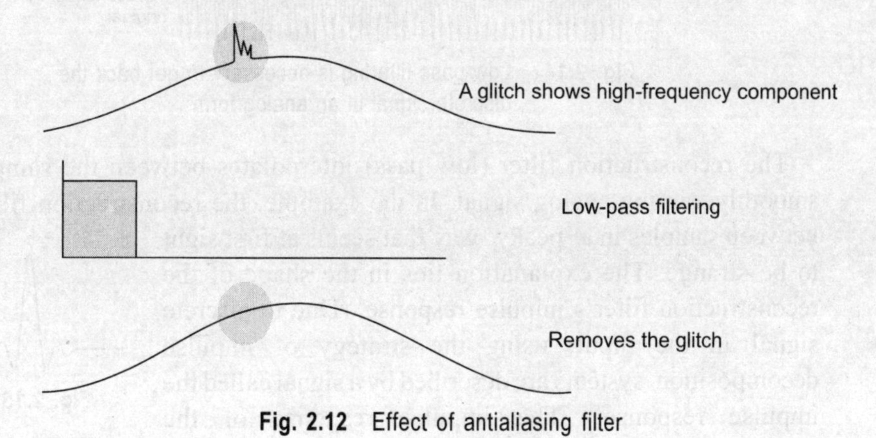

Fig. 2.12 Effect of antialiasing filter

If the highest input frequency is 4 kHz, then the sampling frequency would be set to twice this amount, or 8 kHz. This sampling frequency is the standard used by the voice-digitizing systems.

Note that antialiasing filters must be analog because once the sampling is done then it will be too late to apply it. Nyquist also showed that the sampling theorem does not say that the received signal will look like the original signal, but at least the reconstructed signal will be sufficient to convey the information with Nyquist sampling frequency. Figure 2.13 shows that for a high-frequency signal, even though the Nyquist criterion is used, the signal cannot be reconstructed fully. The reconstructed signal shows some distortion in the reconstructed signal. Figure 2.14 shows that if the sampled signal is passed through a low-pass filter, the signal can be reconstructed.

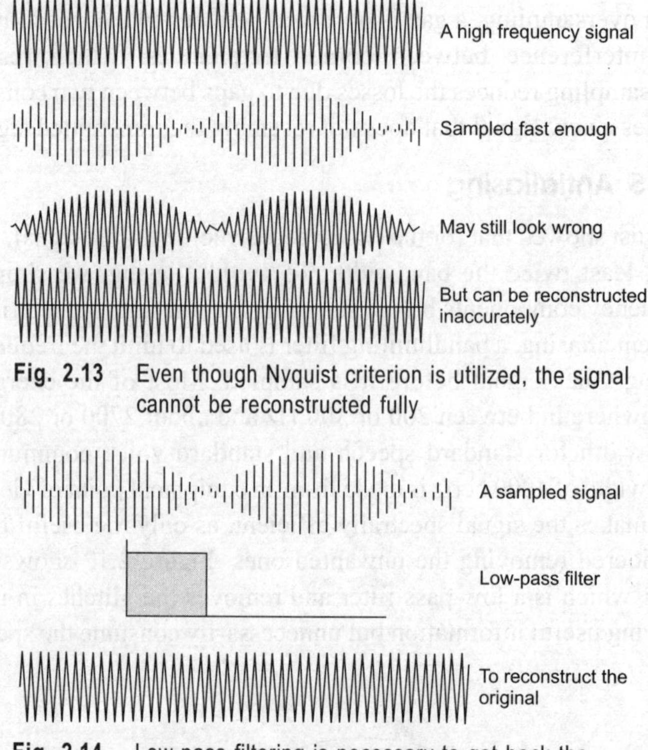

Fig. 2.13 Even though Nyquist criterion is utilized, the signal cannot be reconstructed fully

Fig. 2.14 Low-pass filtering is necessary to get back the discrete signal in an analog form

The reconstruction filter (low pass) interpolates between the samples to make a smoothly varying analog signal. In the example, the reconstruction filter interpolates between samples in a 'peaky' way that seems at first sight to be strange. The explanation lies in the shape of the reconstruction filter's impulse response. (Due to discrete signal in the input, using the strategy of impulse decomposition, systems are described by a signal called the impulse response.) The impulse response of the reconstruction filter has a classic sinc 'sin(x)/x' shape as shown in Fig. 2.15.

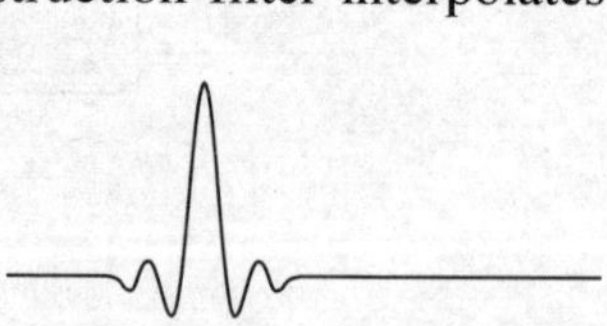

Fig. 2.15 Sinc shape

The stimulus fed to this filter is the series of discrete impulses, which are the samples. Every time an impulse hits the filter, we get 'ringing' and it is the superposition of all these peaky rings that reconstructs the proper signal. If the signal contains frequency components that are close to the Nyquist, then the reconstruction filter has to be very sharp.

2.5.6 Frequency Resolution during Sampling

We only sample the signal for a certain time. We cannot see slow changes in the signal if we do not wait long enough. In fact, we must sample for long enough to detect not only low frequencies in the signal, but also small differences between the frequencies. The length of time for which we are prepared to sample the signal determines our ability to resolve adjacent frequencies, which is called the frequency resolution. We must sample for at least one complete cycle of the lowest frequency we want to resolve.

We can see that we face a forced compromise. We must sample fast to avoid aliasing and for a long time to achieve a good frequency resolution. But sampling fast for a long time means we will have lots of samples and lots of samples means lots of computation, for which we generally do not have time. So, we will have to compromise between *resolving frequency components* of the signal and *being able to see high frequencies.*

2.5.7 Time and Frequency Domain Transformations

This is rather a starting theory. As mentioned earlier, any signal is made up of a number of frequencies, which are sine waves or tones. Jean Baptiste Fourier showed that any signal

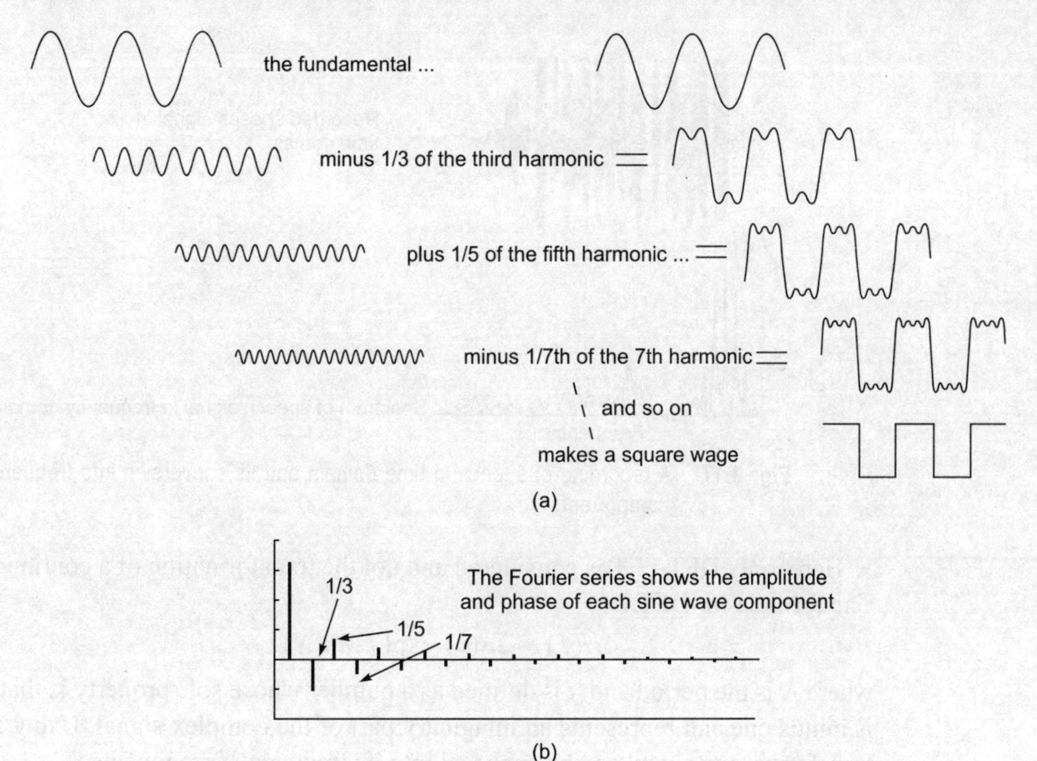

Fig. 2.16 (a) Construction of a square wave in time domain and (b) its Fourier series (frequency domain)

signal or waveform could be made up just by adding together a series of pure tones (sine waves) with appropriate amplitude and phase. Figure 2.16 shows how a square wave (periodic signal) can be made up by adding together pure sine waves at the harmonics of the fundamental frequency. This creates the concept of time and frequency domain, i.e., any signal can be mapped in any domain and converted into another domain. These concepts are very important as frequency domain contents will decide the basic signal bandwidth according to the nature of signal in time domain.

Fourier's theorem assumes that we add sine waves of infinite duration.

Let us summarize some of the transforms and interpret them. The *Fourier transform* (FT) is an equation by which a time domain signal can be mapped into the frequency domain. By inverse FT we do the reverse mapping. In Fig 2.17, one sample speech signal and its frequency domain conversion is shown. Ideally speaking, time to frequency and frequency to time conversions are bidirectional and reversible.

- The Fourier transform is a mathematical formula using integrals for continuous signal.
- *Discrete time Fourier transform* (DTFT) represents non-discretized FT of discrete time domain signal
- The *discrete Fourier transform* (DFT) is a discrete numerical equivalent using sums instead of integrals. DFT is a discretized FT of discrete time domain signal.
- The *fast Fourier transform* (FFT) is just a computationally fast way to calculate the DFT.

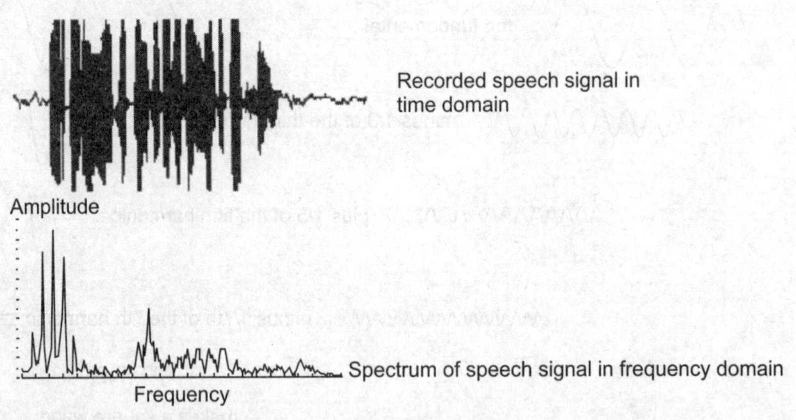

Fig. 2.17 A recording of speech in time domain and its conversion into frequency components

Here, only DFT/FFT is considered and not the transformation of a continuous signal. The formula for DFT is

$$H(f) = \Sigma s(n) \exp(j2\pi n/N) \tag{2.24}$$

where N is the period and j is defined as a number whose sole property is that its square is minus one and represents an imaginary part of the complex signal. Using all Fourier transforms, any signal can be analysed into its frequency components.

These DSP fundamentals are mainly required for OFDM system (Chapter 9).

Example 2.4 Show the parameter mapping in DFT and prove that the duality is maintained between the time and frequency domain.

Solution

When the DFT of a time signal is calculated, the frequency domain results are a function of the time sampling period ΔT_s and the number of samples N as shown in Fig. 2.18. We have $f_s = 1/\Delta T_s$. In the frequency domain, the fundamental frequency of the DFT is equal to $1/N\Delta T_s$ (1/total sample time). Each frequency represented in the DFT is an integer multiple of the fundamental frequency. The maximum frequency that can be represented by a time signal sampled at the rate $1/\Delta T_s$ is $f_{\max} = 1/2\Delta T_s$ as given by the Nyquist sampling theorem. This frequency is located in the centre of the DFT points. All frequencies beyond that point are images of the representative frequencies. The maximum frequency bin of the DFT is equal to the sampling frequency $1/\Delta T_s$ minus one fundamental $1/N\Delta T_s$. The inverse discrete Fourier transform (IDFT) performs the opposite operation to the DFT. It takes a signal defined by frequency components and converts them to a time signal. The parameter mapping is the same as that for the DFT. The time duration of the IDFT time signal is equal to the number of DFT points N times the sampling period ΔT_s. The whole conversion and relationship requirement for DFT and IDFT is shown in Fig. 2.18. Thus, we can say that duality is also maintained during the conversions.

We can move back and forth between the time domain and the frequency domain, without losing information. This statement is true mathematically, but is quite incorrect in any practical sense since we will lose information due to errors in the calculation, or due to deliberately missing out some information that we cannot measure or compute. Understanding the relation between time and frequency domains is useful because of the following factors:

- Some signals are easier to visualize in the frequency domain and some in the time domain.
- Some signals take less information to define in the time domain while some others take less information to be defined in the frequency domain.

For example, a sine wave takes a lot of information to define accurately in the time domain, but in the frequency domain, we need only three data—the frequency, amplitude, and phase.

Short Time Fourier Transform

The Fourier transform assumes that the signal is analysed over all time, an infinite duration, and, in most of the cases, it is assumed to be stationary. This means that there can be no concept of time in the frequency domain and so no concept of a frequency changing with time. Mathematically, frequency and time are orthogonal, you cannot mix one with the other. But we can easily understand that real signals (audio/video) do have frequency components that change with time. The short time Fourier transform (STFT) tries to evaluate the way the frequency content changes with time. These concepts are

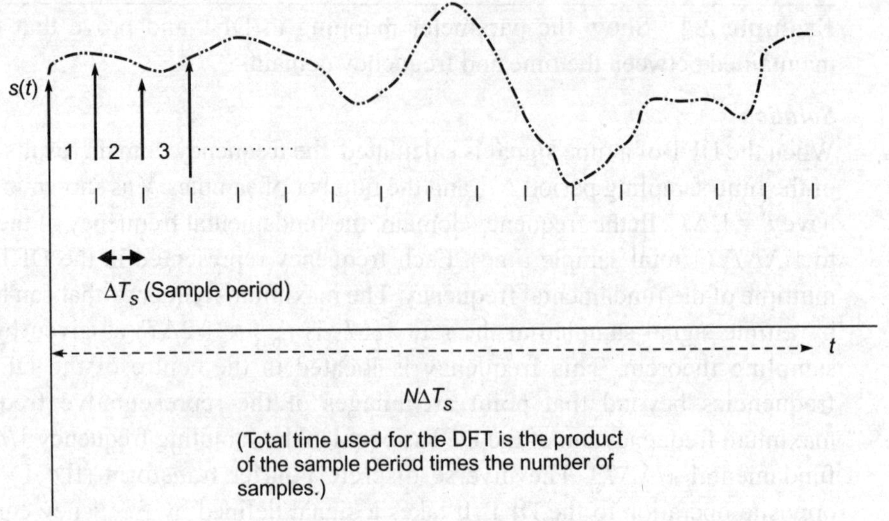

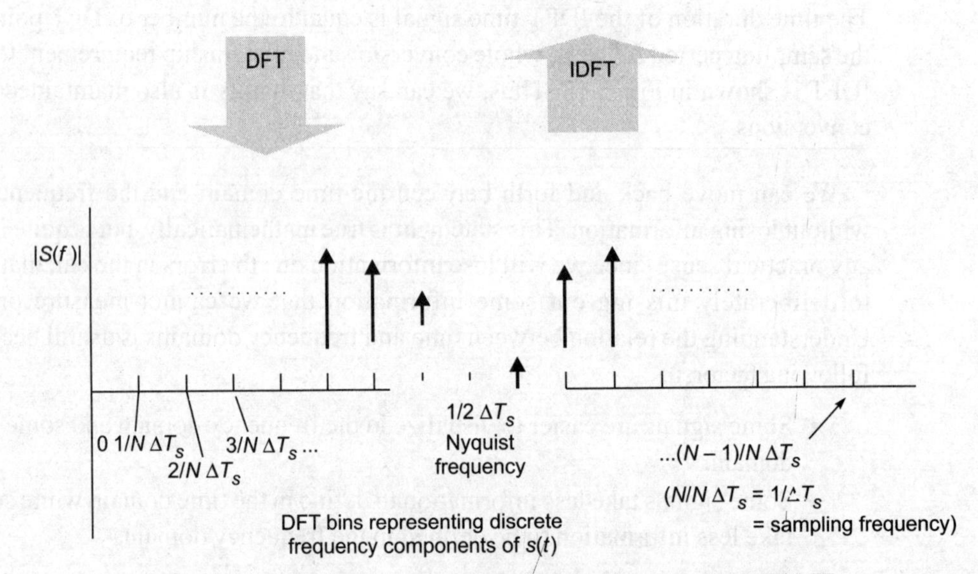

Fig. 2.18 Parameter mapping from time to frequency for DFT/IDFT

also helpful in studying sub band coding and wavelet coding. Figure 2.19 shows a time domain signal chopped into pieces of short duration signals and their independent Fourier transforms.

Each frequency spectrum shows the frequency content during a short time, and so the successive spectra show the evolution of frequency content with time. The spectra can be plotted one behind the other in a 'waterfall' diagram. It is important to realize that the STFT involves accepting a contradiction in terms, because frequency only has a meaning if we use infinitely long sine waves, so we cannot apply Fourier transforms to short pieces of a signal. So, the following constraints should be kept in mind for the frequency analysis of the short time signal.

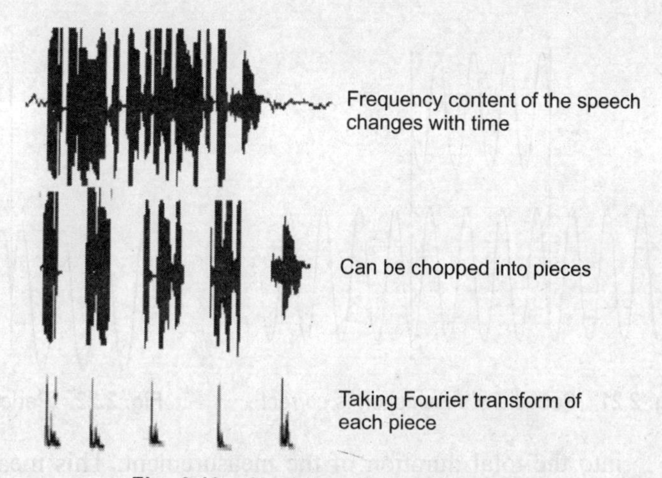

Fig. 2.19 Short time Fourier transform

1. The Fourier transform works on signals of infinite duration. But if we only measure the signal for a short time, we cannot know what happened to the signal before and after the measurement. The Fourier transform has to make an assumption about what happened to the signal before and after the measurement. Figure 2.20 shows the short signals and their FFT conversion. The Fourier transform assumes that any signal can be made by adding a series of sine waves of infinite duration. Sine waves are periodic signals. So, the Fourier transform works as if the data too is periodic for all time.

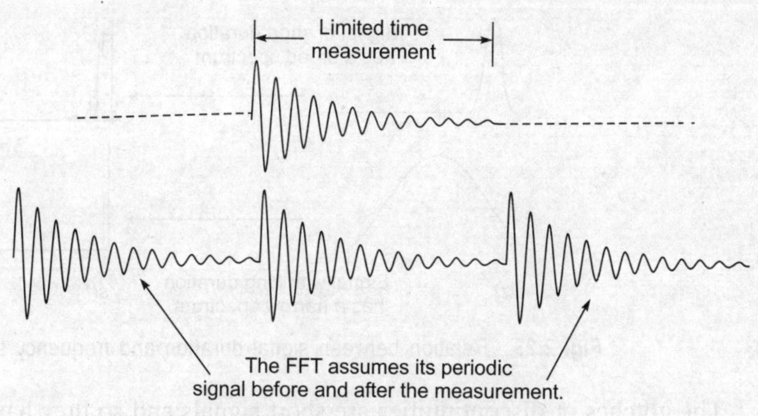

Fig. 2.20 Short signal at FFT stage

The assumption can be correct as shown in Fig. 2.21 and incorrect as shown in Fig. 2.22.

In the first case chosen, it happens that the signal is periodic and that an integral number of cycles fit into the total duration of the measurement. This means that when the Fourier transform assumes the signal repeats, the end of one signal segment connects smoothly with the beginning of the next and the assumed signal happens to be exactly the same as the actual signal. In the second case chosen, it happens that the signal is periodic but that not quite an integral number of cycles fit

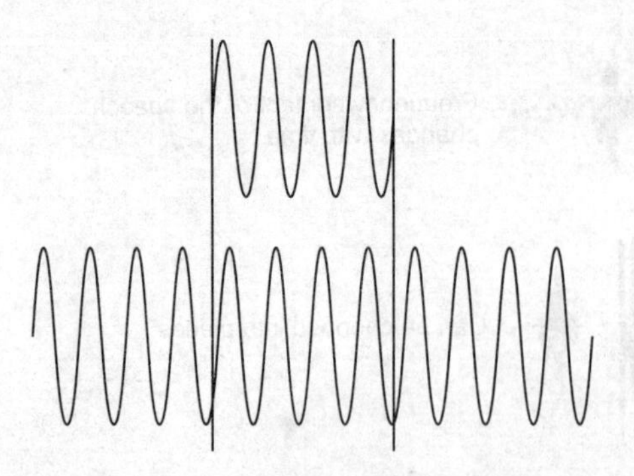

Fig. 2.21 Periodicity assumption is correct

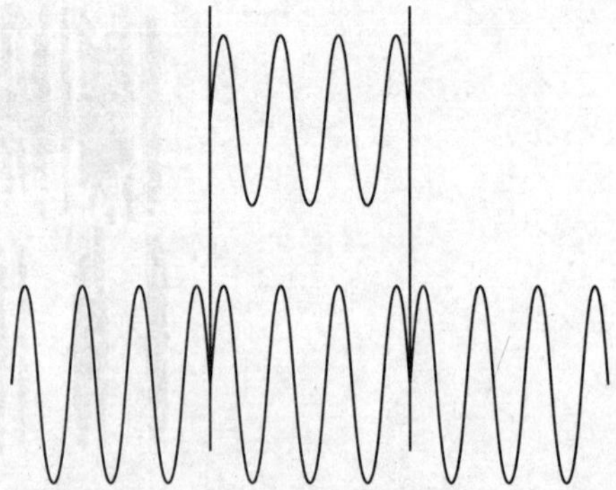

Fig. 2.22 Periodicity assumption is incorrect

into the total duration of the measurement. This means that when the Fourier transform assumes the signal repeats, the end of one signal segment does not connect smoothly with the beginning of the next the assumed signal is similar to the actual signal, but has a little 'discontinuity' at regular intervals. The assumed signal is not the same as the actual signal

2. There is a direct relation between a signal's duration in time and the width of its frequency spectrum as shown in Fig. 2.23. Let us consider the pulse durations/bit durations of the digital signal:
    - Short pulse duration signals have broad frequency spectra.
    - Long pulse duration signals have narrow frequency spectra.

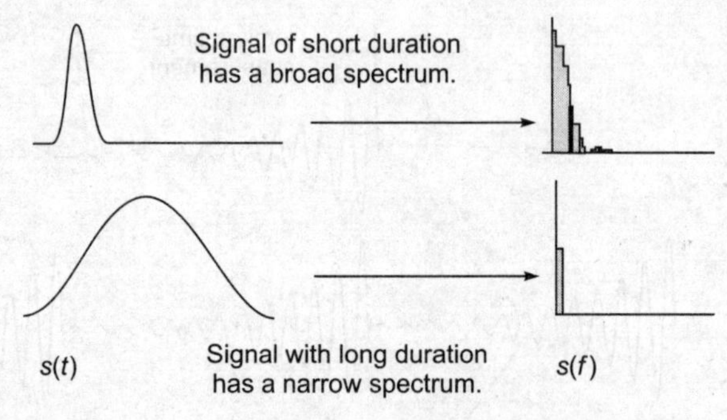

Fig. 2.23 Relation between signal duration and frequency spectrum

The glitches or discontinuities are short signals and so they have a broad frequency spectrum. This broadening is superimposed on the frequency spectrum of the actual signal.

In short, if the signal is periodic, the following cases arise:

- If an integral number of cycles fit into the total duration of the measurement, then when the Fourier transform assumes the signal repeats, the end of one signal segment connects smoothly with the beginning of the next and the assumed signal happens to be exactly the same as the actual signal.
- If an integral number of cycles do not fit into the total duration of the measurement, then when the Fourier transform assumes the signal repeats, the end of one signal

segment does not connect smoothly with the beginning of the next, the assumed signal is similar to the actual signal, but has little glitches at regular intervals.

- if the period does not match the measurement time, the frequency spectrum is incorrect and it is broadened.

This broadening of the frequency spectrum determines the frequency resolution, the ability to resolve (that is, to distinguish between) two adjacent frequency components. Only the one circumstance, where the signal is such that an integral number of cycles exactly fit into the measurement time, gives the expected frequency spectrum. In all other cases, the frequency spectrum is broadened by the discontinuities/fast disturbances at the ends. Matters are made worse because the size of these glitches depends on when the first measurement occurred in the cycle and the broadening will change if the measurement is repeated.

Wavelets

Fourier's theorem assumes that we add sine waves of infinite duration, i.e. Fourier transform fits the data with sine waves. As a consequence, the Fourier transform is good at representing signals that are long and periodic. But the Fourier transform has problems when used with signals that are short and not periodic. Other transforms are possible fitting the data with different sets of functions than sine waves. The trick is to find a transform whose base set of functions look like the signal with which we are dealing. The concept is given in Fig. 2.24. A signal that is not a long, periodic signal but rather a periodic signal with decay over a short time can have wavelet transform.

The signal shown in Fig. 2.24 is not very well matched by the Fourier transform's infinite sine waves. But it might be better matched by a different set of functions, say decaying sine waves. Such functions are called 'wavelets' and can be used in the 'wavelet transform'. Note that the wavelet transform cannot really be used to measure frequency, because frequency has meaning only when applied to infinite sine waves. But, as with the short time Fourier transform, we are always willing to find out the useful features by studying it in detail.

Short duration non-sinusoidal wave whose frequency content changes with time

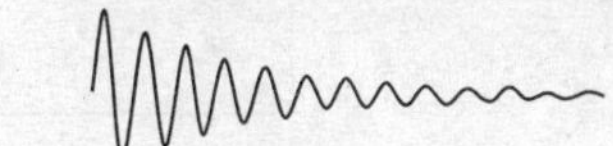

Fig. 2.24 An example of signal for wavelet transform

2.5.8 Correlation

Correlation is explained here as a time-domain process. It is a weighted moving average and can be defined for the continuous time as well as discrete time signals. Following is the mathematical form of representing the correlation for the discrete time signal. For continuous time signals, summation symbol will be replaced by integration.

$$r(n) = \Sigma x(k)\ y(n + k) \tag{2.25}$$

where $x(k)$ and $y(k)$ are two different discrete signals. The depth of correlation is represented by value of correlation function $r(n)$. One signal provides the weighting function. Correlation is required mainly at the receiver side where signal matching is to be done for acquisition of the signal, especially for CDMA system. Matched correlators are used there (Chapter 8) in the rake receiver. Figures 2.25 and 2.26 show how the correlation process is applied.

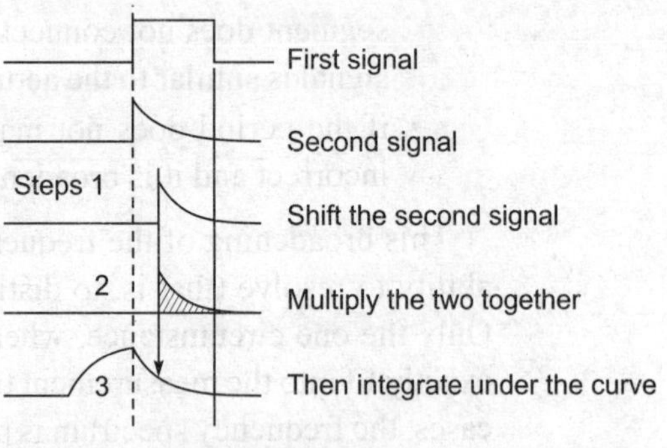

First signal
Second signal
Steps 1
Shift the second signal
2
Multiply the two together
3
Then integrate under the curve

Fig. 2.25 Steps of correlation function calculation

The correlation function calculation is done as follows:

- First, one signal is shifted with respect to the other.
- The amount of the shift is the position of the correlation function point to be calculated.
- Each element of one signal is multiplied by the corresponding element of the other.
- The area under the resulting curve is integrated.

Correlation requires a lot of calculations. If one signal is of length M and the other is of length N, then we need $(N \times M)$ multiplications, to calculate the whole correlation function. Correlation is a maximum when two signals are similar in shape and are in phase (or 'unshifted' with respect to each other). Correlation is a measure of the similarity between two signals as a function of time shift between them.

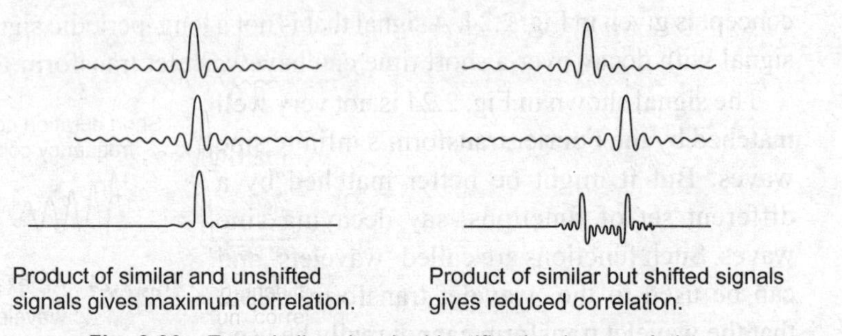

Fig. 2.26 Two similar signals: identification by correlation

When the two signals are similar in shape and unshifted with respect to each other, their product is all positive and the correlation function is largest. This is like constructive interference, where the peaks add and the troughs subtract to emphasize each other. The area under this curve gives the value of the correlation function at point zero, and this is a large value. As one signal is shifted with respect to the other, the signals go out of phase and the peaks no longer coincide. So, the product can have negative-going parts. This is a bit like destructive interference, where the troughs cancel the peaks. The area under this curve gives the value of the correlation function at the value of the shift. The negative-going parts of the curve now cancel some of the positive-going parts and, therefore, the correlation function is smaller.

The breadth of the correlation function, where it has significant value shows for how long the signals remain similar. Correlating a signal with itself is called *autocorrelation*. Correlation with another signal is called *cross correlation*.

Autocorrelation

Autocorrelation is defined as the degree of correspondence between the core (data signal) and the phase-shifted replica of itself. The digital realization of the autocorrelation requires some means to perform the 'multiplication and shift' of the samples of the incoming signal with the stored reference signal. Time-domain processing of autocorrelation is shown here. Different sorts of signals shown in Fig. 2.27 have distinctly different autocorrelation functions. We can use these differences to differentiate between them.

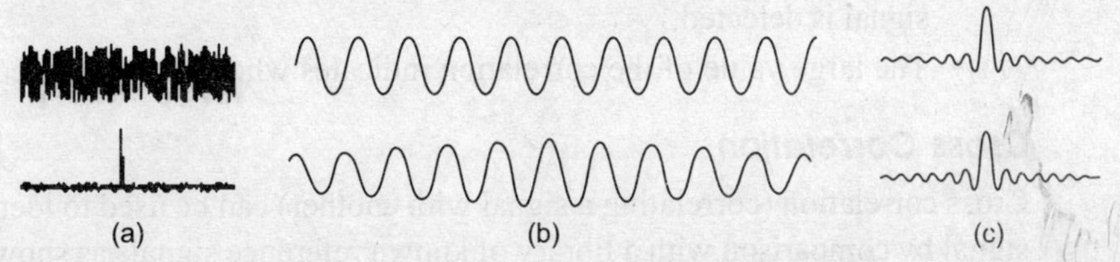

Fig. 2.27 Three different types of signals and their corresponding autocorrelation functions

The following points may be noted:

(a) Random noise is defined to be uncorrelated, which means that it is only similar to itself with no shift at all. Even a shift of one sample either way means that there is no correlation at all and so the correlation function of random noise with itself is a single sharp spike at shift zero.

(b) Periodic signals go in and out of phase as one is shifted with respect to the other. So, they will show strong correlation at any shift where the peaks coincide. The autocorrelation function of a periodic signal is itself a periodic signal, with a period the same as that of the original signal.

(c) Short signals can only be similar to themselves for small values of shift and so their autocorrelation functions are short.

The three types of signals have easily recognizable autocorrelation functions.

Autocorrelation (correlating a signal with itself) can be used to extract a signal from noise. A signal can be extracted from the noise as follows (see Fig. 2.28):

- Random noise has a distinctive 'spike' autocorrelation function.
- A sine wave has a periodic autocorrelation function.
- So, the autocorrelation function of a noisy sine wave is a periodic function with a single spike containing all the noise power.

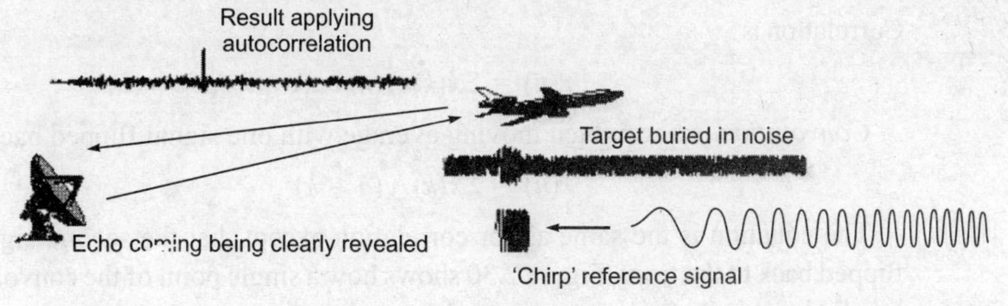

Fig. 2.28 The signal can be located within the noise due to autocorrelation

The separation of signal from noise using autocorrelation works because the autocorrelation function of the noise is easily distinguished from that of the signal. Autocorrelation (correlating a signal with its known replica– it is a special case of cross correlation also) can be used to detect and locate a known reference signal in noise.

If a copy of the known reference signal is correlated with the unknown signal, then

- The correlation will be high when the reference is similar to the unknown signal.
- A large value for correlation shows the degree of confidence that the reference signal is detected.
- The large value of the correlation indicates when the reference signal occurs.

Cross Correlation

Cross correlation (correlating a signal with another) can be used to identify an unknown signal by comparison with a library of known reference signals as shown in Fig. 2.29.

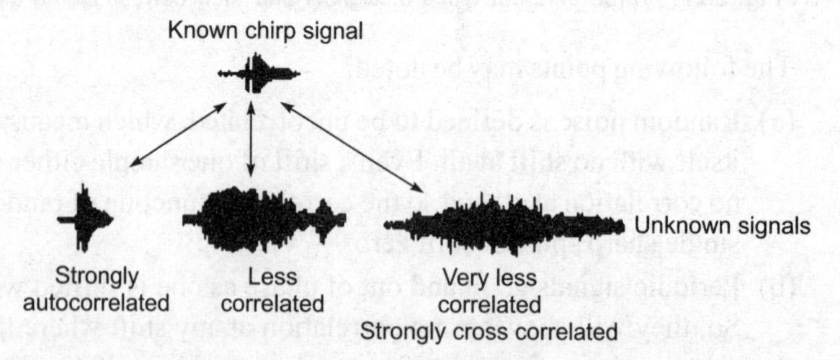

Fig. 2.29 The unknown signal can be identified using cross correlation

If a copy of a known reference signal is correlated with the unknown signals, then

- The cross correlation will be minimum if the reference is similar to the unknown signal.
- The unknown signal is correlated with a number of known reference functions.
- A small value for cross correlation shows the degree of similarity to the reference.
- The largest value for autocorrelation is the most likely match.
- Autocorrelation will be high with required signal while cross correlation will be high with undesired different signals.

2.5.9 Convolution

Correlation is

$$r(n) = \Sigma x(k)\ y(n + k) \tag{2.26}$$

Convolution is a weighted moving average with one signal flipped back to front:

$$r(n) = \Sigma x(k)\ y(n - k) \tag{2.27}$$

The equation is the same as for correlation except that the second signal $y(n - k)$ is flipped back to the front. Figure 2.30 shows how a single point of the convolution function is calculated.

The convolution process can be visualized as follows:

- First, one signal is flipped back to front.
- Then, one signal is shifted with respect to the other.
- The amount of the shift is the position of the convolution function point to be calculated.
- Each element of one signal is multiplied by the corresponding element of the other.
- The area under the resulting curve is integrated.

Convolution is used to find the output of the system for a given input. Convolution requires a lot of calculations. If one signal is of length M and the other is of length N, then we need $(N \times M)$ multiplications, to calculate the whole convolution function.

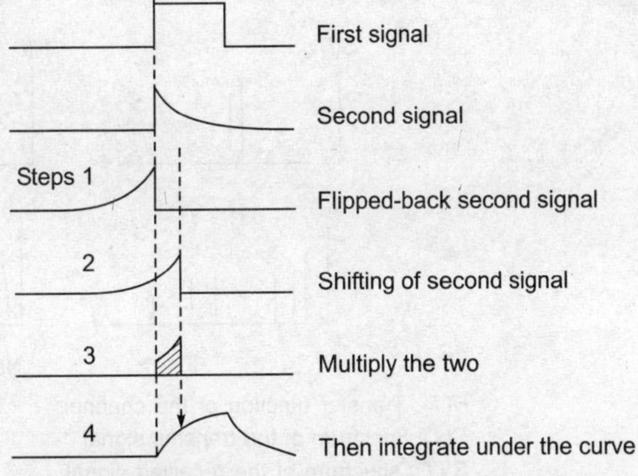

Fig. 2.30 Convolution process

Convolving two time-domain signals is equivalent to multiplying the frequency spectra of those two signals. Correlation is equivalent to multiplying the *complex conjugate* of the frequency spectrum of one signal by the frequency spectrum of the same or another signal. This complex conjugation is not so easily understood. The concept is used in channel estimation.

Convolving a signal with a smooth weighting function can be used to smooth a signal. Figure 2.31 shows how a noisy sine wave can be smoothed by convolving with a rectangular smoothing function—this is just a moving average. The smoothing property leads to the use of convolution for digital filtering. Sending the signal over a channel is a convolution process and by the output of this process we can say that it is a narrowband or wideband system. This process output in terms of channel impulse response is shown in Fig. 2.32. It also shows the time and frequency domain relationships due to convolution.

Wideband is a word with many meanings. One way to define the system is that the modulation bandwidth is considerably larger than the bandwidth of the modulating signal just like FM or SSM. In RF engineering, it describes the relative bandwidth of a certain quantity (e.g., antenna bandwidth) compared to the carrier frequency—if the ratio is not much smaller than unity, the system is called wideband. In digital mobile radio, 'wideband'

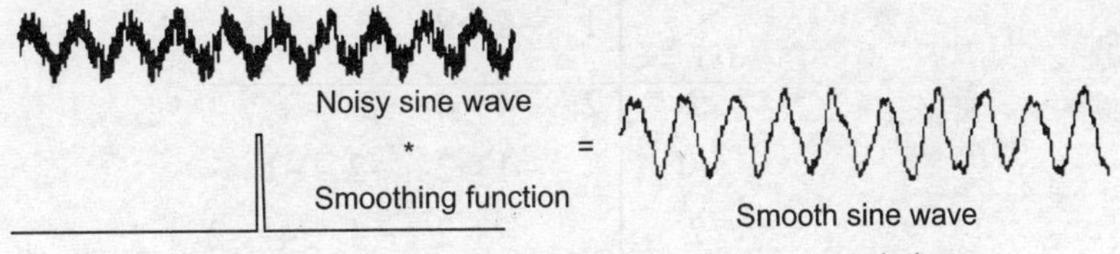

Fig. 2.31 Smoothening of the signal using convolution

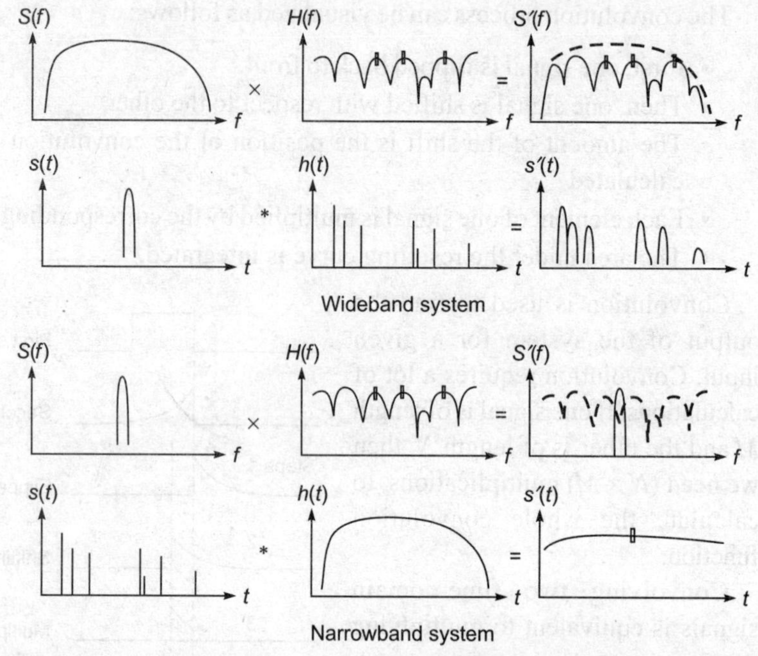

$H(f)$: Transfer function of the channel
$S(f)$: Spectrum of the transmit signal
$S'(f)$: spectrum of the received signal
$h(t)$: Impulse response of the channel
$s(t)$: Waveform for transmission
$s'(t)$: Received time domain signal

Fig. 2.32 (a) Wideband and (b) narrowband system frequency and time domain responses (The symbol * shows convolution process and × shows multiplication process.)

has even more complicated implications: it compares certain properties of the system with properties of the propagation environment.

Example 2.5 If the channel impulse response coefficients are given as $h(n) = [1, -2, 2, 1]$ (refer to Chapter 6 for channel modelling). Let us consider a small period discrete input signal $x(n) = [2, -1, 2]$. If the signal is transmitted through the channel, find out the output signal $y(n)$.

Solution

The signal passing through a channel represents convolution process. Hence

$$y(n) = x(n) * h(n)$$

Now, using multiplication and shift method,

$x(n)$ \ $h(n)$	1	−2	2	1		
2	2	−4	4	2		
−1		−1	2	−2	−1	
2			2	−4	4	2
$y(n)$	2	−5	8	−4	3	2

If M = number of channel coefficients

N = samples in period

then maximum length of $y(n) = M + N - 1 = 4 + 3 - 1 = 6$

(This result is useful during simulation when channel coefficients represent the channel vector, $h(n) = [h_0, h_1, h_2, \ldots, h_k]^T$, where $h_0, \ldots, h_k$ are the channel coefficients.)

Here

$$y(n) = [2, -5, 8, -4, 3, 2]$$

2.5.10 Convolution vs Correlation

Figure 2.33 shows the similarity between convolution and correlation, comparing Figures 2.25 and 2.30.

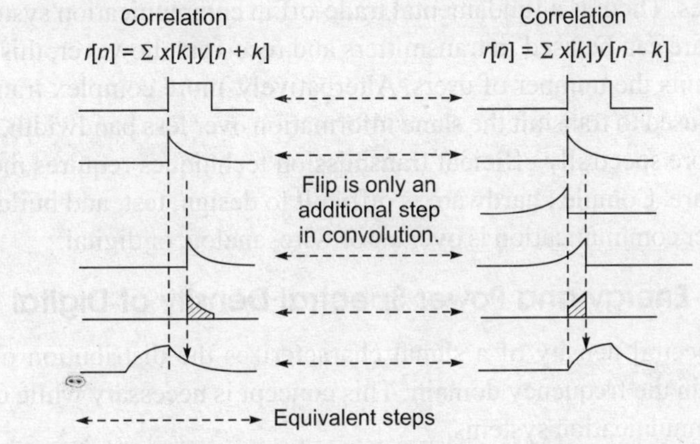

Fig. 2.33 Similarity and difference between convolution and correlation

The preference of convolution over correlation for filtering has to do with how the frequency spectra of the two signals interact. If one (flipping) signal is symmetric, convolution and correlation are identical and flipping that signal back to front does not change it. So, convolution and correlation are the same in this case.

Convolution by multiplying frequency spectra can take advantage of the fast Fourier transform, which is a computationally efficient algorithm. So, convolution in frequency domain can be faster than convolution in the time domain and is called *fast convolution*.

2.6 FUNDAMENTALS OF DIGITAL COMMUNICATION OVER THE CHANNEL

While transmitting a digital signal over the wireless channel, one must process the signal in an appropriate way so that reliable reception of the signal can be achieved. During processing, many DSP fundamentals are applied. Without DSP, digital transmission over the channel is not possible. Hence, some more important concepts are highlighted in brief in the following sections. Modulation is one very important signal processing stage.

2.6.1 Factors that Influence Digital Modulation Selection

Digital communication is possible due to digital modulation schemes (described in Chapter 7). The move to digital modulation provides more information capacity, compatibility with digital data services, better quality communications, and quicker system availability. Developers of communication systems face the following constraints while choosing the modulation scheme:

- Available bandwidth
- Permissible power
- Inherent noise level of the system

The RF spectrum must be shared, yet every day there are more users for that spectrum, as demand for communications services increases. Digital modulation schemes have a greater capacity to convey large amounts of information than the analog modulation schemes. There is a fundamental trade-off in communication systems in general. Simple hardware can be used in transmitters and receivers; however, this uses a lot of spectrum that limits the number of users. Alternatively, more complex transmitters and receivers can be used to transmit the same information over less bandwidth. The transition to more and more spectrally efficient transmission techniques requires more and more complex hardware. Complex hardware is difficult to design, test, and build. This trade-off exists, whether communication is over air or wire, analog, or digital.

2.6.2 Energy and Power Spectral Density of Digital Signals

The spectral density of a signal characterizes the distribution of the signal energy or power in the frequency domain. This concept is necessary while considering filtering in the communication systems.

The total energy of a real-valued energy signal $s(t)$, defined over interval $(-\infty, \infty)$ and can be represented using Parseval's theorem in time as well as frequency domain as follows. Here $s(t)$ is a continuous digital signal in bit form.

$$E_s = \int_{\infty}^{-\infty} s^2(t)dt = \int_{\infty}^{-\infty} |S(f)|^2 df \tag{2.28}$$

where $s(t)$ and $S(f)$ are time and frequency domain representations of the same signal.

Here $|S(f)|^2$ is the waveform energy spectral density (ESD) of the signal $s(t)$ and its integration gives total energy with respect to frequency. The equation states that the energy of the signal is equal to the area under $|S(f)|^2$ versus frequency curve. ESD describes the signal energy per unit bandwidth measured in joules/Hz. There are equal energy contributions from positive as well as negative frequency components.

The average power P_s of a real-valued power signal $s(t)$ is defined in Eq. (2.14). If $s(t)$ is a periodic signal with period T_0, it is classified as a power signal. The expression for the average power of a periodic signal takes the form of Eq. (2.12), where the time average is taken over period T.

Parseval's theorem for real-valued periodic signal takes the form

$$P_s = \frac{1}{T}\int_{-T/2}^{T/2} s^2(t)dt = \sum_{n=-\infty}^{\infty} |c_n|^2 \quad (2.29)$$

where the $|c_n|$ terms are the complex Fourier series coefficients of the periodic signal and hence their values are necessary to be known for the *power spectral density* (PSD). The PSD function $G_s(f)$ of a periodic signal $s(t)$ is a real, even, and non-negative function of frequency that gives the distribution of the power of $s(t)$ in the frequency domain, as follows:

$$G_s(f) = \sum_{n=-\infty}^{\infty} |c_n|^2\ \delta(f - nf_0) \quad (2.30)$$

Equation (2.30) shows the power spectral density of the periodic signal $s(t)$ as a succession of the weighted delta functions with f_0 being the fundamental frequrency. Therefore, the PSD of a periodic signal is a discrete function of frequency. Average normalized power of a real-valued signal can be written as

$$P_s = \int_{-\infty}^{\infty} G_s(f)df = 2\int_{0}^{\infty} G_s(f)df \quad (2.31)$$

Non-periodic signal cannot be represented in the Fourier series; hence PSD can be defined in limiting sense as a truncated version.

2.6.3 Line Coding or Signalling

Digital baseband signal is often converted from the binary form of signal to another form. Sometimes, the extra pulses or transitions may be added in this conversion. Hence, it is considered as coding–line coding technique. It provides particular spectral characteristics of a pulse train. The most common codes for mobile communications are RZ, NRZ, and Manchester encoding. All of these may either be unipolar (with voltage levels 0 or V) or bipolar (with voltage levels +V or –V). The RZ code implies that pulse returns to zero within every bit period. This widens the spectrum but improves the timing synchronization. The NRZ codes, on the other hand, do not return to zero during a bit period, the signal stays at constant level throughout a bit interval. The NRZ codes are more spectrally efficient than the RZ codes but are poor in synchronization capabilities. In unipolar signals, a DC component is present, which may be blocked in some part of the hardware, so normally bipolar NRZ is preferred. The Manchester code is NRZ with special features, maybe without a DC level and ideally suitable for signalling purpose. It has synchronization capabilities. It offers at least one transition per bit interval; hence Manchester codes provide two pulses to represent each binary symbol. Easy clock recovery is possible with Manchester coding. Figure 2.34 shows examples of waveforms of these line codes. There are other forms of line codes also, such as differential Manchester encoding, multi-level coding, etc.

Example 2.6 For the binary pattern 01001110, draw the waveforms of NRZ, RZ, and Manchester coding.

Solution

The waveforms are shown in Fig. 2.34.

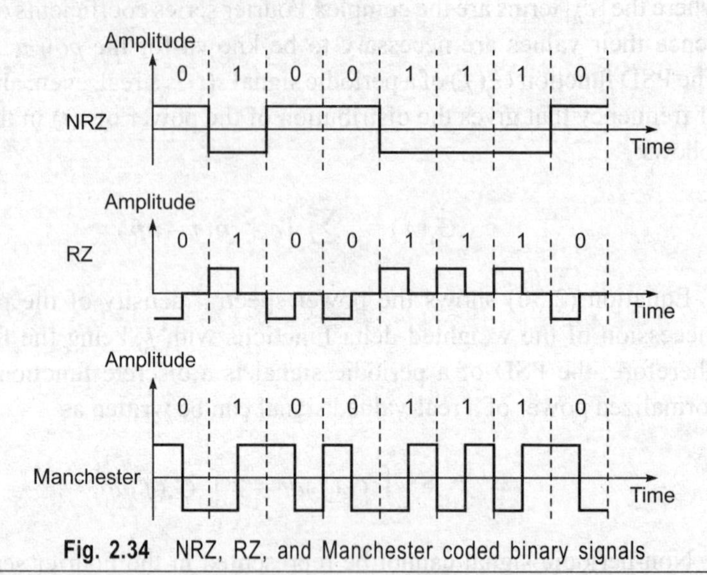

Fig. 2.34 NRZ, RZ, and Manchester coded binary signals

2.6.4 Nyquist Criteria for Zero ISI

Nyquist showed that the theoretical minimum system bandwidth needed in order to detect R_s symbols per second without ISI is $R_s/2$ hertz. The sinc (t/T) shaped pulse is called the *ideal Nyquist pulse*. Its multiple lobes comprise a main lobe and side lobes that are infinitely long. Nyquist established that if each pulse of a received sequence is of the form sinc (t/T), the pulses can be detected without ISI. Figure 2.35 illustrates how ISI is avoided. There are two successive pulses $s(t)$ and $s(t - T)$. Even though $s(t)$ has long tails, the figure shows a tail passing through zero amplitude at the instant $t = T$ when $s(t - T)$ is to be sampled and likewise all tails pass through zero amplitude when any other pulse of the sequence is to be sampled. They satisfy the orthogonality condition.

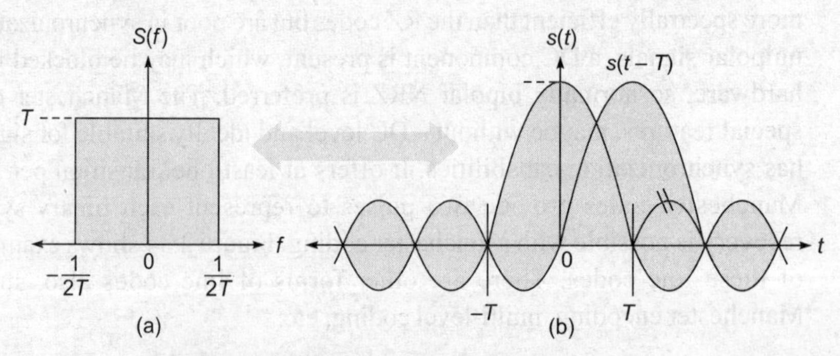

Fig. 2.35 Nyquist pulse detection without ISI

The more compact we make the signalling spectrum, the higher is the allowable data rate or the greater is the number of users that can simultaneously be served. Nyquist has imposed the basic limitation on such bandwidth reduction. If we operate the system at smaller bandwidths, then according to the Nyquist condition, the pulse would become spread in time that would degrade the system BER performance due to increased ISI. A prudent goal is to compress the bandwidth of the data impulses to some reasonably small bandwidth greater than the Nyquist minima. This is accomplished with a Nyquist filter. Without such measure, each pulse extends into next pulse in the entire sequence. Long-time response exhibits large amplitude tails nearest to the main lobe of each pulse. Such tails are undesirable because they contribute zero ISI only when the sampling is performed at exactly the correct sampling time, when the tails are large, small timing errors will result in ISI. Therefore, although a compact spectrum provides optimum bandwidth utilization, it is very susceptible to ISI degradation induced by timing errors. The ISI effect is shown in Fig. 2.36.

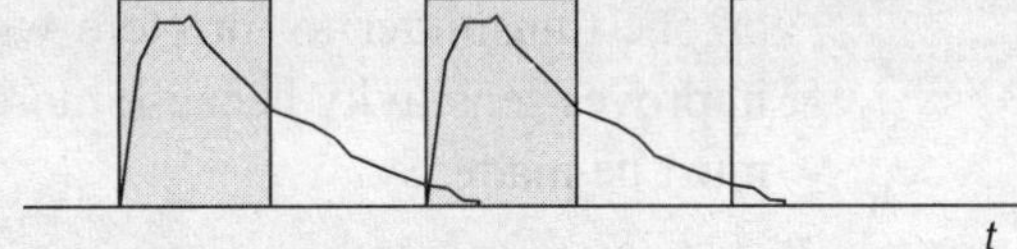

Fig. 2.36 Intersymbol interference between pulses or symbols

The terms *Nyquist filter* and *Nyquist pulse* are often used to describe the general class of filtering and pulse shaping that satisfies zero ISI at the sampling points. The basic transversal filter can be used for pulse shaping for zero ISI; it is explained in Chapter 6. Amongst the class of Nyquist filters, the most popular ones are the raised cosine filter and square root raised cosine filter. Others are Gaussian filter and Chebyshev filter.

2.6.5 Filtering (Pulse Shaping)

There are various filters (and reactive circuit elements such as inductors and capacitors) throughout the system—in the transmitter, channel, and receiver. Some band-pass systems, such as wireless systems, are characterized by fading channels that behave like undesirable filters manifesting signal distortion. When the receiving filter is configured to compensate for the distortion caused by both the transmitter and the channel, it is often referred to as an equalizing filter or a receiving/equalizing filter. Lumping all these filtering effects into one overall equivalent system transfer function,

$$H(f) = H_t(f)\, H_c(f)\, H_r(f) \tag{2.32}$$

where $H_t(f)$ characterizes the transmitting filter shaping the pulse, $H_c(f)$ is the filtering within the channel, and $H_r(f)$ is the receiving/equalizing filter.

Due to the effects of system filtering, the received pulses can overlap one another as shown in Fig. 2.37. The tail of the pulse can 'smear' into adjacent symbol intervals,

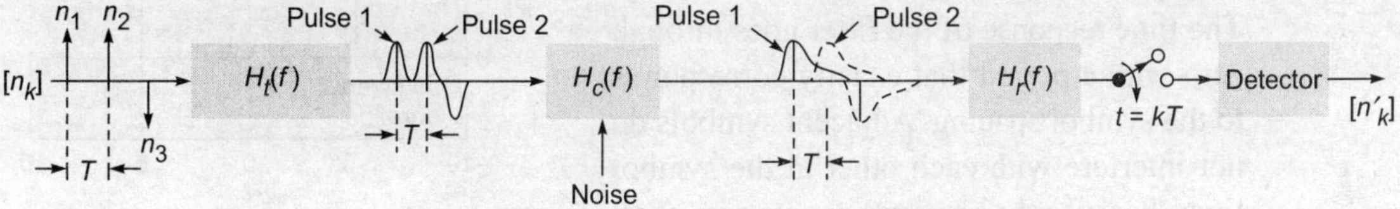

Fig. 2.37 Pulse transmission over a channel

thereby interfering with the detection process and degrading the error performance. Such interference is termed as intersymbol interference as mentioned earlier. Even in the absence of noise, the effects of filtering and channel-induced distortion leads to ISI. Wireless channel problems are better highlighted in Chapter 5. Sometimes, $H_c(f)$ is specified, and the problem remains to determine $H_t(f)$ and $H_r(f)$ such that ISI is minimized at the output of $H_r(f)$. Equalization techniques are described in Chapter 6.

In short, filtering or pulse shaping [determining $H_t(f)$] allows the transmitted bandwidth to be significantly reduced without losing the content of the digital data. This improves the spectral efficiency of the transmission.

Any fast transition in a signal, whether it may be amplitude, phase, or frequency, will require a wide occupied bandwidth. Any technique that helps to slow down these transitions will narrow the occupied bandwidth. Filtering serves to smooth these transitions. Filtering reduces interference because it reduces the tendency of one signal or one transmitter to interfere with another. On the receiver end, reduced bandwidth improves sensitivity because more noise and interference are rejected. Some trade-offs must be made.

1. Some types of filtering cause the trajectory of the signal (the path of transitions between the states) to overshoot in many cases. This overshoot can occur in certain types of filters, such as Nyquist. This overshoot path represents carrier power and phase. For the carrier to take on these values, it requires more output power from the transmitter amplifiers. It requires more power than would be necessary to transmit the actual symbol itself. Carrier power cannot be clipped or limited (to reduce or eliminate the overshoot) without causing the spectrum to spread out again. Since narrowing the spectral occupancy was the reason the filtering was inserted in the first place, it becomes a fine balancing act.
2. Filtering makes the hardware more complex and can make them larger, especially if performed in an analog fashion. Filtering can also create intersymbol interference. This occurs when the signal is filtered enough so that the symbols blur together and each symbol affects those around it. This is determined by the time-domain response or impulse response of the filter.

Pulse Shaping Using a Raised Cosine Filter

Figure 2.38 shows the impulse or time-domain response of a raised cosine filter, one class of Nyquist filter. Nyquist filters have the property that their impulse response rings at the symbol rate. The filter is chosen to ring or have the impulse response of the filter cross through zero, at the symbol clock frequency. The time response of the filter goes through zero with a period that exactly corresponds to the symbol spacing. Adjacent symbols do not interfere with each other at the symbol times because the response equals zero at all symbol times except the centre (desired)

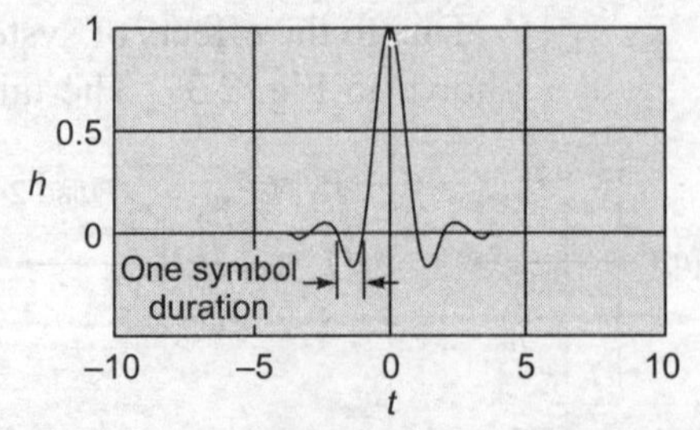

Fig. 2.38 Time domain response of a raised cosine filter

one. Nyquist filters heavily filter the signal without blurring the symbols together at the symbol times. This is important for transmitting information without errors caused by intersymbol interference. Note that the intersymbol interference does exist at all times except the symbol (decision) times.

The sharpness of a raised cosine filter is described by the roll-off factor alpha (α), which decides the slope of the low-pass filter cut-off region. Figure 2.39 gives the response of the raised cosine filter with different α values. The value of α gives a direct measure of the occupied bandwidth of the system and is calculated as

$$\text{Total bandwidth } W = \text{symbol rate} \times (1 + \alpha) \tag{2.33}$$

If the filter had a perfect (ideal) characteristic with sharp transitions/cut off and $\alpha = 0$, the occupied bandwidth for $\alpha = 0$ would be

$$\text{Total bandwidth } W = \text{symbol rate} \times (1 + 0) = \text{symbol rate} \tag{2.34}$$

Figure 2.39 gives the roll-off characteristics of raised cosine filter. With the change in α, the sharpness of the cut-off region or the bandwidth also changes.

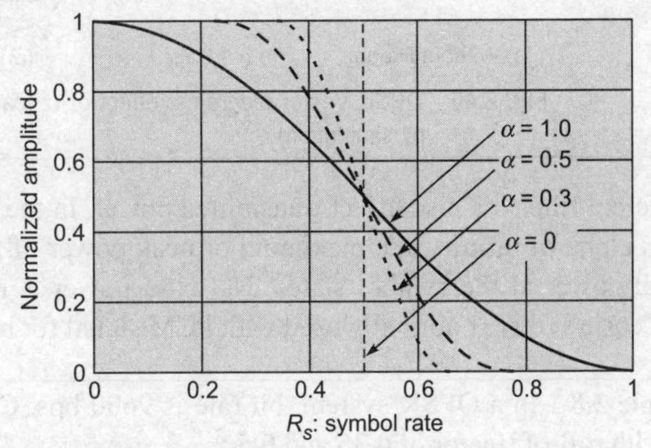

Fig. 2.39 Raised cosine filter response for different α

Ideally, with sharp cut off region the occupied bandwidth would be the same as the symbol rate, but this is not practical. An alpha of zero is impossible to implement. Alpha is sometimes called the 'excess bandwidth factor' as it indicates the amount of occupied bandwidth that will be required in excess of the ideal occupied bandwidth (which would be the same as the symbol rate). At the other extreme, take a broader filter with an alpha of one, which is easier to implement. The occupied bandwidth for $\alpha = 1$ will be

$$\text{Total bandwidth } W = \text{symbol rate} \times (1 + 1) = 2 \times \text{symbol rate} \tag{2.35}$$

Here $\alpha = 1$ uses twice as much bandwidth as for $\alpha = 0$. In practice, it is possible to implement an alpha below 0.2 and make good, compact, and practical radios. Typical values range from 0.35 to 0.5, though some video systems use an alpha as low as 0.11.

Example 2.7 Show that the different filter bandwidths (change in α) show different effects on the vector diagram of modulated signal.

Solution

If the radio has no transmitter filter, as shown in Fig. 2.40(a), the transitions between states are instantaneous. No filtering means $\alpha = \infty$ hypothetically. Transmitting this signal would require infinite bandwidth. Figure 2.40(b) is an example of a signal at $\alpha = 0.7$. Figure 2.40(c) shows the signal at $\alpha = 0.3$. The filters with $\alpha = 0.7$ and 0.3 smooth the transitions and narrow the frequency spectrum required.

Figure 2.40 illustrates the effect of α on the vector diagram. The variation from the theoretical vector diagram is observed due to change in roll-off factor. The modulation scheme will be clear after studying Chapter 7.

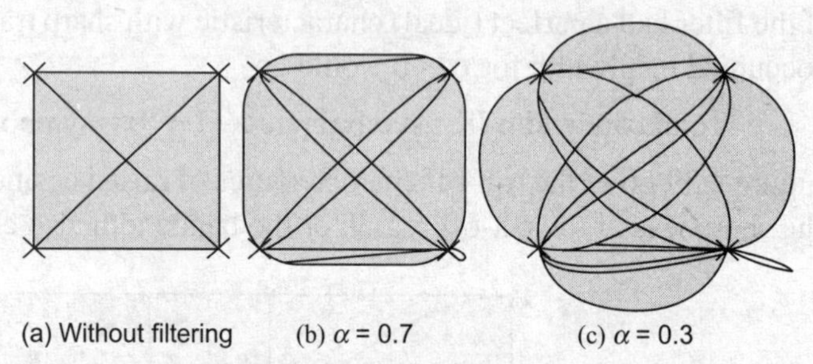

(a) Without filtering (b) $\alpha = 0.7$ (c) $\alpha = 0.3$

Fig. 2.40 QPSK Vector diagrams: effect of α (can be observed by simulation)

Different filter α's also affect transmitted power. In the case of the unfiltered signal, with an alpha of infinity, the maximum or peak power of the carrier is the same as the nominal power at the symbol states. No extra power is required due to the filtering. Raised cosine filter is normally used with PCM signal for baseband transmissions.

Example 2.8 In a QPSK system, bit rate is 9600 bps. Check the bandwidth requirement with roll-off factor of 0.35 and 0.5.

Solution

Total bandwidth W = symbol rate $\times (1 + \alpha)$

For QPSK scheme, symbol rate = 4800 bps

Thus, for $\alpha = 0.35$: W = 4800 symbols/sec $(1 + 0.35)$ = 6480 sps

$\Rightarrow$ occupied bandwidth = 6480 Hz

For $\alpha = 0.5$: W = 4800 symbols/sec $(1 + 0.5)$ = 7200 sps

$\Rightarrow$ occupied bandwidth = 7200 Hz

Square Root Raised Cosine Filter

Usually the filter is split, half being in the transmit path and half in the receiver path. In this case, root Nyquist filters (commonly called root raised cosine) are used in each part, so that their combined response is that of a Nyquist filter. Figure 2.41 shows the block diagram of the square root raised cosine filter.

Sometimes, filtering is desired at both the transmitter and receiver. Filtering in the transmitter reduces the adjacent channel power radiation, while filtering at the receiver reduces the effects of broadband noise and also interference from other transmitters in

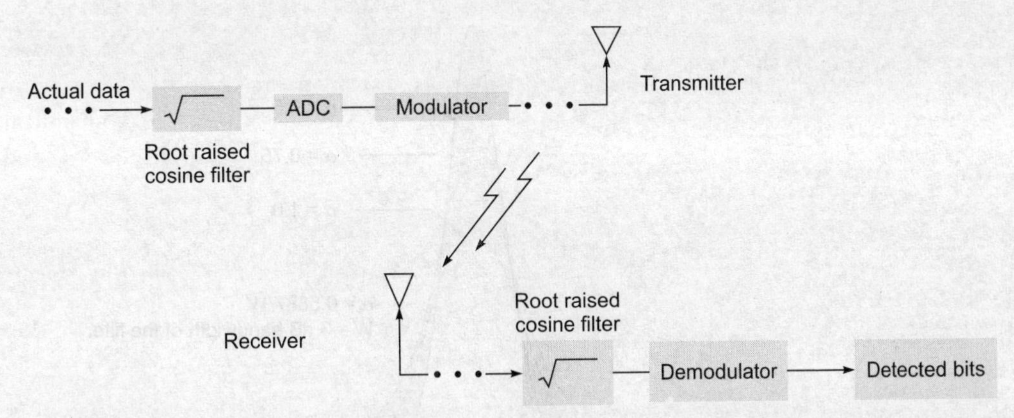

Fig. 2.41 Square root raised cosine filter shown as premodulation filtering and predemodulation filtering

nearby channels. To get zero ISI, both filters are designed until the combined result of the filters and the rest of the system is a full Nyquist filter. Potential differences can cause problems in manufacturing because the transmitter and receiver are often manufactured by different companies. The receiver may be a small hand-held model and the transmitter may be a large cellular base station. If the design is performed correctly, the results are the best data rate, the most efficient radio, and reduced effects of interference and noise. This is why root Nyquist filters are used in receivers and transmitters as

root Nyquist filter × root Nyquist filter = Nyquist filter

Gaussian Pulse Shaping Filter

Gaussian filter is normally used in GMSK and GFSK modulation schemes (Chapter 7). In these schemes, MSK or FSK will be performed after passing the signal through Gaussian filter. The GMSK modulation method is used in GSM (Chapter 13). The GSM signal will have a small blurring of symbols on each of the four states because the Gaussian filter used in GSM does not have zero inter-symbol interference. The phase states vary somewhat causing a blurring of the symbols. Wireless system architects must decide just how much of the inter-symbol interference can be tolerated in a system and combine that with noise and interference.

Gaussian filters are used in GSM because of their advantages in carrier power, occupied bandwidth, and symbol-clock recovery. The Gaussian filter has a Gaussian shape in both the time and frequency domains, and it does not ring like the raised cosine filters. Its effects in the time domain are relatively short and each symbol interacts significantly (or causes ISI) with only the preceding and succeeding symbols. This reduces the tendency for particular sequences of symbols to interact, which makes amplifiers easier to build and more efficient. Constant envelope modulation, as used in GMSK, can use class-C amplifiers, which are most efficient. Figure 2.42 shows the response of a Gaussian filter.

$$h(t) = \frac{\sqrt{\pi}}{\alpha} \exp\left[-\left(\frac{\pi t}{\alpha}\right)^2\right] \tag{2.36}$$

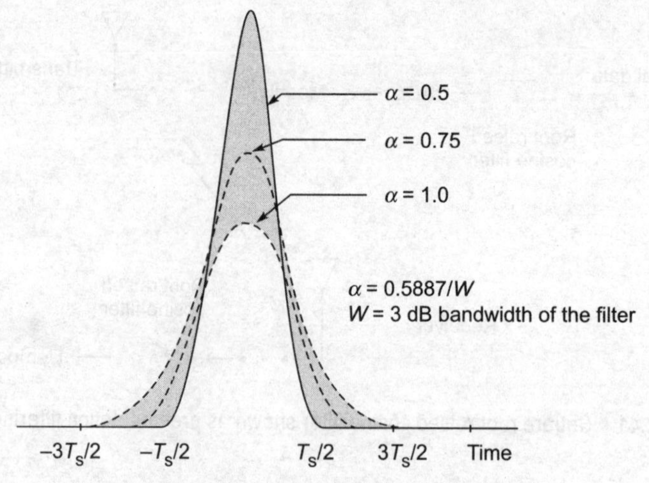

Fig. 2.42 Gaussian filter response

Roll-off factor is very important in raised cosine filter. The corresponding term for a Gaussian filter is *bandwidth time product* (BT). Occupied bandwidth cannot be stated in terms of BT because a Gaussian filter's frequency response does not go identically to zero, as does a raised cosine. Common values for BT are 0.3 to 0.5.

Chebyshev Equiripple FIR Filter

A Chebyshev equiripple FIR (finite impulse response) filter is used for baseband filtering in IS-95 CDMA (Chapter 13). With a channel spacing of 1.25 MHz and a symbol rate of 1.2288 MHz in IS-95 CDMA, it is vital to reduce leakage to adjacent RF channels. This is accomplished by using a filter with a very sharp shape factor using an alpha value of only 0.113. An FIR filter means that the filter's impulse response exists for only a finite number of samples. Equiripple means that there is a 'rippled' magnitude frequency-response envelope of equal maxima and minima in the pass and stop bands. This finite impulse response (FIR) filter uses a much lower order than a Nyquist filter to implement the required shape factor. The IS-95 FIR filter does not have zero ISI. However, ISI in CDMA is not as important as in other formats since the correlation of 64 chips at a time is used to make a symbol decision. This 'coding gain' tends to average out the ISI and minimize its effect. Figure 2.43 shows the filter response.

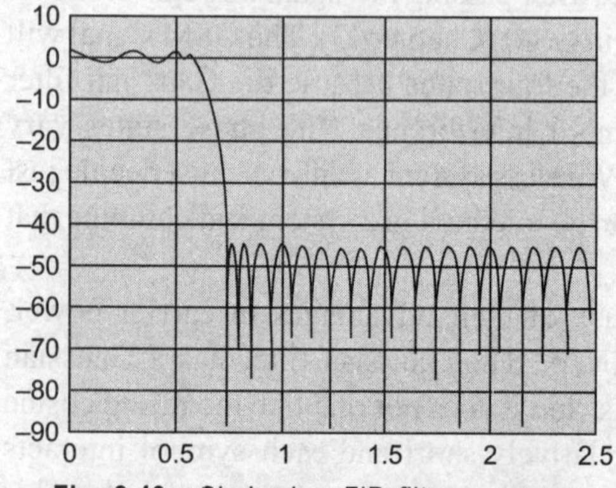

Fig. 2.43 Chebyshev FIR filter response

As with any natural resource, it makes no sense to waste the RF spectrum by using channel bands that are too wide. Therefore, narrower filters are used to reduce the occupied bandwidth of the transmission. Narrower filters with sufficient accuracy and repeatability are more difficult to build. Smaller values of alpha increase the ISI because more symbols can contribute. This tightens the requirements on clock accuracy. These narrower filters also result in more overshoot and, therefore, more peak carrier power.

The power amplifier must then accommodate the higher peak power without distortion. The bigger amplifier causes more heat and electrical interference to be produced since the RF current in the power amplifier will interfere with other circuits. Larger, heavier batteries will be required. The alternative is to have shorter talk time and smaller batteries. In summary, spectral efficiency is highly desirable, but there are penalties in cost, size, weight, complexity, talk time, and reliability.

2.6.6 Windowing Techniques

If we measure the signal only for a short time, the Fourier transform works as if the data were periodic for all time. If not quite an integral number of cycles fit into the total duration of the measurement, then when the Fourier transform assumes the signal repeats, the end of one signal segment does not connect smoothly with the beginning of the next, the assumed signal is similar to the actual signal but has little discontinuity at regular intervals as shown in Fig. 2.44. The discontinuities can be reduced by shaping the signal so that its ends match more smoothly as shown in Fig. 2.44. Since we cannot assume anything about the signal, we need a way to make any signal's ends connect smoothly to each other when repeated.

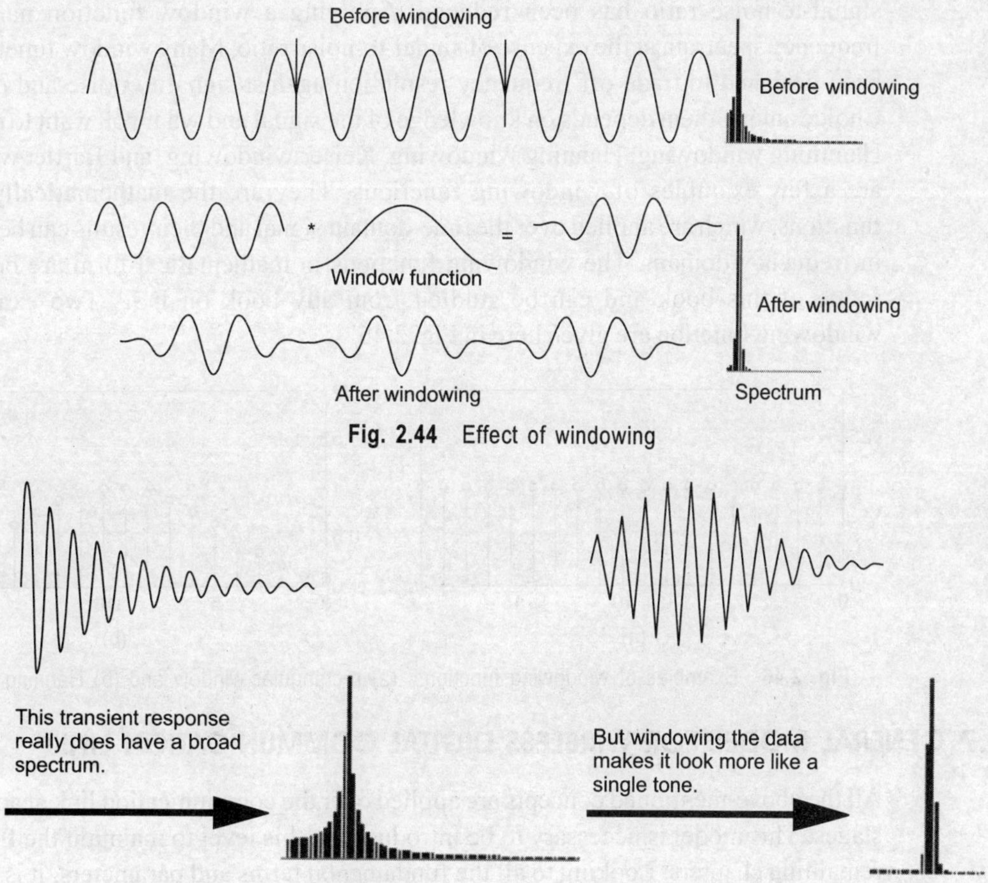

Fig. 2.44 Effect of windowing

Fig. 2.45 The result of applying a window function without proper thought

One way to do this is to multiply the signal by a 'window' function as discussed below.

The easiest way to make sure the ends of a signal match is to force them to be zero at the ends: that way, their value is necessarily the same. Actually, we also want to make sure that the signal is going in the right direction at the ends to match up smoothly. The easiest way to do this is to make sure neither end is going anywhere, that is, the slope of the signal at its ends should also be zero. Mathematically, a window function has the property that its value and all its derivatives are zero at the ends. Multiplying by a window function (called 'windowing') suppresses discontinuities and so avoids the broadening of the frequency spectrum caused by the discontinuities.

Windowing can narrow the spectrum and make it closer to what was expected, but it is important to remember that windowing is really a distortion of the original signal. It adds bit error rate in the performance but improves spectral efficiencies at the same time. Using the windowing function in a system is a compromise.

As per Fig. 2.45, the transient response really does have a broad frequency spectrum but windowing forces it to look more as if it had a narrow frequency spectrum instead. Worse than this, the window function has attenuated the signal at the point where it was largest and so has suppressed a large part of the signal power. This means that the overall signal-to-noise ratio has been reduced. Applying a window function narrows the frequency spectrum at the expense of signal-to-noise ratio. Many window functions have been designed to trade off frequency resolution against signal to noise and distortion. Choice among them depends on knowledge of the signal and what you want to do with it. Hamming windowing, Hanning windowing, Keiser windowing, and Bartlet windowing are a few examples of windowing functions. They are the mathematically defined functions, which are applied over the time-domain signal and their results can be observed in frequency domain. The windowing functions in mathematical form are beyond the scope of this book and can be studied from any book on DSP. Two examples of windowing function are given here in Fig. 2.46.

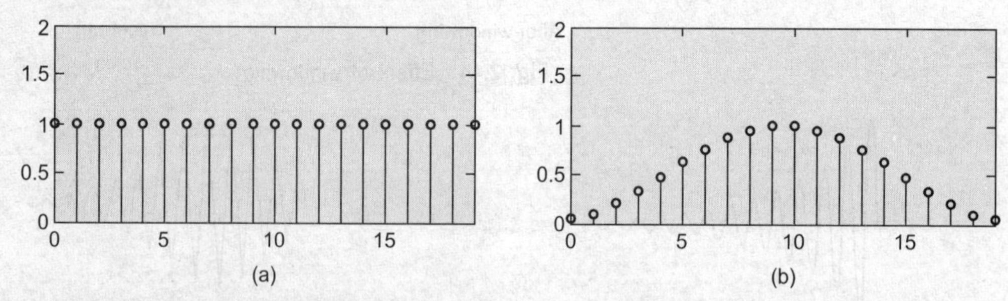

Fig. 2.46 Examples of windowing functions: (a) rectangular window and (b) Hanning window

2.7 GENERAL MODEL FOR WIRELESS DIGITAL COMMUNICATION LINK

All the above-mentioned concepts are applied over the communication link at appropriate stages. This model is necessary to be introduced at this level to maintain the flow of the remaining chapters. Looking to all the fundamental terms and parameters, it is now easy to associate them at the appropriate stages of the given wireless link. Most of the above-mentioned parameters are applied to the transmitting signal to shape it as per the channel and receiver requirements. Whenever one deals with wireless digital communication, it is the study of the whole point-to-point link in depth, covering the fundamentals of each and every block of the link. The main blocks of the complete link is given in Fig. 2.47. Still readers may not be familiar with few of the blocks of the link. Very soon, they will get the introduction about the blocks and their importance. In subsequent chapters, all the blocks will be explained in detail.

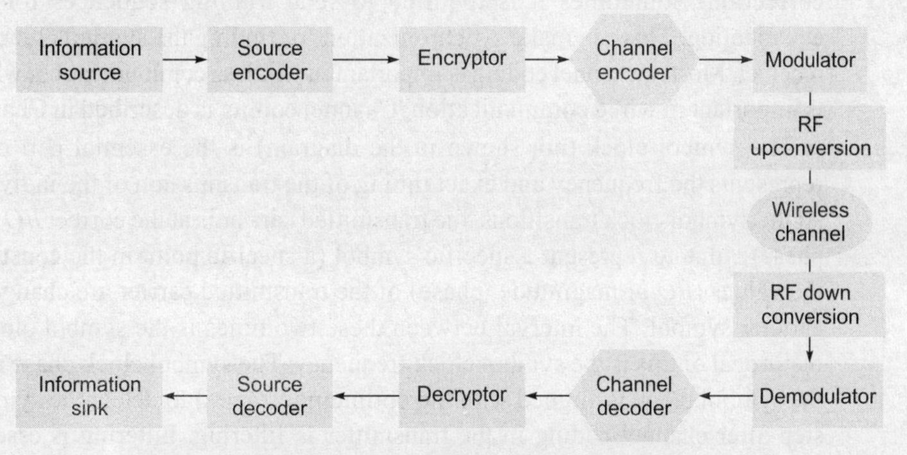

Fig. 2.47 Basic model of wireless digital communication link

Here is a simplified block diagram of a digital communications link. A transmitter begins and ends with an analog signal (except the readily stored digital base). The signal that comes out as multimedia information is analog in nature, which should be first converted into a digital form as usual. Previously, the wireless communication was started with the concept of voice communication, but now any signal can be communicated. For video communication, data requires tremendous storage capacity as well as a high speed of communication and hence source encoding is must. Here the data may be compressed by standard methods and the stored files with standard extensions, such as .jpg, .avi, .mp3, .gif, .tif, and .dat, can be made available for the transmission. The basic communication model is a *systematic assemblage* of *forward path*—information source, source encoder, encrypter, channel encoder, modulator, RF upconversion a *noisy channel* (wireless), and *reverse path*—RF downconversion, demodulator, channel decoder, decrypter, source decoder, and information sink.

The first step is to convert a continuous analog signal into a discrete/digital bit stream. This is called *digitization*. The next step is to add information coding for data compression. Information to be transmitted from source may be human originated (speech) or machine originated (data, image). The source encoder with compression eliminates the 'inherent redundancy' in information (thus compressing) to maximize the transmission rate. Encrypter ensures the secrecy of data. It is described in Chapter 3.

Then channel coding is added. The data must also be protected against perturbations introduced by the noisy channel, which could lead to misinterpretation of the transmitted message at the receiving end. This protection can be achieved through error control strategies: *forward error correction (FEC)*, i.e. using error-correcting codes that are able to correct errors at the receiving end, or *automatic repeat request (ARQ) systems* using error-detecting codes only with no capabilities of correcting it. Channel coding encodes the data in such a way as to minimize the effects of noise and interference in the communications channel. Channel coding adds extra bits to the input data stream even after removed redundant ones by the source coder. Those extra bits are used for error correction. Sometimes it is required to send training sequences for estimation or equalization. This can make synchronization (or finding the symbol clock) easier for the receiver. Mostly, channel coding is important in wireless communication, while line coding is important in wired communication. Channel coding is described in Chapter 4.

The symbol clock (not shown in the diagram) is the essential part of the link that represents the frequency and exact timing of the transmission of the individual symbols. At the symbol clock transitions, the transmitted carrier is at the correct *I*/*Q* (or magnitude/phase) value to represent a specific symbol (a specific point in the constellation). Then the values (*I*/*Q* or magnitude/ phase) of the transmitted carrier are changed to represent another symbol. The interval between these two times is the symbol clock period. The reciprocal of this is the symbol clock frequency. The symbol clock phase is correct when the symbol clock is aligned with the optimum instant(s) to detect the symbols. The next step after channel coding in the transmitter is filtering. Filtering is essential for good bandwidth efficiency. Without filtering, signals would have very fast transitions between states and, therefore, very wide frequency spectra—much wider than is needed for the purpose of sending information. A single filter can be shown for simplicity, but in reality there are two filters—one each for *I* and *Q* channels. This creates a compact and spectrally efficient signal that can be placed on a carrier.

The modulation method should be selected after knowing the channel characteristics and hence channel-related issues and correction for channel effects are described in Chapter 5 and 6. Correction stages are required before demodulation stages to reduce the probability of errors.

The output from the channel coder is then fed into the modulator. Since there are independent *I* and *Q* components in the radio, half of the information can be sent on *I* and the other half on *Q*. This is one reason digital radios work well with this type of digital signal. The *I* and *Q* components are separate. The modulator block generates a signal apposite for the transmission channel. The blocks in the reverse path do the opposite of those in forward path. Modulation techniques are logically divided into three types—pulse modulation, carrier modulation, and spread spectrum techniques. The latest modulation technique, which eliminates most of the problems of wireless channel, is OFDM (orthogonal frequency division multiplexing). In Chapters 7, 8, and 9, the overview about the modulation technologies is given.

After the modulator, the rest of the transmitter looks similar to a typical RF transmitter or a microwave transmitter/receiver pair. The signal is converted up to a higher intermediate frequency (IF), and then further upconverted to a higher radio frequency (RF). Any undesirable signals that were produced by the upconversion are then filtered out. Depending upon requirements the power amplifier, selection is made for power amplification to cover the required distance.

The receiver RF section provides efficient coupling between the antenna and rest of the hardware that utilizes, as effectively as possible the energy abstracted from the radio wave. Also it provides discrimination or selectivity against image and IF signals.

The desired receiver characteristics or issues are as follows.

Sensitivity It is expressed in terms of the voltage that must be applied to the receiver input to give a standard output.

Selectivity It is the characteristic that determines the extent to which the receiver is capable of distinguishing between the desired signal and signal of other frequencies.

Fidelity It represents the variation of the output with the modulation frequency, when the output load impedance is a resistance. At the lower modulation frequencies, it is determined by the low frequency characteristics of the audio frequency amplifier. At the higher modulation frequencies the fidelity is affected by the high frequency characteristics of the audio frequency amplifier.

Noise figure It is a measure of the extent to which the noise appearing in the receiver output in the absence of a signal is greater than the noise that would be present if the receiver was a perfect receiver from the point of view of generating the minimum possible noise. It determines the smallest power that may be received without being drowned out by the noise.

Other than these, automatic gain control (AGC), automatic power control, etc., are some of the issues related to the receiver end.

Learning about the *wireless medium* is essential to understand the reasoning behind specific designs for wireless communication protocols/systems. In particular, the design of the physical and medium access protocols are highly affected by the behavior of the channel that varies substantially in different indoor and outdoor areas. The diversity and complexity of transmission techniques in wireless communications are far more involved than those of wired communications.

The receiver is similar to the transmitter but in reverse. It is more complex to design. The incoming (RF) signal is first downconverted to (IF) and demodulated. The ability to demodulate the signal is hampered by factors including atmospheric noise, competing signals, and multi path or fading. Generally, demodulation involves the following stages:

1. Carrier frequency recovery (carrier lock)
2. Symbol clock recovery (symbol lock)
3. Signal decomposition to *I* and *Q* components
4. Determining *I* and *Q* values for each symbol ('slicing')
5. Decoding and de-interleaving
6. Expansion to original bit stream
7. Digital-to-analog conversion, if required

In more and more systems, however, the signal starts out as digital and stays digital. It is never analog in the sense of a continuous analog signal like audio. The main difference between the transmitter and receiver is the issue of carrier and symbol clock recovery. Both the symbol-clock frequency and phase (or timing) must be correct in the receiver in order to demodulate the bits successfully and recover the transmitted information. A symbol clock could be at the right frequency but at the wrong phase. If the symbol clock was aligned with the transitions between symbols rather than the symbols themselves, demodulation would be unsuccessful. Symbol clocks are usually fixed in frequency and this frequency is accurately known by both the transmitter and receiver. The difficulty is to get them aligned both in phase and timing. There are a variety of techniques and most systems employ two or more. If the signal amplitude varies during modulation, a receiver can measure the variations. The transmitter can send a specific synchronization signal or a predetermined bit sequence such as 10101010101010 to 'train' the receiver's clock. In systems with a pulsed carrier, the symbol clock can be aligned with the power of the carrier turn on. In the transmitter, it is known where the RF carrier and digital data clock are because they are being generated inside the transmitter itself. In the receiver this luxury is not there. The receiver can approximate where the carrier is but has no phase or timing symbol clock information. A difficult task in a receiver design is to create the carrier and symbol-clock recovery algorithms. That task can be made easier by the channel coding performed in the transmitter.

Mobile telephony, mobile Internet services, wireless LAN, etc., are a few of the applications that are protocol based; the basic physical layer considerations of these are described here. Mobile communications and systems are based on the technologies used to establish the wireless digital communication link.

Summary

- The bit rate defines the rate at which the information is passed, while the baud (or signalling) rate defines the number of symbols per second, each symbol represents n bits and has M signal states, where $M = 2^n$. This is called M-ary signalling.
- Different types of signal can be categorized and they can be identified by unique mathematics.
- Using the transforms, any time domain signal can be analysed into its frequency components. For every signal, 'the signal defines the spectrum and the spectrum defines the signal,' i.e. they are unique and opposite conversions ideally but may not be so practically.
- Bandwidth gives important information about the signal in frequency domain.
- The SNR and bandwidth are exchangeable and to be compromised always.
- Modern communication system hardware cannot eliminate digital signal processors.
- *A*-to-*D* and *D*-to-*A* conversions of signals are always lossy, same as sampling. They are achieved with constraints.
- A signal should be sampled at minimum Nyquist frequency or at a higher frequency to reconstruct the signal back at the receiver; otherwise aliasing can be observed.
- Antialiasing filters must be analog and should be used before the sampling stage.
- Autocorrelation is established with the signal and its phase shifted replica, while cross correlation is established between two different signals.

- Correlation is a weighted moving average and convolution is a weighted moving average with one signal flipped back to front. If one signal is symmetric then convolution and correlation are same.
- The FT, DTFT, STFT, and DFT are mathematically related to each other in this sequence as shown in Fig. 2.48.
- All channels will act as a low-pass filter (except for a few cases in wireless) and spreads a pulse due to one or other reasons.
- Nyquist pulse is sinc-shaped signal.
- Line coding is applied to digital baseband for desired spectral characteristics.
- Pulse shaping can be employed to remove unnecessary spectral spread or to improve spectral efficiency.
- Windowing can improve the spectral efficiency at the cost of BER.
- Wireless link transmitter employs source coding, channel coding, modulation, and upconversion and opposite blocks are present at the receiver side.

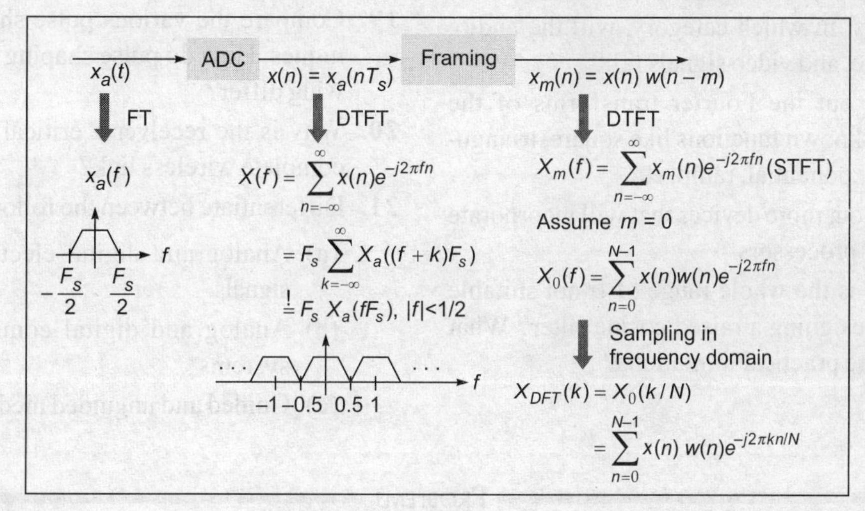

Fig. 2.48

REVIEW QUESTIONS

1. Write short notes on the following terms: Information, transmitter, types of channel, types of noise, receiver, modulation, carrier, bandwidth, SNR
2. A channel bandwidth is 250 kHz. What kind of information signals can be transmitted over it? Why should system bandwidth be higher than the signal bandwidth?
3. What are the various commercial ranges for various wireless applications? Few commercial ranges are used for multi-applications. Which factors are considered to derive reliable communication in these situations?
4. With reference to Fig. 2.2, find out the theoretical range of bit interval for which bit occurrence can be detected and establish the relation with the system bandwidth.
5. What is the relationship between the fundamental frequency and the period of a signal?

6. Shannon and Nyquist formulas of channel capacity places upper limit on the bit rate of a channel. Are they related? How?
7. What are the key factors that affect channel capacity? Explain how.
8. Explain the SNR-bandwidth trade-off.
9. Prove that relationship between SNR and bandwidth expansion factor is non-linear.
10. Are the signal spectrum and the signal bandwidth same? Why?
11. List out the various types of signal for communication and draw their waveforms. In which category, will the audio, image, and video signals fall?
12. Find out the Fourier transforms of the well-known functions like square, triangular, exponential, ramp, etc.
13. Find out more devices that will incorporate DSP processors.
14. Why is the whole range of α not suitable for designing a raised cosine filter? What are the practical limitations?
15. Where should the windowing stage in the overall wireless link be placed and why?
16. Why does a channel act as a filter? Why is convolution process incorporated to find out the output of a wireless channel? Will the channel always act as a filter?
17. What is the basic difference between DTFT and DFT? Why are they considered separately?
18. What is the importance of short-time Fourier transform? Find out the applications where the concept of STFT is applied.
19. Compare the various pulse shaping techniques. How do pulse shaping and windowing differ?
20. Why is the receiver a critical part of the complete wireless link?
21. Differentiate between the following terms:
    (a) Analog and digital electromagnetic signals
    (b) Analog and digital communication systems
    (c) Guided and unguided media

PROBLEMS

1. If the bit rate is to be maintained to 10 Mbps, what modifications should be made in the system to cope up with the SNR variations between 10 dB to 20 dB?
2. If square pulses, each of duration 0.05 μs, are to be transmitted at a carrier frequency of 100 MHz, what will be the shape of the spectrum? According to this spectrum, find out the (a) null to null (significant energy) bandwidth, (b) fractional power containment bandwidth, (c) bounded power spectral density, and (d) absolute bandwidth.

   Hint:

   Fractional power containment bandwidth: According to FCC rules, the occupied bandwidth is the band that levels exactly 0.5% of the signal power above the upper band limit and exactly 0.5% of the signal power below the lower band limit. Thus, 99% of the signal power is inside the occupied band.

   Bounded power spectral density: Typical attenuation level might be 35 or 50 dB.

   Absolute bandwidth: It is the interval between frequencies beyond which the spectrum is zero. However, for all realizable waveforms, absolute bandwidth is infinite.
3. Energies of signals $g_1(t)$ and $g_2(t)$ are E_{g1} and E_{g2}, respectively.

(a) Show that in general the energy of the signal $g_1(t) + g_2(t)$ is not $Eg_1 + Eg_2$.

(b) Under what condition, is the energy of $g_1(t) + g_2(t)$ equal to $Eg_1 + Eg_2$?

(c) Can the energy of the signal $g_1(t) + g_2(t)$ be zero? If so, under what condition(s)?

4. Determine the energy spectral density of the square pulse $s(t) = \text{rect}(t/T)$, where $\text{rect}(t/T)$ equals 1, for $-T/2 \le t \le T/2$ and equals zero elsewhere. Calculate the normalized energy E_s in the pulse.

5. A signal $g(t)$ is given by the expression $g(t) = \sin(At)/\pi t$. Determine the Nyquist frequency for sampling this signal.

6. What is the least sampling rate required to sample the signal $f(t) = \sin^3(\omega_0 t)$? Show graphically the effect caused by a reduction of the sampling rate, falling below the Nyquist rate.

7. The spectral range of a modulated signal extends from 1.0 MHz to 1.2 MHz. Find out the minimum sampling rate and maximum sampling time.

8. Determine the signal $g(t)$ whose Fourier transform is as shown in Fig. 2.49.

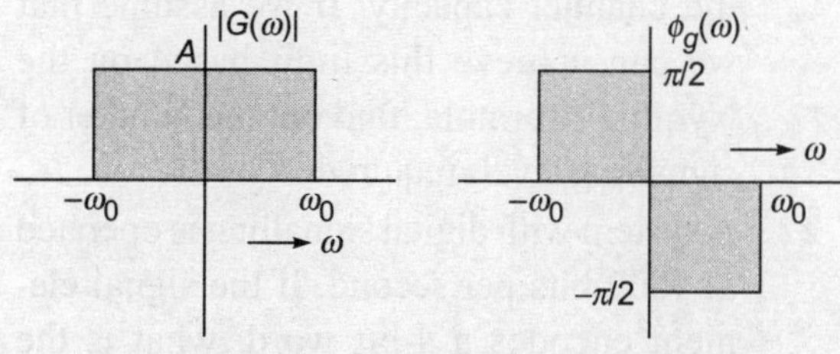

Fig. 2.49 Fourier transform for Problem 8

9. Fourier transforms of two real signals $s_1(t)$ and $s_2(t)$ are $S_1(\omega)$ and $S_2(\omega)$, respectively.

(a) Show that if $s_1(t)$ and $s_2(t)$ are non-overlapping, then the signals $s_1(t)$ and $s_2(t)$ are orthogonal, that is

$$\int_{-\infty}^{\infty} s_1(t) s_2(t) = 0.$$

(b) Show that if spectra $S_1(\omega)$ and $S_2(\omega)$ are orthogonal, that is $\int_{-\infty}^{\infty} S_1(\omega) S_2^*(\omega) = 0$, then $s_1(t)$ and $s_2(t)$ are also orthogonal.

10. Find the autocorrelation function $s(t) = A \cos(2\pi f_c t + \theta)$ in terms of its period $T_c = 1/f_c$. Using this result, find out the autocorrelation function of waveform $s(t) = 5 \cos 5t + 10 \cos 10t$.

11. Evaluate the convolution integral $\sin t\, u(t) * u(t)$.

12. If $g(t) * f(t) = c(t)$, then show that $g(t - T_1) * f(t - T_2) = c(t - T_1 - T_2)$.

13. For the given signal $s(t)$, as shown in Fig. 2.50, draw $s(2t)$ and $s(t/2)$.

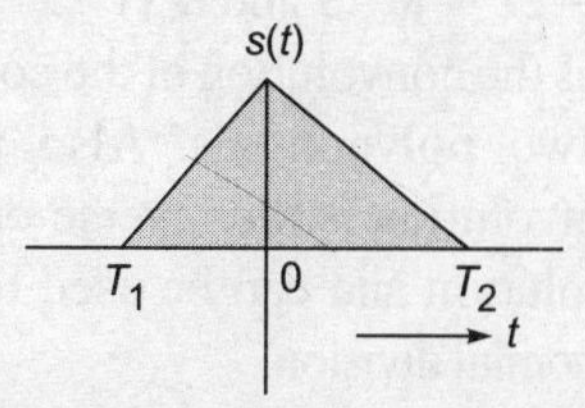

Fig. 2.50 Signal for Problem 13

14. The input x and output y of a certain non-linear channel are related as $y = x + 0.22x^3$. The input signal $x(t)$ is a sum of two modulated signals as shown below:

$$x(t) = x_1(t) \cos \omega_1 t + x_2(t) \cos \omega_2 t$$

where the $X_1(\omega)$ and $X_2(\omega)$ are as shown in Fig. 2.51.

Here $\omega_1 = 2\pi(100 \times 10^3)$

and $\omega_2 = 2\pi(110 \times 10^3)$

(a) Sketch the spectra of the input signal $x(t)$ and output signal $y(t)$.

(b) Can signals $x_1(t)$ and $x_2(t)$ be recovered without distortion and interference from the output $y(t)$?

15. Show that an arbitrary function $s(n)$ can be represented by a sum of even function $s_e(n)$ and odd function $s_o(n)$, i.e.

$$s(n) = s_e(n) + s_o(n)$$

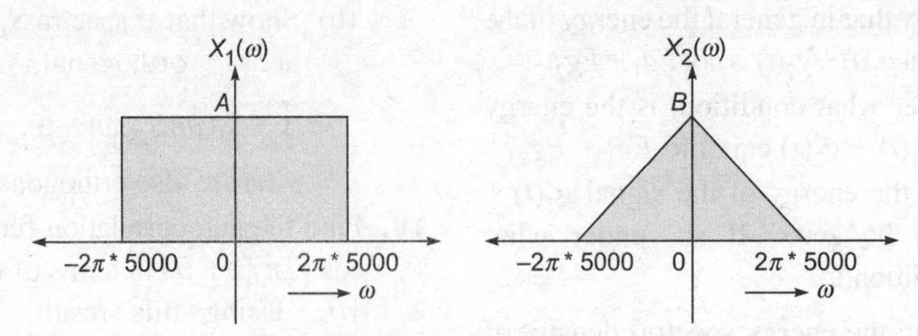

Fig. 2.51 Spectra for Problem 14

16. If $x(n)$ and $h(n)$ both are equal to [–1, 0, 1]. Find out the convolution between $x(n)$ and $h(n)$ and also between $h(n)$ and $x(n)$. Is the order of convolution important? Why?
17. Prove that the product of two polynomials $h(x) = 2x^2 - x - 3$ and $u(x) = x^4 + 2x^3 + 4x^2 + 2x$ is the convolution of the coefficient of the two polynomials. Also prove that deconvolution is the inverse operation of convolution and can be used to carry out polynomial division.
18. If the correlation is to find out between two signals $x(n)$ = [–1, 0, 1] and $y(n)$ = [–1, 0, 1], what is the expected result? If the order of correlation is changed, will we get any difference in the result?
19. Given a sequence of samples $s(n)$, an impulse train $s(t)$ can be generated such that

$$s(t) = \sum_{n=-\infty}^{\infty} s(n)\delta(t - nT_s).$$

If this is the input to a continuous time filter with frequency response as shown in Fig. 2.52 and the corresponding impulse response is $h(t)$, find out the impulse response and reconstructed output signal.

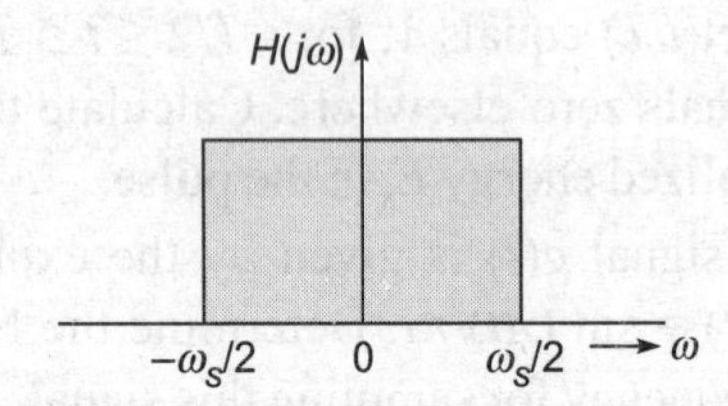

Fig. 2.52 Filter response for Problem 19

20. In a multilevel signalling, if the number of discrete signal or voltage levels is 8 in a MODEM and the bandwidth is of 4 kHz, find out the channel capacity. If the data rate is increased by increasing the number of signalling elements, for a given bandwidth, what will be the expected changes? Comment on it.
21. A channel bandwidth is 2 MHz and SNR = 25 dB. Using Shannon's formula, find out the channel capacity. If we assume that we can achieve this limit based on the Nyquist's formula, find out the number of signalling levels required.
22. A system with digital signalling is operated at 4800 bits per second. If the signal element encodes a 4-bit word, what is the minimum required bandwidth?
23. Room temperature is specified as T = 17°C or 290 K. At this temperature, find out the thermal noise power density. If the bandwidth of the receiver is 10 MHz, find out the thermal noise level at the receiver output in dBW.

Multiple Choice Questions

1. If the transmission bandwidth is W and available channel bandwidth is $W_{channel}$, what should be the condition that will allow fruitful reception?
   (a) $W = W_{channel}$
   (b) $W < W_{channel}$
   (c) $W > W_{channel}$
   (d) All of the above
2. If the bit rate of the data is 1 Mbps, what should be the bandwidth occupied by that rectangular wave?
   (a) 1 MHz (b) 0.1 MHz
   (c) 0.5 MHz (d) 2 MHz
3. Real audio/video signal is a/an
   (a) energy signal
   (b) power signal
   (c) deterministic signal
   (d) periodic signal
4. Unit ramp signal is
   (a) an energy signal
   (b) a power signal
   (c) a periodic signal
   (d) none of the above
5. Which of the following measures cannot be effective in reducing the noise?
   (a) Reduction in signalling rate
   (b) Increase in channel bandwidth
   (c) Increase in transmitter power
   (d) Use of redundancy
6. Thermal noise power is proportional to
   (a) bandwidth only
   (b) bandwidth and temperature
   (c) temperature only
   (d) bandwidth, temperature, and resistance
7. Which of the following statements is correct?
   (a) Flicker noise is a 'white noise process'.
   (b) The spectral density of thermal noise is considered 'pink' in nature.
   (c) Flicker noise is most serious at low frequencies.
   (d) In a cascade stage, noise figure is least dependent on the first stage.
8. Shot noise is caused by
   (a) random variations in the arrival of electrons at the collector
   (b) random variations in the arrival of holes at the collector
   (c) random variations in the arrival of both electrons and holes at the collector
   (d) none of the above
9. The number of spectral components when two sine waves are multiplied is
   (a) 1 (b) infinite
   (c) 2 (d) 4
10. A narrowband signal occupying a frequency range of nf_0 to $(n + 1) f_0$ is to be sampled. Consider n as an integer. The minimum sampling rate is
    (a) $2\ nf_0$ (b) $2(n + 1)f_0$
    (c) $2f_0$ (d) none of these
11. The signal that shows the aliasing effect must have a sampling rate that is
    (a) less than signal bandwidth
    (b) greater than Nyquist rate
    (c) equal to Nyquist rate
    (d) all of the above
12. In pulse modulation systems, the number of samples required to ensure no loss of information is given by
    (a) Fourier transform
    (b) Nyquist theorem
    (c) Parseval's theorem
    (d) Shannon's theorem
13. The autocorrelation function of a signal $s(t)$ at $t_0 = 0$ is equal to

(a) zero
(b) average voltage of the signal
(c) infinite
(d) average power of the signal

14. The autocorrelation function for the function $x(t) = V \sin \omega t$ is given by
(a) $V^2 \cos \omega t$
(b) $1/2(V^2 \cos \omega t)$
(c) $V^2 \cos^2 \omega t$
(d) $2V^2 \cos \omega t$

15. Which of the following processes is used to identify how the spectrum of two signals interacts?
(a) Correlation (b) DFT
(c) Convolution (d) STFT

16. Windowing process will
(a) increase the spectrum efficiency
(b) decrease the out of band components
(c) shape the spectrum
(d) do all of the above

17. By which of the following processes, one can detect the required signal from all the coexisting ones?
(a) Cross correlation
(b) Convolution
(c) Autocorrelation
(d) Multiplexing

18. By which of the following processes, one can reject an undesired signal from the channel?
(a) Cross correlation
(b) Convolution
(c) Autocorrelation
(d) Windowing

19. A signal of maximum of 10 kHz is sampled at Nyquist rate. The time interval between two successive samples is
(a) 50 μs (b) 100 μs
(c) 1000 μs (d) 5 μs

20. The channel capacity C of a bandlimited Gaussian channel is defined as
(a) $W_{\text{channel}} \log_2 (1 + \text{SNR})$
(b) $(1/W_{\text{channel}}) \log_2 (1 + \text{SNR})$
(c) $W_{\text{channel}} \log_2 (\text{SNR})$
(d) $(1/W_{\text{channel}}) \log_2 (\text{SNR})$

21. Two orthogonal signals $s_1(t)$ and $s_2(t)$ satisfy the relation
(a) $\int_0^T s_1(t)\, s_2(t)\, dt = 1$
(b) $\int_0^T s_1(t)\, s_2(t) dt = \infty$
(c) $\int_0^T s_1(t)\, s_2(t) dt = 0$
(d) $\int_0^T s_1(t) s_2(t) dt = \pi$

22. The autocorrelation function of a continuous time signal $s(t)$ is defined as
(a) $r(t_0) = \dfrac{1}{T}\int_0^T s(t)s(t + t_0)dt$
(b) $r(t_0) = \dfrac{1}{T}\int_0^T s(t - t_0)s(t + t_0)dt$
(c) $r(t_0) = \dfrac{1}{T}\int_0^T s(t)s(t - t_0)dt$
(d) $r(t_0) = \dfrac{1}{T}\int_{-T}^T s(t)s(t + t_0)dt$

23. The autocorrelation function is
(a) an odd function in t_0
(b) an even function in t_0
(c) an exponential function in t_0
(d) a linear function in t_0

24. An ergodic process in communication is present if many random signals have
(a) identical time averages
(b) identical time and ensemble averages
(c) identical ensemble averages
(d) none of the above

25. In communication receivers the fidelity is provided by
(a) mixer stage
(b) audio stage

(c) IF stage
(d) detector stage

26. A receiver has poor IF selectivity. It will, therefore, also has poor
(a) sensitivity
(b) double spotting
(c) diversity reception
(d) blocking

27. Noise figure is used as a figure of merit of
(a) oscillator (b) modulator
(c) amplifier (d) isolator

28. Average thermal noise is
(a) constant as long as the bandwidth is fixed
(b) dependent on channel loading
(c) due to amplitude non-linearity in the system
(d) not related to microwave communication at all

29. The selectivity of most receivers is determined largely by
(a) sensitivity
(b) antenna direction
(c) characteristics of IF section
(d) all of the above

30. Which one of the following is not a useful quantity for comparing the noise performance of receivers?
(a) Noise figure
(b) Equivalent noise resistance
(c) Input noise voltage
(d) Noise temperature

chapter 3

Source Coding Techniques

Key Topics

- Wireless real-time communication: voice, audio, image, and video
- Basic properties of digital speech/audio signal, image signal, and video signal
- Quantization techniques in general: uniform, non-uniform, adaptive, and vector
- Pulse code modulation (PCM), differential PCM, and adaptive differential PCM
- Information sources and entropy
- Information source coding fundamentals (with example of Huffman coding)
- Voice coders
- Linear predictive coding
- Methods for coding in frequency domain: sub-band coding and transform coding (discrete cosine transform and wavelet transform)
- Encrypters

Chapter Outline

In wireless digital communication, the information source may be real-time audio or video signal; hence, digitization is always required. Digitization is a lossy process but can be achieved successfully with some constraints. For that, appropriate sampling, quantization, and digitization methods should be followed. At the same time, the generated database is tremendously large, which can not be transmitted with the desired high speed. Instead, if the source coding and compression to the database is applied, speedy communication can be achieved. Source coding is also entropy coding. While sending the information to the other end of the system, minimum energy should be consumed for useful information only. This chapter gives most of the source coding methods utilized in the present wireless systems. Coding may be done by removing redundancy from the information so that useful information can be optimally preserved and a minimum database is transmitted. However, channel coding stage again adds redundant bits for error-handling purpose at the cost of reduction in information transmission efficiency.

3.1 WIRELESS REAL-TIME COMMUNICATION: VOICE, AUDIO, IMAGE, AND VIDEO

Most of the wireless applications are voice/audio/image/video communication, which in general, is called multimedia communication. Multimedia information is real-time, probabilistic (non-deterministic), continuous analog, and energy signal in time domain. As far as frequency domain is concerned, audio bandwidth (20 Hz to 20 kHz) is comparatively less than image and video bandwidth (which is in MHz). Multimedia best suits the human being's perception and communicating behaviour as well as the way of acting, which provides the communication capabilities and information sharing irrespective of location and time. Integration and interaction among different multimedia types create new challenging opportunities. The concept is represented in Fig. 3.1.

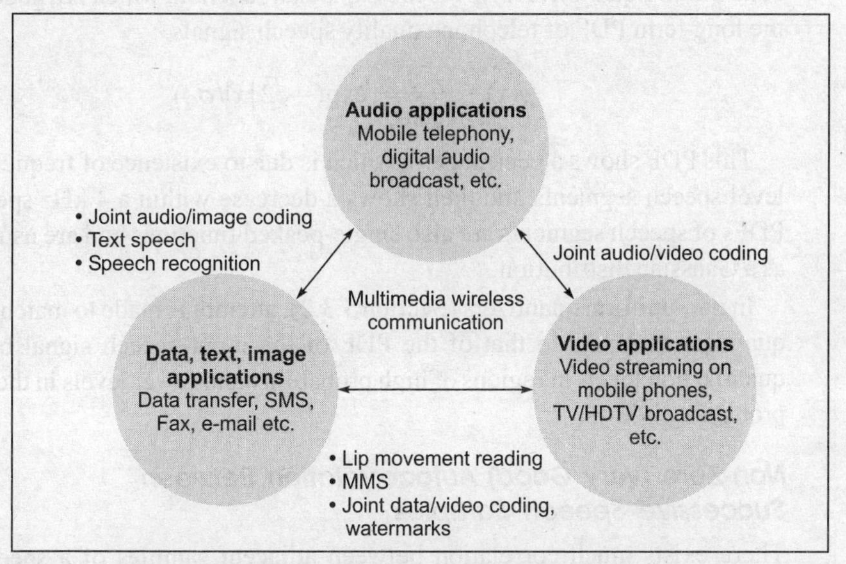

Fig. 3.1 Multimedia communication and a few media interactions in existing wireless communication systems (applied as well as on wired lines)

The first step for a multimedia transmitter is the *source coding.* Without source coding, the digitized information can be transmitted, but the speed and efficiency of information transmission may be poor. Source coding helps to reduce energy of total transmissions and also improves speed of useful data transfer. Source coders may be of three types: waveform coders, parametric coders and hybrid coders. To begin with the source coding concept, one must know the signal to be communicated and its properties. Audio signal is the simplest of all and so the description can be started with the study of the properties of speech signal.

3.1.1 Basic Properties of Speech Signal used for Quantization and Coding

Speech waveforms have a number of useful properties that can be exploited while designing the coders. The most basic property of the speech waveform is that they can be band limited and this property is exploited by all the speech coders. Finite bandwidth means that it can be time discretized at a finite rate and reconstructed completely from its samples, provided that the sampling rate follows the Nyquist criteria. Some other properties of speech waveform are discussed below.

Non-uniform Probability Distribution of Speech Amplitude

The PDF of a speech signal is in general characterized by the following factors:

- Very high probability of non-zero amplitudes

- Significant probability of very high amplitudes
- Monotonically decreasing function of amplitudes between these extremes

The exact distribution is, however, dependent upon the recording conditions and input bandwidth. Equation (3.1) gives the Laplacian function, which is a good approximation to the long-term PDF of telephone quality speech signals.

$$p(x) = \frac{1}{\sqrt{2\sigma_x}} \exp(-\sqrt{2}|x|/\sigma_x) \tag{3.1}$$

This PDF shows a peak at zero, which is due to existence of frequent pauses and low-level speech segments and then shows a decrease within a 4 kHz spectrum. Short time PDFs of speech segments are also single-peaked functions and are usually approximated as a Gaussian distribution.

In non-uniform quantizers (Section 3.3.2), attempt is made to match the distribution of quantization levels to that of the PDF of the input speech signal by allocating more quantization levels in regions of high probability and fewer levels in the regions where the probability is low.

Non-Zero (Very Good) Autocorrelation Between Successive Speech Samples

There exists much correlation between adjacent samples of a speech segment. This implies that in every sample of speech, there is a large component that is easily predicted from the value of the previous samples with small random error. All differential and predictive coding schemes are based on exploiting this property. Following the autocorrelation function gives the quantitative measure of the closeness or similarity between successive samples of speech signal as a function of their time separation index.

$$C_{\text{auto}}(k) = \frac{1}{N} \sum_{n=0}^{N-k-1} s(n)s(n+k) \tag{3.2}$$

Here $s(k)$ represents the kth speech sample. Normally, autocorrelation function is normalized to the variance of the speech signal and hence is restricted to have values in the range [–1, 1] with $C_{\text{auto}}(0)$, which is due to correlation of a sample with itself. Typical signals have an adjacent sample correlation $C_{\text{auto}}(1) = 0.8$ to 0.9, at the most.

Non-Flat Nature of Speech Spectra

The non-flat nature of power spectral density of speech makes it possible to obtain significant compression by coding speech in the frequency domain (sub-band coding). It is achieved by dividing the total band into sub-bands and finding significant frequencies. The non-flat nature of the PSD is due to frequency domain manifestation of the non-zero autocorrelation property. Typical long-term averaged PSDs of speech show that high-frequency components contribute very little to the total speech energy. This indicates that coding speech separately in different frequency bands can lead to a significant coding gain. It should be noted that the high-frequency components, though insignificant in energy are very important carriers of speech information and hence need to be adequately represented in the coding system.

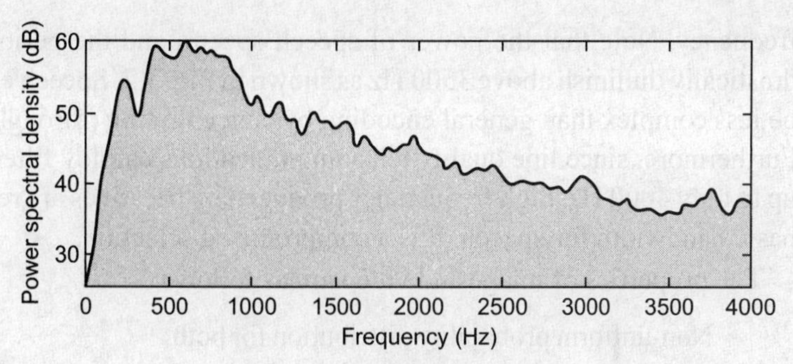

Fig. 3.2 PSD of speech signal

A qualitative measure of the theoretical maximum coding gain that can be obtained by exploiting the non-flat characteristics of the speech spectra is given by the *spectral flatness measure* (SFM). It is defined as

SFM = Arithmetic mean of the samples/geometric mean of the samples (3.3)

where PSD is taken at uniform intervals in frequency.

$$\text{SFM} = \frac{\left[\dfrac{1}{N}\sum_{k=1}^{N} S_k^2\right]}{\left[\prod_{k=1}^{N} S_k^2\right]^{1/N}} \tag{3.4}$$

where S_k is the kth frequency sample of the PSD of the speech signal. Typically, speech signals have long-term SFM value of 8 and short-term SFM value varying widely between 2 and 500.

Existence of Voiced and Unvoiced Speech Segments

It will be explained in Section 3.9.1, where the human speech generation is described.

Quasiperiodicity of Voiced Speech Signals

Both the above features can be observed clearly from the sample speech signal shown in Fig. 3.3.

Fig. 3.3 Sample speech signal in analog form

The spectrum for speech (combined voiced and unvoiced sounds) has a total bandwidth of approximately 7000 Hz with an average energy at about 3100 Hz. The auditory canal optimizes speech detection by acting as a resonant cavity at this average frequency. Note that the power of speech spectra and the periodic nature of formants drastically diminish above 3500 Hz as shown in Fig. 3.2. Speech encoding algorithms can be less complex than general encoding by concentrating (through filters) on this region. Furthermore, since line quality telecommunications employ filters that pass frequencies up to only 4000 Hz, high frequencies produced by fricatives are removed. This forms one basic bandwidth for speech. It is a compromised selection.

The properties of image and video are as follows:

- Non-uniform probability distribution for both
- Non-flat nature of image spectra for both
- A good correlation between successive samples (or pixels) in images and between successive image shots in video.

3.1.2 Digital Baseband/Sources

Digital Speech/Audio

Digital speech/audio signal is a one-dimensional signal, same as analog audio but discrete in nature. It is achieved by sampling analog input and then by A to D conversion. The fundamentals are discussed in Section 3.3 while explaining pulse code modulation. A Nyquist criterion is the fundamental that supports selecting the sampling frequency for discretization of the analog speech/audio as described in Chapter 2.

Digital Image

Digital images are recorded or captured directly from digital cameras or indirectly by scanning a photograph with a scanner. The image is divided into a matrix or array of small picture elements called *pixels* [Fig. 3.4(a)]. Again, pixels are nothing but samples of the image and decide the resolution of the image. Each pixel is represented by a numerical value and its position in 2D array. The numerical value can be converted into a code. The advantage of digital images is that they can be processed, in many ways, by computer systems.

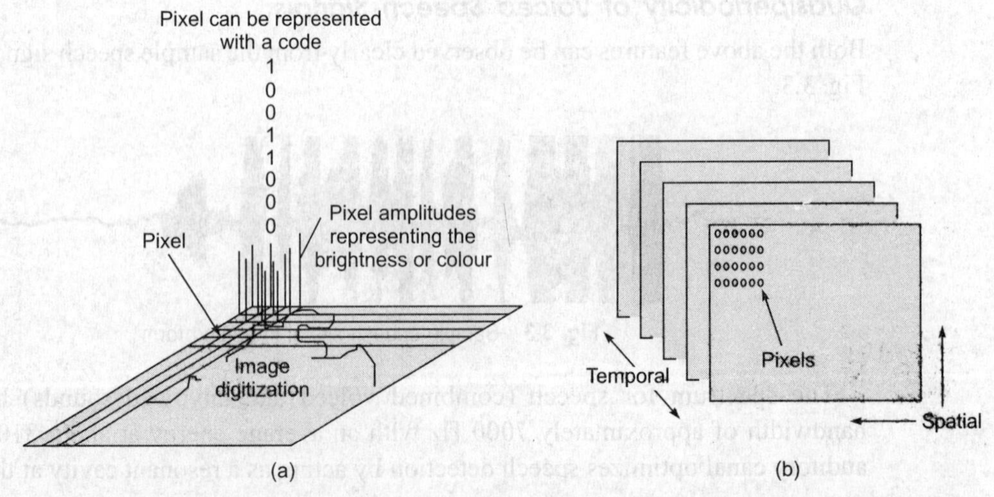

Fig. 3.4 (a) Image converted into matrix of pixels and (b) multiple images transmitted in video

Mathematically, a *digital image* can be defined as a two-dimensional function $f(x,y)$, where x and y are the spatial (image plane) coordinates representing a pixel and amplitude of $f(x, y)$ at any pair of coordinates (x, y) is called *intensity* or *grey level* of the image at that point. In colour images each pixel is associated with a colour. Collection of pixels is arranged as a two-dimensional matrix (spatial representation) form, the size of which decides the image resolution. The image can be represented in frequency domain also. In general, the pixel value is related to the brightness or colour, which can be understood when the digital image is converted into an analog image for display and viewing.

The numerical size (number of bits) of an image is the product of the following two factors:

1. *The number of pixels in an image* This is found by multiplying the number of pixels in a row (pixel length) and the total number of rows created in the image due to resolution of the image.
2. *The pixel depth (bits per pixel)* This is usually in the range of 8–16 bits or 1–2 bytes per pixel.

The significance of numerical size is that the larger the image (numerically), the more memory and disk storage space is required and more time for processing and distribution of images is required.

Example 3.1 When an image is digitized, it contains 256 different kinds of pixels. Find out the pixel depth. If the 256×256 resolution is used, what will be the size of the storage required for one digital image?

Solution

There are 256 different kinds of pixels, which means that they have 256 different quantized levels while converting the image into digital form.

Now $2^8 = 256$

It means that the pixel depth will be 8 bits/pixel. (It is similar to 8 bits/sample.)

Thus, total $256 \times 256 \times 8 = 524288$ bits/image will be required for the storage. In other words, 65536 bytes will be occupied.

Digital Video

It is nothing but multiframe images continuously generated in sequence [Fig. 3.4(b)], converted into digital form. It can be achieved from digital video cameras, stored movies on the computer, or digitized by sampling the analog video signal using Nyquist rate. It is a multidimensional function.

Digital video is not like normal analog video used by everyday televisions. To understand how digital video works, it is best to think of it as a sequence of non interlaced images, each of which is a two-dimensional frame of pixels. Present-day analog television systems such as the National Television Standards Committee (NTSC) used in North America and Japan and Phase Alternate by Line (PAL) used in Western Europe, employ line interlacing. Systems that use frame-wise line interlacing alternately scan odd and even lines of the video, which can produce problems when analog video is digitized. This issue complicates any discussion of digital video and the compression process, and so is best left aside for now.

According to fundamentals of colour images, associated with each pixel are two values, *luminance* and *chrominance*. The luminance is a value proportional to the pixel's intensity. The chrominance is a value that represents the colour of the pixel and there are a number of representations to choose from. There are different colour models. Any colour can be synthesized by an appropriate mixture of three properly chosen primary colours. Red, green, and blue (RGB) are usually chosen for the primary colours.

Digital video can be characterized by the following variables.

1. *Frame rate* It is the number of frames displayed per second. The illusion of motion can be experienced at frame rates as low as 12 frames per second.
2. *Frame dimensions* It is the width and height of the image expressed in the number of pixels. Digital video comparable to television, requires dimensions of around 640×480 pixels.
3. *Pixel depth* It is the number of bits per pixel. In some cases, it might be possible to separate the bits dedicated to luminance from those used for chrominance. In others, all the bits might be used to reference one of a range of colours from a known palette.

Table 3.1 Possible values of parameters for typical applications of digital video

Application	Frame Rate	Dimensions	Pixel Depth
Multimedia (computer)	15	320×240	16
Entertainment TV	25	640×480	16
Surveillance	5	640×480	12
Video telephony	10	320×240	12
HDTV	25	1920×1080	24

A few statements can be made to summarize this section:

1. Speech, audio, image, and video signals are of stochastic nature.
2. For source coding, the above signals may be available in two different ways:
    - In the form of bits, 1's and 0's, which can be compressed directly (entropy coding) using algorithms.
    - In form of real waveforms, which must be digitized and then compressed (waveform + entropy coding) using signal properties.

    Different methods of source coding will use appropriate form out of these two.
3. Source coding may be done in time or frequency domain.

3.2 GENERAL DIAGRAM OF SOURCE CODING STAGE

As shown in Fig. 3.5, information source output is obtained through measurement and observation stage. Signal processing generally involves two tasks as shown in the figure.

First, it is a vehicle for obtaining a general representation of a speech, audio, image, or video signal in either waveform or parametric form. Second, signal processing serves the function of aiding in the processing of transforming the signal representation into alternate forms, which are less general in nature but more appropriate to the coding method. In order to put the signal into a more convenient form for some new transform-based efficient source-coding techniques, such as DCT and wavelet, appropriate signal processing is applied. The last step in the process is the extraction of the useful message information. This step may be performed either by human listeners or automatically by machines. On the basis of useful information, the source coding and compression is applied. Finally, the source-coded data is available, which goes to error handling/channel coding stage.

As illustrated in Fig. 3.5, source coding results in entropy coding.

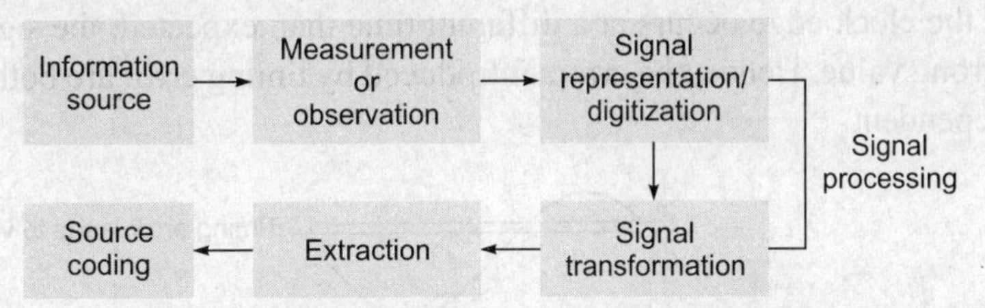

Fig. 3.5 General block diagram of source coding steps (All the stages may not be always required.)

Firstly, it is necessary to know the method of digitization of the signal for waveform coders. In Chapter 2, the sampling process in practical aspects is explained along with its limitations. After sampling and A-to-D conversion, signal reconstruction is always a lossy process. This is caused due to quantization error. Before we study digitization methods, such as PCM, let us first study the types of quantization processes, so that the best method can be adopted for the digitization of the information signal and the conversion may be near lossless. The PCM, DM, DPCM, and ADPCM are nothing but waveform coders.

3.3 QUANTIZATION TECHNIQUES

Quantization is the process of mapping a continuous range of amplitudes of a signal into a finite set of discrete amplitudes. When the signal is converted into a digital form, the precision is limited by the number of converted bits. The smoothly varying analog signal can only be represented as a 'stepped' waveform due to the limited precision as shown in Fig. 3.6. The errors introduced by digitization are both non-linear and signal dependent. We cannot calculate their effects using normal mathematics. The errors cannot be reduced by simple means and that even if we could calculate their effect, we would have to do so separately for every type of signal we expect. Figure 3.6 shows quantization error as if it were a source of random noise. Because of this, quantization error is called quantization noise. In simplified form,

$$\text{Quantization error } q_e = \text{actual value of the sample} - \text{quantized value of that sample} \tag{3.5}$$

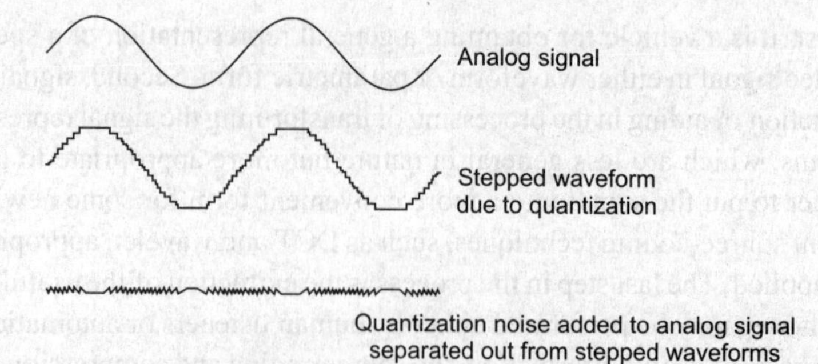

Fig. 3.6 Quantization noise

Figure 3.7 shows a real-time signal, which is held on the rising edge of a clock signal. If the clock edge occurs at a different time than expected, the signal will be held at the wrong value. Hence, the errors introduced by timing error are both non-linear and signal dependent.

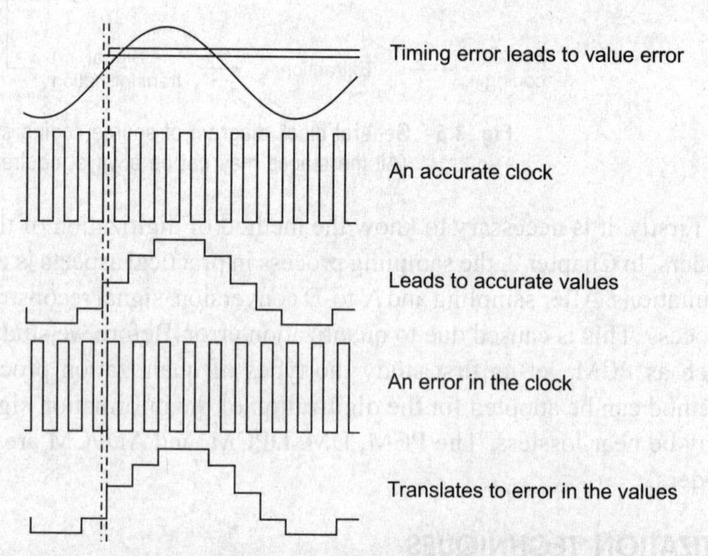

Fig. 3.7 Uncertainty in the clock timing leads to errors in the reconstructed signal

In practice, conversion of analog signal into digital form by quantization is always a lossy affair. Its quality of output is always dependent upon the maximum frequency content of the signal and the sampling rate applied to it. To convert discrete natural samples into binary codes the discrete level allocation to samples is required. That is why, the noise-like effect in the signal artificially. Quantization can be thought of as a process that removes relevancies in a signal, and their operation is irreversible.

Uniform quantization, non-uniform quantization, adaptive quantization, and vector (statistical data based) quantization are four approaches for the quantization. The first two methods are conventional and simple. Adaptive is the improvement over the conventional one and statistical quantization methods can improve on adaptive techniques. Generally the same numbers of quantization levels are used in adaptive and statistical quantization so there is no savings there. However, statistical methods consider the distribution of the data in each section when determining the breakdown of the quantization levels. Therefore, the improvement lies in better definition of each quantization level. Sometimes quantization becomes an integral part of the speech coders or source coders.

3.3.1 Uniform Quantization

As the input signal samples enter the quantization phase, each input sample is assigned a quantization interval that is closest to its amplitude height. If all quantization intervals are equally spaced throughout the dynamic range of the input analog signal, it is called *uniform quantization*. Each quantization interval is assigned a discrete value in the form of a binary code word.

Figure 3.8 shows an example of uniform quantization. In uniform quantization, the total range is divided into M discrete levels uniformly (equal step size). A quantizer that uses n bits can have $M = 2^n$ discrete amplitude levels. The distortion introduced due to quantization operation is directly proportional to the square of the step size, which in turn is inversely proportional to the number of levels for given amplitude range. The mean square error is the measure for the distortion introduced by the quantization process.

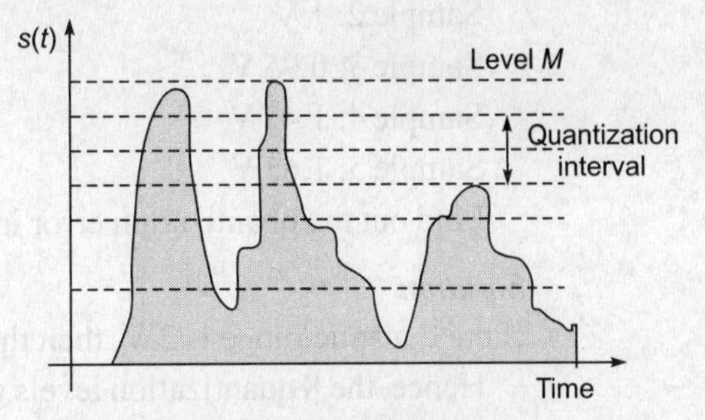

Fig. 3.8 Uniform quantization

$$E_{\text{ms}} = E[(s(t) - s_q(t))^2] = \frac{1}{T}\int_0^T [s_q(t) - s(t)]^2 dt \tag{3.6}$$

where $s(t)$ is the analog speech signal and $s_q(t)$ is the quantized speech signal. For calculating the quantization error in discrete form, each sample is considered separately, that is the difference between natural amplitude and the quantized amplitude of the sample will give quantization error [Eq. (3.5)] and all such sample errors must be squared, summed up and averaged for calculating the mean square error. This is the additive quantization noise that can never be removed a hundred percent. The performance of a quantization process is measured by another term, *signal-to-quantization noise ratio* (SQNR). Quantization noise is equivalent to the random noise that impacts the signal-to-noise ratio of original signal. S/N_q is a measure of signal strength relative to background noise. The ratio is usually measured in decibels (dB). If the incoming signal strength in microvolts is V_s, and the noise level, also in microvolts, is V_n, then the signal-to-noise ratio, S/N_q (SQNR), in decibels is given by the formula

$$S/N_q = 20 \log_{10}(V_s/V_n) \tag{3.7}$$

The higher the SQNR, the better is the voice quality. Quantization noise reduces the SQNR of a signal. Therefore, an increase in quantization noise degrades the quality of a voice signal.

One way to reduce quantization noise is to increase the number of quantization intervals. The difference between the amplitude of the input signal and the quantized signal decreases as the quantization intervals are increased (increase in the number of discrete levels decreases the quantization noise). However, the amount of code words would also have to be increased in proportion to the increase in quantization intervals. This process would introduce additional problems dealing with the capacity of a PCM system to handle more code words.

Example 3.2 Following are the readings for the measurement of quantization error in five consecutive samples. The number of quantization levels in the dynamic range of 2 volts is 8.

Sample 1: 1.2 V
Sample 2: 1 V
Sample 3: 0.95 V
Sample 4: 1.41 V
Sample 5: 1.65 V

Find out the quantization error in terms of its mean square value.

Solution

If the dynamic range is 2 V, then the smallest step size will be 2/8 = 0.25 V.

Hence, the 8 quantization levels will be 0.25, 0.5, 0.75, 1.0, 1.25, 1.50, 1.75, 2.0.

The measured samples will be assigned the following quantization values:

Sample 1: 1.25 V quantization error = –0.05 V
Sample 2: 1.0 V quantization error = 0
Sample 3: 1.0 V quantization error = –0.05 V
Sample 4: 1.5 V quantization error = 0.09 V
Sample 5: 1.75 V quantization error = 0.1 V

$$\begin{aligned}\text{Mean square error} &= [(-0.05)^2 + 0 + (-0.05)^2 + (0.09)^2 + (0.1)^2]/5 \\ &= [0.0025 + 0.0025 + 0.0081 + 0.01]/5 \\ &= 0.0231/5 \\ &= 0.00462\end{aligned}$$

$$\text{RMS error} = 0.0678 \text{ V}$$

The low signals will have a small SQNR and high signals will have a large SQNR for the same number of quantization levels. Most voice signals generated are of the low kind. To improve voice quality at lower signal levels, uniform quantization was replaced by a non-uniform quantization process.

3.3.2 Non-Uniform Quantization

When the step size is fixed and uniform, it decides the fixed resolution of that quantizer stage. If the speech signal is weak or it is necessary to maintain the smallest amplitude information, which is having an amplitude less than the uniform step size, the method fails.

Here beginning from the smallest possible step size at the lowest amplitude level, step size should be increased gradually, thus non-uniform distribution of levels. This is a more efficient method than the previous one (see Fig. 3.9).

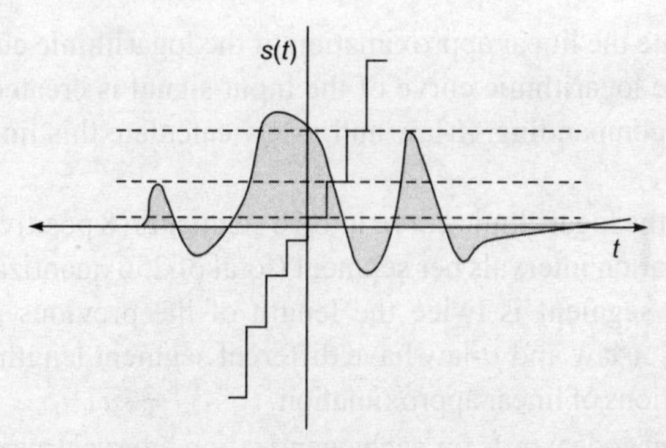

Fig. 3.9 Non-uniform quantization

Non-uniform quantizers distribute the levels in accordance with the PDF of the input waveform. For an input with the PDF $p(x)$, the mean square error is given by

$$E_{\text{ms}} = E[(s(t) - s_q(t))^2] = \int_{-\infty}^{\infty} [s_q(t) - s(t)]^2 \, p(x)dt \tag{3.8}$$

This is the same relationship as mentioned in Eq. (3.6) with necessary addition of PDF function. From the above equation, it is clear that the total distortion can be reduced by decreasing the quantization noise, $[s(t) - s_q(t)]^2$. This means that quantization levels need to be concentrated in amplitude regions of high probability and those that are at the lower amplitude levels. This suggests that there must be lower step sizes at the lower amplitude regions and then increasing at higher voltage levels.

There are many ways to design non-uniform quantizers basically by companding.

Companding refers to the process of first *com*pressing an analog signal at the source, and then ex*panding* this signal back to its original size when it reaches its destination. There are many algorithms for compression and expansion, but a simple and robust method, which is used in commercial telephony is the logarithmic quantizer. During the companding process, input analog signal samples are compressed into logarithmic segments and then each segment is quantized and coded using uniform quantization. In other words, the larger amplitude signals are compressed more than the smaller amplitude signals, causing the quantization noise to increase as the signal amplitude increases. A logarithmic increase in quantization noise throughout the dynamic range of an input sample signal will keep the SQNR constant throughout this dynamic range. The ITU-T standards for companding are called *A*-law and μ-law. The *A*-law standard is used by European countries and the μ-law is used by North America and Japan. The *A*-law and μ-law are audio compression schemes (codecs) defined by the Consultative Committee for International Telephony and Telegraphy (CCITT).

Example 3.3 Give procedure for A-law and μ-law standard companding processes for 256 quantization levels.

Solution

1. Calculate the linear approximation of the logarithmic curve of a sample input signal. The logarithmic curve of the input signal is created during the compression part of companding. A-law and μ-law calculate this linear approximation differently.
2. Divide the logarithmic curve into 16 segments (8 positive and 8 negative), with 16 quantization intervals per segment (total of 256 quantization intervals). Each successive segment is twice the length of the previous segment (logarithmic increase). A-law and μ-law have different segment lengths because of the different calculations of linear approximation.
3. Use 8-bit code words for each quantization interval (maximum of 256 code words, one for each quantization interval). The first bit of code word represents the polarity of quantization interval; bits 2, 3, and 4 represent segment number, and the last four bits indicate the quantization interval within the segment. Eight-bit code words allow for a bit rate of 64 kbps, calculated by multiplying the sampling rate (twice the maximum input frequency) by the size of the code word (2×4 kHz $\times 8$ bits = 64 kbps).

The A-law and μ-law PCM produce acceptable voice quality, which is called *toll quality*.

Limiting the linear sample values to twelve magnitude bits, the A-law compression is defined by Eq. (3.9), where A is the compression parameter ($A = 87.7$ in Europe) and x is the normalized (normalized to 1) integer to be compressed.

$$F(x) = \begin{cases} \dfrac{A \times |x|}{1 + \ln(A)} & 0 \le |x| < \dfrac{1}{A} \\ \dfrac{\operatorname{sgn}(x) \times (1 + \ln(A|x|))}{1 + \ln(A)} & \dfrac{1}{A} \le |x| \le 1 \end{cases} \tag{3.9}$$

Limiting the linear sample values to thirteen magnitude bits, the μ-law compression is defined by Eq. (3.10), where μ is the compression parameter ($\mu = 255$ in the US and Japan) and x is the normalized integer to be compressed.

$$F(x) = \frac{\operatorname{sgn}(x) \times \ln(1 + \mu|x|)}{\ln(1 + \mu)} \qquad 0 \le |x| \le 1 \tag{3.10}$$

The A-law standard is primarily used by Europe and the rest of the world, while the μ-law is used by North America and Japan. Here

$$\operatorname{sgn}(x) = \begin{cases} +1, \text{ for } x \ge 0 \\ -1, \text{ for } x < 0 \end{cases} \tag{3.11}$$

Similarities between A-law and μ-law

- Both provide linear approximations of logarithmic input/output relationship.

- Both are implemented using eight-bit code words (256 levels, one for each quantization interval). Eight-bit code words allow for a bit rate of 64 kbps. This is calculated by multiplying the sampling rate (twice the maximum input frequency) by the size of the code word (2×4 kHz $\times 8$ bits = 64 kbps).
- Both break a dynamic range into a total of 16 segments (piecewise approximated companding curve):
  - There are eight positive and eight negative segments.
  - Each segment is twice the length of the preceding one.
  - Uniform quantization is used within each segment.
- Both use a similar approach to coding the eight-bit word.

Differences between A-law and μ-law

- Different linear approximations lead to different lengths and slopes.
- The numerical assignment of the bit positions in the eight-bit code word to segments and the quantization levels within segments are different.
- The A-law provides a greater dynamic range than the μ-law.
- The μ-law provides better signal/distortion performance for low-level signals than the A-law. The μ-law has a better SQNR at the lower signal levels than the A-law.
- The A-law requires 12 bits for a uniform PCM equivalent. The μ-law requires 13 bits for a uniform PCM equivalent.
- An international connection needs to use the A-law and μ to A conversion is the responsibility of the μ-law country.

3.3.3 Adaptive Quantization

Adaptive quantization is an upgraded version of non-uniform quantization. There is a difference between the long-term and short-term PDFs of speech waveform because it is non-stationary stochastic process. The time-varying random nature of speech signal results into a dynamic range of approximately 35– 40 dB or more. Similarly, image and video signals are also of random and time-varying nature. An efficient way to accommodate large dynamic range is to adopt a time-varying quantization technique. It means that when the input signal is of low peak-to-peak amplitude, keeping the same number of intervals with a small step size, as may be in the case of large peak-to-peak amplitude signal with larger step size. Adaptive quantization scheme is an important scheme, which is suitable for audio, image as well as for very low bit rate video coders. In recent years, many adaptive quantization schemes have been proposed. Broadly, these schemes can be divided into two categories: backward and forward adaptive quantization schemes.

Backward Adaptation

The basis to adjust quantization step of backward adaptive quantization schemes is the fullness of buffer and the bit rate of channel. This kind of method can make the output bit rate of the encoder fit the channel effectively and the decoded audio quality is also good but it causes much fluctuation of the decoded image/video quality when the bit rate is low. Backward adaptation uses the past decoded speech to estimate the linear prediction (LP) model but the following issues are also important due to which the method is not popular nowadays:

- LP parameters are not transmitted directly because of the following reasons:
    - LP model is poorly trained on delayed data.
    - LP paramters are hard to quantize and consume considerable bandwidth.
    - LP model has simpler bitstream format.
- Longer model order is possible and is not limited by quantization error.
- Decoder needs more power to generate the LP model.
- Backward adaptation technique works best for low-delay systems.

The method can be used in linear predictive coders, which are economical with 2.4 kbps bit rate and lossy.

Forward Adaptation

Forward adaptive quantization schemes adjust quantization step by predicting the output bit amount or iterative trial. Here statistical concepts or blind methods can be followed. Although the computation cost of forward adaptive quantization schemes is higher than that of backward schemes, the decoded image quality in forward schemes is better than that in backward schemes because they take considerations of signal contents (properties). Figure 3.10 shows a typical example of forward adaptive quantizer.

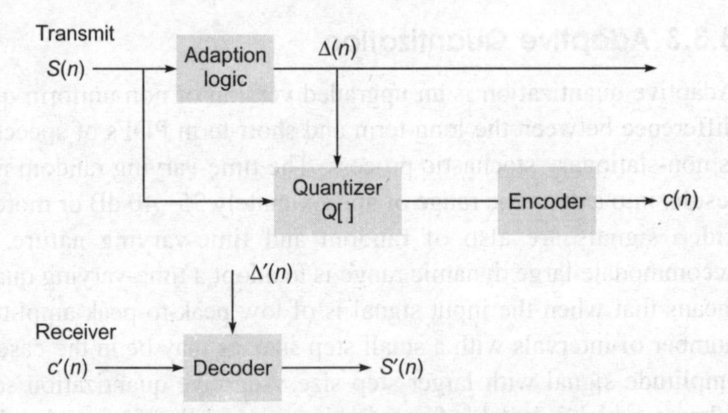

Fig. 3.10 Typical forward adaptive quantizer

There are many new techniques derived for adaptive quantization. Comparing with traditional backward schemes, the new adaptive quantization schemes take considerations of not only the status of buffer but also the information contents and human characteristics for perceptions of multimedia. Comparing with traditional forward schemes, new schemes take advantage of fuzzy theory to avoid complex computation and iterative trial. With these adaptive quantization schemes, we can keep the decoded information quality consistent with good audio-visual effects also.

3.3.4 Vector Quantization

Till now the methods studied are scalar quantization, in which the scalar value is selected from a finite list of possible values to represent an input sample, which is close (in some sense) to the sample it is representing. In vector quantization, a vector is selected from a finite list of possible vectors to represent an input vector of samples. The selected vector is chosen to be close (in some sense) to the vector it is representing. *Vector quantization* (VQ) is a lossy data compression method based on the principle of block coding. It is a fixed-to-fixed length algorithm. In the earlier days, the design of a vector quantizer was considered to be a challenging problem due to the need for multidimensional integration. In 1980, Linde, Buzo, and Gray (LBG) proposed a VQ design algorithm based on a training sequence. The use of a training sequence bypasses the need for multidimensional integration. A VQ that is designed using this algorithm is referred to in the literature as an LBG-VQ and is nothing more than an approximator. The idea is similar to that of 'rounding-off' (say, to the nearest integer).

If we talk in terms of waveforms, Shannon's rate distortion theorem states that there exists a mapping from a source waveform to output code words such that for a given distortion D, $R(D)$ bits per sample are sufficient to reconstruct the waveform with an average distortion arbitrarily close to D. Therefore, the actual rate R has to be greater than $R(D)$. This function is called rate-distortion function, represents a fundamental limit on the achievable rate for a given distortion. Scalar quantizers do not achieve performance close to this information theoretical limit, thus vector quantization becomes important. Shannon predicted that better performance can be achieved by coding many samples at a time instead of one sample at a time because of better approximation or averaging.

Vector quantization is an example of competitive learning. Normally, a basic competitive learning network has one layer of input nodes and one layer of output nodes. Binary valued outputs are often (but not always) used. There are as many output nodes as there are classes. The goal here is to have the network 'discover' the structure in the data by finding how the data is clustered. The results can be used for data encoding and compression.

Each input vector (samples) can be visualized as a point in an N-dimensional space. The quantizer is defined by a partition of this space into a set of non-overlapping volumes. These volumes are called intervals, polygons, and polytops respectively for 1-, 2-, and N-dimensional vector spaces. The task of the vector quantizer is to determine the interval, polygon, or volume, in which input vector is located. The output of the optimal quantizer is the vector identifying the centroid of that volume for N-dimensional case. Just like in scalar quantizers, the mean square error is a function of the boundary locations for the partition and the multidimensional PDF of the input vector. The mentioned terms will be clarified with the following examples. Quantization in 1-D and 2-D cases is given in Example 3.4.

Example 3.4 With the help of numerical example, explain the difference between one-dimensional VQ and two-dimensional VQ.

Solution

Typical numerical values are selected for the explanation here.

For one-dimensional VQ:

Every number less than –2 is approximated by –3.

Every number between –2 and 0 are approximated by –1.

Every number between 0 and 2 are approximated by +1.

Every number greater than 2 is approximated by +3.

Thus, four quantization levels are chosen. Note that the approximate values are uniquely represented by 2 bits. This is a one-dimensional, two-bit VQ. It has a rate of 2 bits/dimension [Fig. 3.11(a)].

For two-dimensional VQ:

In contrast to one-dimensional VQ, here, every pair of numbers falling in a particular region (polygon) is approximated by a centroid (•) (shown in Fig. 3.11) associated with that region. For example, if there are 16 regions, each of them can be uniquely represented by 4 bits. Thus, this is a two-dimensional, four-bit VQ. Its rate is also 2 bits/dimension [shown in Fig. 3.11(b)].

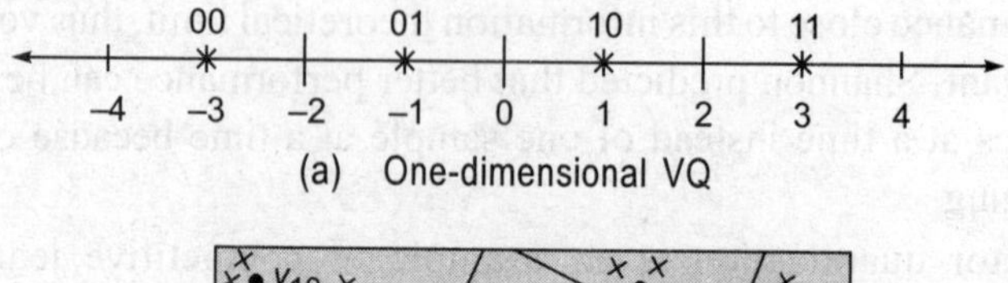

(a) One-dimensional VQ

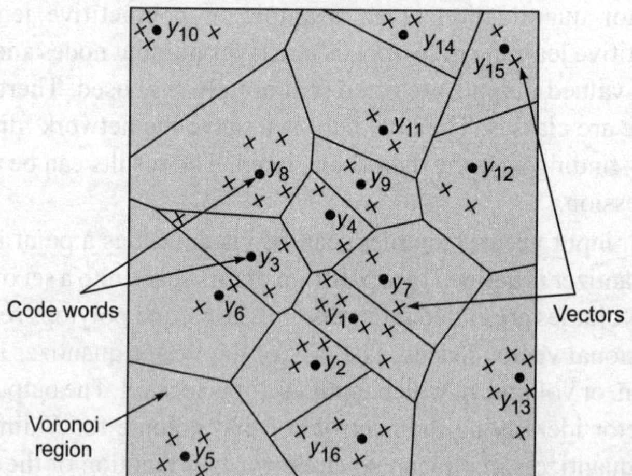

(b) Concept of encoding region, input vectors, and code words (Input vectors are marked with crosses (x), code words are marked with dots, and the Voronoi regions are separated with boundary lines.)

Fig. 3.11

In the above example, the stars are called *vectors* and the regions are called *encoding regions*, which are nothing but polygons as mentioned previously. The regions are also called the *clusters*. Dots represent the code vectors. The set of all code vectors is called the *code book* and the set of all encoding regions is called the *partition* of the space. Code book has an address too. Each code word has an address. We only transmit the addresses and not the code. The source encoder and the decoder using vector quantization are shown in Fig. 3.12. In Fig. 3.12, such addresses are represented with indices.

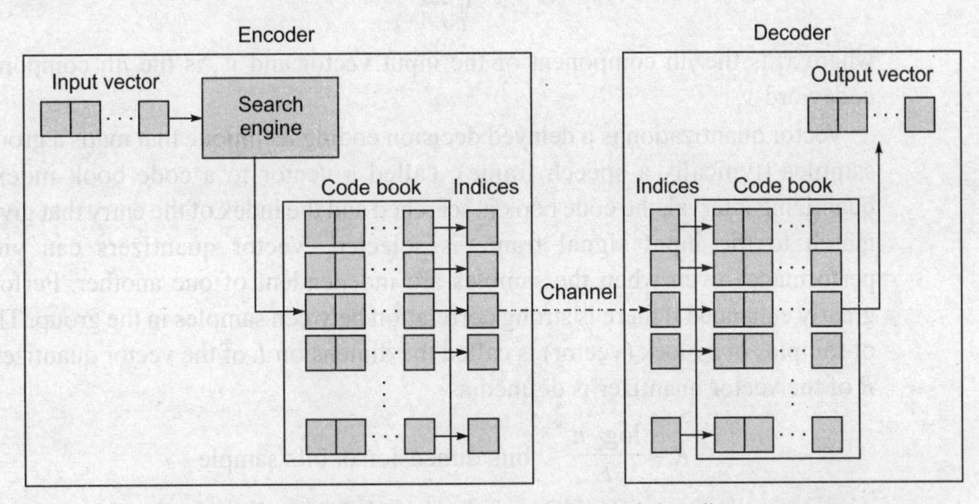

Fig. 3.12 Vector quantized source encoding and decoding processes

From Fig. 3.11(b), let us understand the vector quantization mathematically, which can be extended for *N*-dimensional case. Say, a vector quantizer maps *k*-dimensional vectors in the vector space R^k into a finite set of vectors $Y = \{y_i: i = 1, 2, ..., N_c\}$. Hence, N_c clusters will be there and corresponding N_c vectors will be generated. Each vector y_i is called a code vector or a *code word* and the set of all the code words is listed out in a *code book Y*. Associated with each code word y_i, there is a nearest neighbouring region, called *Voronoi* region, and it is defined by

$$V_i = \{x \in R^k : \|x - y_i\| \leq \|x - y_j\|, \text{ for all } j \neq i\} \tag{3.12}$$

An input belongs to cluster *i* if *i* is the index of the closest prototype (closest in the sense of the normal Euclidean distance). This has the effect of dividing up the input space into Voronoi regions [Figures 3.11(b) and 3.13]. For each training pattern, the reference vector that is closest to it is determined.

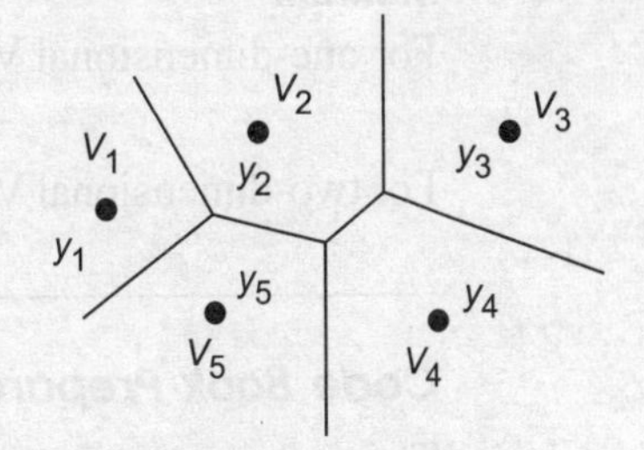

Fig. 3.13 Voronoi regions and code vectors for 2-D VQ mathematics

Method of Clustering

Divide the data into a number of clusters such that the inputs in the same cluster are in some sense similar.

The set of Voronoi regions V_i's partition the entire space R^k such that

$$\bigcup_{i=1}^{N_c} V_i = R^k \text{ and } \bigcap_{i=1}^{N_c} V_i = \phi \quad \text{for all } i \neq j \tag{3.13}$$

The representative code word is determined to be the closest in Euclidean distance from the input vector x. The Euclidean distance is defined by

$$d(x, y_i) = \sqrt{\sum_{j=1}^{k} (x_j - y_{ij})^2} \tag{3.14}$$

where x_j is the jth component of the input vector and y_{ij} is the jth component of the codeword y_i.

Vector quantization is a delayed decision coding technique that maps a group of input samples (typically a speech frame), called a vector to a code book index. In each quantizing interval, the code book is searched and the index of the entry that gives the best match to the input signal frame is selected. Vector quantizers can yield better performance even when the samples are independent of one another. Performance is greatly enhanced if there is strong correlation between samples in the group. The number of samples in a block (vector) is called the dimension L of the vector quantizer. The rate R of the vector quantizer is defined as

$$R = \frac{\log_2 n}{L} \quad \text{bits/dimension or bits/sample} \tag{3.15}$$

where n is the size of the VQ code book and R may also take fractional values. All the quantization principles used in scalar quantization apply to vector quantization as a straight forward extension. Instead of quantization levels we have quantization vectors and distortion is measured as a squared Euclidian distance between the quantization vector and the input vector. VQ is known to be most efficient at a very low bit rate ($R = 0.5$ bits/sample or less).

Example 3.5 The speech frame with 24 samples is to be quantized using vector quantization method and 8 bits/sample will be used for representing the quantized samples. Find out the rate of quantization for 1 dimensional VQ and for 2 dimensional VQ.

Solution

For one-dimensional VQ, the rate of quantization is

$$R = \log_2 8/1 = 3 \text{ bits/dimension}$$

For two-dimensional VQ,

$$R = \log_2 8/2 = 3/2 = 1.5 \text{ bits/dimension}$$

Code Book Preparation and Its Use

There are two main tasks in the vector quantization. First is the code design task, which deals with problem of performing the multidimensional volume quantization (partition) and selecting the allowable output sequences. The second task is that of using the code, and deals with searching for the particular volume with this partition that corresponds (according to some fidelity criteria) to the best description of the source. The form of the algorithm selected to control the complexity of encoding and decoding may couple both the tasks—the partition and the search. The standard vector coding methods are code book, tree, and trellis-coding algorithms.

Code books are nothing but look-up table algorithms. A list of code words is stored in the code book memory. Each pattern is identified by an address or pointer index. The algorithm tries to find the best match from the code book and transmits the pointer index. The tree and trellis coders are sequential coders. Here the code words are not selected independently but a node steering structure is used. Hence, tree graph suffers form exponential memory growth as the depth of the tree increases. The trellis graph reduces the dimensionality problem by the simultaneous tracking of contender paths with an associated path weight metric called intensity, with the use of finite state trellis.

The code vectors stored in the code book, tree, or trellis are the likely or typical vectors. The task of identification of likely code vectors is called *populating* the code. The methods of determining the code population are classically deterministic, stochastic and iterative.

Given an input vector and a populated code book, tree, or trellis, the coder algorithm must conduct a search to determine the best matching contender vector. Coder performance improves for the larger dimensional spaces but complexity increases at the same rate. An exhaustive search over a large dimension improves the coder performance, but is prohibitively time consuming also. Alternatively, non-exhaustive, sub-optimum search schemes can be used, with acceptably degraded performance.

VQ and Data Compression

Vector quantization can be used for (lossy) data compression. If we are sending information over a phone line, instead of sending the data/code directly, we

- initially send the code book vectors and
- for each input, send the index of the class that the input belongs.

For a large amount of data, this can be a significant reduction. Basic concept of data compression is described in Section 3.8. If code vectors y_i, have $i = 1$ to 64, then it takes only 6 bits to encode the index. If the data itself consists of floating point numbers (4 bytes, i.e. 32 bits), there is an 80% data reduction [$100 \times (1 - 6/32)$] will be achieved.

3.3.5 New Quantization Methods

Some new quantization methods have come up due to the vast development in digital signal processing. They are transform-based quantization methods, which are associated with the coding as well as the compression parts. Here they are introduced in brief. The details are given as independent topics afterwards. Taking example of an image, during quantization, each pixel of the transformed digital image is mapped to a discrete number. Each integer in the range of numbers used in the mapping symbolizes a colour.

In Haar wavelet quantization, if the image to be quantized has $N \times M$ pixels and the new colour scale is to have $M = 2^n$ numbers, then the number of bits required to store the quantized image is $N \times M \times n$. If the images used are 256×256 pixels, in order to uniformly quantize the image, we would need $256 \times 256 \times n$ bits. In the case of transformed digital image, all of the energy resides in a particular section of the array. Adaptive quantization methods can take advantage of the fore knowledge of smooth regions by using fewer quantization levels in these areas. The fewer the colour levels used, the smaller the number of bits required. Likewise, neighbourhoods of the digital image with high energy should be represented with more quantization levels. More bits will be used in these regions, but if the neighbourhoods are small, then the overall reduction in bits will be greater than that achieved by uniform quantization.

If the image is DCT transformed, it is divided into $m \times m$ blocks. These blocks are then encoded individually. The blocks on the top left corner would be encoded with more bits to keep the important information or energies. As we move away from the upper left-hand corner, the blocks are encoded with fewer and fewer bits. Eventually hitting the bottom right corner, the blocks are encoded with few if any bits. This is the usual DCT quantization method.

Both these methods are described afterwards at the end of this chapter.

3.4 PULSE CODE MODULATION (PCM)

In analog transmissions, analog signals become weak because of transmission loss and noise. Amplifying analog signals also amplifies noise, and eventually they become too noisy to use. Digital signals, having only 'one-bit' and 'zero-bit' states, are more easily separated from noise and can be amplified without corruption. Over time, it has become obvious that digital coding is more immune to noise corruption on long-distance connections, and the world's communications systems have converted to a digital transmission format called *pulse code modulation* (PCM). The PCM is a waveform coding method defined in the ITU-T G.711 specification. The PCM block diagram is given in Fig. 3.14.

Pulse code modulation is formed from an analog signal by operations of antialiasing filtering sampling, quantization, A/D conversion, and encoding. The input analog signal,

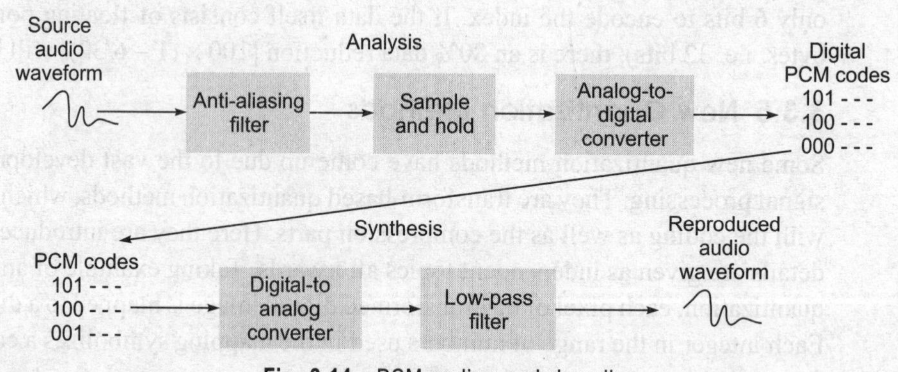

Fig. 3.14 PCM coding and decoding

denoted as $s(t)$ in Fig. 3.15. is continuous, in both time and amplitude and random in amplitude. Pulse code modulation is a process that begins by low-pass filtering the analog signal to ensure that no frequencies above f_{max} are present; such a filter is called an *antialiasing filter*. The next step is to sample the filtered analog input. A clock signal generates pulses at Nyquist sampling rate of at least 2 f_{max}, which are used by a sampler to produce 2 f_{max} samples of the analog signal per second. The sampling operation will keep the signal still continuous in amplitude but discrete in time. Such signals are often referred to as sampled data signals. Sampling concept is introduced in Chapter 2, however, proof of *sampling theorem* can be referred from any basic book on communication theory or on digital signal processing.

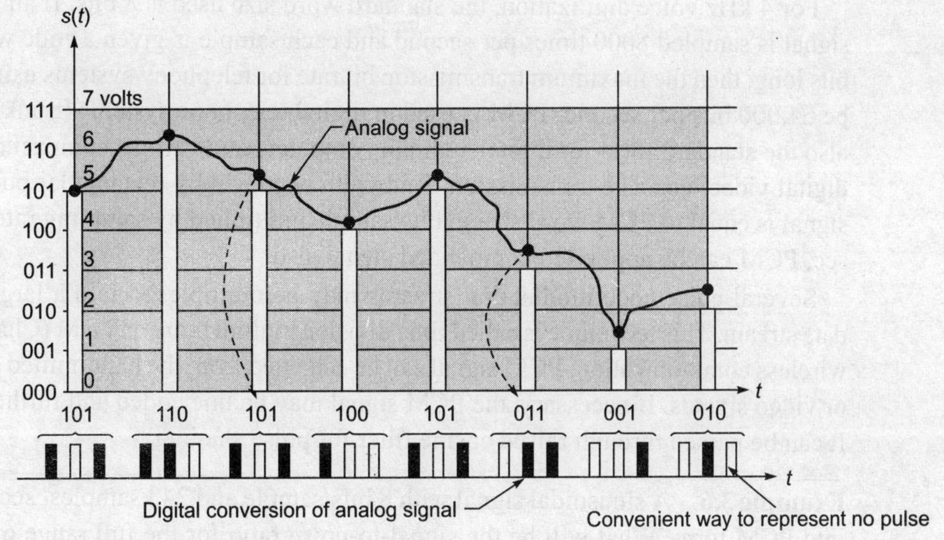

Fig. 3.15 Analog signal converted into PCM signal by discrete level assigned to the sample and A/D conversion

Sampling is followed by quantizing and rounding off each sampled values. The result of sampling and quantizing–called digitization–is a series of varying-amplitude pulses at sampling rate. Such a Pulse Amplitude Modulated (PAM) signal, discrete in time and amplitude, is shown in Fig. 3.15. This is the stage where errors are introduced at each step of this process. This signal is basically a step-modulated AM signal subject to degradation by noise as in any AM signal. To convert these step pulses to a digital signal, the pulses are grouped according to the grouping of the bits in the binary numbers and are then transmitted in rapid sequence. We can verify that the values at the sampling moments are 5, 6, 5, 4, 5, 3, 1, 2, etc. The integers are translated into a binary representation; the PCM data encoding would look like this in binary: 101, 110, 101, 100, 101, 011, 001, 010, etc. After PCM, digital signal can be either return-to-zero (RZ), non-return-to-zero (NRZ), or any other line coding format. For an NRZ system to be synchronized using in-band information, there must not be long sequences of identical symbols, such as 1's or 0's, so data *scrambling* is applied. Scrambling is nothing but data randomization.

The PCM is a lossy conversion. It may be noted that there are two sources of impairment:

- Rounding the analog signal to the nearest integer value: quantization error/noise.
- The frequency range of the analog signal is higher than half the sampling rate: aliasing error.

As the transmitted on-off pulses are of equal amplitude, they are immune to additive noise in the sense that only the presence or absence of a pulse needs to be determined at the receiver, this process is reversed at the receiver, which converts binary data into analog levels, decoding the received binary number into a staircase analog signal which then smoothed by passing it through a low-pass filter with bandwidth f_{max}.

For 4 kHz voice digitization, the standard word size used is 8 bits. If an input analog signal is sampled 8000 times per second and each sample is given a code word that is 8 bits long, then the maximum transmission bit rate for telephony systems using PCM will be 64,000 bits per second. PCM is used in digital telephone systems (trunk lines) and is also the standard form for digital audio in computers and various compact disc formats, digital video, etc. The transmission bandwidth occupied by rectangular pulses of PCM signal is equal to PCM word size in bits/sample multiplied by sampling rate in samples/sec. PCM can be applied to companded signal also.

Several pulse code modulation streams may be multiplexed into a larger aggregate data stream. This technique is called time division multiplexing or TDM (Chapter 10). For wireless communication, PCM signal can be achieved from the bandlimited audio, image or video signals. If necessary, the PCM signal may be line coded and further processed. It can be passed through raised cosine filter for pulse shaping.

Example 3.6 A sinusoidal signal with 8 bits/sample and 24 ksamples/ sec is converted into PCM form. What will be the signal-to-noise ratio for the full range of sinusoid. If companding is applied before PCM conversion, what will be the new SNR value with $\mu = 100$?

Solution

For uniformly quantized sinusoid, for fine quantization, the quantized signal power can be approximated to the second moment of the clean signal $s(t)$.

So, $$\text{Quantized signal power} = \overline{s^2(t)}$$

The quantization noise power is given by $\dfrac{s_p^2}{3M^2}$, where M is the number of quantization levels and s_p^2 is the peak quantization level of the uniform quantizer.

Also $\left[\dfrac{s_p^2}{\overline{s^2(t)}}\right]^{1/2}$ is known as the crest factor.

Now $$\text{SNR} = 3M^2 \frac{\overline{s^2(t)}}{s_p^2} = 3(2^n)^2 \frac{\overline{s^2(t)}}{s_p^2}$$

$$\Rightarrow \qquad \frac{\overline{s^2(t)}}{s_p^2} = 1/2 \text{ for sinusoid}$$

$$\Rightarrow \qquad \text{SNR} = 3(2^8)^2 \times 1/2$$
$$= 3(256)^2 \times 1/2$$
$$= 98{,}304$$
$$= 49.93 \text{ dB}$$

For companded sinusoid,

$$\text{SNR} = \frac{3M^2}{[\ln(1+\mu)]^2} = \frac{3(2^8)^2}{[\ln(1+100)]^2} = \frac{196608}{21.3} = 9230.423 = 39.65 \text{ dB}$$

Example 3.7 Find out the bandwidth expansion factor for the PCM signal with sync pulses and with rectangular pulses, whose basic bandwidth is 4 kHz and the number of bits per sample is 8.

Solution

According to Nyquist criteria, the sampling rate is 8 ksamples/sec.

For sync pulses:

Transmission bandwidth = 8 × 8/2 = 64/2 = 32 kHz

Bandwidth expansion factor = 32/4 = 8

For rectangular pulses:

Transmission bandwidth = 8 × 8 = 64 kHz

Bandwidth expansion factor = 64/4 = 16

The SNR can be improved by trading the bandwidth because SNR increases with transmission bandwidth of PCM signal.

3.5 DELTA MODULATION

The pulse-code modulation is an absolute coding method and uses no data compression. In playback, the data bits representing the absolute values of each successive signal sample are sent to a full-resolution D/A converter and reproduced at the same rate at which they were recorded: 96,000 bps in, 96,000 bps out.

On the other hand, voice waveforms contain much redundant data. Long periods of silence are interspersed with sounds that vary in pitch slowly. If some time is taken to analyse the A/D samples, it will be noticed that the changes are, for the most part, gradual and the variations in the signal between adjacent samples are a limited portion of the full dynamic range. One method of reducing the data rate used in PCM voice reproduction is called delta modulation. This process assumes that the input signal's waveform has a fairly uniform and predictable slope (rate of rising and falling). Rather than storing an 8- or 12-bit quantity for each sample, a delta modulator stores only a single bit. When the computer samples the input signal from the A/D converter, it compares the current reading to the preceding sample. If the amplitude of the new sample is greater, then the computer stores a bit value of 1. Conversely, if the new sample is less than the previous one, then a 0 will be stored.

Each sample of the source waveform is tested to see if its amplitude is higher or lower (within the resolution of a fixed quantization value Δr) than that of the previous sample. If the amplitude is higher, the single-bit delta-modulated encoding value is set to 1; if lower, the encoding value is set to 0 as shown in Fig. 3.16.

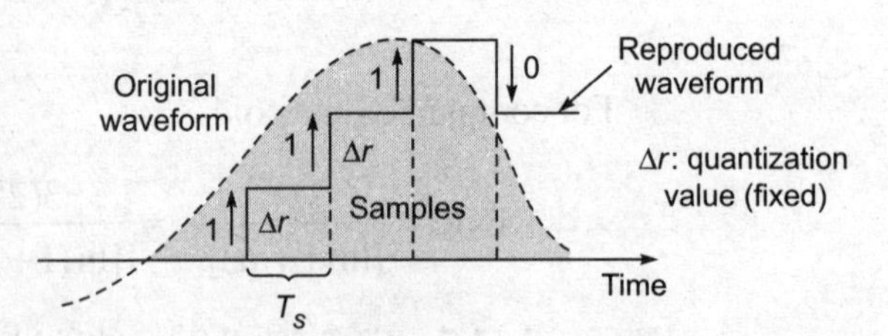

Fig. 3.16 Waveform sampling, coding, and reproduction concept for delta modulation

Reproduction of the waveform is accomplished by sending the stored bits in sequence to the output, where their values are integrated. But, like other techniques, delta modulation has limitations, one of them the familiar sampling-rate restriction. Because only a single bit changes between samples, the rate at which samples are taken must be sufficiently fast such that no significant information is lost from the input signal. When the source waveform changes too rapidly, the fixed quantization value may be too small to express the full change in the input; this slope overload causes a *compliance error* or when there is little change in the input waveform (at the extreme, a DC signal), vertical deflection in the quantization value results in *granular noise* in the output due to small signal amlitude. Slope overload and granular noise are shown in Fig. 3.17.

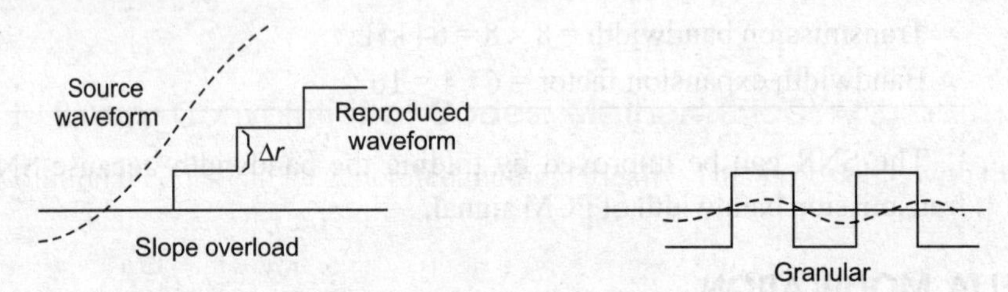

Fig. 3.17 Two potential problems occurring in delta modulation

Furthermore, if the slope of the input waveform varies a lot, the reproduced waveform may be audibly distorted. So using delta modulation may not reduce the data rate much, although there are many different variant schemes and it is difficult to predict which is optimal in a given situation. Delta modulation is popular not for commercial use but it forms the basis for DPCM and ADPCM when it is combined with PCM, which gives the appreciable performance.

3.6 DIFFERENTIAL PULSE CODE MODULATION METHODS OF CODING

In conventional PCM, the analog signal is processed before being digitized. Once the signal is digitized, the PCM signal is usually subjected to further processing (e.g., data compression). Some modified forms of PCM combine signal processing with coding. These simple techniques have been largely rendered obsolete by modern transform-based signal compression techniques. There are two modifications to the conventional PCM.

1. *Differential (or delta) pulse-code modulation (DPCM)*, which encodes the PCM values as differences between the current and the previous values. For audio, this type of encoding reduces the number of bits required per sample by about 25% compared to PCM.
2. *Adaptive DPCM (ADPCM)*, which is a variant of DPCM that varies the size of the quantization step, to allow further reduction of the required bandwidth for a given SQNR.

3.6.1 DPCM

In telephony, a standard audio signal for a single phone call is encoded as 8000 analog samples per second, of 8 bits each, giving a 64 kbps digital signal known as DS0. The default encoding on a DS0 is either μ-law PCM or A-law PCM. These are logarithmic compression systems where a 12- or 13-bit linear PCM sample number is mapped into an 8 bit value as discussed earlier.

DPCM was designed to calculate the difference between consecutive sample and then transmit this small difference signal instead of the entire input sample signal. Since the difference between input samples is less than an entire input sample, the number of bits required for transmission is reduced. Using DPCM can reduce the bit rate of voice transmission down to 48 kbps. Figure 3.18 shows the DPCM diagram.

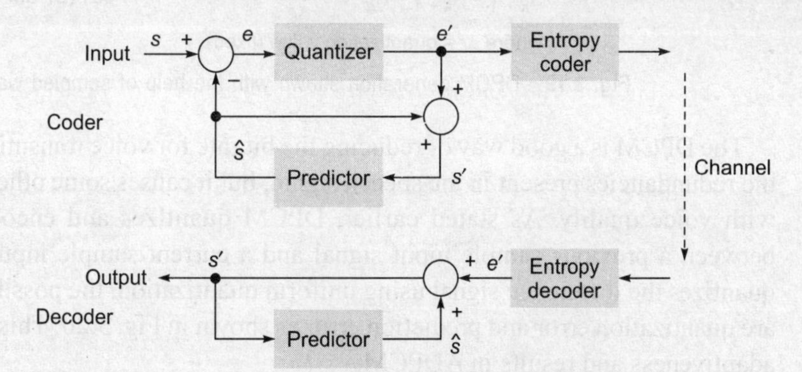

Fig. 3.18 Block diagram of DPCM

The DPCM works exactly like PCM in which quantization process is applied to difference value of the samples and then coding of difference value is done (that is why it is called differential PCM). How does DPCM calculate the difference between the current sample signal and a previous sample? The input signal is sampled at a constant Nyquist sampling frequency resulting in PAM signal. At this stage, the DPCM process takes over. The sampled input signal is stored in what is called a predictor. The predictor takes the stored sample signal and sends it through a differentiator. The differentiator compares the previous sample signal with the current sample signal and sends this difference to the quantizing and coding phase of PCM. (This phase could be uniform quantizing or companding with A-law or μ-law.) After quantizing and coding, the difference signal is transmitted to its final destination. At the receiving end of the network, everything is reversed: First the difference signal is decoded and dequantized. Then this difference signal is added to a sample signal stored in a predictor and sent to a low-pass filter that reconstructs the original input signal.

Figure 3.19 shows the calculation of DPCM values for the final transmission. The three consecutive samples are shown ($k-1$, k, and $k+1$), which have amplitudes nearer to each other and due to correlation between consecutive samples the next one can be predicted, which gives the difference signal to be coded. It is clear that for representing the absolute levels three different PCM codes are required while for DPCM equivalently only two differential values are coded. Thus, reduction in data is achieved.

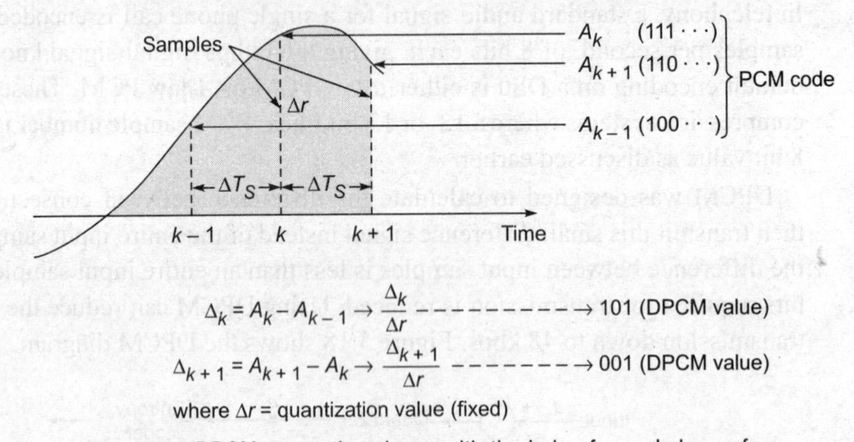

Fig. 3.19 DPCM generation shown with the help of sampled waveform

The DPCM is a good way of reducing the bit rate for voice transmission by exploiting the redundancies present in the speech signal, but it causes some other problems dealing with voice quality. As stated earlier, DPCM quantizes and encodes the difference between a previous sample input signal and a current sample input signal. If DPCM quantizes the difference signal using uniform quantization, the possible sources of error are quantization error and prediction error as shown in Fig. 3.20. This needs some sort of adaptiveness and results in ADPCM.

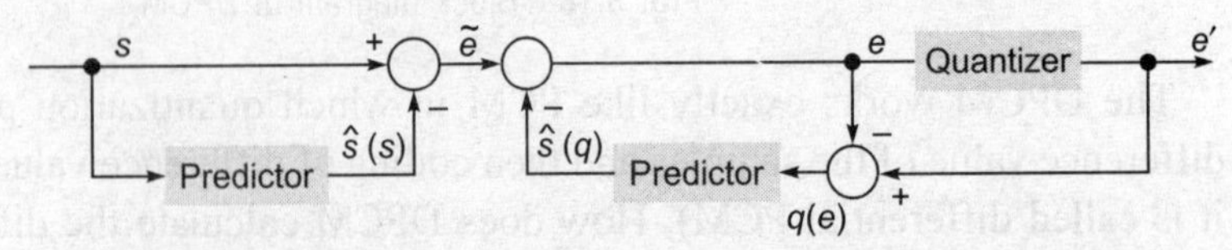

Fig. 3.20 Quantization error feedback in DPCM coder and prediction error

The DPCM exhibits some of the same limitations as simple delta modulation but to a lesser degree. Only when the difference between samples is greater than the maximum DPCM encoding value, then the distortion (called a compliance error) occurs. The only solution is to reduce the input bandwidth or raise the sampling frequency.

3.6.2 ADPCM

The ADPCM codec is also a waveform codec, which quantizes the difference between the consecutive speech signal samples or a prediction has been made of the speech signal as in DPCM but quantization is adaptive. In practice, ADPCM encoders are implemented using signal prediction techniques accommodating the forward adaptive quantizers, instead of simply encoding the difference between adjacent samples. A linear predictor is used to predict the current sample using previous ones. The difference between the predicted and actual sample called the prediction error is then encoded for transmission. Prediction is based on the knowledge of the autocorrelation properties of speech or fuzzy logic may be applied.

If the prediction is accurate then the difference between the real and predicted speech samples will have a lower variance than the real speech samples, and will be accurately quantized with fewer bits than would be needed to quantize the original speech samples. At the decoder the quantized difference signal is added to the predicted signal to give the reconstructed speech signal or recovers an approximation to the original speech signal by essentially integrating the quantized adjacent sample differences. Since the *quantization error variance* for a given number of bits/sample is directly proportional to the *input signal variance*, the reduction obtained in the quantizer input variance reduces the *reconstruction error variance* for a given value of bits/sample. Figure 3.21 shows the ADPCM generation. The difference in the method can be compared with Fig. 3.19 showing DPCM generation.

The performance of the DPCM codec is aided by using adaptive prediction and quantization. How does ADPCM adapt these quantization levels? If the difference signal

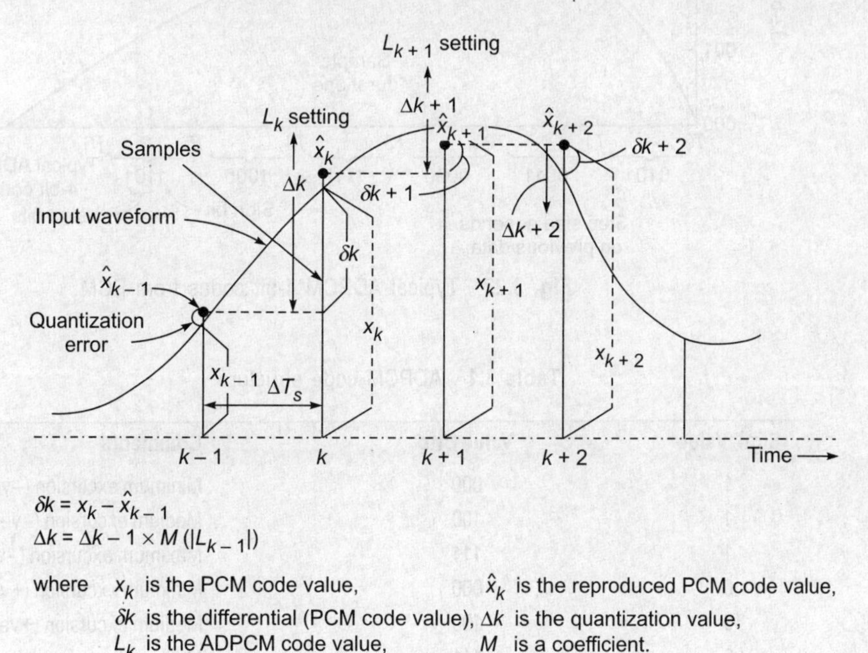

Fig. 3.21 ADPCM generation concept

is low, ADPCM lowers the size of the quantization levels. If the difference signal is high, ADPCM decreases the size of the quantization levels. So, ADPCM adapts the quantization level to the size of the input difference signal, generating an SNR that is uniform throughout the dynamic range of the difference signal. In ADPCM, each sample's encoding is derived by a procedure that includes the following steps (refer to Fig. 3.21):

- A differential-k (δk) is obtained by subtracting the previous PCM-code value from the current value.
- The quantization value (Δk) is obtained by multiplying the previous quantization value times a coefficient times the absolute value of the previous ADPCM-code value.
- The PCM-valued differential-k is then expressed in terms of the quantization value (Δk) and encoded in four bits.

Using the above procedure, typical values of ADPCM are derived from the PCM values and are illustrated in Fig. 3.22 and Table 3.1.

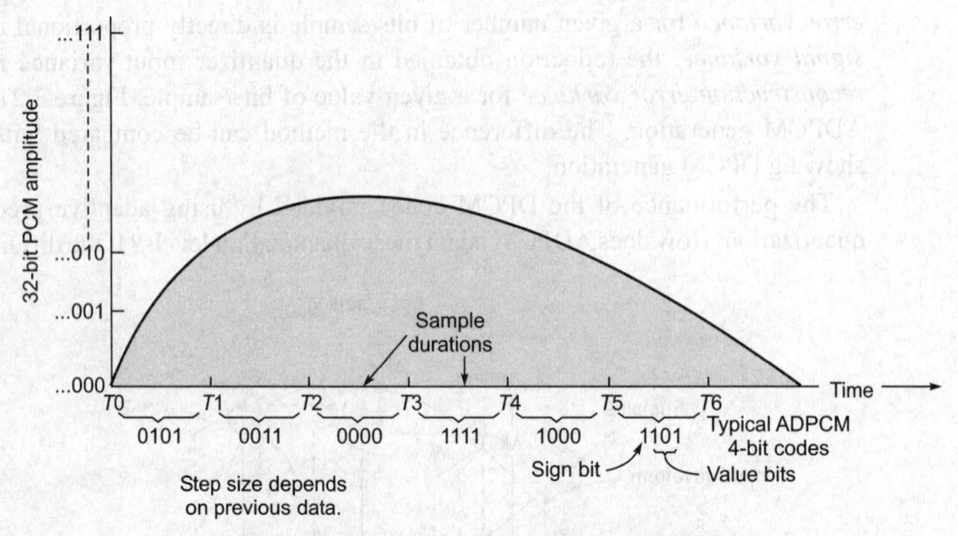

Fig. 3.22 Typical ADPCM 4-bit codes from PCM

Table 3.1 ADPCM code structure

Sign Value	Value Bits	Comments
1	000	Minimum excursion (–ve direction)
1	100	Medium excursion (–ve direction)
1	111	Maximum excursion (–ve direction)
0	000	Minimum excursion (+ve direction)
0	100	Medium excursion (+ve direction)
0	111	Maximum excursion (+ve direction)

The ADPCM condenses 12-bit PCM samples into only 3 or 4 bits. Using ADPCM, we can reduce the bit rate of voice transmission down to 32 kbps, half the bit rate of A-law or μ-law PCM. The ADPCM produces 'toll quality' voice just like A-law or μ-law PCM.

The ADPCM encoder makes best use of the available dynamic range of four bits by varying its step size in an adaptive manner. The step size of the quantizer depends on the dynamic range of the input, which is speaker dependent and varies with time. The adaption is in practice achieved by normalizing the input signals via a scaling factor derived from a prediction of the dynamic range of the current input. This prediction is obtained from two components: A fast component for signals with rapid amplitude fluctuations and a slow component for signals that vary more slowly. The two components are weighted to give a single-quantization scaling factor.

The ADPCM can be applied to image and video signals also in the same manner. In images, there is correlation between nearer pixels and in video there is a correlation between two consecutive frames. In the mid 1980s the CCITT standardized a 32 kbps ADPCM for speech signal, known as G-721, which gave reconstructed speech almost as good as the 64 kbps PCM codec. Later in recommendations G-726 and G-727, codec operating at 40 kbps, 32 kbps, 24 kbps and 16 kbps were standardized. There are many algorithms for ADPCM but the CCITT standard G-721 ADPCM algorithm for 32 kbps speech coding is used in cordless telephone systems like CT2 and DECT. Where circuit costs are high and loss of voice quality is acceptable, it sometimes makes sense to compress the voice signal even further. An ADPCM algorithm is used to map a series of 8 to 12 bit PCM samples into a series of 4 bit ADPCM samples. In this way, the capacity of the line is almost doubled. (Later it was found that even further compression was possible and additional standards were published. Some of these international standards describe systems and ideas that are covered by privately owned patents and thus use of these standards requires payments to the patent holders). Some ADPCM techniques are used in voice over IP communications.

3.7 INFORMATION SOURCES AND ENTROPY

As the digital baseband signal is achieved from the analog sources, the database may not be completely useful and some redundant information may be present. Hence, the actual outcome may be different than the actual source-data. When we deal with sources, the amount of useful information must be measured so that data compression can be applied by removing redundant information and the best possible speed of useful data communication can be achieved and for that, it is necessary to study information theory. At the same time, channel capacity is also limited according to Shannon who discovered long before that information transmission, noise, and communications channel capacity have a very special relationship. By long tradition, engineers have used the term *bit* to describe both the symbol and information content. Representing in the form of a statement:

If the discrete channel has the capacity C and a discrete source has the entropy H (information content) and if $H \leq C$, then there exists a coding system such that the output of the source can be transmitted over the channel with an arbitrarily small possibility of errors.

For the sake of our discussion, we shall consider entropy per second to be bits per second of information transmitted. The word 'discrete' refers to the information transmitted in symbols, i.e. digital information. Shannon also gave the remarkable consideration that if our information source is sending the data at the rate what the communication channel can handle, then we can add some extra bits to the data stream to push the error rate down to an arbitrarily low level. This is the concept of channel coding and described in detail in Chapter 4. Thus, source coding must be an efficient technique with good compression. Reduced source data makes the channel coding possible. But again a trade-off! Communication delay is increased due to channel coding. The cost of using channel coding to protect the information is a reduction in data rate or an expansion in bandwidth. However, for every situation, there are enough choices of channel coding such that there exists some satisfactory compromise between delay and error performance.

Another point of consideration that an information source has a *probability distribution*, i.e. a set of probabilities assigned to a set of outcomes (containing the useful information). This reflects the fact that the information contained in an outcome is determined not only by the outcome but also by how uncertain it is. An almost certain outcome contains little information. A measure of the information contained in an outcome was introduced by Hartley in 1927. He defined the *information* (sometimes called *self-information*) contained in an outcome ω_i as

$$I(\omega_i) = \log_2 \frac{1}{P\{\omega_i\}} = -\log_2(P\{\omega_i\}) \tag{3.16}$$

This measure satisfies our requirement that the information contained in an outcome is proportional to its uncertainty. If $P\{\omega_i\} = 1$, then $I\{\omega_i\} = 0$, telling us that a certain event contains no information. The definition also satisfies the requirement that the total information in independent events should be added, which is shown in Eq. (3.17). Clearly, our rain forecast for two days contains twice as much information as for one day. From Eq. (3.16), for two independent outcomes ω_i and ω_j,

$$\begin{aligned} I(\omega_i \text{ and } \omega_j) &= \log_2 \left[\frac{1}{P\{\omega_i \text{ and } \omega_j\}}\right] \\ &= \log_2 \left[\frac{1}{P\{\omega_i\}\, P\{\omega_j\}}\right] \\ &= \log_2 \left[\frac{1}{P\{\omega_i\}}\right] + \log_2 \left[\frac{1}{P\{\omega_j\}}\right] \\ &= I(\omega_i) + I(\omega_j) \end{aligned} \tag{3.17}$$

Hartley's measure defines the information in a single outcome. The measure *entropy* $H(X)$, sometimes *absolute entropy*, defines the information content of the source X as a whole. It is the mean information provided by the source per source output or symbol. We have

$$H(X) = \Sigma P\{\omega_i\} I\{\omega_i\} = \Sigma -P\{\omega_i\} \log_2(P\{\omega_i\}) \tag{3.18}$$

A *binary symmetric source* (BSS) is a source with two outputs whose probabilities are p and $(1 - p)$ respectively ($P = \{p, (1 - p)\}$). The rain forecast discussed is a BSS. Binary signal with output '0' or '1' is a BSS. The entropy of the source is

$$H(X) = -p\log_2(p) - (1 - p)\log_2(1 - p) \tag{3.19}$$

The function of Fig. 3.23 takes the value zero when $p = 0$. When one outcome is certain, so is the other, and the entropy is zero. As p increases, so too does the entropy, until it reaches a maximum when $p = 1-p = 1/2$. When p is greater than 1/2, the curve declines symmetrically and reaches zero, reached when $p = 1$. We conclude that the average information in the BSS is maximized when both outcomes are equally likely. The entropy is measuring the average uncertainty of the source. (The term *entropy* is borrowed from thermodynamics. There too, it is a measure of the uncertainty or disorder of a system.)

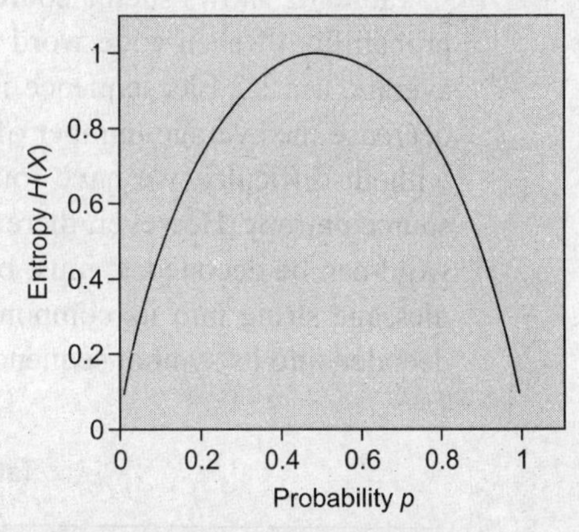

Fig. 3.23 Entropy of the binary symmetric source

When $p = 1/2$, then $H(X) = 1$. The unit of entropy is bits/symbol. An equally probable BSS has an entropy, or average information content per symbol of 1 bit per symbol. Sometimes there may be more than one bit per symbol. Hence, more information can be read at a time. This is the necessity of high-speed communications.

A BSS whose outputs are 1 or 0 has an output we describe as a bit. The entropy of the source is also measured in bits, so that we might say that the equi-probable ($p = 1/2$) BSS has an information rate of 1 bit/bit. The *numerator* bit refers to the information content. The *denominator* bit refers to the symbol 1 or 0. We can avoid this by writing it as 1 bit/symbol. When $p \neq 1/2$, the BSS information rate falls. When, $p = 0.1$, $H(X) = 0.47$ bits/symbol. This means that on average each symbol (1 or 0) of source output is providing 0.47 bits of information. For a BSS, the entropy is maximized when both outcomes are equally likely. This property is generally true. If an information source X has J symbols, its maximum entropy is $\log_2(J)$ and this is obtained when all J outcomes are equally likely. Thus, for a J symbol source.

$$0 \leq H(X) \leq \log_2(J) \tag{3.20}$$

3.8 INFORMATION SOURCE CODING FUNDAMENTALS (WITH HUFFMAN CODING EXAMPLE)

It seems intuitively reasonable that an information source of entropy $H(X)$ needs on average only H binary bits to represent each symbol. However, consider the rain forecast. Suppose the probability of rain is 0.1 and that of no rain 0.9. We have already noted that this BSS has entropy of 0.47 bits/symbol. Suppose we identify rain with 1 and no rain with zero, this representation uses 1 bit per symbol.

The replacement of the symbols (rain/no-rain) with a binary representation is termed *source coding*. In any coding operation, we replace the symbol with a *code word*. The purpose of source coding is to reduce the number of bits required to convey the information provided by the information source. Source coding uses the sequences. By this, we mean that code words are not simply associated to a single outcome, but to a sequence of outcomes. We can group the outcomes according to their probability and assign binary code words to these grouped outcomes according to their probability.

Table 3.2 shows such a source-coding scheme called variable length coding and the probability of each code word occurring. It is easy to compute that this code will on average use 1.2 bits/sequence instead of 3 bits/sequence. This example shows how to decrease the average number of bits per symbol on the basis of probability. Moreover, without difficulty, we have found a code that has an average bit usage less than the source entropy. However, there is a difficulty with the code in Table 3.2. Before a code word can be decoded, it must be parsed. Parsing describes the activity of breaking the message string into its component code words. After parsing, each code word can be decoded into its symbol sequence.

Table 3.2 Variable length coding

Probability of Occurrence of a Code	No. of Bits/Code
0.729	0
0.081	1
0.081	01
0.081	10
0.009	11
0.009	00
0.009	000
0.001	111

The code in Table 3.3, however, is an instantaneously parseable code. It satisfies the prefix condition.

Table 3.3 Instantly parseable variable length coding

Sequence	Probability of Occurrence of a Code	Code Word	Letter for Identification in the Tree
000	0.729	1	A
001	0.081	011	B
010	0.081	010	C
011	0.081	001	D
100	0.009	00011	E
101	0.009	00010	F
110	0.009	00001	G
111	0.001	00000	H

The code in Table 3.3 uses on average 1.568 (~ 1.6) bits per sequence. This is a 47% improvement on identifying each symbol with bits. In fact this variable length code is the *Huffman code* for the sequence set. The code for each sequence is found by generating the *Huffman code tree* for the sequence. A Huffman code tree is an unbalanced binary tree. The derivation of the Huffman code tree is shown in Fig. 3.24 and the tree itself is shown in Fig. 3.25. In both these figures, the letters A to H have been used in place of the sequences in Table 3.3 to make them easier to read.

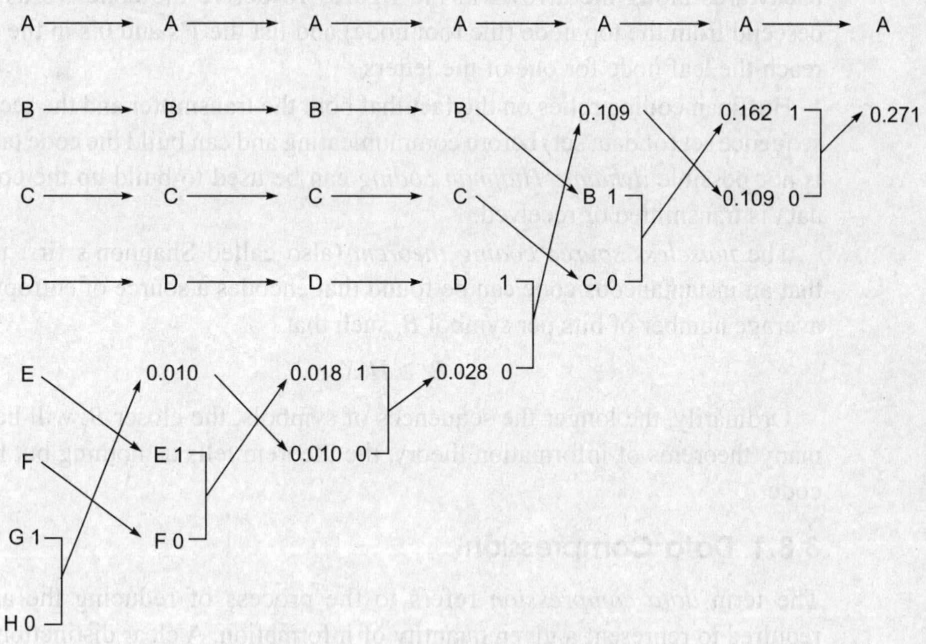

Fig. 3.24 Derivation of a Huffman code tree

In Fig. 3.25, the sequences are ordered with respect to the probability of the sequence occurring, highest probability at the top of the list. (The probabilities are effectively used as *weights* in the process to be described.) The tree is derived bottom up, in terms of

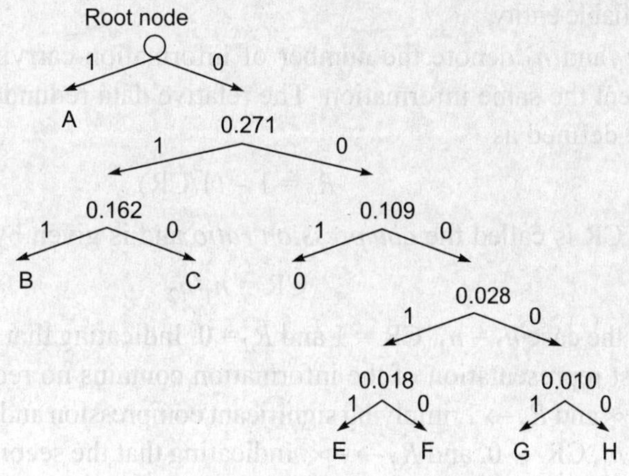

Fig. 3.25 A Huffman code tree

branch nodes and *leaf nodes* by combining weights and removing leaf nodes in progressive stages. As shown in Fig. 3.25, the two lowest leaf nodes G and H have their weights added, and the topmost node is labelled with a 1 and the lower one with a 0. The next stages are represented by that weight and the list is rewritten in order of the weights. The two lowest leaf nodes are now *E* and *F*, and they are labelled 1 and 0, respectively, and their weights are added to be taken on to the next stage. This continues until only two nodes remain. The Huffman tree shown in Fig. 3.25 is then produced by following backwards along the arrows in the figure. To derive the code words from the tree, descend from the top node (the root node) and list the 1's and 0's in the order until you reach the leaf node for one of the letters.

Huffman coding relies on the fact that both the transmitter and the receiver know the sequence set (or data set) before communicating and can build the code table. Where this is not possible *dynamic Huffman coding* can be used to build up the code table as the data is transmitted or received.

The *noiseless source coding theorem* (also called Shannon's first theorem) states that an instantaneous code can be found that encodes a source of entropy $H(X)$ with an average number of bits per symbol B_s such that

$$B_s \geq H(X) \tag{3.21}$$

Ordinarily, the longer the sequences of symbols, the closer B_s will be to $H(X)$. Like many theorems of information theory, the theorem tells us nothing but how to find the code.

3.8.1 Data Compression

The term *data compression* refers to the process of reducing the amount of data required to represent a given quantity of information. A clear distinction must be made between *data* and *information*. They are not synonymous. Data are the means by which information is conveyed. A story (information) can be prepared by using the different words (data). There may be some redundant data that may not have relevance or significance in that story and can be removed from the total database. Data redundancy is the critical issue in data compression. It is not an abstract concept but mathematically quantifiable entity.

If n_1 and n_2 denote the number of information-carrying units in two data sets that represent the same information. The relative data redundancy R_d of the first data set n_1 can be defined as

$$R_d = 1 - (1/\text{CR}) \tag{3.22}$$

where CR is called the *compression ratio* and is given by

$$\text{CR} = n_1/n_2 \tag{3.23}$$

For the case $n_2 = n_1$, CR = 1 and $R_d = 0$. Indicating that, relative to the second data set the first representation of the information contains no redundant data. When $n_2 << n_1$, CR $\rightarrow \infty$ and $R_d \rightarrow 1$. implying significant compression and highly redundant data. Finally, $n_2 >> n_1$, CR $\rightarrow 0$, and $R_d \rightarrow -\infty$. Indicating that the second data sets contains more data than the original information. This is the case of data expansion instead of compression. A practical compression ratio such as 10 means that the first data set has 10 information-carrying units or bits for every one unit or bit in the second or compressed data set. The corresponding redundancy of 0.9 implies that 90% of the data in the first data set is redundant.

Example 3.8 An 8 level image has the grey level distribution (corresponding bit patterns) shown in Table 3.4. There are two methods employed to encode the 8 possible grey levels. In one method, equal bits are used and in the other method, the number of bits/level is variable. What should be the average length of bits according to Table 3.4? Which is the more efficient technique according to the concept of information theory?

Table 3.4 Example data for the image representing variable length coding

L_k = No. of Possible Levels	$P(L_k)$ = Probability of Occurrence of *k*th Level	Coding Method 1	Bits Used per Code for Method 1	Coding Method 2	Bits Used per Code for Method 2
$L_0 = 0$	0.18	000	3	11	2
$L_1 = 1/7$	0.25	001	3	01	2
$L_2 = 2/7$	0.21	010	3	10	2
$L_3 = 3/7$	0.17	011	3	001	3
$L_4 = 4/7$	0.08	100	3	0001	4
$L_5 = 5/7$	0.06	101	3	00001	5
$L_6 = 6/7$	0.03	110	3	000001	6
$L_7 = 1$	0.02	111	3	000000	6

Solution

Average number of bits for the coding method 1 = 3 bits (as they are equal)

Average number of bits for the coding method 2

$$= 2(0.18) + 2(0.25) + 2(0.21) + 3(0.17) + 4(0.08) + 5(0.06) + 6(0.03) + 6(0.02)$$
$$= 2.71 \text{ bits}$$

The resulting compression ratio, CR = 3/2.71 = 1.107

Thus, approximately 10% of the data resulting from the use of code 1 is redundant. Hence, the second method of variable length coding is more efficient technique. The exact level of redundancy can be determined by

$$R_d = 1 - (1/1.107) = 1 - 0.903 = 0.097$$

3.9 SPEECH CODING TECHNIQUES

Today wireless communication is primarily of voice/speech, especially for mobile telephony. Speech coding is required for transmission of compressed speech. Coder is a hardware circuit (chip) or a software routine that converts the spoken word into digital code and vice versa. We have seen previously the waveform coders using PCM/

ADPCM techniques. They are relatively simple, better adaptive capability and better speech quality and widely used at the 16–64 kbps range.

Other types of coders are based on parametric coding. These encoders take advantage of predictable elements in human speech, i.e. specialized for human voice characteristics. By analysing vocal tract sounds, a recipe for rebuilding the sound at the other end is sent rather than the sound waves themselves. As a result, the speech codec is able to achieve a much higher compression ratio, which yields a smaller amount of digital data for transmission. However, if music is encoded with a speech codec, it will not sound as good when decoded at the other end. Speech/voice coders are called *vocoders*. Several low data rate encoders are described here with an assessment of their subjective quality.

Speech signal may be processed before coding, for example noise may be removed by DSP techniques. Noise reduction techniques are conceptually very similar, regardless of the signal being processed; however, knowledge of the characteristics of an expected signal can mean that the implementation of these techniques varies greatly, depending on the type of signal. Similarly, few frequency domain operations like up sampling, down sampling, removal of certain harmonics, etc., can be performed over the speech signal, depending on the requirements before speech coding.

The speech coders may be classified on the basis of bit rate as follows:

- Medium rate: 8–16 kbits /s
- Low rate: below 8 kbits/s and down to 2.4 kbits/s
- Very low rate: below 2.4 kbits/s

3.9.1 Voice Coders/Vocoder

Vocoders and voders are actually devices that allow machines to manipulate human speech and create similar sounds. The vocoder is an electronic device that analyses the human voice, breaking it up into its component parts. These parts can then be manipulated, transmitted over wires, or stored to serve a number of purposes. The voder is a similar device and sometimes the two are combined in one unit but, strictly speaking, a voder synthesizes a human voice. More commonly, however, it is used to create music or motion picture effects.

Now voders and vocoders are considered to be the same. A vocoder, in general, is a speech analyser and synthesizer.

Theory of Vocoders

A cross-sectional view of the human speech organ is shown in Fig. 3.26.

Speech is produced by a cooperation of lungs, glottis (with vocal cords), and articulation tract (mouth and nose cavity). The human voice consists of the sounds generated by the opening and closing of the glottis by the vocal cords. For the production of voiced sounds, the lungs press air through the epiglottis, the vocal cords vibrate, and they interrupt the air stream and produce a quasi-periodic pressure waveform with many harmonics. This basic sound is then filtered by the nose and throat (a complicated resonant piping system) to produce differences in harmonic content (formants) in a controlled way, creating the wide variety of sounds used in speech. The pressure impulses are commonly called *pitch impulses* and the frequency of the pressure signal is the *pitch frequency* or *fundamental frequency.* In Fig. 3.27, a typical impulse sequence (sound pressure function) produced by the vocal cords for a voiced sound is shown. It is the part of the voice signal that defines the speech melody. When we speak with a constant pitch frequency, the speech sounds monotonous, but in normal cases a permanent change of the frequency ensues. Figure 3.28 depicts how the pitch frequency varies with time.

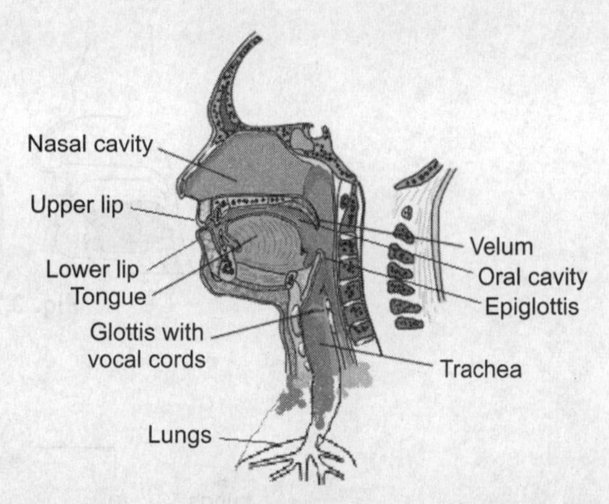

Fig. 3.26 Cross section of the human speech organ

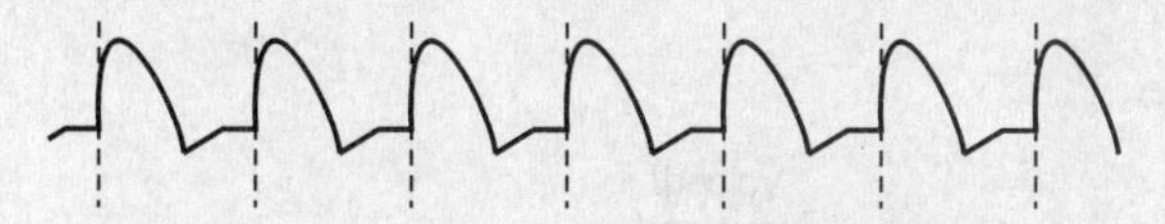

Fig. 3.27 Typical pitch impulse sequence

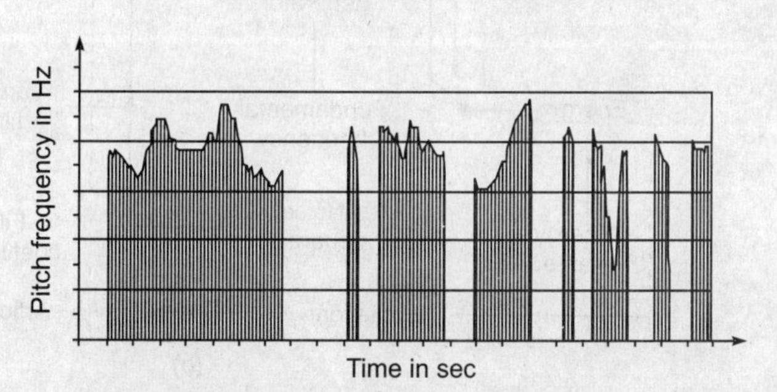

Fig. 3.28 Variation of the pitch frequency also called pitch frequency contour

The pitch impulses stimulate the air in the mouth and for certain sounds also the nasal cavity. When the cavities resonate, they radiate a sound wave that is the speech signal. Both cavities act as resonators with characteristic resonance frequencies, called *formant frequencies*. Since the mouth cavity can be greatly changed, we are able to pronounce many different sounds. There is another set of sounds, known as the unvoiced and plosive sounds, which are not modified by the mouth in the same fashion. In the case of unvoiced sounds, the excitation of the vocal tract is more noise-like.

These characteristics are used to analyse and generate the synthetic sound as follows. Figure 3.29 demonstrates the production of the sounds. The different shapes and positions of the articulation organs are obvious. The above-mentioned speech-generation process is modelled in the form of a block diagram as shown in Fig. 3.30(a).

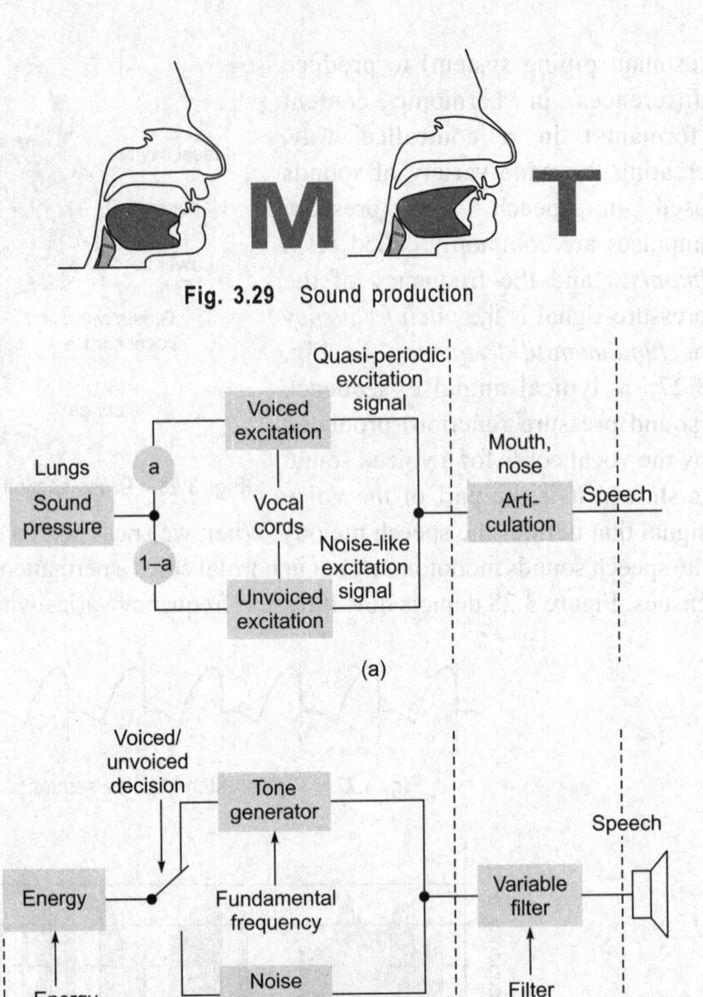

Fig. 3.29 Sound production

Fig. 3.30 (a) Human speech production modelling and (b) equivalent synthetic speech production blocks

The components of the human speech organ, namely the excitation and the vocal tract parameters, are computed. The components are then fed into the synthesis part of a vocoder, which finally generates a synthesized speech signal. The user can replay the signal and compare it with the reference speech signal. For the reconstructed signal also, the pitch frequency contour is graphically presented and the user can directly manipulate this contour.

The vocoder examines speech by finding this basic carrier wave, which is at the *fundamental frequency*, and measuring how its spectral characteristics are changed over time by recording someone speaking. This results in a series of numbers representing these modified frequencies at any particular time as the user speaks. In doing so, the vocoder dramatically reduces the amount of information needed to store speech, from a complete recording to a series of numbers. To recreate speech, the vocoder simply reverses the process, creating the fundamental frequency in an oscillator, and then passing it through a stage that filters the frequency content based on the originally recorded series of numbers. The user can manipulate the fundamental frequency contour, the number of prediction coefficients, the signal energy, etc., and he or she can then hear the result of these manipulations.

Modern Vocoders' Applications

Vocoders may be used in musical instruments, television and films, robots or, talking computers. Table 3.5 lists some of the existing standards/systems and speech coders used.

Table 3.5 ITU-T Vocoder standards for speech processing and audio processing codecs

Standards/Systems	Speech Coder
ITU-T G.729	8 kbps CS-ACELP based codec
ITU-T G.729A	Annex A to the G.729 standard–reduced complexity 8 kbps CS-ACELP based codec
ITU-T G.729B/G.729AB	Annex B to the G.729 standard–adds silence compression scheme to G.729
ITU-T G.728	16 kbps LD-CELP based codec
ITU-T G.726	40, 30, 24, 16 kbps ADPCM based codec
ITU-T G.723.1	6.3 and 5.3 kbps, MP-MLQ, and ACELP based codec
ITU-T G.722.2 (AMR-WB)	ITU-T adaptive multi rate – wide band codec
ITU-T G.722.1	24 and 32 kbps MLT based codec
ITU-T G.722	64, 56, 48 kbps SB-ADPCM based codec
ITU-T G.711	64, 56, 48 kbps *A*-law and μ-law, PCM-based codec
GSM-AMR (AMR-NB)	3GPP adaptive multi rate – narrow band codec

Quality and bit rate variations can be plotted for waveform as well as parametric vocoders as shown in Fig. 3.31. Parametric vocoders use different approaches for coding the voice signal or they treat the different parameters for coding purpose; so they can be

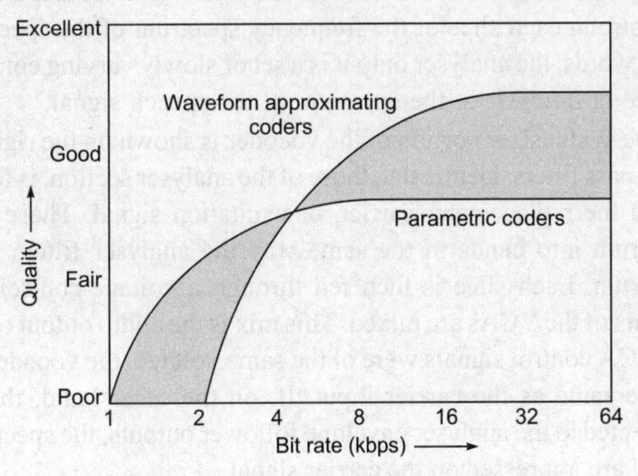

Fig. 3.31 Quality vs bit rate for speech coders

further classified as explained in the following topics. In fact, speech coding system development continues to this day to be a vigorous area of research and development. Channel, formant, cepstrum as well as voice excited vocoders do not have much applications in mobiles but they are the good examples of how the speech properties are utilized.

Channel Vocoder

Channel vocoders were originally developed for signal coding purposes, towards reducing the amount of data that would be needed to be transmitted over communication channels. Thereafter, channel vocoders (and the functionally equivalent, but computationally quite different, phase vocoder) have found application in music production.

Channel vocoders are the simplest, frequency-domain vocoders that determine the envelope of the speech signal for a number of frequency bands and then sample, encode, and multiplex these samples with the encoded outputs of the other filters. Here, just like frequency band allocation for different channels, the range of frequency is divided. Hence, the name channel vocoder. It was the first among the analysis-synthesis systems of speech demonstrated practically. The sampling is done synchronously every 10 to 30 ms. Along with the energy information about each band, the voiced/unvoiced decision and the pitch frequency for voiced speech are also transmitted.

The channel vocoder uses a bank of filters or digital signal processors to divide the signal into several sub-bands as shown in Fig. 3.32. It is the source filter model that represents the statistical characteristics of the speech signal. The left side of diagram is the analyser portion of the device. A speech signal is fed through a series of band-pass filters. (The power levels are transmitted together with a signal that represents a model of the vocal tract).

The centre frequencies *F1, F2, F3,...,Fn* of the filters are spaced one-quarter to one-half octave apart; together, the filter bands cover most of the audio spectrum. Thus, the filter band slices up the spectrum of the speech signal. Each slice then goes to an envelope follower (ENV), the output of which is a control voltage that is proportional to the strength of that slice. The envelope follower outputs are control signals. They tell us how strong each slice of the frequency spectrum of the speech signal is at any time. In other words, the analyser output is a set of slowly varying control voltages that constitute a *code* or *analysis* of the spectrum of the speech signal.

The synthesizer portion of the vocoder is shown on the right side of Fig. 3.32. A set of band-pass filters, identical to those of the analyser section, is fed by a second audio signal, called the replacement, carrier, or excitation signal. These filters slice up the carrier spectrum into bands in the same way the analyser filters slice up the speech signal spectrum. Each slice is then fed through a voltage-controlled amplifier (VCA). The outputs of the VCAs are mixed. This mix is the audio output of the basic vocoder. If all of the VCA control signals were of the same voltage, the vocoder output would in principle be the same as the carrier input. If, on the other hand, the VCA control inputs are connected to the analyser envelope follower outputs, the spectral variations of the speech signal are impressed on the carrier signal.

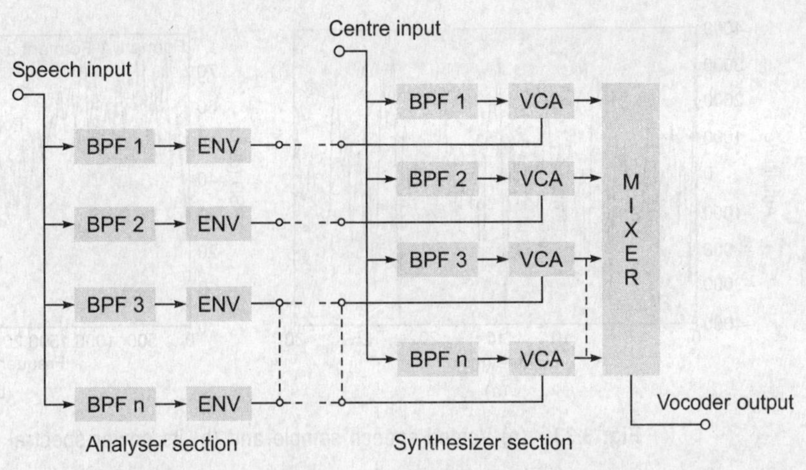

Fig. 3.32 Basic analysis synthesis function in a channel vocoder

These vocoders typically operate between 1 and 2 kbps. Even though these coders are efficient, they produce a synthetic quality and, therefore, are not generally used in commercial mobile telephone systems.

Formant Vocoder

It is also a parametric vocoder. As mentioned earlier, the two types of speech sounds, voiced and unvoiced, produce different sounds and spectra due to their differences in sound formation. With voiced speech, air pressure from the lungs forces the normally closed vocal cords to open and vibrate. The vibrational frequencies (pitch) vary from about 50 to 400 Hz (depending on the person's age and sex) and form resonance in the vocal track at odd harmonics. These resonance peaks are called formants and can be seen in the voiced speech in Figures 3.33(a) and (b).

Unvoiced sounds, called fricatives (e.g., *s*, *f*, *sh*) are formed by forcing air through an opening (hence the term, derived from the word *friction*). Fricatives do not vibrate the vocal cords and therefore do not produce as much periodicity as seen in the formant structure in voiced speech; unvoiced sounds appear more noise-like [see Figures 3.34(a) and (b)]. Time-domain samples lose periodicity and the power spectral density does not display the clear resonant peaks that are found in voiced sounds.

Formant vocoders are like channel vocoders but the transmitted parameters is the formant information—the positions of the peaks (formants) of the spectral envelope [Fig. 3.33(b)], instead of sending samples of the entire power spectrum envelope, which is the case of channel vocoder. Since speech signal information is primarily contained in the formants, a vocoder that can predict the position and bandwidths of the formants could achieve high quality at very low bit rates. These typically operate in the range of 1000 bit/s. Theoretically, the bit rate of the formant vocoder is lower than the channel vocoder because it uses fewer control signals. Typically, a formant vocoder must be able to identify at least three formants for representing the speech sound. Formant vocoders must control the intensities of the formants.

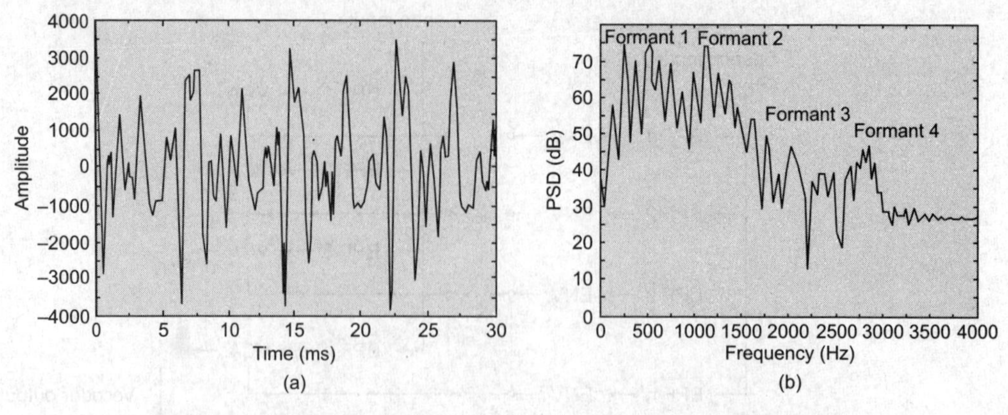

Fig. 3.33 (a) Voiced speech sample and (b) its power spectral density

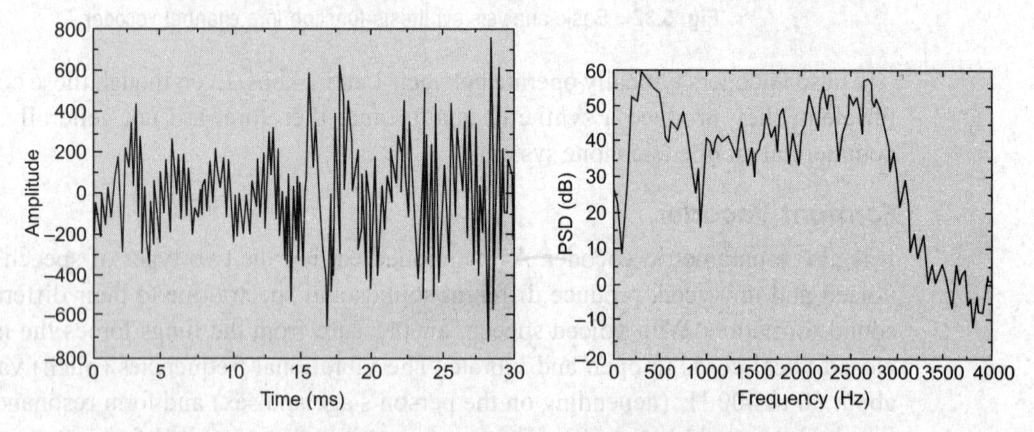

Fig. 3.34 (a) Unvoiced speech sample and (b) its power spectral density

The lower information rate and harder to get right (greater distortion) are the two main drawbacks of the system. There are difficulties in computing the location of formants accurately, the formants are difficult to predict and also formant transitions from human speech. Due to these reasons formant vocoders have not been very successful.

Cepstrum Vocoder

One more parametric coder, the ceptrum vocoder, separates the excitation and vocal tract spectrum by inverse Fourier transform of the log magnitude spectrum to produce the cepstrum of the signal. The low-frequency coefficients in the cepstrum correspond to the vocal tract spectral envelope, with the high-frequency excitation coefficients forming a periodic pulse train at multiples of the sampling period. Linear filtering is performed to separate the vocal tract cepstral coefficients from the excitation coefficients. In the receiver, the vocal tract cepstral coefficients are Fourier transformed to produce the vocal tract impulse response. By convolving this impulse response with a synthetic excitation signal (random noise or periodic pulse train), the original speech is reconstructed. It forms the concept of sub-band coding.

Voice Excited Vocoder

Pitch excited and voice excited vocoders category is somewhat hybrid. Voice excited vocoders eliminate the need for pitch extraction and voicing detection operations. This system uses a hybrid combination of PCM transmission for the low frequency band of speech, combined with channel vocoding for higher frequency bands. A pitch signal is generated at the synthesizer by rectifying, band-pass filtering, and clipping the baseband signal, thus creating a spectrally flat signal with energy at pitch harmonics. Voice excited vocoders have been designed for operation at 7200 bps to 9600 bps and their quality is typically superior to that obtained by the traditional pitch excited vocoders.

Linear Prediction Based Vocoders

Linear predictive coders form the basis for the hybrid coders. The human speech production is illustrated by a simple model in Fig. 3.26. Here the lungs are replaced by a DC source, the vocal cords by an impulse generator and the articulation tract by a linear filter system. A noise generator produces the unvoiced excitation. In practice, all sounds have a mixed excitation, which means that the excitation consists of voiced and unvoiced portions. Of course, the relation of these portions varies strongly with the sound being generated.

Based on this model, a further simplification can be made. We use a 'hard' switch, which only selects between voiced and unvoiced excitation. The filter, representing the articulation tract, is a simple recursive digital filter; its resonance behaviour (frequency response) is defined by a set of filter coefficients. Since the computation of the coefficients is based on the mathematical optimization procedure of *linear prediction coding* they are called *linear prediction coding coefficients* or *LPC coefficients* and the complete model is the so-called *LPC vocoder*. In practice, the LPC vocoder is used for speech telephony. It's great advantage, is the very low bit rate needed for speech transmission (about 3 kbit/s) compared to PCM (64 kbit/s).

A great advantage of the LPC vocoder is the manipulation facilities and the narrow analogy to the human speech production. Since the main parameters of the speech production, namely the pitch and the articulation characteristics, expressed by the LPC coefficients, are directly accessible; the audible voice characteristics can be widely influenced. For example, the transformation of a male voice into the voice of a female or a child is very easy. Also the number of filter coefficients can be varied to influence the sound characteristics, above all, the formant characteristics.

Linear predictive encoders are the most popular today and are used mainly in digital personal communications services. The LPC algorithm assumes that each speech sample is a linear combination of previous samples. Speech is sampled, stored and analysed. Coefficients, calculated from the samples are transmitted and processed in the receiver. With long-term correlation from samples, the receiver accurately processes and categorizes voiced and unvoiced sounds. The LPC family use pulses from an excitation pulse generator to drive filters whose coefficients are set to match the speech sample. The excitation pulse generator differentiates the various types of LP coders discussed below. LP filters are simple to implement and simulate filtering and acoustic pulses produced in the mouth and throat. An LPC coder development is shown in Figures 3.35 to 3.38.

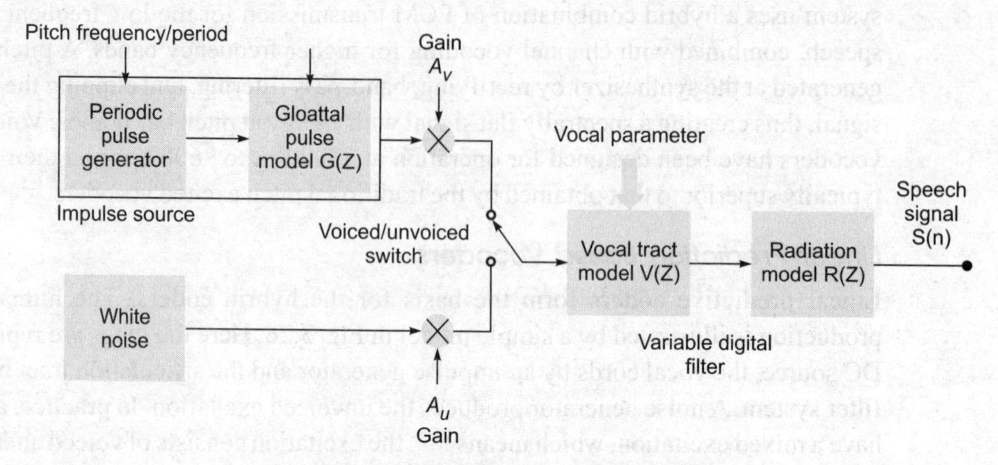

Fig. 3.35 Synthetic speech generation model

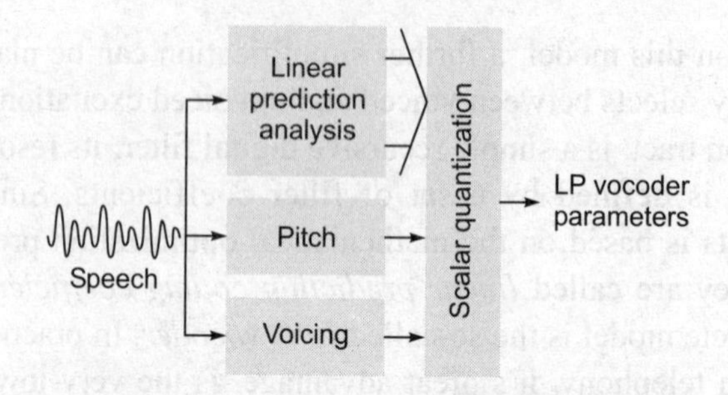

Fig. 3.36 Vocoder parameter extraction

Regular Pulse Excited (RPE) LPC Coder

The excitation source signal is modelled as either a periodic impulse train for voiced speech like vowel sounds, or a random noise for unvoiced speech like consonants. The RPE, MPE, CELP are the hybrid coders based on excitation. The regular pulse excited (RPE) coder analyses the signal to determine if it is voiced or unvoiced. After determining the period for voiced sounds, the periodicity is encoded and the coefficient is transmitted. When the signal changes from voiced to unvoiced, a code is transmitted that stops the receiver from generating periodic pulses and starts generating random pulses to correspond to the noise-like nature of fricatives. The RPE is used in the GSM (Chapter 13) full rate vocoder (Fig. 3.39).

The GSM vocoder on the network side is in the transcoder and rate adapter unit (TRAU) that transcodes data from 16 kbit/s to 64 kbit/s. Phase one of the GSM specification defines full rate coders; phase two improves capacity by supporting half rate CELP coders at comparable quality (but requires more processing capability).

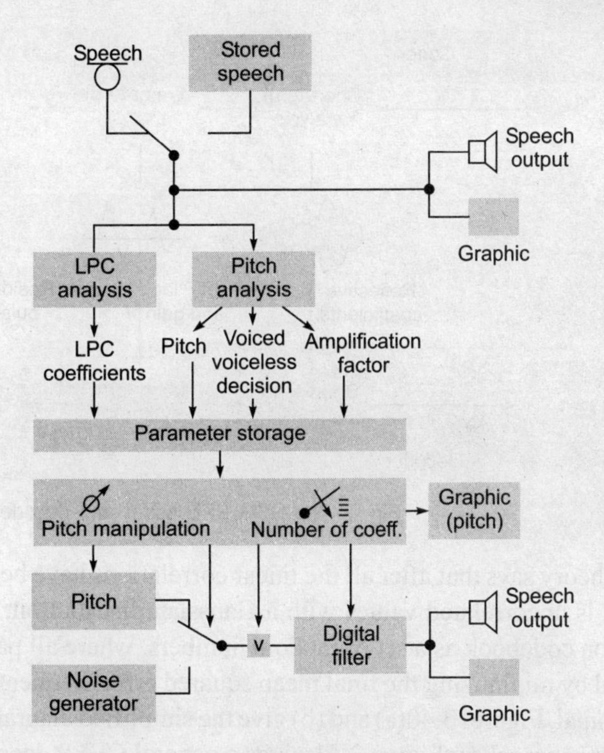

Fig. 3.37 Combining the concept of Figures 3.35 and 3.36 for LPC

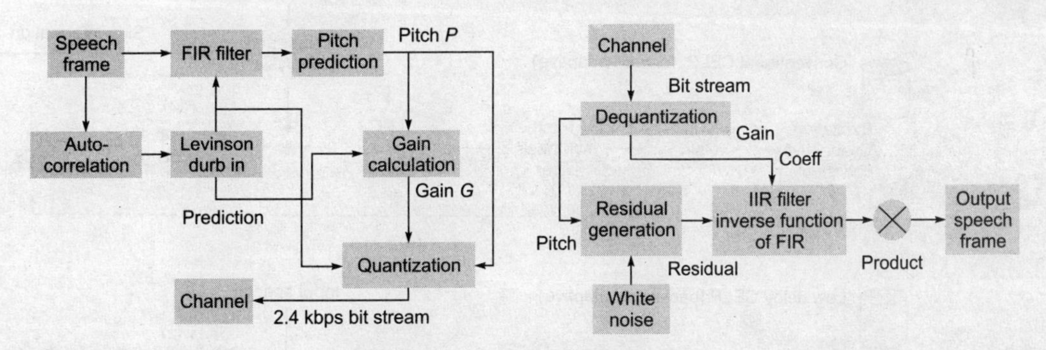

Fig. 3.38 LPC encoder and decoder using Levinson–Durbin recursion

Code Book Excited Linear Predictive (CELP) coder

Code excited linear prediction (CELP) coders are state-of-the-art in telephony coding. This coder is optimized by using a code book (look up table prepared by vector quantization method) to find the best match for the signal. First of all, vector quantizes the LP residual, typically 40 samples (5ms), using 1024 codebook entries. The coding requires more computations than decoding (need for codebook searches). This method reduces the processing complexity and the required data transmission rate is achieved. Resultant bit rate is about 4 kbps.

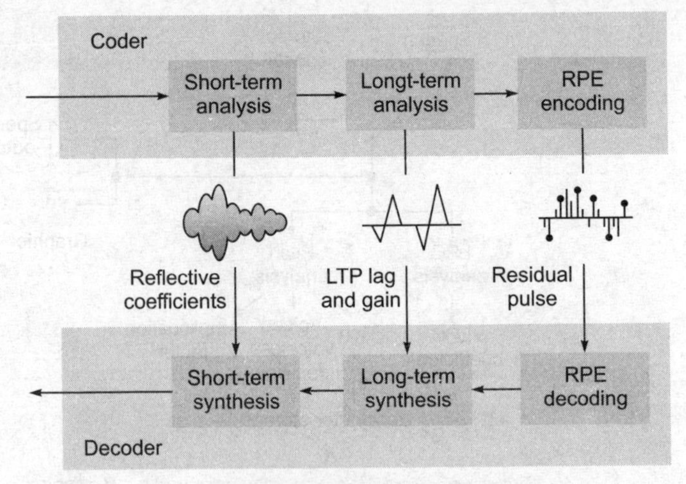

Fig. 3.39 GSM RPE coder and decoder

The theory says that after all the linear correlations have been removed, the prediction residual is uncorrelated values with a Gaussian distribution, therefore can generate the excitation codebook as a set of random numbers, where all parameters of the system are obtained by minimizing the final mean squared error. Sequential parameter estimation is non-optimal. Figures 3.40(a) and (b) give the simplified diagram of CELP coders with two different approaches. Figure 3.41 gives a general CELP decoder diagram.

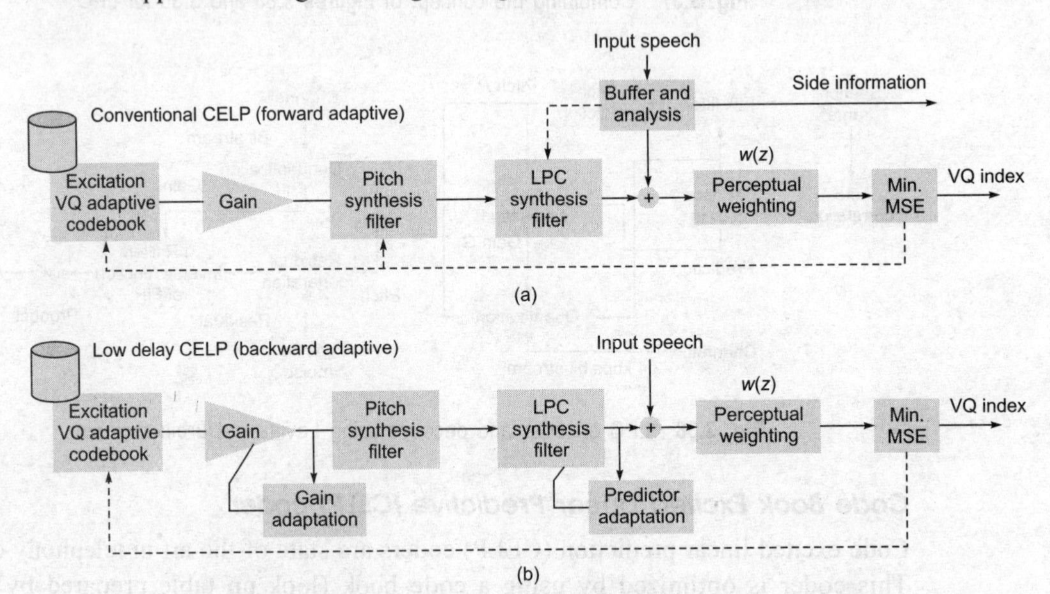

Fig. 3.40 (a) Forward and (b) backward adaptive systems for CELP

Most digital cellular systems use CELP (or CELP-based) coders. Improved CELP models include the vector sum excited linear prediction (VSELP) and the algebraic code excited linear prediction (ACELP). The VSELP, used in D-AMPS (North American digital cellular, IS-54), GSM half rate and PDC (Japan), simplifies the code book arrangement so that frequently occurring speech combinations are organized close together. ACELP coders do not require fixed code books at both the transmitter and receiver but optimize codes by using a series of nested loops. These coders are used in the GSM enhanced full rate (EFR), D-AMPS enhanced and U.S. PCS (GSM) 1900 EFR systems.

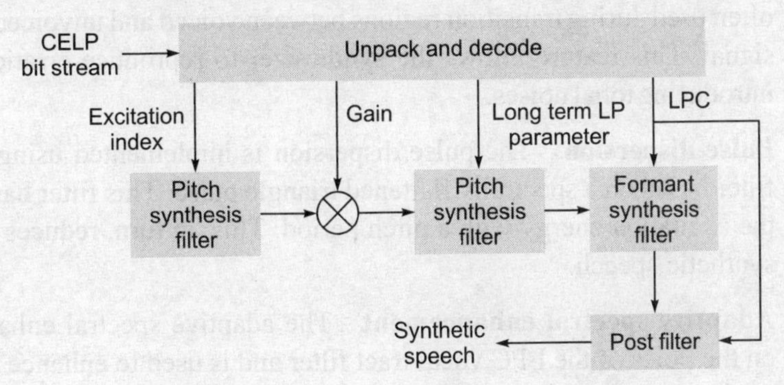

Fig. 3.41 A general, simplified CELP decoder

Mixed Excitation LPC (MELPC)

This is suggested as a standard for GSM and explained in the G series recommendation. The mixed excitation linear predictive coder (MELPC) algorithm is the new 2400 bps federal standard hybrid speech coder. The United States Department of Defense (DoD) Digital Voice Processing Consortium (DVPC) selected it after an extensive testing programme for many years.

The selection test concentrated on four areas: intelligibility, voice quality, talker recognizability, and communicability. The selection criteria also included hardware parameters such as processing power, memory usage, and delay. MELPC was selected as the best of the seven candidates. MELPC is robust in difficult background noise environments, such as those frequently encountered in commercial and military communication systems. It is very efficient in its computational requirements. This translates into relatively low power consumption, an important consideration for portable systems.

Traditional pitch-excited LPC vocoders use either a periodic pulse train or white noise as the excitation for an all-pole synthesis filter producing intelligible speech at very low bit rates, but they sometimes sound mechanical or buzzy and are prone to annoying thumps and tonal noises. These problems arise from the inability of a simple pulse train to reproduce all kinds of voiced speech. The MELPC vocoder uses a mixed excitation model that can produce more natural sounding speech because it can represent a richer ensemble of possible speech characteristics.

The MELPC vocoder is based on the traditional LPC parametric model, but also includes four additional features. These are discussed below.

Mixed excitation The mixed excitation is implemented using a multi band mixing model. This model can simulate frequency dependent voicing strength using a novel adaptive filtering structure based on a fixed filterbank. The primary effect of this multi band mixed excitation is to reduce the buzz usually associated with LPC vocoders, especially in broadband acoustic noise.

Aperiodic pulses When the input speech is voiced, the MELPC vocoder can synthesize speech using either periodic or aperiodic pulses. Aperiodic pulses are most often used during transition regions between voiced and unvoiced segments of the speech signal. This feature allows the synthesizer to reproduce erratic glottal pulses without introducing tonal noises.

Pulse dispersion The pulse dispersion is implemented using fixed pulse dispersion filter based on a spectrally flattened triangle pulse. This filter has the effect of spreading the excitation energy with a pitch period. This, in turn, reduces the harsh quality of the synthetic speech.

Adaptive spectral enhancement The adaptive spectral enhancement filter is based on the poles of the LPC vocal tract filter and is used to enhance the formant structure in the synthetic speech. This filter improves the match between synthetic and natural band pass waveforms, and introduces a more natural quality to the speech output.

Multipulse Excited LPC

Multipulse excited LPC are like LP coders but the excitation is modelled as a number of pulses. Pulse position is found by an exhaustive search to minimize mean squared error. About one pulse per millisecond is enough to generate high-quality speech. The method is not so popular. Typical bit rate is 9600 bps.

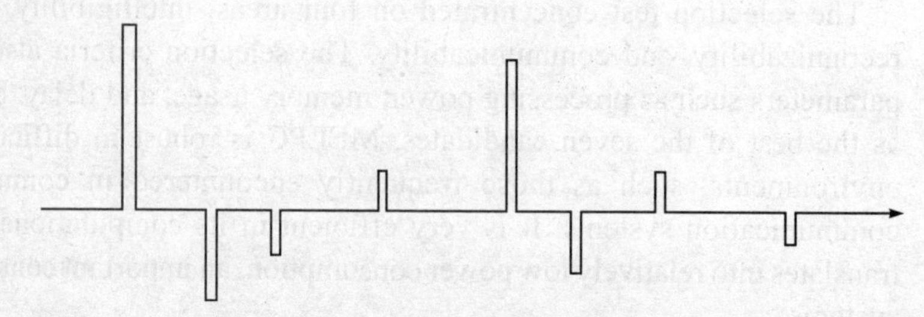

Fig. 3.42 Typical multipulse excitation sequence

Excitation by a single pulse per pitch period produces audible distortion, no matter how well the pulse is positioned. This is shown by Atal (1986). He suggested that if more than one pulse, typically 8 per pitch period, are used and if individual pulse position and amplitudes are adjusted sequentially as shown in Fig. 3.42, spectrally weighted mean square error can be minimized. This technique is called multipulse excited LPC. It gives better results not only because of the prediction residual is better approximated by several pulses per pitch period, but also because of the multipulse algorithm does not require pitch detection. The number of pulses used can be reduced, in particular for high pitched voices, by interpolating the linear filter with a pitch loop in the synthesizer.

QCELP Vocoders

The two basic speech coding methods for data rates between 4.8 kbps and 16 kbps are *analysis-and-synthesis* (AaS), and *analysis-by-synthesis* (AbS) as we saw previously.

In the AaS approach, an analyser in the transmitter analyses the original speech, and extracts a set of parameters that represent some kind of source filter model. These parameters are then transmitted out to the receiver, where a synthesizer reconstructs the speech based on the received parameters. In this system, the distortion throughout the whole coding process is difficult to check and control, because the analyser and the synthesizer are located separately in the transmitter and receiver.

In the AbS approach, an analyser and a local synthesizer are introduced in the transmitter. The synthesized speech is now available in the transmitter for analysis. A trial and error procedure, similar to a closed loop, determines the optimum parameters in the transmitter. In the receiver, these parameters reconstruct the synthetic speech, which match the real signal with minimum perceptual error. Compared to the AaS approach, the AbS approach is capable of producing higher quality speech at low data rates, but the encoder in the transmitter is more complex.

The basic CELP algorithm is one of the AbS methods widely used in the low bit rate of speech coding. QCELP is also a CELP algorithm, but it differs from the traditional CELP in that it dynamically adjusts the encoded data rate based on speech signal energy, background noise, and other speech characteristics. Therefore, the average data rate of the compressed speech is significantly reduced, while the voice quality is not affected. The QCELP vocoder consists of an encoder and decoder. When using a general DSP chip to implement a QCELP vocoder, approximately 20 to 25 MIPS are needed, of which 90% of these MIPS are for the encoder and the remaining 10% for the decoder. QCELP vocoders are mainly used in CDMA-based systems (Chapter 13).

A summary of systems and vocoders used is presented in Table 3.6.

Table 3.6 Different vocoders in different systems

System Type	System Name	Reference	Codec Type	Codec Rate (kbps)	Forward Error Correction (kbps)	Codec Algorithmic Delay (mS)	Est. Processing
Cellular TDMA	GSM/DCS/PCS Full Rate	ETSI/ETS 300580 ANSI J-STD-007	RPE-LTP	13	9.8	40	2.5 Mips
Cellular TDMA	GSM/DCS/PCS Half Rate	ETSI/ETS 300581	VSELP	5.6	5.8	40	17.5 Mips
Cellular TDMA	GSM EFR US PCS 1900 EFR	ETSI/ETS 300723 ANSI J-STD-007A	ACELP	12.2	10.6	40	15.4 WMops
Cellular	D-AMPS	TIA/EIA	VSELP	8.0	5.0	28	22 Wmops

Contd.

Table 3.6 continued

System Type	System Name	Reference	Codec Type	Codec Rate (kbps)	Forward Error Correction (kbps)	Codec Algorithmic Delay (mS)	Est. Processing
TDMA	Full Rate	IS-85					
Cellular	D-AMPS	TIA/EIA	ACELP	7.4	5.6	25	14 Mops
TDMA	Enhanced	IS-641					
Cellular	PDC	RCR-STD	VSELP	6.7	4.5	Unavailable	7.8 Mops
TDMA	Full rate	–27					
Cellular	PDC	RCR-STD	PSI-CELP	3.45	2.15	Unavailable	18.7 Mops
TDMA	Half rate	–27					
Cellular	Composite	TIA/EIA	CELP-like	7.2	3.2	26	11 Mips
CDMA	CDMA/TDMA	IS-661					
Cellular	CDMA	TIA/EIA	RCELP	8/4/0.8	19.2	30	20 Mops
CDMA		IS-127					
Cellular	W-CDMA	USA	ADPCM	32	0	Unavailable	Unavailable
CDMA		Unavailable					
Digital cordless	CT2	ITU-T-G 726	ADPCM	32	0	0.25	10 Mips
Digital cordless	DECT	ITU-T-G 726	ADPCM	32	0	0.25	10 Mips
Digital cordless	PHS	ITU-T-G 726	ADPCM	32	0	Unavailable	1.0 Mops

3.10 METHODS FOR SOURCE CODING IN FREQUENCY DOMAIN

Frequency domain speech/image coders take the advantage of the speech perception and generation models. But the algorithms will not totally be dependent on the models used. sub-band coding and transform coding are the two main methods. In sub-band coding speech signal spectrum may be divided into many smaller sub-bands and encodes each sub-band separately according to some criteria. A transform coder codes short time transform of a windowed sequence of samples and encodes them with number of bits proportional to its perceptual significance. Discrete cosine transform (DCT) and wavelet transform are the two main techniques described here.

3.10.1 Sub-band Coding

A variety of techniques have been developed to efficiently represent the speech signals in digital form for either storage or transmission purpose. Basic principle is that most of the energy of speech signal is contained in the lower frequencies, so one should encode the lower frequency band with more number of bits than the high frequency band. sub-band coding is the method in which speech signal is subdivided into several frequency bands and each band is digitally encoded separately.

If we think in other way, sub-band coding is a method of controlling and distributing quantization noise across the signal spectrum. Due to non-linear quantization process, distortion products are typically broad in spectrum. The human ear does not detect the quantization noise equally well at all the frequencies. It is therefore, possible to achieve considerable improvement in the quality by coding the signal into narrower bands.

Each sub-band is sampled at the Nyquist rate. Decimation (The process of reducing the sampling rate by a factor D is called decimation) by a factor of 2 is performed after frequency subdivision. By allocating a different number of bits per sample to the signal in the four sub-bands, reduction in the bit rate of digitized speech signal can be achieved; hence, effectively coding with compression is achieved.

Band splitting can be done in many ways. Two approaches are suggested below.

Approach 1 Divide the entire speech band into unequal bands using perceptual criteria that contribute equally to the articulation index. The method is suggested by Crochiere, et al. (1976). An example is shown in Fig. 3.43.

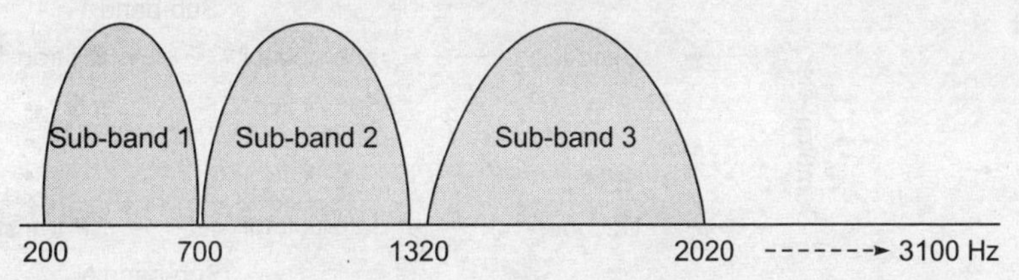

Fig. 3.43 Split of spectrum into unequal bands using perceptual criteria

Approach 2 Splitting the speech band would be to divide it into equal width sub-bands and assign to each sub-band, a number of bits proportional to perceptual significance while encoding them.

Instead of partitioning into equal width bands, octave band splitting, matching the first approach is often employed. As the human ear has an exponentially decreasing sensitivity to frequency, this kind of splitting is more in tune with the perception process.

Octave Band Splitting

An example of a frequency subdivision is given here Let us assume that the speech signal is sampled at a rate f_s samples per second. The first frequency subdivision splits the signal spectrum into two equal width bands:

- A low-pass signal ($0 \le f \le f_s/4$)
- A high-pass signal ($f_s/4 \le f \le f_s/2$)

The second frequency subdivision splits the low-pass signal from the first stage into two equal bands again:

- A low-pass signal ($0 \le f \le f_s/8$)
- A high-pass signal ($f_s/8 \le f \le f_s/4$)

As the third stage is accommodated, the third frequency subdivision splits the low-pass signal from the second stage into two equal bandwidth signals. Thus, the signal is subdivided into four frequency bands, covering three octaves.

There are various methods for generating and processing sub-band signals. One way is to make a low-pass translation of the sub-band signal to zero frequency by modulation process equivalent to SSB modulation. This type of translation makes it possible to reduce sampling rate and possesses other benefits from coding low-pass signals. An LP translator-based sub-band coder/decoder is shown in Fig. 3.44. Decimation and interpolation processes make the coding process possible.

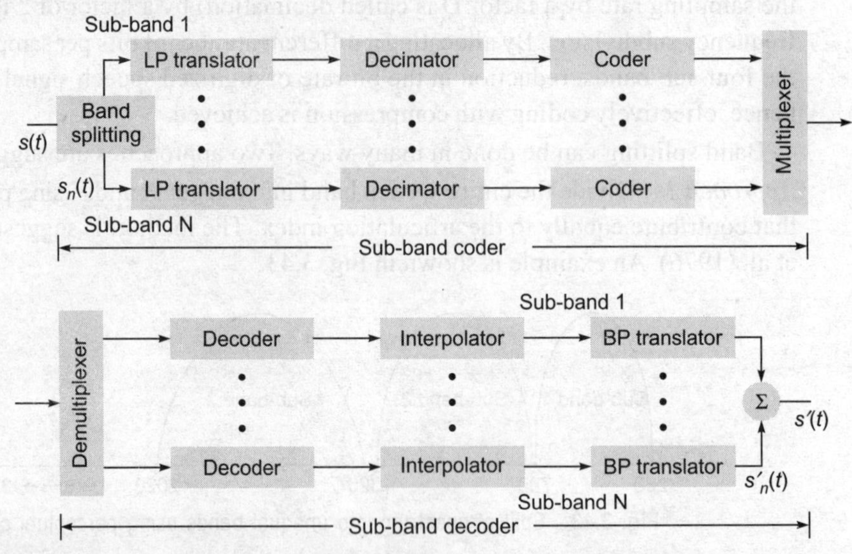

Fig. 3.44 Block diagram of sub-band coder/decoder

The input signal is filtered and translated to sub-bands. Each band has a different width. Hence sampling rate is decided accordingly. The resulting signal $s_n(t)$ (n stands for nth channel of sub-bands) is modulated by a cosine wave and filtered using *low-pass filter* (LPF). This LP translated signal undergoes the decimation process and is then digitally encoded. Finally, all channels are multiplexed. At the receiver, the data is demultiplexed into separate channels, decoded and band-pass translated to give the final estimate of $s'_n(t)$.

Filter design is particularly important in achieving good performance in sub-band coding. Aliasing resulting from decimation of the sub-band signals must be negligible. We cannot use sharp cut off filters as they are practically unrealizable. So the practical solution is to use quadrature mirror filter (QMF). This filter has the frequency response characteristics as shown in Fig. 3.45.

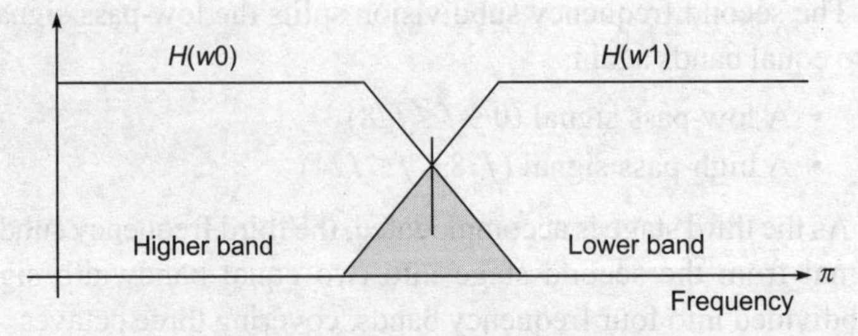

Fig. 3.45 Quadrature mirror filter

The synthesis method for the sub-band encoded speech signal is basically the reverse of the encoding process. The signals in adjacent low-pass and high-pass frequency bands are interpolated, filtered and combined as shown in Fig. 3.44. A pair of QMF is used in the signal synthesis for each octave of the signal.

Sub-band coding is also an effective method to achieve data compression in image signal processing. By combining the sub-band coding with vector quantization for each sub-band signal, Safranek, et al. (1988) has obtained coded images with approx.1/2 bit/ pixel, compared with 8 bits/pixel for the uncoded image. In general sub-band coding of signals is an effective method for achieving bandwidth compression in a digital representation of the signal, when the signal energy is concentrated in a particular region of the frequency band.

Sub-band coding can be used for coding the speech at bit rates of around 9.6 kbps to 32 kbps. In this range, speech quality is approximately equivalent to that of ADPCM at an equivalent bit rate. In addition, its complexity and relative speech quality at low bit rates make it particularly advantageous for coding below about 16 kbps. Due to increased complexity of sub-band coding when compared to other higher bit rate techniques does not warrant its use at bit rate greater than 20 to 25 kbps.

In short, sub-band coding has the following features:

- It exploits the frequency sensitivity of the auditory system.
- It splits the signal into sub-bands using band-pass filters.
- It codes each sub-band at an appropriate resolution, e.g., 4 bits per sample in the lower sub-bands and 2 bits per sample in the upper sub-bands.
- It can also exploit auditory masking and use fewer bits if a neighboring sub-band is much louder.
- It is basis for the MPEG audio standard (5:1 compression of CD quality audio with no perceptual degradation).

3.10.2 Transform Coding

Discrete Cosine Transform and Coding

Discrete cosine transform (DCT) is closely related to the discrete or fast Fourier transform. It is an energy compression technique. It plays a role in coding audio signals and images (e.g., in the widely used standard JPEG compression). All transform operations are mathematically related. The one-dimensional discrete cosine transform is defined by the mathematics as follows:

$$t(k) = c(k) \sum_{n=0}^{N-1} s(n) \cos \frac{\pi(2n+1)k}{2N} \tag{3.24}$$

where s is the array of N original values, t is the array of N transformed values, and the coefficients c are given by

$$c(0) = \sqrt{1/N}\,, \quad c(k) = \sqrt{2/N} \quad \text{for } 1 \leq k \leq N-1 \tag{3.25}$$

The discrete cosine transform in two dimensions (for images), for a square matrix, can be written as

$$t(i, j) = c(i, j) \sum_{n=0}^{N-1} \sum_{m=0}^{N-1} s(m, n) \cos \frac{\pi(2m+1)i}{2N} \cos \frac{\pi(2n+1)j}{2N} \quad (3.26)$$

with an analogous notation for N, s, t, and the $c(i, j)$ given by $c(0, j) = 1/N$, $c(i, 0) = 1/N$, and $c(i, j) = 2/N$ for both i and $j \neq 0$.

The DCT has an inverse, defined by

$$s(n) = \sum_{k=0}^{N-1} c(k)t(k) \cos \frac{\pi(2n+1)k}{2N} \quad (3.27)$$

for the one-dimensional case, and

$$s(m, n) = \sum_{i=0}^{N-1} \sum_{j=0}^{N-1} c(i, j)t(i, j) \cos \frac{\pi(2m+1)i}{2N} \cos \frac{\pi(2n+1)j}{2N} \quad (3.28)$$

for two dimensions.

Most of the practical transform coding schemes vary the bit allocation among different coefficients adaptively from frame to frame while keeping the total number of bits constant. This dynamic bit allocation is controlled by time-varying statistics, which have to be transmitted as a side information.

Considering typically for image, the DCT-based encoder can be thought of as essentially a compression of a stream of 8×8 blocks of image samples. Each 8×8 block makes its way through each processing step, and yields the output in a compressed form into the data stream. Because adjacent image pixels are highly correlated, the 'forward' DCT processing step lays the foundation for achieving data compression by concentrating most of the signal in the lower spatial frequencies. For a typical 8×8 sample block from a typical source image, most of the spatial frequencies have zero or near-zero amplitude and need not be encoded. In principle, the DCT introduces no loss to the source image samples; it merely transforms them to a domain in which they can be more efficiently encoded.

After output from the FDCT, each of the 64 DCT coefficients is uniformly quantized in conjunction with a carefully designed 64-element *quantization table* (QT). At the decoder, the quantized values are multiplied by the corresponding QT elements to recover the original unquantized values. After quantization, all of the quantized coefficients are ordered into the 'zig-zag' pattern as shown in Fig. 3.46. This ordering helps to facilitate entropy encoding by placing low-frequency non-zero coefficients before high-frequency coefficients. The DC coefficient, which contains a significant fraction of the total image energy, is differentially encoded.

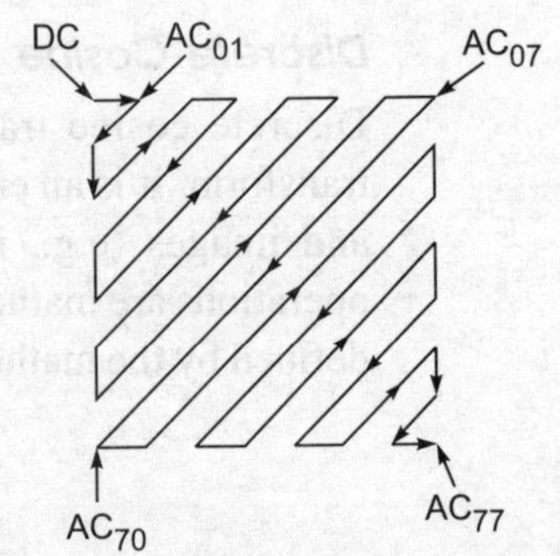

Fig. 3.46 Zig zag pattern of sequence

Entropy coding achieves additional compression losslessly by encoding the quantized DCT coefficients more compactly based on their statistical characteristics. The JPEG proposal specifies both Huffman coding and arithmetic coding. The baseline sequential codec uses Huffman coding, but codecs with both methods are specified for all modes of operation. Arithmetic coding, though more complex, normally achieves 5–10% better compression than Huffman coding.

Wavelet Transform

Along with speech compression, image and video compression is also equally important nowadays. Use of wavelet transform in speech applications is less. Certain ideas of the wavelet theory appeared quite a long time ago. For example, early in 1910 A. Haar published the full orthonormal system of basis functions with local definition domain (which are now known as Haar wavelets). The first record of wavelets was in the literature on digital processing and analysis of the seismic signals (the works by A. Grossman and J. Morlet). In the recent years, there even appeared a separate scientific area to deal with wavelet analysis and the wavelet transformation theory. Unlike the Fourier transform whose basis functions are sinusoids (composition of some combination of sine and cosine signals), wavelet transforms are based on small waves, called wavelets of varying frequency and limited duration. Fourier analysis, using the Fourier transform, is a powerful tool for analysing the components of a stationary signal (a stationary signal is a signal that repeats). The Fourier transform is less useful in analysing non-stationary data, where there is no repetition within the region sampled. Wavelet transforms (of which there are, at least formally, an infinite number) allow the components of a non-stationary signal to be analysed. Wavelets also allow filters to be constructed for stationary and non-stationary signals.

Wavelet transform is a major research topic presently. Some introduction is given in Chapter 2. Wavelets are extensively used for the purposes of filtration and preprocessing data, analysis and predictions, image recognition, as well as for processing and synthesizing various signals like speech or medical signals, for images compressing and processing, training neural networks and so on.

Just like quadrature mirror filtering from digital speech recognition, sub-band coding from signal processing and pyramidal image processing (Fig. 3.47), wavelet transform is also based on multi-resolution theory, which is described in the next topic. Many different types of wavelets are defined and all can exhibit different features.

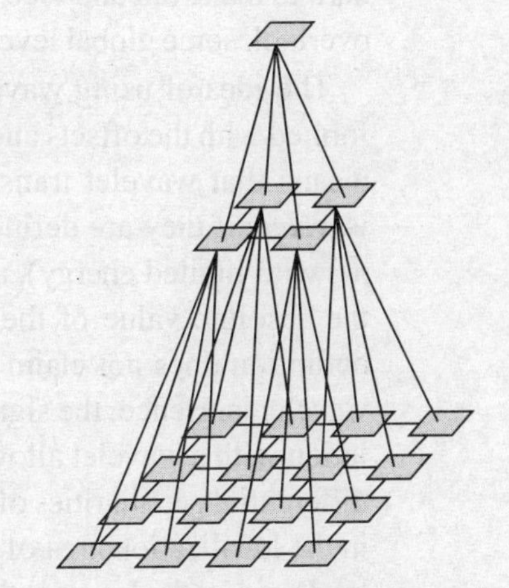

Fig. 3.47 Pyramidal image processing

The mathematics of wavelets is much larger than that of the Fourier transform. In fact, the mathematics of wavelets encompasses the Fourier transform. Wavelets can be symmetrical, asymmetrical or non-symmetrical. Wavelets are also grouped into those having compact domain and those not having it. Some functions have analytical form; others have the fast algorithm for calculating the wavelet transform associated with them.

Wavelets and Multi-resolution Analysis

Let us consider quite a typical problem. Suppose we have a signal (which can be whatever from sensor readings to digitized voice or image). The idea behind the multi-resolution analysis (MRA) is that the signal is looked at very closely—first under a microscope, then with a magnifying lens. After that, we make a couple of steps aside, and finally take a look at it from afar, as shown in Fig. 3.48.

Fig. 3.48 An example of multi-resolution analysis of an image

What do we get from all this? First, by consecutively refining the signal, we can reveal its local peculiarities (emphasis in speech or distinctive details of an image) and range them according to their intensiveness. Second, this demonstrates how the dynamics of the signal changes depending on the zoom. While sudden small changes are visualized, interactions of such events on a small scale that develop into large-scale events are pretty hard to make out and vice versa, while concentrating on small details only, we can easily overlook some global level events.

The idea of using wavelets for the MRA is that the signal is expanded by the basis formed with the offsets and non-uniformly scaled copies of the function prototype (which means that wavelet transform is essentially fractal). Such basis functions are called *wavelets* if they are defined in the space $L^2(R)$ (the space of complex-valued functions $f(t)$ with limited energy), are oscillating about the abscissa axis, and converge to zero as the absolute value of the argument increases. We have to note at this point that this definition does not claim to be full and accurate, but it is just a sort of sketch of what wavelet is. Hence, the signal convolution with a wavelet allows finding differential peculiarities of the signal in the localization area of the wavelet. Besides, the larger is the scale of the wavelet the wider portion of the signal influences the result of the convolution. An example of a Mexican hat wavelet is given in Fig. 3.49.

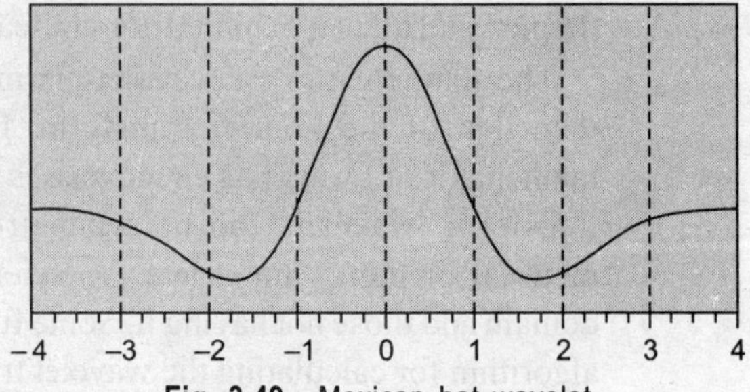

Fig. 3.49 Mexican hat wavelet

According to the uncertainty principle, the better the function is concentrated in time the more it is smeared in the frequency domain. When the function is re-scaled, the product of time and frequency ranges remains constant and represents the area of the cell in the time-and-frequency (phase) plane. The advantage of wavelet transform is that it covers the phase plane with cells of equal area but of a different shape (Fig. 3.50). Thanks to this, the signal's low-frequency components can be localized in the frequency domain (dominant harmonics) and high-frequency ones—in the time domain (sudden changes, peaks, etc.). Moreover, wavelet analysis allows investigating behaviour of fractal functions, i.e. the ones that have no derivatives in any point.

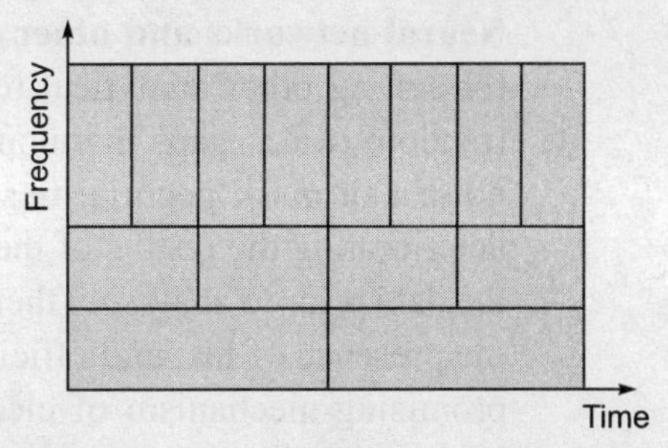

Fig. 3.50 Phase plane of wavelet transform

Recently, the wavelet transform theory has been developing at an incredible pace. Such areas have appeared and developed as orthogonal and biorthogonal wavelets, discrete wavelets, multi-wavelets, wavelet packages, etc.

Practical Use of Wavelet Transform

Processing experimental findings Since wavelets showed up merely as a mechanism of experimental data processing, their use as such is still very attractive. Wavelet transform provides the most visual and informative picture of experiment results. It allows clearing input data from noise and random distortions and even gives clues about peculiarities in data or prompts to the ways of further processing and analysing the data. Besides, wavelets are good for analysing transient signals that appear in medicine, stock market analysis and other areas.

Image processing The peculiarity of the human vision is that we concentrate attention on significant details of an image and cut off unnecessary information. With wavelet transform we can smooth or emphasize some details of an image, zoom in it or out, single out important details and even improve its quality!

Data compression A characteristic feature of orthogonal multiscale analysis is that for smooth enough data, values of details obtained after the transform are for the most part close to zero. Therefore, they are easily compressed with ordinary statistical methods. A great advantage of the wavelet transform is that it does not introduce additional redundancy in the input data, and the signal can be fully restored using just the same filters. Besides, lossy compression can be easily implemented, thanks to the fact that as a result of the transformation, details are separated from the main signal. It is quite simple then, to discard the details on those scales where they are insignificant. Suffice it to say that an image processed with wavelets can be compressed from one-third to one-tenth without any significant information loss (and up to 1/300th with tolerable loss). As an example we should note that wavelet transform underlies the data compression standard MPEG-4.

Neural networks and other data analysis mechanisms When training a neural net (or setting other analytical tools), data noisiness or a large amount of special cases (random peaks, gaps, harmonic distortion, etc.) present quite serious problems. Such noise can mask peculiarities in data or pretend to replace them, thus significantly deteriorating the results of the network learning. Therefore it is recommended to clean the data prior to analysing them. Because of the above-mentioned reasons and thanks to the presence of fast and efficient algorithms, wavelets seem to be very convenient and promising mechanism of cleaning and pre-processing data for use in statistical and business applications, AI systems, etc.

Systems of data transmission and digital signal processing Thanks to high performance of the algorithms and their noise stability, wavelet transform can be a powerful tool in the areas where other methods of data analysis have been traditionally used, like Fourier transform. Because it is possible to use existing methods of transform processing and thanks to particular qualities of the wavelet transform in the time-and-frequency domain, capabilities of such systems can be significantly expanded and enlarged.

Wavelet analysis is a technique to transform an array of M samples from their actual numerical values to an array of M wavelet coefficients. Each wavelet coefficient represents the correlation between the wavelet function at a particular size and a particular location within the data array. By varying the size of the wavelet function (usually in powers of two) and shifting the wavelet so it covers the entire array, you can build up a picture of the overall match between the wavelet function and your data array.

Since the wavelet functions are compact, the wavelet coefficients only measure the variations around a small region of the data array. This property makes wavelet analysis very useful for signal or image processing; this 'localized' nature of the wavelet transform allows to easily pick out features in the data, such as spikes (for example, noise or discontinuities), discrete objects (in, for example, astronomical images or satellite photos), and edges of objects.

The localization also implies that a wavelet coefficient at one location is not affected by the coefficients at another location in the data. This makes it possible to remove 'noise' of all different scales from a signal, simply by discarding the lowest wavelet coefficients. [All of the wavelet algorithms must be applied to a data set (a time series or a signal) with a power of two number of elements (e.g., 256 elements = 2^8).]

In the ordered wavelet transforms, the result consists of a wavelet scaling function value (also known as a smooth value or a low-pass filter value), followed by bands of wavelet function values (wavelet coefficients), in increasing frequency. The sizes of these wavelet coefficient bands are ordered in increasing powers of two. If there are M elements in the data set (where M is a power of two), the coefficient bands following the scaling value will have sizes $2^0, 2^1, 2^2, ..., M/2$.

The above-discussed concepts will be more clear with the following examples.

Example 3.10 The discrete function $f[m]$ has $M = 4$ samples,

$$f[0] = 1, \quad f[1] = 4, \quad f[2] = -3, \quad f[3] = 0$$

Calculate the approximation coefficient as well as detail coefficients and the forward and inverse discrete wavelet transform using Haar scaling.

The $M = 4$ point discrete Haar scaling and wavelet functions are used:

$$\begin{bmatrix} 1 & 1 & 1 & 1 \\ 1 & 1 & -1 & -1 \\ \sqrt{2} & -\sqrt{2} & 0 & 0 \\ 0 & 0 & \sqrt{2} & -\sqrt{2} \end{bmatrix} \begin{matrix} \varphi_{0,0}[m] \\ \psi_{0,0}[m] \\ \psi_{1,0}[m] \\ \psi_{1,1}[m] \end{matrix}$$

Solution

The coefficient for V_0:

$$W_\varphi[0, 0] = \frac{1}{2}\sum_{m=0}^{3} f[m]\varphi_{0,0}[m] = \frac{1}{2}[1 \cdot 1 + 4 \cdot 1 - 3 \cdot 1 + 0 \cdot 1] = 1$$

The coefficient for W_0:

$$W_\psi[0, 0] = \frac{1}{2}\sum_{m=0}^{3} f[m]\psi_{0,0}[m] = \frac{1}{2}[1 \cdot 1 + 4 \cdot 1 - 3 \cdot (-1) + 0 \cdot (-1)] = 4$$

The two coefficients for W_1:

$$W_\psi[0, 0] = \frac{1}{2}\sum_{m=0}^{3} f[m]\psi_{1,0}[m] = \frac{1}{2}[1 \cdot \sqrt{2} + 4 \cdot (-\sqrt{2}) - 3.0 + 0.0] = -1.5\sqrt{2}$$

$$W_\psi[1, 1] = \frac{1}{2}\sum_{m=0}^{3} f[m]\psi_{1,1}[m] = \frac{1}{2}[1.0 + 4.0 - 3 \cdot \sqrt{2} + 0 \cdot (-\sqrt{2})] = -1.5\sqrt{2}$$

In matrix form, we have

$$\begin{bmatrix} 1 \\ 4 \\ -1.5\sqrt{2} \\ -1.5\sqrt{2} \end{bmatrix} = \frac{1}{2}\begin{bmatrix} 1 & 1 & 1 & 1 \\ 1 & 1 & -1 & -1 \\ \sqrt{2} & -\sqrt{2} & 0 & 0 \\ 0 & 0 & \sqrt{2} & -\sqrt{2} \end{bmatrix}\begin{bmatrix} 1 \\ 4 \\ -3 \\ 0 \end{bmatrix}$$

Now the function $f[m]$ can be expressed as a linear combination of these basis functions:

$$f[m] = \frac{1}{2}[W_\varphi[0, 0]\varphi_{0,0}[m] + W_\psi[0, 0]\psi_{0,0}[m] + W_\psi[1, 0]\psi_{1,0}[m] + W_\psi[1, 1]\psi_{1,1}[m]]$$

$$f[0] = \frac{1}{2}[W_\varphi[0, 0]\varphi_{0,0}[0] + W_\psi[0, 0]\psi_{0,0}[0] + W_\psi[1, 0]\psi_{1,0}[0] + W_\psi[1, 1]\psi_{1,1}[0]]$$

$$= \frac{1}{2}[1 \cdot 1 + 4 \cdot 1 - 1.5\sqrt{2} \cdot \sqrt{2} - 1.5\sqrt{2} \cdot 0] = 1 \quad \text{and so on}$$

or in matrix form:

$$\begin{bmatrix} 1 \\ 4 \\ -3 \\ 0 \end{bmatrix} = \frac{1}{2}\begin{bmatrix} 1 & 1 & \sqrt{2} & 0 \\ 1 & 1 & -\sqrt{2} & 0 \\ 1 & -1 & 0 & \sqrt{2} \\ 1 & -1 & 0 & -\sqrt{2} \end{bmatrix}\begin{bmatrix} 1 \\ 4 \\ -1.5\sqrt{2} \\ -1.5\sqrt{2} \end{bmatrix}$$

Example 3.11 Give the scheme of finding DWT of a signal $x(n)$ by passing it through 3 level filter banks of quadratic mirror filters.

Solution

Let us first draw the three levels of the filter bank.

At each level in Fig. 3.51, the signal is decomposed into low and high frequencies. Due to the decomposition process the input signal must be a multiple of 2^n where n is the number of levels. For example, a signal with 16 samples, frequency range 0 to f_n, and 3 levels of decomposition, 4 output scales are produced:

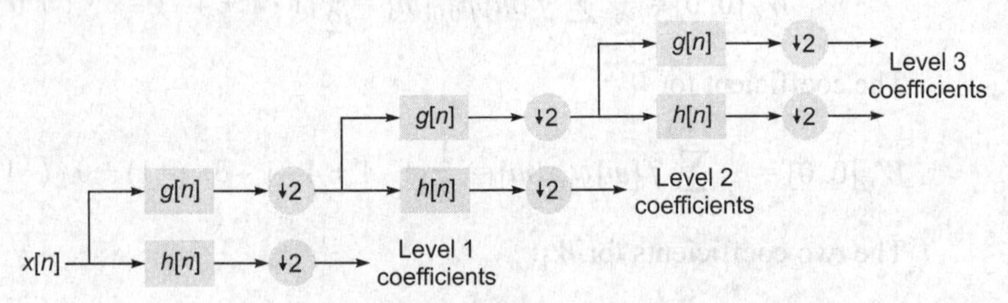

Fig. 3.51 A three-level filter bank

Table 3.7 Decomposition levels

Level	Frequencies	Samples
3	0 to $f_n/8$	4
	$f_n/8$ to $f_n/4$	4
2	$f_n/4$ to $f_n/2$	8
1	$f_n/2$ to f_n	16

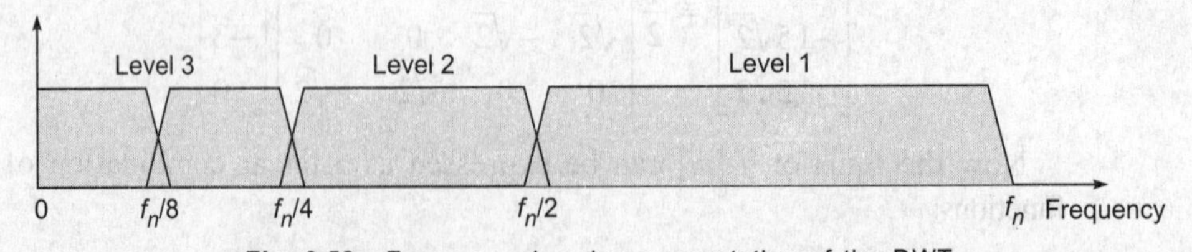

Fig. 3.52 Frequency domain representation of the DWT

3.11 ENCRYPTION/DECRYPTION

Source-coded information may be given to the encryptor stage before passing to the channel coding part. Cryptography is the art or science of secret writing by encryption or more exactly of storing information (for a shorter or longer period of time) in a form that allows it to be revealed to those you wish to see it by decryption, yet hides it from all others. A cryptosystem is a method to accomplish this. Cryptanalysis is the practice of defeating such attempts to hide information. Cryptology includes both cryptography and cryptanalysis.

The source-coded information to be hidden is called 'plaintext'. The hidden information is called 'ciphertext', which may be given to channel coding stage. Encryption process may or may not increase the size of the data but it simply converts it into another secret text. Encryption is any procedure to convert plaintext into ciphertext by means of an encryption engine (again, generally a computer program) whose operation is fixed and determinate (the encryption method) but which functions in practice in a way that is dependent on a piece of information (the encryption key), which has a major effect on the output of the encryption process. Decryption is any procedure to convert ciphertext into plaintext. A cryptosystem is designed such that decryption can be accomplished only under certain conditions, which generally means only by persons in possession of both a decryption engine (these days, generally a computer program) and a particular piece of information, called the decryption key, which is supplied to the decryption engine in the process of decryption.

The result of using the decryption method and the decryption key to decrypt ciphertext produced by using the encryption method and the encryption key should always be the same as the original plaintext (except perhaps for some insignificant differences). In this process, the encryption key and the decryption key may or may not be the same. When they are, the cryptosystem is called a 'symmetric key' system; when they are not it is called an 'asymmetric key' system. The most widely known instance of a symmetric cryptosystem is DES (the so-called data encryption standard). The most widely known instance of an asymmetric key cryptosystem is PGP ('Pretty Good Privacy').

An encryption algorithm (a precise specification of the steps to be taken when encrypting plaintext and when decrypting the resulting ciphertext) is known as an 'asymmetric algorithm' if the encryption and decryption keys that it uses are different; otherwise it is a 'symmetric algorithm'. There are many reasons for using encryption and the cryptosystem that one should use is the one best suited for one's particular purpose and that satisfies the requirements of security, reliability, and ease of use.

- Ease of use is easy to understand.
- Reliability means that the cryptosystem, when used as its designer intended it to be used, will always reveal exactly the information hidden when it is needed (in other words, that the ciphertext will always be recoverable and the recovered data will be the same as to the original plaintext).
- Security means that the cryptosystem will in fact keep the information hidden from all but those persons intended to see it despite the attempts of others to crack the system.

A cryptosystem that is easy to use should allow keyboard entry of a string of 10 to 60 characters, and easy to remember words. Spaces should not be significant, and upper and lower case should be equivalent. Reliability is the quality next easiest to test for. If it is not possible to provide a formal proof that the decryption of the encryption of the plaintext is always identical to the plaintext, it is at least possible to write software to perform multiple encryptions and decryptions with many different keys to test for reliability.

Finally, there is the question of security. The security of a cryptosystem is always relative to the task it is intended to accomplish and the conditions under which it will be used. A theoretically secure system becomes insecure if used by people who write their encryption keys on pieces of paper, which they stick to their computer terminals. In general a cryptosystem can never be shown to be completely secure in practice.

Summary

- In audio, image and video, the real-time signals will follow the envelope of stochastic nature.
- Speech signal has its own property, which can be exploited in source coding.
- Information signal should be source coded so that final minimum digital base is created with maximum information.
- For digitization, the analog signal must be converted into a digital form by sampling and quantization, which is a lossy process.
- Quantization process also adds the quantization noise. Good quantization techniques can make it possible to recover data back at the receiver. It may be uniform, non-uniform, adaptive, and vector.
- Vector quantization is the statistical base approach and used in modern coder designs.
- PCM, DM, and DPCM-ADPCM are the methods for digitizing the analog waveforms.
- Information theory correlates the entropy of the information and capacity of the channel for reliable and fruitful transmission.
- Coding methods give compression in the data size. Coding redundancy and compression ratio is the measure for amount of compression.
- There are various low bit rate speech coding techniques, are known as vocoders. These methods take the advantage of the speech signal properties.
- The human voice generation mechanism forms the basis of synthetic speech generation model. Linear predictive coding method has given birth to the latest voice coding methods like, RPELP, CELP, MELP, QCELP, etc.
- Sub-band coding is the frequency domain coding method. It uses the speech properties and divides the bands accordingly to achieve the compression effect also.
- Transform coding methods are discrete cosine transform and wavelet transform. Wavelet transform method is very much suitable for coding of images, for which multi-resolution analysis approach is used.(It is still a research topic and is less suitable for real-time communication due to heavy processing.)

Review Questions

1. What is the reason one has to quantize the sampled signal?
2. Observe the real-time signal waveforms and list out their general characteristics. What is the nature of a speech signal?
3. List out different quantization methods and compare them. If we have to choose any one method for high bit rate communication, which will be most suitable?
4. How does the vector quantization method differ from conventional methods?

5. Explain how the concept of vector quantization is equally applied in case of intervals, polygons, or polytops. What will be the code vector in each case?
6. How can a vector quantization exhibit compression also?
7. What are the drawbacks of PCM that can be removed by differential methods? What are the drawbacks of differential PCM methods?
8. What are the waveform coding methods by which information transmission efficiency can be improved?
9. A binary source $S = \{s_1, s_2\}$ has $P = \{p, (1 - p)\}$. Plot the entropy of the source versus p, as p varies from 0 to 1 and comment on the result.
10. What is the difference between waveform coding and source coding? Give one example of each.
11. Why are vocoders designed separately and conventional PCM is not used?
12. List out various vocoders and the speech properties exploited by them.
13. How do DCT and DFT differ?
14. How can you achieve compression by sub-band coding?
15. What do you mean by multi-resolution analysis? How is this method used in wavelet transform?

PROBLEMS

1. Find out the number of quantization levels for sending 8 bits per sample. If more bits are used for the encoding purpose, what will be the effect on the bandwidth? If fewer bits are used, comment on the reconstruction of the signal.
2. Consider an audio signal with spectral components in the range of 300 Hz to 3 kHz. Assume that a sampling rate of 8 kHz will be used to generate a PCM signal.
   (i) For S/N = 30 dB, what is the number of uniform quantization levels needed?
   (ii) What is the required data rate?
3. The bandwidth of a TV video plus audio signal is 5 MHz. If the signal is quantized at 512 levels, determine the data rate of the resulting PCM signal. Assume that the signal is sampled at a rate of 20% above the Nyquist rate.
4. In a PCM conversion, an analog signal is sampled at Nyquist sampling rate $1/T_s$, quantized using L quantization levels and converted into binary digital form for sending over the channel. Prove that the bit duration T of the transmitted signal must satisfy the following condition:

$$T \le T_s/\log_2 L$$

   Also comment on this.
5. A voice signal bandwidth is between 300 Hz and 3.1 kHz, considering significant frequency components. If 8 ksamples/sec is used to generate PCM signal and if peak signal power to average quantization noise power requirement is 32 dB.
   (a) What are the minimum number of bits/sample and minimum number of uniform quantization levels required?
   (b) What will be the final bit rate?
   (c) What will be the transmission bandwidth to make the PCM signal detectable?
6. In an eight-bit uniform quantizer with a dynamic range of 2 V with span of –1 V to 1 V, what may be the step size? Find out the SQNR for the sinusoidal signal that spans the full dynamic range.
7. A signal bandlimited to 100 kHz is sampled at a rate of 50% higher than the Nyquist

rate and quantized, so that 8 bits per sample can be assigned. A μ-law quantizer with $\mu = 255$ is used. Determine SQNR. If the SQNR found is to be increased at least by 10 dB, how will you achieve the same?

8. Consider a source $S = \{s_1, s_2, s_3, s_4\}$ with $P = \{1/2, 1/4, 1/8, 1/8\}$. Find out self-information of each message and entropy of source S.

9. Given the messages x_1, x_2, x_3, x_4, x_5, and x_6 with respective of probabilities of 0.4, 0.2, 0.2, 0.1, 0.06, and 0.04, construct the Huffman code. Determine the efficiency and redundancy of the code.

10. A source emits eight messages with probabilities 0.5, 0.25, 0.1, 0.07, 0.05, 0.02, 0.005, and 0.0005. Find out the entropy of the source. Obtain the Huffman code and find the average length of the code. Determine the efficiency and redundancy of the code.

11. Following is the data given:

$$S = \{s1, s2, s3, s4, s5, s6, s7, s8, s9, s10\}$$

$$P(S) = \{0.18, 0.17, 0.16, 0.15, 0.10, 0.08, 0.05, 0.05, 0.04, 0.02\}$$

The encoding alphabet is $\{0,1,2,3\}$.

Apply Huffman coding and determine coding gain and redundancy.

12. In a delta modulation system, the input is 8 kHz sine wave with 1 V peak-to-peak amplitude. The sampling rate is 10 times the Nyquist rate. Find out the step size required to prevent from the slope overload condition and to minimize granular noise? What is PSD for granular noise? Find out the SNR of this DM system.

MULTIPLE CHOICE QUESTIONS

1. If p_k is the probability of a message being received or transmitted, the amount of information I_k associated in bits is given by

(a) $I_k = \log_2 p_k$
(b) $I_k = \log_2 (1/p_k)$
(c) $I_k = 2\log_2 p_k$
(d) $I_k = \log_{10} p_k$

2. Entropy is basically a measure of

(a) rate of information
(b) average information
(c) disorder of information
(d) probability of information

3. A PCM receiver can see

(a) quantization noise
(b) channel noise
(c) interference noise
(d) all of the above

4. In PCM for q quantizing levels, the number of pulses p in a code group is given by

(a) $\log_{10} q$ (b) $\log_2 q$
(c) $\ln q$ (d) $2 \log_2 q$

5. The principal merit of PCM system is its

(a) lower bandwidth
(b) lower noise
(c) lower power requirement
(d) lower cost

6. A compander is used in communication systems to

(a) compress the bandwidth
(b) improve the frequency response
(c) reduce the channel noise
(d) improve signal-to-noise ratio

7. The channel capacity is exactly equal to the

(a) noise rate in channel
(b) bandwidth of demand
(c) amount of information per second
(d) available bandwidth

8. Which of the following systems is digital?

(a) PPM (b) PWM
(c) PCM (d) PFM

9. Which of the following is unlikely to happen when the quantizing noise is decreased in PCM?
(a) Increase in the bandwidth
(b) Increase in the number of standard levels
(c) Increase in the channel noise
(d) All of the above

10. The signal-to-quantization noise ratio in a PCM system depends upon
(a) the sampling rate
(b) the message signal bandwidth
(c) the number of quantization level
(d) all of the above

11. Which of the following systems is not digital?
(a) Differential PCM (b) DM
(c) ADPCM (d) PAM

12. In which of the following methods of source coding, maximum compression is achieved?
(a) *A*-law PCM (b) μ-law PCM
(c) DM (d) ADPCM

13. Which of the following errors is/are likely to occur in the differential PCM schemes?
(a) Prediction error
(b) Compliance error
(c) Quantization error
(d) All of the above

14. For successful information transmission over a channel with capacity C, the source entropy must be
(a) less than or equal to C
(b) equal to C
(c) greater than C
(d) less than C

15. If a source outcome has probability of 1, the information content in that is
(a) one (b) zero
(c) one-half (d) none

16. The average information in the binary symmetric source is maximized when both outcomes with probabilities p and $1-p$ are
(a) 1 and 0, respectively
(b) zero
(c) equally likely to occur
(d) maximum

17. A Huffman tree algorithm is used for
(a) redundancy removal
(b) coding gain
(c) variable length coding
(d) none of the above

18. Which of the following vocoders is based on filter banks?
(a) Formant (b) Cepstrum
(c) Channel (d) LP

19. Which of the following vocoders requires past record of speech samples?
(a) Formant (b) Cepstrum
(c) Channel (d) LP

20. Which of the following is the time-domain source coding method?
(a) DCT (b) Sub-band
(c) Wavelet (d) LPC

21. In which of the following methods, multiresolution analysis method is used?
(a) Sub-band coding
(b) Wavelet transform
(c) QMF filtering
(d) All of the above

22. Which of the following vocoders is based on speech properties?
(a) Formant (b) LPC
(c) CELP (d) QCELP

23. Which of the following vocoders is based on linear predictive vocoder?
(a) Channel (b) Formant
(c) Voice excited (d) MELP

24. Fourier transform is suitable for ______ signals while wavelet transform is suitable for _____ signals.
(a) Stationary, non-stationary
(b) Stationary, stationary as well as non-stationary
(c) Stationary as well as non-stationary, stationary
(d) Non-stationary, stationary

chapter 4

Channel Coding Techniques

Key Topics

- Fundamentals of channel coding and decoding
- Shannon's limit
- Trade-offs in error detection/correction
- Channel capacity and coding gain
- Error-detecting and error-correcting capabilities
- Various block codes, such as Hamming, BCH, and RS
- Hard and soft coding and decoding methods
- Convolutional coding and Viterbi decoding
- Interleaver, puncture coding, and turbo coding

Chapter Outline

The task of source coding (described in the previous chapter) is to represent the source information with the minimum number of symbols by removing the redundant bits systematically. When a code is transmitted over a channel in the presence of noise, errors are likely to occur. On the other hand, channel coding represents the source information over the channel in a manner that minimizes the error probability in decoding by adding the redundant bits systematically. Channel coding is more important over the wireless channel. It reduces the bit error rate in the final reception. Hence, the quality of reception improves. With this aim, the various channel coding schemes are highlighted in this chapter. Other related terms important for performance comparison for different channel coding schemes are also discussed. Channel coding in general can be done by error-detecting or error-correcting codes. How error-correcting codes are useful for the wireless digital communication system is also dealt with in the chapter. Basically, coding methods are based on mathematical or logical operations and also its corresponding decoding.

4.1 FUNDAMENTALS OF CHANNEL CODING AND DECODING

Coding theory deals with transmission of data over noisy channels by adopting various source and channel coding/decoding schemes. Coding theory (in general) was originated by Claude Shannon in 1948 with the publication of his paper, A Mathematical Theory of Communication. He also worked on entropy and source coding, capacity of the channel, etc., which we have studied in the last chapter. Afterwards, Huffman and Hamming also worked on this theory. In the last fifty years, due to growth of digital communication, coding theory has grown into a discipline intersecting mathematics and engineering with applications to almost every area of communication, such as satellite, cellular telephony, Internet, CD recording, and data storage.

The existence of noise as well as the variation in propagation characteristics of different communication channels makes the correct handling of errors vital. *Channel coding* refers to the class of signal transformations designed to improve communication performance by enabling the transmitted signals to better withstand the effects of various channel impairments, such as noise, interference, and fading (Chapter 5). Some of the ways in which errors are handled include error detection, error correction, data acknowledgment, and data resends. It can be partitioned into two areas (same as in source coding):

1. Waveform (or signal) coding
2. Structured sequences (or structured redundancy) coding

Waveform coding deals with the transformation of waveforms into 'better waveforms' to make the detection process less subject to errors. This approach is mainly suitable for signalling purposes, for example, *m*-ary signalling, antipodal signalling, orthogonal signalling, trellis coded modulation, etc. Line coding is a kind of waveform coding method for handling channel impairments, synchronization between transmitter and receiver, maintaining power levels, etc.

Structured sequences deal with transforming data sequences into 'better sequences' having structured redundancy (redundant bits addition). The redundant bits can be used for the detection and correction of errors. For example, CRC, block coding, convolutional coding, turbo coding, etc.

There are two basic approaches of adding structured redundancy used for controlling errors—automatic repeat request (ARQ) and forward error correction (FEC).

The automatic repeat request or error detection, correction (sometimes) and retransmission, utilizes parity bits or redundant bits for checksum added to the data to detect an error. These errors may be corrected sometimes when single bit error or error bit positions can be identified. The receiving terminal does not attempt to correct the error always; rather it requests the transmitter to retransmit the data. It is clearly noticeable that a two-way link is required for such a dialogue between transmitter and receiver. It controls the error by allowing retransmissions rather than error correction every time. This is possible only in packet based systems. For real-time transmissions like multimedia, retransmissions are not possible and hence structured redundancy coding is a better option.

Techniques for error detection, such as performing a cyclic redundancy check (CRC) on the data, can guarantee detection of errors with a vastly greater certainty than a typical system normal BER. This detection process can then be used to send an acknowledgement (ACK) or a failed/not acknowledgement (NAK) back to the transmitting node to request retransmission. Typically, a system based on a combination of error detection and acknowledgments can perform well in a stationary environment where data latency is not a problem. The principal advantages of this type of system are the simplicity and the high data throughput relative to the data rate for good SNR conditions. This type of system does not, however, tend to work well in poor SNR conditions due to the large number of re-sends necessary. In order to improve the typical BER of a mobile communications systems error detection and correction is usually employed. This involves the transmission of some additional data in order to allow the detection, identification, and correction of errors.

The *forward error correction* uses redundant bits added systematically to detect and correct the error independently. These schemes do not send ACK/NAK signals and are more important in the wireless environment. In this case, only a one-way link will suffice for proper communication. Error handling is totally done by the receiver in this case if the appropriate method according to application is selected.

In short, there are two types of codes in general:

1. Error-detecting codes, e.g., parity check, LRC-VRC check, CRC check, etc.
2. Error-correcting codes: These are generally used over wireless medium/time dispersive channel. The examples are block codes (like Hamming code, Hadamard code, Golay code, cyclic code, BCH code, Reed–Soloman code, etc.), convolutional codes and turbo codes.

For using the above two schemes, the following basic concept is applied in common. On the transmitter side, an encoder adds redundancy to the data. Then at the receiver, a decoder is able to exploit the redundancy in such a way that a reasonable number of channel errors can be detected/corrected. A binary encoder takes in k bits at a time and produces an output (or code word) of n bits, where $n > k$. While there are 2^n possible sequences of n bits, only a small subset of them, 2^k to be exact, will be valid code words. The ratio k/n is called the *code rate.* Lower rate codes can generally correct more channel errors than higher rate codes and thus are more *energy efficient*. However, higher rate codes are more *bandwidth efficient* than the lower rate codes because the amount of overhead (in the form of parity bits) is lower. Thus, selection of the code rate involves a trade-off between energy efficiency and bandwidth efficiency.

Error-correcting codes that are mainly used in wireless communication are broadly categorized as follows:

1. Block codes:
   (i) Hamming code
   (ii) Bose–Chaudhury–Hocquenghem (BCH) codes
   (iii) Reed–Solomon codes and so on
2. Convolutional codes
3. Turbo codes:
   (i) Block/product turbo codes (BTC or PTC)
   (ii) Convolutional turbo codes (CTC)

Block codes are based rigorously on finite field arithmetic and abstract algebra. Finite set of elements on which two binary operations, addition and multiplication, are defined. Block codes as mentioned previously accept a block of k information bits and produce a block of n coded bits mostly by using generator matrix. By predetermined rules, $(n - k)$

redundant bits are added to the k information bits to form the n coded bits. Commonly, these codes are referred to as (n, k) block codes. The same basic concept is applied to different block coding schemes.

There are many differences between block codes and convolutional codes. Convolutional codes are one of the most widely used channel codes in practical communication systems. These codes are developed with a separate strong mathematical structure and are primarily used for real-time error correction. Convolutional codes convert the entire data stream into single codeword. The encoded bits depend not only on the current k input bits but also on past input bits. The main decoding strategy for convolutional codes is based on the widely used Viterbi algorithm.

As a result of the wide acceptance of convolutional codes, there have been many advances to extend and improve this basic coding scheme. This advancement resulted in two new coding schemes, namely, trellis coded modulation (TCM) and turbo codes. TCM adds redundancy by combining coding and modulation into a single operation (as the name implies) and we categorize it as waveform coder. The unique advantage of TCM is that there is no reduction in data rate or expansion in bandwidth as required by most of the other coding schemes, while recently a near channel capacity error correcting codes, called turbo codes, were introduced. This error correcting code is able to transmit information across the channel with an arbitrary low (approaching zero) bit error rate.

It has been shown that a turbo code can achieve performance within 1 dB of channel capacity. Random coding of long block lengths may also perform close to channel capacity, but this code is very hard to decode. Without a doubt, the performance of a turbo code is partly due to the random interleaver used to give the turbo code a 'random' appearance. However, one big advantage of a turbo code is that there is enough code structure (from the convolutional codes) to decode it efficiently. There are two primary decoding strategies for turbo codes. They are based on a maximum a posteriori (MAP) algorithm and a soft output Viterbi algorithm (SOVA). Regardless of which algorithm is implemented, the turbo decoder requires the use of two (same algorithm) component decoders that operate in an iterative manner. The SOVA is much less complex than MAP and it provides comparable performance results. Furthermore, SOVA is an extension of the Viterbi algorithm and thus has an implementation advantage over MAP.

All these channel coding techniques will be discussed in this chapter after introducing general concepts for channel coding.

4.1.1 Channel Capacity

After the channel coding stage, the modulated signal with the final transmission bandwidth has to be exposed to the channel and hence channel capacity is an important topic in this chapter. In Chapters 2 and 3, the channel capacity is introduced as per requirement. Here it is explained in a better way so as to understand its practical significance. Most communication engineers always try to minimize the transmission bandwidth. The trend has been to use narrower bandwidths because narrow bandwidths permit more communication channels to be packed into a defined frequency band and also the advantage of good SNR.

Claude Shannon worked a lot on the fundamental information transmission capacity of a communication channel. He showed that 'there exists a limiting value of E_b/N_o below which there can be no error-free communication at any information rate'. This limiting value of E_b/N_o is called the *Shannon limit*. This theory is applicable to both wired as well as wireless channels.

As described in Chapter 2, Shannon showed that the system capacity C (maximum data rate in bits per second) of a channel perturbed by AWGN (explained in Chapter 5) is a function of the average received signal power S, the average white Gaussian noise power N, and the bandwidth W. The only options available to increase a channel's capacity are to increase either bandwidth (W) or the signal-to-noise ratio (S/N). The capacity relationship (popularly known as *Shannon–Hartley theorem*) can be stated as

$$C = W \log_2\left(1 + \frac{S}{N}\right) \tag{4.1}$$

When W is in hertz and the logarithm is taken to the base 2, as shown, the capacity C is given in bits/s. It is theoretically possible to transmit information over such a channel at any rate R, where $R \leq C$, with an arbitrarily small error probability by using a sufficiently complicated coding scheme. For an information rate $R > C$, it is not possible to find a code that can achieve an arbitrarily small error probability. An increase in the SNR requires an increase in transmitted power as the noise within the channel is beyond our control. Thus, we can either trade power or bandwidth to achieve a specified channel data rate. Because of the logarithmic relationship, increasing the power output is often unrealistic. However, if frequency allocation constraints permit, the bandwidth can be increased. An appreciable increase in data capacity or SNR (for a fixed data rate) can then be achieved.

One of the most famous results of information theory is Shannon's channel coding theorem, which says that, for a given channel, there exists a code that will permit the error-free transmission across the channel at a rate R, provided $R \leq C$ the channel capacity. Equality is achieved only when the SNR is infinite. (Proof is not given here.)

Description of how the capacity is calculated will not be given here. However, an example is instructive, which is also described in Chapter 3 in terms of entropy or information. Like the binary symmetric source, the *binary symmetric channel* is a channel with a binary input and output. Associated with each output are a probability p that the output is correct and a probability $(1 - p)$ that it is not. For such a channel, the channel capacity turns out to be

$$C = 1 + p\log_2(p) + (1 - p)\log_2(1 - p) \tag{4.2}$$

Here, p is the bit error probability. If $p = 0$ then $C = 1$. If $p = 0.5$, then $C = 0$ (Fig. 4.1). Thus, if there is equal probability of receiving a 1 or 0, irrespective of the signal sent, the channel is completely unreliable and no message can be sent across it.

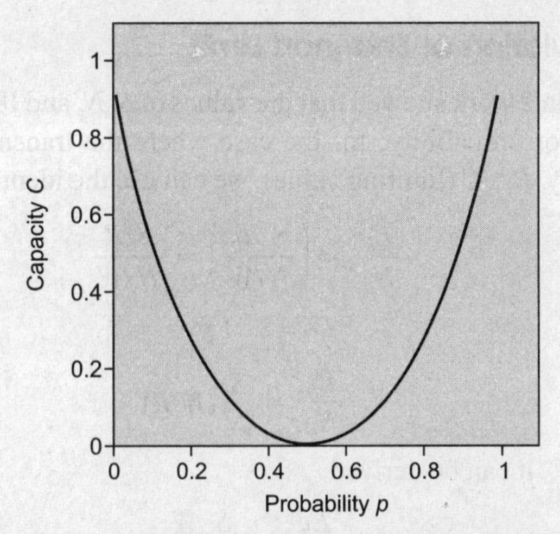

Fig. 4.1 Capacity of a binary symmetric channel

The graphs in Figures 3.23 and 4.1 are exactly opposite. So defined–the channel capacity is a non-dimensional number. We normally quote the capacity as a rate in bits/s. To do this, we relate each output to a change in the signal. For a channel of bandwidth W, we can transmit at most $2W$ changes per second. Thus, the maximum capacity in bps is $2W$. For the binary channel, we have

$$C = 2W[1 + p\log_2(p) + (1 - p)\log_2(1 - p)] \tag{4.3}$$

Thus, for the binary channel the maximum bit rate is $2W$. We note that the capacity is always less than the maximum bit rate. The final data rate R, or information rate describes the rate of transfer of data bits across the channel. In theory, we can write

$$W \geq C \geq R \tag{4.4}$$

[Instead of $H(X)$, R is used here purposely (think why).]

As a matter of practical fact,

$$W > C > R \tag{4.5}$$

If the source is optimally coded, we can rephrase the channel coding theorem as follows: A source of information with entropy $H(X)$ can be transmitted error free over a channel provided $H(X) \leq C$.

All the modulations are described in Chapter 7. If we consider the case of a binary channel, we should note that QPSK uses half the bandwidth of BPSK for the same bit rate. We might suppose that for the same bandwidth, QPSK would have twice the capacity, but this is not so. The BPSK modulation is far from optimum in terms of bandwidth use. The QPSK makes better use of the bandwidth. The increase in bit rate provided by QPSK does not reflect an increase in capacity; merely a better use of bandwidth.

The capacity of the binary channel is much less than that calculated from the Hartley–Shannon law [Eq. (4.1)]. Why so? The answer is that the equation applies to the systems whose outputs may take any values. We use systems obeying the equation because they are technically convenient, not because they are desirable.

4.1.2 Calculation of Shannon Limit

Though Shannon's work showed that the values of S, N, and W set a limit on transmission rate, not on error probability; for the case where the transmission bit rate is equal to channel capacity, $R = C$ (limiting value), we can use the identity

$$\frac{E_b}{N_o} = \left(\frac{S \cdot T_b}{N/W}\right) = \frac{S/R}{N/W} \tag{4.6}$$

Simplifying,

$$\frac{E_b}{N_o} = \frac{S}{N}(W/R) \tag{4.7}$$

Putting $R = C$, it can be derived

$$\frac{E_b}{N_o} = \frac{S \cdot W}{(N \cdot C)} \tag{4.8}$$

Now using Equations (4.1) and (4.8),

$$\frac{C}{W} = \log_2\left[1 + \frac{E_b}{N_o}\left(\frac{C}{W}\right)\right] \tag{4.9}$$

where C/W denotes *bandwidth efficiency*. Now if $C/W \to 0$, using identity

$$\lim_{x \to 0}(1 + x)^{1/x} = e$$

we get

$$\frac{E_b}{N_o} = -1.6 \text{ dB} \tag{4.10}$$

Equation (4.10) is popularly known as *Shannon limit,* below which there can be no error-free communication at any information rate.

In Fig. 4.2, the Shannon limit is shown in undashed bit error probability P_B versus E_b/N_o curve. The curve is discontinuous going from a value of $P_B = 0.5$ to $P_B = 0$ at $E_b/N_o = -1.6$ dB. For a bit error probability of 10^{-5}, the optimum uncoded binary phase-shift keying (BPSK) modulation requires an E_b/N_o of typically 9.6 dB. Therefore, for this case, Shannon's work promised the existence of a theoretical performance improvement of 11.2 dB [9.6 – (–1.6) = 11.2 dB] over the performance of an optimum uncoded binary modulation through the use of coding techniques. Today, most of that promised improvement (as much as 10 dB) is realizable with turbo codes.

Remarkably, Shannon showed that capacity could be achieved with bit error probability approaching zero by a *completely random code*, (i.e. a randomly chosen mapping set of code words) only when the block length n approaches infinity. However, random codes are not practically feasible. In order to be able to encode and decode with reasonable complexity, codes must possesses some sort of structure. Unfortunately, *structured codes perform considerably worse than random codes*. This is the basis of the *coding paradox*. It was also surprising that for practical purposes, a capacity limit was applied that was a few decibels lower than Shannon's, called the *cut-off rate bound*. Finally, turbo codes have become popular for achieving Shannon's limit.

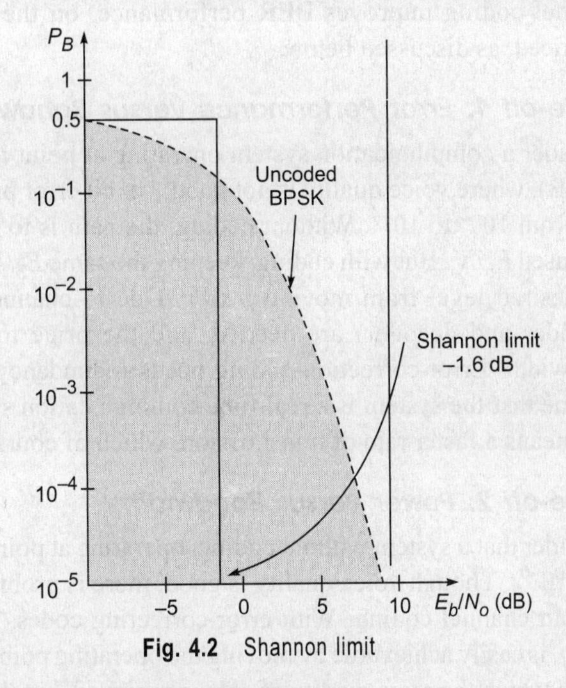

Fig. 4.2 Shannon limit

4.1.3 Why to Use Error-Correcting Codes

Error-correcting codes can be regarded as a vehicle for effecting various system trade-offs. Figure 4.3 compares two curves depicting bit error performance (P_B) versus bit energy-to-noise density ratio (E_b/N_o) (same as SNR, E_b/N_o is normally a parameter for digital signal representing signal strength). One curve represents a typical modulation scheme without coding. The second curve represents the same modulation with coding.

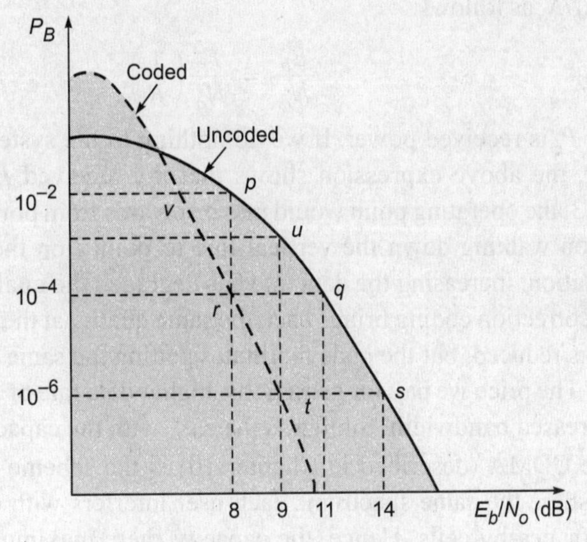

Fig. 4.3 Comparison of typical coded versus uncoded error performance

Though the channel coding is incorporated in the wireless links to improve BER, there exist some trade-offs especially in terms of transmission bandwidth. While on one side, channel coding improves BER performance, on the other side, something is to be sacrificed, as discussed below.

Trade-off 1: Error Performance versus Bandwidth

Consider a communication system operating at point *t* in Fig. 4.3 ($P_B = 10^{-2}$ and E_b/N_o = 8 dB), where voice quality is not good, i.e. bit error probability (P_B) has to be reduced, say, from 10^{-2} to 10^{-4}. Without coding, the path is to be followed through *p* to *q* with increased E_b/N_o. But with coding, keeping the same E_b/N_o, bit error rate can be reduced at the desired level from moving *p* to *r*. Due to channel coding, the new components (encoder and decoder) are needed, and the price to be paid is more transmission bandwidth. Error-correction coding needs redundancy to be added to the data. If we assume that the system is a real-time communication system, the addition of redundant bits means a faster rate of transmission, which of course means more bandwidth.

Trade-off 2: Power versus Bandwidth

Consider that a system without coding, operating at point *s* in Fig. 4.3 (E_b/N_o = 14 dB and $P_B = 10^{-6}$). Though voice quality is good, there is problem with reliability of the system without channel coding. With error-correcting codes, reduction in power level, i.e. in E_b/N_o, is easily achievable by moving the operating point from *s* to *t*. Thus, the trade-off is one in which the same quality of data is achieved, but the coding allows saving in power or E_b/N_o. Again the cost to be paid is nothing but more bandwidth requirement.

Trade-off 3: Data rate versus Bandwidth

Now consider that the system is operating at the same point *s*. Assume that there is no problem with the data quality and no particular need for reducing power. But the requirement is for increased data rate. Now there is well-known relation of data rate *R* with E_b/N_o as follows:

$$\frac{E_b}{N_o} = \frac{P_r}{N_o}\left(\frac{1}{R}\right) \tag{4.11}$$

where P_r is received power. If we do nothing to the system, except increasing the data rate *R*, the above expression shows that the received E_b/N_o would decrease, and in Fig. 4.3, the operating point would move upwards from point *s* to, say, some point *u*. Now, envision walking down the vertical line to point *t* on the curve that represents coded modulation. Increasing the data rate has degraded the quality of the data. But, the use of error-correction coding brings back the same quality at the same power level (P_r/N_o). The E_b/N_o is reduced, but the code facilitates getting the same error probability with a lower E_b/N_o. The price we pay for getting this higher data rate or greater capacity is nothing but an increased bandwidth. Similar is the case with the capacity and bandwidth trade-off.

The CDMA (described in Chapter 10) is the scheme for cellular telephony, where users share the same spectrum. Each user interfers with each of the other users in the same or nearby cells. Hence, the capacity (here maximum users per cell) is inversely proportional to E_b/N_o. It means a lower E_b/N_o and, therefore, more capacity; the codes achieve a reduction in each user's power, which in turn allows for more number of users.

One thing must be noted that in each of the above trade-off examples, a 'traditional' code involving redundant bits and faster signalling for real-time communication system has been assumed; hence, in each case, the cost was *expanded bandwidth*. However, trellis-coded modulation does not require faster signalling or expanded bandwidth for real-time systems.

Note It is clear that for non-real-time communication systems, error-correction coding can be used with a somewhat different trade-off. In this case, it is possible to obtain an improved bit-error probability or reduced power (similar to trade-off 1 or 2 above) but, by paying the price of *delay*, instead of *bandwidth*.

Can all the inevitable nonsense errors incorporated in the channel be detected and corrected? We shall see in the next sections. Error-correction coding for real-time signals is more sophisticated than simple error-detection coding. Error detection aims at detecting and locating errors in transmission and once located, the correction is simply an inversion of the bit. Error-correction coding requires lower rate codes than error detection. It is, therefore, uncommon in terrestrial communication, where better performance is usually obtained with error detection and retransmission. However, in satellite communications, sometimes error-correction coding is used because if retransmissions are used, the propagation delay often means that many frames are transmitted before an instruction for retransmission is received. This can make the task of data handling very complex.

4.1.4 Coding Gain

For a given bit-error probability, coding gain is defined as the reduction in E_b/N_o that can be realized through the use of the code. Coding gain is generally expressed in dB, such as

$$G(\text{dB}) = \left(\frac{E_b}{N_o}\right)_{\text{uncoded}} (\text{dB}) - \left(\frac{E_b}{N_o}\right)_{\text{coded}} (\text{dB}) \tag{4.12}$$

4.1.5 Error Detecting and Correcting Capability

All of the error patterns can never be correctly decoded and corrected. There are some restrictions and conditions. Before investigating into that, let us define a few important terms related to coding theory.

Hamming Weight

The Hamming weight $W(U)$ of a code word U is defined to be the number of non-zero elements in U. For a binary vector, this is equivalent to the number of 1's in the vector. For example, if $U = \underline{1}\,0\,0\,\underline{1}\,0\,\underline{1}\,0\,\underline{1}$, then $W(U) = 4$.

Hamming Distance

It is apparent that channel coding requires the use of redundancy. If all possible outputs of the channel correspond uniquely to a source input, there is no possibility of detecting errors in the transmission. To detect, and possibly correct errors, the channel code sequence must be longer than the source sequence. The rate of a channel code is the average ratio of the source sequence length to the channel code length as mentioned previously. Thus, rate of a channel code is always less than 1.

A good channel code is designed so that if a few errors occur in transmission, the output can still be identified with the correct input. This is possible because, although incorrect, the output is sufficiently similar to the input to be recognizable. The idea of similarity is made more firm by the definition of a *Hamming distance.* Let x and y be two binary sequences of the same length. The Hamming distance between these two codes is the number of symbols or bits that disagree. Suppose the code x is transmitted over the channel. Due to errors, y is received. The decoder will assign to y the code x that minimizes the Hamming distance between x and y.

If the transmitter sends 10000 but there is a single bit error and the receiver gets 10001, it can be seen that the 'nearest' codeword is in fact 10000 and so the correct codeword is found. It can be shown that to detect n bit errors, a coding scheme requires the use of codewords with a Hamming distance of at least $n + 1$. It can also be shown that to correct n bit errors requires a coding scheme with at least a Hamming distance of $2n + 1$ between the codewords. By designing a good code, we try to ensure that the Hamming distance between possible codewords x is larger than the Hamming distance arising from errors. The Hamming distance between two codewords U and V, denoted by $d(U, V)$, is defined to be the number of elements in which they differ.

Example 4.1 If $U = 1\,0\,0\,1\,0\,0\,0\,1\,0\,1$
$V = 1\,0\,1\,1\,0\,0\,1\,1\,0\,1$

then find out the Hamming distance between two code words.

Solution

While comparing both the sequences, the bit positions that differ in terms of bits are underlined and highlighted:

$$U = 1\,0\,\mathbf{0}\,1\,0\,0\,\mathbf{0}\,1\,0\,1$$
$$V = 1\,0\,\underline{\mathbf{1}}\,1\,0\,0\,\underline{\mathbf{1}}\,1\,0\,1$$

Thus, the Hamming distance is

$$d(U, V) = 2$$

Hamming distance between two code words is equal to the Hamming weight of their sum. In mathematical notation,

$$d(U, V) = W(U + V) \tag{4.13}$$

Example 4.2 Prove Eq. (4.13).

Solution

Summing up U and V means exclusive OR operation.

$$\begin{array}{r} U = 1001000101 \\ V = 1011001101 \\ \hline 0010001000 \end{array}$$

Now, Hamming weight of this sum is equal to the number of 1's, which is 2.

Thus, $d(U, V) = 2$

Minimum Distance of a Linear Code in General

This will indirectly give the code word selection criteria. Using the property of linear codes, i.e. if U and V are code words, then $Z = U + V$ must also be a code word. Hence, the distance between two code words is equal to the weight of a third code word; that is

$$d(U, V) = W(U + V) = W(Z) \tag{4.14}$$

There are a number of possible values of U and V. Thus, the minimum distance of a linear code can be ascertained without examining the distance between all combinations of codeword pairs. We only need to examine the weight of each code word (excluding the all-zeros codeword) in the sub-space; the minimum weight of code word in a sub-space corresponds to the minimum distance $d_{\min}$ between two code words in sub-space.

$$d_{\min} = \text{minimum weight}$$

Equivalently, the minimum distance, $d_{\min}$, corresponds to the smallest of the set of distances between the all-zeros codeword and all other codewords. If the code used is binary, the distance is known as Hamming distance. Now, *error-correcting capability* t_c of a code is defined as the maximum number of guaranteed correctable errors per codeword and it is written as

$$t_c = \left\lfloor \frac{d_{\min} - 1}{2} \right\rfloor \tag{4.15}$$

where $\lfloor x \rfloor$ means the largest integer not to exceed x. The *error-detecting capability* of a code t_d is defined as the maximum number of guaranteed detectable errors per codeword prior to correction and it is written as

$$t_d = d_{\min} - 1 \tag{4.16}$$

So, the number of errors that can be detected and corrected depends on the minimum Hamming distance (or *free distance*).

4.1.6 Statistical Concepts for Decoding

Now readers should be acquainted with the statistical concepts behind decoding. The concepts are also necessary to understand Viterbi and turbo decoding properly.

Maximum Likelihood Decoding

If all message sequences are equally likely (equi-probable), *maximum likelihood decoding* is one that achieves the minimum probability of error by choosing the maximum out of all possible likelihood functions $P(Y|S^{(m)})$, for all $S^{(m)}$, where Y is the received sequence and $S^{(m)}$ is one of the possible transmitted sequences.

Maximum likelihood decoder is an optimal decoder that chooses $S^{(m)}$ if

$$P(Y|S^{(m)}) = \max P(Y|S^{(m)}) \quad \text{over all } S^{(m)} \tag{4.17}$$

In consideration of simple case, let us say that there are only two equally likely possible sequences, $s_1(t)$ or $s_2(t)$, that might have been transmitted. Therefore, from the received signal, we need to decide whether $s_1(t)$ was transmitted [if $p\,(y/s_1) > p\,(y/s_2)$] or to decide that $s_2(t)$ was transmitted (maximum likelihood decisions).

Hard versus Soft Decision Decoding

Consider a two-level signal for digital system. One bit can be transmitted per symbol, with, say, a '0' being sent as –1V and a '1' as +1V. At a receiver, assuming that the gain is correct, we should expect to receive a signal always in the vicinity of either –1V or +1V, depending on whether a '0' or a '1' was transmitted, the departure from the exact values ±1V being caused by the inevitable noise added in transmission.

A simple receiver might operate according to the rule that negative signals should be decoded as '0' and positive ones as '1'. This is an example of a *hard decision*, with 0 V as the decision boundary. This simple receiver would make no mistakes unless there were noise effects. But noise usually has a continuous distribution such as Gaussian and may occasionally have a large amplitude, although with lower probability than for smaller values. Thus, if say +0.5 V is received, it most probably means that a '1' was transmitted, but there is a smaller yet still finite probability that actually '0' was sent. Common sense suggests that if a large amplitude signal is received we can be more confident in the hard decision than when the amplitude is small. It is just like a comparator to reset the correct logic levels. For multilevel transmission, hard decision making is a quantization process.

This quantization process is exploited in *soft decision* Viterbi decoders (which are mainly important for convolutional and turbo codes-refer contents). These maintain a history of many possible transmitted sequences, building up a view of their relative and finally selecting the value '0' or '1' for each bit according to which the *maximum likelihood* is decided. For convenience, a Viterbi decoder adds *log-likelihood* (rather than multiplying probabilities) to accumulate the likelihood of each possible sequence. It can be shown that in the case of BPSK or QPSK the appropriate log likelihood measure or *metric* of the certainty of each decision is indeed simply proportional to the distance from the decision boundary on constellation. (The slope of this linear relationship itself also depends directly on the signal-to-noise ratio.) Thus, the Viterbi decoder is fed with a soft decision comprising both the hard decision (the sign of the signal) together with a measure of the amplitude of the received signal.

With other rectangular-constellation modulation systems, such as 16-QAM or 64-QAM, each axis carries more than one bit. At the receiver, a soft decision can be made separately for each received bit. The metric functions are now more complicated than for QPSK, being different for each bit, but the principle of the decoder exploiting knowledge of the expected reliability of each bit remains same.

Consider that a binary signal transmitted over a symbol interval $(0, T)$ is represented by $s_1(t)$ for binary one and $s_0(t)$ for binary zero. The received signal is $y(t) = s_m(t) + n(t)$, where $n(t)$ is zero mean Gaussian noise process. So at the receiver side decision is to be made on the basis of comparing $y(t)$ to a threshold. The conditional probabilities of $y(t)$, $p(Y|s_1)$ and $p(Y|s_0)$ are shown in Figure 4.4, labelled likelihood of s_1 and likelihood of s_0. The demodulator of the wireless link converts the set of time random variables $\{y(t)\}$ into a code sequence Y, and passes it on to the decoder. In short, the demodulator output can be configured in a variety of ways.

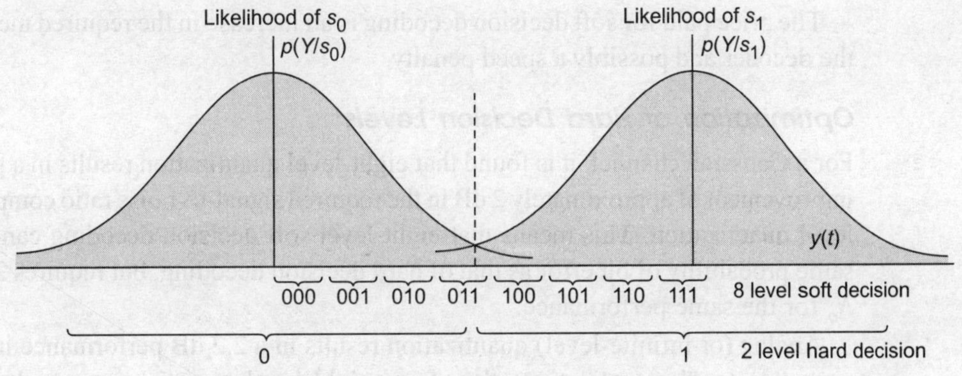

Fig. 4.4 Hard and soft decoding decisions

Demodulator output quantized to two levels (zero or one) ⇒ Hard decision In this case, output of the demodulator is quantized to two levels, zero and one, and fed into the decoder (same as threshold detection using comparator or 1-bit ADC). Since the decoder operates on the firm/hard decisions made by the demodulator, the decoding is called *hard decision decoding*.

Demodulator output quantized to more than two levels ⇒ Soft decision The demodulator can also be configured to feed the decoder with a quantized value of $y(t)$ greater than two levels. Such an implementation furnishes the decoder with more information than is provided in the hard decision case. When the quantization level of the demodulator output is greater than two (same as n-bit ADC), the decoding is called *soft decision decoding*.

Eight levels (3-bits) of quantization are illustrated on the abscissa of Fig. 4.4, when the demodulator sends a hard binary decision; it sends it as a single binary symbol. When the demodulator sends a soft binary decision, quantized to three bit, i.e., eight levels, it sends the decoder a 3-bit word describing an interval along $y(t)$. In effect, sending such a 3-bit word in place of a single binary symbol is equivalent to sending the decoder a measure of confidence along with the code symbol decision. Referring to Fig. 4.4, if the demodulator sends 111 to the decoder, this is tantamount to declaring the code symbol to be a 1 with very high confidence, while sending a 100 is tantamount to declaring the code symbol to be a 1 with very low confidence. It should be clear that ultimately, every message decision out of the decoder must be a hard decision. The idea behind the demodulator not making the hard decisions and sending more data (soft decisions) to the decoder can be thought of as an interim step to provide the decoder with more information, which the decoder then uses for recovering the message sequence (with better error performance than it could in the case of hard decision decoding). In Fig. 4.4, the eight-level soft-decision metric often shows –7, –5, –3, –1, 1, 3, 5, 7. Such a designation lends itself to a simple interpretation of the soft decision; the sign of the metric represents a decision (zero or one) and the magnitude of the metric represents the confidence level of that decision.

So it can be inferred that

Soft decision = hard decision + confidence level

The price paid for soft decision decoding is an increase in the required memory size at the decoder and possibly a speed penalty.

Optimization of Hard Decision Levels

For a Gaussian channel, it is found that eight-level quantization results in a performance improvement of approximately 2 dB in the required signal-to-noise ratio compared to two-level quantization. This means that eight-level soft decision decoding can provide the same probability of bit error as that of hard decision decoding, but requires 2 dB less E_b/N_o for the same performance.

Analog (or infinite-level) quantization results in a 2.2 dB performance improvement over two-level quantization; therefore, eight-level quantization results in loss of approximately 0.2 dB compared to infinitely fine level quantization. For this reason, quantization to more than eight levels can yield little performance improvement. Thus, *eight levels are optimum as the number of hard decision levels for soft decoding.*

A Posteriori Probability (APP) and A Priori Probability

For an AWGN channel, *a posteriori probability* (APP) of decision of data bit can be thought of as a 'refinement' of the prior knowledge about the data, brought about by examining the received signal s (which is a 'continuous valued random variable' or a 'test statistic' that is obtained at the output of a demodulator or some other signal processor).

The most useful form of Bayes' theorem expresses the APP of a decision in terms of a continuous-valued random variable s as

$$\text{APP} = P(d = i|s) = \frac{P(s|d = i) \cdot p(d = i)}{p(s)} \tag{4.18}$$

and
$$p(s) = \sum_{i=1}^{M} p(s|d = i)P(d = i) \tag{4.19}$$

where

APP represents the probability density function of transmitted data at receiver side to be $d = i$ from a set of possible M classes,

$d = i$ represents data d belonging to the ith signal class from a set of M classes,

$P(s|d = i)$ represents the probability density function of a received continuous valued data-plus-noise signal s, conditioned on the signal class is $d = i$ out of M classes, and

$p(d = i)$, called a priori probability, is the probability of occurrence of the ith signal class, which is known prior to transmission/reception experiment (may be training sequence known at the receiving end, which one can easily estimate).

Note P denotes probability density function and p denotes probability.

Maximum A Posteriori (MAP) Rule

The MAP rule is useful for MAP algorithm for turbo decoding. The general expression for the MAP rule in terms of APP for two levels ±1 can be expressed as

$$P(d=+1|s) \underset{H_2}{\overset{H_1}{\gtrless}} P(d=-1|s) \tag{4.20}$$

Equation (4.20) states that one should choose the hypothesis H_1, i.e., $(d=+1)$ if the APP, $P(d=+1|s)$, is greater than the APP, $P(d=-1|s)$. Otherwise, one should choose hypothesis H_2, that is $(d=-1)$.

Using the Bayes' theorem, the APPs in Eq. (4.20) can be replaced by their equivalent expressions, yielding

$$p(s|d=+1)\cdot P(d=+1) \underset{H_2}{\overset{H_1}{\gtrless}} p(s|d=-1)\cdot P(d=-1) \tag{4.21}$$

$$\frac{p(s|d=+1)}{p(s|d=-1)} \underset{H_2}{\overset{H_1}{\gtrless}} \frac{p(d=-1)}{p(d=+1)} \tag{4.22}$$

or

$$\frac{p(s|d=+1)\cdot P(d=+1)}{p(s|d=-1)\cdot P(d=-1)} \underset{H_2}{\overset{H_1}{\gtrless}} 1 \tag{4.23}$$

where PDF appearing on the both sides of the inequality in Eq. (4.22) has been manipulated. Equation (4.22) is generally expressed in terms of a ratio, yielding the so called likelihood ratio test, as follows.

Log Likelihood Ratio (LLR): Soft Output of Decoder

By taking logarithm of the likelihood ratio developed in Eq. (4.20) through Eq. (4.23), we obtain a useful metric called the log-likelihood ratio (LLR). It is a real number representing a soft decision out of a detector/decoder, designated by

$$\text{LLR} = L(d|s) = \log\left[\frac{P(d=+1|s)}{P(d=-1|s)}\right] \underset{H_2}{\overset{H_1}{\gtrless}} \log(1) \tag{4.24}$$

Using Bayes' theorem,

$$L(d|s) = \log\left[\frac{p(s|d=+1)}{p(d=-1)}\right] + \log\left[\frac{P(d=+1)}{P(d=-1)}\right] \underset{H_2}{\overset{H_1}{\gtrless}} 0 \tag{4.25}$$

or,

$$L(d|s) = L(s|d) + L(d) \tag{4.26}$$

or,

$$L'(\hat{d}) = L_c(s) + L(d) \tag{4.27}$$

where

$L(s|d) = L_c(s)$ = the LLR test statistic s obtained by measurements of the *channel output* s under the alternate conditions that $d = +1$ or $d = -1$ may have been transmitted

$L(d)$ = the priori LLR of the data bit d

Till now, decision-making benefits of the decoder itself from the parity check bits during the decoding process have not been considered. So considering that, it can be shown, for a systematic code, the LLR (soft output) out of the decoder is equal to

$$L(\hat{d}) = L'(\hat{d}) + L_e(\hat{d}) \tag{4.28}$$

or,

$$L(\hat{d}) = L_c(s) + L(d) + L_e(\hat{d}) \tag{4.29}$$

So soft output (LLR) of a systematic decoder can be represented as having three LLR elements:

1. A channel measurement: $L_c(s)$
2. A priori knowledge of data: $L(d)$
3. An extrinsic LLR stemming solely from decoder: $L_e(\hat{d})$

As three terms are statistically independent, the effective expression can be obtained as summation of the three individual terms as shown above.

The soft decoder output $L(\hat{d})$ is a real number that provides a hard decision as well as the reliability of that decision. The sign of $L(\hat{d})$ denotes the hard decision—that is, for positive values of $L(\hat{d})$ decide that $d = +1$ and for negative values that $d = -1$. The magnitude of $L(\hat{d})$ denotes the reliability of that decision. Often the value of $L_e(\hat{d})$ due to the decoding has the same sign as $L_c(S) + L(d)$ and, therefore, acts to improve the reliability of $L(\hat{d})$.

4.2 CHANNEL CODING SCHEMES

4.2.1 Error-Detection Codes

The simplest error detection mechanism is a repetition code. Repeat of the sequence means that instead of sending 0, 1, ..., we shall send 00,11, ..., etc. This is an inefficient way. The theoretical limitations of coding are placed by the results of information theory. These results are frustrating in that they offer little clue as to how the coding should be performed. Error detection coding is designed to permit the detection of errors. Once detected, the receiver may ask for a retransmission of the erroneous bits, or it may simply inform the recipient that the transmission was corrupted.

As mentioned previously, practical codes are normally (n, k) binary block codes. For block codes, the rate of the code is the ratio k/n, and the redundancy of the code is $1 - (k/n)$. Here data will be in the same order sequence and additional bit or bits will be added to it. Our ability to detect errors depends on the rate. A low rate has a high detection probability, but a high redundancy. The receiver will assign to the received code word the pre-assigned code word that minimizes the Hamming distance between the two words. If we wish to identify any pattern of m or less errors, the Hamming distance between the pre-assigned code words must be $m + 1$ or greater.

A very common block code is the *single parity check code*. This code appends to each k data bits an additional bit, whose value is taken to make the $n = k + 1$ word even (or odd). Such a choice is said to have even (odd) parity. With even (odd) parity, a single bit error will make the received word odd (even). The pre-assigned codewords are always even (odd) and hence are separated by a Hamming distance of 2 or more. A single error in a binary code can be detected without much loss of efficiency. For a given data bit, we count the number of times the digit 1 appears in the string, and then append an extra check bit (digit) to the string, which would make this number even. If the number of 1's is odd, the check digit should be 1, changing the data 0010 to codeword 00101, which is a (5, 4) parity check code. For the data 0011, the check digit will be 0, making it 00110.

Now, a single error in any position will make the parity odd. When receiving a codeword, we do a parity check. We count once again the numbers of 1's. If this number is odd we know an error occurred. If the number is even, we simply remove the last digit and decode the original codeword. Here, no two errors can be detected. More generally, no even number of errors can be detected and every odd number of errors will be detected. Such a code has $n - 1$ message bits and 1 check bit. It is a $(n, n - 1)$ block code. The redundancy of this code is

$$\frac{n}{n-1} = 1 + \frac{1}{n-1} \tag{4.30}$$

The information rate of a (5, 4) parity check code is 4/5 = 0.80. The information rate for the (8, 4) block repetition code is 4/8 = 0.50. Since both are used to detect errors in codes with four message bits and both detect only a single error, parity check is clearly more efficient.

Example 4.3 illustrates how the addition of a parity bit can improve error performance.

Example 4.3 Show that addition of parity bits in the data bits $k = 8$ improves the bit error performance, where bit error rate $p = 10^{-4}$ without addition of parity bits.

Solution

For no addition of parity bits:

Probability of single bit error = p
Probability of no error in single bit = $(1 - p)$
Probability of no error in 8 bits = $(1 - p)^8$
Probability of unseen error in 8 bits = $1 - (1 - p)^8 = 7.9 \times 10^{-4}$
So, the probability of a transmission with an error is 7.9×10^{-4}.

With the addition of a parity error bit, we can detect any single bit error.

So,

Probability of no error in single bit = $(1 - p)$
Probability of no error in 9 bits = $(1 - p)^9$
Probability of single error in 9 bits = $9(p)\,(1 - p)^8$
Probability of unseen error in 9 bits = $1 - (1 - p)^9 - 9(p)\,(1 - p)^8 = 3.6 \times 10^{-7}$

As can be seen, the addition of a parity bit has reduced the uncorrected error rate by three orders of magnitude.

Single parity bits are common in asynchronous, character-oriented transmission. Where synchronous transmission is used, additional parity symbols are added, that checks not only the parity of each 8 bit *row*, but also the parity of each 8 bit *column*. The column is formed by listing each successive 8 bit word one beneath the other. This type of parity checking is called *block sum checking* or *longitudinal redundancy check (LRC) + vertical redundancy check (VRC),* and it can correct any single 2 bit error in the transmitted block of rows and columns. However, there are some combinations of errors that will go undetected in such a scheme (Fig. 4.5).

	P1	B6	B5	B4	B3	B2	B1	B0
	0	0	0	0	0	0	0	0
	1	0	1	0	1	0	0	0
	0	1	0*	0	0	1*	1	0
	0	0	1	0	0	0	0	0
	1	0	1	0	1	1	0	1
	0	1	0	0	0	0	0	0
	1	1	1*	0	0	0*	1	1
	1	0	0	0	0	0	1	1
P2	1	1	0	0	0	0	0	1

P1: odd parity for rows
P2: even parity for columns
*: mark undetected error combination

Fig. 4.5 Example of block sum check showing undetected errors

Parity checking in this way provides good protection against single and multiple bit errors when the probability of the errors is independent. However, in many circumstances, errors occur in groups, or bursts. Parity checking of the kind just described then provides little protection. In these circumstances, a polynomial code is used.

The mechanism of *polynomial codes* is not covered in detail in this book however any book related to data link control can be referred for this. They are also block codes. Polynomial codes work on each frame (for frame based data). Additional digits are added to the end of each frame. These digits depend on the contents of the frame. The number of added digits depends on the length of the expected error burst. Typically 16 or 32 digits are added. The computed digits are called the *frame check sequence* (FCS) or *cyclic redundancy check* (CRC). Before transmission, each frame is divided by a generator polynomial. The remainder of this division is added to the frame. On reception, the division is repeated. Since the remainder has been added, the result should be zero. A non-zero result indicates that an error has occurred. These codes are sometimes categorized under error correction coding as it can correct single bit error.

A polynomial code can detect any error burst of length less than or equal to the length of the generator polynomial. The technique requires the addition of hardware to perform the division. However, with modern integrated circuitry, this hardware is now available inexpensively. CRC error checking is quite common for both wired as well as wireless systems. Most of the above codes are suitable for nonreal-time data and used to decide the retransmissions.

4.2.2 Error Correction Codes

Real-time transmission often precludes retransmission. It is necessary to get it right first time. In these special circumstances, the additional bandwidth required for the redundant check bits is an acceptable price. Let us study the important channel codes (error correcting codes) in the next section under three main categories: block codes, covolutional codes, and turbo codes.

4.3 BLOCK CODES

Properties used: Linear, systematic, or cyclic.

A (n, k) block code is said to be 'linear' block code if it satisfies the condition as mentioned. Let $C1$ and $C2$ be any two n-bit codewords belonging to a set of (n, k) block code. If $C1 \oplus C2$ is also an n-bit codeword belonging to the same set of (n, k) block code, then such block code is called *linear block code*. This linear block code is said to be 'systematic' if the k-message bits appear either at the beginning of the codeword or at the end of the codeword.

A (n, k) linear block code C is said to be a 'cyclic' code if every cyclic shift of the code is also a code vector of C. For example if $C1$ = 0111001 be a code vector of C. If $C2$ = 1011100 (the last '1' of $C1$ is shifted into first position) is also a code vector of C, then it is called a cyclic code. Cyclic codes have two advantages over linear codes. First is, encoding circuits can be easily implemented with simple shift register with feedback connections and some basic gates. Second, the cyclic codes have fair amount of mathematical structure (algebraic structure) that makes possible to design codes with useful error-correcting properties.

4.3.1 Hamming Codes

A Hamming code is a block code capable of identifying and correcting any single-bit error occurring within the block. It is identified as an (n, k) Hamming code. Hamming codes employ modulo-2 arithmetic (Ex-OR). This operation has a truth table as shown in Table 4.1.

The theory of Hamming codes is beyond the scope of the book. However, Hamming code has the following parameters with conditions to be satisfied.

Code length: $$n \le 2^{n-k} - 1 \quad (4.31)$$

Number of message bits: $$k \le n - \log_2(n + 1) \quad (4.32)$$

Table 4.1 Truth table for exclusive OR

A	B	A ⊕ B
0	0	0
0	1	1
1	0	1
1	1	0

Number of parity check bits = $n - k$ (4.33)

Error-correcting capability: $t_c = \left\lfloor \frac{d_{\min} - 1}{2} \right\rfloor$ as in Eq. (4.15)

Example 4.4 For the code word bit pattern 1 0 0 1 1 0 1, find out a Hamming code to detect and correct for single-bit errors assuming each codeword contains an ASCII character, i.e. a seven-bit data field.

Solution

Such a coding scheme requires four check bits since, with this scheme, the check bits occupy all bit positions that are powers of 2. Such a code is thus known as an (11, 7) block code with a rate of 7/11 and a redundancy of (1 – 7/11). For example, the bit positions of the value 1001101 are:

Bit Position	11	10	9	8	7	6	5	4	3	2	1
Bit value	1	0	0	x	1	1	0	x	1	x	x

The four bit positions marked with x are used for the check bits, which are derived as follows. The four-bit binary numbers corresponding to those bit positions having a binary 1 are added together using Modulo-2 arithmetic and the four check bits are then the four-bit sum:

11 = 1011
7 = 0111
6 = 0110
3 = 0011
1001

The transmitted code word is now:

Bit position	11	10	9	8	7	6	5	4	3	2	1
Bit value	1	0	0	1	1	1	0	0	1	0	1

Similarly, at the receiver, the four-bit binary numbers corresponding to those bit positions having a binary 1, including the check bits, are again added together and, if no errors have occurred, the Modulo-2 sum should be zero:

11 = 1101
8 = 1100

$$
\begin{aligned}
7 &= 0111 \\
6 &= 0110 \\
3 &= 0011 \\
1 &= 0001 \\
&\ 0000
\end{aligned}
$$

Now consider a single-bit error, say bit 11 is corrupted from 1 to 0. The new modulo-2 sum would now be

$$
\begin{aligned}
8 &= 1100 \\
7 &= 0111 \\
6 &= 0110 \\
3 &= 0011 \\
1 &= 0001 \\
&\ 1011
\end{aligned}
$$

Firstly, the sum is non-zero, which indicates an error, and secondly the modulo-2 sum, equivalent to decimal 11, indicates that bit 11 is the erroneous bit. The latter would, therefore, be inverted to obtain the corrected code word and hence data bits.

Hamming codes suffer from the same difficulty as parity codes. They offer protection against single-bit errors. They offer little protection against burst errors. Convolutional codes are designed to deal with these circumstances. Convolutional codes are different from previous codes we have examined in that they work in a statistical sense. By this, we mean that we cannot say that, for example, every single-bit error will be corrected. We can only say that, on an average, the use of the convolutional code will improve the error rate.

4.3.2 Bose–Chaudhuri–Hocquenghem (BCH) Codes

One of the major considerations in the design of optimum codes is to make the block size (n) smallest for a given message block of k so as to obtain a desired value of $d_{\min}$ or for a given n and k, one may wish to design codes with largest $d_{\min}$. One of the most important and powerful classes of linear block codes are BCH codes, whose properties are discussed below.

For any integer $m \geq 3$, $t_c < (2^m - 1)/2$, there exits a BCH code with the following parameters.

Block length:	$n = 2^m - 1$	(4.34)
Number of message bits:	$k \geq n - mt_c$	(4.35)
Minimum distance:	$d_{\min} \geq 2t_c + 1$	(4.36)

Clearly BCH codes are t_c-random error correcting codes. The major advantage of these codes lie in the flexibility while choosing the code parameters like block length and code rate.

Let us describe the single error-correcting BCH codes. In fact, it can be proven that single error-correcting BCH code is isomorphic to a Hamming code, which is a single error-correcting linear code. Further, we will assume that the code length is $n = 2^m - 1$. We want to construct codes of length n over a finite field Φ. Start by factoring $(x^n - 1)$ into irreducible over Φ using the following rule:

$$f(x) = f_1(x) f_2(x) \ \dots \ f_k(x) \tag{4.37}$$

If $(x^n - 1)$ is not separable in Φ, extend Φ to Φ' and write

$$x^n - 1 = (x - 1)\,(x - \alpha^1)\,(x - \alpha^2)\,(x - \alpha^3) \dots (x - \alpha^{n-1}) \tag{4.38}$$

where α^n are the n distinct roots of unity in the extension of Φ. Each of the powers of α is also a root of some $f_i(x)$. Then for each α^i, define $q_i(x)$ to be the polynomial $f_k(x)$ such that $f_k(\alpha) = 0$. Note that $q_i(x)$ are the minimal polynomials for α^i and that they need not be distinct. This follows since α^i and α^j may have the same minimal polynomial.

A BCH code of length n with designed distance d is a code with generating polynomial

$$g(x) = \text{LCM}\ [q_{k+1}(x), q_{k+2}(x), \dots, q_{k+d-1}(x)] \tag{4.39}$$

for some integer k.

Example 4.5 Find a generating polynomial for BCH code if $n = 4$.

Solution

We start by factoring $x^4 - 1$ to get

$$x^4 - 1 = (x - 1)\,(x - 2)\,(x - 3)\,(x - 4)$$

Then of all the roots {1, 2, 3, 4} we must determine which is a primitive 4th root of unity. That means there is some element such that $\alpha^4 = 1$ and $\alpha^k \neq 1$ for $1 \leq k < 4$. The root 2 satisfies this condition. Let $\alpha = 2$. Thus, the $q_i(x)$ are labelled as follows:

$$q_0(x) = x - 1,\ q_1(x) = x - 2,\ q_2(x) = x - 4,\ q_3(x) = x - 3$$

since

$$q_0(\alpha^0) = 0, \quad q_1(\alpha^1) = 0, \quad q_2(\alpha^2) = 0, \quad q_3(\alpha^3) = 0$$

Now suppose we were looking for a code of distance 3. We can choose k arbitrarily, and so assuming $k = 0$, the generating polynomial is

$$g(x) = \text{LCM}\ [q_{0+1}(x), q_{0+2}(x)] = (x - 2)\,(x - 4)$$

or

$$g(x) = x^2 + 4x + 3$$

Then the first row of the generating matrix is the row vector (3, 4, 1, 0).

Because we have defined a BCH code to be cyclic, we get the complete generating matrix from cyclic shifts of the generating code vector

$$\begin{pmatrix} 3 & 4 & 1 & 0 \\ 0 & 3 & 4 & 1 \end{pmatrix}$$

which is a cyclic [4, 2] code.

In Example 4.5, the weight of the first row vector was 3, and since it is a cyclic code, we know that the minimum weight is exactly 3. So the result tells us that the exact minimum weight = 1. One of the advantages of this type of code is that there is a relatively simple way to decode.

The parameters of some useful BCH codes are given in Table 4.2.

Table 4.2 Parameters of some BCH codes

n	k	t_c	Generator Polynomial Coefficients
7	4	1	1 011
15	7	2	111 010 001
15	5	3	10 100 110 111

4.3.3 Reed–Solomon Codes

Reed–Solomon codes are a specific type of BCH code. The encoder for and RS code differs from binary encoders in that it operates on multiple bits rather than individual bits. The t_c-random error-correcting RS codes have the parameters we wish to explore here.

Block length: $n = 2^m - 1$ (4.40)

Message size: k symbols

Parity check size: $r = (n - k) = 2t_c$ symbols (4.41)

Minimum distance: $d_{\min} = 2t_c + 1$ symbols (4.42)

No. of correctable symbols in error: $t_c = (n - k)/2$ (4.43)

The encoder for an RS (n, k) code on m-bit symbols, divides the incoming binary data stream into blocks, each of $k \times m$ bits long. Each block is treated as k symbols with each symbol having 8-bits. The encoding algorithm expands a block of k symbols by adding $(n - k)$ redundant symbols, when m is an integer power of 2. For bytes the popular value of $m = 8$. Eight-bit RS codes are extremely powerful. For any (n, k) linear block code $d_{\min} < (n - k + 1)$, but for RS code $d_{\min} = n - k + 1$. It means the minimum distance is always equal to the design distance of the code. So, the code is also called 'maximum distance separable' code.

Example 4.6 For (31, 15) Reed–Solomon code, answer the following questions:

1. How many bits per symbol of the code are there?
2. What is the block length in terms of bits?
3. What is the minimum distance of the code?
4. How many symbols in error can the code correct?
5. What is the length of an in-phase burst that the code can correct?

Solution

We have $n = 2^m - 1 = 31$

$\Rightarrow$ $2^m = 32$

So, the number of bits per symbol, $m = 5$ bits
Block length = $31 \times 5 = 155$ bits
Now, the minimum distance of the code,

$$d_{min} = n - k + 1$$
$$= 31 - 15 + 1$$
$$= 17$$

We have the number of correctable symbols in error,

$$t_c = (n - k)/2$$
$$= (31 - 15)/2$$
$$= 8$$

The length of an in-phase burst that the code can correct, $m = 5$

4.4 CONVOLUTIONAL CODES

Convolutional codes are often used to improve the performance of wireless links. A convolutional encoder is called so because it performs a *convolution* of the input stream with the encoder's *impulse responses*: A convolutional encoder is a discrete linear time-invariant system. Every output of an encoder can be described by its own transfer function, which is closely related to a generator polynomial. Convolutional codes are of importance as they are used in most of latest mobile networks. Convolutional codes are more powerful for error correction than block codes. In this section, convolutional codes are explained in detail.

4.4.1 Binary Convolutional Codes: Mathematical Approach

Convolutional codes can be generated mathematically. This is possible with the help of a generator matrix. The generator matrix for a convolutional code can define the code and it is semi-infinite, since the length of the input is semi-infinite. Because of this, it is not a convenient method. Still for understanding of the method or even to design the hardware, mathematics is necessary. In general, a code rate $R = k/n$, $k \leq n$, convolutional encoder input (information sequence) is a sequence of binary k-tuples,

$$u = \ldots, u_{-1}, u_0, u_1, u_2, \ldots \quad (4.44)$$

where $$u_i = (u_i^{(1)} \ldots, u_i^{(k)}) \quad (4.45)$$

The output (code sequence) is a sequence of binary n-tuples,

$$v = \ldots, v_{-1}, v_0, v_1, v_2, \ldots \quad (4.46)$$

where $$v_i = (v_i^{(1)}, \ldots, v_i^{(n)}) \quad (4.47)$$

Generator sequences specify convolutional code completely by the associated generator matrix. Encoded convolutional code is produced by matrix multiplication of the input and the generator matrix. The sequences must start at a finite (positive or negative) time and may or may not end. The relation between the information sequences and the code sequences is determined by the equation

$$v = u^*G \quad (4.48)$$

where

$$G = \begin{pmatrix} G_0 & G_1 & \dots & G_m & & & \\ & G_0 & G_1 & \dots & G_m & & \\ & & G_0 & G_1 & \dots & G_m & \\ & & & \ddots & \ddots & & \ddots \end{pmatrix} \quad (4.49)$$

is the semi-infinite generator matrix and where the sub-matrices G_i, $0 \le i \le m$, are binary $k \times n$ matrices. The arithmetic in it is carried out over the binary field and the parts left blank in the generator matrix $\boldsymbol{G}$ are assumed to be filled in with zeros.

The right-hand side of Eq. (4.48) defines a discrete time convolution between $\boldsymbol{u}$ and $\boldsymbol{G}$ = ($G0$ $G1$ … Gm), hence, the name convolutional codes. As in many other situations where convolutions appear it is convenient to express the sequences in some sort of transform. In information theory and coding theory it is common to use the delay operator D, the D-transform. The information and code sequences become

$$u(D) = \dots + u_{-1}D^{-1} + u_0 + u_1 D + u_2 D^2 + \cdots \quad (4.50)$$

and
$$v(D) = \dots, + v_{-1}D^{-1} + v_0 + v_1 D + v_2 D^2 + \cdots \quad (4.51)$$

They are related through the equation

$$v(D) = u(D)\ G(D) \quad (4.52)$$

where
$$G(D) = G_0 + G_1 D + \cdots + G_m D^m \quad (4.53)$$

is the generator matrix. According to the formula for the convolution, encoder outputs are formed by *modulo-2 discrete convolutions*:

$$v^{(1)} = u * g^{(1)},\ v^{(2)} = u * g^{(2)},\ \dots,\ v^{(j)} = u * g^{(j)} \quad (4.54)$$

where u is the information sequence.

Therefore, the lth bit of the jth output branch is

$$v_l^{(j)} = \sum_{i=0}^{m} u_{l-i}\, g_l^{(j)} = u_l\, g_0^{(j)} + u_{l-1}\, g_1^{(j)} + \cdots + u_{l-m}\, g_m^{(j)} \quad (4.55)$$

Here, L is the number of registers used to generate the code sequence, $m = L + 1$, $u_{l-i} \ne 0$, $l < i$

Example 4.7 An information source is $u = [1, 0, 1, 1, 1]$. Assume the generator matrix as

$$\begin{pmatrix} \mathbf{g}^{(1)} = [1\ 0\ 1\ 1] \\ \mathbf{g}^{(2)} = [1\ 1\ 1\ 1] \end{pmatrix}$$

Find the convolutional codes.

Solution

We have $\boldsymbol{v} = \boldsymbol{u} * \boldsymbol{G}$

So, the following equations result:

$$v_l^{(1)} = u_l + 0 + u_{l-2} + u_{l-3}$$
$$v_l^{(2)} = u_l + u_{l-1} + u_{l-2} + u_{l-3}$$

Hence the encoder output becomes

$$v = [v_0^{(1)} \, v_0^{(2)} \, v_1^{(1)} \, v_1^{(2)} \, v_2^{(1)} \, v_2^{(2)} \ldots]$$

The information sequence $\boldsymbol{u} = (1\ 0\ 1\ 1\ 1)$. Then the output sequences are

$$\mathbf{v}^{(1)} = (1\ 0\ 1\ 1\ 1) * (1\ 0\ 1\ 1) = (1\ 0\ 0\ 0\ 0\ 0\ 0\ 1)$$
$$\mathbf{v}^{(2)} = (1\ 0\ 1\ 1\ 1) * (1\ 1\ 1\ 1) = (1\ 1\ 0\ 1\ 1\ 1\ 0\ 1)$$

and the codeword is

$$\mathbf{v} = (1\,1, 0\,1, 0\,0, 0\,1, 0\,1, 0\,1, 0\,0, 1\,1)$$

If the generator sequences $\mathbf{g}^{(1)}$ and $\mathbf{g}^{(2)}$ are interlaced and then arranged in the matrix

$$\mathbf{G} = \begin{bmatrix} g_0^{(1)} g_0^{(2)} & g_1^{(1)} g_1^{(2)} & g_2^{(1)} g_2^{(2)} & \cdots & g_m^{(1)} g_m^{(2)} & & \\ & g_0^{(1)} g_0^{(2)} & g_1^{(1)} g_1^{(2)} & \cdots & g_{m-1}^{(1)} g_{m-1}^{(2)} & g_m^{(1)} g_m^{(2)} & \\ & & g_0^{(1)} g_0^{(2)} & \cdots & g_{m-2}^{(1)} g_{m-2}^{(2)} & g_{m-1}^{(1)} g_{m-1}^{(2)} & g_m^{(1)} g_m^{(2)} \\ & & \ddots & & & & \ddots \end{bmatrix}$$

where the blank areas are all zeros. The encoding equations can be rewritten in matrix form as

$$\mathbf{v} = \mathbf{uG}$$

where all operations are modulo-2. **G** is called the *generator matrix* of the code. Note that each row of **G** is identical to the preceding row but shifted $n = 2$ places to the right, and that **G** is a semi-infinite matrix, corresponding to the fact that the information sequence **u** is of arbitrary length. If **u** has finite length L, then **G** has L rows and $2(m + L)$ columns, and **v** has length $2(m + L)$.

If $\mathbf{u} = (1\ 0\ 1\ 1\ 1)$, then

$$\mathbf{v} = \mathbf{uG}$$

$$= (1\ 0\ 1\ 1\ 1)\begin{bmatrix} 1\,1 & 0\,1 & 1\,1 & 1\,1 & & & & & \\ & 1\,1 & 0\,1 & 1\,1 & 1\,1 & & & & \\ & & 1\,1 & 0\,1 & 1\,1 & 1\,1 & & & \\ & & & 1\,1 & 0\,1 & 1\,1 & 1\,1 & & \\ & & & & 1\,1 & 0\,1 & 1\,1 & 1\,1 \end{bmatrix}$$

$$= (1\,1, 0\,1, 0\,0, 0\,1, 0\,1, 0\,1, 0\,0, 1\,1),$$

So far, the described codes are time invariant with fixed generator matrix. Another type is the time-varying convolutional codes, which are beyond the scope of this book. Convolutional code generation by hardware is described next.

4.4.2 Convolutional Encoder Hardware

Convolutional codes are used in applications that require good performance with low implementation cost. They operate on code streams (not in blocks). A continuous sequence of information bits is mapped into a continuous sequence of encoder output bits.

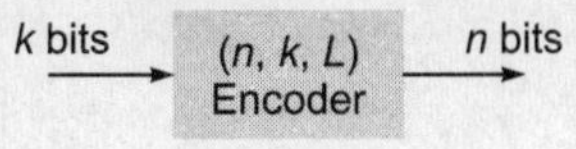

This mapping is highly structured/systematic so that the decoding can be possible. Decoding methods are different than the block codes. It is found that convolutional codes can have a larger coding gain than that can be achieved through block coding with the same complexity. Convolutional codes have a memory that utilizes previous bits to encode or decode the following bits (block codes are memoryless as mentioned earlier). Convolutional codes achieve good performance by expanding their memory depth.

k bits → (*n*, *k*, *L*) Encoder → *n* bits

Fig. 4.6 Simple conceptual diagram of convolutional encoder

Convolutional codes are denoted by (n, k, L), where n is the number of output bits (coded), k is the number of input bits (uncoded), and L is code (or encoder) memory depth, which represents the number of register stages.

$$\text{Constraint length, } C = (L + 1) \tag{4.56}$$

It is defined as the number of encoded bits a message bit can influence. The constraint length concept is shown in Fig. 4.7.

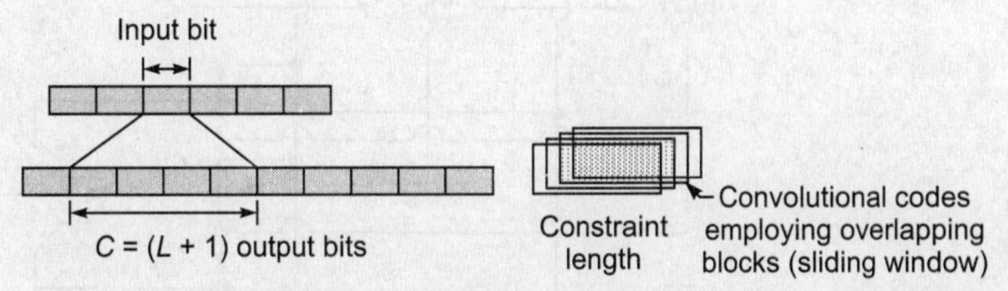

Fig. 4.7 Constraint length concept

Note To provide the extra bits needed for error control, an output rate greater than the message bit rate could be achieved by connecting two or more MOD-2 summers to the register and interleaving the encoded bits via a switch. In this situation, each message bit influences a span of $n(L + 1)$ successive output bits, which is the constraint length measured in terms of encoded output bits.

Convolutional encoder is a *finite state machine* (FSM) processing information bits in a serial manner. Thus, the generated code is a function of the input and the state of the FSM. Hence, convolutional code generation can be represented using a state diagram also.

To convolutionally encode the data, start with L memory elements shift register, each element holding one input bit. Pass the information sequence through a finite state shift register. Unless otherwise specified, all memory registers start with a value of 0. The encoder has modulo-2 adders, and n generator polynomials—one for each adder. Figure 4.8 shows a typical convolutional coder. An input bit m_j is fed into the rightmost register. Using the generator polynomials and the existing values in the remaining registers, the encoder outputs n bits. Now the bit shift, all register values to the left (m_j moves to m_{j-1}, m_{j-1} moves to m_{j-2}) and wait for the next input bit. If there are no remaining input bits, the encoder continues output until all registers have returned to the zero state. Similar explanation can be given for Fig. 4.9.

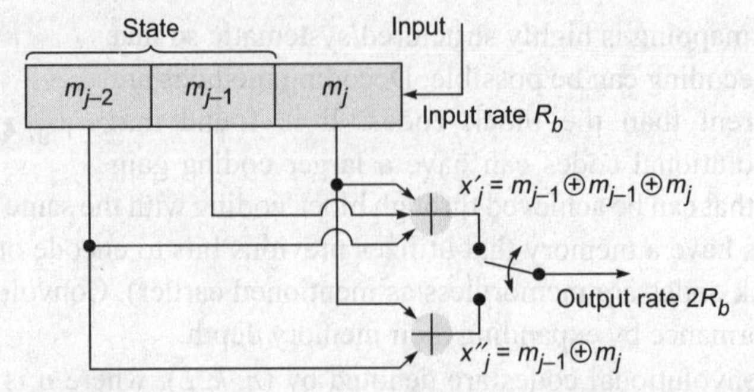

Fig. 4.8 Typical convolutional encoder ($x_{out} = x'_1\, x''_1\, x'_2\, x''_2\, x'_3\, x''_3 \ldots$ is the convolution coded output ready for modulation.)

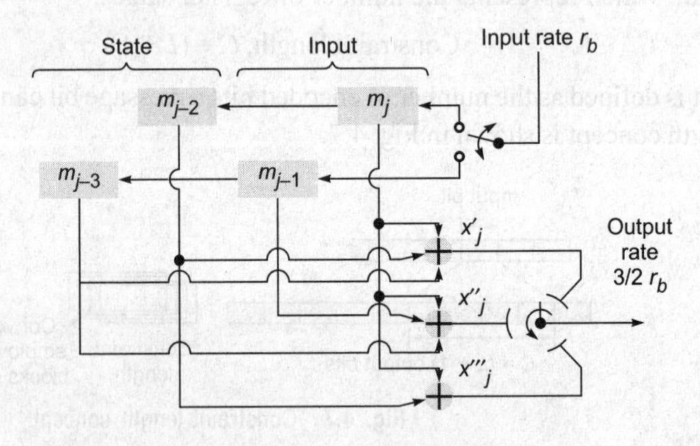

Fig. 4.9 Typical rate 3/2 convolutional encoder

Example 4.8 For $(n, k, L) = (2, 1, 2)$ convolutional encoder, find out the constraint length.

Solution

Each message bit influences a span of $C = (L + 1) = 3$ successive output bits = constraint length C. Thus, for generation of 2-bit output, we require 2 shift registers and one bit input.

4.4.3 Trellis Diagram and Convolutional Code

Trellis diagram can be used for the decoding of convolutional codes. A typical tree is given in Fig. 4.10. Also, it is explained further in section 7.1.5.

The example discussed in Section 4.4.4 will explain how to use the trellis diagram.

4.4.4 Viterbi Decoding Algorithm (Viterbi, 1967)

Viterbi algorithm is mainly used for convolutional decoding. The Viterbi decoding algorithm was discovered and analysed by Viterbi in 1967. In 1969, Omura demonstrated that the Viterbi algorithm is in fact, a maximum likelihood. However, it reduces the

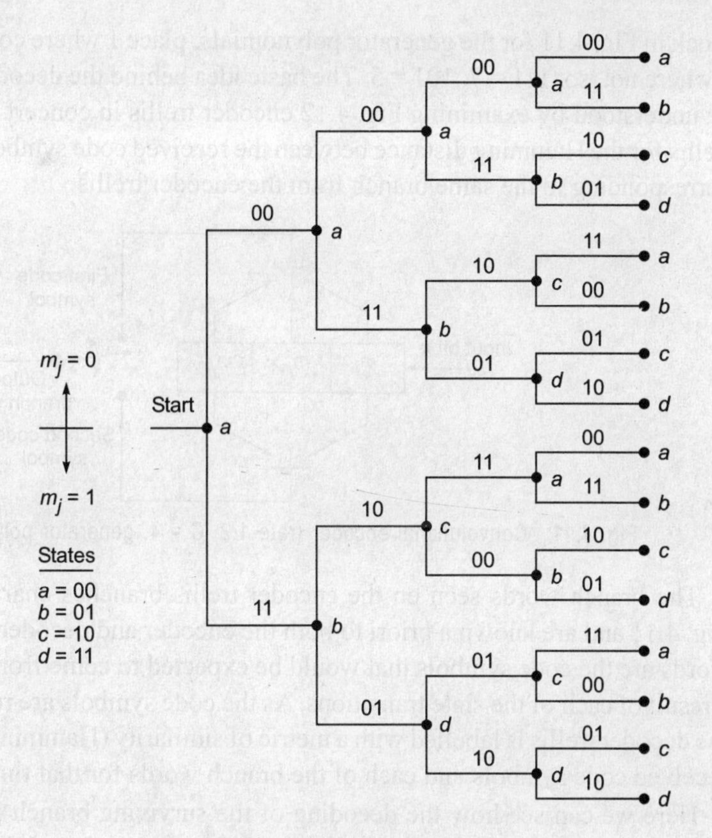

Fig. 4.10 Typical trellis diagram

computational load by taking advantage of the special structure in the code trellis. The advantage of Viterbi decoding, compared with brute force decoding, is that the complexity of a Viterbi decoder is not a function of the number of symbols in the codeword sequence. The algorithm involves calculating a measure of similarity, or distance, between the received signal, at time t_i, and the entire trellis path entering each state at time t_i.

The Viterbi algorithm removes from consideration, those trellis paths that could not possibly be candidates for the maximum likelihood choice. When two paths enter the same state, the one having the best metric is chosen; this path is called the *surviving path*. This selection of surviving paths is performed for all the states. The decoder continues in this way to advance deeper into the trellis, making decisions by eliminating the least likely paths. The early rejection of the unlikely paths reduces the decoding complexity. Note that the goal of selecting the optimum path can be expressed, equivalently, as choosing the codeword with the maximum likelihood metric, or as choosing the code word with the minimum distance metric.

Hard Decision Viterbi Convolutional Decoding for (2, 1, 3) coder: An Example

For simplicity, Hamming distance is a proper distance measure to consider here. The encoder for this example is shown in Fig. 4.11. See the connection with each memory block in Fig. 4.11 for the generator polynomials; place 1 where connection is present and 0 where not, so, 111 = 7, 101 = 5. The basic idea behind the decoding procedure can best be understood by examining Fig. 4.12 encoder trellis in concert with Fig. 4.13 decoder trellis for the Hamming distance between the received code symbols and the branch word corresponding to the same branch from the encoder trellis.

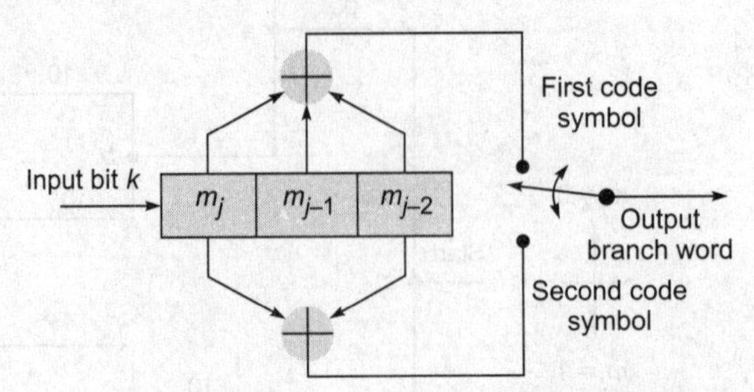

Fig. 4.11 Convolutional encoder (rate 1/2, C = 4, generator polynomials = [7, 5])

The branch words seen on the encoder trellis branches characterize the encoder in Fig. 4.11 and are known a priori to both the encoder and decoder. These encoder branch words are the code symbols that would be expected to come from the encoder output as a result of each of the state transitions. As the code symbols are received, each branch of the decoder trellis is labelled with a metric of similarity (Hamming distance) between the received code symbols and each of the branch words for that time interval.

Here we can see how the decoding of the surviving branch is facilitated by having drawn the trellis branches with solid lines for the input 0's and dashed lines for the input 1's. Note that the bit cannot be decoded until the path metric computation has proceeded to a much greater depth into trellis. For a typical decoder implementation, this represents a decoding delay, which can be as much as five times the constraint length in bits. The decoder trellis in Fig. 4.13 shows

A message sequence, m = 11011 …
The corresponding code word sequence, V = 11 01 01 00 01 …
A noise corrupted received sequence, Z = 11 01 01 10 01 …

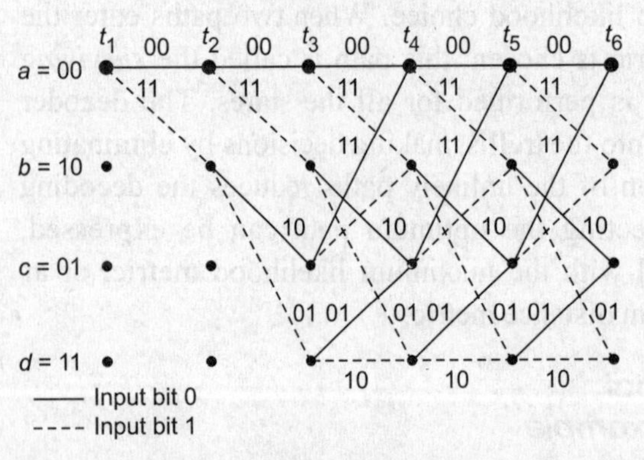

Fig. 4.12 Encoder trellis diagram (rate 1/2, C = 4)

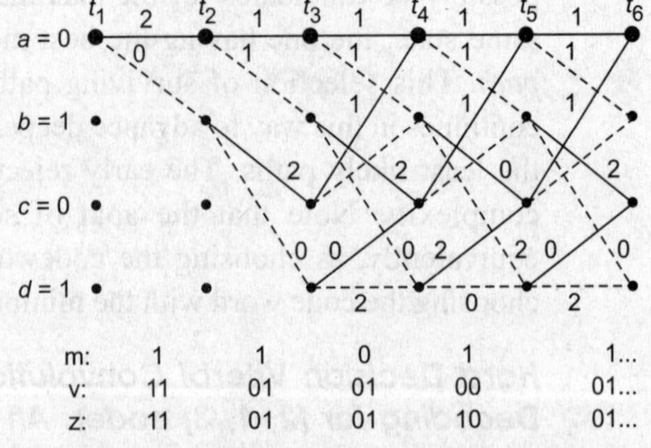

Fig. 4.13 Decoder trellis diagram (rate 1/2, C = 4)

Labelling procedure is done like this. From the received sequence Z, we see that the code symbol received at time t_1 is 11. In order to label the decoder branches at (departing) time t_i with the appropriate Hamming distance metric, we look at Fig. 4.12 encoder trellis. Here we see that a state 00 $\rightarrow$ 00 transition yields an output branch word of 00. But we received 11. Therefore, on the decoder trellis we label the state 00 $\rightarrow$ 00 transition with Hamming distance between them, namely 2. Looking at the encoder trellis again, we see that a state 00 $\rightarrow$ 10 transition yields an output branch word of 11, which corresponds exactly with the code symbols we received at time t_1. Therefore, on the decoder trellis, we label the state 00 $\rightarrow$ 10 transition with Hamming distance 0 and so on.

Thus, the metric entered on a decoder trellis branch represents the difference (distance) between what was received and what should have been received, had the branch word been associated with that branch been transmitted. In effect, these metrics describe a correlation-like measure between a received branch word and each of the candidate branch words. We continue labelling the decoder trellis branches in this way as the symbols are received at each time t_i. Thus the most likely (minimum distance) path through the trellis can be found that gives the decoded output.

If any two paths in the trellis merge to a single state, one of them can always be eliminated in the search for an optimal path.

A typical example of Fig. 4.14 shows two paths merging at time t_5 to state 00. Let us define *cumulative Hamming path metric* of a given path at time t_i as *the sum of the branch Hamming distance metrics along the path up to time* t_i. In Fig. 4.14, the upper path has metric 4; the lower has metric 1. The upper path cannot be a portion of the *optimum path* because the lower path, which enters the same state, has *lower metric*. This observation holds true because of the *Markov* (*chain process*) *nature* of the encoder state: the present state summarizes the encoder history in the sense that previous states cannot affect future states or future output branches.

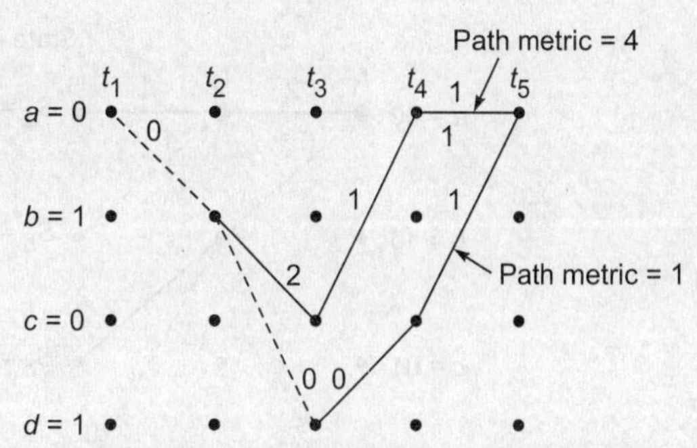

Fig. 4.14 A case of two merging paths

At each time t_i, there are $2^C - 1$ states in trellis, where C is the constraint length, and each state can be entered by means of two paths. Viterbi decoding consists of computing the metrics for the two paths entering each state and eliminating one of them. This computation is done for each of the $2^C - 1$ states or nodes at time t_i, then the decoder moves to time t_{i+1} and repeats the process. At a given time, the *winning path metric* for each state is designated as the *state metric* for that state at that time.

With reference to Figures 4.12 and 4.13, the first few steps in our decoding example are as follows.

- Assume that the decoder knows the correct initial state of trellis. (This assumption is not necessary in practice but simplifies the explanation.)

- At time t_1, the received code symbols are 11 as in Fig. 4.15. From the state 00 the only possible transitions are to state 00 or state 10, as shown in Fig. 4.15. State 00 → 00 transition has branch metric 2; state 00 → 10 transition has branch metric 0. At time t_2 there are two possible branches leaving each state metric $S_a = 2$ and $S_b = 0$, as shown in Fig. 4.15.

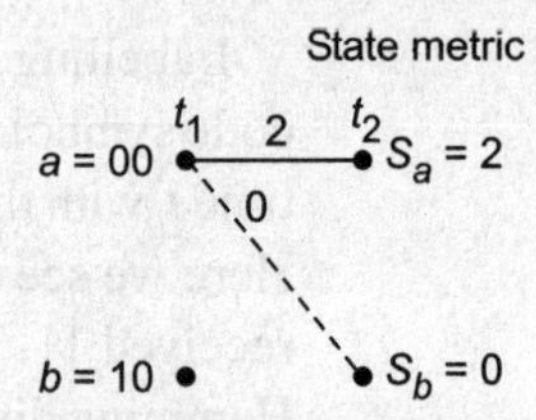

Fig. 4.15 Survivors at t_2

- At time t_3 in Fig. 4.16, there are again two branches diverging from each state. As a result, there are two paths entering each state at time t_4. One path entering each state can be eliminated, namely, the one having the larger cumulative path metric. The cumulative metrics of these branches are labelled as state metrics $S_a = 3$, $S_b = 3$, $S_c = 2$, and $S_d = 0$ corresponding to the terminating state. If metrics of the two entering paths be of equal value, one path is chosen for elimination by using an arbitrary rule.The surviving path into each state is shown in Fig. 4.17. At this point in the decoding process, there is only a single surviving path, termed the *common stem*, between times t_1 to t_2. Therefore, the decoder can now decide that the state transition which occurred between t_1 and t_2 was 00 → 10. Figure 4.18 shows the survivors at time t_4.

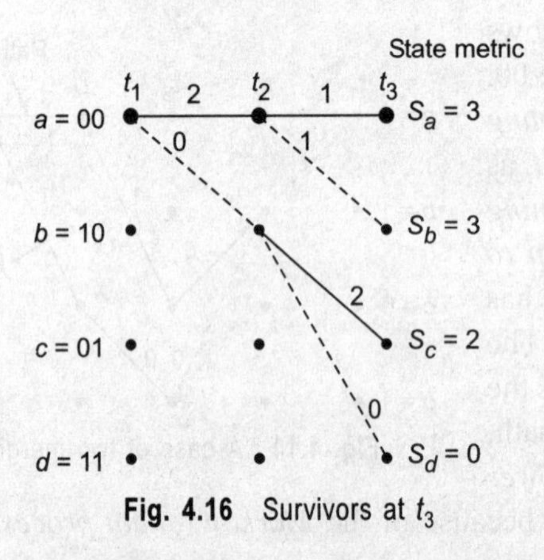

Fig. 4.16 Survivors at t_3

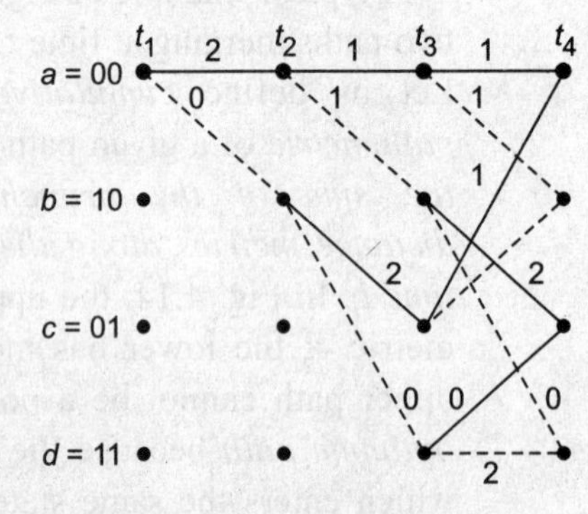

Fig. 4.17 Metric comparisons at t_4

- Figure 4.19 shows the next step in the decoding process. Again, at time t_5, there are two paths entering each state, and one of each pair can be eliminated. Figure 4.20 shows the survivors at time t_5. (Note that in our example, we cannot yet make a decision on the second input data bit because there still are two paths leaving the state 10 node at time t_2.)
- At time t_6, in Fig. 4.21, we again see the pattern of reemerging paths. In Fig. 4.22, we see the survivors at time t_6. Also in Fig. 4.22, the decoder output one has the second decoded bit, corresponding to the single surviving path between path t_2 and t_3. The decoder continues in this way to advance deeper into the trellis and to make decisions on the input data bits by eliminating all paths but one.

Pruning the trellis, because paths remerge, guarantees that there are never more paths than states. From the above example, it can be verified that after each pruning in Figures 4.19 to 4.22, there are only 4 paths. Compare this to 'brute force' maximum likelihood

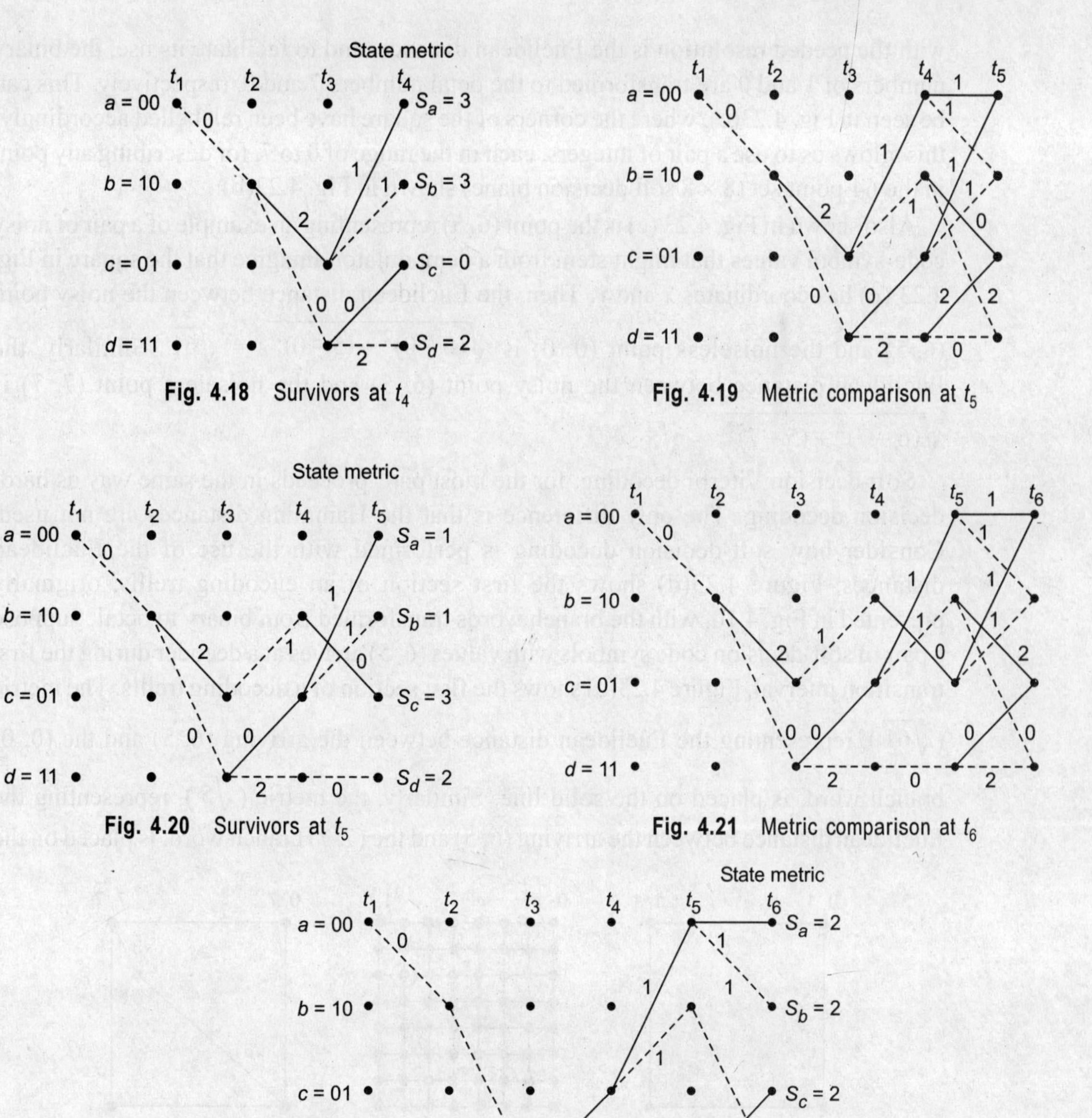

Fig. 4.18 Survivors at t_4

Fig. 4.19 Metric comparison at t_5

Fig. 4.20 Survivors at t_5

Fig. 4.21 Metric comparison at t_6

Fig. 4.22 Survivors at t_6

sequence estimation without using the Viterbi algorithm. In that case, the number of possible paths (representing possible sequences) is an exponential function of sequence length. For a binary codeword sequence that has length of L branch words, there are 2^L possible sequences.

Concept of soft decision Viterbi algorithm The primary difference between hard-decision and soft-decision Viterbi decoding algorithm is that the soft-decision algorithm cannot use Hamming distance metric because of its limited resolution. A distance metric with the needed resolution is the Euclidean distance, and to facilitate its use, the binary numbers of 1 and 0 are transformed to the octal numbers 7 and 0, respectively. This can be seen in Fig. 4.23(c), where the corners of the square have been relabelled accordingly; this allows us to use a pair of integers, each in the range of 0 to 7, for describing any point in the 64-point set (8×8 soft decision plane) shown in Fig. 4.23(b).

Also shown in Fig. 4.23 (c) is the point (6, 5) representing an example of a pair of noisy code-symbol values that might stem from a demodulator. Imagine that the square in Fig. 4.23 (c) has coordinates x and y. Then, the Euclidean distance between the noisy point (6, 5) and the noiseless point (0, 0) is $\sqrt{(6-0)^2+(5-0)^2} = \sqrt{61}$. Similarly, the Euclidean distance between the noisy point (6, 5) and the noiseless point (7, 7) is $\sqrt{(6-7)^2+(5-7)^2} = \sqrt{5}$.

Soft-decision Viterbi decoding, for the most part, proceeds in the same way as hard-decision decoding. The only difference is that the Hamming distances are not used. Consider how soft-decision decoding is performed with the use of the Euclidean distances. Figure 4.23(d) shows the first section of an encoding trellis, originally presented in Fig. 4.10, with the branch words transformed from binary to octal. Suppose a pair of soft-decision code symbols with values (6, 5) arrives at a decoder during the first transition interval. Figure 4.23(e) shows the first section of a decoding trellis. The metric ($\sqrt{61}$), representing the Euclidean distance between the arriving (6, 5) and the (0, 0) branch word, is placed on the solid line. Similarly, the metric ($\sqrt{5}$), representing the Euclidean distance between the arriving (6, 5) and the (7, 7) branch word, is placed on the

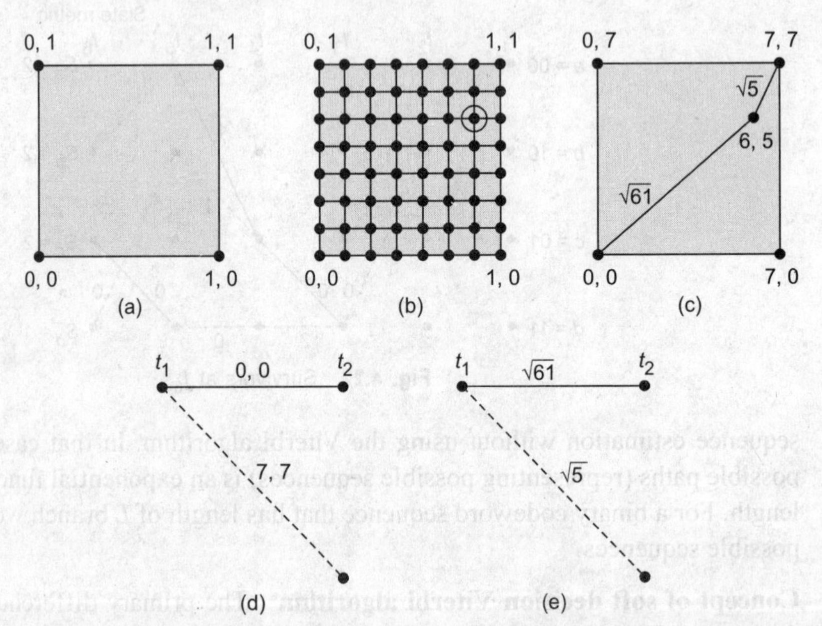

Fig. 4.23 (a) Hard-decision plane, (b) 8-level by 8-level soft-decision plane, (c) example of soft code symbols, (d) encoding trellis section, and (e) decoding trellis section

dashed line. The rest of the task, pruning the trellis in search of a common stem, proceeds in the same way as hard-decision decoding.

4.5 INTERLEAVER

In a memoryless channel, the errors can be characterized as single randomly distributed bit errors whose occurrences are independent from bit to bit. Whereas a channel that has memory exhibits mutually dependent signal transmission impairments. A channel that exhibits multipath fading (Chapter 5), where signals arrive at the receiver over two or more paths of different lengths, is an example of a channel with memory. The effect is that the signals can arrive out of phase with each other, and the cumulative received signal is distorted. Wireless mobile communication channels, as well as ionospheric and tropospheric propagation channels, suffer from such phenomenon. All of these time-correlated impairments result in statistical dependence among successive symbol transmissions. That is, the disturbance tends to cause errors that occur in *bursts*, instead of as *isolated events*.

Under the assumption that channel has memory, the errors no longer can be characterized as single randomly distributed bit errors whose occurrence is independent from bit to bit. Most of the channel coding techniques are designed to combat random independent errors. The result of a channel having memory on such coded signals is to cause degradation in error performance. Some coding techniques for channels with memory have been proposed, but the greatest problem with such coding is the difficulty in obtaining accurate models of the often time-varying statistics of such channels. One technique, which only requires knowledge of the duration or span of the channel memory, not its exact statistical characterization, is the use of *time diversity* or *interleaving*.

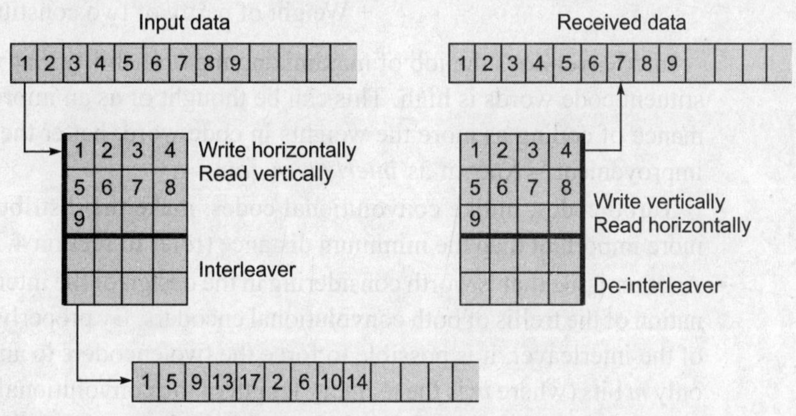

Fig. 4.24 Interleaving and de-interleaving operation

As shown in Fig. 4.24, an interleaver is a device that arranges the ordering of sequence of symbols in a deterministic manner. For a typical case, it can be thought of as a storage memory, where data are read in row-wise and read out column-wise. There may be other algorithms also. Associated with the interleaver is a de-interleaver that applies the inverse permutation to restore the original sequence. The most critical part in the design of a turbo code is the interleaver design. Two main issues in the interleaver design are the *interleaver size* and the *interleaver map*. The size of the interleaver plays an important role in the *trade-off between performance and time* (*delay*) since both of them are directly proportional to the size. On the other hand, the map of the interleaver plays an important role in setting the code performance.

Conventionally, interleaver is used to spread out the errors occurring in a burst. Separating the symbols in time effectively transforms a channel with memory to a memoryless one, and thereby enables the random error correcting codes to be useful in a burst noise channel.

For turbo codes, the interleaver has more functions (this will be clear after studying turbo codes).

- Interleaver is used to feed the encoders with permutations so that the generated redundancy sequences can be assumed to be independent. The validity of the assumption that the generated redundant sequences are independent, is a function of the particular interleaver used. This will exclude a number of interleavers, which generate regular sequences such as cyclic shift.
- Second key role of the interleaver is to shape the weight distribution of the code, which ultimately, controls its performance. This is so because the interleaver will decide which word of the second encoder will be concatenated with the current word of the first encoder, and hence what weight the complete code word will have. So the aim of the designer is to produce (by manipulating the weights of the second redundancy part through interleaver mapping) whole code words with the overall weights as large as possible. In turbo code,

$$\text{Weight of code word} = \text{Weight of input data} + \text{Weight of parity of two constituent code} \quad (4.57)$$

Interleaver does the job of maximizing the probability that weight of two constituent code words is high. This can be thought of as an improvement in performance of coding as more the weights in code word, better the linear code.[1] This improvement is known as *interleaver gain*.

Turbo codes, unlike convolutional codes, make the distribution of the weight more important than the minimum distance (refer to section 4.7 on Turbo Codes).

- Another issue that is worth considering in the design of the interleaver is the termination of the trellis of both convolutional encoders. By properly designing the map of the interleaver, it is possible to force the two encoders to an all-zero state with only m bits (where m is the memory length of the convolutional encoder assuming that the same convolutional code is used in both encoders).

The interleaver shuffles the code symbols over a span of several block lengths (for block codes) or several constraint lengths (for convolutional codes). Several types of interleavers are commonly used, including *block*, *convolutional*, and *pseudo-random* interleavers. A typical interleaver used in OFDM system (Chapter 9) is shown in Fig. 4.25.

[1] Turbo code is nothing but modified linear block code as both RSC encoders are linear.

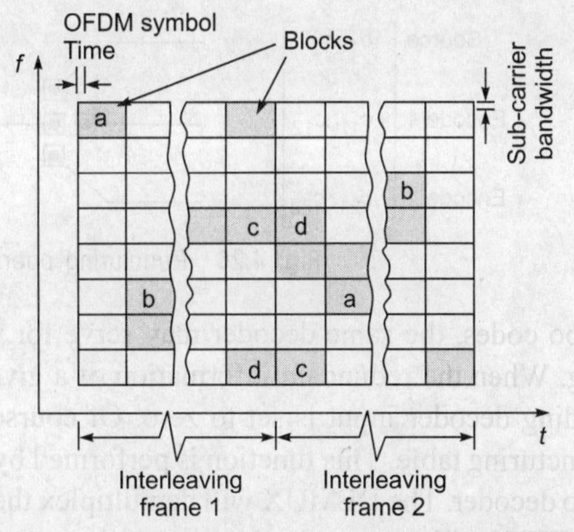

Fig. 4.25 A typical example of burst error handling by block interleaving in OFDM

Data transmission in this OFDM system concept is performed by blocks of several sub-carriers in several consecutive OFDM symbols. Which block of sub-carriers and which OFDM symbols are used for transmission is determined by a frequency hopping pattern (Chapter 8). This pattern is repetitive with some period. We refer to this period as an interleaving frame. Coding and interleaving is performed over all blocks of data symbols transmitted during such an interleaving frame. Thus, one way of interpreting the OFDM concept is by the time and frequency grids shown in Fig. 4.25, where each transmission block (rectangle) is orthogonal to all other transmission blocks in the grid. The size of a transmission block depends on the choice of parameters for the OFDM system. This may eliminate the probability of the whole block of frame in error. The concept is just writing vertically and reading horizontally associated with randomness.

4.6 CODE PUNCTURING

Puncturing is the process of deleting some bits from the codeword according to a puncturing matrix. The puncturing matrix (P) consists of 0's and 1's where the zero represents an omitted bit and the one represents an emitted bit. It is usually used to increase the rate of a given code. Puncturing can be applied to both block and convolutional codes.

An example of the puncturing matrix to go from rate 1/3 to rate 1/2 is given by

$$P = \begin{bmatrix} 1 & 1 \\ 1 & 0 \\ 0 & 1 \end{bmatrix}$$

This matrix implies that the systematic bit is always transmitted while every other bits coming from two encoders are alternatively omitted, explained in Fig. 4.26 with more clarity. So puncturing can be thought of as MUX operation along with redundancy removal. In Fig. 4.26, $\{b_1, b_2, \ldots\}$ are systematic bits and $\{C_{11}, C_{12} \ldots\}$ $\{C_{21}, C_{22} \ldots\}$ are coming from two encoders, but finally transmitted bits are $\{b_1, C_{11}\}\{b_2, C_{22}\}\ldots$ as per the puncturing matrix P. So, now instead of 6 bits, 4 bits will be transmitted finally.

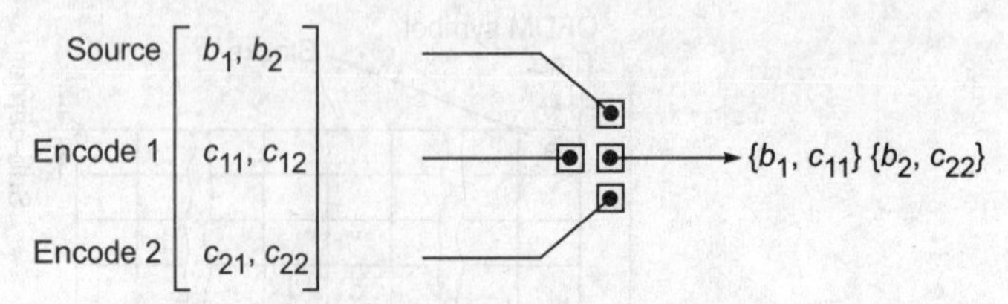

Fig. 4.26 Puncturing operation

For turbo codes, the same decoder may serve for various coding rates by means of puncturing. When the redundant information of a given encoder is not transmitted, the corresponding decoder input is set to zero. Of course, the decoder needs to know the current puncturing table. This function is performed by the DEMUX/INSERTION block in the turbo decoder. The DEMUX will demultiplex the stream between the decoders and the INSERTION will insert an analog zero if the corresponding bit is omitted. When the code is punctured, the branch metric corresponding to the punctured bits need not to be computed.

Determining the best puncturing pattern for turbo codes is still an open problem.

4.7 TURBO CODES

Turbo codes were first introduced in 1993 by Berrou, Glavieux, and Thitimajshima, where a scheme was desired that achieves a bit-error probability of 10^{-5}, using a rate 1/2 code over an additive white Gaussian noise (AWGN) channel and BPSK modulation at an E_b/N_o of 0.7 dB.

The turbo codes are constructed by using two or more component codes on different interleaved versions of the same information sequence. Whereas for conventional codes, the final step at the decoder yields hard-decision decoded bits, for a *concatenated scheme*, such as a turbo code, to work properly, the decoding algorithm should not limit itself to passing hard decisions among the decoders. To best exploit the information learned from each decoder, the decoding algorithm must be based on soft decisions rather than hard decisions. For a system with two component codes, the concept behind turbo decoding is to pass soft decisions from the output of one decoder to the input of the other decoder, and to *iterate* this process several times through *feedback* so as to produce more reliable decisions.

According to Fig. 4.27, the principle of concatenated code is to feed the output of one encoder (outer encoder) to another encoder and so on. The final encoder before the

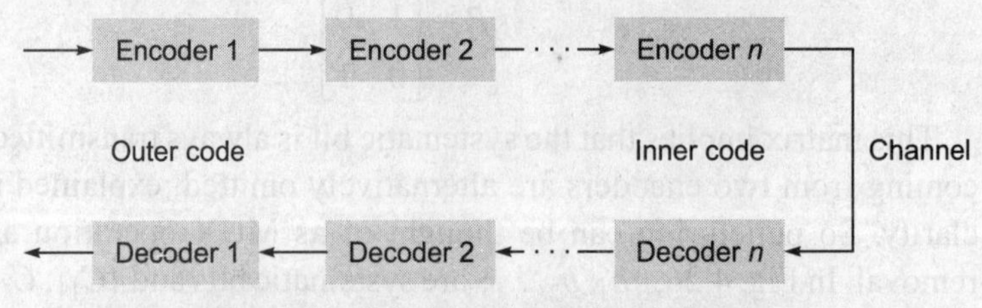

Fig. 4.27 Principle of concatenated codes

channel is known as the inner encoder. The resulting composite code is clearly much more complex than any individual code. However, it can be decoded step by step as shown. We simply apply each of the component decoders in turn, from the inner to the outer. This scheme suffers from many drawbacks, the most significant of which is error propagation. If decoding error occurs in a codeword, it usually results in a number of data errors. When these are passed on the next decoder they may overwhelm the ability of that code to correct the errors. The performance of the outer decoder might be improved if these errors where distributed between a number of separate codewords. This can be achieved using an interleaver/de-interleaver.

The interleaver may be placed between the outer and inner encoders of a concatenated code that uses two component codes and the de-interleaver between inner and outer decoders. Then, provided the rows of the interleaver are at least as long as the outer code words and the columns at least as long as the inner data blocks, each data bit of an inner codeword falls into a different outer codeword. Hence, the outer code is able to correct at least one error, it can always cope with single decoding error in the inner code.

4.7.1 Block/Product Turbo Codes (BTC or PTC)

Consider the two-dimensional code (product code) depicted in Fig. 4.28. The configuration can be described as a data array made up of k_1 rows and k_2 columns. The k_1 rows contain codewords made up of k_2 information data bits and $(n_2 - k_2)$ redundant parity bits. Thus, each of the k_1 rows represents a codeword from an (n_2, k_2) code. Similarly, the k_2 columns contain codewords made up of k_1 information data bits and $(n_1 - k_1)$ redundant parity bits. Thus, each of the k_2 columns represents a codeword from an (n_1, k_1) code. The various portions of the structure are labelled d for data, p_h for horizontal parity (along the rows), and p_v for vertical parity (along the column). In effect, the block of $(k_1 \times k_2)$ information data bits is encoded with two codes—a horizontal code and a vertical code.

Additionally, in Fig. 4.28, there are blocks labelled L_{eh} and L_{ev} containing the extrinsic LLRs values learned from the horizontal and vertical decoding steps, respectively. Error-correction codes generally provide some improved performance. The extrinsic LLRs

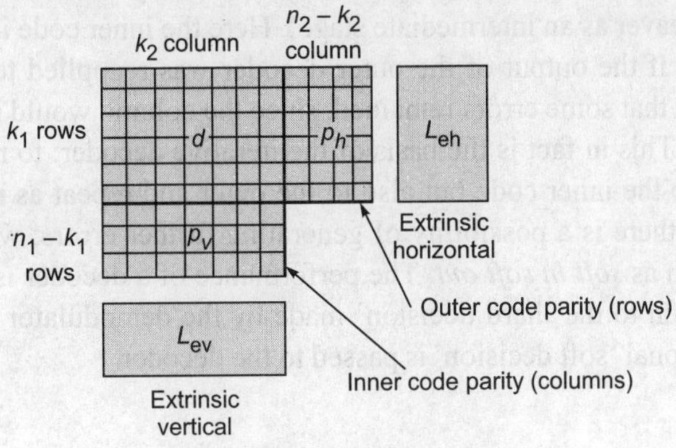

Fig. 4.28 Product turbo code

represent a measure of that improvement. Note that this product code is a simple example of a concatenated code explained in Fig. 4.27. Its structure encompasses two separate encoding steps—horizontal and vertical.

Recall that the final decoding decision for each bit and its reliability hinges on the value of $L(\hat{d})$, as shown in Equations (4.28) and (4.29). With these equations in mind, an algorithm yielding the extrinsic LLRs (horizontal and vertical) and the final $L(\hat{d})$ can be calculated as discussed in section 4.7.2.

4.7.2 Iterative Decoding Algorithm

It is the feedback (for several iterations) that has given rise to the term *turbo*, since the original inventors linked the process similar to turbo-charged engine, in which a part of power at the output is fed back to the input to boost the performance of the whole system. Thus, the term *turbo* should really be applied to the *decoder structure* than the code themselves. Following are the mathematical steps for iterative decoding.

1. Set the a priori LLR $L(d) = 0$ (unless the priori probabilities of the data bits are other than equally likely).
2. Decode horizontally, and using Eq. (4.29), obtain the horizontal extrinsic LLR:
$$L_{\mathrm{eh}}(\hat{d}) = L(\hat{d}) - L_c(s) - L(d)$$
3. Set $L(d) = L_{\mathrm{eh}}(\hat{d})$ for the vertical decoding of step 4.
4. Decode vertically, and using Eq. (4.29), obtain the vertical extrinsic LLR:
$$L_{\mathrm{ev}}(\hat{d}) = L(\hat{d}) - L_c(s) - L(d)$$
5. Set $L(d) = L_{\mathrm{ev}}(\hat{d})$ for the step 2 horizontal decoding.
6. After enough iterations (i.e., repetitions of steps 2 through 5) to yield a reliable decision, go to step 7.
7. The soft output is
$$L(\hat{d}) = L_c(s) + L_{\mathrm{eh}}(\hat{d}) + L_{\mathrm{ev}}(\hat{d}) \tag{4.58}$$

The innovation is in the decoding technique, iterative decoding—the turbo principle. The conventional decoding technique for array codes is shown in Fig. 4.27 (with the interleaver as an intermediate stage). Here the inner code is decoded first rather than the outer. If the output of the outer decoder was reapplied to the inner decoder, it would detect that some errors remained, since the column would not be codewords of the inner code. This in fact is the basis of the iterative decoder: to reapply the decoded word not just to the inner code but also to the outer and repeat as many times as necessary. But since there is a possibility of generating further errors, we require another component known as *soft in soft out*. The performance of a decoder is significantly enhanced, if in addition to the 'hard decision' made by the demodulator on the current symbol, some additional 'soft decision' is passed to the decoder.

4.7.3 Convolutional Turbo Codes (CTC)

Brief Introduction of NSC and RSC Code

For classical *non-systematic convolutional* (NSC), consider a binary rate $R = 1/2$ convolutional encoder with constraint length C and memory $M = C - 1$, as shown in Fig. 4.29. The input to the encoder at time k is a bit d_k and the corresponding codeword C_k is the binary couple (X_k, Y_k) with

$$X_k = \sum_{i-0}^{C-1} g_{1i}\, d_{k-i} \quad \text{mod. 2} \quad g_{1i} = 0, 1 \tag{4.59a}$$

$$Y_k = \sum_{i-0}^{C-1} g_{2i}\, d_{k-i} \quad \text{mod. 2} \quad g_{2i} = 0, 1 \tag{4.59b}$$

where G_1: $\{g_{1i}\}$, G_2: $\{g_{2i}\}$ are the two encoder generators, generally expressed in octal form.

The BER of a classical NSC code is lower than that of a classical systematic code with the same memory M at a large SNR. At a low SNR, it is in general the other way around. The new class of *recursive systematic convolutional* (RSC) codes at any SNR for high code rates.

A binary rate $R = 1/2$, RSC is obtained from an NSC code by using a *feedback loop* (because due to feedback loop, its impulse response is infinite like a recursive system and hence the name 'recursive' and original bits appear in the encoder output and thus the name 'systematic') and setting one of the two outputs X_k or Y_k equal to the input bit data d_k. For an RSC code, the shift register (memory) input is no longer the bit d_k but is a new binary variable a_k. If $X_k = d_k$ (respectively $Y_k = d_k$), the output Y_k (respectively X_k) is equal to Eq. (4.60) by substituting a_k for d_k and the variable a_k is recursively calculated as follows:

$$a_k = d_k + \sum_{i=1}^{k-1} \gamma_i\, a_{k-i} \quad \text{mod. 2} \tag{4.60}$$

where γ_i is respectively equal to g_{1i} if $X_k = d_k$ and to g_{2i} if $Y_k = d_k$. Equation (4.60) can be rewritten as

$$d_k = \sum_{i=0}^{k-1} \gamma_i\, a_{k-i} \quad \text{mod. 2} \tag{4.61}$$

One RSC encoder with memory $M = 4$ obtained from an NSC encoder defined by generators $G_1 = 37$, $G_2 = 21$ is depicted in Figures 4.29 and 4.30.

It can be shown that the trellis structure is identical for the RSC code and the NSC code and these two codes have same *free distance* d_f (same as Hamming distance). However, the two output sequences $\{X_k\}$ and $\{Y_k\}$ do not correspond to the same input sequence $\{d_k\}$ for RSC and NSC codes. This is the main difference between the two codes.

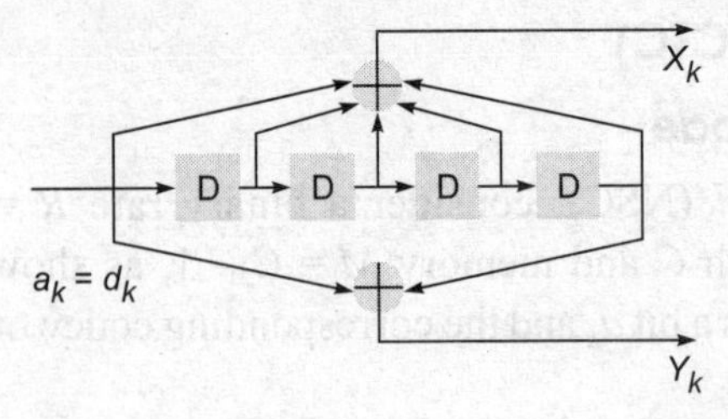

Fig. 4.29 Classical non-systematic code

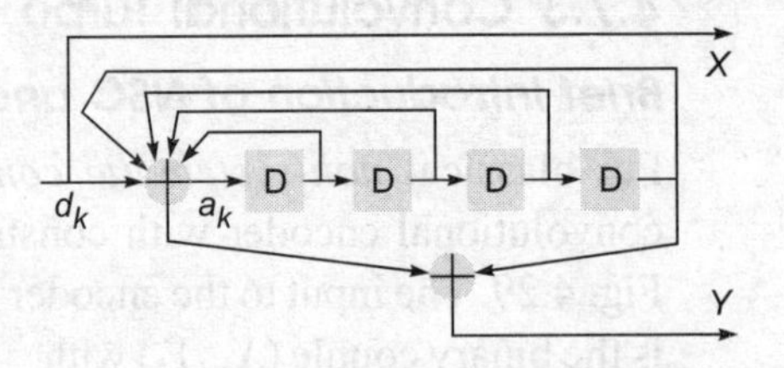

Fig. 4.30 Recursive systematic code

Main reasons behind choosing RSC encoders for turbo codes are as follows:

1. It gives systematic bit (i.e. original bit), which directly appears in the encoder output.
2. It has infinite impulse response, which is the requirement for turbo code, as in IIR system. An arbitrary input will cause a 'good' (high-weight) output with high probability, though some inputs will cause bad (low-weight) outputs.

Parallel Concatenated (RSC) Convolutional Codes (PCCC)

The goal of designing turbo codes is to choose the best component codes by maximizing the effective free distance of the code. At large value of E_b/N_o, this is tantamount to maximizing the minimum weight codeword. However, at low value of E_b/N_o (the region of greatest interest), the optimizing the weight distribution of the codewords is more important than maximizing the minimum weight codeword.

With RSC codes, a new concatenation scheme, called parallel concatenation, can be used. In Fig. 4.31, an example of two identical RSC codes with parallel concatenation is shown. Both elementary encoder (RSC1 and RSC2) inputs use the same bit d_k but according to a different sequence, due to the presence of an interleaver (pseudo-random interleaver is more popular). Intuitively, it is aimed to avoid pairing low-weight codewords from one encoder with low-weight codewords from the other encoder. Many such pairings can be avoided by proper design of the interleaver. For an input bit sequence $\{d_k\}$, encoder outputs X_k and Y_k at time k are respectively equal to d_k (for systematic encoder) and to encoder RSC1 output Y_{1k}, or to encoder RSC2 output Y_{2k}. If the coded outputs (Y_{1k}, Y_{2k}) of encoders RSC1 and RSC2 are used respectively n_1 times and n_2 times and so on, the encoder RSC1 rate R_1 and encoder RSC2 rate R_2 are, respectively, equal to

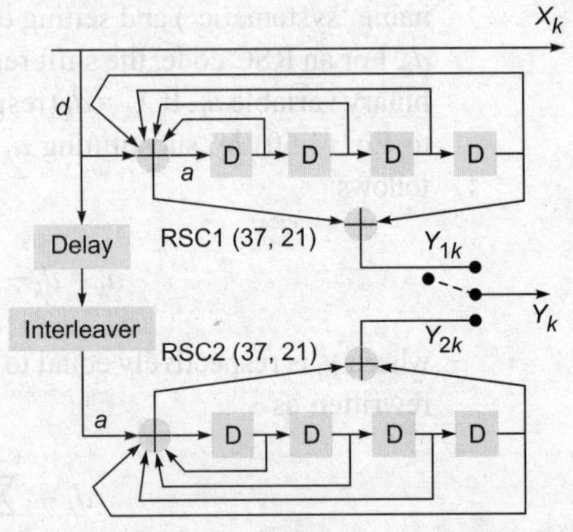

Fig. 4.31 Recursive systematic code with parallel concatenation

$$R_1 = \frac{n_1 + n_2}{2n_1 + n_2} \quad \text{and} \quad R_2 = \frac{n_1 + n_2}{2n_2 + n_1}$$

Thus, the *systematic bits* (i.e., original message bits) and the two sets of *parity-check bits* generated by the two encoders constitute the output of the turbo encoder.

Serial Concatenated Convolutional Codes (SCCC)

The concept of SCCC is already described previously and a block diagram is given in Fig. 4.32. The characteristic for PCCC schemes is to perform better than SCCC schemes at low SNRs. However, increasing the SNR, SCCC schemes outperform PCCC schemes. The cross-point depends on the interleaver size and interleaver design.

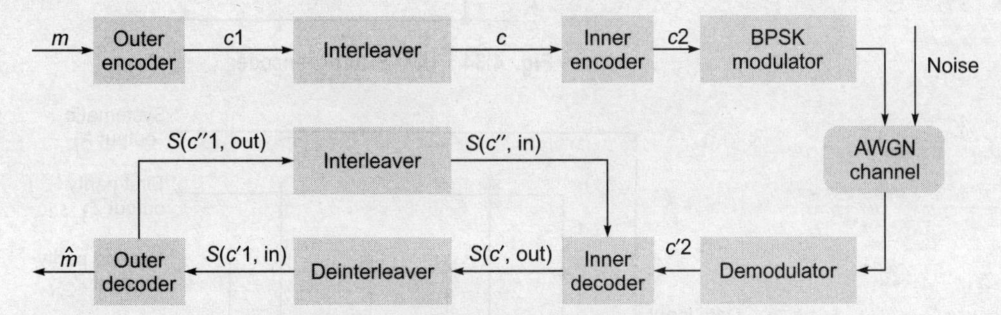

Fig. 4.32 System model for iterative decoding SCCC

HYBRID Concatenated Convolutional Codes (HCCC)

The name itself is self-explanatory. The HCCC contains both SCCC and PCCC and have advantages of both the SCCC and PCCC. Thus, HCCC outperforms both SCCC and PCCC at the cost of more complexity in the structure. In Fig. 4.33, such an example is given in which INT stands for interleaver.

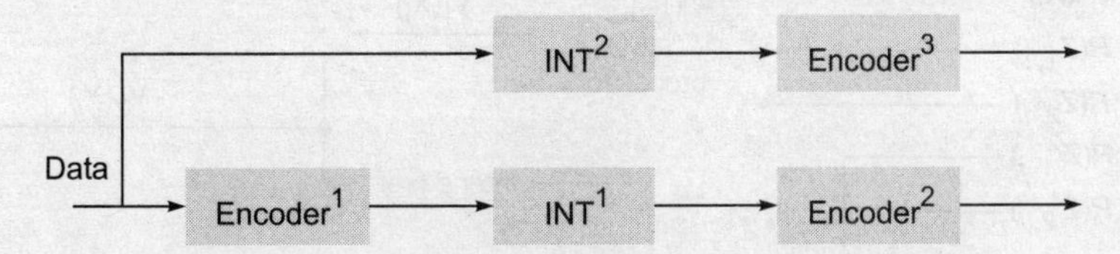

Fig. 4.33 Hybrid concatenated convolutional encoder

Various Turbo Encoders/Decoders for Latest Systems[2]

Figures 4.34 to 4.36 show different turbo encoders/decoders for UMTS and CDMA2000.

Algorithm used in decoding convolutional codes can be modified to be used in decoding turbo codes. In the original paper on turbo codes, a modified Bahl et al. algorithm (also known as *BCJR algorithm* as it was originally invented by Bahl, Cocke, Jelinek, and Raviv) was proposed for the turbo decoding stage. This algorithm is based on MAP rules. The problem with this algorithm is the inherent complexity and time delay. MAP algorithm was originally developed to minimize the bit error probability instead of the sequence error probability. The algorithm, although optimal, seems less attractive due to the increased

[2] It is an additional topic for the awareness of the latest scenario and can be omitted without breaking continuity of the study.

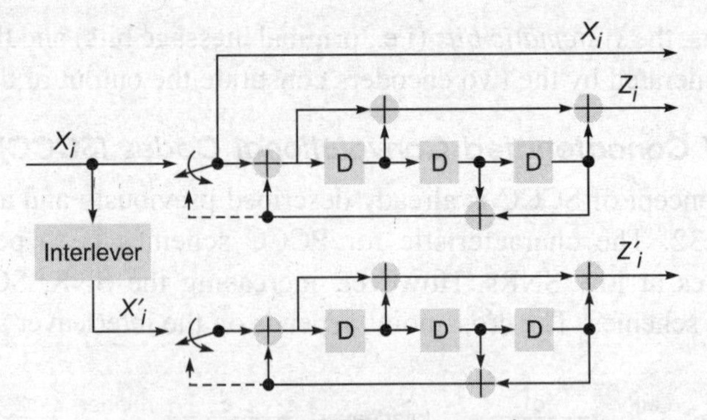

Fig. 4.34 UMTS turbo encoder

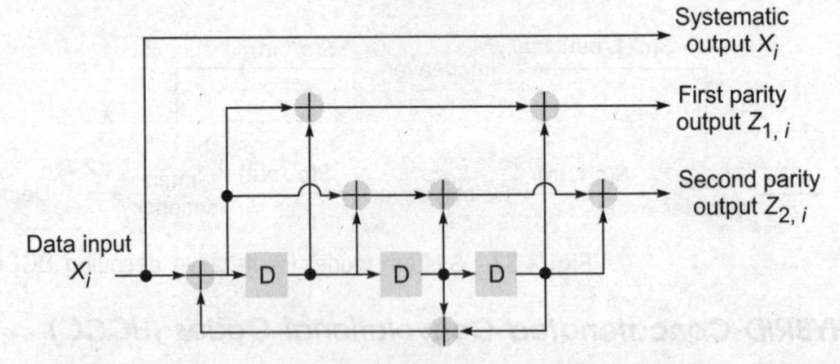

Fig. 4.35 Rate 1/3 RSC encoders used by CDMA2000 turbo code

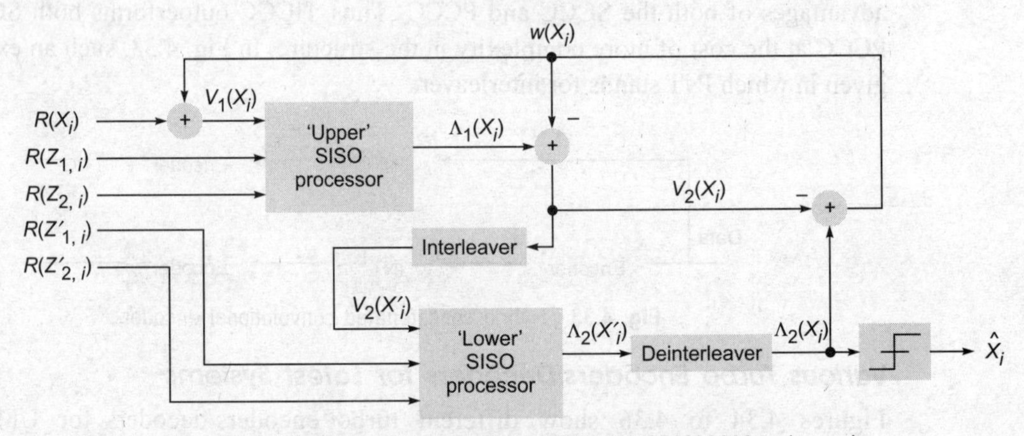

Fig. 4.36 Architecture for decoding UMTS and CDMA2000 turbo code

complexity. Efforts have been made to reduce the decoding complexity of the BCJR algorithm. Max-log-MAP (Log-MAP) and SOVA (soft output viterbi algorithm), a modified version of Viterbi algorithm, which uses soft output decision instead of hard decision, are the two top sub optimal decoding algorithms for bit error rate performance of turbo code.

Comparison of log MAP and SOVA MAP algorithm is the maximum likelihood decoding algorithm that minimizes the probability of bit error, whereas SOVA (or Viterbi, in case convolutional coding) is the maximum likelihood decoding algorithm that minimizes the probability of word or sequence error. This subtle difference can also be observed while implementing these algorithms. In SOVA, we add branch metrics to state metrics. Then we compare and select the minimum distance (maximum likelihood) in order to form the next metric. This process is called *add–compare–select*. On the other hand, in MAP algorithm, we multiply (add in logarithmic domain) state metrics by branch metrics. Then, instead of comparing them, we sum them to form the next forward (or reverse) state metric. The MAP algorithm suffers from its inherent complexity and time delay compared to SOVA. The SOVA suffers from two degradations. First, the reliability information of the decoder output is too optimistic especially for bad channels. Secondly, the assumption that the decoders' decisions are uncorrelated is not completely true. An improved decoding with SOVA, called normalized SOVA, that provides a remedy to these two problems, is also introduced.

Although turbo codes have the potential to offer unprecedented power efficiencies, they have some peculiarities that should be taken into consideration.

Error floor is a phenomenon of flattening out of BER curve at higher SNR (Fig. 4.37), which hinders the ability of a turbo code to achieve extremely small bit-error rates. The error floor is mainly due to the presence of a few low-weight codewords. At low SNR, these codewords are insignificant, but as SNR increases, they begin to dominate the performance of the code. For this, sometimes it is preferred to use convolutional code for high SNR values.

The error floor effect can be combated in several ways. First, one is to use two different RSC encoders. One RSC encoder is optimized to perform well at low SNR, while the other is optimized to reduce the error floor. The resulting asymmetric turbo code provides a reasonable combination of performance at both low and high SNRs. Unfortunately, although the error floor has been reduced, it is still present. Second method to reduce the error floor is to arrange the two constituent encoders in serial concatenation. Such serially concatenated convolutional codes offer excellent performance at high SNR, as the error floor is virtually eliminated (actually, it is pushed down to BER $\approx 10^{-10}$). However, performance at low SNR is considerably worse than it is for parallel concatenated convolutional codes. An alternative to choose between SCCCs and PCCCs is to use hybrid turbo codes, which combine features of each type of code. However, the error floor depends on the specific interleaver and can be lowered by improving the interleaver construction instead of using pseudo-random interleavers.

Though several talks can be delivered in favour of turbo codes, its decoding algorithm is no doubt more complex than any other FECs. One easy way to reduce complexity is simply to halt the decoder iterations once the entire frame has been completely corrected. This will prevent over-iteration, which corresponds to wasted hardware clock cycles. Other methods, such as implementing the entire decoder in analog circuitry rather than in digital hardware and sliding window algorithms for saving in memory requirement, are also approached.

Turbo-coded systems typically experience significant latency due to interleaving and iterative decoding. For example, let us think of an interleaver used in turbo code uses 65,536 bits. Since this number of bits must inevitably be stored in the interleaver in the encoder and/or decoder at any given time, this means that there is latency. Thus, for an information rate of (say) 8 kbps (appropriate for speech transmission), there is delay of 65536/8 = 8192 ms or more than 8 s. This delay is quite unacceptable in a telephone system, since it would be highly disruptive of any conversation. In fact, the delay limitation for speech services on wireless links usually takes 40 ms, which corresponds to a latency of 320. Thus, if a turbo code is to be used for speech services at this data rate, then this interleaver can no longer be used. However, to reduce latency one can think of a low interleaver size, i.e. low latency but there will be some performance penalty like departing from Shannon limit (as turbo code is approaching to Shannon limit only when latency is unlimited) and error floor discussed earlier.

4.8 POPULARITY OF TURBO CODES

The features of turbo codes that make it popular are discussed below.

Randomly Structured Code

Turbo code is randomly structured code. According to Shannon, it is only the *random code* that approaches Shannon limit but is difficult to decode whereas the *structured code* possesses the reverse characteristics. Turbo code is an optimal one, which both possesses *randomness* through interleaver, thus approaching Shannon limit and *structure* through concatenation, thus easily decodable.

Good Linear Code with Thin Distance Spectrum

Turbo code is a good linear code. A 'good' linear code is one that has mostly high-weight (Hamming weight) codewords. The job of producing mostly high Hamming weight is done by pseudo-random interleaver. Coding theorists say that turbo codes have a 'thin distance spectrum', which is a function that describes the number of code words each of possible non-zero weight, is thin in the sense that there are not very low weight codewords present. For information, distance properties of the turbo product codes are similar to those of the product of constituent codes.

Most Power Efficient Code

Turbo code is no doubt a power efficient code with high correction capability. For example, (a) for a frame size of 250 × 250 = 65,536 bits, we can achieve a BER = 10^{-5} over AWGN channel at only $E_b/N_o = 0.7$ dB close to Shannon limit and (b) for Rayleigh fading channel, a BER = 10^{-5} can be achieved at $E_b/N_o = 4.3$ dB, which represents a gain of 2.3 dB as compared to classical convolutional code with similar complexity.

4.8.1 Applications of Turbo Codes

The practical applications of turbo codes extend from mobile communications to deep space explorations. Here, a few of them will be explained.

Deep Space Exploration

Turbo code is essential to maintain communication with spacecraft exploring the solar system, because over the vast distances the signal power involved is at a premium. An improvement of a small fraction of a decibel can make a difference of millions of miles to the operational range of a mission. The Jet Propulsion Laboratory (JPL), which carries out research for NASA, was among the first to realize the potential of turbo codes, and as a result turbo codes were used in the Pathfinder mission.

Mobile Communication

Turbo codes are one of the options for FEC coding in the UMTS third generation mobile radio standard. A great deal of development has been carried out here, especially on the design of interleavers of different lengths, for application both to speech services, where latency must be minimized, and to data services, which must provide very low BER. Turbo code is an optional feature in WiMAX.

Terrestrial Digital Video Broadcast (DVB-T) Standard

Recently turbo codes have been incorporated into the standard that will incorporate a return channel in digital broadcast systems.

4.9 BER CONSIDERATIONS FOR PERFORMANCE ANALYSIS OF CONVOLUTIONAL AND TURBO CODES

The performance of turbo coding systems is characterized by two distinct regions—*turbo cliff* region, where the bit-error rate drops within a fraction of a dB of SNR to a very low values, and *error floor/flare* region, which is characterized by a slow decrease of error rate with increasing SNR (Fig. 4.37).

In turbo codes based on various parameters or various changes made in the input or in the system design, the performance differs. How does it vary is discussed below.

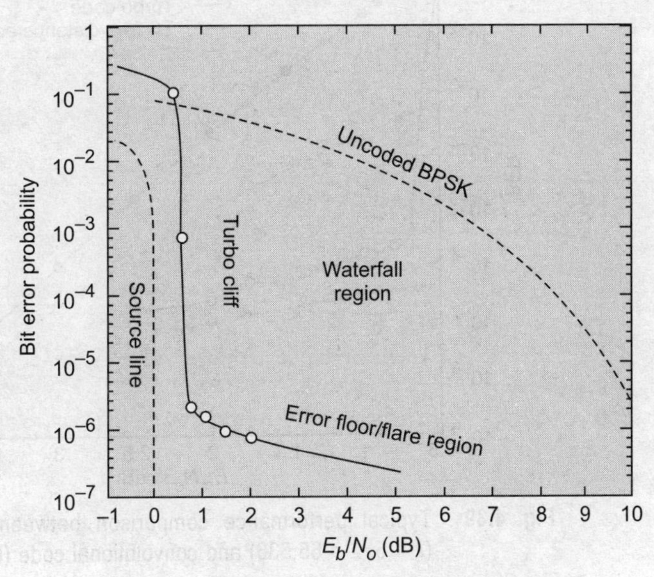

Fig. 4.37 Typical performance of turbo coding

Number of Iterations

Obviously, the number of iterations increases the performance while on the other hand increasing latency period (Fig. 4.38). So, there should be a compromise between the latency period and the performance gain. Normally, the number of iterations is 18 but in most of the cases, 6 will suffice.

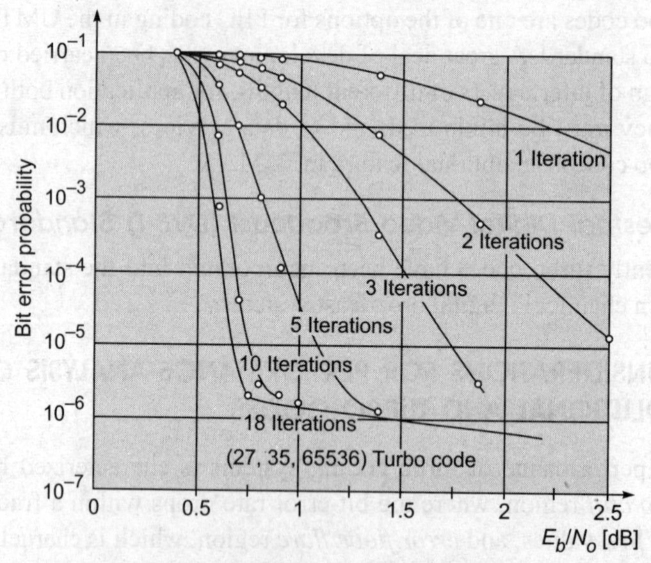

Fig. 4.38 Effect of the number of iterations

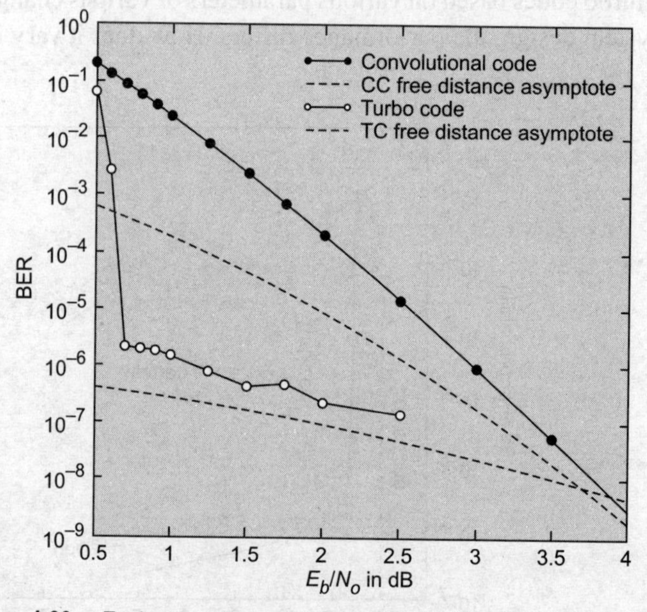

Fig. 4.39 Typical performance comparison between turbo code (k = 5, L = 65,536) and convolutional code (C = 15)

Block Size

It is observed that as the block size is increased there is a steady increase in performance gain. If block size is increased then the size of interleaver must be correspondingly increased. Since a non-uniform random interleaver is used, a large interleaver size will improve the randomization between adjacent bits (the more is randomness, the more approaching towards Shannon limits) and also less weight codewords are hardly to occur.

Interleaver Design

It is observed that pseudo-random interleaver provides better performance than the block interleaver. The larger the interleaver size, the more performance gain due to more randomness and the more latency. So there is trade-off between latency and performance. It is shown that the larger the interleaver, the more iterations are necessary for the MAP algorithm to converge in waterfall region. At low SNRs and high SNRs, little iteration will suffice.

4.10 TRELLIS CODED MODULATION

Trellis coded modulation (TCM) is a technique that combines both coding (convolutional/turbo) and modulation (M-PSK) to achieve significant coding gains without compromising bandwidth efficiency. The TCM scheme employs redundant non-binary modulation in combination with a finite state encoder that decides the selection of modulation signals to generate coded signal sequences. It uses signal state expansion to provide redundancy for coding and to design coding and signal mapping functions jointly so as to maximize directly the free distance (minimum Euclidean distance) between the coded signals. At the receiver, the signals are decoded by a soft decision maximum likelihood sequence decoder.

The general TCM encoder implements TCM by convolutionally encoding the binary input signal and mapping the result to an arbitrary signal constellation. The points in the signal constellation are listed in set partitioned order in the signal constellation parameter. This parameter is a complex vector whose length, M, equals the number of possible output symbols from the convolutional encoder, i.e. $\log_2 M$ is equal to n for a rate k/n convolutional code.

The general TCM decoder uses the Viterbi algorithm to decode a trellis coded modulation signal that was previously modulated using an arbitrary signal constellation. The trellis structure and signal constellation parameters in this block should match those in the general TCM encoder block, to ensure proper decoding. In particular, the signal constellation parameter must be in set partitioned order.

The advantage of TCM is considerable coding gain of 5 to 6 dB that can be obtained without any bandwidth expansion or reduction in the effective information rate. Rectangular QAM TCM is also one more possibility.

Summary

- By channel coding, we add redundant bits in the data in a systematic way.
- Channel coding improves BER performance in general.
- There are two types of channel coding—waveform and structured sequence coding.
- There are two basic approaches of adding structured redundancy—automatic repeat request and forward error correction. Hence, we may have error-detecting codes and error-correcting codes.
- Selection of the code rate involves trade-off between energy efficiency and bandwidth efficiency.
- Transmission bandwidth is always tried to reduce: $W > C > R$.
- Minimum distance between the codes gives the number of bits in error.
- Parity check, CRC, etc., are error-detecting codes and mostly used with retransmissions.
- Block codes, convolutional codes, and turbo codes are error-correcting codes and do not allow retransmissions. These are mainly used with real-time systems.
- Convolutional codes are suitable for the channel with memory.
- Turbo codes are concatenated codes and undergo iterative decoding.
- Turbo codes can approach Shannon limit.
- Interleaver and puncture codes play an important role in turbo coding. Interleaver increases the randomness in the code. Puncturing is used to increase the code rate.

Review Questions

1. Why is information theory applied to both source as well as channel coding? What do you mean by removed redundancy in case of source coding? What do you mean by added redundancy in case of channel coding?
2. How does data and information differ?
3. Describe the methods for controlling the errors. How can the error control be achieved with the error-detection schemes?
4. What types of errors may occur in the data? What may be the reasons behind it?
5. Prove the error-detection and error-correction capabilities of the linear block code.
6. State and prove Shannon–Hartley law. State the significance of Shannon limit. Discuss the overall work of Shannon for which he introduced a number of laws.
7. What do you mean by rate 1/2, rate 1/3 convolutional coder?
8. How can you represent the trellis diagram in terms of a state diagram?
9. What is the limit up to which the redundancy can be added for error handling to the source coded data to be transmitted?
10. Find out the various algorithms for decoding of convolutional code?
11. What will be the role of the number of iterations in turbo codes?
12. How can we say that interleaver plays a trade-off between performance and time delay?
13. How can we incorporate the interleaver in a concatenated code generation and decoding of it? What will be the advantage of this?
14. Find out the turbo coding methods/diagrams in which puncture codes are incorporated.

15. Compare the BER performance of convolution-coded and turbo-coded BPSK modulators. Also comment on the limitations of turbo coding.

16. Why is turbo coding method considered to be the most efficient method?

17. Explain in detail cyclic redundancy with an example. Also explain why CRC check is often used with convolutional coding.

18. Simulate the algorithm for soft decision making for the decoding purpose. List out the decoding algorithms in which the soft decision making is used.

PROBLEMS

1. If the channel coding scheme is incorporated using block code such that it is to transmit three times the binary information symbols, what may be the code rate and redundancy added?

2. Find out a generator polynomial $g(x)$ for a (7, 4) cyclic code and code vectors for the following data vectors: 1010, 1111, 0001, and 1000.

3. Consider the rate 4/7 (7, 4, 3) Hamming codes with the parity check matrix

$$H = \begin{bmatrix} 1 & 0 & 1 & 0 & 1 & 0 & 1 \\ 0 & 1 & 1 & 0 & 0 & 1 & 1 \\ 0 & 0 & 0 & 1 & 1 & 1 & 1 \end{bmatrix}$$

Find out the set of code words for all the combinations of message bits starting from 0000 to 1111.

4. Consider the (3, 1, 2) convolutional code with impulse responses $g^{(1)} = (110), g^{(2)} = (101)$, and $g^{(3)} = (111)$

(a) Draw the encoder block diagram.

(b) Find out the generator matrix.

(c) Find out a code word corresponding to the information sequence (11101).

(d) Draw the code tree.

5. In order to construct a (15,7) BCH code for the coefficients given in Table 4.2, what will be the generator polynomial? Show all the steps.

6. In the example of hard decision Viterbi decoding discussed in Section 4.4.4, if message bit is 110110 and coded bits are 11 10 01 11 10 01 as well as the received bits are 11 10 01 11 00 01, find the trellis and the decoding path on it.

MULTIPLE CHOICE QUESTIONS

1. By channel coding, the information transmission efficiency

(a) increases

(b) reduces

(c) remains same

(d) sometimes increases, sometimes reduces

2. Error correction coding method is also called

(a) automatic repeat request (ARQ)

(b) forword error correction (FEC)

(c) error-detection coding

(d) error control

3. Which of the following is a error-detection scheme?

(a) Turbo coding

(b) Convolutional coding

(c) Cyclic coding

(d) Parity check

4. Which of the following is a error-correction scheme?

(a) Parity check

(b) Reed–Solomon coding
(c) Longitudinal redundancy check
(d) Cyclic redundancy check

5. Code rate is nothing but
(a) nk (b) n/k
(c) k/n (d) $n - k$

6. Hamming weight in the sequence 1110100 is
(a) 3 (b) 1
(c) 4 (d) 7

7. Error-detecting capability of a code t_d is defined as
(a) $t_d = d_{\min} - 1$
(b) $t_d = (d_{\min} - 1)/2$
(c) $t_d = 1 - d_{\min}$
(d) $t_d = d_{\min} - 1/2$

8. Which of the following codes can approach Shannon limit?
(a) Convolutional code
(b) Turbo code
(c) Hamming code
(d) Block code

9. Quantization process is used in
(a) ML decoding
(b) soft decision decoding
(c) Viterbi decoding
(d) hard decision decoding

10. Which of the following stages adds randomness in the turbo code?
(a) Interleaver
(b) Code puncturing
(c) Inner encoder
(d) Convolutional coding

11. Which of the following changes the code rate?
(a) Interleaver
(b) Code puncturing
(c) Hamming bits
(d) Convolutional coding

chapter 5

Radio Propagation over Wireless Channel

Key Topics

- Properties and problems associated with wireless channel
- Radio propagation in atmospheric layers
- Free space propagation model
- Additive white Gaussian noise (AWGN)
- Multipath fading effects
- Delay spread: small-scale fading and large-scale fading
- Coherence bandwidth: flat fading and frequency-selective fading
- Doppler effect: fast fading and slow fading
- Shadowing
- Outage and outage probability

Chapter Outline

Radio propagation is very much dependent upon the site/location of the communication distance between transmitter, receiver, etc., especially for wireless channel. The performance of the same system can vary significantly depending on the terrain, type of propagation, weather, frequency of operation, velocity of mobile terminal, interference sources, and other dynamic factors. The channel considerations for long- and short-distance communication differ. Accurate characterization of the radio channel through important parameters, a mathematical model, and simulation of the channel model is important for predicting signal coverage, achievable data rates, BER performance, specific performance attributes of alternative signalling, and reception schemes as well as for analysis of interference from different systems and determination of the optimum location for installing base station antennas or earth stations for applications like mobile satellite communication. All the above channel effects are described in this chapter and channel models, diversity techniques, channel estimation, and equalization methods are described in the next chapter.

5.1 GENERAL CONSIDERATIONS ABOUT RADIO WAVES AND WIRELESS CHANNEL

Wireless channel is a dielectric unguided medium that can be analysed in different ways as follows:

1. By using the fundamentals of electromagnetic wave theory, such as phase velocity, phase propagation constant, amplitude, frequency, and phase.

2. By using ray theory, refractive index dependence, reflection, refraction, diffraction of electromagnetic waves (rays), etc.
3. By using DSP fundamentals, such as channel transfer function, spectrum, channel impulse response, and convolution with transmitting signal. The channel can act as a low-pass filter in certain conditions.
4. The channel characteristic is random (channel processes are assumed to be stochastic processes) and depends upon the situation; hence, the probability theory and concept of PDF can be applied to wireless channels. The PDF gives the behavioural model for the channel.

A radio signal exists in the form of an electromagnetic wave over a wireless channel, which is radiated through an antenna. Since the radiation and propagation of radio waves cannot be seen, the discussion here is based on the accepted theories and the predictions and measurements applied. The theory of electromagnetic radiation was proposed by a British physicist, James Clerk Maxwell, in 1857 and finalized in 1873. It is not possible to cover the mathematics of Maxwell's equations here. Electromagnetic waves are nothing but the energy propagated through free space at the velocity of light, 3×10^8 metres per second. The energy level decreases with the distance and it can have both vertical and horizontal components.

The electromagnetic waves [Fig. 5.1(a)] have two elements. They are made from electric and magnetic field components, which are inseparable. The planes of the fields are at right angles to each other and to the direction in which the wave is travelling.

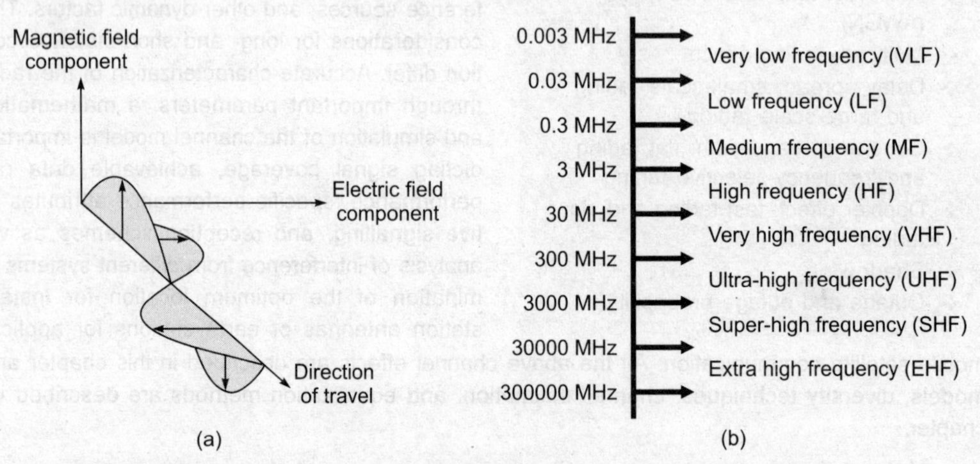

Fig. 5.1 (a) An electromagnetic wave and (b) radio frequency ranges

The electric component of the wave results from the voltage changes, which occur as the antenna element is excited by the alternating waveform. The lines of force in the electric field run along the same axis as the antenna but spread out as they move away from it. This electric field is measured in terms of the change of potential over a given distance, e.g., volts per metre, and this is known as the *field strength*. This measure is often used in measuring the intensity of an electromagnetic wave at a particular point. The other component, namely the magnetic field, is at right angles to the electric field and hence at right angles to the plane of the antenna. It is generated as a result of the current flow in the antenna.

Light wave (LASER) communication by optical rays falling in the 10^{14} Hz range can be considered as wireless communication. Infrared communication is also categorized under wireless communication, but these systems do not require conventional metallic antenna-based radiation. They require solid state optical device-based radiators. Some obstacles may interrupt the light rays, which can be considered as limitation of such communication systems. Let us skip these special cases for now and consider the radio frequency ranges. Figure 5.1(b) gives an idea about the radio frequency (RF) ranges suitable for wireless communication, of the whole electromagnetic spectrum. Any radio wave will follow the propagation mechanisms as follows:

Reflection The propagating wave impinges on an object that is large compared to the wavelength and reflects, e.g., the surface of the earth, buildings, walls, etc.

Diffraction The radio path between the transmitter and the receiver is obstructed by a surface with sharp irregular edges, e.g., waves bend around the obstacle.

Scattering The propagating wave impinges on the objects smaller than the wavelength of the propagating wave and gets scattered, e.g., lamp posts, foliage, street signs, and particles in the air.

Refraction Due to variation of refractive index of the atmospheric layers, the EM wave bends (in cases other than satellite communication).

These effects will become significant on the basis of the distance between the transmitter and the receiver and the number of objects in the surroundings.

The *polarization* of an electromagnetic wave indicates the plane in which it is vibrating. It often has a significant effect on the way the wave propagates. While it is important to match the polarization of the transmitting and receiving antennas, the choice of polarization is also important for the signal propagation and for diversity mitigation (Chapter 6). As an electromagnetic wave consists of an electric field and a magnetic field vibrating at right angles to each other, however, it is necessary to adopt a convention to determine the polarization of the signal. For this purpose, the plane of the electric field is used.

Vertical and horizontal polarizations are the most straightforward forms and they fall into a category known as *linear polarization*. Here the wave can be thought of vibrating in one plane. It is also possible to generate waveforms that have circular or elliptical polarization. *Circular polarization* can be visualized by imagining a signal propagating from an antenna that is rotating. The tip of the electric field vector can be seen to trace out a helix or corkscrew as it travels away from the antenna. Circular polarization can be either right- or left-handed depending upon the direction of rotation as seen from the transmitting antenna. *Elliptical polarization* occurs when there is a combination of both linear and circular polarizations. The question may arise: Why is the polarization needed for propagation?

For many terrestrial applications, it is found that once a signal has been transmitted, its polarization will remain broadly the same. However, reflections from objects in the path can change the polarization. As the received signal is the sum of the direct signal and a number of reflected signals, the overall polarization of the signal can change slightly although it usually remains the same. When reflections take place from the ionosphere, greater changes may occur.

In some applications, there are performance differences between horizontal and vertical polarization. For example, medium wave broadcast stations generally use vertical polarization because ground wave propagation over the earth is considerably better using vertical polarization, whereas horizontal polarization shows a marginal improvement for long-distance communications using the ionosphere. Circular polarization is sometimes used for satellite communications as there are some advantages in terms of propagation and in overcoming the fading caused if the satellite is changing its orientation.

Wireless communication systems exist in a variety of different forms and consequently different channel characteristics are to be considered, but mainly two differentiations can be made as follows.

Firstly, microwave point-to-point links or satellite communication links are established for very large distance communication and hence no matter of reflections or multipaths. These effects are negligible as the distance is too large and the direct signal is very strong. A majority of these links are considered as pure line of sight (LOS). These are typically designed with slim margins as the channel will vary little over time, with the exception of possibly increased loss from rain (absorption), scattering due to solid particles or Faraday rotation, or simply attenuation with distance. The dynamic range required at the receiver will consequently be small. Hence, to analyse such a system and link power budget estimation, one can start with the simple free space propagation model and then gradually add the losses in it for the actual scenario.

We can model this type of channel as time-invariant non-dispersive channel.

Secondly, multipath links are usually described by resultant of both of the following:

- **Line-of-sight (LOS) component** The direct path between the transmitter (TX) and the receiver (RX)
- **Non-line-of-sight (NLOS) component** The path due to the reflection from reflectors, scatterers, or diffractors

If only NLOS component is present, then it is a Rayleigh fading channel and if LOS component is present along with NLOS, then it becomes Rician fading channel (Chapter 6). Mobile communication systems operating in the VHF and UHF bands are typically multipath based and, therefore, consist of the sum of signals from different paths. Reflections, refraction, and diffraction are more pronounced in these systems. This scenario is a little bit complex. Potentially, this will result in multipath fading effect resulted from considerable amplitude variations with time and/or distance. Physical objects will also cause 'shadowing' resulting in a loss of signal, for instance, a vehicle entering a tunnel or travelling behind a building. Additionally, due to the varying proximity between two nodes, a considerabiy larger dynamic range will be required at the receiver together with an increased link budget margin. We can model this type of channel as a time-variant, time-dispersive channel.

It is clear that these two communication systems will require different properties from a modulation scheme to transmit the information.

Attenuation is the drop in the signal power when transmitting from one point to another. It is caused by the transmission path length, obstructions in the signal path, and multipath effects. This effect can be observed along with small-scale and large-scale fading, slow and fast fading, or flat and frequency-selective fading, where the drop in the signal strength with distance is observed. Any object that obstructs the line-of-sight signal from the transmitter to the receiver can cause attenuation. Wooden objects, glass, bricks, etc., have different attenuation values when an electromagnetic ray penetrates through it and so attenuation will not be uniform.

Shadowing of the signal can occur whenever there is a covering obstruction between the transmitter and the receiver. Shadowing can be observed in long- as well as short-distance communication. It is generally caused by buildings and hills and is the most important environmental attenuation factor. Shadowing is most severe in heavily built-up areas and is caused by buildings. Hills can cause a major problem due to the large shadow they produce. Ground wave propagation described next is also affected due to shadowing. Radio signals diffract off the boundaries of obstructions, thus preventing total shadowing of the signals behind hills and buildings. However, the amount of diffraction is dependent on the radio frequency used. With low frequencies, diffraction is more than that with high-frequency signals. Thus, high-frequency signals, especially, ultra-high frequencies, and microwave signals require line of sight for adequate signal strength. To overcome the problem of shadowing, transmitters are usually elevated as high as possible to minimize the number of obstructions. In Fig. 5.4, the car can be considered in the shadow portion.

Important relationships for a dielectric medium like wireless channel:

$E = E_0 \sin(\omega t - \beta z)$:	Representation of an electromagnetc wave with its peak amplitude, frequency, and phase
$\omega = 2\pi f$:	Angular frequency
$\lambda = c/f$:	Wavelength
$n = c/v_{em}$:	Refractive index of the medium
$k = 2\pi/\lambda$:	Media propagation constant (1 for free space)
$\beta = nk$:	Wave propagation constant
c:	Speed of light or EM waves in free space
v_{em}:	Velocity of an EM wave in the dielectric medium

5.1.1 Radio Propagation and the Atmosphere

The way that radio signals propagate or travel from the radio transmitter to the radio receiver is of great importance when planning a radio network or system. This is governed to a great degree by the regions of the atmosphere through which they pass. Without the action of the atmosphere it would not be possible for radio signals to travel around the globe on the short wave bands, or travel greater than only the line-of-sight distance at higher frequencies. In view of the importance of the atmosphere, an overview of its make-up is given here. The atmosphere can be split up into a variety of different layers according to their properties. As different aspects of science look at different properties, there is no single nomenclature for the layers.

- The lowest layer is the troposphere that extends to a height of 10 km.
- Above this, at altitudes between 10 and 50 km, the stratosphere is found. This contains the ozone layer at a height of around 20 km.
- Above the stratosphere, there is the mesosphere extending from an altitude of 50 km to 80 km.
- Above this is the thermosphere, where temperatures rise dramatically.
- Ionosphere is in between the mesosphere and the thermosphere, which again contains four sub-layers.

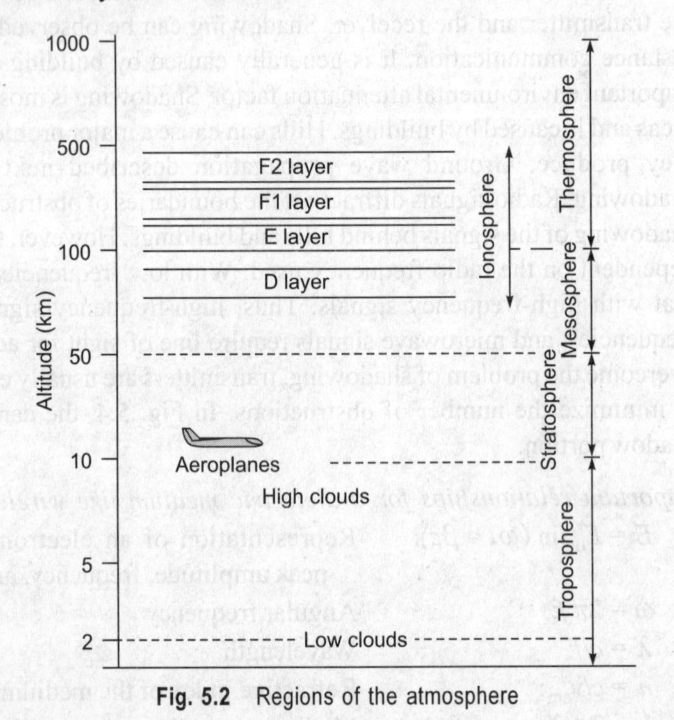

Fig. 5.2 Regions of the atmosphere

The layers that are important for long-distance communication are described here in brief (refer to Fig. 5.2).

Troposphere

The lowest of the layers of the atmosphere is the troposphere. This extends from ground level to an altitude of 10 km. It is within this region that the effects that govern our weather occur. To give an idea of the altitudes involved, it is found that low clouds occur at altitudes of up to 2 km, whereas medium level clouds extend to about 4 km. The highest clouds are found at altitudes up to 10 km, whereas modern jet airliners fly above this at altitudes of up to 15 km.

Within the troposphere, there is generally a steady fall in temperature with height and this has a distinct bearing on some propagation modes that occur in this region. The fall in the temperature continues in the troposphere until the tropopause is reached. This is the area where the temperature gradient levels out and then the temperature starts to rise. At this point, the temperature is around −50°C. The refractive index of the air in the troposphere plays a dominant role in radio signal propagation. This depends on the temperature, pressure, and humidity. When radio signals are affected, this often occurs at altitudes up to 2 km.

Ionosphere

The ionosphere is a region where there is a very high level of free electrons and ions. It is found that the free electrons affect radio waves. Although there are low levels of ions and electrons at all altitudes, the number starts to rise noticeably at an altitude of around 30 km. However, it is not until an altitude of approximately 60 km is reached that it rises to a degree sufficient to have a major effect on radio signals.

In general, the ionosphere is very complicated. It involves radiation from the sun striking the molecules in the upper atmosphere. This radiation is so intense that when it strikes the gas molecules, some electrons are given sufficient energy to leave the molecular structure. This leaves a molecule with a deficit of one electron, which is called an ion. It causes free electron. As might be expected, the most common molecules to be ionized are nitrogen and oxygen.

Most of the ionization is caused by radiation in the form of ultraviolet light. At very high altitudes, the gases are very thin and only low levels of ionization are created. As the radiation penetrates further into the atmosphere, the density of the gases increases and, accordingly, the number of molecules being ionized increases. However, when molecules are ionized, the energy in the radiation is reduced, and even though the gas density is higher at lower altitudes, the degree of ionization becomes less because of the reduction of the level of ultraviolet light. At the lower levels of the ionosphere, where the intensity of the ultraviolet light has been reduced, most of the ionization is caused by X-rays and cosmic rays, which are able to penetrate further into the atmosphere.

Because of the above reason, the ionosphere is thought of consisting of a number of distinct layers. Each layer overlaps the others with the whole of the ionosphere having some level of ionization. The layers are best thought of as peaks in the level of ionization. These layers are given designations D, E, F1, and F2.

D layer The D layer is the lowest of the layers of the ionosphere. It exists at altitudes around 60 to 90 km. It is present during the day when radiation is received from the sun. However, the density of the air at this altitude means that ions and electrons recombine relatively quickly. This means that after sunset, electron levels fall and the layer effectively disappears. This layer is typically produced as the result of X-rays and cosmic ray ionization. It is found that this layer tends to attenuate signals that pass through it.

E layer The next layer beyond the D layer is called the E layer. This exists at an altitude of between 100 and 125 km. Instead of acting chiefly as an attenuator, this layer reflects radio signals, although they still undergo some attenuation. In view of its altitude and the density of the air, electrons and positive ions recombine relatively quickly. This occurs at a rate of about four times that in the F layers, which are higher up, where the air is less dense. This means that after nightfall, the layer virtually disappears although there is still some residual ionization. There are a number of methods by which the ionization in this layer is generated. It depends on various factors, including the altitude within the layer, the state of the sun, and the latitude. However, X-rays and ultraviolet rays produce a large amount of the ionization light, especially that with very short wavelengths.

F layer The F layer is the most important region for long distance HF communications. During the day it splits into two separate layers. These are called the F_1 and F_2 layers, the F_1 layer being the lower of the two. At night these two layers merge to give one layer called the F layer as shown in Fig. 5.9 later on. The altitudes of the layers vary considerably with the time of day, season and the state of the sun. Typically in summer the F_1 layer may be around 300 km with the F_2 layer at about 400 km or even higher. In winter these figures may be reduced to about 200 km and 300 km. Then at night the F layer is generally around 250 to 300 km. Like the D and E layers, the level of ionization falls at night, but in view of the much lower air density, the ions and electrons combine much more slowly and the F layer decays much less. Accordingly, it is able to support communications, although changes are experienced because of the lessening of the ionization levels. The figures for the altitude of the F layers are far more variable than those for the lower layers. They change greatly with the time of day, the season, and the state of the sun. As a result, the figures that are given must only be taken as an approximate guide. Most of the ionization in this region of the ionosphere is caused by ultraviolet light, both in the middle of the UV spectrum and the portions with very short wavelengths.

5.2 BASIC PROPAGATION MECHANISMS

Radio signals are affected in many ways by objects in their path and by the media through which they travel. The properties of the path by which the radio signals will propagate govern the level and quality of the received signal. It is, therefore, very important to know the likely radio propagation characteristics that are likely to prevail.

The distances over which radio signals may propagate vary considerably. For some applications, only a short range may be needed. For example, a Wi-Fi link may only need to be established over a distance of a few metres. On the other hand, a short-wave broadcast station or a satellite link would need the signals to travel over much greater distances. Even for these two examples of the short-wave broadcast station and the satellite link, the radio propagation characteristics would be completely different–the signals reaching their final destinations having been affected in very different ways by the media through which the signals have travelled.

5.2.1 Radio Propagation Categories for Long-Distance Case

There are a number of categories into which different types of long-distance radio propagation mechanisms can be placed. These relate to the effects of the media through which the signals propagate.

Free space propagation Here, the radio signals travel in free space and away from other objects that influence the way in which they travel. If the wavelength is fixed, it is only the distance from the source that affects the way in which the field strength reduces.

This type of radio propagation is mainly encountered with signals travelling to and from satellites. This is rather a hypothetical case.

Ground wave propagation (below 2 MHz) When signals travel via the ground, they are modified by the ground or terrain over which they travel. They also tend to follow the earth's curvature. Signals heard on the medium waveband during the day use this form of propagation.

Ionospheric propagation or sky wave propagation (2 to 30 MHz) Here the radio signals are modified and influenced by the action of the free electrons in the upper reaches of the earth's atmosphere in the ionosphere. This form of radio propagation is used by stations on the short wavebands for their signals to be heard around the globe.

Tropospheric propagation Here the signals are influenced by the variations of refractive index in the troposphere just above the earth's surface. Tropospheric radio propagation is often the means by which the signals at VHF and above (maybe TV signals) are heard over extended distances.

Most of the above applications are broadcast applications. The AWGN is observed in these communications as it is due to temperature variations. Even sometimes the shadowing effect is also observed. All these propagation methods are described afterwards.

A special case is line-of-sight propagation above 30 MHz as shown in Fig. 5.3.

During line-of-sight transmission, the following significant impairments are observed:

- Attenuation and attenuation distortions
- Atmospheric absorption

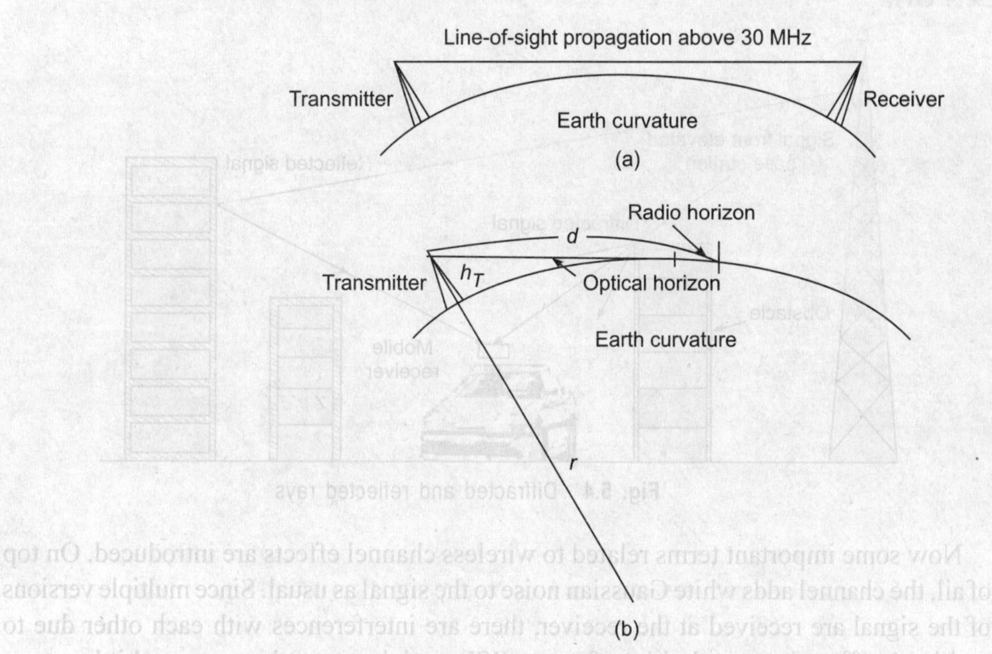

Fig. 5.3 (a) Line-of-sight propagation mode and (b) limitations on antenna height due to optical horizon and radio horizon (due to radio horizon slightly more coverage is achieved)

- Free space loss
- Noise
- Refraction
- Mulipath

In addition to these categories, many short-range radio or wireless systems have radio propagation scenarios that do not fit neatly into these categories because they may have indoor models also. However, for outdoor models, groundwave concepts can be correlated. Wi-Fi and cellular systems, for example, need to have their radio propagation models generated for office, or urban situations like for buildings, vegetations, vehicles, etc. Under these circumstances, the analysis approach is modified by multiple reflections, refractions, and diffractions with multipath (NLOS). Despite these complications, it is still possible to generate rough guidelines and models for these radio propagation scenarios.

5.2.2 Short-Distance Mobile Communication Case

In an ideal radio channel condition, the received signal would consist of only a single direct path signal, which would be a perfect reconstruction of the transmitted signal. However, in a real channel, the signal is modified during transmission through the channel. For NLOS multipath, the received signal consists of a combination of attenuated, reflected, refracted, and diffracted replicas of the transmitted signal. Consider the situation depicted in Fig. 5.4, where a receiver in a moving automobile receives a signal from a single transmitter, which has propagated along two paths. One propagation path is a direct path (maybe slightly diffracted) from the transmitter to mobile. The second path is due to a reflection from a building. A majority of such systems work on above 800 MHz, up to 2–11 GHz.

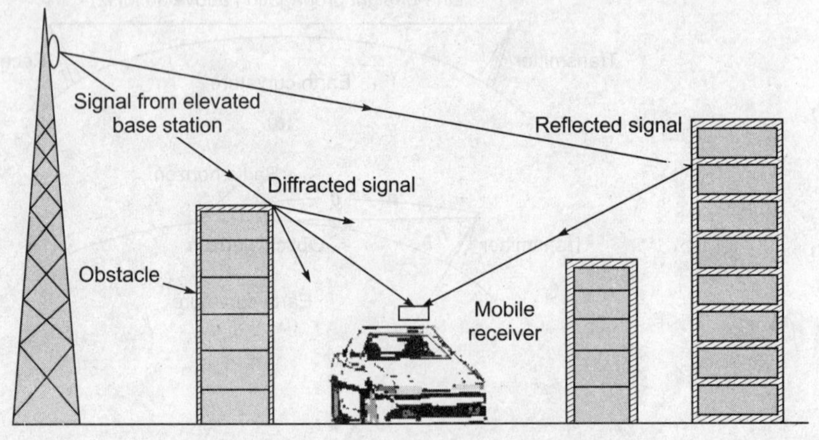

Fig. 5.4 Diffracted and reflected rays

Now some important terms related to wireless channel effects are introduced. On top of all, the channel adds white Gaussian noise to the signal as usual. Since multiple versions of the signal are received at the receiver, there are interferences with each other due to multipath effect, intersymbol interference (ISI) or delay spread occurs, and it becomes very hard to extract the original information without complicated *equalizers* (described in the next chapter) or some modulation schemes designed to combat ISI (Chapter 9). The common representation of the multipath channel is the *channel impulse response* (CIR) of the channel, which is the response at the receiver if a single impulse is transmitted. It can be calculated by *channel estimation* procedures, which is described in the next chapter. Mobility of users and hence mobile receivers or mobility of surrounding objects can cause a shift in the received carrier frequency, which is called *Doppler effect*. Figure 5.5 gives the typical scenario of the multipath/NLOS communication.

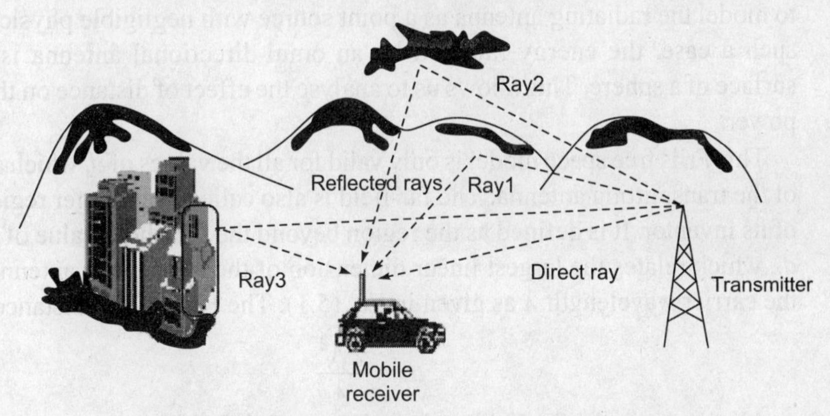

Fig. 5.5 Multipath effect between transmitter and receiver

5.3 FREE SPACE PROPAGATION MODEL

Free space is the space that does not interfere with the normal radiation and propagation of the radio waves. Thus, it does not have any magnetic or gravitational fields, solid bodies, and ionized particles. Apart from the fact that free space is unlikely to exist anywhere (due to ideal conditions), it certainly does not exist near the earth. The concept of free space is used to study the wave propagation in a simplified manner and then applying the same conditions to different categories of long-distance communication with actual scenario. Any power escaping into free space is governed by the characteristics of free space. If such power is released purposely, it is said to have been radiated and it then

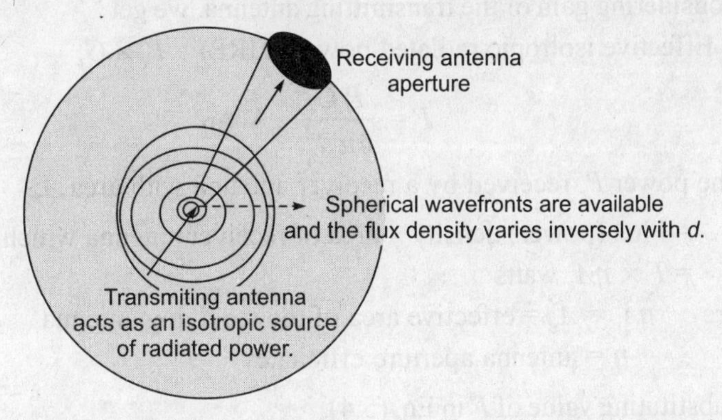

Fig. 5.6 Transmit antenna modelled as a point source (Transmit power is spread over the surface area of a hypothetical sphere at a distance d. The receiver antenna has an aperture A_r, illustrated over area $4\pi d^2$.)

form of an electromagnetic wave. This action can also be related to the term *power density*, which is radiated power per unit area. Power density reduces with the distance.

The free space propagation model assumes a transmit antenna and a receiver antenna to be located in an empty environment. Neither absorbing obstacles nor reflecting surfaces are there. In particular, the influence of the earth's surface is assumed to be entirely absent. For propagation distances d much larger than the antenna size, the far field of the electromagnetic wave dominates all other components, that is, we are allowed to model the radiating antenna as a point source with negligible physical dimensions. In such a case, the energy radiated by an omni-directional antenna is spread over the surface of a sphere. This allows us to analyse the effect of distance on the received signal power.

This Friis free space model is only valid for all the values of d, which are in the far-field of the transmitting antenna. The far-field is also called Fraunhofer region after the name of its inventor. It is defined as the region beyond the threshold value of far-field distance d_f, which relates the largest linear dimension of the transmitter antenna aperture A_t and the carrier wavelength λ as given in Eq. (5.1). The Fraunhofer distance is given by

$$d_f = \frac{2A_t^2}{\lambda} \tag{5.1}$$

Also, to be in the far-field region, $d_f >> A_t$ and $d_f >> \lambda$.

As with most large-scale radio wave propagation models, the free space model predicts that the received power decays as a function of the distance between transmitter and receiver raised to some power (a power law function). It is seen that the power density is inversely proportional to the square of distance from the source. This is the inverse square law, which applies universally to all forms of radiation in free space. Considering the isotropic source of transmitted power (which may be a half-wave dipole antenna), we get

Flux density at distance d,

$$F = \frac{P_t}{4\pi d^2} \text{ W/m}^2 \tag{5.2}$$

Considering gain of the transmitting antenna, we get

Effective isotropic radiated power (EIRP) = $P_t \times G_t$

or

$$F = \frac{P_t G_t}{4\pi d^2} \text{ W/m}^2 \tag{5.3}$$

The power P_r received by a receiver antenna with area A_r

= Flux (power) density × area of receiver antenna which receives flux density

= $F \times \eta A_r$ watts (5.4)

where $\eta A_r = A_e$ = effective area of the receiving antenna

η = antenna aperture efficiency

Substituting value of F in Eq. (5.4),

$$P_r = \frac{P_t G_t A_e}{4\pi d^2} \tag{5.5}$$

Thus, the received power is proportional to EIRP of the transmitting antenna and does not depend on the frequency.

Now, the receiving antenna gain,

$$G_r = \frac{4\pi A_e}{\lambda^2} \tag{5.6}$$

Using relationship of Eq. (5.6) and substituting for A_e, Eq. (5.5) will become

$$P_r = \frac{\lambda^2}{(4\pi d)^2} G_t P_t G_r \tag{5.7}$$

The wavelength λ (in metres) is

$$\frac{c}{f_c} = \frac{2\pi c}{\omega_c} \tag{5.8}$$

where c = velocity of light

f_c = carrier frequency in Hz

ω_c = angular frequency

Now, the path loss in general is defined as

$$P_{LdB} = 10 \log \frac{P_t}{P_r} \tag{5.9}$$

$$= 10 \log \left[\frac{(4\pi d)^2}{\lambda^2}\right] \quad \text{(for unity gain conditions)} \tag{5.10}$$

$$= -10 \log \frac{\lambda^2}{(4\pi d)^2} \tag{5.11}$$

If the wavelength is constant, loss simply depends upon the distance d between isotropic transmitter and receiver. Considering the real systems, according to Friis free space equation,

$$P_r = \frac{P_t G_t G_r \lambda^2}{(4\pi d)^2 L} \tag{5.12}$$

where L is the system loss factor, which is not related to propagation. L may be greater than or equal to 1. These miscellaneous losses are usually due to transmission line attenuation, filter losses, antenna losses, etc. $L = 1$ means no losses in the system hardware.

In practice, *effective radiated power* (ERP) is used instead of EIRP to denote the maximum radiated power. Also antenna gains are given in units of dBi (dB gain with respect to an isotropic antenna) or dBd (dB gain with respect to a half-wave dipole).

Example 5.1 A satellite link is established between an earth station and a satellite transponder for the RF frequency of 4 GHz. For the earth station transmitter, the transmitted power is 1 kW, and the transmitter and receiver antenna gains are 0 dB. The free space distance is 30,000 m. Find out the received power at the transponder. Assume $L = 1$.

Solution

$f_c = 4$ GHz $\Rightarrow \lambda = c/f = (3 \times 10^8 \text{ m/s})/(4 \times 10^9 \text{ Hz}) = 3/40$ metre

Now,
$$P_r = \frac{P_t G_t G_r \lambda^2}{(4\pi d)^2 L}$$

$$P_r = \frac{1000 \times 1 \times 1 \times (3/40)^2}{(4\pi\, 30000)^2}$$

$$= \frac{5.625}{142{,}122{,}303{,}375}$$

$$= 3.96 \times 10^{-11} \text{ W} = -104 \text{ dB}$$

5.3.1 20 log *d* Path Loss Law

As the propagation distance increases, the radiated energy is spread over the surface of a sphere of radius *d*. So, the power received decreases proportional to d^{-2}. For the unity gain antennas, the received power expressed in dB is

$$P_{r\text{dB}} = p_{t\text{dB}} - 20 \log \frac{d}{\lambda/4\pi} \tag{5.13}$$

This is called path loss law, where $p_{t\text{dB}}$ is the transmitted power in dB. The characteristic plot is given in Fig. 5.7. Furthermore, Friis equation does not hold for $d = 0$. This will be useful to understand large-scale propagation models using a close-in distance d_0 as a known received power reference point. The received power at any distance $d > d_0$ may be related to P_r at d_0 as follows:

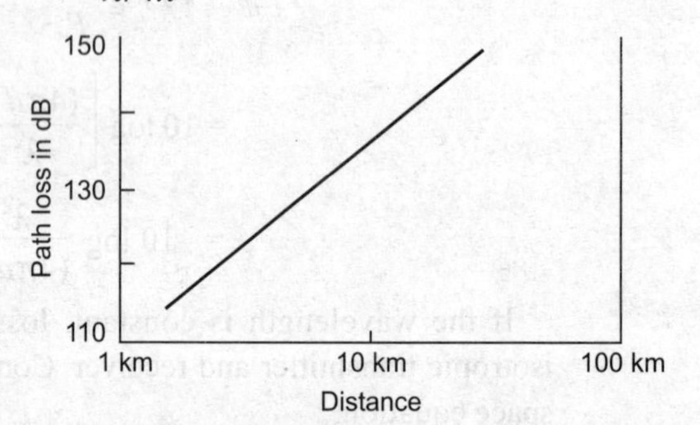

Fig. 5.7 Typical average path loss in dB versus distance

$$P_r(d) = P_r(d_0)\left(\frac{d_0}{d}\right)^2 \quad \text{for } d \ge d_0 \ge d_f \tag{5.14}$$

In mobile radio systems, it is common to find that P_r may have large variations of amplitude values over a typical coverage area of several square kilometers. As the dynamic range of the received power levels is large, often dBm or dBW units are used to express the received power levels. Equation (5.14) can be expressed in dBm or dBW units, simply by taking the logarithm of both sides and multiplying by 10. If P_r is represented in units of dBm, the received power is given by

$$P_r(d)\text{dBm} = 10 \log\left[\frac{P_r(d_0)}{0.001\text{W}}\right] + 20 \log\left(\frac{d_0}{d}\right) \quad \text{for } d \ge d_0 \ge d_f \tag{5.15}$$

Here, $P_r(d_0)$ is in units of watts.

The reference distance (d_0) for practical systems using low-gain antennas in the 1–1.5 GHz region is typically chosen to be 1 m for indoor environments and 100 m or 1 km for outdoor environments.

Path loss causes attenuation between the transmitter power amplifier and receiver front end. Some other effects are listed below, which should be considered during the link design:

- Losses in the antenna feeder (0–4 dB).
- Losses in transmit filters, particularly if the antenna radiates signal of multiple transmitters (0–3 dB) and antenna directionality gain (0–12 dB).
- Losses in duplex filter.
- Fade margins to anticipate for multipath (9–19 dB) and shadow losses (~5 dB).
- Penetration losses if the receiver is indoors, typically about 10 dB for 900 MHz signals.

5.3.2 Field Strength

While cellular telephone operators mostly calculate received power, broadcasters plan the coverage area of their transmitters using the (CCIR-recommended) *electric field strength E* (e-field) at the location of the receiver. The conversion formula is

$$E = \sqrt{\frac{120\pi P_r}{A_e}} \tag{5.16}$$

5.4 GROUND WAVE PROPAGATION

Ground wave propagation is particularly important on the LF and MF portions of the radio spectrum. It is used to provide relatively local coverage, especially by radio broadcast stations that require covering a particular locality. It can also be considered for land mobile telecommunication.

Ground wave radio signal propagation is ideal for relatively short-distance propagation on these frequencies during the daytime. Sky wave ionospheric propagation is not possible during the day because of the attenuation of the signals on these frequencies caused by the D region in the ionosphere. In view of this, stations need to rely on the ground wave propagation to achieve their coverage.

A ground wave signal is made up from a number of constituents. If the antennas are in the line-of-sight, then there will be a *direct wave* as well as a *reflected wave*. As the name suggests, the direct signal is one that travels directly between two antennas and is not affected by the locality. There will also be a reflected signal as the transmission will be reflected by a number of objects, including the earth's surface and any hills or large buildings. In addition to this, there is a *surface wave*. This tends to follow the curvature of the earth and enables coverage to be achieved beyond the horizon. It is the sum of all these components that is known as the ground wave.

Beyond the horizon, the direct and reflected waves are blocked by the curvature of the earth, and the signal is purely made up from the diffracted surface wave. It is for this reason that surface wave is commonly called ground wave propagation.

5.4.1 Surface Wave

The signal spreads out from the transmitter along the surface of the earth. Instead of just travelling in a straight line, the signals tend to follow the curvature of the earth as shown in Fig. 5.8. This is because currents are induced in the surface of the earth and this action slows down the wavefront in this region, causing the wavefront to tilt downwards towards the earth. With the wavefront tilted in this direction, it is able to curve around the earth and be received well beyond the horizon.

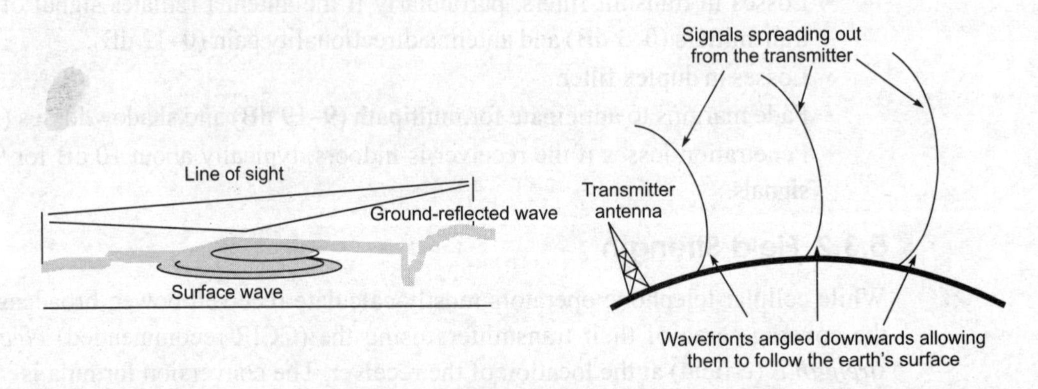

Fig. 5.8 Surface wave propagation

5.4.2 Effect of Frequency in Ground Wave Propagation

As the wavefront of the ground wave travels along the earth's surface, it is attenuated. The degree of attenuation is dependent upon a variety of factors. Frequency is one of the major determining factors as losses rise with increase in the frequency. As a result, it makes this form of propagation impracticable above the bottom end of the HF portion of the spectrum (3 MHz). Typically, a signal at 3.0 MHz will suffer an attenuation that may be in the range of 20 to 60 dB more than one at 0.5 MHz. In view of this, it can be seen why even high power HF broadcast stations may only be audible for a few miles from the transmitting site via the ground wave.

5.4.3 Effect of Ground

The surface wave is also very dependent upon the nature of the ground over which the signal travels. Ground conductivity, terrain roughness, and the dielectric constant, all affect the signal attenuation. In addition to this, the ground penetration varies, becoming greater at lower frequencies, and this means that it is not just the surface conductivity that is of interest. At the higher frequencies, the ground penetration is not of great importance, but at lower frequencies, the penetration means that the ground strata down to 100 metres may have an effect.

Despite all these variables, it is found that terrain with good conductivity gives the best result. Thus, soil type and the moisture content are of importance. Salty sea water is the best, and rich agricultural or marshy land is also good. Dry sandy terrain and city centres are by far the worst. This means that sea paths are optimum, although even these are subject to variations due to the roughness of the sea, resulting on path losses being slightly dependent upon the weather. It should also be noted that in view of the fact that signal penetration has an effect, the water table may have an effect, dependent upon the frequency in use.

5.4.4 Effect of Polarization

The type of antenna has a major effect. Vertical polarization is subject to considerably less attenuation than horizontally polarized signals. In some cases, the difference can be of several tens of decibels. It is for this reason that medium-wave broadcast stations use vertical antennas, even if they have to be made physically short by adding inductive loading.

At distances that are typically towards the edge of the ground wave coverage area, some sky wave signal may also be present, especially at night when the D layer attenuation is reduced. This may serve to reinforce or cancel the overall signal resulting in figures, which will differ from those that may be expected.

5.5 IONOSPHERIC PROPAGATION

Ionospheric conditions are already described. Figure 5.9 gives the summarized view of ionospheric conditions.

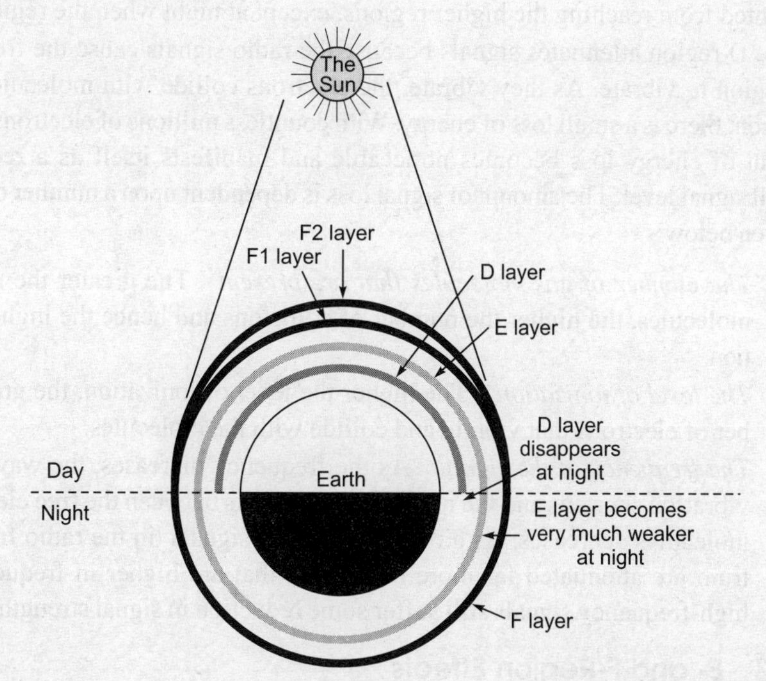

Fig. 5.9 A simplified view of the layers in the ionosphere over the period of a day

Radio signals in the medium and short wave bands travel by two basic means. The first is known as a ground wave (covered in the previous section) and the second a sky wave using the ionosphere. The following topic will describe sky wave propagation.

5.5.1 Sky Waves

When using ionospheric propagation, the radio signals leave the earth's surface and travel towards the ionosphere, where some of these are returned to the earth. These radio signals are termed sky waves for obvious reason. If they are returned to the earth, the ionosphere can simply be viewed as a vast reflecting surface encompassing the earth that enables the signals to travel over much greater distances than would otherwise be possible. Naturally, this is a great simplification because the frequency, time of day, and many other parameters govern the reflection or, more correctly, the refraction of signals back to the earth. There are, in fact, a number of layers or more correctly regions within the ionosphere, and these act in different ways as described below.

5.5.2 D-Region Effects

When a sky wave leaves the earth's surface and travels upwards, the first region of interest it reaches in the ionosphere is the D region. This region attenuates the signals as they pass through it. The level of attenuation depends on the frequency; low frequencies are attenuated more than the higher ones. In fact, it is found that the attenuation varies as the inverse square of the frequency, i.e. doubling the frequency reduces the level of attenuation by a factor of four. This means that low-frequency signals are often prevented from reaching the higher regions, except at night when the region disappears.

The D region attenuates signals because the radio signals cause the free electrons in the region to vibrate. As they vibrate, the electrons collide with molecules and at each collision, there is a small loss of energy. With countless millions of electrons vibrating, the amount of energy loss becomes noticeable and manifests itself as a reduction in the overall signal level. The amount of signal loss is dependent upon a number of factors, such as given below:

1. *The number of gas molecules that are present* The greater the number of gas molecules, the higher the number of collisions and hence the higher the attenuation.
2. *The level of ionization* The higher the level of ionization, the greater the number of electrons that vibrate and collide with the molecules.
3. *The frequency of the signal* As the frequency increases, the wavelength of the vibration shortens and the number of collisions between the free electrons and gas molecules decreases. As a result, the lower signals in the radio frequency spectrum are attenuated far more than those that are higher in frequency. Even so, high-frequency signals still suffer some reduction in signal strength.

5.5.3 E- and F-Region Effects

Once a signal passes through the D region, it travels on and reaches first the E region and next the F region. At the altitude where these regions are found, the air density is much lesser and this means that when the free electrons are excited by radio signals and vibrate, fewer collisions occur. As a result, the way in which these regions act is somewhat different. The electrons are again set in motion by the radio signal, but they tend to re-radiate it. As the signal is travelling in an area where the density of electrons is increasing, as it progresses further and further into the region, the signal gets refracted away from the area of higher electron density. In the case of HF signals, this refraction is often sufficient to bend them back to the earth. In effect, it appears that the region has 'reflected' the signal as shown in Fig. 5.10.

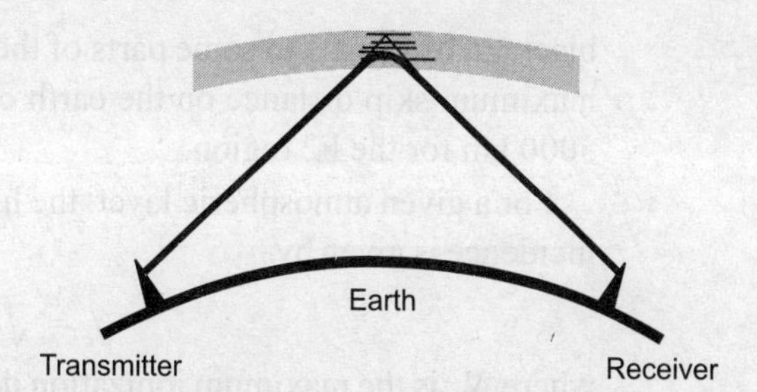

Fig. 5.10 Refractions of a radio signal as it enters an ionized region

The tendency for this reflection is dependent upon the frequency and the angle of incidence. As the frequency increases, it is found that the amount of refraction decreases until a frequency is reached where the signals pass through the region and onto the next. Eventually, a point is reached where the signal passes through all the regions and into outer space.

5.5.4 Effect of Frequencies

To gain a better idea of the characteristics of ionospheric propagation, it is worth viewing what happens to a radio signal if the frequency is increased across the frequency spectrum. Starting with a signal in the medium-wave broadcast band, during the daytime, signals on these frequencies can only propagate using the ground wave. Any signals of this frequency that reach the D region are absorbed. However, at night, as the D region disappears, signals reach the other regions and may be heard over much greater distances.

If the frequency of the signal is increased in the short-wave range, a point is reached where the signals start to penetrate the D region and reach the E region. Here, it is reflected and will pass back through the D region and return to the earth a considerable distance away from the transmitter. As the frequency is increased further, the signal is refracted less and less by the E region and eventually it passes right through. It then reaches the F1 region and here it may be reflected passing back through the D and E regions to reach the earth again. As the F1 region is higher than the E region, the distance reached will be greater than that for an E region reflection. Finally, as the frequency rises still further, the signal will eventually pass through the F1 region and onto the F2 region. This is the highest of the regions in the ionosphere and the distances reached using this are the greatest. The whole scenario is shown in Fig. 5.11. The sky wave communication suffers from a

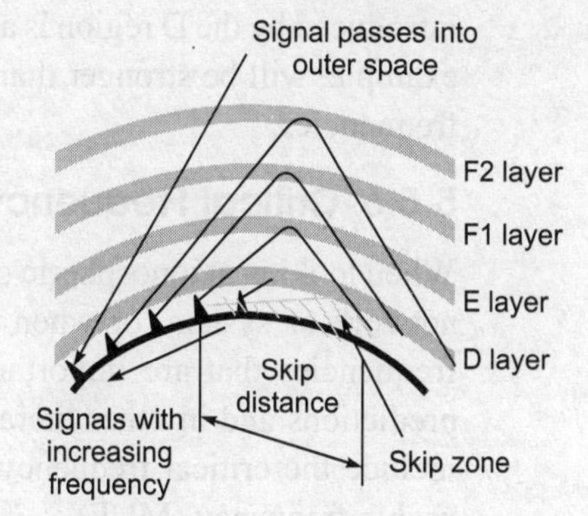

Fig. 5.11 Signals reflected by the E and F regions

blackout of signals to some parts of the earth known as *skip zones*. As a rough guide, the maximum skip distance on the earth curvature for the E region is around 2500 km and 5000 km for the F2 region.

For a given atmospheric layer, the highest frequency that is reflected back for vertical incidence is given by

$$f_c = \sqrt{81 N_m}$$

where N_m is the maximum ionization density in the layer.

5.5.5 Multiple Hops

In the ionospheric propagation, certain frequency signals get reflected towards the ground. Once the signals are returned to the earth from the ionosphere, they are reflected back upwards by the earth's surface, and again they are able to undergo another 'reflection' by the ionosphere. Naturally, the signal is reduced in strength at each reflection, and it is also found that different areas of the earth reflect radio signals differently. As might be anticipated, the surface of the sea is a very good reflector, whereas desert areas are very poor. As shown in Fig. 5.12, multiple hops of signal will be formed due to these multiple reflections.

It is not just the earth's surface that introduces losses into the signal path. In fact, the major cause of loss is the D region, even for frequencies high up into the HF range of the spectrum. One of the reasons for this is that the signal has to pass through the D region twice for every reflection by the ionosphere. This means that to get the best signal strengths, it is necessary to get the minimum number of hops to be used. This is generally achieved using frequencies close to the maximum frequencies that can support communications using ionospheric propagation and thereby using the highest regions in the ionosphere. In addition to this, the level of attenuation introduced by the D region is also reduced. This means that a radio signal on 20 MHz, for example, will be stronger than one on 10 MHz if propagation can be supported at both frequencies.

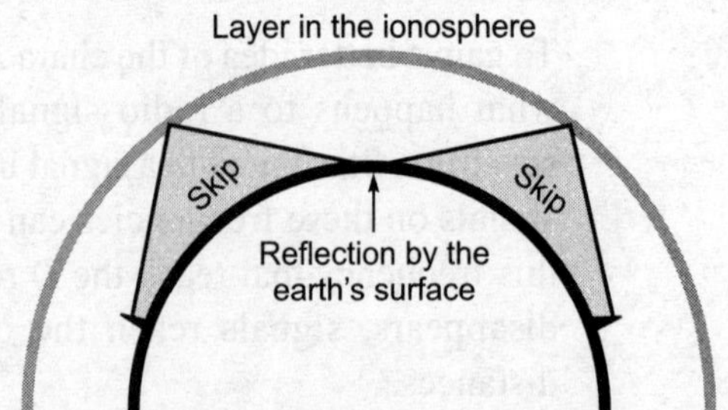

Fig. 5.12 Formation of multiple hops due to reflection by the earth's surface

5.5.6 Critical Frequency, MUF, and LUF

When looking at ionospheric or short-wave radio signal propagation for planning a radio network or system, or when predicting the propagation conditions, there are several frequencies that are important and are often mentioned in radio signal propagation predictions and in other literature associated with signal propagation. The frequencies include the critical frequency, the lowest usable frequency (LUF), and the maximum usable frequency (MUF).

Critical Frequency

It is important to give an indication of the state of the ionosphere. It is obtained by sending a signal pulse directly upwards. This is reflected back and can be received by a receiver on the same site as the transmitter. The pulse may be reflected back to the earth, and the time is measured to give an indication of the height of the layer. As the frequency is increased, a point is reached where the signal will pass right through the layer and onto the next one or onto outer space. The frequency at which this occurs is called the *critical frequency*. The equipment used to measure the critical frequency is called an *ionosonde*. In many respects, it resembles a small radar set, but for the HF bands. Using these sets, a plot of the reflections against frequency can be generated.

LUF

As the frequency of a transmission is reduced, further reflections from the ionosphere may be needed, and the losses from the D layer increase. These two effects mean that there is a frequency below which the communication between two stations will be lost. In fact, the *LUF* is defined as the frequency below which the signal falls below the minimum strength required for satisfactory reception.

From the above discussion, it can be seen that the LUF is dependent upon the stations at either end of the path. Their antennas, receivers, transmitter powers, the level of noise in the vicinity, and so forth, all affect the LUF. The type of modulation used also has an effect, because some types of modulation can be detected at lower strengths than others. In other words, the LUF is the practical limit below which the communication cannot be maintained between two particular stations.

If it is necessary to use a frequency below the LUF, then as a rough guide a gain of 10 dB must be made to decrease the LUF by 2 MHz. This can be achieved by methods like increasing the transmitter powers and improving the antennas.

MUF

When a signal is transmitted over a given path, there is the maximum frequency that can be used. The maximum frequency results from the fact that as the signal frequency increases, it will pass through more layers and eventually travel into outer space. As it passes through one layer, the communication may be lost because the signal then propagates over a greater distance than is required. Also when the signal passes through all the layers, the communication will be lost.

The frequency at which the communication just starts to fail is known as the *MUF*. It is generally three to five times the critical frequency, depending upon the layer being used and the angle of incidence. It is given by $f_c \sec \theta$, where f_c is the critical frequency and θ is the angle of incidence at the ionosphere measured with respect to the normal.

In general, the higher frequency is the better selection. When using the higher frequencies, it is necessary to ensure that the communications are still reliable. In view of the ever-changing state of the ionosphere, a general rule of thumb is to use a frequency that is about 20% below the MUF. This should ensure that the signal remains below the MUF despite the short-term changes. However, it should be remembered that the MUF will change significantly according to the time of day, and, therefore, it will be necessary to alter the frequency periodically to take account of this.

5.6 TROPOSPHERIC PROPAGATION

For frequencies at VHF and above, different modes of propagation prevail. Although some ionospheric modes may be experienced, the main effects are caused by changes in the troposphere. On frequencies above 30 MHz, it is found that the troposphere has an increasing effect on radio signals and they are able to travel over greater distances than would be suggested by line-of-sight calculations. Radio signals may be detected over distances of 500 or even 1000 km and more. This is normally done by a form of tropospheric enhancement, often called 'tropo' for short. At times, signals may even be trapped in an elevated duct in a form of radio signal propagation known as *tropospheric ducting* or *duct propagation*.

The way in which signals travel at frequencies of VHF and above is of great importance for those looking at radio coverage of systems like cellular telecommunications, mobile radio, and other wireless systems as well as other users, including radio hams. It might be thought that most radio communications at VHF and above follow a line-of-sight path. This is not strictly true and it is found that even under normal conditions, radio signals are able to travel or propagate over distances that are greater than the line of sight.

The reason for increase in the distance travelled by the radio signals is that they are refracted by small changes that exist in the earth's atmosphere close to the ground. It is found that the refractive index of the air close to the ground is very slightly higher than that of the air higher up. As a result, the radio signals are bent towards the area of higher refractive index, which is closer to the ground. It thereby extends the range of the radio signals.

The refractive index of the atmosphere varies according to a variety of factors. Temperature, atmospheric pressure, and water vapour pressure, all influence its value. Even small changes in these variables can make a significant difference because radio signals can be refracted over whole of the signal path and this may extend for many kilometres.

5.6.1 Enhanced Conditions for Tropospheric Propagation

Under certain conditions, the radio propagation conditions provided by the troposphere are such that signals travel over even greater distances. This form of 'lift' in conditions is less pronounced in the lower portions of the VHF spectrum but is more apparent on some of the higher frequencies. Under some conditions, radio signals may be travelling over distances of 2000 to 3000 kilometres. This can give rise to significant levels of interference for periods of time. These extended distances result from much greater changes in the values of refractive index over the signal path. This enables the signal to achieve a greater degree of bending and, as a result, to follow the curvature of the earth over greater distances.

It is also possible for tropospheric ducts to occur above the earth's surface. These elevated tropospheric ducts occur when a mass of air with a high refractive index has a mass of air with a lower refractive index underneath and above it as a result of the movement of air that can occur under some conditions. When these conditions occur, the signals may be confined within the elevated area of air with the high refractive index and they cannot escape and return to the earth. As a result, they may travel for several hundred miles and receive comparatively low levels of attenuation. They may also not be audible to stations underneath the duct and, in this way, create a skip or dead zone similar to that experienced with HF ionospheric propagation.

There is a strong link between weather conditions and radio propagation conditions and coverage. Normally, the temperature of the air closest to the earth's surface is higher than that at a greater altitude. This effect tends to reduce the air density gradient (and hence the refractive index gradient) as air with a higher temperature is less dense. However, under some circumstances, what is termed as a temperature inversion occurs. This happens when the hot air close to the earth rises allowing the colder denser air to come close to the earth. When this occurs, it gives rise to a greater change in refractive index with height.

When signals are propagated over extended distances as a result of enhanced tropospheric propagation conditions, the signals are normally subject to slow deep fading. This is caused by the fact that the signals are received via a number of different paths. As the winds in the atmosphere move the air around, the different paths will change over a period of time. Accordingly, the signals appearing at the receiver will fall in and out of phase with each other as a result of the different and changing path lengths and, as a result, the strength of the overall received signal will change.

5.6.2 Troposcatter Propagation

One useful form of communications for applications where path lengths of around 800 km are needed is *tropospheric scatter* or *troposcatter*. It is a reliable form of communications that can be used regardless of the prevailing tropospheric conditions. Although reliable, when using troposcatter, the signal strengths are normally very low. Accordingly, troposcatter links require high powers, high antenna gains, and sensitive receivers.

Troposcatter is often used for commercial applications, normally on frequencies above 500 MHz for over the horizon links. It is ideal for remote telemetry or other links where low- to medium-rate data needs to be carried. Where viable, troposcatter provides a means of communication that is much cheaper than using satellites. The whole scenario is shown in Fig. 5.13.

As the name implies, troposcatter uses the troposphere as the region that affects the radio signals being transmitted, returning them to the earth so that they can be received by the distant receiver. Troposcatter relies on the fact that there are areas of slightly different dielectric constant in the atmosphere at an altitude of between 2 and 5 kilometres. Even dust in the atmosphere at these heights adds to the reflection of the signal. A transmitter launches a high-power signal, most of which passes through the

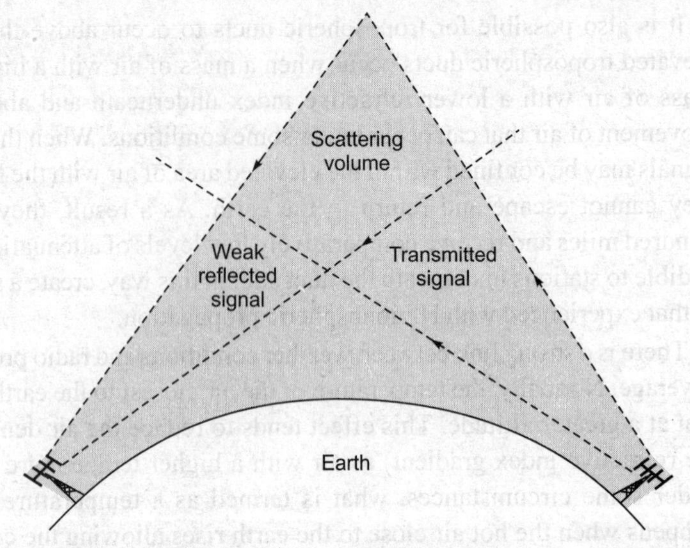

Fig. 5.13 Principle of troposcatter

atmosphere into outer space. However, a small amount is scattered when it passes through this area of the troposphere and passes back to the earth at a distant point. As might be expected, only a small amount of the signal is scattered back to the earth and, as a result, path losses are very high. Additionally, the angles through which signals can be reflected are normally small.

The area within which the scattering takes place is called the *scatter volume* and its size is dependent upon the gain of the antennas used at either end. In view of the fact that scattering takes place over a large volume, the received signal will have travelled over a vast number of individual paths, each with a slightly different path length. As they all take a slightly different time to reach the receiver, this has the effect of blurring the overall received signal and this makes high-speed data transmissions difficult.

It is also found that there are large short-term variations in the signal as a result of turbulence and changes in the scatter volume. As a result, commercial troposcatter propagation systems use multiple diversity systems. This is achieved by using vertically and horizontally polarized antennas as well as different scatter volumes (angle diversity) and different frequencies (frequency diversity). Control of these systems is normally undertaken by computers. In this way, troposcatter systems can run automatically giving high degrees of reliability.

5.7 CHANNEL NOISE AND LOSSES

Both channel noise and losses are unavoidable parameters of any communication system. Free space losses are calculated previously in the chapter. A few others are described as follows. Here a specific case of macrocellular environment/ground wave propagation is considered for the calculation of losses. Similar approach can be developed for the calculation of losses for other types of propagation. The AWGN is always present in any type of channel model.

5.7.1 AWGN

In reality, transmission is always corrupted by noise whatever may be the type of channel assumed. The simplest mathematical model of the radio channel is the additive white Gaussian noise (AWGN) channel. It is a very good model for the physical reality as long as the thermal noise at the receiver is the only source of disturbance. Nevertheless, because of its simplicity, it is often used to model human-made noise or multi-user interference. The AWGN channel model can be characterized as follows.

- The noise is *additive*. The received signal equals the transmit signal plus some noise, where the noise is statistically independent of the signal. The noise $w(t)$ is an additive random disturbance in the useful signal $s(t)$, i.e. the received signal is given by

$$r(t) = s(t) + w(t) \tag{5.17}$$

- The noise is *white*. The power spectral density is flat or constant power spectral density. So, the autocorrelation of the noise in time domain is zero for any non-zero time offset. The one-sided PSD is usually denoted by N_0. Thus, $N_0/2$ is the two-sided PSD and WN_0 is the noise inside the noise bandwidth W [see Fig. 5.14 (a)]. For thermal resistor noise, $N_0 = kT_0$, where k is the Boltzmann constant ($\approx 1.38 \times 10^{-23}$J/K) and T_0 is the absolute temperature. The unit of N_0 is W/Hz, which is the same as the unit J for the energy. Usually, N_0 is written as dBm/Hz.

  For $T_0 = 290$ K, $\quad N_0 \approx -174$ dBm/Hz

  However, this is only the ideal physical limit for an ideal receiver. In practice, some decibels according to the so-called *noise figure* (which is related to the antenna temperature also) have to be added. Typically, N_0 will be a value slightly above –170 dBm/Hz.

- The noise samples have a *Gaussian distribution*. The Gaussian probability density function with variance σ^2 is given by

$$p(x) = \frac{1}{\sqrt{2\pi\sigma^2}} e^{-(x-m)^2/2\sigma^2} \tag{5.18}$$

The noise is a stationary and zero-mean Gaussian random process, which means that the output of every (linear) noise measurement is a zero-mean Gaussian random variable that does not depend on the time instant when the measurement is done [Fig. 5.14 (c)].

One must keep in mind that the AWGN model is a mathematical function, because it implies that the total power (i.e. the PSD integrated over all frequencies) is infinite. Thus, a time sample of the white noise has infinite average power, which is certainly not a physically reasonable property. It is known from statistical physics that the thermal noise density decreases exponentially at very high frequencies. But to understand the physical situation in communications engineering, it is better to keep in mind that every receiver limits the bandwidth as well as every physical noise measurement. So, it makes sense to think of the noise process to be white, but it cannot be sampled directly without an input device. Each input device filters the noise and leads to a finite power. Instead of two-

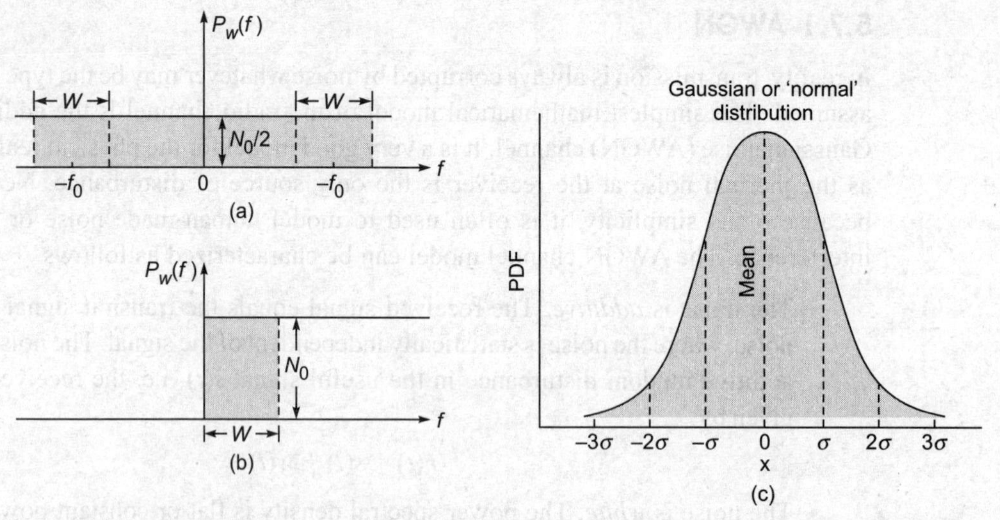

Fig. 5.14 Noise represented with (a) two-sided PSD, (b) equivalent one-sided PSD, and (c) typical Gaussian distribution with zero mean and σ standard deviation

sided PSD, if noise over W is represented with equivalent single band, it will look like Fig. 5.14(b).

The mean of the multiplication of two noise samples (autocorrelation) is

$$E\{w(t_1)w(t_2)\} = \frac{N_0}{2}\,\delta(t_1 - t_2) \tag{5.19}$$

where $\delta(t)$ represents the unit impulse at a time t.

5.7.2 Ground Reflection Loss

There are three components of a ground wave—line-of-sight wave, ground reflected wave, and surface wave (Fig. 5.15). The surface wave has negligible amplitudes. The relative amplitude of the surface wave is very small for most cases of mobile UHF communication (<< 1). Its contribution is relevant only at a few wavelengths above the ground. So, mostly a *two-ray model* is considered as shown in Fig. 5.16. [This model has been found to be reasonably accurate for calculation of large-scale signal strength over distances of several kilometres, especially for mobile radio systems using tall antenna towers (heights > 50 m) as well as for line-of-sight microcellular channels for urban area.]

If we consider the effect of the earth's surface, the expressions for the received signal become more complicated than in case of free space propagation. The main effect is that

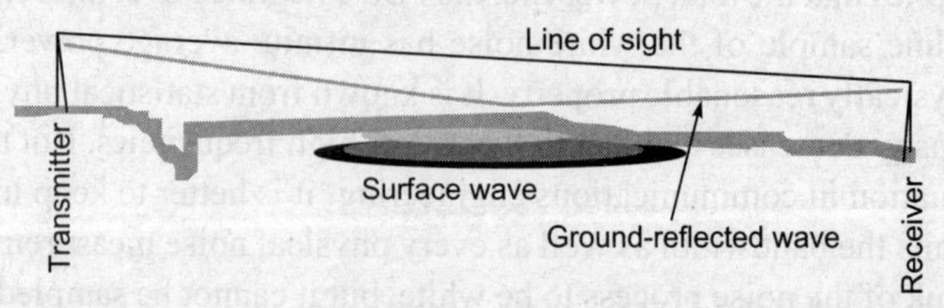

Fig. 5.15 Three components of the ground wave propagation (microwave line of sight)

signals reflected off the earth's surface may (partially) cancel the line-of-sight wave. As per Fig. 5.16,

- Transmit antenna height = h_T
- Receive antenna height = h_R
- Path length of line-of-sight wave = d
- Reflection component = R

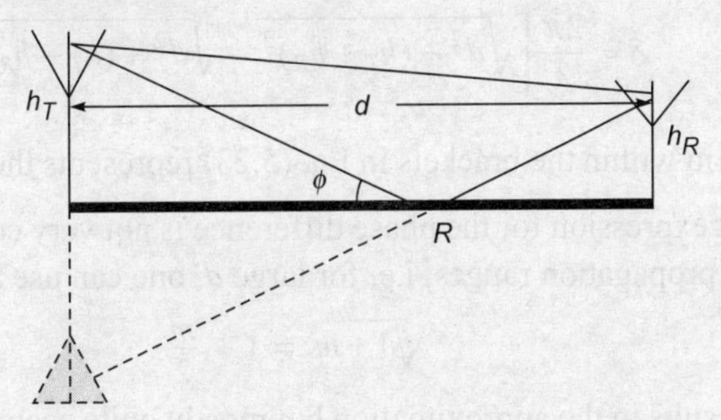

Fig. 5.16 Two-ray model for UHF propagation over a plane reflecting earth

(The length can be found by taking the mirror image of the transmitting antenna and extending the line to it from the point of reflection.) For isotropic antennas above a plane earth, the received electric field strength E is

$$E = E_0(1 + R_c e^{j\delta} + (1 - R_c) \cdot e^{j\delta} + \cdots) \tag{5.20}$$

where R_c = ground reflection coefficient
E_0 = field strength for propagation in free space
δ = phase difference between direct and reflected rays

This expression can be interpreted as the complex sum of a direct line-of-sight wave, a ground-reflected wave, and a surface wave. The phasor sum of the first and second terms is known as the *space wave*. Referring to Fig. 5.16, for a horizontally polarized wave incident on the surface of a perfectly smooth earth,

$$R_c = \frac{\sin\phi - \sqrt{(\varepsilon_r - jx) - \cos^2\phi}}{\sin\phi + \sqrt{(\varepsilon_r - jx) - \cos^2\phi}} \tag{5.21}$$

where ε_r is the relative dielectric constant of the earth, ϕ is the angle of incidence between the radio ray and the earth's surface, and

$$x = \sigma_g/(\omega_c\varepsilon_0) \approx 18 \times 10^9\, \sigma_g/f_c$$

with σ_g being the conductivity of the ground, ω_c the angular carrier frequency, and ε_0 the dielectric constant of vacuum.

For vertical polarization,

$$R_c = \frac{(\varepsilon_r - jx)\sin\phi - \sqrt{(\varepsilon_r - jx) - \cos^2\phi}}{(\varepsilon_r - jx)\sin\phi + \sqrt{(\varepsilon_r - jx) - \cos^2\phi}} \tag{5.22}$$

The value of δ in Eq. (5.20) can be found as follows.

A line-of-sight wave and a ground-reflected wave arrive at the receive antenna. These two rays are out of phase due to differences in path length and due to phase shifts at the reflection area. The phase difference between the direct and the ground-reflected wave can be found from the two-ray approximation considering only a line of sight and a ground reflection. Denoting the transmit and receive antenna heights as h_T and h_R, respectively, the phase difference can be expressed as

$$\delta = \frac{2\pi}{\lambda}\left[\sqrt{d^2 + (h_T + h_R)^2} - \sqrt{d^2 + (h_T - h_R)^2}\right] \tag{5.23}$$

The term within the brackets in Eq. (5.23) represents the path difference $\Delta \cong \frac{2h_T h_R}{d}$. The exact expression for the phase difference is not very convenient for further analysis. For large propagation ranges, i.e. for large d, one can use an approximation based on

$$\sqrt{1+m} \cong 1 + \frac{m}{2} \tag{5.24}$$

This results in the approximation but mostly quite accurate phase difference.

$$\delta \cong \frac{4\pi}{\lambda}\frac{h_T h_R}{d} \tag{5.25}$$

This equation and an approximation for the reflection coefficient give an often used two-ray model for plane earth propagation. For large d ($d >> 5h_T h_R$), the reflection coefficient tends to –1. So, the received signal power becomes

$$P_r = \frac{\lambda^2}{(4\pi d)^2}\left[2\sin\frac{2\pi}{\lambda}\frac{h_T h_R}{d}\right]^2 G_t P_t G_r \tag{5.26}$$

if

$$d = \frac{4}{\lambda} h_T h_R \tag{5.27}$$

then the distance acts as a turnover point. For this distance, the argument of the sine becomes equal to $\pi/4$. It must be noted that as d becomes large, the path difference Δ becomes very small and the amplitudes of the line of sight as well as ground-reflected signals become virtually nearly identical and differ only in phase.

Example 5.2 For carrier frequencies around 1 GHz, the base station height is 30 metre and the mobile antenna height is 2 metre. Find out the turnover distance when the signal is transmitted from the base station to the mobile.

Solution

$h_T = 30$ metre

$h_R = 2$ metre

Wavelength, $\lambda = c/f = (3 \times 10^8 \text{ m/sec})/(1 \times 10^9 \text{ Hz})$

$d = (4/3) \times 10 \times 30 \times 2$

$= 800$ metre

For 900 MHz, the wavelength is about 720 metre. Hence, for macrocellular systems, the distances beyond the turnover distance are more relevant.

The full path loss expression shows an interference pattern of the line of sight and the ground-reflected waves for relatively short ranges and a rapid decay of the signal power beyond the turnover distance. For propagation distances substantially beyond the turnover point, the path loss tends to the fourth power distance law as per Egli's model[1]:

$$P_r \to \frac{(h_T h_R)^2}{d^4} P_t G_t G_r \tag{5.28}$$

By experiments, it is confirmed that in macrocellular links over smooth, plane terrain, the received signal power (expressed in dB) decreases with 40 log d. Also, a 6 dB/octave height gain is experienced—doubling the antenna height increases the received power by a factor of 4. These two effects are correctly predicted by the two-ray model.

However, in contrast to the theoretical plane earth loss, Egli, in 1957, measured a significant increase of the path loss with the carrier frequency f_c for ranges $1 < d < 50$ km. He proposed the semi-empirical model:

$$P_r = \left(\frac{40\text{ MHz}}{f_c}\right)^2 \frac{(h_T h_R)^2}{d^4} P_t G_t G_r \tag{5.29}$$

i.e. he introduced a frequency-dependent empirical correction $(40\text{ MHz}/f_c)^2$ for carrier frequencies 30 MHz $< f_c <$ 1 GHz.

For communication at short range, this formula loses its accuracy because the reflection coefficient is not necessarily close to –1. For $d << 4h_T h_R/\lambda$, free space propagation is more appropriate, but a number of significant reflections against the earth's surface must be anticipated. However, these are often hard to distinguish from reflections against obstacles in the vicinity of the antenna site. Moreover, in streets with high buildings, guided propagation may occur.

5.7.3 Diffraction Loss

If the direct line-of-sight is obstructed by a single knife-edge type of obstacle with height h_m (Fig. 5.17), then the diffraction parameter v can be defined as follows:

$$v = h_m \left(\sqrt{\frac{2}{\lambda}\left(\frac{1}{d_T} + \frac{1}{d_R}\right)}\right) \tag{5.30}$$

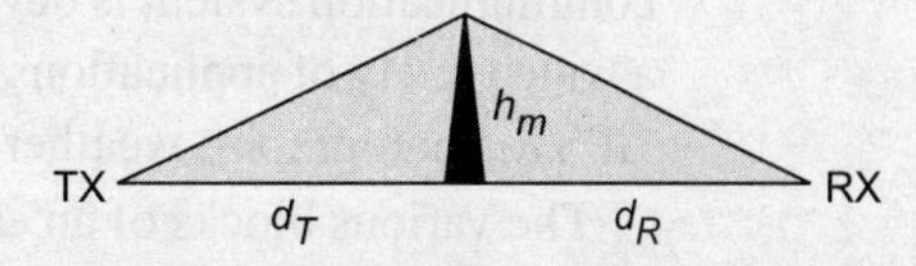

Fig. 5.17 Path profile model for (single) knife-edge diffraction

where d_T and d_R are the terminal distances from the knife-edge. The diffraction loss, additional to the free space loss and expressed in dB, can be closely approximated by

[1] Egli's model is an implementation of John Egli's methodology published in October 1957 in the Proceedings of the IRE. This model is derived from measured data to a distance of approximately 50 miles and over gently rolling terrain with average hill heights of approximately 50 feet. It relies on frequency, distance, and heights of transmitting and receiving antennas.

$$P_{\text{diffr dB}} = 0 \quad \text{if } v < 0,$$
$$P_{\text{diffr dB}} = 6 + 9v + 1.27v^2 \quad \text{if } 0 < v < 2.4$$
$$P_{\text{diffr dB}} = 13 + 20 \log v \quad \text{if } v > 2.4$$

The attenuation over rounded obstacles is usually higher than $P_{\text{diffr dB}}$ in the above formula.

5.7.4 Total Path Loss

The previously presented methods for ground reflection loss and diffraction losses suggest an interpretation of the path profile: Obstacles occur as straight vertical lines while horizontal planes cause reflections, that is the propagation path is seen as a collection of horizontal and vertical elements. Accurate computation of the path loss over non-line-of-sight paths with ground reflections is a complicated task and does not allow such simplifications.

Many measurements of propagation losses for paths with combined diffraction and ground reflection losses indicate that knife-edge type of obstacles significantly reduce ground wave losses. Blomquist suggested two methods to find the total loss:

$$P_{\text{total dB}} = P_{\text{fs dB}} + \sqrt{P_{\text{gr dB}}^2 + P_{\text{diffr dB}}^2} \tag{5.31}$$

and the empirical formula

$$P_{\text{total dB}} = P_{\text{fsdB}} + P_{\text{grdB}} + P_{\text{difrdB}} \tag{5.32}$$

where $P_{\text{fs dB}}$ is the free space loss, $P_{\text{gr dB}}$ is the ground reflection loss, and $P_{\text{diffr dB}}$ is the multiple knife-edge diffraction loss in dB values.

5.8 SATELLITE LINK

A satellite communication link consists of a transmitting earth station, a transponder, and a receiving earth station. Corresponding to uplink for one earth station there is a downlink for another and vice versa. The free space propagation model can be directly applied to a satellite link and can also be used for its budgetary analysis. Satellite communication is of two types— geostationary and orbital. Again orbits may be low, medium, or high. Satellite communication system is developed based on Kepler's three laws. Satellites are used for a wide variety of applications from satellite TV broadcasting and navigation in the case of GPS to photography, weather monitoring, and many more applications.

The various blocks of an earth station are shown in Fig. 5.18. As the earth stations are capable of transmitting signals at a high power level, they can withstand the path losses. So, uplinks are at higher frequency than the downlinks. The transponder reproduces the signal and sends it back to the earth with a down-conversion. At the transponder, solar cells are used for power generation and hence the downlink is at lower frequency in order to reduce frequency-dependent path losses.

The *circulator* is used to make sure that the transmit signals go out through the dish and not back into the receive chain. It also makes sure that the received signals come from the dish into the receiver chain and not into the transmitter chain. It works much like a roundabout in principle. This is often referred to as an orthomode transducer (OMT) and is built into the feed assembly.

The Rx or *receive filter* is usually a waveguide filter that tightly controls the frequencies allowed into the receiver chain. This has the effect of reducing the unwanted noise from space and preventing interference from outside of the receiver band of frequencies.

The Tx or *transmit filter* is usually a waveguide filter that tightly controls the frequencies allowed into antenna. This has the effect of reducing the unwanted signals being accidentally transmitted onto the satellite and prevent interference from outside of the transmit band of frequencies.

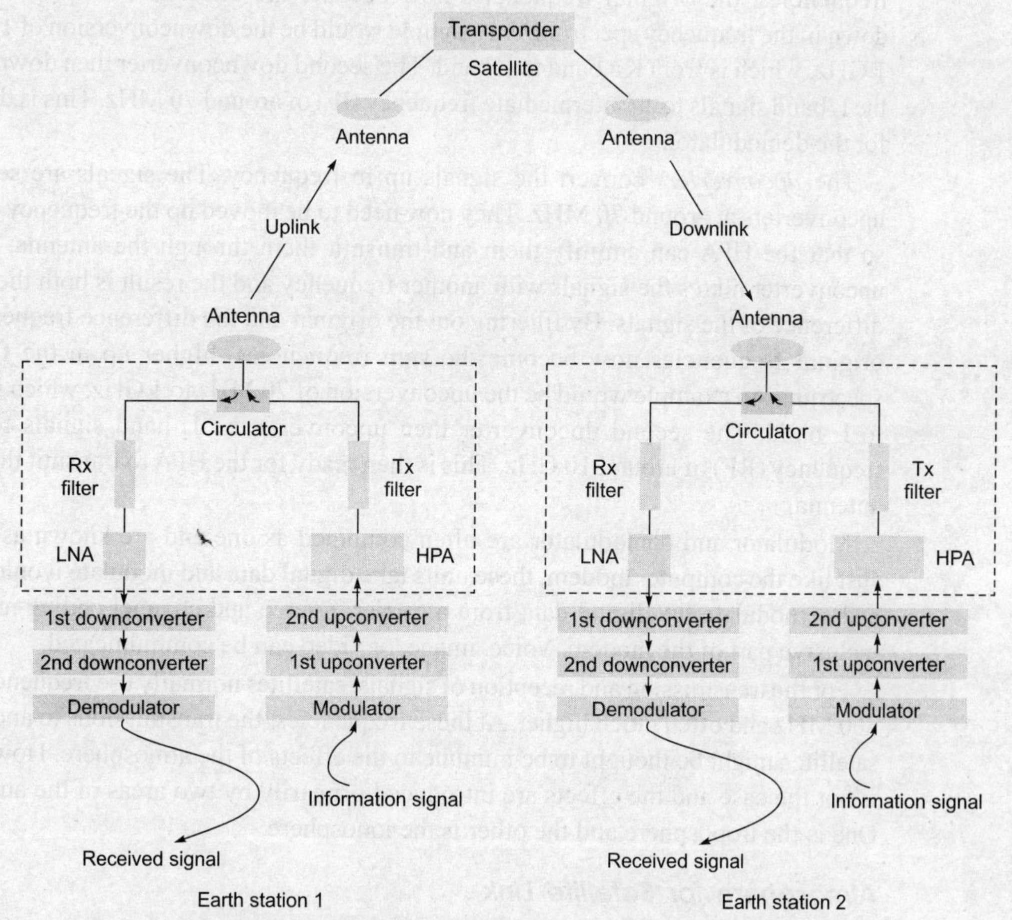

Fig. 5.18 A typical bidirectional satellite link for video communication

The low noise amplifier (LNA), sometimes known as an LNB on the receive-only terminals, is a very good amplifier that has the job of amplifying the small signals picked up by the antenna without amplifying the noise. Various kinds of LNAs exist but they all do the same thing—provide enough signal level to demodulate the data from the carrier.

The high power amplifier (HPA), otherwise known as a TWTA (travelling wave tube amplifier) or an SSHPA (solid state high power amplifier), has one job. It amplifies a specific band of frequencies by a large amount, sufficiently large to enable the antenna to beam them up to the satellite. These can range in power from a few watts up to over 1000 watts. The largest HPA has to be cooled using liquid nitrogen and resembles electron microscopes. The smallest HPA looks more like a lump of metal bolted to a small heat sink.

The *downconverters* convert signals down in frequency. The signals arrive at the dish at anything from 10 to 40 GHz and are then filtered and amplified. They now need to be moved down the frequency spectrum so that the equipment can be made cheaper and easier. The first downconverter mixes the signals with another frequency and the result is both the sum and difference of the signals. By filtering out the original and the sum frequencies, the original frequencies now become the difference frequencies, lower down in the frequency spectrum. An example would be the downconversion of 10 GHz to 1 GHz, which is from Ku band to L band. The second downconverter then downconverts the L band signals to an intermediate frequency (IF) of around 70 MHz. This is then ready for the demodulator.

The *upconverters* convert the signals up in frequency. The signals are sent to the upconverters at around 70 MHz. They now need to be moved up the frequency spectrum so that the HPA can amplify them and transmit them through the antenna. The first upconverter mixes the signals with another frequency and the result is both the sum and difference of the signals. By filtering out the original and the difference frequencies, the original frequencies now become the sum frequencies, higher up in the frequency spectrum. An example would be the upconversion of 70 MHz to 1 GHz, which is from IF to L band. The second upconverter then upconverts the L band signals to a radio frequency (RF) of around 10 GHz. This is then ready for the HPA to transmit through the antenna.

Modulator and demodulator are often combined as one and are known as *modems*. Just like the computer modem, these units take digital data and modulate it onto a carrier and demodulate the digital data from a carrier. Source and channel coding may be the inclusive part of the modem. Voice, image, or video can be communicated.

For the transmission and reception of signals, satellites normally use frequencies above 500 MHz and often much higher. At these frequencies, the transmissions to and from the satellites might be thought to be immune to the effects of the atmosphere. However, this is not the case and the effects are introduced primarily by two areas of the atmosphere. One is the troposphere and the other is the ionosphere.

Atmosphere for Satellite Link

The atmosphere can be divided into several areas. It is found that the temperature varies according to the height. Initially, the temperature falls until altitudes of around 10 km are reached. At this point, the temperature is around –50 or –60°C. It is around this point that the temperature starts rising again. The region below this inflexion point is the troposphere.

The second area that affects radio signals is the ionosphere. This is a region of the atmosphere that starts at altitudes of around 50 km and extends to more than 400 km. Beyond the ionosphere, the signals can be considered to be in free space. The region between the upper reaches of the troposphere and the ionosphere is often termed *inner free space*. This region too has a little effect. The troposphere and ionosphere have refractive indices that differ from unity. The refractive index of troposphere is greater than unity and that of ionosphere is less than unity. As a result, refraction and absorption occur.

A further effect that is introduced by the ionosphere is known as *Faraday rotation*, which results from the fact that the ionosphere is a magneto-ionic region. The Faraday rotation of a signal causes different elements of a signal to travel in different ways, particularly rotating the plane of polarization. This can create some problems with reception.

Another effect introduced by the ionosphere (already described previously) is *ionospheric scintillations*. These scintillations manifest themselves as a variety of variations of the amplitude, phase, and polarization angle. They can also change the angle of arrival of the signals. These variations change over a period of between one to fifteen seconds, and they can affect signals well into the microwave region. The variations are caused primarily by the variations in electron density arising in the E region, often as a result of sporadic E, as well as in the F layer, where a spreading effect is the cause. The level of scintillation is dependent upon a number of factors, including the location of the earth station and the state of the ionosphere as a result of the location, the sunspot cycle, the level of geomagnetic activity, latitude, and local time of day. The scintillations are more intense in equatorial regions, falling with increasing latitude away from the equator but then rising at high latitudes. The effects are also found to decrease with increasing frequency and generally not noticeable above frequencies of 1–2 GHz. As such, they are not applicable to many direct broadcast television signals, although they may affect GPS and some communication satellites.

There are a number of effects that the troposphere introduces, including signal bending as a result of refraction, scintillation, and attenuation. The signal refraction in the troposphere is in the opposite sense to that in the ionosphere. This is because the refractive index in the troposphere is greater than unity and is also frequency independent. The signal refraction gives them a greater range than would be expected as a result of the direct geometric line of sight. Tropospheric ducting and extended range effects that are experienced by terrestrial VHF and UHF communications may also be experienced when low angles of elevation are used.

The scintillations induced by the troposphere are often greater than those seen as a result of the ionosphere. They occur as a result of the turbulence in the atmosphere where areas of differing refractive index move around as a result of the wind or convection currents. The degree to which the scintillations occur is dependent upon the angle of inclination, and above angles of around 15 degrees, the effect can normally be ignored. At the angles within 5 to 10 degrees, the changes can often be around 6 dB at frequencies of around 5–6 GHz.

Frequency changes as a result of the Doppler shift (explained in Section 5.9.2) principle may be in evidence with signals from some satellites. Satellites in low earth orbits move very quickly and, as a result, a Doppler frequency shift is apparent in many cases. With the satellite moving towards the earth station, the frequency appears higher than nominal and then as it moves away, the apparent frequency falls. The degree of shift is dependent upon a number of factors, including the speed of the satellite (more correctly, its speed relative to the earth station) and the frequencies in use. Shifts of the order of 10 kHz may be experienced. As most satellites operate in a cross-mode configuration, the Doppler shift is not just applicable to the band on which the signal is received but also to the cumulative effect of the uplink and downlink transmissions. In many instances, the effects will subtract because of the way the satellite mixing process is configured. Such effects can even be observed in mobile satellite communication.

5.9 MULTIPATH EFFECT/FADING IN LAND MOBILE SYSTEMS

Because there are obstacles and reflectors in the wireless propagation channel, the transmitted signal arrives at the receiver from various directions over a multiplicity of paths. Such a phenomenon is called *multipath*. The multipath effect is mostly there in wireless communications but depends upon the distance and surroundings. (Some of the issues were discussed previously.) However, specifically for very short distance communication systems like indoor/outdoor communication and urban area systems, multipath is a serious problem because of the dynamic environment. It is an unpredictable set of reflections and/or direct waves, each with its own degree of attenuation and delay as shown in Fig. 5.4.

The multipath will cause amplitude and phase fluctuations and time delay in the received signals. We can use diverse schemes to combat multipath. The results of the multipath effect are also discussed in Chapter 15, where the basic concepts for MIMO are described showing that MIMO exploits multipath.

The major effect of multipath is fading.

The communications quality between a base station transmitter and a mobile (or stationary) receiver depends on a number of factors, including the general quality of the propagation channel through which the signal passes. In wireless applications, especially cellular communications, the propagation path is terrestrial air with a significant number of human-made and natural objects that get in the way. As the transmitted signal gets absorbed by the atmosphere and reflected off buildings and trees, it experiences fluctuations in its amplitude as well as phase. Wireless channels, in short, exhibit highly irregular amplitude or signal strength behaviour. It is the time variation of the received signal power caused by changes in the transmission medium or path. This phenomenon is generally referred to as *fading*. The fading, essentially caused by the reception of multiple reflections of the transmitted signal (illustrated in Fig. 5.19), is a key inherent problem of wireless channels, which unfortunately cannot be avoided.

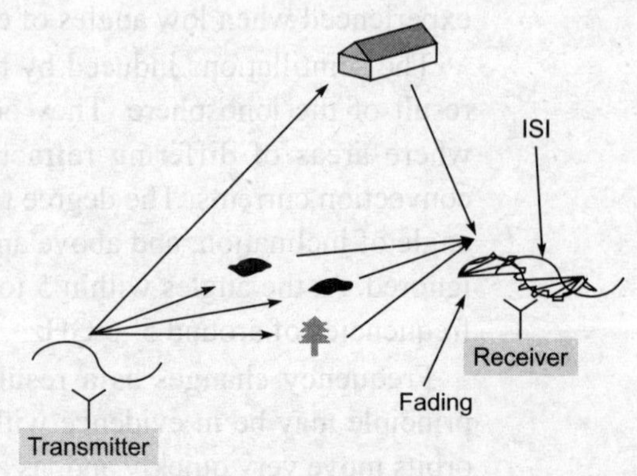

Fig. 5.19 Multipath fading effect

As shown in Fig. 5.20, the mechanism of fading is normally broken down into two different categories based on the position of the receiver relative to the transmitter:

- Large-scale fading for radio propagation over long distances
- Small-scale fading due to time-varying reflections from the surroundings near the receive antenna

However, both the effects are overlapped and distinct identification of both at the receiver is difficult. From the plot of signal strength versus distance, one can approximate them.

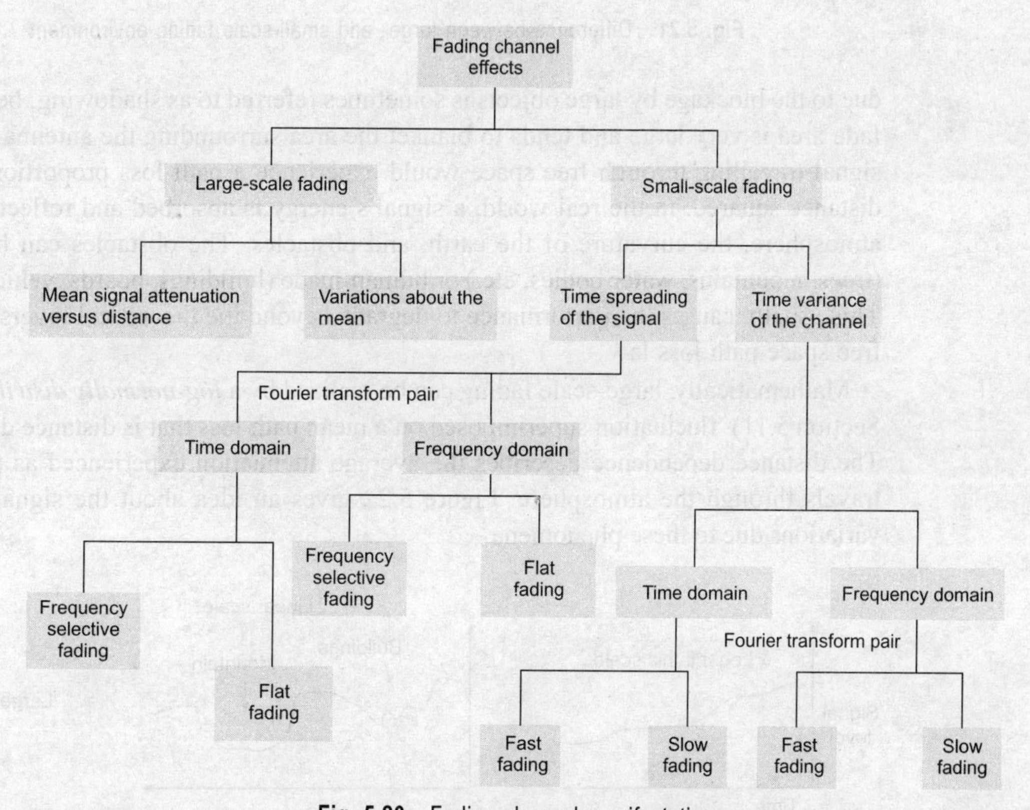

Fig. 5.20 Fading channel manifestation

Figure 5.21 differentiates between the small- and large-scale fading effects and the reasons for these effects.

5.9.1 Large-scale Fading

It is due to the following reasons:

(i) Attenuation in free space: Power degrades with $1/d^2$.

(ii) Shadows: Signals are blocked by obstructing structures.

The large-scale fading essentially represents the average attenuation of a wireless signal as it travels a long distance (several hundred wavelengths or more). Degradation

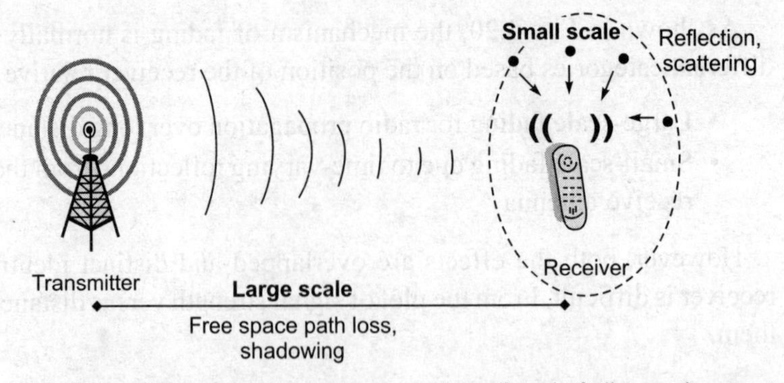

Fig. 5.21 Difference between large- and small-scale fading environment

due to the blockage by large objects is sometimes referred to as shadowing, because the fade area is very large and tends to blanket the area surrounding the antenna. An ideal signal travelling through free space would experience a path loss proportional to the distance squared. In the real world, a signal's energy is absorbed and reflected by the atmosphere, the curvature of the earth, and obstacles. The obstacles can be natural (trees, mountains, water bodies, etc.) or human-made (buildings, boards, vehicles, etc.). This usually causes the performance to degrade beyond the theoretical inverse squared free space path loss law.

Mathematically, large-scale fading can be realized by a *log-normally distributed* (see Section 5.11) fluctuation superimposed on a mean path loss that is distance dependent. The distance dependence describes the average attenuation experienced as the signal travels through the atmosphere. Figure 5.22 gives an idea about the signal strength variations due to these phenomena.

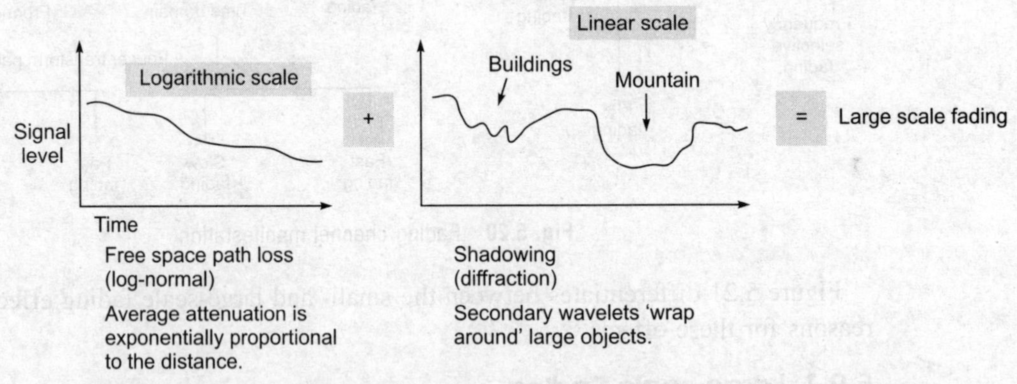

Fig. 5.22 Effect of attenuation and shadowing individually results in large-scale fading

5.9.2 Small-scale Fading

It is due to the following reasons:

- Random frequency modulation due to varying Doppler shifts on different multipath signals.

- Time dispersion (echoes) caused by multipath propagation delays due to nearby objects.
- Even when the mobile is stationary, the received signals may fade due to movement of surrounding objects.

In addition to the path loss over large distances, the receiver antenna will also experience fluctuations in signal levels that vary significantly over small distances (on the order of one to tens of wavelengths). The signal being transmitted from the base station can take different paths to the receiver due to reflection, diffraction, and local scattering. Different paths have different lengths associated with them, which causes the receiver to see multiple copies of the signal at different times of arrival. Also, the signal can shift in phase as it is reflected and scattered off local objects. All of these signals at different power levels and phase converge on the receiver antenna with constructive or destructive interference. As the antenna moves through space, it will experience peaks and valleys of signal strength as these interfering wavelets add and subtract at the receiver. Another adverse effect of motion is Doppler shift. As the receiver antenna moves in relation to a fixed transmitter, the incoming signal will modulate in frequency according to the direction of movement. The copies of the signal that arrive via paths directly in front of the moving receiver will seem to have a higher frequency, while the copies of the signal arriving via paths behind the moving receiver will seem to have a lower frequency. This will also cause small-scale fading.

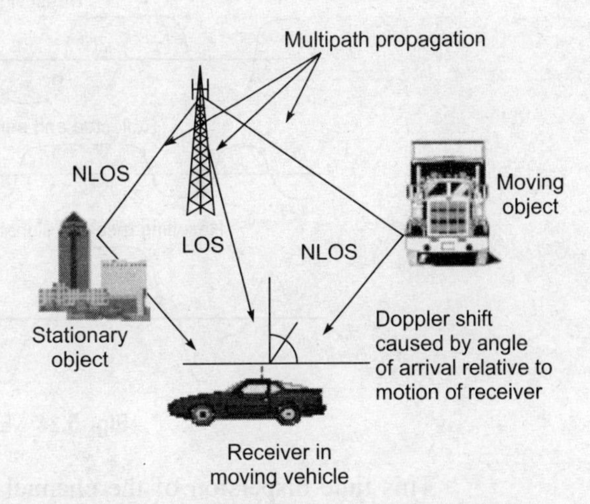

Fig. 5.23 Various reasons of small-scale fading

Time spreading/delay spread and time variance/Doppler effect are the two main effects observed on the signal due to small-scale fading and leads to the concept of coherence bandwidth and coherence time. Contribution of large-scale fading is also unavoidable in the resultant signal.

After a general idea of what causes fading, let us describe the effects observed on the signal. Because small-scale fading is less predictable and more potentially destructive than large-scale fading, most of the discussion here will focus on small-scale fading. In the following discussion, it is shown how small-scale fading is observed on the signal and how delay spread and Doppler shift alter the transmitted signal and make it more difficult for the receiver to accurately detect the altered signal.

Delay Spread Effect

Delay spread effect is mainly due to small-scale fading. Because multiple reflections of the transmitted signal may arrive at the receiver at different times and all get added constructively or destructively, this can result in intersymbol interference or bits 'crashing/smearing' into one another, as shown in Fig. 5.24, which the receiver cannot sort out.

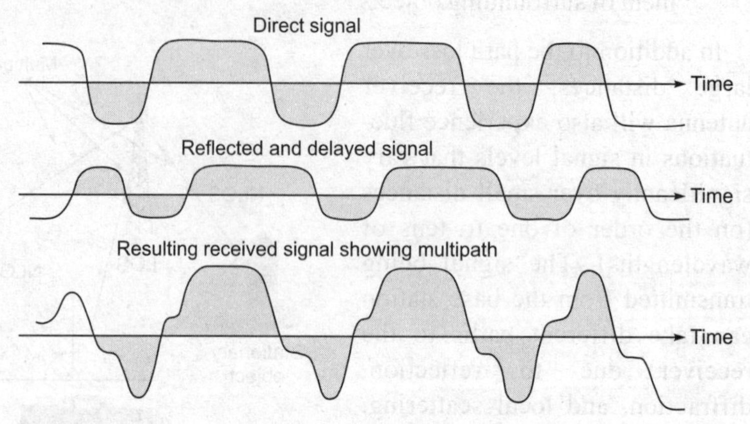

Fig. 5.24 Effect of delay spread

This time dispersion of the channel is called *multipath delay spread*, which is an important parameter to access the performance capabilities of wireless systems. A common measure of multipath delay spread is the *root mean square* (*RMS*) *delay spread*. There is some finite delay between the time when the antenna receives the first copy of the signal on the shortest path and when it receives the last copy of the same signal on the longest path. The maximum excess delay time is represented by T_m. (The RMS value of delay spread, T_{rms}, is more commonly used in practice than the maximum delay time but is a bit more mathematically complex.)

Because of multipath reflections, the channel impulse response of a wireless channel looks like a series of impulses with decreasing amplitudes as shown in Fig. 5.25. In practice, the number of impulses that can be distinguished is very large and depends on the time resolution of the communication or measurement system.

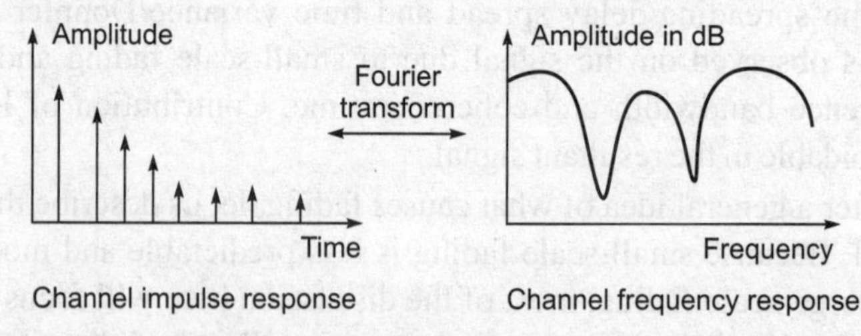

Fig. 5.25 Example of impulse response and frequency transfer function of a multipath channel

In system evaluations, we typically prefer to address a class of channels with properties that are likely to be encountered, rather than one specific impulse response. Therefore, we define the average (local-mean) power that is received with an excess delay that falls within the interval $(T, T + d\tau)$. Such characterization for all T gives the delay profile of the channel. The delay profile determines the frequency dispersion, that is the extent to which the channel fading at two different frequencies f_1 and f_2 is correlated.

Important Definitions Related to Delay Spread Effect

We will start with defining the maximum delay time spread and the RMS delay spread.

- The maximum delay time spread T_m is the total time interval, during which reflections with significant energy arrive.
- The RMS delay spread, T_{rms}, is the standard deviation (or root-mean-square) value of the delay of reflections, weighted proportional to the energy in the reflected waves.

For a digital signal with a high bit rate, this dispersion is experienced as frequency selective fading (which will be considered in the consecutive topics) and ISI. No serious ISI is likely to occur if the symbol duration is longer than, say, ten times the RMS delay spread.

Power delay profile The *delay profile* is the expected power variation per unit of time received with a certain excess delay. It is obtained by averaging a large set of impulse responses. In Figures 5.26, 5.27, 5.28, and 5.30, delay profiles are shown with various environmental conditions.

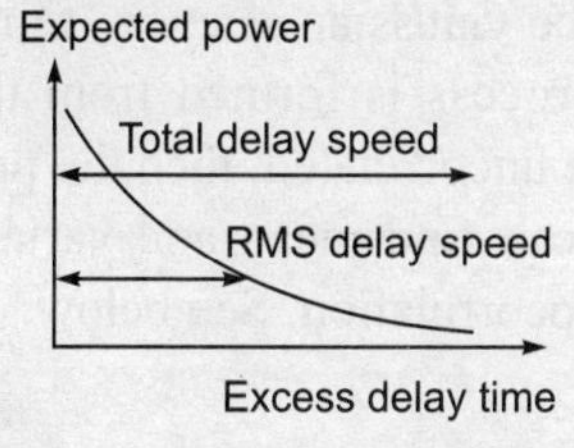

Fig. 5.26 Typical delay profile: exponential

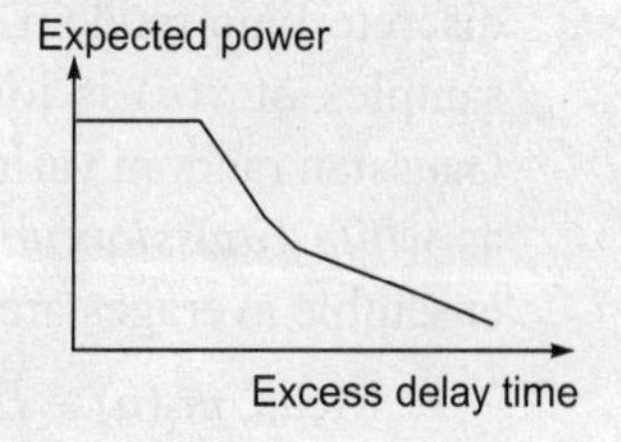

Fig. 5.27 Typical indoor delay profile

In an indoor environment (Fig. 5.27), early reflections often arrive with almost identical power. This gives a fairly flat profile up to some point and a tail of weaker reflections with larger excess delay. From the delay profile, one can compute the correlation of the fading at different carrier frequencies as mentioned earlier.

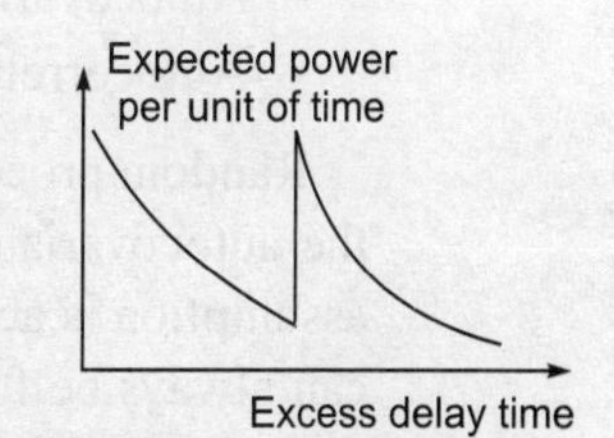

Fig. 5.28 Typical 'bad urban' delay profile with tall buildings

Correlation of fading and autocovariance (With a case study using statistical signal modelling concepts) The received signal is random because of noise and channel effects. Now, the discrete time signal analysis of random processes is easy. A discrete time random process is a mapping from the sample space Ω into the set of discrete time signal $x(n)$. Thus, it is a collection or *ensemble* of discrete time signals. A sequence of coefficients $x(n)$ forms the discrete time random process. It is also indexed sequence of random variables … $x(-2), x(-1), x(0), x(1), x(2)$… . Normally, each random variable in the sequence has its own PDF. In discrete time random processes, rather than probability density function, the mean and autocorrelation of a process are of interest. The same concepts are applied to wireless channel also.

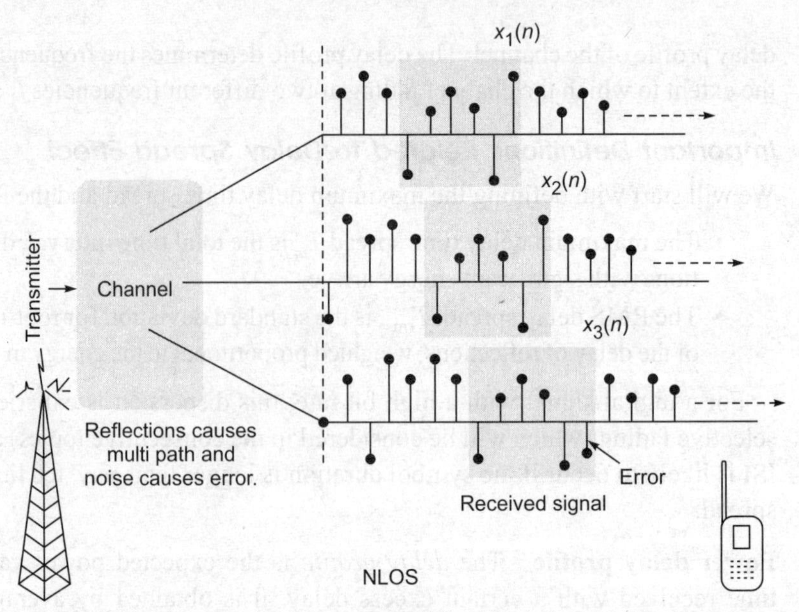

Fig. 5.29 Autocorrelation among multiple reflected rays at the receiver

The direct and reflected rays add up but some autocorrelation exists among them. A discrete time random process $x(n)$ is said to be Gaussian if every finite collection of samples of $x(n)$ is jointly Gaussian. If the process is formed from the sequence of Gaussian random variables $x(n)$ and if they are uncorrelated, then the process is known as *white Gaussian noise*. For each $x(n)$, we can find mean and variance. Two other ensemble averages are autocovariance and autocorrelation. See below:

- Mean, $m_x(n) = E\{x(n)\}$
- Variance, $\sigma_x^2 = E\{|x(n) - m_x(n)|^2\}$
- Autocovariance, $c_x(k, l) = E\{[x(k) - m_x(k)]\,[x(l) - m_x(l)]^*\}$
- Autocorrelation, $r_x(k, l) = E\{x(k)\, x^*(l)\}$

Random processes will always be assumed to have zero mean for simplicity, so that the autocovariance and autocorrelation sequences may be used interchangeably. This assumption is acceptable because if $x(n)$ has non-zero mean, then a zero-mean process can always be formed.

As in the case of random variables, the autocorrelation and autocovariance functions provide information about the degree of linear dependence between two random variables $x(k)$ and $x(l)$. For example, if

$$c_x(k, l) = 0$$

for $k \neq 0$, then the random variables $x(k)$ and $x(l)$ are uncorrelated and knowledge of one does not help in the estimation of the other using a linear estimator.

A random process $x(n)$, due to channel impulse response $h(n)$, is said to be, in wide sense, stationary if the following three conditions are satisfied:

1. The mean of the process is a constant $m_x(n) = m_x$.

2. The autocorrelation $r_x(k, l)$ depends only on the difference $k - l$.
3. The variance of the process is finite.

Now, from the delay profile, one can compute the *correlation of fading* at different carrier frequencies. The transform of the delay profile gives the autocorrelation of the complex amplitudes of sinusoidal signals at frequencies f_1 and f_2 as given in Eq. (5.33).

$$\text{Delay profile} \leftrightarrow E[x(f_1)\, x^*(f_2)] \tag{5.33}$$

The received signal is a random process. Random processes (due to channel) will always be assumed to have zero mean, so that autocovariance and autocorrelation terms are used interchangeably. After some algebraic manipulations, relationship can be derived to express the autocorrelation or autocovariance of the amplitude versus frequency separation, $f_1 - f_2$. In Eq. (5.33), if $x(f_1) = R_1$ and $x(f_2) = R_2$, then the normalized autocovariance of the amplitudes R_1 and R_2 of two carriers, one at f_1 and another at f_2, is

$$C = \frac{E\{R_1 R_2\} - E\{R_1\}\, E\{R_2\}}{s(R_1)\, s(R_2)} \tag{5.34}$$

where $s(R_1)$ = standard deviation of R_1 and $s(R_2)$ = standard deviation of R_2.

Coherence bandwidth For a reliable communication, without using adaptive equalization or other anti-multipath techniques, the transmitted data rate may be much smaller than the inverse of the RMS delay spread, called *coherence bandwidth.*

We shall see in this chapter that narrow band transmission uses radio signals that see flat fading. The channel may be considered relatively constant over the transmit bandwidth. This criterion is found to be satisfied if the transmission bandwidth does not substantially exceed the *coherence* bandwidth B_c of the channel. This is the bandwidth over which the channel transfer function remains virtually constant.

One can define *narrowband* transmission in the time domain, considering the inter-arrival times of multipath reflections and the time scale of variations in the signal caused by modulation: A signal sees a narrow band channel if the bit duration is sufficiently larger than the inter-arrival time of reflected waves. In such cases, the intersymbol interference is small.

Formally, the coherence bandwidth is the bandwidth for which the autocovariance of the signal amplitudes at two extreme frequencies reduces from 1 to 0.5. For a Rayleigh fading channel with an exponential delay profile, one finds

$$B_c = 1/(2\pi\, T_{rms}) \tag{5.35a}$$

where T_{rms} is the rms delay spread

or sometimes $$B_c \cong 1/(5\, T_{rms}) \tag{5.35b}$$

Example 5.3 In a measurement of power delay profile, the maximum excess delay is 50 ns. Assuming exponentially decaying profile and Rayleigh fading channel, find out the maximum transmission bandwidth for which the data can be transferred with minimum ISI.

Solution

The RMS delay spread is approximately maximum excess delay/$\sqrt{2}$. So, RMS delay spread = 0.707 × 50 = 35.35 ns.

The maximum transmission bandwidth for minimum ISI is nothing but coherence bandwidth

$$B_c = 1/(2\pi T_{\text{rms}})$$
$$= 1/ 2\pi \times 35.35 \times 10^{-9}$$
$$= 0.00450225 \times 10^{9}$$
$$\sim 4.5 \text{ MHz}$$

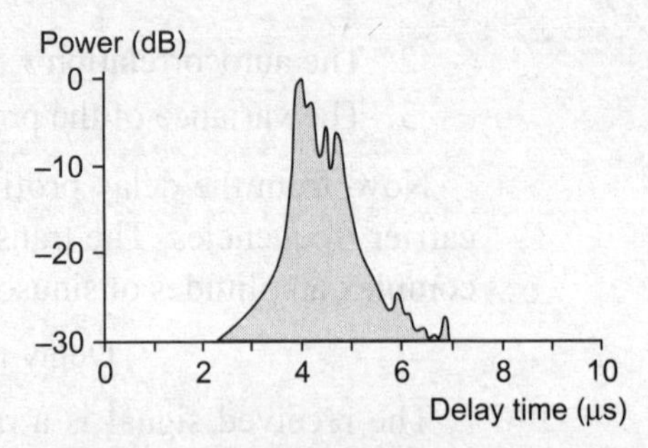

Fig. 5.30 Typical power delay profile for delay spread = 1.2 msec and coherence bandwidth B_c = 1.3 MHz

Resolvable paths A wideband signal with symbol duration T_s [or a direct sequence (DS)-CDMA signal with chip time t_{chip}] can 'resolve' the time dispersion of the channel with an accuracy of about T_s. For DS-CDMA, the number of resolvable paths is

$$N_p = \text{round}\left(\frac{T_{\text{delay}}}{t_{\text{chip}}} + 1\right) \tag{5.36}$$

where round (x) is the largest integer value smaller than x and T_{delay} is the total length of the delay profile. A DS-CDMA rake receiver can exploit N_p-fold path diversity which can be.

Delay Spread and ISI

Table 5.1 shows the typical delay spread for various environments. The maximum delay spread in an outdoor environment is approximately 20 μs. Thus, significant intersymbol interference can occur at bit rates as low as 25 kbps.

Table 5.1 Typical delay spreads

Environment or Cause	Delay Spread	Maximum Path Length Difference
Indoor (room)	40 nsec–200 nsec	12 m–60 m
Outdoor	1 μ sec–20 μsec	300 m–6 km

As such, delay spread and ISI have similar effects. However, there is a difference. The ISI is caused by the reception of a small number of reflections from remote objects (as opposed to the large number of reflections from nearby objects that causes fading). The ISI causes the receiver to receive the original signal, overlapped by some delayed versions of the signal. Traditionally, different types of equalizers were used to reject ISI. The OFDM modulation provides a fairly strong and simple ISI rejection mechanism using cyclic prefix (Chapter 9). Intersymbol interference may be measured by eye patterns. In ASK-like modulations, it is possible to (ideally) remove the interference between different symbols using a filter satisfying the Nyquist ISI criterion. The ISI can cause significant time errors in high bit rate systems, especially while using *time division multiple access* (TDMA). Figure 5.31 shows the effect of intersymbol interference on the received signal. As the transmitted bit rate is increased, the amount of intersymbol interference also increases. The effect starts to become very significant when the delay spread is greater than ~50% of the bit time.

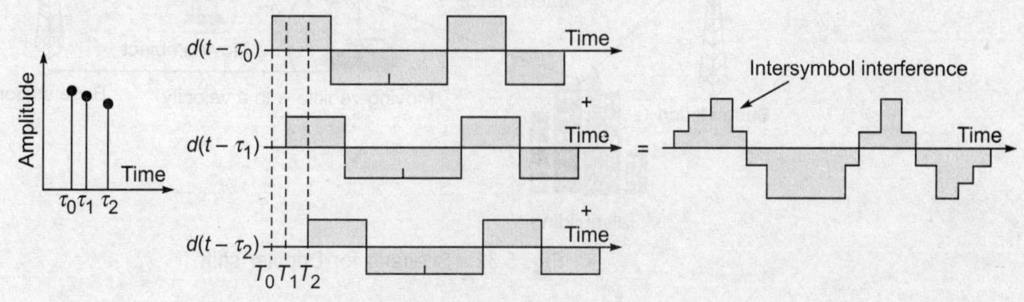

Fig. 5.31 ISI due to delay spread effect

The delay spread restricts the maximum allowable symbol transmission rate. The intersymbol interference can be minimized in several ways. One method is to reduce the symbol rate by reducing the data rate for each channel (i.e. splitting the bandwidth into more channels using frequency division multiplexing or OFDM). Another is to use a coding scheme that is tolerant to intersymbol interference, such as CDMA.

Table 5.2 lists out some measures to eliminate delay spread.

Table 5.2 Measures to eliminate delay spread

System	Measure to Eliminate Delay Spread
Analog	• Narrowband transmission
GSM	• Adaptive channel equalization • Channel estimation training sequence
DECT	• Use the handset only in small cells with small delay spreads • Diversity and channel selection can help a little bit (pick a channel where late reflections are in a fade)
IS95 CDMA	• Rake receiver separately recovers signals over paths with excessive delays. CDMA array processing can further improve performance, because it also exploits angle spreads.
Digital audio/video broadcasting	• OFDM multicarrier modulation: The radio channel is split into many narrowband sub channels with orthogonality. There is no ISI.
MIMO	• Exploits multipath diversity. Uses multiple transmitting and receiving antennas.

Doppler Shift/Spread

Figures 5.32 and 5.33 show the scenario of Doppler effect. Along with the multiple reflected signals, the receiver undergoes one more effect of Doppler spread due to its own mobility, especially vehicular mobility. The Doppler spread is the width of the received spectrum when a single tone is transmitted and it is related to the rate at which fading

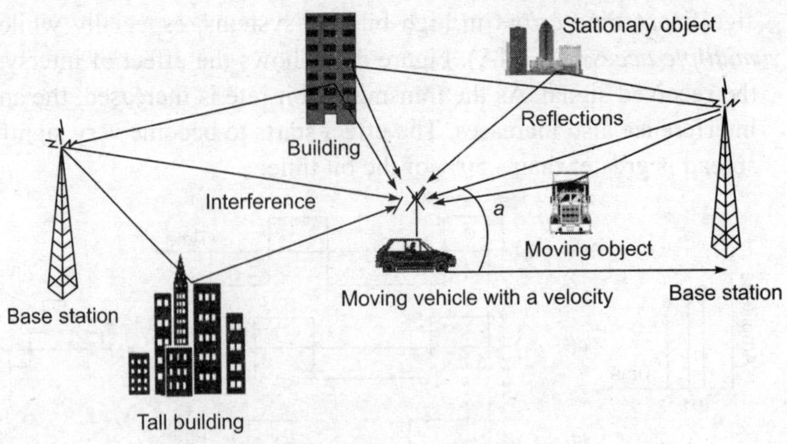

Fig. 5.32 Scenario for Doppler shift

occurs. The Doppler spread is important in determining the minimum adaptation rate for an adaptive receiver. Motion of a receiving antenna produces the Doppler shifts of incoming received waves. We consider a signal received over a multipath channel, with many incoming waves. Let the nth reflected wave with amplitude c_n and phase ϕ_n arrive with an angle α_n relative to the direction of the motion of the antenna.

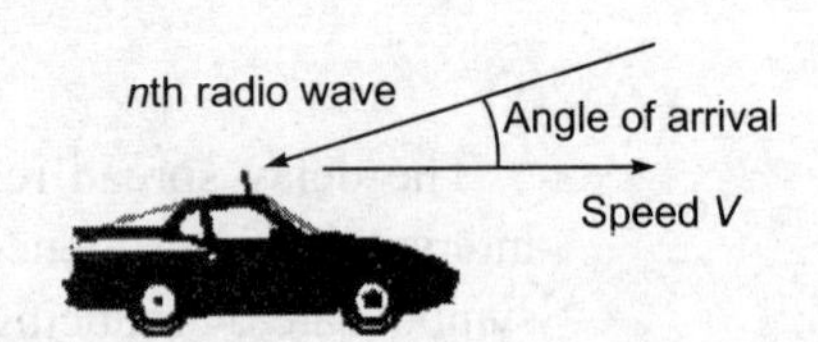

Fig. 5.33 Jake's model for Doppler effect

The Doppler shift of each wave is

$$\delta f_n = \frac{v}{\lambda} \cos \alpha_n \tag{5.37}$$

where v is the speed of the receiving antenna.

The maximum Doppler shift f_d occurs for a wave coming from the direction opposite to the direction the antenna is moving. It has a frequency shift, with f_c being the carrier frequency and c the velocity of light,

$$f_d = \frac{v}{c} f_c \tag{5.38}$$

Such motion of the antenna leads to (time-varying) phase shifts of individual reflected waves. It is not so much this minor shift that bothers radio system designers as a receiver oscillator can easily compensate for it. Rather, the problem is that many waves arrive with different shifts. Thus, their relative phases change all the time and, so, it affects the amplitude of the resulting composite signal. So, the Doppler effects determine the rate at which the amplitude of the resulting composite signal changes.

If the same effect is applied to multicarrier transmission (Chapter 9), all the orthogonal tones placed near each other will be received with a spread. It means that all the carriers will be received with offsets. This will be the worst situation because the carriers will not remain orthogonal and as they are placed near each other, the limiting conditions will be generated. Hence, in multicarrier environment, more attention is required about the allowable Doppler spread and subcarrier spacing.

Example 5.4 A vehicle receives a 910 MHz transmission while travelling at a constant velocity for 15 s. The average fade duration for a Rayleigh fading signal level 10 dB below the RMS level is 1 ms. How far does the vehicle travel during 15 s time duration? Assume that the local mean remains constant during travel.

Solution

The average fade duration is defined as the average period of time for which the received signal is below the specified (RMS here) level and can be expressed as

$$\tau_f = \frac{e^{l^2} - 1}{l\, f_d \sqrt{2\pi}}$$

where l is the minimum signal strength below the specified (RMS) level due to fade.

We have $l = -10 \text{ dB} = 0.316$

$$\Rightarrow \qquad \tau_f = \frac{e^{(0.316)^2} - 1}{0.316 \times f_d \times \sqrt{2\pi}} = 1 \text{ ms} \quad \Rightarrow \; f_d = 132.8 \text{ Hz}$$

Now,

$$v = f_d \times \frac{c}{f_c} = 132.8 \times \frac{3 \times 10^8}{910 \times 10^6} = 43.78 \text{ m/s}$$

$$\Rightarrow \qquad \text{Total distance} = 15 \times 43.78 = 656.7 \text{ m}$$

Doppler Power Spectrum

The models behind Rayleigh or Rician fading (Chapter 6) assume that many waves arrive each with its own random angle of arrival (thus, with its own Doppler shift), which is uniformly distributed within [0, 2π], independently of other waves. This allows us to compute a probability density function of the frequency of incoming waves. Moreover, we can obtain the Doppler spectrum of the received signal.

This leads to the U-shaped power spectrum for isotropic scattering as shown in Figures 5.34 and 5.35.

$$S(f) = \frac{1}{4\pi f_d} \frac{1}{\sqrt{1 - \dfrac{(f - f_c)^2}{f_d^2}}} \qquad (5.39)$$

where we assume a unity local mean power.

If a sinusoidal signal is transmitted (represented by a central line in the frequency domain in Fig. 5.34) after transmission over a fading channel, we will receive a power spectrum that is spread around the single-frequency tone. The frequency range where the power spectrum is non-zero defines the Doppler spread. The Doppler spread is relevant, for instance, to compute threshold crossing rates and average fade durations.

Excess Delay and Doppler Spread

There are two different forms of multipath scattering according to the excess time delay of the given channel tap—small excess and large excess. Their corresponding effects on the Doppler spectrum are discussed below.

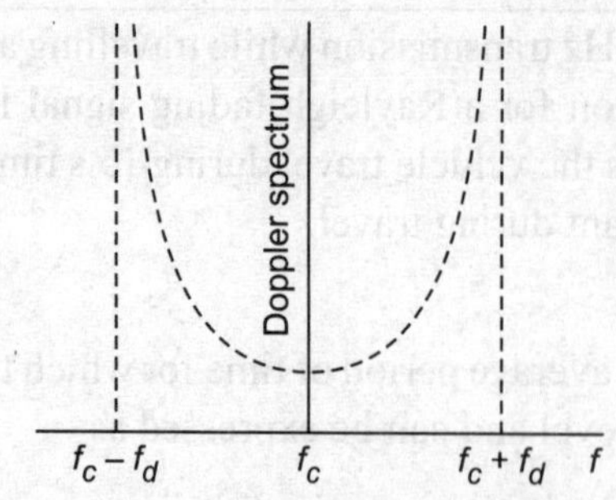

Fig. 5.34 Power density spectrum of a sine wave suffering from a Dopple spread (ideally)

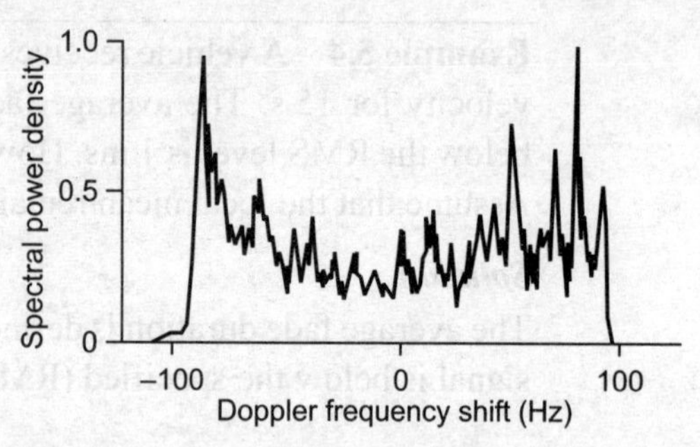

Fig. 5.35 Doppler spread at 1800 MHz = 60.3 Hz (practically achieved)

Table 5.3 list out some measures to tackle Doppler spread.

Table 5.3 How do systems handle Doppler spreads

System	Countermeasure
Analog	• Doppler causes random FM modulation that may be audible; carrier frequency is low enough to avoid problems.
GSM	• Channel bit rate is well above Doppler spread. • TDMA during each bit/burst transmission, the channel is fairly constant. • Receiver training/updating during each transmission burst. • Feedback frequency correction.
DECT	• Intended for pedestrian use: o Only small Doppler spreads are to be anticipated. o Bit rate is very large compared to Doppler spread.
IS95 Cellular CDMA	• Downlink: Pilot signal for synchronization and channel estimation. • Uplink: Continuous tracking of each signal.
Wireless LAN's	• Mobility is slow and thus Doppler spread is only a few hertz.

Small excess time delays The channel tap may be modelled as the accumulation of multipath components received from the scatterers close to the mobile. This gives rise to the classical Doppler power spectrum of the received multipath components.

Large excess time delays The classical Doppler model does not provide a satisfactory geometric model for this type of scattering. Instead, multipath energy is more likely to have a narrow Doppler spread, having arisen from the reflections of isolated obstacles, such as buildings or hills. The instantaneous variation of signal power in space for a channel depends on the angles of arrival of the multipath components.

5.10 FADING EFFECTS TO THE SIGNAL AND FREQUENCY COMPONENTS

The fading effects to the frequency components are related to bit or symbol transmission rate and time spreading of those pulses. When the transmitted data rate is much less than the coherence bandwidth, the wireless channel is referred to as the *flat channel* or *narrowband channel* and the effect through which the signal undergoes is called *flat fading*. In flat fading, all the frequency components of the transmitted signal will be affected equally and received with almost equal power level. When the transmitted data rate is closely equal to or larger than the coherence bandwidth, such a channel is called the *frequency selective channel* or *wideband channel* and the fading effect in this situation is called *frequency selective fading*. In this case, one can find the different power levels of different frequency components. The whole scenario is shown in Fig. 5.36.

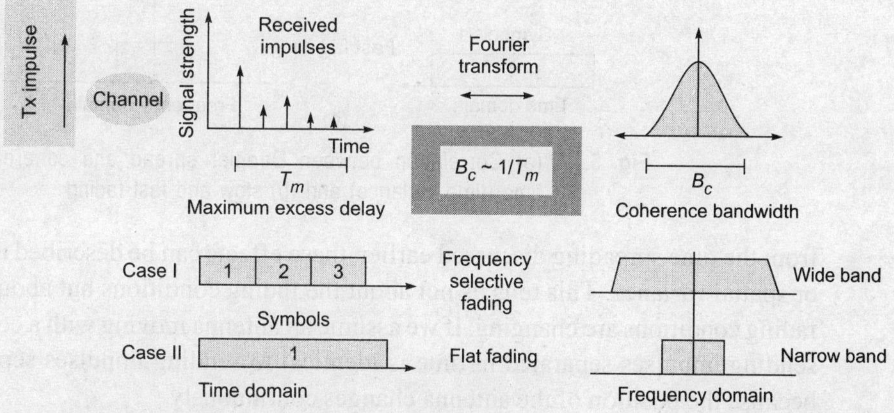

Fig. 5.36 Correlation between delay spread and coherence bandwidth (time spreading) as well as definition of flat fading and frequency selective fading

When observed in the frequency domain, the delay spread manifests itself as a sort of frequency notch filter. The coherence bandwidth B_c is the range of frequencies within which the signal impairments of the channel do not vary significantly. In other words, if two signals are sent that are more than B_c apart from each other in frequency, they will experience different channel conditions. The coherence bandwidth B_c is inversely proportional to T_m.

If the time taken to transmit one symbol T_s is longer than the maximum delay spread T_m ($T_s > T_m$), the path is said to exhibit flat fading, which means that all the multipath components of the transmitted symbol are received within one symbol period. Since signalling rate is inversely proportional to the symbol period, this is equivalent to the case where the symbol bandwidth W is less than the coherence bandwidth ($W < B_c$). On the other hand, frequency selective fading occurs when $T_s < T_m$ or when $W > B_c$. In this case, different spectral components of the signal will be altered by the channel in different ways. As a result, a wideband signal can experience a large variation in received power over its bandwidth.

After studying the time spreading effect on the signal, let us describe the time variance. The relationship of coherence time and Doppler spread is shown in Fig. 5.37.

The multipath effects seen previously can affect a receiver that is stationary, since a transmitted signal will always experience reflection when encountering objects. For receivers that change their location relative to the transmitter, there are other factors in addition to multipath reflection that affect the signal's amplitude and phase. To distinguish

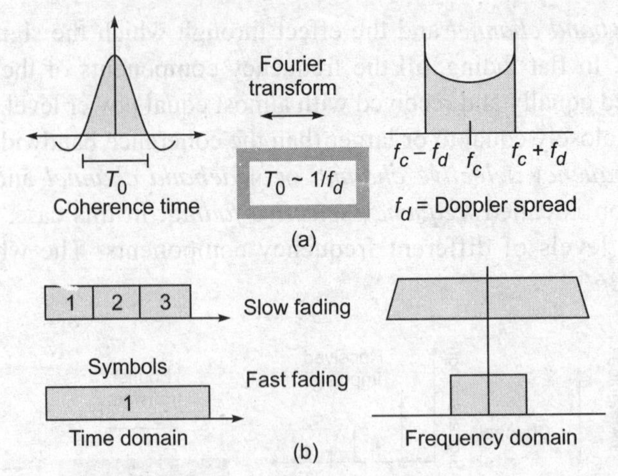

Fig. 5.37 (a) Correlation between Doppler spread and coherence time (time variance) and (b) slow and fast fading

from the time spreading discussed earlier, these effects can be described as time variance or spatial variance. This tells us not about the fading conditions but about how rapid the fading conditions are changing. If we assume an antenna moving with a constant velocity, sending impulses separated in time is identical to sending impulses separated in space because the position of the antenna changes continuously.

One important thing to know while sending a signal through a varying channel is the period of time we have before the conditions change. This is called the coherence time, denoted by T_0. In more technical terms, T_0 is the time period over which the channel's impulse response is highly correlated. If we look at the time variance in the frequency domain, we see that a receiver antenna in constant motion will experience shifts in frequency that are dependent on the angle of arrival of the incoming signal and on the speed of motion. Instead of the signal being spread in time, we shall see it being spread in frequency. This Doppler spread, which is represented by f_d, is reciprocally related to the coherence time T_0.

If the symbol period is longer than the coherence time ($T_s > T_0$), the channel exhibits *fast fading*, because the fading conditions are coming and going faster than the symbols are being transmitted. This is the same thing as saying that the user bandwidth is smaller than the Doppler spread ($W < f_d$). If $T_s < T_0$ (or $W > f_d$), the channel exhibits *slow fading*, meaning that the channel conditions are stable and predictable during the time that the symbol is transmitted. Note that for most cellular applications, the user bandwidth is generally of order of magnitude greater than the frequency spread. So, we usually see slow fading effects. Here is a word of caution: Since the symbol frequency is inversely related to the symbol period, changing the signalling rate to compensate for frequency-selective fading can also alter its performance.

Example 5.5 For a typical wireless system, in laboratory measurements, it is found that the delay over the channel between a steady transmitter and a receiver is 50 ms to receive the first component of the signal. The maximum excess delay observed is 65 µs.

The transmission bandwidth is 10 MHz with RF carrier frequency 2.4 GHz. Check whether the frequency selective or flat fading will occur over the channel. If now, the mobile receiver is in vehicular mobility condition with the speed of 50 km/hr, check whether the fast fading or slow fading will occur.

Solution

Bandwidth: $W = 10$ MHz, $T_m = 65$ μs

Coherence bandwidth: $B_c = 1/T_m = 1/(65 \times 10^{-6}) = 15.4$ kHz

Now, as $W > B_c$, frequency-selective fading will occur.

The vehicle moves with 50 km/hr = 13.88 m/s

Doppler spread: $f_d = \frac{v}{c} f_c = (13.88/3 \times 10^8)\,(2.4 \times 10^9) = 111$ Hz

Now, as $W > f_d$, slow fading will be exhibited.

5.10.1 Modelling of Flat Fading Channel

It is a fundamental requirement to know the properties of the flat fading channels before going into details of the complicated channel models. Rayleigh, Rician, and Nakagami models are described in detail in the next chapter. However, a few equations can be utilized to have an idea about how the flat fading can be incorporated in the signal while modelling it mathematically. In a flat fading channel, the received signal $r(t)$ is obtained by the addition of the transmitted signal $s(t)$ multiplied by a time-varying attenuation $\alpha(t)$ and the noise contribution $w(t)$:

$$r(t) = \alpha(t)s(t) + w(t) \tag{5.40}$$

The time variation of the attenuation is known as fading, as studied previously. The time function of attenuation variation is difficult to represent mathematically as it is a matter of probability and depends upon the channel environment. For such cases, it is better to represent the probability distribution function for α. It has been shown experimentally and also argued theoretically that it usually follows a Rayleigh distribution.

$$\text{pdf}(\alpha) = (\alpha/\sigma^2) \times \exp[-\alpha^2/2\sigma^2] \quad \text{for } 0 < \alpha < \infty \tag{5.41}$$

where σ^2 is the variance of the underlying Gaussian process. The condition for this to be valid is that there are many statistically independent scatterers and no single scatterer makes a dominant contribution. If there is one dominant contribution (usually a line-of-sight component), the distribution of α is a Rice distributed variable characterized by

$$\text{pdf}(\alpha) = (\alpha/\sigma^2) \times \exp[-(\alpha^2 + A^2)/2\sigma^2] \times I_o(\alpha A/\sigma^2) \text{ for } 0 < \alpha < \infty \tag{5.42}$$

where $I_o(x)$ is the modified Bessel function of first kind, zero order. The parameter A is the amplitude of the dominant component. The Rice parameter K is defined as $A^2/2\sigma^2$.

The multipath propagation also introduces a phase shift. If the amplitude is Rayleigh fading, the phase shift is statistically independent of the amplitude distribution and is uniformly distributed between 0 and 2π. If the amplitude fading is Rician, the joint PDF of amplitude and phase ψ is

$$\text{pdf}_{\alpha,\psi} = (\alpha/2\pi\,\sigma^2) \times \exp[-(\alpha^2 + A^2 - 2\alpha A \cos\psi)/2\sigma^2] \tag{5.43}$$

An alternative amplitude distribution, which has gained popularity, especially for the evaluation of measurements, is Nakagami m-distribution given by

$$\text{pdf}_\alpha(\alpha) = \frac{2}{\Gamma(m)}\left(\frac{m}{\Omega}\right)^m \alpha^{2m-1} \exp\left(-\frac{m}{\Omega}\alpha^2\right) \text{ for } \alpha \geq 0 \text{ and } m \geq 1/2 \tag{5.44}$$

For $m = 1$, this distribution reduces to the Rayleigh distribution. The parameter Ω is the mean square value $\Omega = \overline{\alpha^2}$ and the parameter m is given by

$$m = \frac{\Omega^2}{\overline{(\alpha^2 - \Omega)^2}} \tag{5.45}$$

Nakagami and Rice distributions are quite similar and each can be approximately converted to the other for $m \geq 1$:

$$m = \frac{(K+1)^2}{(2K+1)} \tag{5.46}$$

and

$$K = \frac{\sqrt{m^2 - m}}{m - \sqrt{m^2 - m}} \tag{5.47}$$

The movement of the mobile station leads to a frequency shift of the arriving waves (Doppler effect). If the sinusoidal wave of frequency f_c is transmitted, the spectrum of the received signal is

$$Y(f) \propto [\,pdf_\gamma(\gamma)G(\gamma) + pdf_\gamma(-\gamma)G(-\gamma)\frac{1}{\sqrt{\left(\frac{v}{c}f_c\right)^2 - (f - f_c)^2}} \tag{5.48}$$

where f is the variable for the frequency, $(v/c)\,f_c$ is the extreme new frequency due to Doppler effect as explained previously other than f_c and it varies in the range $-(v/c)\,f_c < f < (v/c)\,f_c$ and 0 elsewhere, where v is the velocity or speed of movement, γ is the angle between the direction of incidence of the move and the direction of movement, and $G(\gamma)$ is the antenna pattern. For the case when the waves are all incident horizontally, uniformly distributed in azimuth, and the antenna has a uniform pattern in azimuth, we get

$$v(f) \propto \frac{1}{\sqrt{\left(\frac{v}{c}f_c\right)^2 - (f - f_c)^2}} \tag{5.49}$$

The Rayleigh or Rice-fading is also known as small-scale fading since it describes the variation of the amplitude within an area of about ten wavelengths. Over a large scale, it has been shown experimentally that the small-scale averaged amplitude obeys a log-normal distribution. Free space propagation model and the related equations can be applied to calculate the path losses.

5.10.2 Comparison of Flat and Frequency-Selective Fading

Practically, in any radio transmission, the channel spectral response is not flat. It has dips or fades in the response due to the reflections causing cancellation of certain frequencies at the receiver. Reflections off nearby objects (e.g., ground, buildings, and trees) can lead to multipath signals of similar signal power as the direct signal. This can result in deep nulls in the received signal power due to destructive interference.

For narrow bandwidth transmissions, if the null in the frequency response occurs at the transmission frequency, then the entire signal can be lost. This can be partly overcome in two ways. One method is to transmit a wide bandwidth signal or spread spectrum as CDMA (Chapter 8). In this way, any dips in the spectrum result in will a small loss of signal power rather than a complete loss. Another method is to split the transmission up into many small bandwidth carriers, as is done in a COFDM/OFDM transmission (as described in Chapter 9). The original signal is spread over a wide bandwidth and so nulls in the spectrum are likely to affect only a small number of carriers rather than the entire signal. The information in the lost carriers can be recovered by using forward error correction techniques. Thus, the frequency selective fading sometimes becomes advantageous.

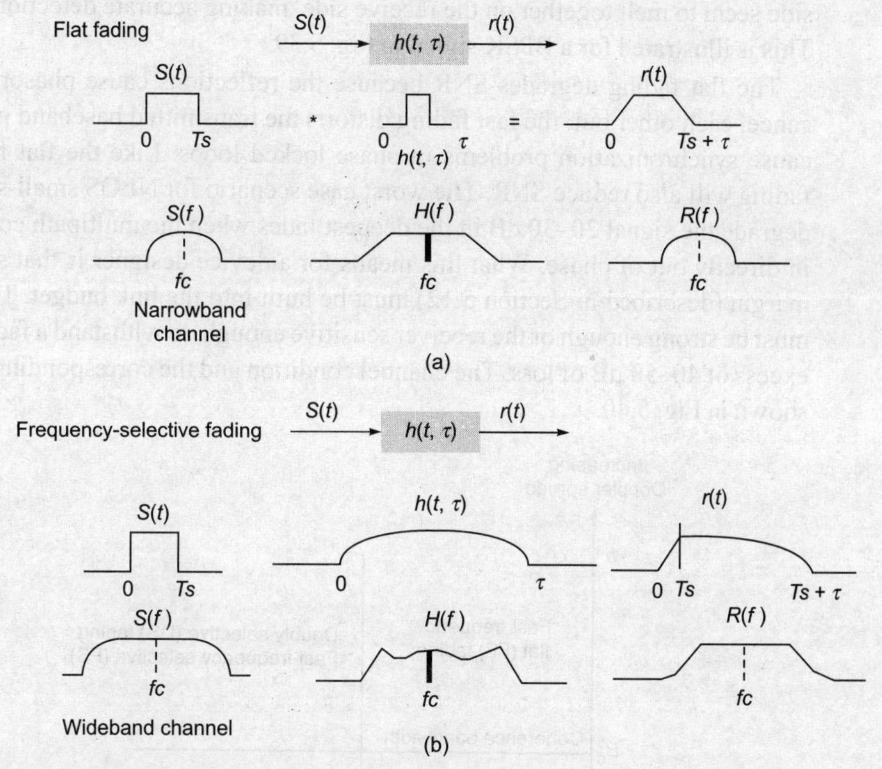

Fig. 5.38 Channel responses: (a) flat fading response and (b) frequency-selective response

5.10.3 How Does Fading Affect a Device

The fading has the following effects on a device:

- Reduced signal-to-noise ratio (SNR)
- Intersymbol interference
- Must account for total fading margin in link budget analysis

The large-scale direct path fading results mostly in the attenuation of the overall signal level. Path loss is heavily dependent on the distance. The effect on the device is to reduce SNR by lowering the received signal power. Shadowing and large-scale reflections can be represented as deviations from the mean path loss mentioned above. These variations, when measured on a log scale, tend to have a normal (or Gaussian) distribution. It will typically degrade the signal 6–10 dB, in addition to the direct path attenuation mentioned above.

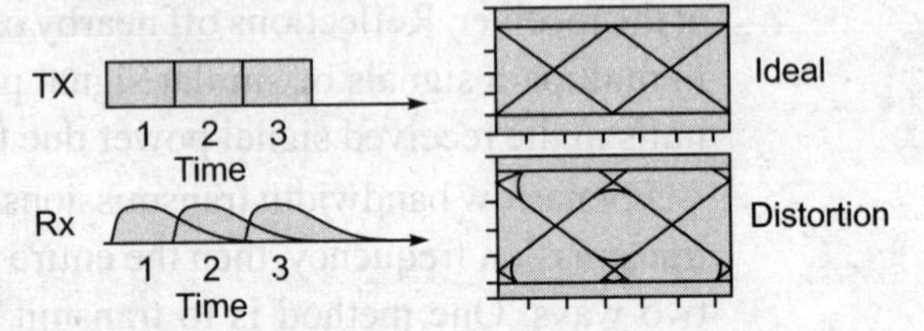

Fig. 5.39 The effect of fading shows distortions in the eye diagram (Chapter 7)

The small-scale fading by multipath and Doppler is the most potentially destructive fading and can be observed at the receive antenna in a number of ways. The frequency selective fading causes intersymbol interference, where adjacent symbols on the transmit side seem to melt together on the receive side, making accurate detection more difficult. This is illustrated for a BPSK signal in Fig. 5.39.

The flat fading degrades SNR because the reflections cause phasor components to cancel each other out. the fast fading distorts the transmitted baseband pulse, which can cause synchronization problems in phase locked loops. Like the flat fading, the slow fading will also reduce SNR. The worst case scenario for NLOS small-scale fading will degrade the signal 20–30 dB in the deepest fades when the multipath components come in directly out of phase. What this means for a device designer is that sufficient fading margin (described in Section 5.12) must be built into the link budget. The signal power must be strong enough or the receiver sensitive enough, to withstand a fading condition in excess of 40–50 dB of loss. The channel condition and the corresponding fading effect is shown in Fig. 5.40.

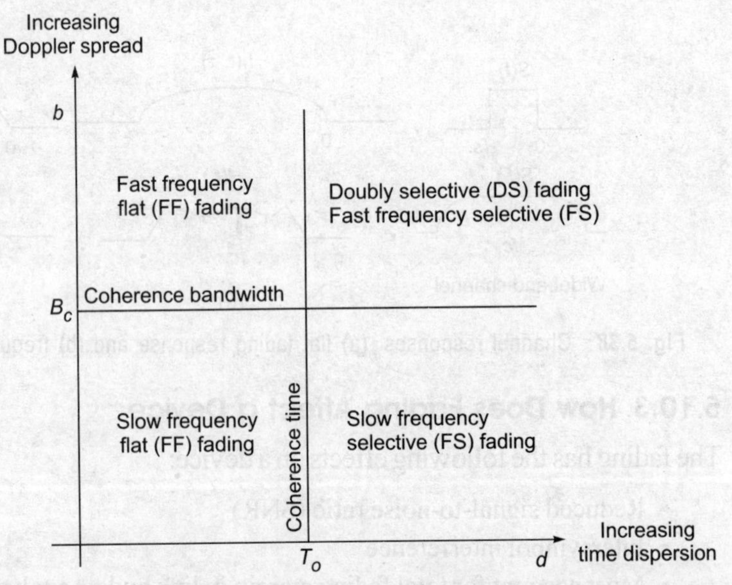

Fig. 5.40 Types of channels and their effects

The various ways to reduce the fading effects are summarized below:

- Signal conditioning limits on bandwidth and bit rate (mainly controlled by information theory—Chapters 2, 3, and 4)
- Error correction-interleaving and coding (mainly controlled by channel coding—Chapter 4)
- Device design
  - Adaptive equalizer (Chapter 6), for example, GSM
  - Rake receiver (Chapter 8), for example, CDMA
- Modulation format (Chapters 7 and 9), for example, OFDM and UWB

The best performance of a wireless link will only be realized in the absence of any type of channel impairment. The presence of Gaussian noise prevents the wireless channel from being completely clear, but there are a number of techniques that can be employed in the design of wireless devices to reduce the effects of fading. These techniques reduce the error probability from the worst case, Rayleigh curve, and bring it closer to the best case, AWGN curve. As we have seen earlier, fading can take a number of different forms. The frequency selective or fast fading affects the bit error rate probability significantly, while the flat and slow fading affect it less significantly. When designing a wireless link to withstand fading degradation, it is important to first determine what kind of fading is evident in the channel. When this is known, the bit rate used can be chosen to reduce the avoidable errors.

Equalization is a commonly used technique to fight ISI caused by the frequency selective fading. The process invokes a filter that has an impulse response opposite to that of the propagating channel. Thus, the combination of the transmission path plus the receive filter yields a flat linear response. The GSM uses adaptive equalization to mitigate distortion. The CDMA technologies use a rake receiver to mitigate ISI. It involves using special filters (called fingers) that sniff out the spread-out signal components, collect them, and add them coherently by delaying the earlier arriving paths longer than the later arriving paths. Interleaving and coding can be used to reduce the E_b/N_o required for accurate detection. Coding provides redundancy by sending systematically added bits for error detection. Interleaving provides robustness to the link by spreading out errors in time, avoiding large amounts of contiguous data loss. Then there are transmission technologies whose signalling properties avoid the most common effects of fading. Ultra-wideband transmits pulses of such short duration that they are not affected by delay spread in a channel. Orthogonal frequency division multiplexing combats frequency selective fading by breaking the carrier signal into subcarriers with lower bit rates and thereby longer symbol durations.

5.11 SHADOWING

Shadowing is the effect that the received signal power fluctuates due to the objects obstructing the propagation path between the transmitter and receiver. These fluctuations are experienced on local-mean powers, i.e. short-term averages to remove fluctuations due to multipath fading.

Experiments reported by Egli in 1957 showed that for paths longer than a few hundred metres, the received (local-mean) power fluctuates with a 'log-normal' distribution about the area-mean power. By 'log-normal' is meant that the local-mean power expressed in logarithmic values, such as dB or neper, has a normal (i.e. Gaussian) distribution.

We distinguish between local means and area means below:

- *Local means* Average over about 40λ, to remove multipath fading
- *Area means* Average over tens or hundreds of metres, to remove multipath fading and shadowing

The mean received power $P_{\log}$, expressed in logarithmic units (neper), is defined as the natural logarithm of the local-mean power $\overline{p}$ over the area-mean power $\overline{\overline{p}}$. Thus,

$$\overline{P}_{\log} = \ln \frac{\overline{p}}{\overline{\overline{p}}} \tag{5.50}$$

It has the normal probability density function

$$f_{\overline{P}_{\log}}(\overline{P}_{\log}) = \frac{1}{\sqrt{2\pi\sigma}} e^{-\left\{\frac{1}{2\sigma^2}\overline{P}^2_{\log}\right\}} \tag{5.51}$$

where σ is the *logarithmic standard deviation* in natural units. If we convert nepers to watts, the log-normal distribution of received (local-mean) power is found as.

$$f_{\overline{p}}(\overline{p}) = \frac{1}{\sqrt{2\pi\sigma\,\overline{p}}} e^{-\left\{\frac{1}{2\sigma^2}\left(\ln\frac{\overline{p}}{\overline{\overline{p}}}\right)^2\right\}} \tag{5.52}$$

Here the factor '1/local-mean power' occurs due to the conversion of the PDF of $P_{\log}$ to local-mean power. There may be large as well as small area shadowing, which can be modelled as two independent Markovian processes.

The shadowing makes practical cell planning complicated. To fully predict local shadow attenuation, up-to-date and highly detailed terrain databases are needed. If one extends the distinction between large area and small area shadowing, the definition of shadowing covers any statistical fluctuation of the received local-mean power about a certain area-mean power, with the latter determined by (predictable) large-scale mechanisms. Multipath propagation is separated from shadow fluctuations by considering the local-mean powers. That is, the standard deviation of the shadowing will depend on the geographical resolution of the estimate of the area-mean power. A propagation model that ignores specific terrain data produces about 12 dB of shadowing. On the other hand, the prediction methods using topographical databases with unlimited resolution can, at least in theory, achieve a standard deviation of 0 dB. Thus, the standard deviation is a measure of the impreciseness of the terrain description. If, for generic system studies, the (large-scale) path loss is taken of simple form depending only on the distance but not on details of the path profile, the standard deviation will necessarily be large. On the other hand, for the planning of a practical network in a certain (known) environment, the accuracy of the arge-scale propagation model may be refined. This may allow a spectrally more efficient planning if the cellular layout is optimized for the propagation environment.

With shadowing, the interference power accumulates more rapidly than proportional to the number of signals. The accumulation of multiple signals with shadowing is a relevant issue in the planning of cellular networks.

Table 5.5 shows some methods to tackle shadowing.

Table 5.5 How do systems handle shadowing

DECT	• Base station location
IS95	• Power control. (Power control in IS95 is also needed to achieve sufficient performance of the CDMA receiver.) • Base station locations
Digital audio broadcasting	• Single frequency networks (Shadow fades are filled in by signals from other co-channel transmitters.)

5.12 SIGNAL OUTAGES IN FADING CHANNELS AND FADING MARGIN

The mobile Rayleigh or Rician radio channel is characterized by rapidly changing channel characteristics. As typically, a certain minimum (threshold) signal level is needed for acceptable communication performance, the received signal will experience periods of

- sufficient signal strength or 'non-fade intervals' and
- insufficient signal strength or 'fades'.

During the fades, the user experiences a signal *outage*. The term outage can be understood by Fig. 5.41.

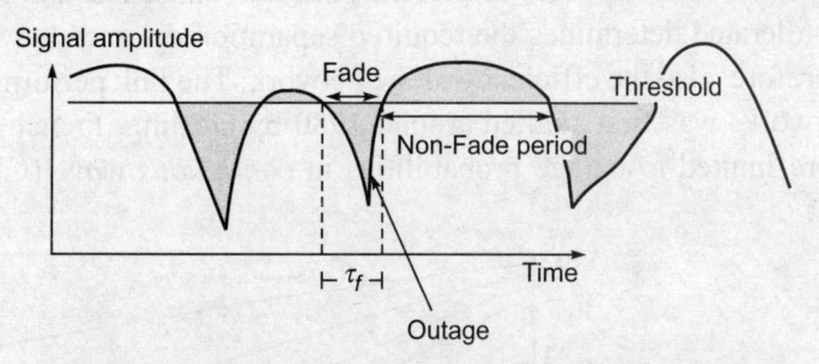

Fig. 5.41 Signal outages occur during channel fades

Fade margin (Fig. 5.42) is the ratio of the average received power over some threshold power needed for reliable communication. The signal outage probability is fairly simple to compute if one knows the probability distribution of the fading (e.g. Rayleigh or Rician/Nakagami) and outage occurs if the signal drops below the noise power level. If noise is also significant, an outage occurs when the (carrier) signal-to-noise plus interference ratio $C/(I + N)$ drops below the threshold z. In mobile telephony, the quality of service is often expressed in terms of the probability of outage experienced by subscribers near the boundary of the base station service area. Because of limited spectrum availability, radio networks become increasingly limited by mutual interference between the users. Therefore, the outage probability is usually determined in terms of the

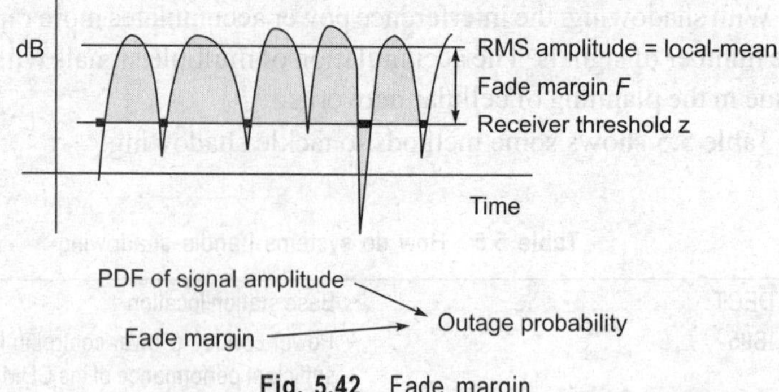

Fig. 5.42 Fade margin

probability that the signal-to-joint-interference ratio drops below a minimum required ratio z.

The choice of the value of z, in general, depends on the required quality of service and may, therefore, be somewhat arbitrary. The 'outage criterion' can, for instance, be a certain figure of merit subjectively determined by a representative panel of listeners or, in a digital system, an instantaneous bit error rate or digital word erasure rate.

The link budget calculation provides an estimate of the margin for how deep a signal fade can be before the receiver loses the signal. The calculation becomes much more complicated if the (multiple) interfering signals also exhibit fading. Moreover, shadowing affects the outage probability also because it depends upon the fade margin as shown in Fig. 5.43.

A crucial aspect in the evaluation and planning of radio networks is the computation of the effect of co-channel interference in radio links. The amount of interference that can be tolerated determines the required separation distance between co-channel cells and, therefore, also the efficiency of the network. The link performance of cellular telephone networks was first studied around 1980 by Gosling, French, and Cox. Initial analyses were limited to outage probabilities in *continuous wave* (CW) voice communication,

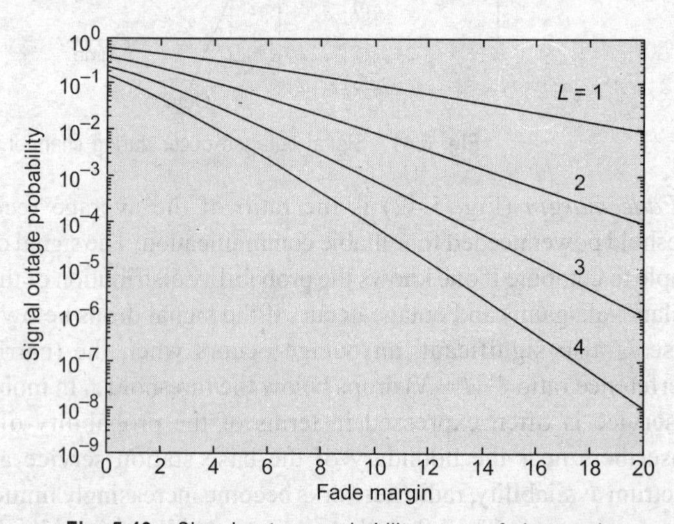

Fig. 5.43 Signal outage probability versus fade margin

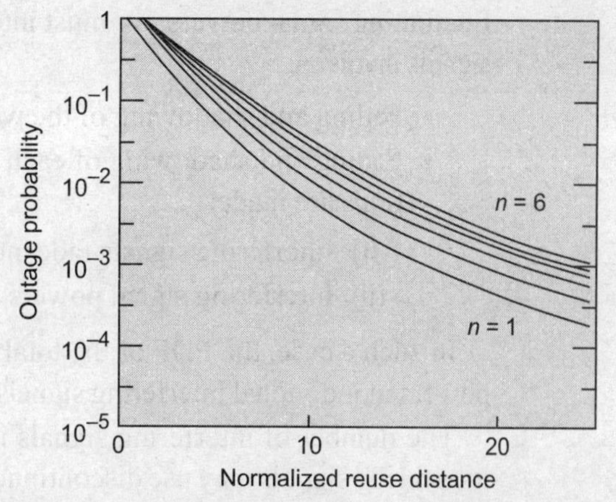

Fig. 5.44 Signal outage probability versus normalized reuse distance for n = 1,..., 6 interfering signals. [Shadowing 6 dB, receiver threshold 10 dB, and plane earth loss (40 log d)]

taking into account the path loss and flat Rayleigh fading. In the 1980's, the technique for computing outage probabilities was refined step by step, considering among other things shadowing, multiple interfering signals cumulating coherently or more realistically incoherently, the modulation technique and error correction method, and more recently the presence of a dominant line-of-sight propagation path, as it occurs in microcellular networks. The stochastic occupation of nearby co-channel cells according to the traffic laws by Erlang was included in some studies.

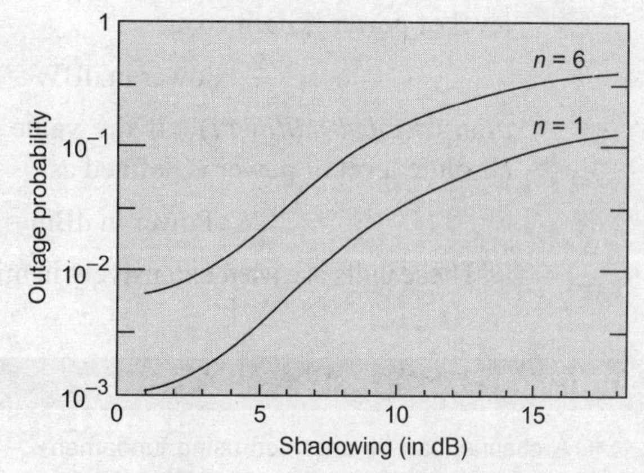

Fig. 5.45 Outage probability versus degree of shadow fading for 1 and 6 interfering signals. [Shadowing 6 dB, receiver threshold 10 dB, reuse distance equal to 10 times the cell radius, and plane earth loss (40 log d)]

Any cellular operator will use topographical databases to estimate outage probabilities in the area covered by the telephone service, but the results for idealized hexagonal cell layouts are nonetheless illustrative for the effect of the reuse distance and shadowing.

Evidently, the operator can reduce the outage probability by choosing larger reuse distances as shown in Fig. 5.44. However, this requires large cluster sizes, which results in poor spectrum efficiency. As shown in Fig. 5.45, with the rise in degree of shadow fading, the outage probability increases. The degree of shadowing is dependent upon the reuse distance and the number of interfering signals.

For a crude propagation model that does not include terrain data, the standard deviation of the error is about 12 dB. In such a case, either unacceptably large outage probabilities would be experienced or the cluster size would have to be unacceptably large. Hence, improved cell planning is required and virtually no operator plans his her services considering idealized hexagonal cells. Mostly, topographical databases are used to estimate the area-mean signal power and interference powers as accurately as possible.

Computation of the outage probability requires that we find the probability that the signal-to-interference ratio drops below a certain threshold. As all (fading) signals have fluctuating signal powers, we must integrate over the probability density functions of all signals involved:

- Fading and shadowing of the wanted signal
- Fading and shadowing of each interfering signal. Mostly, the following assumptions are made:
  - (i) Interfering signals fade independently and independent of the wanted signal.
  - (ii) Interfering signal powers accumulate incoherently, i.e. we may add powers.

In such a case, the PDF of the total interference power is the convolution of the PDF powers of individual interfering signals.

The number of interfering signals is six in a typical hexagonal reuse pattern, but the interfering signals may use discontinuous voice transmission. In such a case, the terminal switches of the carrier during speech pauses to reduce the amount of interference caused to other users (see chapter 11 for cellular concepts).

Important Unit Conversions

dBW (decibel-watt) If the value of 1 W is selected as a reference, then the absolute level of power is defined as

$$\text{Power in dBW} = 10 \log_{10} (\text{power in W/1 W})$$

dBm (decibel-milliwatt) If the value of 1 mW is selected as a reference, then the absolute level of power is defined as

$$\text{Power in dBm} = 10 \log_{10} (\text{power in W/1 mW})$$

These units are used extensively in microwave applications.

Summary

- A channel can be analysed using fundamentals of EM waves, fundamentals of DSP, and probability theory concept.
- Radio frequency ranges from 0.003 MHz to 3,00,000 MHz.
- Radio waves propagate with reflection, diffraction, scattering, and refraction.
- Polarization of the EM wave is important in radio wave propagation and reception.
- Long-distance and short-distance wireless link considerations differ a lot.
- The atmosphere has various layers with different characteristics.
- Four important types of radio propagation are free space, ground wave, ionospheric, and tropospheric.
- The AWGN is significant over any type of channel. Channel losses can be considered according to the situation.
- A satellite link is a typical wireless link with the largest propagation delay.
- For short-distance communication, a direct path signal and a number of reflected rays are considered, which cause ISI effect.
- Multipath causes fading, which is observed in terms of delay spread and Doppler shift.
- Large-scale fading is mainly due to attenuation and shadowing.
- Power delay profile gives CIR and maximum excess delay, which decides coherence bandwidth.
- Coherence bandwidth decides frequency selective or flat fading.

- Doppler spectrum decides coherence time.
- Coherence time decides fast or slow fading.
- Shadowing effect can be considered over small area or large area.
- Reduction in signal strength due to fading gives signal outage and it reduces by increasing fade margin.

REVIEW QUESTIONS

1. What is the importance of polarization of EM waves?
2. What is the mathematical relationship between the wavelength and the frequency of a sinusoidal EM wave?
3. What are the channel problems to be taken into account precisely for long- and short-distance propagation? What are the common problems faced by both types of propagation?
4. Show that for a reference transmitter with EIRP of 1 kW in free space, the usable field strength $E_u = \sqrt{\dfrac{30 P_t G_t}{d}}$.
5. Show that for ground wave propagation, the reflection coefficient tends to −1 for the angles close to 0. Verify that for horizontal polarization, $R_c > 0.9$ for $\phi < 10$ degrees. Also verify that for vertical polarization, $R_c > 0.5$ for $\phi < 5$ degrees, and $R_c > 0.9$ for $\phi < 1$ degree. Here R_c is the absolute value.
6. Why is it required to have distinct uplink and downlink frequencies in satellite communication?
7. What is attenuation? Does it degrade the signal? How?
8. List out the applications that use microwave communication. Find out the advantages and disadvantages of microwave communication.
9. What is the difference between ionospheric propagation and tropospheric propogation?
10. In which applications, will the multihop transmission be useful? How?
11. Describe the nature of additive white Gaussian noise.
12. What is the difference between mean excess delay and maximum excess delay for the multipath power delay profile?
13. Define the terms: delay spread, Doppler shift, coherence bandwidth, power delay profile.
14. Differentiate between the following terms:
    (a) Large-scale fading and small-scale fading
    (b) Fast fading and slow fading
    (c) Frequency-selective fading and flat fading
    (d) Time spreading and time variance
15. How does the time variance become significant in deciding fast or slow fading?
16. Define the terms 'fading margin' and 'outage probability'. Mention their significance in cellular planning.
17. Write a note on channel impulse responses for flat and frequency-selective fading.
18. Comment on the observations of Doppler shift when a moving vehicle moves towards a transmitter and away from the transmitter.
19. Discuss the effect of path loss on the performance of a cellular radio network.
20. Comment on the mathematical modelling of the wireless channel, which can take into account all possible effects observed over the channel. Can the power delay

profile help in the modelling? If so, then explain how this profile becomes useful.

21. What are the fundamental concepts of DSP?

22. How does shadowing affect the reception? Discuss various situations of shadowing. How does the shadowing in hilly area differ from that in the crowded urban area?

PROBLEMS

1. Show that the path loss L between two isotropic antennas ($G_r = 1$, $G_t = 1$) can be expressed as $L = -32.44 - 20 \log f_c/1\text{MHz} - 20 \log d/1\text{km}$, where the loss is found in dB.

2. Assume a receiver located 10 km away from a 50 W transmitter. The carrier frequency is 6 GHz. Free space propagation is assumed.
   (a) Find out the power received at the receiver.
   (b) Find out the magnitude of the e-field at the receiver antenna.
   (c) Find out the RMS voltage applied to the receiver input, assuming that the receiving antenna has a real impedance of 50 Ω and is matched to the receiver. (**Hint:** $v_{\text{rms}} = \sqrt{4P_r R}$ V)

3. A transmitter produces 55 W of power. If this power is fed to a unity gain antenna with 890 MHz carrier frequency, determine the received power in dBm and dBW. The free space distance is 200 metre between the transmitting and receiving antennas. Also find out the received powers at distances of 500 metres, 1 km, and 10 km. Will we get linear relationship between the received power and the distance? Comment on the results.

4. Consider a base station antenna with $h_T =$ 30 metres and a mobile antenna with $h_R =$ 2 metres. The mobile is $d = 1000$ metres separated from the base station. The carrier frequency is 1 GHz.
   (a) Compute length of the line of sight.
   (b) Compute length of the ground reflected wave.
   (c) Compute the phase difference due to path length differences.
   (d) Can we apply Egli's formula here? Explain.

5. In a cellular system, considering a cell, a mobile receiver is 2 km away from the base station using $\lambda/4$ monopole antenna with a gain of 2 dB. The e-field, at a distance of 1 km from the transmitter, is measured to be 10^{-3} V/m and the carrier frequency used is 910 MHz. Calculate the length L and the effective aperture of the receiving antenna.

6. In Problem 5, if two-ray ground reflection model is assumed, calculate the phase difference between the direct and reflected rays as well as the received power at the mobile if the height of the transmitting antenna is 72 m and the receiving mobile antenna height is 1.6 m above the ground. Gain of base station antenna can be assumed as unity.

7. Find the far-field distance for an antenna with maximum dimension of 1 m and operating frequency of 900 MHz.

8. Find whether flat or frequency-selective fading will occur in the following conditions. The rms delay spread is 1.3 ms and the transmission bandwidth is 1.5 MHz. What is the allowable symbol rate to allow only the flat fading condition for BPSK?

9. A transmitter carrier frequency is 890 MHz. For a vehicle moving at 30 km/hr, calculate the received carrier frequency if the mobile is
   (a) moving toward the transmitter,
   (b) moving away from the transmitter, and
   (c) moving in a direction that is perpendicular to the direction of arrival of the transmitted signal.

10. The RMS delay spread and mean excess delay are defined from a power delay profile, which is the spatial average of consecutive impulse response measurements collected and averaged over a local area. The power delay profile of a typical outdoor channel is given in Fig. 5.46.

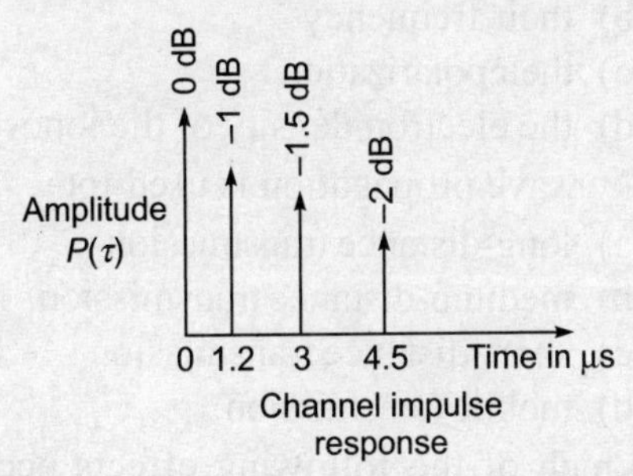

Fig. 5.46 Power delay profile for Problem 10

Calculate the RMS delay spread over the channel. If BPSK modulation is used, what is the maximum bit rate that can be sent over the channel without equalization? Estimate the coherence bandwidth of the channel.

Hint: Use the following equations:

(i) Mean excess delay,

$$\bar{\tau} = \frac{\sum_k P(\tau_k)\tau_k}{\sum_k P(\tau_k)}$$

(ii) RMS delay spread,

$$T_{\text{rms}} = \sqrt{\overline{\tau^2} - (\bar{\tau})^2}$$

(iii) $$\overline{\tau^2} = \frac{\sum_k P(\tau_k)\tau_k^2}{\sum_k P(\tau_k)}.$$

Further, note that RMS delay spread/symbol time should be less than or equal to 0.1. This condition will be helpful to decide allowable bit rate RMS delay spread is of the order of microsecond in outdoor channels and nanosecond in indoor radio channels.

11. Determine the isotropic free space loss at 4 GHz for the shortest path to a synchronous satellite from the earth at 35,864 km.

Multiple Choice Questions

1. The electric field of an electromagnetic wave at a point in free space is in the positive Y direction and the magnetic field is in the positive X direction. The direction of power flow will be in the
   (a) $+X$ direction (b) $+Y$ direction
   (c) $+Z$ direction (d) $-Z$ direction

2. In isotropic propagation, the maximum usable frequency (MUF) equals
   (a) $f_c \cos\theta$ (b) $\sqrt{f_c \cos\theta}$
   (c) $f_c \sec\theta$ (d) $\sqrt{f_c \sec\theta}$

   where f_c is the critical frequency and θ is the angle of incidence at the ionosphere measured with respect to the normal.

3. The skip distance is
   (a) independent of the frequency
   (b) independent of the state of ionization
   (c) independent of the transmitting power
   (d) dependent on the transmitting power

4. In the sky wave propagation, the skip distance is used
   (a) so as not to exceed the critical frequency
   (b) to avoid the Faraday effect
   (c) to prevent sky wave and upper ray interference
   (d) to obey tilting
5. The critical frequency of an ionospheric layer depends upon
   (a) only the height
   (b) only the electron density
   (c) both the height and electron density and nothing else
   (d) the height, electron density, and angle of incidence
6. Polarization of EM waves is due to
   (a) reflection
   (b) transverse nature of EM waves
   (c) longitudinal nature of EM waves
   (d) spherical wavefronts of EM waves
7. The virtual height of an ionospheric layer is
   (a) more than the height a wave actually reaches
   (b) less than the height a wave actually reaches
   (c) same as the height a wave actually reaches
   (d) none of the above
8. The ionosphere roughly extends from
   (a) 50 km to several earth radii
   (b) 50 km to 80 km
   (c) 50 km to 400 km
   (d) 50 km to 150 km
9. The total noise of a satellite earth station receiving system consists of
   (a) sky noise
   (b) parametric amplifier noise
   (c) antenna and feeder noise
   (d) all of the above
10. The ground wave eventually disappears as one moves from the transmitter because of
    (a) surface attenuation
    (b) loss of the line of sight
    (c) diffraction
    (d) tilting
11. If the reflected wave has to travel one-half wavelength more than the direct wave, the two waves will
    (a) arrive at the receiving antenna in phase
    (b) arrive at the receiving antenna out of phase
    (c) arrive at the receiving antenna in exactly the same time
    (d) not arrive
12. In the atmosphere, the absorption of the radio waves depends on
    (a) their distance from the transmitter
    (b) their frequency
    (c) their polarization
    (d) the electron density of the ionosphere
13. Sky wave propagation is used for
    (a) long-distance transmission
    (b) medium-distance transmission
    (c) short-distance transmission
    (d) mobile transmission
14. Which of the following effects occurs in tropospheric scatter propagation?
    (a) Faraday effect
    (b) Fading
    (c) Super-refraction
    (d) Atmospheric storm
15. Spectral density of white noise
    (a) varies with frequency
    (b) is constant for all frequency ranges
    (c) varies with bandwidth
    (d) is constant for limited range of frequencies
16. If η is the positive frequency power density, the power spectrum density of white noise $\delta(\omega)$ is equal to
    (a) $1/\eta$ (b) $\eta/2$
    (c) η (d) 2η

17. Gaussian probability density is defined as
    (a) $p(x) = \frac{1}{\sqrt{2\pi\sigma^2}} e^{-(x-m)^2/2\sigma^2}$
    (b) $p(x) = \frac{1}{2\pi\sigma^2} e^{-(x-m)^2/2\sigma^2}$
    (c) $p(x) = \sqrt{2\pi\sigma^2}\, e^{-(x-m)^2/2\sigma^2}$
    (d) $p(x) = 2\,\pi\sigma^2\, e^{-(x-m)^2/2\sigma^2}$
18. If x and y are two independent Gaussian random variables, each with average value zero and with variance σ^2, the joint density function is
    (a) $p(x, y) = p(x) + p(y)$
    (b) $p(x, y) = p(x) - p(y)$
    (c) $p(x, y) = p(x) \times p(y)$
    (d) $p(x, y) = p(x)/p(y)$
19. A geostationary satellite completes one orbit in
    (a) 1 hour (b) 6 hours
    (c) 24 hours (d) 7 days
20. A geostationary satellite and the earth have the same
    (a) acceleration
    (b) momentum
    (c) velocity
    (d) angular velocity
21. A transponder comprises
    (a) transmitter alone
    (b) receiver alone
    (c) transmitter and receiver
    (d) transmitter receiver and antenna
22. Which of the following happens when the transmission bandwidth is greater than the coherence bandwidth?
    (a) Frequency-selective fading
    (b) Flat fading
    (c) Large-scale fading
    (d) Small-scale fading
23. Signal outages may occur due to
    (a) destructive addition of the reflected rays
    (b) shadowing
    (c) fading
    (d) all of the above
24. What happens when the coherence time is less than the symbol duration?
    (a) Fast fading
    (b) Slow fading
    (c) Slow and flat fading
    (d) Fast and frequency-selective fading
25. Doppler effect is nothing but
    (a) a tone received with delay
    (b) a tone received with reduced amplitude
    (c) a tone received with a spectrum spread
    (d) motion of the mobile
26. Power delay profile can be used for
    (a) receiving the resultant phase of the signal
    (b) measurement of channel impulse response
    (c) velocity estimation
    (d) coherence time
27. Which of the following theories is/are applied to land mobile systems?
    (a) Troposcatter propagation
    (b) Free space propagation path loss law
    (c) Ground wave propagation
    (d) All of the above
28. What makes the wireless channel unpredictable?
    (a) Shadowing
    (b) Doppler spread
    (c) Multipath
    (d) All of the above

chapter 6

Channel Models, Diversity, Equalization, and Channel Estimation Techniques

Key Topics

- Considerations for modelling a wireless channel
- Realization of channels by digital filter
- Wideband channel model
- Phase error calculation by mathematical model
- Stochastic channel modelling
- Rayleigh fading channel model
- Rician channel model
- Nakagami channel model
- Comparison of Rayleigh, Rician, and Nakagami models
- Diversity techniques and space diversity combining
- Equalizers
- Channel estimation issues

Chapter Outline

After studying the various channel effects, one must be able to model a channel in order to analyse the effects of the channel on the transmitted signal and to predict the reception of the signal. Modelling of a channel is also required for the simulation purpose to check the performances in various selected conditions. Rayleigh, Rician, and Nakagami models are some popular models. Modelling a channel means correlating mathematics with the channel statistics. The earlier part of the chapter represents such models, while the latter deals with important techniques to eliminate the channel effects. Due to the channel effects, the received signal is degraded. By using some techniques, these channel effects can be reduced up to certain extent. This can improve the BER performance. There are a few techniques by which the receiver can combat the channel problems. They are diversity, equalization, and channel estimation methods. These all are described in detail in the chapter.

6.1 INTRODUCTION TO CHANNEL MODELLING

Perfect channel modelling is always required to study or analyse a wireless system. For that, the properties of the channels must be listed out and one or the other must be considered at a time for the modelling purpose. Based on the assumptions for the channel, the performance of the system will vary. Some important properties of a channel are as follows:

- Channels may be time varying or static. (Multipath effect makes the channel time varying and, depending upon the constructive or destructive interference, the quality of the received signal will vary.) Effect of mobility is that the channel varies with the user's location and time, which results in rapid fluctuations of received power. The slower you move, the lesser variations will be observed.

- Channels may be time dispersive or non-dispersive. (Due to dispersion, pulse spreading will be observed, which will result in ISI effect.)
- Channels may be linear or non-linear.
- All channels act as a low-pass filter under certain conditions as they show pulse spreading effect.
- Channel may be fast fading or slow fading, frequency selective or flat fading.

Depending upon the scenario or requirement of the system, mobile systems must follow the appropriate model for the analysis purpose. Two main categories of models for this purpose are as follows:

- Outdoor propagation model (e.g., Longley–Rice model, Durkin's Okumura model, and Hata model)[1]
- Indoor propagation model

For most of the channel modelling considerations as well as the equalization and channel estimation problems, the concept of digital signal processing is applied. Hence, reader must be thorough with the basics of DSP (discussed in Chapter 2) as well as the concept of digital filters, unit impulse response, etc.

The channel for mobile system applications is characterized by *multipath reception*—The signal offered to the receiver contains not only a direct line-of-sight radio wave but also a large number of reflected radio waves. Even worse in urban centres, the line of sight is often blocked by obstacles, and a collection of differently delayed waves is all what is received by a mobile antenna. The reflected waves interfere with the direct wave, which causes significant degradation of the performance of the link. If the antenna moves, the channel varies with location and time, because the relative phases of the reflected waves change. This leads to fading—time variations of the received amplitude and phase (as described in Chapter 5).

There are many constraints in modelling a channel. There may be different environment for different situation and different weather. There may be different objects in a room, or different surrounding conditions. Hence, it is very difficult to predict the number of reflected rays and whether the constructive (add-in phase) or destructive interference will occur. Because of this, the channel model is based on probability and its behaviour is represented by the probability density function (PDF).

Most of the conventional digital modulation techniques are sensitive to intersymbol interference unless the channel symbol rate is small compared to the delay spread of the channel. On the other hand, a narrowband signal with bit durations much longer than the delay spread may vanish completely in fade. A signal received at a frequency and location where reflected waves cancel each other, is heavily attenuated and may thus suffer large bit error rates.

1. These models can be self-studied.

A typical characteristic is observed on the wireless channel within fading and non-fading environment. In Fig. 6.1, the bit error probability for fading and non-fading channels is shown. In the non-fade channel, if the *C*/*I* ratio is increased slightly, there will be considerable drop in bit errors, i.e. the BER decreases rapidly.

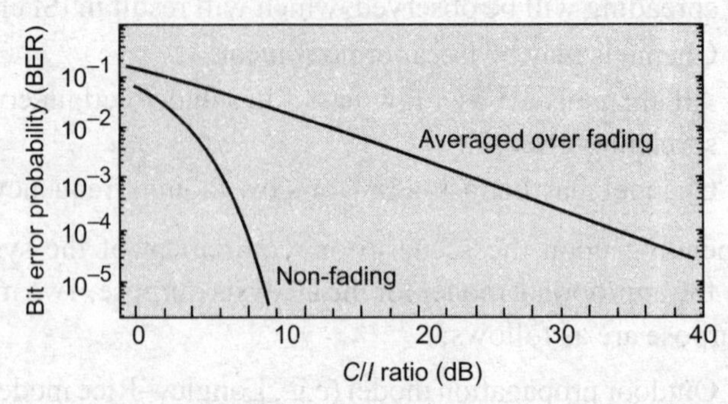

Fig. 6.1 Channel response over fading and non-fading environment

In a fading channel, the received signal is very weak and many bit errors occur. This phenomenon remains present even if the (average) signal-to-noise ratio is large. So, the BER only improves very slowly and with a fixed slope if plotted on a log-log scale. (Diversity or error correction can help to make the slope steeper and hence improves performance.) A few effects due to multipath reception are listed out in Table 6.1 in a simplified manner.

Table 6.1 Effects due to multipath reception

Application	Effect
1. Fast moving user	Rapid fluctuations of the signal amplitude and phase
2. Wideband (digital) signal	Dispersion and intersymbol interference
3. Analog television signal	'Ghost' images (shifted slightly to the right)
4. Multi-carrier signal	Different attenuation at different (sub)carriers and at different locations
5. Stationary user of a narrowband system	Good reception at some locations and frequencies; poor reception at other locations and frequencies.
6. Satellite positioning system	Strong delayed reflections may cause a severe miscalculation of the distance between user and satellite. This can result in a wrong estimate of the position.

As shown in Fig. 6.2, the model of many randomly phased sinusoids with variations in amplitudes appears to describe the wireless radio channel appropriately and to allow calculation of outage probabilities, fade durations, and many other critical parameters of wireless links. It greatly facilitated the development systems that can reliably communicate despite the anomalies and unpredictability of the mobile communication channel. However, the number of sinusoids will be limited due to inconvenience in adding all.

Although channel fading is experienced as an unpredictable and stochastic phenomenon to the user's device or the system planner, powerful models have been

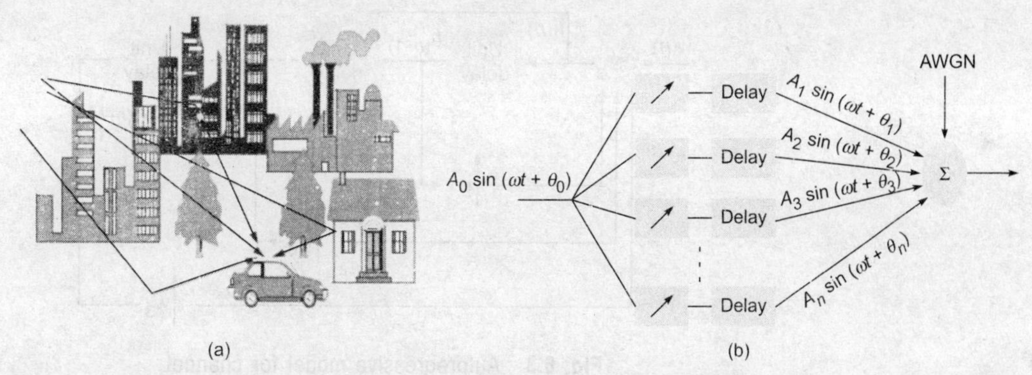

Fig. 6.2 Multipath reception scenario shown in (a) can be modelled as summation of signal and its phase delayed versions as shown in (b)

developed that can accurately predict average system performance. Countermeasures can be used to avoid system failure even if the channel exhibits fades at particular frequencies of particular locations. Such countermeasures are mostly used at the receiver side and are covered in Sections 6.9 to 6.12.

6.2 REPRESENTATION OF A DISCRETE CHANNEL BY FILTER

Wireless channel for mobile radio communications is a challenging medium that requires careful system design for reliable transmission. Conceptually, on a stationary channel, if an impulse is transmitted, then at the receiver, multiple delayed versions of impulses will be received at different instants of time due to multipath, which are non-correlated and with reducing amplitudes with time. (Power delay profiles studied in Chapter 5 represents the same concepts.) They are just like delayed samples and hence difference equation with coefficient values can be used to represent this concept, which is the equation for an FIR filter:

$$r(n) = w_0 s(n) + w_1 s(n-1) + w_2 s(n-2) + \cdots \tag{6.1}$$

In Eq. (6.1), the coefficients w's are the complex channel impulse response or fading coefficients. For the time-varying channels, this whole scenario becomes dynamic. The equation forms the basis for an equalizer design. The channel model for time-varying channel can be approximated by autoregressive (AR) model (Fig. 6.3):

$$h(n+1) = w_1\, h(n) + w_2\, h(n-1) + \ldots + w_n\, h(n-N) + w(n) \tag{6.2}$$

where w's are the weights/coefficient values to estimate the channel.

The concept will be better highlighted while discussing adaptive channel equalization and channel estimation where weights are to be adjusted. In this case, the received time domain signal $r(n)$ is a function of the transmitted signal $s(n)$, the estimated channel transfer function and AWGN $w(n)$. It can be expressed as

$$r(n) = s(n) * h_e(n) + w(n) \tag{6.3}$$

where '*' denotes the convolution and $h_e(n)$ is the estimated channel impulse response. The whole time domain scenario can be mapped into frequency domain representation.

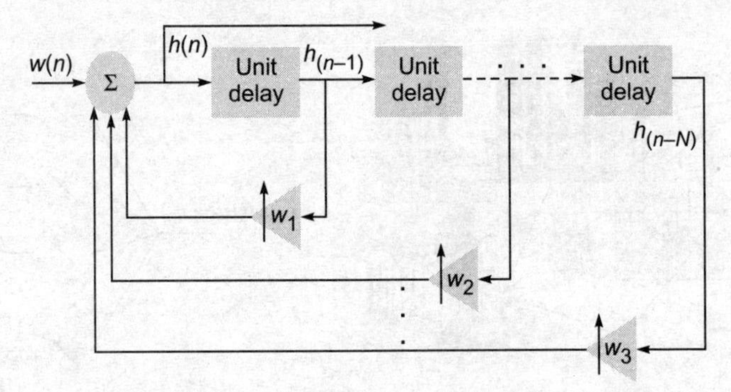

Fig. 6.3 Autoregressive model for channel

6.3 STOCHASTIC/STATISTICAL CHANNEL MODELLING CONSIDERATIONS

In principle, three different domains determine the radio signal transmission. These are as follows:

- Physical conditions selected or operational scenario
- Dispersion phenomena of wave propagation
- Transceiver characteristics

Table 6.2 shows a generic scheme of all the effects and parameters that are to be considered with regard to the set-up of the stochastic radio channel model (SCRM).

Table 6.2 Effects to be considered for stochastic radio channel model

Operation Scenario	Multipath Propagation	Dispersion Short-term Fluctuations	Long-term Fluctuations	Transceiver Characteristics
• Frequency range • Bandwidth • Type of environment	• Number of paths For each path: • Mean power • Delay • Doppler shift • Incidence direction • Scattering function • Polarization	For each path: • Fast fading	For each path: • Path loss • Shadowing • Transitions • Delay drift • Direction drift	• Trajectory • Velocity • Antenna configuration

Of course, the deterioration of the transmission quality is also strongly dependent on the signal processing, that is modulation, coding, detection, etc.

The *operation scenario* implies some fundamental data like frequency range, system bandwidth, and environments, (urban, rural, indoor, etc.), which even impress the general character of a wireless transmission link due to the relationship between the topographical features and the wavelength or the time resolution. *Dispersion* in frequency, time, direction, and polarization is a crucial aspect of radio communication. We have to distinguish between the multipath propagation, where each path suffers from a multiplicity of well-known effects, and the temporal fluctuations, which can be split up into short-term or small-scale fluctuations (fast fading) and long-term or large-scale fluctuations, comprising gradual and sudden changes of path parameters mainly because of movements either of the mobile terminals (MT) or reflectors and the scatters, respectively. *Transceiver characteristics* contributing to the SCRM are the parameters that describe the MT movement and the antenna configuration with its radiation pattern and diversity properties.

Concluding from Table 6.2, the consideration of all aspects leads to a rather high computational complexity of the SRCM. The dedication of the SRCM to a certain class of operational scenario, i.e. the delivery of indoor services with a given maximum bit rate (bandwidth) at a prescribed frequency range, is a first step toward a simplified model approach. Moreover, the identification of a few types of environments allows a remarkable complexity reduction. Thus, four important indoor situations can be chosen: small rooms, large rooms, factory halls, and corridors.

Obviously, the allocated frequency range plays a key role with respect to the path loss, while the room dimensions influence directly the delay spread. Referring to the uncertainty relation, the system bandwidth is inversely proportional to the time resolution and, therefore, determines the necessary sampling rate of the model. The evaluation of future generations of indoor systems equipped with smart antennas and intended for unrestricted mobile operation will require full complexity channel modelling.

A stochastic radio channel model has been developed in order to simulate realistic channel impulse responses (CIR) according to a wide range of possible physical situations within a given category of environments. If the channel is estimated on the basis of channel statistics, it is called the *blind method* of channel estimation. The complexity of this method is very high.

The transceiver characteristics as well as nearly all dispersion effects, i.e. the different phenomena of multipath propagation and short- and long-term fluctuations, are implemented in the SRCM. Under certain conditions, some effects that are expected to have only little impact to the system performance can be neglected. This may result in a considerable reduction of the model complexity. Further improvements of the SRCM, especially with regard to the consideration of terminal movements along arbitrary trajectories with varying velocity, are envisaged.

Among the different algorithms for the calculation of the SRCM parameters from the measured data, the space altering generalized expectation maximization (SAGE) algorithm reveals itself to be a powerful tool. Due to advanced semiconductor signal processing devices, even an online multiple parameter extraction (that is, complex amplitude, delay, incidence direction, and Doppler shift of the impinging wave components) seems to be possible in the near future. Thus, the SAGE algorithm will also be a promising signal processing scheme for the next generation of wireless communication systems using smart antennas. Figure 6.4 shows some processing stages for the algorithm in which fading and AWGN can be added artificially to measure the channel effects.

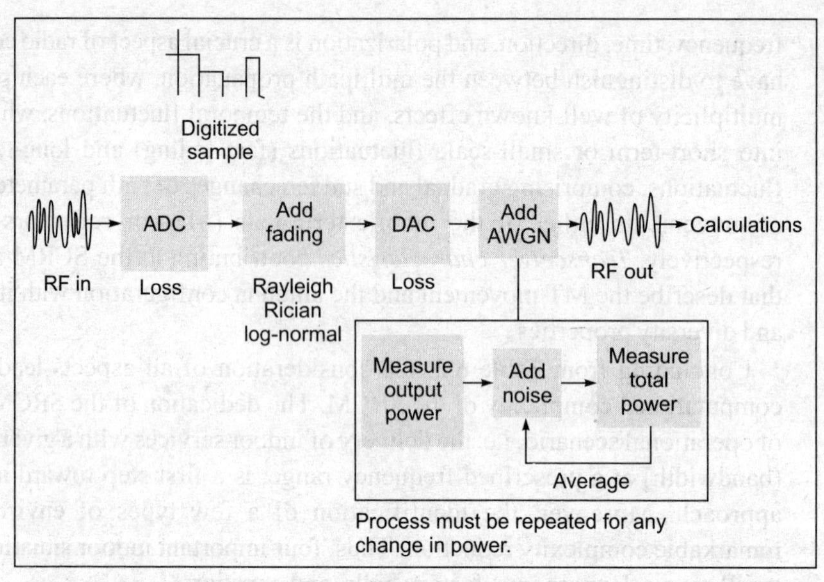

Fig. 6.4 Processing stages for SAGE algorithm to calculate channel parameters

6.4 WIDEBAND TIME-DISPERSIVE CHANNEL MODELLING CONSIDERATIONS

Most of the latest communication systems are digital and wideband, like CDMA and OFDM described in Chapters 8 and 9. At the same time, wireless channel is a dielectric medium and hence the refractive index of the channel, phase velocity, wave, and media propagation constants, etc., are the key parameters. So, over a wide range of frequencies in the transmission bandwidth, frequency-dependent performance is obtained.

Research in this area has been going on for more than 25 years and has concentrated mainly on two areas (modelling may be done mostly by using measurements):

1. Results are collected from extensive measurement campaigns. Wideband measurements are much more complicated than simple field strength (i.e. narrowband) measurements.
2. From the measurements, the channel models are derived, which should fulfill the following two criteria:
   - They must be simple enough to allow an analytical computation of basic system performance.
   - They must be very close to the physical reality; in other words, the performance computed by these models must be close to the performance measured in actually existing mobile radio channels.

These requirements are contradictory. So, models of different complexity and accuracy have been developed. Furthermore, some models might be more suitable for certain systems than others. For example, we can see that a so-called delay-based model is well suited for unequalized systems, but it is usually not for CDMA systems.

A further field of research, which is an intermediate step between measurement and modelling, is information condensation. As a wideband measurement campaign gives a lot of data, for further interpretation and processing, a few characteristic parameters must be extracted. This will reduce the amount of data while keeping the loss of information as less as possible. Stanford University Interim (SUI) 1 to 6 channels and ITU pedestrian A, B as well as ITU vehicular A, B channels are now used for wideband system simulations.

6.5 RAYLEIGH FADING MODEL

When the waves of multipath signals are out of phase, a reduction of the signal strength at the receiver can occur. This may result in deep fading. The basic model of Rayleigh fading assumes a received multipath signal to consist of a (theoretically infinite) large number of reflected waves with independent and identically distributed in-phase and quadrature amplitudes. This model has played a major role in our understanding of mobile propagation. The model was first proposed in a comment paper written by Lord Rayleigh in 1889. Signal amplitude (in dB) versus time graph for an antenna moving at constant velocity exhibiting Rayleigh fading is shown in Fig. 6.5. Note that the deep fades occur

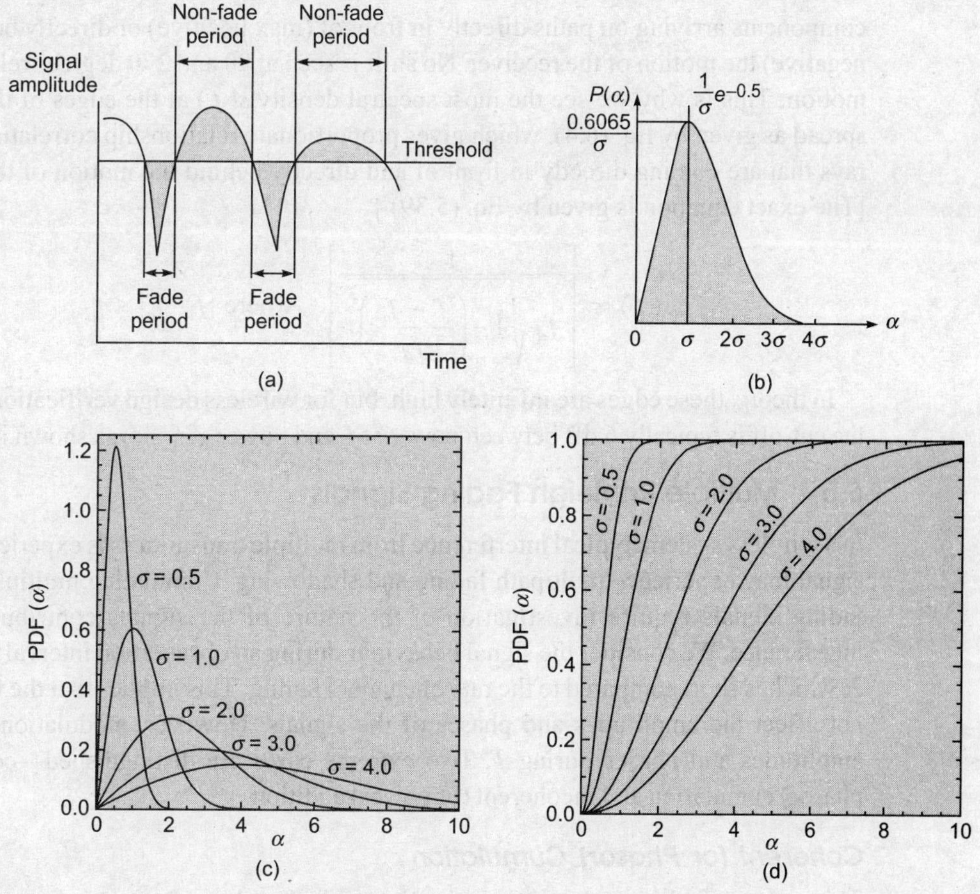

Fig. 6.5 (a) A sample of a Rayleigh fading signal, (b) Rayleigh PDF, (c) variations in Rayleigh PDF with change in standard deviation, and (d) variations in Rayleigh CDF with change in standard deviation or mean μ

occasionally. Although fading is a random process, deep fades have a tendency to occur approximately every half a wavelength of motion.

The Rayleigh distribution is a good model for channel propagation, where there is no strong line-of-sight path from the transmitter to the receiver. This can be used to represent the channel conditions seen on a busy street in a city, where the base station is hidden behind a building several blocks away and the arriving signal is bouncing off many scattering objects in the local area. In the time domain, Rayleigh fading looks like periodic peaks of 10 dB or less interspersed between deep troughs of 40 dB or more. These deep fades (nulls in signal power) will typically occur at separations of half a wavelength. The dense scatterer model, which explains cellular communications propagation, states that the amplitude of multipath rays will follow a Rayleigh distribution, while the angle of arrival (multipath phase) follows a uniform distribution. The PDF of Rayleigh distribution shown in Fig. 6.5(b) is given in Chapter 5, which is covered while explaining the nature of flat fading.

The Doppler shift of each incident ray is given by Eq. (5.37). The max spread f_d is determined by the speed of the vehicle and is only experienced by the spectral components arriving on paths directly in front of (max positive) or directly behind (max negative) the motion of the receiver. No shift is seen at 90 and 270 degrees relative to the motion. This is why we see the most spectral density $s(f)$ at the edges of the Doppler spread as given by Eq. (6.4), which gives proportionate relationship correlating with the rays that are coming directly in front of and directly behind the motion of the antenna. [The exact equation is given by Eq. (5.39).]

$$s(f) \propto \left[\frac{1}{f_d\sqrt{1-\left(\dfrac{f-f_c}{f_d}\right)^2}}\right] \quad \text{where } |f-f_c| < f_d \tag{6.4}$$

In theory, these edges are infinitely high, but for wireless design verification purposes, the cut-off is typically 6 dB between power at f_c and power at $f_c \pm f_d$ as shown in Fig. 5.34.

6.5.1 Multiple Rayleigh Fading Signals

In a wireless system, typical interference from multiple transmitters is experienced. Each signal may experience multipath fading and shadowing. Cumulating multiple Rayleigh fading signals require investigation of the nature of the signals contributing to the interference. We consider the signal behaviour during an observation interval of duration T, which is short compared to the rate of channel fading. This implies that the fading does not affect the amplitudes and phases of the signals. However, modulation can affect amplitudes and phases during T. Two extreme cases are distinguished—coherent (or phasor) cumulation and incoherent (or power) addition.

Coherent (or Phasor) Cumulation

This occurs if, during the observation interval, the phase fluctuations caused by the modulating signals are sufficiently small and the carrier frequencies of the signals are exactly equal. The joint signal behaves as a Rayleigh phasor, with Gaussian in-phase and quadrature phase components. The instantaneous power is exponentially distributed. The local-mean power is equal to the sum of the local-mean powers of the individual signals as shown in Fig. 6.6.

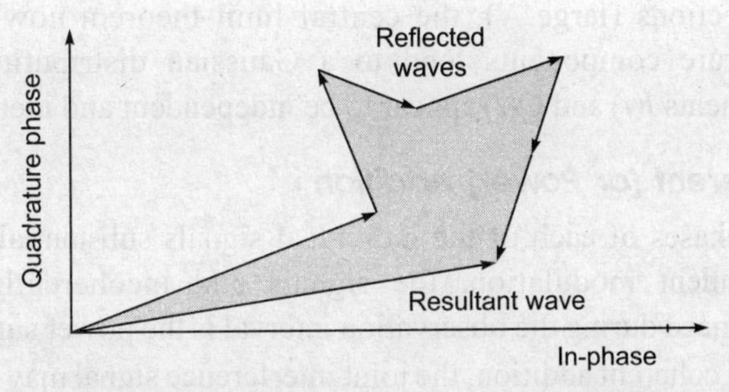

Fig. 6.6 Phasor diagram of a set of scattered waves resulting in a Rayleigh fading envelope

To understand Fig. 6.6 in a better way, consider a case of an unmodulated carrier, the transmitted signal has the form

$$s(t) = \cos(\omega_c t + \psi) \tag{6.5}$$

Considering the Doppler effect in Fig. 5.32, the received unmodulated carrier $r(t)$ can be expressed as

$$r(t) = \sum_{n=1}^{N} \alpha_n \cos(2\pi f_c t + \psi + \phi_n + 2\pi\Delta f_n t) \tag{6.6}$$

An in-phase quadrature representation of the form

$$r(t) = I(t)\cos\omega_c t - Q(t)\cos\omega_c t \tag{6.7}$$

can be found with the in-phase component, including the Doppler effect, as

$$I(t) = \sum_{n=1}^{N} \alpha_n \cos\left(\frac{2\pi v f_c t}{c}\cos\alpha_n + \psi + \phi_n\right) \tag{6.8a}$$

and the quadrature phase component can be found as

$$Q(t) = \sum_{n=1}^{N} \alpha_n \sin\left(\frac{2\pi v f_c t}{c}\cos\alpha_n + \psi + \phi_n\right) \tag{6.8b}$$

Coherent addition can occur only if phase modulation with a very small deviation is applied, or if the observation interval is taken short with respect to the rate of modulation. This occurs, for instance, in digital systems if the joint interference signal is studied during one bit interval or during the lock-in of a carrier-recovery loop in a synchronous detector.

Let us consider a stationary user, i.e. $v = 0$. An in-phase and quadrature representation reduces to

$$I(t) = \sum_{n=1}^{N} \alpha_n \cos(\psi + \phi_n) \tag{6.9a}$$

$$Q(t) = \sum_{n=1}^{N} \alpha_n \sin(\psi + \phi_n) \tag{6.9b}$$

Thus, both the in-phase and quadrature components, $I(t)$ and $Q(t)$, can be interpreted as the sum of many (independent) small contributions. Each contribution is due to a particular reflection, with its own amplitude α_n and phase. For sufficiently large number of reflections (large N), the central limit theorem now says that the in-phase and quadrature components tend to a Gaussian distribution of their amplitude. The components $I(t)$ and $Q(t)$ appear to be independent and identically distributed.

Incoherent (or Power) Addition

If the phases of each of the individual signals substantially fluctuate due to mutually independent modulation, the signals add incoherently. The interference power experienced during the observation interval is the power sum of the individual signals.

With coherent addition, the joint interference signal may exhibit deep fades, caused by mutually cancelling phasors from the signals. This cannot continue for a sustained period due to the phase variations caused by angle modulation of each signal or by slightly different carrier frequencies due to Doppler shifts and free-running oscillators. With incoherent cumulation, the joint interference signal behaves as a band-limited Gaussian noise source if the number of components is sufficiently large. Moreover, any fade of one of the signals is likely to be hidden by other interfering signals. Hence, the joint interference signal tends to exhibit less multipath fluctuations per unit of time than the signal from one individual interferer.

From the above description, it can be noted that if the antenna speed is set to zero, channel fluctuations no longer occur. Fading is due to motion of the antenna. An exception occurs if the reflecting objects move. In a vehicular cellular phone system, the user is likely to move out of a fade, but in a wireless LAN, a terminal may, by accident, be placed permanently in a fade where no reliable coverage is available.

6.5.2 Multiple Incoherent Rayleigh Fading Signals with Equal Mean Power

If the interference is caused by the power sum of n Rayleigh fading signals with identical local-mean power, the PDF of the joint interference power is the nth convolution of the exponential distribution of the power of an individual interfering signal. The PDF of the joint interference power caused by interfering signals with different local-mean powers can be approximated by a gamma distribution. It may not be fully appropriate to speak of the envelope of such a joint interference signal, but if one defines the amplitude to be proportional to the square root of the power, then one finds that the amplitude has a Nakagami distribution, which will be studied afterwards in the chapter.

6.5.3 PDF of Rayleigh Signal Amplitude

In order to obtain the PDF of the signal amplitude r of a Rayleigh fading signal, we observe the random processes of the in-phase and quadrature components, $I(t)$ and $Q(t)$, respectively, at one particular instant t_0.

If the number of received waves N becomes very large and all the waves are independent and identically distributed, the central limit theorem says that $I(t_0)$ and $Q(t_0)$

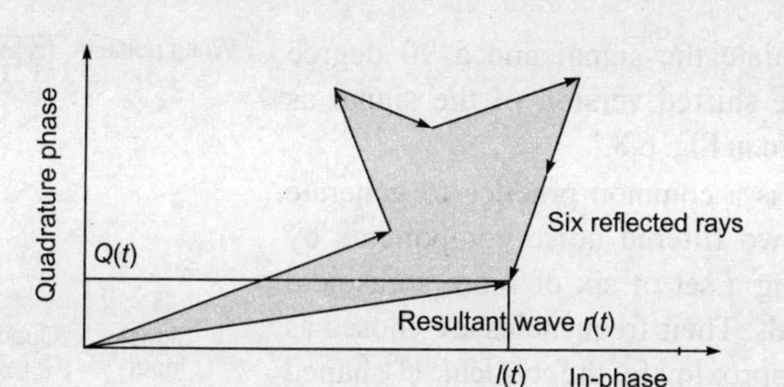

Fig. 6.7 A received signal consisting of $N = 6$ reflected waves (The resulting signal amplitude r consists of an in-phase component I and a quadrature component Q. If the antenna moves, the relative phases of the reflected waves change over time. So, r, I, and Q become functions of time t.)

at a particular instant t_0 are (zero-mean) Gaussian random variables, each with variance σ^2. Then Lord Rayleigh argued that the received signal is of the form mentioned in Eq. (5.40) and has a Rayleigh amplitude $\alpha(t)$, which is found from $\sqrt{I^2(t)+Q^2(t)}$, and a uniform phase $q(t)$ between 0 and 2π. The probability density of the amplitude is described by the 'Rayleigh' PDF represented in Eq. (5.41).

A few more relationships for Rayleigh fading distribution are given below.

1. The probability that the envelope of the received signal does not exceed a specified (threshold) value J is given by the corresponding CDF

$$P(J) = \int_0^J p(\alpha)d\alpha = 1 - \exp\left(-\frac{J^2}{2\sigma^2}\right) \tag{6.10}$$

2. The mean value of Rayleigh distribution is given by

$$E[\alpha] = \sigma\sqrt{\frac{\pi}{2}} = 1.2533\sigma \tag{6.11a}$$

3. The variance of Rayleigh distribution is given by

$$\sigma_\alpha^2 = E[\alpha^2] - E^2[\alpha] = 0.4292\sigma^2 \tag{6.11b}$$

4. The RMS value of the envelope is the square root of the mean square, i.e. $\sqrt{2}\sigma$, where σ is the standard deviation and median value of α is 1.177σ.

An important application of the probability density function is the calculation of outage probabilities, that is the probability that the signal strength drops below a certain threshold level.

If the set of reflected waves is dominated by one strong component, Rician fading is a more appropriate model.

6.5.4 Rayleigh Fading Simulator: An Example

Narrowband Rayleigh fading is modelled often as a random process that multiplies the radio signal by a complex-value Gaussian random function (for I and Q components). The spectrum of this random function is determined by the Doppler spread of the channel. Thus, one can generate two appropriately filtered Gaussian noise signals and use these to modulate the signal and a 90 degree phase shifted version of the signal as shown in Fig. 6.8.

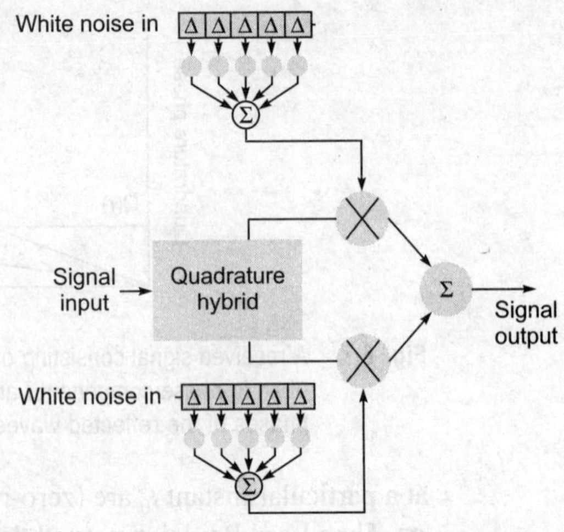

Fig. 6.8 Block diagram of a narrowband Rayleigh fading simulator (in baseband form)

It is a common practice to generate the two filtered noise components by adding a set of six or more sinusoidal signals. Their frequencies are chosen as to approximate the typical U-shaped Doppler spectrum as shown in Fig. 6.9. The N frequency components are taken at

$$f_i = f_d \cos [2\pi i/2(2N+1)] \quad (6.12)$$

where $i = 1, 2, \ldots, N$.

This specific set of frequencies is chosen to approximate the U-shaped Doppler spectrum. All amplitudes are taken equal to unity. One component at the maximum Doppler shift is also added but at an amplitude of $1/\sqrt{2}$, i.e. at about 0.707. One more simulator is designed by Dr. Kemilo Feher and is given in his book on *Digital Communication*.

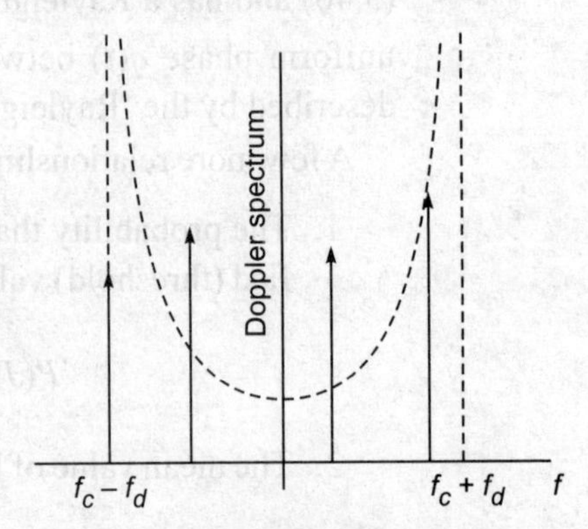

Fig. 6.9 N frequency components are chosen from the Doppler spectrum for simulation

As the demand for mobile communication increases, systems have to be more efficient and cell sizes smaller and smaller as described in Chapter 11. To describe microcellular propagation, the Rayleigh model lacked the effect of a dominant line-of-sight component, and Rician model appeared to be more appropriate.

6.6 RICIAN FADING MODEL

The Rician fading model is similar to that for Rayleigh fading, except that in Rician fading, a strong dominant component is present. This dominant component can, for instance, be the line-of-sight wave. The refined Rician models also consider the following points:

- The dominant wave can be a phasor sum of two or more dominant signals, e.g. the line of sight and a ground reflection. This combined signal is then mostly treated as a deterministic (fully predictable) process.
- The dominant wave can also be subject to shadow attenuation. This is a popular assumption in the modelling of satellite channels.

Besides the dominant component, the mobile antenna receives a large number of reflected and scattered waves as shown in Fig. 6.10.

Again consider the Doppler shift theory explained previously. The max spread f_d is determined by the speed of the vehicle and is only experienced by the spectral

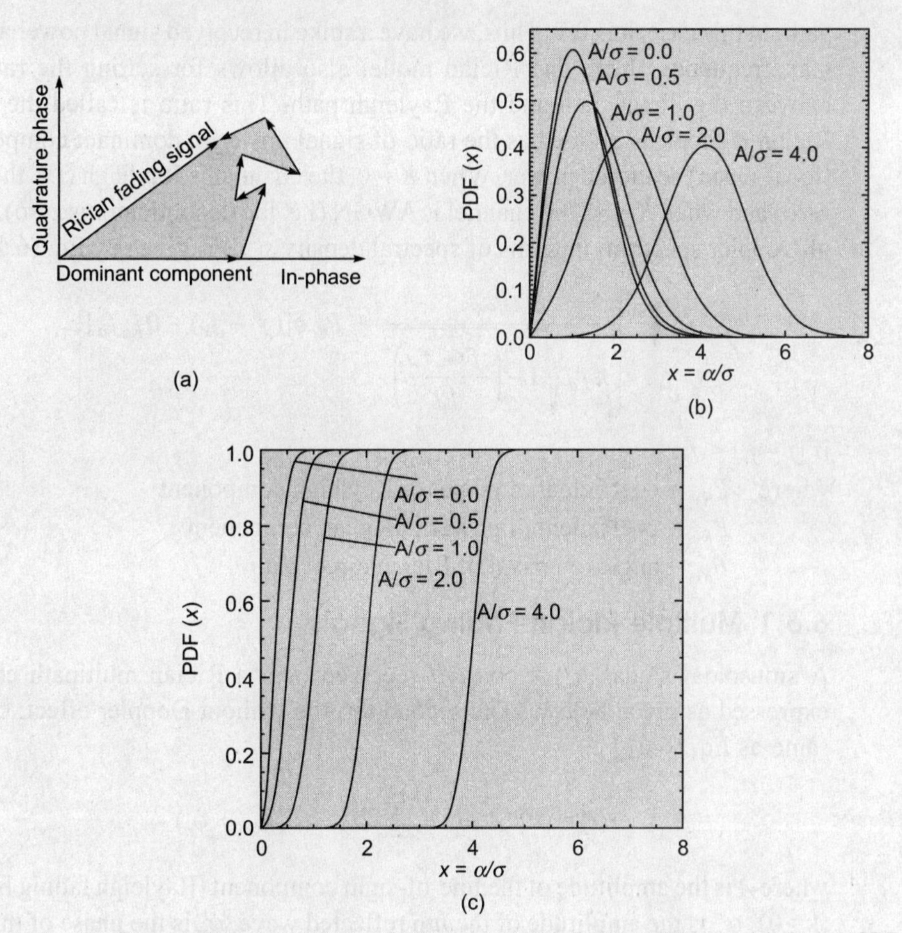

Fig. 6.10 (a) Phasor diagram of Rician fading signal, (b) Rician PDF, and (c) Rician CDF with change in standard deviation

components arriving on paths directly in front of (max positive) or directly behind (max negative) the motion of the receiver because this is where the cosine of the angle of arrival is 1. At any other angle, the cosine is less than 1. No shift is seen at 90 and 270 degrees relative to the motion. This is why we see the most spectral density at the edges of the Doppler spread, which correlates with the rays that are coming in directly in front of and directly behind the motion of the antenna.

The Rayleigh fading is considered a worst-case scenario here. In rural environments, where the multipath profile includes a few reflected paths combined with a strong line-of-sight path, the spectral power follows a Rician distribution. The angle of arrival of the direct ray as well as the ratio of the powers between the direct ray and the multipath rays, determine how much effect the energy from the direct path has on the normal Rayleigh model. Observing this effect in the frequency domain, what you see is a spike in the power corresponding to the frequency shift attributed to the direct ray. As an example, in the GSM specifications, the angle of arrival of the direct path is set to 45 degrees. We take the cosine of the angle to see how much statistical weight we give to that particular path. In this case, it is 0.7. Thus, we have a spike in received signal power at 0.7 times the max frequency shift. The Rician model also allows for setting the ratio of powers between the direct path and the Rayleigh path. This ratio is called the K-factor. The Rician K-factor is defined as the ratio of signal power in dominant component over the (local-mean) scattered power. When $K = 0$, the channel is Rayleigh (i.e. the numerator is zero) and when $K = \infty$, the channel is AWGN (i.e. the denominator is zero). The variation of Doppler spectrum in terms of spectral density $s(f)$ is given by Eq. (6.13).

$$S(f) \propto \left\{ \frac{P_{Ry}}{\pi f_d \sqrt{1-\left(\frac{f-f_c}{f_d}\right)^2}} + P_{Rc}\delta[(f-f_c)-\theta_{Rc} f_d] \right\} \tag{6.13}$$

if $|f - f_c| < f_d$

where P_{Ry} = coefficient of power of Rayleigh component
P_{Rc} = coefficient of power of Rician component
θ_{Rc} = angle of arrival of Rician direct path

6.6.1 Multiple Rician Fading Signals

A sinusoidal signal $s(t) = \cos \omega_c t$ received over a Rician multipath channel can be expressed as given below. [The second term is without Doppler effect; otherwise, it is same as Eq. (6.6).]

$$r(t) = A \cos \omega_c t + \sum_{n=1}^{N} \alpha_n \cos (\omega_c t + \phi_n) \tag{6.14}$$

where A is the amplitude of the line-of-sight component (Rayleigh fading is recovered for $A = 0$), α_n is the amplitude of the nth reflected wave, ϕ_n is the phase of the nth reflected wave, and $n = 1, ..., N$ identify the reflected, scattered waves.

Examples of Rician fading are found in

- Microcellular channels
- Vehicle-to-vehicle communication
- Indoor propagation
- Satellite channels

Similar to the case of Rayleigh fading, the in-phase and quadrature phase components of the received signal are independent and identically distributed jointly Gaussian random variables. However, in Rician fading, the mean value of (at least) one component is non-zero due to a deterministic strong wave.

For a large fade margin $F >> 1$, the probability that the instantaneous power drops below the noise threshold z tends to

$$\text{Outage probability} = \frac{(1+K)\exp(-K)}{F} \tag{6.15}$$

where the fade margin F is defined as the local mean power minus the threshold z.

6.6.2 PDF of Rician Signal Amplitude

The derivation of the probability density function of amplitude in Rician fading [Fig. 6.10(b)] is more involved than in Rayleigh fading, and a Bessel function occurs in the mathematical expression. It has been proposed to approximate this expression by the model for Nakagami fading. However, the behaviour of Nakagami and Rician fading in deep fades is essentially different. Approximations that focus on the behaviour near the mean value will divert by orders of magnitude in predicting the probability of deep fades.

The derivation is similar to the derivation for Rayleigh fading. In order to obtain the probability density of the signal amplitude α, we observe the random processes $I(t)$ and $Q(t)$ at one particular instant t_0. If the number of scattered waves is sufficiently large and the waves are independent and identically distributed, the central limit theorem says that $I(t_0)$ and $Q(t_0)$ are Gaussian but, due to the deterministic dominant term, are no longer zero mean. The Rician PDF is given by Eq. (5.42).The transformation of variables shows that the amplitude and the phase have the joint PDF represented by Eq. (5.43).

In the expression for the received signal, the power in the line of sight equals $A^2/2$. In indoor channels with an unobstructed line of sight between the transmit and receive antennas, the K-factor is between, say, 4 and 12 dB. Rayleigh fading is recovered for $K = 0$ ($-\infty$ dB). The total local-mean power $\overline{p}$ is the sum of the power in the line of sight and the local-mean scattered power. The local-mean scattered power equals $\overline{p}/(K + 1)$. The amplitude of the line of sight is $A = \sqrt{[2K\overline{p}/(K+1)]}$.

6.7 NAKAGAMI FADING MODEL

Besides Rayleigh and Rician fading, a refined model for the PDF of a signal amplitude exposed to mobile fading has been suggested. The distribution of the amplitude and signal power can be used to find probabilities on signal outages. The following points should be noted:

- If the envelope is Nakagami distributed, the corresponding instantaneous power is gamma distributed.
- The parameter m is called the *shape factor* of the Nakagami or the gamma distribution.
- In the special case $m = 1$, Rayleigh fading is recovered, with an exponentially distributed instantaneous power.
- For $m > 1$, the fluctuations of the signal strength reduce as compared to Rayleigh fading.

The Nakagami fading model was initially proposed because it matched empirical results for short-wave ionospheric propagation. In current wireless communication, the main role of the Nakagami model can be summarized as follows.

- It describes the amplitude of received signal after maximum ratio diversity combining (described in Section 6.11.4). After k-branch maximum ratio combining (MRC) with Rayleigh fading signals, the resulting signal is Nakagami with $m = k$.

The MRC combining of m-Nakagami fading signals in k-branches gives a Nakagami signal with a shape factor mk.

- The sum of multiple independent and identically distributed (IID) Rayleigh fading signals has Nakagami-distributed signal amplitude. This is particularly relevant to model interference from multiple sources in a cellular system.
- The Nakagami distribution matches some empirical data better than other models.
- The Nakagami fading occurs for multipath scattering with relatively large delay time spreads, with different clusters of reflected waves. Within any one cluster, the phases of individual reflected waves are random, but the delay times are approximately equal for all waves. As a result, the envelope of each cumulated cluster signal is Rayleigh distributed. The average time delay is assumed to differ significantly between clusters. If the delay times also significantly exceed the bit time of a digital link, the different clusters produce serious intersymbol interference. So, the multipath self-interference then approximates the case of co-channel interference by multiple incoherent Rayleigh fading signals.
- The Rician and the Nakagami models behave approximately equivalently near their mean value. This observation has been used in many papers to advocate the Nakagami model as an approximation for situations where a Rician model would be more appropriate. While this may be accurate for the main body of the probability density, it becomes highly inaccurate for the tails. As bit errors or outages mainly occur during deep fades, these performance measures are mainly determined by the tail of the probability density function (for probability to receive a low power).

6.8 COMPARISON OF RAYLEIGH, RICIAN, AND NAKAGAMI FADING MODELS

The Rician and Nakagami fadings are two generalizations of the model for the Rayleigh fading. In literature, often a Nakagami model is used for analytical simplicity in cases where Rician fading would be a more appropriate model. In contrast to common belief, the Nakagami model is not an appropriate approximation for Rician fading. It has an essentially different behaviour for deep fades such that results on outage probabilities or error rates can differ in the order of magnitude.

To describe microcellular propagation, the Rayleigh model lacked the effect of a dominant line-of-sight component and the Rician model appeared to be more appropriate. For analytical and numerical evaluation of system performance, the expressions for Rician fading are less convenient, mainly due to the occurrence of a Bessel function in the Rician probability density function of the received signal amplitude. Approximations by a Nakagami distribution, with simpler mathematical expressions, have become popular.

In the analysis of outage probabilities or error rates, it is the behaviour of the model for signals in deep fades that has the determining effect. As the behaviours of the probability density functions for amplitudes near zero differ significantly, approximations based on the behaviour near the mean are inappropriate.

The Rician and Nakagami models have a fundamentally different density for deep fades. Modelling a Rician fading signal by a Nakagami distribution of the amplitude leads to overly optimistic results, and discrepancies can be in many orders of magnitude. However, the Nakagami model is sometimes used to approximate the PDF of the power of a Rician fading signal. Matching the first and second moments of the Rician and Nakagami PDFs, we get

$$m = \frac{K^2 + 2K + 1}{2K + 1} \tag{6.16}$$

as in Eq. (5.46), which tends to $m = K/2$ for large K. In the special case that the dominant component is zero ($K = 0$) or $m = 1$, Rayleigh fading occurs with an exponentially distributed power.

The results are strikingly different for m larger than one. As the relation between K and m was based merely on the first and second moments, it is likely to be most accurate for the values close to the mean. Outage probabilities, however, highly depend on the tail of the PDF for small power of the wanted signal. The probability of deep fades differs for these two models. So, an approximation of the PDF of a Rician fading wanted signal by a Nakagami PDF can be highly inaccurate: Results differ even in the first-order behaviour. For Rician fading, the slope of the outage probability versus C/I is the same as for Rayleigh fading. For Nakagami fading, the slope is steeper, similar to that of m-branch diversity reception of a Rayleigh fading signal.

6.9 OPTIMAL SIGNAL DETECTION IN AWGN LTI CHANNEL USING MATCHED FILTER

The theory for signal transmission over AWGN linear time invariant (LTI) channels is very well developed and covered in many books. Many fundamental theorems in signal detection theory have been developed. However, the theory of the matched filter receiver is of particular interest.

The signal is multiplied by a locally stored reference copy and integrated over time (correlation). The matched filter (Fig. 6.11) correlates the incoming signal with a locally stored reference copy of the transmit waveform. The matched filter maximizes the signal-to-noise ratio for a known signal. It can be shown to be the optimal detector if

- the channel produces AWGN,
- the channel is LTI, and
- exact time reference is available (the signal amplitude as a function of time is precisely known).

If the noise is non-white, the matched filter can still be applied. In this case, one can pre-filter the incoming signal to make the noise component white. This is called a *whitening filter*. Evidently, this also filters the wanted signal. Therefore, the filtered incoming signal is not matched to (i.e. correlated with) the reference transmit signal but is matched to a reference signal fed through the filter that is identical to the whitening filter. Thus, two copies of the whitening filter are needed. It can be shown that one can build a detector that mathematically is equivalent but uses only one filter.

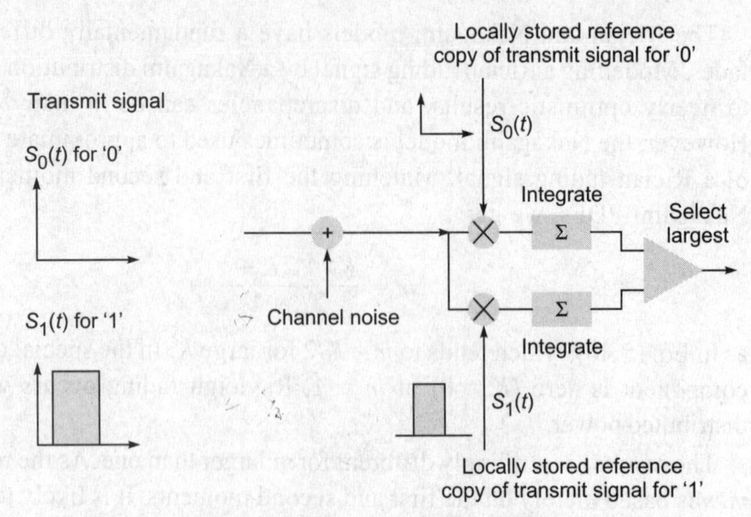

Fig. 6.11 Possible implementation of a matched filter receiver

If the channel is dispersive, the matched filter concept can still be used, but one must process the incoming signal with a locally generated copy of the expected waveform after transmission over the channel. That is, the receiver must estimate the channel impulse response and apply this to the reference signal waveform. The incoming signal is correlated with a reference waveform, which is dispersed in the same manner as the channel disperses the radio signal.

A complication is that such dispersion causes intersymbol interference. Theoretically, it is no longer optimum to detect the received symbols one by one. The maximum likelihood (ML) receiver correlates the incoming sequence with dispersed sequences of potentially transmitted waveforms, containing multiple successive bits.

6.10 DIVERSITY TECHNIQUES

In telecommunications, a diversity scheme refers to a method for improving the reliability of a message signal by utilizing two or more communication channels with different characteristics. Diversity plays an important role in combating fading and co-channel interference and avoiding error bursts. Multiple versions of the same signal may be transmitted and/or received and combined in the receiver. Alternatively, a redundant forward error-correction code may be added and different parts of the message may be transmitted over different channels.

In other words, a diversity scheme is a method that is used to develop information from several signals transmitted over independent fading paths. This means that the diversity method requires that a number of transmission paths be available, all carrying the same message but having independent fading statistics. The mean signal strengths of the paths should also be approximately the same. The basic requirement of the independent fading is that the received signals are uncorrelated. Therefore, the success of diversity schemes depends on the degree to which the signals on the different diversity branches are uncorrelated. The general basic model of diversity concept is given in Fig. 6.12, where different independent fading paths are shown with different channel impulse responses

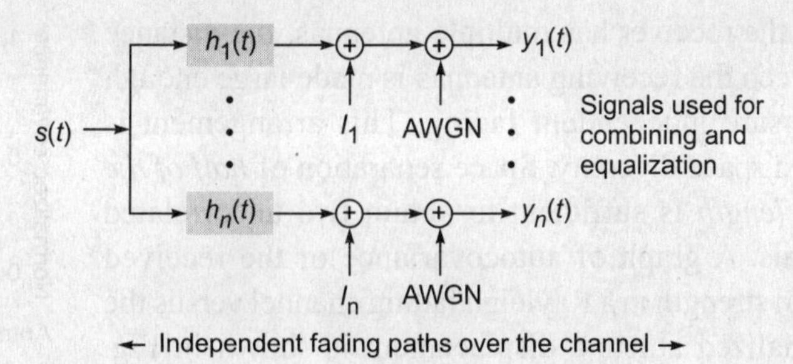

Fig. 6.12 Basic concept of diversity

and the distortion in the signal is observed due to the addition of interfering signal and white Gaussian noise. Then the multiple received signals are used to find out the strong signal, which can be further equalized and demodulated to receive the digital signal with minimum bit error rate.

Combining is the process to extract the main transmitted signal with minimum channel effects. This process is based on digital signal processing techniques in most of the cases. Properly combining the multiple signals will greatly reduce severity of fading and improve reliability of the transmission, because deep fades seldom occur simultaneously during the same time intervals on two or more paths.

Depending upon the type of fading, the diversity techniques have been categorized as microscopic diversity techniques and macroscopic diversity techniques.

Microscopic Diversity Techniques

These techniques are used in small-scale fading environment. They exploit the rapidly changing signal to avoid deep fades, for example, in case of small-scale fading it is known that if the two antennas are separated by a fraction of a metre, one may receive a null while the other receives a strong signal. By selecting the best signal at all times, a receiver can mitigate small-scale fading effects.

Macroscopic Diversity Techniques

These techniques are used in large-scale fading techniques. Here, we generally select the base station that is not shadowed when others are. The mobile can substantially improve the average signal-to-noise ratio on the forward link.

There are a number of diversity schemes described in the literature, and we will briefly consider some basic diversity schemes here.

Space Diversity

Space diversity is a macroscopic diversity technique. This is used at the transmitter or the receiver. The signal is transferred over several different propagation paths. In the case of wired transmission, this can be achieved by transmitting via multiple wires. In the case of wireless transmission, it can be achieved by antenna diversity using multiple transmitter antennas (transmit diversity) and/or multiple receiving antennas (diversity reception). In the latter case, a diversity combining technique is applied before further signal processing takes place.

If the receiver has multiple antennas, the distance between the receiving antennas is made large enough to ensure independent fading. This arrangement is called space diversity. Space separation of *half of the wavelength* is sufficient to obtain two uncorrelated signals. A graph of autocovariance of the received signal strength in a Rayleigh fading channel versus the normalized antenna displacement is shown in Fig. 6.13. Based on space diversity, various types of systems are possible. Space diversity and such systems are better highlighted in Chapter 15 on MIMO Systems.

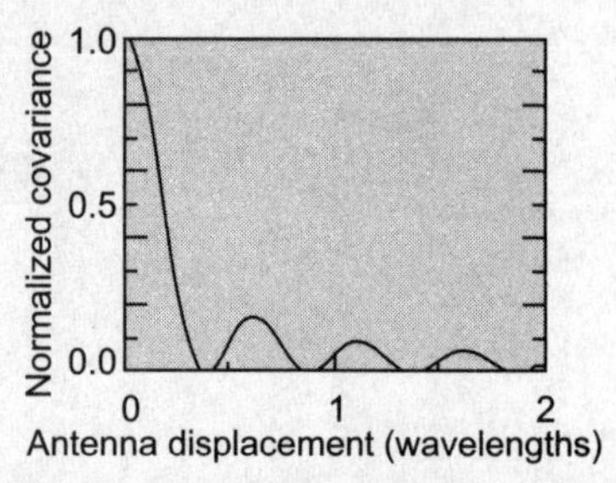

Fig. 6.13 Autocorrelation coefficient versus space separation

Site-based macrodiversity is a special form of space diversity. In site diversity, the receiving antennas are located at different receiver sites. For instance, in land mobile radio, where vehicle-mounted and hand-held radios communicate with a base station radio over a single frequency, space diversity is achieved by having several receivers at different sites. Signals from within a cell may be received at the different corners of the hexagonal area. The advantage is that not only the multipath fading attenuation is independent at each branch but the shadowing and path losses are also uncorrelated to some extent.

Polarization Diversity

Polarization diversity exploits the fact that obstacles scatter waves differently depending on their polarization. It is hoped that in (at least) one of the branches, the received waves do not cancel out each other, resulting in a relatively strong signal.

Antennas can transmit either a horizontal polarized wave or a vertical polarized wave. When both waves are transmitted simultaneously, received signals will exhibit uncorrelated fading statistics. This scheme can be considered as a special case of space diversity because separate antennas are used. However, only two diversity branches are available, since there are only two orthogonal polarizations.

Angle Diversity

In angle diversity, directional antennas receive only a fraction of all scattered energy. Since the received signal arrives at the antenna via several paths, each with a different angle of arrival, the signal component can be isolated by using directional antennas. Each directional antenna will isolate a different angular component. Hence, the signals received from different directional antennas pointing at different angles are uncorrelated.

Frequency Diversity

Here information is transmitted on more than one carrier frequency. The reason is that the frequencies separated by more than the coherence bandwidth of the channel will not experience the same fading. This is often employed in microwave line-of-sight links that carry several channels in a FDM mode. In practice, a 1:*N* protection switching is provided by a radio licence, wherein one frequency is nominally idle but is available on a standby basis to provide the frequency diversity switching for any one of the N other carriers. When diversity is needed, the appropriate traffic is simply switched to the backup frequency.

This scheme is exploited in wideband systems. Instead of transmitting the same signal using a different carrier, the signal is transferred using several frequency channels or spread over a wide spectrum that is affected by frequency-selective fading. Examples are OFDM multicarrier modulation (Chapter 9) and frequency hopping spread spectrum techniques (Chapter 8). Signals with different carrier frequencies far apart with each other are possibly independent. The carrier frequencies must be separated enough so that the fading associated with the different frequencies are uncorrelated. Thus, we can say that in an OFDM system, wide frequency-selective fading channel is converted into narrow flat fading channels due to multiple subcarrier channels (refer to Chapter 9). For frequency separations of more than several times the coherence bandwidth (refer to delay spread effect), the signal fading would be essentially uncorrelated, that is the difference in the carrier frequency should be more than the coherence bandwidth to achieve effective diversity. In frequency diversity, digital cellular systems can use slow frequency hopping (SFH) for diversity reasons: Each block of bits is transmitted at a different carrier frequency.

Time Diversity

Multiple versions of the same signal are transmitted at different time instants. Alternatively, a redundant forward error-correction code is added and the message is spread in time by means of bit interleaving before it is transmitted. Thus, error bursts are avoided, which simplifies the error correction. When the same data is sent over the channel at different time instants, the received signals can be uncorrelated if the time separations are large enough.

The time difference between the two transmissions should be large compared to the time the mobile antenna takes to move half a wavelength. The required time separation is at least as great as the reciprocal of the fading bandwidth, which is two times the speed of the mobile station divided by the wavelength. Hence, the time separation is inversely proportional to the speed of the mobile station. For the stationary mobile station, time diversity is useless. This is in contrast to all of the other diversity types discussed earlier because they are independent of the movement of the mobile station.

As a handy selection, we can repeatedly transmit information at time spacing that exceeds the coherence time of the channel, so that multiple repetitions of the signal will be received with independent fading conditions, thereby providing for diversity.

The systems with stationary antennas, like indoor wireless communication, time diversity will be less effective as the channel characteristics does not change very much with time. However, time diversity may be helpful if uncorrelated interference signals are experienced during successive attempts.

Joint Diversities

The increased mobility of users in cellular communication often results in fast fading or large Doppler spreads and the degradation of system performance is expected. The use of joint time-frequency diversity techniques provides significant performance improvement over the existing systems in the single user receiver design. Like the time-frequency diversity for the wireless channel characteristics, one can create the diversity dimension artificially, such as space-time diversity for multi-user detection in CDMA systems. Diversity gains can be achieved by using multiple elements for antenna arrays processing and adaptively combining space-time-frequency diversity signals.

Similarly, space and frequency diversity is exhibited in MIMO OFDM system. However, due to heavy computational complexity, the trade-off between the performance and complexity becomes an important issue when applying joint diversity techniques.

6.11 DIVERSITY COMBINING TECHNIQUES

Space diversity reception/combining methods can be classified into four types:

1. Selection diversity
2. Threshold diversity
3. Maximal ratio combining diversity
4. Equal gain diversity

6.11.1 Pure Selection Combining (SC)

The simplest combining scheme is selection combining, which is based on the principle of selecting the best signal (the largest energy or SNR) among all of the signals received from different branches.

In this scheme, m modulators are used to provide m diversity branches whose gains are adjusted to provide the same SNR for each branch. The receiver branch having the highest instantaneous SNR is then connected to the demodulator. The main drawback of this system is that it cannot function on a truly instantaneous basis. The conceptual diagram is shown in Fig. 6.14.

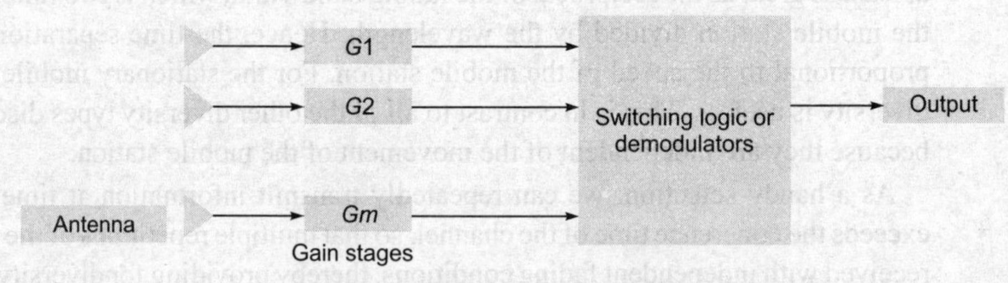

Fig. 6.14 Selection combining diagram

6.11.2 Threshold Combining (TC)

It is a special form of selection diversity, which is less expensive to implement. Here, a limited number of signals are considered for the extraction purpose using the threshold level. Scanning and feedback mechanism is used here. In this scheme, instead of always using the best of M signals, the M signals are scanned in a fixed sequence until one is found to be above the predetermined threshold. This signal is then received until it falls below the threshold and the scanning process is again initiated. The receiver switches to another signal when the current signal drops below a predefined threshold and so on. The constraint on this scheme is that the statistics obtained by this method are inferior to those obtained using the other methods. The threshold method is illustrated in Fig. 6.15.

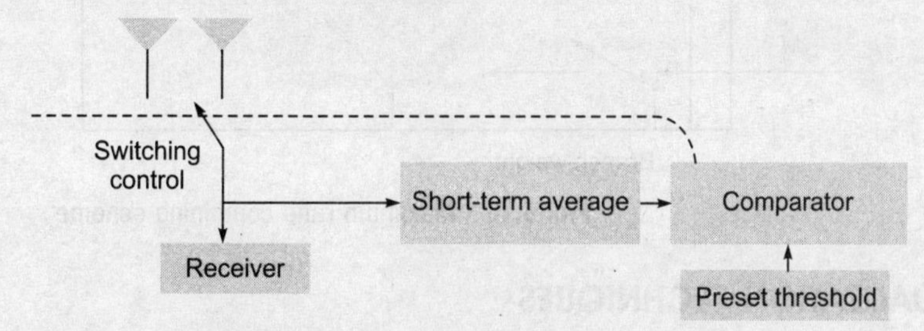

Fig. 6.15 Threshold diversity combining

6.11.3 Equal Gain Combining (EGC)

Equal gain combining (EGC) is better than selection diversity and almost as good as the MRC described next but is less complex in terms of signal processing. The idea will be clear from Fig. 6.16, which is given for the MRC case. In EGC, the branch weights are all set to unity (eliminated), that is the adaptively controlled amplifiers/attenuators are not needed. The signals from each branch are co-phased to provide equal gain combining. This allows the receiver to exploit signals that are simultaneously received on each branch. Thus, EGC is simpler to implement than MRC. Moreover, no channel amplitude estimation is needed. Though it is simpler than MRC, it gives worse results than MRC. However, the average SNR improvement of EGC is typically about 1 dB, worse than with MRC, but still much better than without diversity.

6.11.4 Maximum Ratio Combining (MRC)

For noise-limited systems without interference, MRC results in the best signal-to-noise ratio. Here all the incoming signals from all the M branches are weighted according to their individual signal voltage to noise power ratios and then added. All the individual signals must be co-phased before being added. This requires an individual receiver circuitry and phasing circuit for each antenna element. It produces an acceptable SNR compared to other techniques. It has tremendous signal processing and complex hardware. Figure 6.16 illustrates this scheme.

The received signals are weighted with respect to their SNR and then added. The resulting SNR yields $\sum_{k=1}^{M} \text{SNR}_k$, where SNR_k is SNR of the kth received signal.

There is one more method, interference rejection combining, which is used in smart antennas.

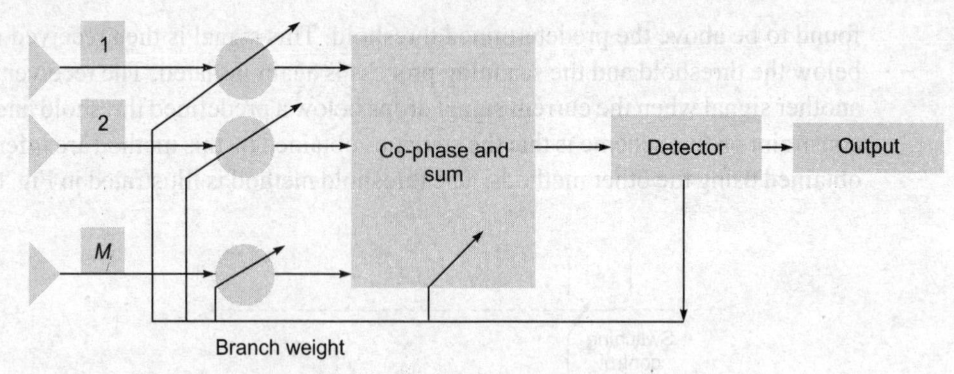

Fig. 6.16 Maximum ratio combining scheme

6.12 EQUALIZATION TECHNIQUES

Equalization is a method for the recovery of the distorted signal, when it is transmitted through the channel, received by a convolution process, and observed in additive noise. The process of recovery of a signal, that is convolved with the impulse response of a communication channel is known as *deconvolution* or *equalization*. For this, we must design a stage in the receiver whose transfer function will be exactly opposite to the channel transfer function. Figure 6.17 represents the basic concept of equalization.

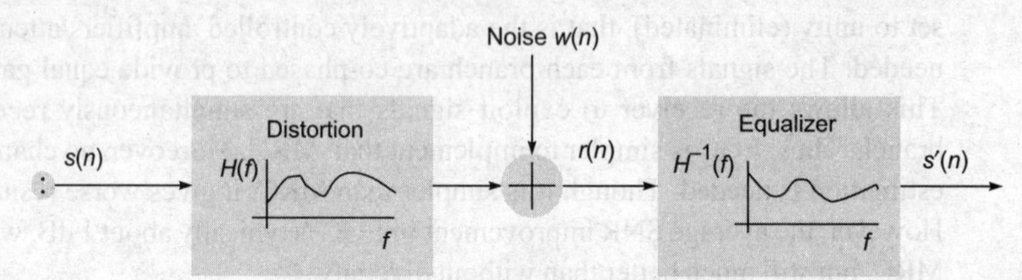

Fig. 6.17 Basic concept of equalization

The concept can be represented mathematically in discrete from with the equations as follows.

$$r(n) = h(n)*s(n) + w(n) \tag{6.17}$$

If noise is removed by noise elimination filter and if the following operation is performed, the original signal can be detected with the removal of channel effects and phase ambiguities.

$$s'(n) = h^{-1}(n)*r(n) \tag{6.18}$$

From Eq. (6.18), it is clear that one must design inverse channel filter for the equalization purpose. Thus, the equalization problem is relatively simple when the channel response is known and invertible and when the channel output is not noisy. However, in most practical cases, the channel response is unknown, time varying, and non-linear and may also be non-invertible. Also, the AWGN is observed in the output.

The simplest equalizer is the matched filter correlator. The use of matched filter in channel equalization is given in Fig. 6.18. Afterwards, the equalized signal can be demodulated.

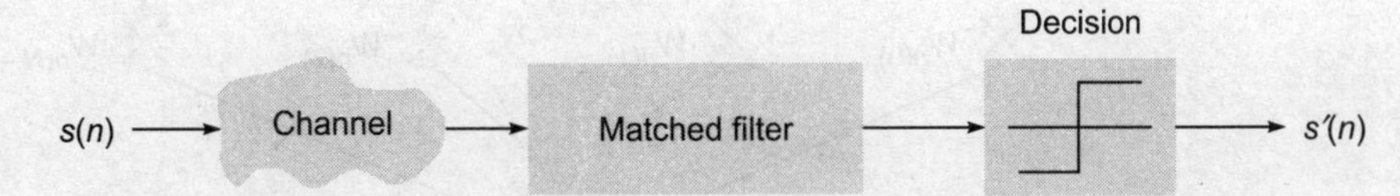

Fig. 6.18 Channel equalization concept in a digital communication system using matched filter and decision making

The equalization error signal or convolution noise is generated due to white Gaussian noise present in the signal. The *equalization error* is defined as the difference between the channel equalizer output and the desired signal. If the channel distorts the pulse shape, the matched filter will no longer be matched perfectly, intersymbol interference may increase, and the system performance will degrade. The least square error (LSE), minimum mean square error (MMSE), linear mean square (LMS), or some other methods may be used for checking the minimum error or maximum matching condition.

Digital communication systems can provide schemes for equalizer training periods, during which a training pseudo-noise (PN) sequence, also known at the receiver, is transmitted. A synchronized version of the PN sequence is generated at the receiver, where the channel input and output signals are used for channel estimation based equalizer by using some adaptive algorithm. The obvious drawback of using training periods for channel equalization is that power, time, and bandwidth are consumed for equalization process.

6.12.1 Transversal Filter

Before there were truly discrete, digital filters, there were continuous time analog filters (such as delay lines) that were 'tapped' at discrete points. The signal input appeared in delayed versions; it transversed in time (i.e. 'went across') as the delay line taps are viewed or accessed at one instant of time. Thus, they were identified as transversal filters. After digital filters, the task of equalization has become easier and is made adaptive. Digital filters can be realized with FIR filter for the causal systems in which only past components are considered for estimating the present component.

From an analysis point of view, we can just substitute 'FIR' for 'transversal'. A transversal filter and an FIR filter can be viewed as the same thing. The frequency response is periodic in the inverse of the (assumed regular) tap separation in time. A transversal equalizer is a transversal filter that usually has adjustable weights/coefficients in the filter such that the equalization can be adjusted for best performance. The transversal filter acting as a basis for adaptive filter is described in detail as follows. The block diagram and the concept behind this equalization will be clear from Fig. 6.19, where $y(n)$'s (arbitrary constants) represent the delayed versions of discrete signal and w's represent the weights for the coefficient adjustment.

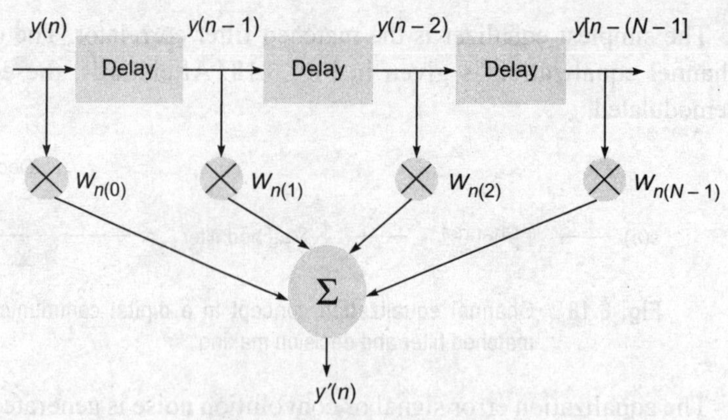

Fig. 6.19 Transversal filter

In N-tap transversal filter, the value of N is determined by practical considerations. An FIR filter is chosen because of its stability. Equations (6.1) and (6.2) represent the suitability of selection of an FIR filter. The use of the transversal structure allows relatively straightforward construction of the filter to model a channel. As the input, coefficients, and output of the filter are all assumed to be complex valued, the natural choice for the property measurement is the modulus or instantaneous amplitude. If $y'(n)$ is the complex-valued filter output, then $|y(n)|$ denotes the amplitude. The convergence error $e(n)$ can be defined as follows:

$$e(n) = |y(n)| - A \tag{6.19}$$

where A is the amplitude in the absence of signal degradations.

The error $e(n)$ should be zero when the envelope has the proper value, and non-zero otherwise. The error carries sign information to indicate the direction in which the envelope is in error. The adaptive algorithm is defined by specifying a performance/cost/fitness function based on the error $e(n)$ and then developing a procedure that adjusts the filter impulse response so as to minimize or maximize that performance function.

$$y'(n) = \sum_{i=0}^{i=N-1} W_{n(i)} y(n-i) \tag{6.20}$$

Conventional transversal filters require a multiplying element for each tap. As multiplication is a time-consuming process, a multiplier-less transversal filter can be designed, which can carry out the multiplication in the logarithmic domain to save time. That is, all the multiplications are replaced by additions. Such a filter is shown in Fig. 6.20.

6.12.2 Adaptive Equalizer

A transversal filter is often inserted in front of the matched filter correlator to compensate for the channel variations as shown in Fig. 6.21, where W represents the weights or the coefficients of the filters that can be varied in accordance with the channel conditions so that an exact approximation of the channel can be possible. Also, we must note that the filter is usually linear—although modifying the coefficients very rapidly can cause the filter to become 'non-linear' during the change.

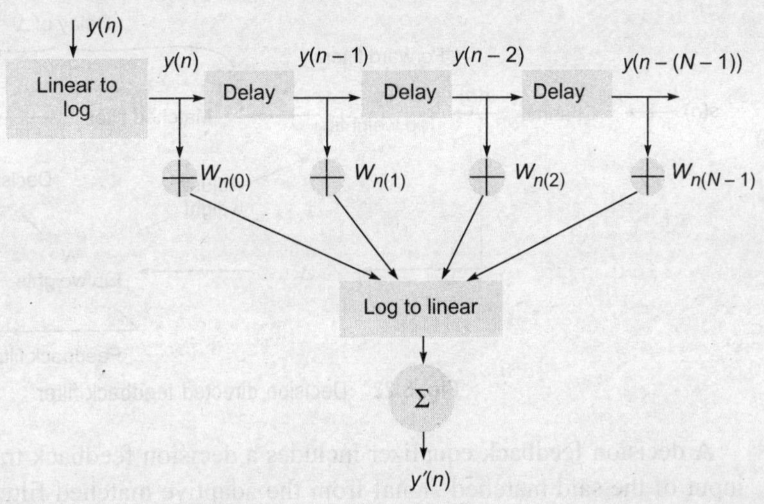

Fig. 6.20 Multiplier-less transversal filter (multiplications are replaced by additions)

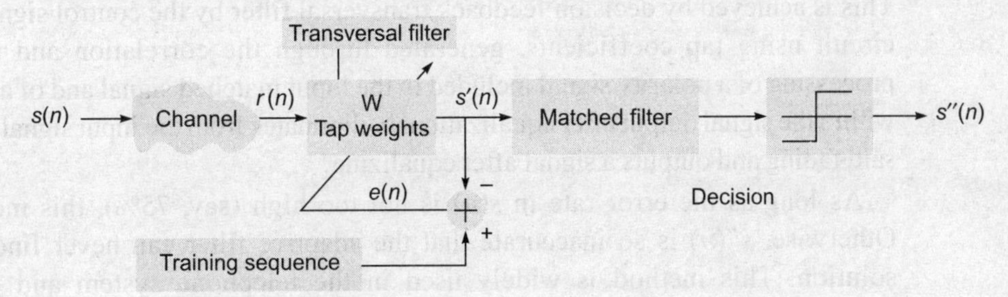

Fig. 6.21 Adaptive equalization using adaptive matched filter

This is, of course, unrealizable unless we have access to the original transmitted signal, here the training sequence. The training sequence can be used in this manner. There is a periodical broadcast of a known training signal. The adaptation is switched on only when the training signal is being broadcast and thus $s(n)$ is known.

An adaptive matched filter (which includes a transversal filter for inputting the said digital baseband signal and a control signal generating circuit for supplying tap coefficients to the transversal filter) makes symmetric and asymmetric impulse responses within the said input digital baseband signal due to the said fading in the said propagation paths and outputs of a matched signal. This is achieved by having control over the transversal filter by the control signal generating circuit using tap coefficients generated through correlation and time-average processing of a polarity signal indicating polarity of the digital baseband signal and of a polarity signal indicating polarity of the output signal of the transversal filter.

6.12.3 Decision Directed Feedback Equalizer

Another option is decision directed feedback (Fig. 6.22). If the overall system is working well, then the output s″(n) should almost always equal $s(n)$. We can thus use our received digital communication signal as the desired signal, since it has been cleaned of noise (we hope) by the non-linear threshold device.

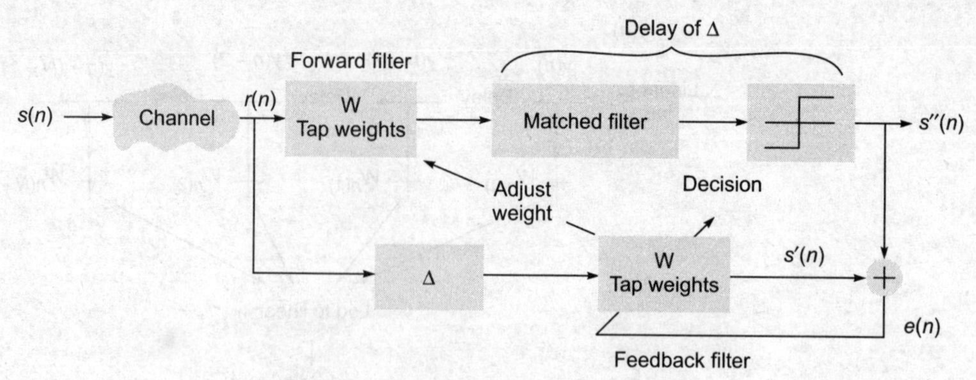

Fig. 6.22 Decision directed feedback filter

A decision feedback equalizer includes a decision feedback transversal filter for the input of the said matched signal from the adaptive matched filter and a control signal generating circuit for supplying tap coefficients to the decision feedback transversal filter. This is achieved by decision feedback transversal filter by the control signal generating circuit using tap coefficients, generated through the correlation and time-average processing of a polarity signal included in the input matched signal and of an error signal within the signal output after equalization. It eliminates from the input signal the ISI due to said fading and outputs a signal after equalizing.

As long as the error rate in $s(n)$ is not too high (say, 75%), this method works. Otherwise, $s''(n)$ is so inaccurate that the adaptive filter can never find the Wiener solution. This method is widely used in the telephone system and other digital communication networks.

To avoid the transmission power and bandwidth consumption, it is preferable to have a blind equalization scheme that can operate without access to the channel input. It uses the Bayesian estimation algorithm. Furthermore, in some applications, such as blurred images, all that is available is the distorted signal and the only restoration method applicable is blind equalization. The blind equalization is feasible only if some statistical knowledge of the channel input, and perhaps that of the channel, is available.

The blind equalization involves two stages: channel identification and deconvolution of the input signal and the channel response. Here the channel identification or channel estimation is done by using some statistical models.

6.13 CHANNEL ESTIMATION METHODS

To send the transmitter signal to the receiver with the features to combat the channel problems, some characteristics of the channel must be estimated, maybe in terms of delay between the transmitter and the receiver or in terms of the channel impulse response. Channel estimation is nothing but estimating channel impulse response at the receiver. The received training signal and transmitted training signal are used for estimation of the CIR. Afterwards, it is utilized for the detection of the required signal in equalized form.

One method of blind estimation is described previously. However, many useful estimation algorithms are available based on the application.

We have studied the adaptive equalization in the previous section. A training sequence based adaptive equalization method can be converted into a training sequence (also called pilot symbols) based channel estimation method. A comparison between two main channel estimation methods is given in Table 6.3. However, the hidden Markov model based on the states or the linear prediction model based on past record can also be used to estimate the present condition of the channel. Memory consumption is more in these cases.

Table 6.3 Comparison between channel estimation methods

Training Sequence Based Method (CSI)	Blind Method (No CSI)
1. Sequences known to the receiver are embedded into the frame and sent over the channel.	1. No training sequence required but uses certain underlying mathematical properties of the data being sent.
2. Easy and most popular but consumes more bandwidth.	2. Excellent for applications where bandwidth is scarce.
3. Not too computationally intense.	3. Extremely computationally intensive.
4. Its major drawback is that it is wasteful of the information bandwidth.	4. Its major drawback is that it is hard to implement on real-time systems.
5. Suitable for fast fading environment.	5. Not suitable for fast fading environment.

Channel estimation methods are widely used in CDMA as well as OFDM systems. Pilot-based channel estimation method in OFDM is described in Chapter 9. Kalman filter can be used as an estimation filter. The pilot information represents the channel state information (CSI) as it undergoes the same channel effects as the data. The pilot information can be added in time or frequency domain or in both.

Summary

- Channel modelling is required to simulate a channel during performance analysis while analysing wireless systems.
- Assumptions for the channel in the modelling will decide the performance of the system. With change in assumptions, the performance will change.
- A channel is time varying, dispersive, nonlinear, and fading type when it undergoes multipath.
- For long- and short-distance communication channels, assumptions must be different.
- Channel behaviour is random in nature and hence is represented by its PDF.
- A multipath channel can be represented by a digital filter equation due to direct and/or multiple reflected delayed components received at the receiver.
- A time-varying channel can be approximated by autoregressive model.
- Wideband channels exhibit frequency selective fading.
- Rayleigh fading model has multiple reflected rays added at the receiver while Rician model has a strong direct ray with multiple reflected rays. Nakagami fading occurs when large delay spreads are there.
- Detection of a signal over AWGN LTI channel can be done optimally using matched filter (simple equalization concept). In case of complicated channel assumptions, better equalization techniques are required.
- There are two types of diversity techniques: macroscopic and microscopic.
- Diversity techniques are used to combat the fading effects over the channel, which may be

based on space, time, frequency, angle, polarization, site, etc.

- Combining is the process to extract the main transmitted signal with minimum channel effects. This process is based on digital signal processing techniques in most of the cases.
- Maximum ratio combining gives the best performance but is complex in hardware.
- Equalization techniques are required for the phase compensations, i.e. indirectly the channel compensations.
- Adaptive equalizers and decision feedback equalizers are more useful in case of fast fading channels, where continuous corrections are required for the equalizations. However, they must be fast enough.
- Channel estimation means estimation of channel impulse response. This response is used for the detection of transmitted signal.
- Channel estimation can be done by using training sequences or using channel statistics based blind methods.

REVIEW QUESTIONS

1. Why do we require channel modelling? What are the difficulties one might face while modelling a channel?
2. What are the different characteristics of a channel taken into considerations while modelling mathematically? Correlate this with the mathematical models developed for the various types of systems defined in the field of digital signal processing (e.g., linear time-invariant system).
3. What will be the channel considerations for long- and short-distance channel modelling?
4. Find out the assumptions for the channel in case of indoor propagation model and outdoor propagation model. Find out the models suitable for indoor and outdoor mobile applications.
5. 'The wideband signal transmission is a serious problem in wireless communication.' Justify this statement.
6. Compare the Rayleigh, Rician, and Nakagami fading models. Why do we use Rayleigh fading model in case of short-distance communication?
7. Justify that diversity techniques definitely help in combating the mulipath channel.
8. Why do we require combining methods in space diversity? Do we require combining methods in case of frequency, time, or polarization diversity? Why?
9. What are the types of diversity techniques used in case of cellular communication?
10. What is the difference between channel equalization and channel estimation?
11. What do you mean by weights in case of adaptive equalizers? On what basis, they are varied?
12. How is the training sequence useful in case of channel equalization as well as estimation?
13. Discuss the working of a matched filter and justify its suitability for the equalizers.
14. What complexity will be there in case of blind estimation in comparison with training sequence based methods? How is the training sequence method disadvantageous?

PROBLEMS

1. A pure sinusoidal wave of frequency of 100 MHz with zero phase is transmitted through a channel. If it undergoes different delays over three different paths and if the delay over each path creates the phase difference of $\pi/8$, $\pi/5$, and $\pi/6$, respectively, what will be the phase of the resultant signal at the receiving end? Will it be constructive addition?
2. For a Rayleigh channel model, if the PDF has the standard deviation for a particular amplitude as 0.125, find out the mean and variance of the PDF.
3. If the typical Rayleigh fading envelope at 900 MHz, with the receiver speed of 120 km/hr, is between +7 dB and –25 dB about RMS value of the envelope, get the approximate PDF.
4. Design a three-tap linear transversal equalizer for the received pulse $r(t)$, where

   $r(0) = 1, r(1) = 0.3, r(-1) = -0.3$
   $r(2) = 0.1, r(-2) = 0.2$
   $r(3) = -0.03, r(-3) = -0.02$

   Also find out the setting of the coefficient values.
5. For an indoor mobile communication system, find out the outage probability in case of Rician fading if the local mean power is 15 dB and the receiver noise threshold level is 6 dB. The Rician factor is 5. Also prove the relationship that you use.

MULTIPLE CHOICE QUESTIONS

1. In space diversity reception system, we combine the outputs of several receivers at the output of
   (a) RF amplifier
   (b) detector
   (c) IF amplifier
   (d) audio frequency amplifier
2. Direct (LOS) component is not considered in
   (a) Rician fading
   (b) Nakagami fading
   (c) Rayleigh fading
   (d) all of the above
3. Which of the following models is/are a good approximation for cellular mobile communication with vehicular mobility?
   (a) Rician model
   (b) Nakagami model
   (c) Rayleigh model
   (d) All of the above
4. *K*-factor is the ratio of powers between
   (a) the direct path and the reflected path
   (b) the direct paths of two base stations
   (c) the direct path and the Rayleigh paths
   (d) the scattered paths and direct path
5. When *K*-factor is zero, the model becomes
   (a) Rician model
   (b) Nakagami model
   (c) Rayleigh model
   (d) Hata model
6. For indoor propagation, the most suitable models is/are
   (a) Rician model
   (b) Nakagami model
   (c) Rayleigh model
   (d) all of the above
7. Matching the first and second moments of the Rician and Nakagami PDFs, the shape factor of the Nakagami model and the Rician factor can be related as

(a) $m = \dfrac{K^2 + K + 1}{2K + 1}$

(b) $m = \dfrac{K^2 + 2K + 1}{2K + 1}$

(c) $K = \dfrac{m^2 + m + 1}{2m + 1}$

(d) $m(K + 1) = K^2 + 2K + 1$

8. The best efficient combining method is
   (a) maximum ratio
   (b) equal gain
   (c) pure selection
   (d) threshold based

9. The most important block in the equalizer design is
   (a) integrator
   (b) comparator
   (c) FIR filter
   (d) matched filter

10. Which of the following methods is hard to implement over real-time systems?
   (a) Pilot based channel estimation
   (b) Adaptive equalization
   (c) Blind channel estimation
   (d) None of the above

11. Spread spectrum systems employ a form of diversity called
   (a) frequency diversity
   (b) radio diversity
   (c) spatial diversity
   (d) bandwidth diversity

chapter 7

Single Carrier Digital Modulation Techniques

Key Topics

- Digital modulation performance parameters
- Coherent and non-coherent systems
- Polar and constellation diagrams
- Eye diagram
- Trellis diagram
- Constant envelope modulation schemes: BPSK, QPSK, M-PSK, FSK, MSK, and GMSK
- Variable envelope modulation schemes: ASK, QAM, and M-QAM
- Differential modulation schemes: DQPSK and DQAM
- IQ offset modulation scheme: OQPSK
- Bandwidth and spectrum efficiency
- Transmission power related issues
- Phase error calculation model

Chapter Outline

Modulation is required to make the baseband signal suitable for transmission over a channel as described in Chapter 2. In this chapter, we will concentrate on single carrier based digital modulation schemes. Among these schemes, a few are power efficient and others are spectrally efficient. The general feature of these schemes is the phase mapping in the *IQ* vector plane representing the typical constellations. The system performance is different for different modulation schemes. Hence, on the basis of application, environment/channel and SNR condition, the suitable scheme must be selected. Different diagrams, such as eye diagram, polar/vector diagram, constellation diagram, and trellis diagram, are explained in the chapter as they are important in the performance analysis and comparison of the modulation schemes. The spectrum efficiency and power considerations are also described in the chapter. Here, for all the schemes, the input signal will be in digital form and the output signal will be in analog form.

7.1 DIGITAL MODULATION AND PERFORMANCE PARAMETERS

Let us start with the revision of the fundamentals. Digital information signal in time domain can be viewed as a sequence of pulses depending upon the bit pattern of 1 and 0. The bit rate is simply inverse of the bit period. These signals have equivalent representations in frequency domain, where the energy of the signal is spread across a set of frequencies. This representation is called the *power spectrum* or simply *spectrum*. The signal bandwidth is a measure of the width of the spectrum. The bandwidth and bit rate of a digital signal are related but not exactly the same. The relation between the two depends on the type of modulation used. The ratio of the bit rate to the available bandwidth is called *spectral efficiency* and is represented in the unit bits/s/Hz.

Typically, the objective of a digital communication system is to transport digital data between two or more nodes. In wireless digital communications, this is usually achieved by adjusting physical characteristic of a sinusoidal carrier, that is frequency, phase, amplitude, or a combination thereof. The adjustment of carrier characteristic is performed in real systems with a modulator at the transmitting end to impose the physical change to the carrier and a demodulator at the receiving end to detect such changes and thereafter to get the data back.

Along with the modulator and demodulator stages, developers of communications systems face the following constraints:

- Available bandwidth
- Permissible power
- Inherent noise level of the system

In general, the move to digital modulation provides more information capacity, compatibility with digital data services, higher data security, better quality communications, and quicker system availability. Also, the choice of modulation scheme can satisfy one or more of the above-mentioned constraints.

The choice of digital modulation scheme will significantly affect the characteristics, performance, and resulting physical realization of a communication system. There is no universal 'best' choice of scheme, but depending on the physical characteristics of the channel, required level of performance and target hardware trade-offs, some will prove a better fit than others. Consideration must be given to the required data rate, acceptable level of latency, available bandwidth, anticipated link budget and target hardware cost, size, and current consumption.

The physical characteristics of the time-varying channel undergoes fast changes due to multipath and will typically significantly affect the choice of optimum system rather than long-haul links. The objective of this chapter is to review the key characteristics and salient features of the main digital modulation schemes used, including consideration of the receiver and transmitter requirements. Simulation can be used to compare the performance and trade-offs of different modulation schemes (MSK, M-PSK, M-QAM, etc.), including analysis of key parameters, such as occupied bandwidth and bit error rate in the presence of AWGN.

To understand and compare different modulation schemes and efficiencies, it is important to first understand the difference between bit rate and symbol rate, which is provided in Section 2.4 (Chapter 2). Rather than bits, symbols are considered for the digital modulation schemes because symbols can be easily converted into modulo values for implementation purpose and also the symbols are actually mapped into constellation diagram (Section 7.1.3). The symbol clock represents the frequency and exact timing of the transmission of the individual symbols. At the symbol clock transitions, the transmitted carrier is at the correct *I*/*Q* (or magnitude/phase) value to represent a specific symbol (a specific point in the constellation).

7.1.1 Coherent and Non-coherent Systems

The terms 'coherent' and 'incoherent' are frequently used when discussing the generation and reception of digital modulation. When linked to the process of modulation, the term *coherence* relates to the ability of the modulator to control the phase of the signal and not just the frequency. The example of frequency shift keying (FSK) given in Fig. 7.1 shows that FSK can be generated both coherently with an *IQ* modulator and incoherently with simply a voltage controlled oscillator (VCO) and a digital voltage source.

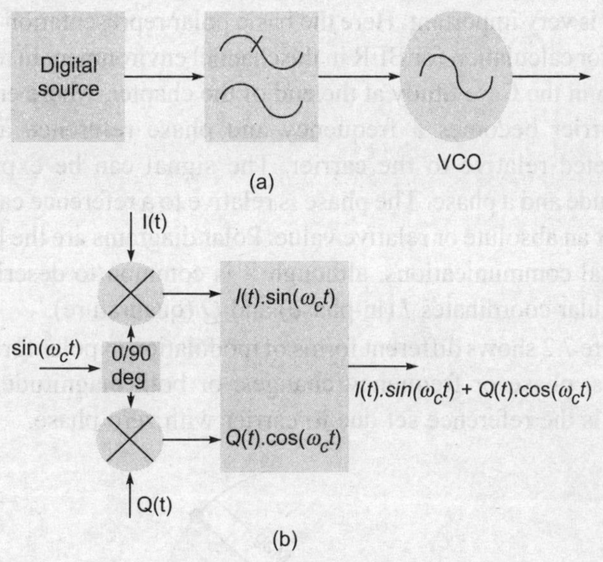

Fig. 7.1 (a) Non-coherent and (b) coherent generation of FSK modulated signal

With the system as shown in Fig. 7.1(a), the instantaneous frequency of the output waveform is determined by the modulator (within a tolerance set by the VCO and data amplitude, etc.), but the instantaneous phase of the signal is not controlled and can have any value. Alternatively, the coherent generation of the modulation is achieved as shown in Fig. 7.1(b). Here the phase of the signal is controlled, rather than the frequency.

The modulator shown above offers the possibility to shape the resultant carrier phase trajectory at baseband either with analog filtering or digital signal processing and a DAC. This can be used to generate both constant amplitude and amplitude modulated signals. Use of the term *coherent* with respect to the act of demodulation refers to a system that makes a demodulation decision based on the received signal phase and not frequency. The additional 'information' due to FEC results in an improved BER performance. The high level of digital integration now possible in semiconductor devices has made digitally based coherent demodulators common in mobile communications systems.

To understand the coherent system, first it is necessary to study polar representation of the signal and then its *IQ* components. For the phase control based systems, at the receiver end, on the basis of received phase, hard or soft decision making is to be performed incorporated to form the bit sequence again before channel decoding is to be performed.

This issue is covered in Chapter 4 and further details are given in the Case Study at the end of this chapter.

7.1.2 Polar Representation and *IQ* Diagrams

Any signal can be represented by its amplitude, frequency, and phase. Carrier is such an analog signal and in digital modulation schemes, amplitude and phase of the carrier become very important once the frequency of the carrier is chosen. Amplitude is important to identify logic levels and phase for synchronization. A simple way to view amplitude and phase is the polar diagram. Especially to find out the bit error rate, a polar display is very important. Here the basic polar representation is explained. Mathematical model for calculation for BER in the channel environment in terms of polar representation is given in the Case Study at the end of the chapter, where error handling is considered. The carrier becomes a frequency and phase reference and the received signal is interpreted relative to the carrier. The signal can be expressed in polar form as a magnitude and a phase. The phase is relative to a reference carrier signal. The magnitude is either an absolute or relative value. Polar diagrams are the basis of many displays used in digital communications, although it is common to describe the signal vector by its rectangular coordinates *I* (in-phase) and *Q* (quadrature).

Figure 7.2 shows different forms of modulation in polar form. There may be magnitude changes, phase or frequency changes, or both magnitude and phase changes. Zero degree is the reference set due to carrier with zero phase.

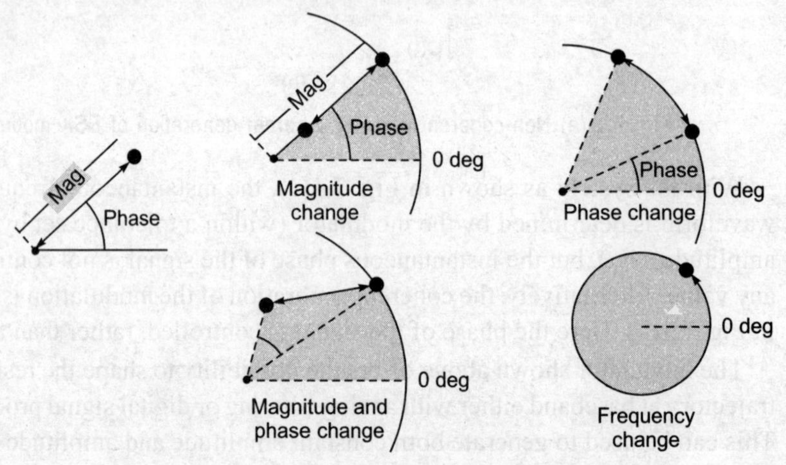

Fig. 7.2 Signal changes or modifications in polar form

In digital communications, modulated signal is often expressed in terms of *I* and *Q* components. This is a rectangular representation of the polar diagram. On a polar diagram, the *I* axis lies on the zero degree phase reference and the *Q* axis is rotated by 90 degrees. The signal vector's projection onto the *I* axis is its '*I*' component and the projection onto the *Q* axis is its '*Q*' component. This representation is very important to understand all shift keying techniques, as well as the OFDM technique (discussed in Chapter 9), where modulation phase mapping is used. These *I* and *Q* components are orthogonal and do not interfere with each other.

The I/Q diagrams are particularly useful because they mirror the way most digital communications signals are created using an I/Q modulator. In the transmitter, I and Q signals are mixed with the same local oscillator (LO) frequency with a 90 degree phase shifter placed in one of the LO paths. Signals in I and Q channels that are separated by 90 degrees are also known as being orthogonal to each other or in quadrature. They are two independent components of the signal. When recombined, they give a composite output signal or complex signal whose magnitude is found by following the trigonometry rules. The main advantage of I/Q modulation is the symmetric ease of combining independent signal components into a single composite signal and later splitting such a composite signal into its independent components. The example for this is given in Figures 7.4 and 7.5.

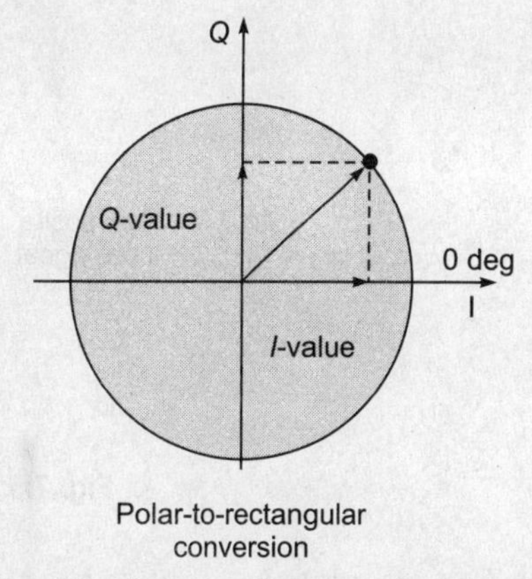

Fig. 7.3 *IQ* vector diagram representation

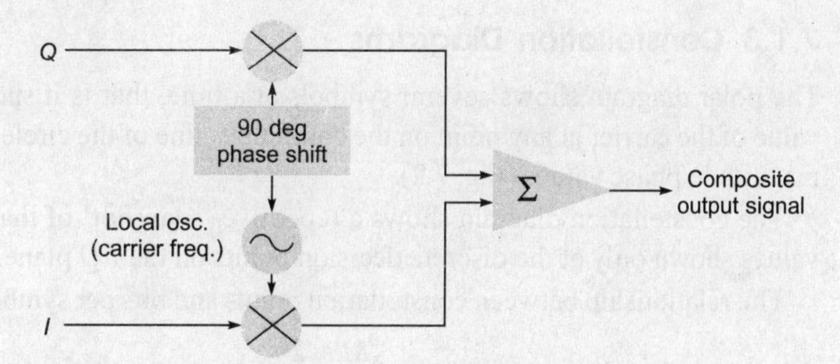

Fig. 7.4 Concept of combining *I* and *Q* channels to generate a composite signal (QPSK case)

The composite signal with magnitude and phase (or I and Q) information arrives at the receiver input. The input signal is mixed with the local oscillator signal at the carrier frequency in two forms as shown in Fig. 7.5. One is at an arbitrary zero phase. The other has a 90 degree phase shift. The composite input signal is thus broken into an in-phase I and a quadrature Q. These two components of the signal are independent and orthogonal. One can be changed without affecting the other. Normally, information cannot be plotted in a polar format and reinterpreted as rectangular (Cartesian) values without doing a polar-to-rectangular conversion. This conversion is exactly what is done by the in-phase and quadrature mixing processes in a digital radio. A local oscillator, phase shifter, and two mixers can perform the conversion accurately and efficiently.

Most digital modulation techniques map the data to a number of discrete points on the I/Q plane. These are known as *constellation points*. As the signal moves from one point to another, simultaneous amplitude and phase modulation usually results. To accomplish this with an amplitude modulator and a phase modulator is difficult and complex. It is also impossible with a conventional phase modulator. The signal may, in principal, circle the

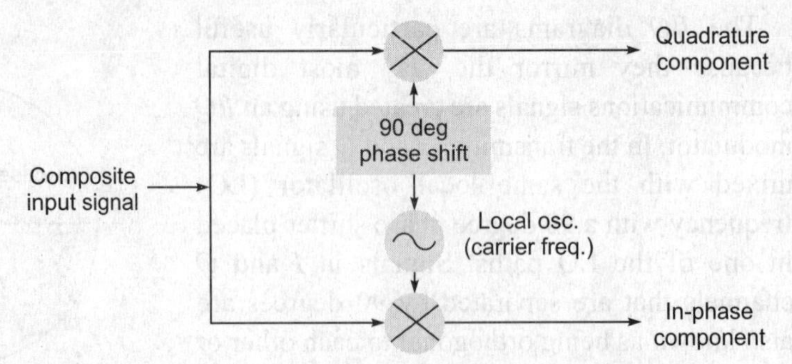

Fig. 7.5 Concept of splitting *I* and *Q* channels at the receiver

origin in one direction forever, necessitating infinite phase shifting capability. Alternatively, simultaneous AM and phase modulation is easy with an *I*/*Q* modulator.

The symbol clock represents the frequency and exact timing of the transmission of the individual symbols. At the symbol clock transitions, the transmitted carrier is at the correct *I*/*Q* (or magnitude/phase) value to represent a specific symbol (a specific point in the constellation).

7.1.3 Constellation Diagrams

The polar diagram shows several symbols at a time, that is it shows the instantaneous value of the carrier at any point on the continuous line of the circle, represented as *I*/*Q* or magnitude/phase values (Fig. 7.2).

The constellation diagram shows a repetitive 'snapshot' of that same burst, with the values shown only at the discrete decision points on the *I*/*Q* plane.

The relationship between constellation points and bits per symbol is

$$M = 2^n$$

where M = number of constellation points and n = bits/symbol

or

$$n = \log_2 (M) \tag{7.1}$$

Constellation diagrams can also be considered as signal space diagram represented in terms of basis functions, which is one dimensional for ASK and two dimensional for M-PSK schemes. In general, M-ary PSK constellation is circular except in the special case of $M = 2$. The constellation diagram displays phase errors as well as amplitude errors, at the decision points. The transitions between the decision points affect the transmitted bandwidth. Constellation diagrams indirectly provide insight into varying power levels, the effects of filtering and phenomena, such as intersymbol interference. Constellation diagrams are shown along with different modulation schemes in Figures 7.9 to 7.11 and various other figures.

7.1.4 Eye Diagrams

Another way to view a digitally modulated signal is with an eye diagram. Separate eye diagrams can be generated, one for the *I*-channel data and another for the *Q*-channel data. Eye diagrams display I and Q magnitude versus time in an infinite persistence mode, with retraces. In Fig. 7.6, I and Q transitions are shown separately and an 'eye' (or eyes) is formed at the symbol decision times. QPSK has four distinct I/Q states, one in each quadrant. There are only two levels for I and two levels for Q. This forms a single eye for each I and Q. Other schemes use more levels and create more nodes in time through which the traces pass. The second example is a 16QAM signal, which has four levels forming three distinct 'eyes'. The eye is open at each symbol. A 'good' signal has wide open eyes with compact crossover points.

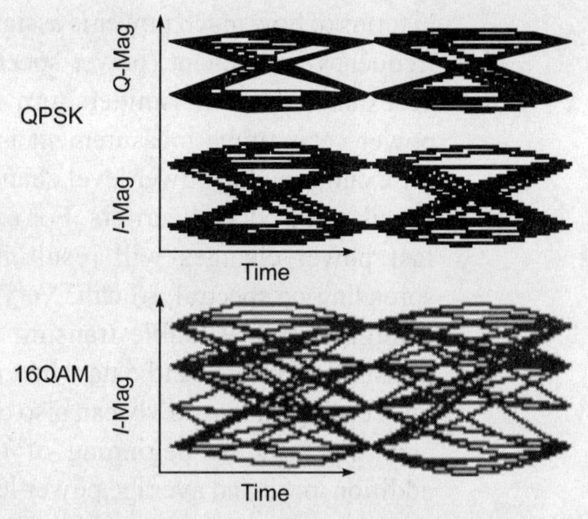

Fig. 7.6 An example of eye diagram

7.1.5 Trellis Diagrams

Figure 7.7 shows a 'trellis' diagram. It resembles a garden trellis and hence the name. We have already studied trellis diagrams for convolution decoding purpose in Chapter 4. Actually, the diagram shows the possible paths for the decoding, indirectly representing different phases vs time, of which the best path is to be recovered following an algorithm so that the minimum phase errors can be observed at the receiver side.

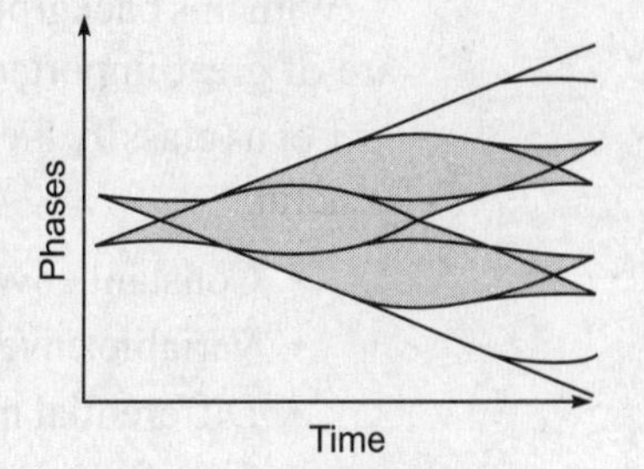

Fig. 7.7 An example of trellis diagram

In general, the trellis diagram shows time on the X-axis and phase on the Y-axis. This allows the examination of the phase transitions with different symbols. If a long series of binary ones is sent, the result is a series of positive phase transitions of certain degrees per symbol. If a long series of binary zeros is sent, there is a constant declining phase of 90 degrees per symbol. Typically, there are intermediate transmissions with random data. When troubleshooting, trellis diagrams are useful in isolating missing transitions, missing codes, or a blind spot in I/Q modulator or mapping algorithm.

There are many different ways of looking towards a digitally modulated signal, such as given below:

1. Time and frequency domain view
2. Power and frequency domain view
3. Frequency vs time view
4. Power vs time view

Signals can be analysed either in time domain or in frequency domain. From one domain, they can be converted into another and vice versa. Signals can also be analysed in terms of how much power is assigned to which frequency component (power spectral density). To examine how transmitters turn on and off, a power versus time measurement is very useful for examining the power level changes involved in pulsed or bursted carriers. For example, very fast power changes will result in frequency spreading or spectral growth. Very slow power changes waste valuable transmit time, as the transmitter cannot send data when it is not fully on. Turning on too slowly can also cause high bit error rates at the beginning of the burst. In addition, peak and average power levels must be well understood, since asking for excessive power from an amplifier can lead to compression or peak power clipping. These phenomena distort the modulated signal and usually lead to spectral growth as well. The *crest factor* is an important parameter and defined as peak to average power ratio.

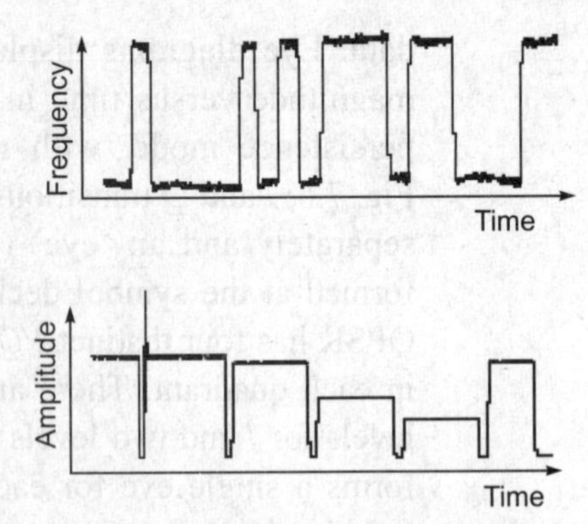

Fig. 7.8 Examples of frequency and amplitude variations with time

Figure 7.8 shows variation of frequency and amplitude with time.

With this background, we can start the study of various digital modulation schemes that are of great importance in the present scenario.

Let us classify the digital modulation schemes in the following four main categories in general:

- Constant envelope modulation, for example M-PSK, FSK, MSK, and GMSK
- Variable envelope modulation, for example ASK and M-QAM
- Differential modulation, for example DPSK and DQAM
- *I/Q* offset modulation, for example OQPSK and OKQPSK

7.2 CONSTANT ENVELOPE MODULATION SCHEMES

In constant-envelope modulation, the amplitude of the carrier is constant, regardless of the variation in the modulating signal. It is a power-efficient scheme that allows efficient class-C amplifiers to be used without introducing degradation in the spectral occupancy of the transmitted signal. However, constant-envelope modulation techniques occupy a larger bandwidth than schemes in which amplitude of the transmitted signal varies with the modulating digital signal. In systems where bandwidth efficiency is more important than power efficiency, constant envelope modulation is not well suited.

The M-ary PSK schemes and FSK are constant envelope type. The MSK is a special type of FSK where the peak-to-peak frequency deviation is equal to half the bit rate. The GMSK is a derivative of MSK, where the bandwidth required is further reduced by passing the modulating waveform through a Gaussian filter. The Gaussian filter minimizes the instantaneous frequency variations over time. All the mentioned schemes are described in the following sections.

7.2.1 Binary Phase Shift Keying (BPSK)

One of the simplest forms of digital modulation is the *binary* or *bi-phase shift keying* (BPSK). One application where this scheme exists is deep space telemetry. The BPSK is simply achieved by multiplying data with a carrier $A \sin \omega_c t$, where data has two possible levels +1 and – 1. The phase of a constant amplitude carrier signal moves between zero and 180 degrees or $+/-\pi/2$.

On I and Q diagrams, the I state has two different values. There are two possible locations in the state diagram. So, a binary 1 or 0 can be sent. The symbol rate is one bit per symbol. The BPSK demonstrates better performance than the ASK and FSK. Figure 7.9 shows BPSK waveform and constellation diagram.

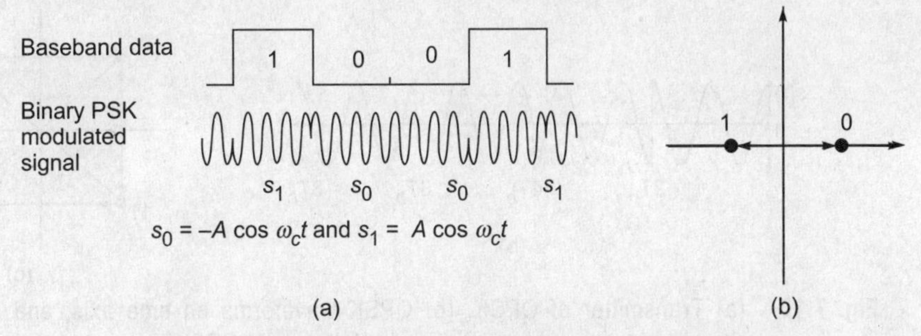

Fig. 7.9 (a) BPSK waveform and (b) constellation diagram

7.2.2 Quadrature Phase Shift Keying (QPSK)

Another common type of phase modulation is quadrature phase shift keying (QPSK). Transmitter and receiver diagram for QPSK is given in Figures 7.4 and 7.5, along with the explanation of I and Q vectors and transmitter diagram is repeated within a better representation in Fig. 7.10(a). Actually, BPSK can be expanded to an M-ary scheme, employing multiple phases and amplitudes as different states and QPSK is a case with $M = 4$. For any PSK schemes, the symbols are mapped into phases. For QPSK, 2 bits/ symbol is used to be mapped into four phases.

The QPSK is used extensively in applications, including CDMA, cellular service, wireless local loop, iridium (a voice/data satellite system), and DVB-S (digital video broadcasting-satellite). Quadrature means that the signal shifts between phase states that are separated by 90 degrees. The signal shifts in increments of 90 degrees from 45 to 135, – 45, or –135 degrees. These points are chosen as they can be easily implemented using I/Q modulator. It is a more bandwidth-efficient type of modulation than BPSK, potentially twice as efficient. For modulation, two carrier waves are used for I and Q channels to be modulated, both are of same frequency but orthogonal to each other, i.e. one will be a sine wave and the other will be the cosine wave of the same frequency.

Variety of schemes can be developed using the basic concept of QPSK. The IQ diagrams for such schemes are given in Fig. 7.11. Other than conventional QPSK, the modulation schemes are described in the subsequent sections. Conventional QPSK has transitions through zero (i.e. $\pm\pi$ or 180° phase transition). Sudden phase reversals can

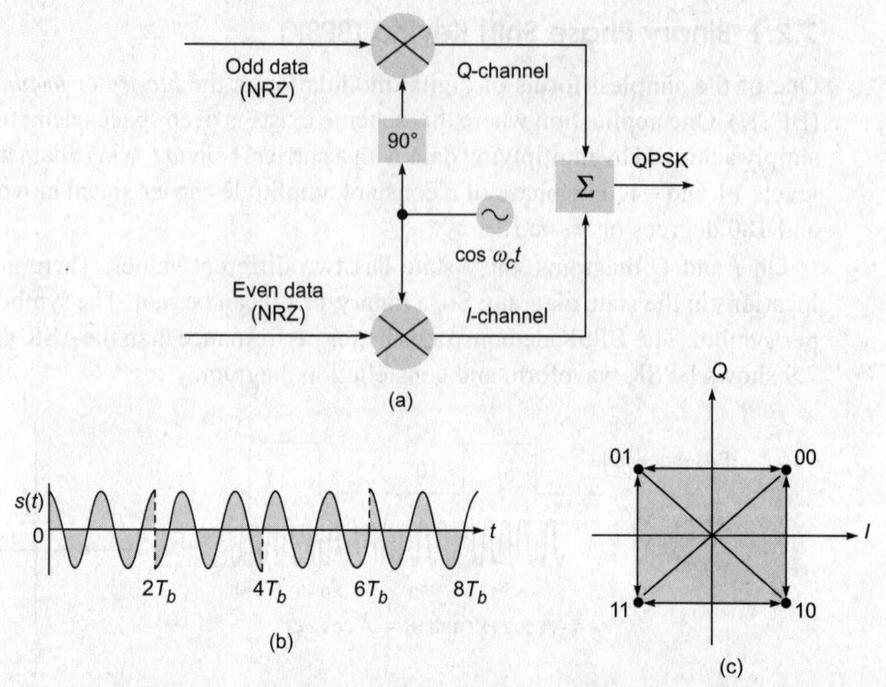

Fig. 7.10 (a) Transmitter of QPSK, (b) QPSK waveforms on time axis, and (c) two bits/symbol representing one phase on the constellation of QPSK

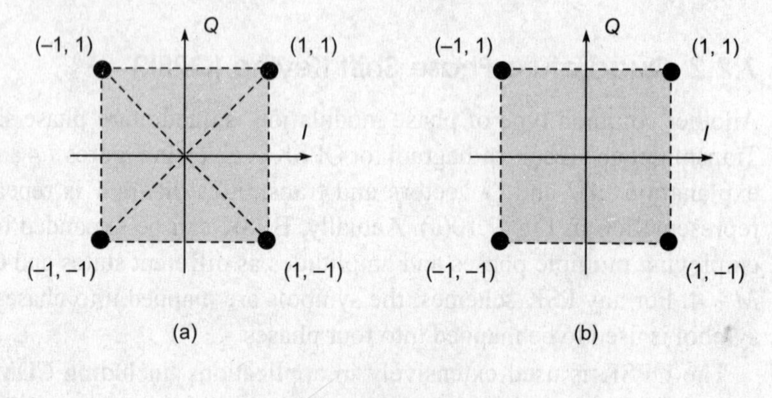

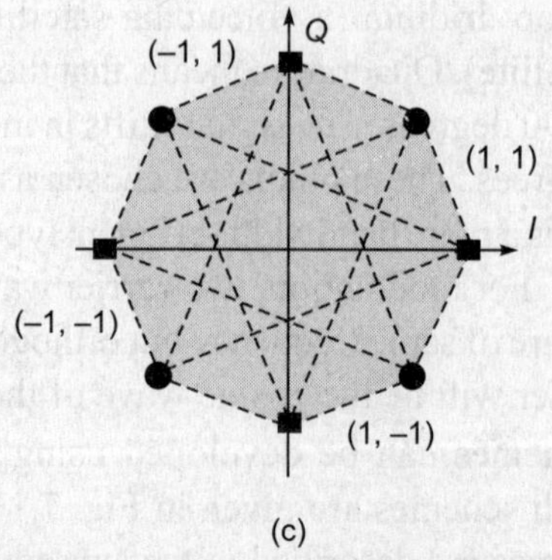

Fig. 7.11 Various modulation schemes based on QPSK but with different constellation (–1 represents bit '0' and 1 represents bit '1'): (a) conventional QPSK, (b) offset QPSK, and (c) $\pi/4$ QPSK

throw the amplifiers into saturation. As shown in Fig.7.10(b), the phase reversals cause the envelope to go to zero momentarily. This may cause non-linearities in the amplifier circuitry. This can be prevented using linear amplifiers but they are more expensive and power consuming. A solution to this problem is the use of OQPSK.

The OQPSK modulation is such that phase transitions about the origin are avoided. The scheme is used in IS-95 handsets. On offset QPSK, the transitions on I and Q channels are staggered. In this case, transitions are, therefore, limited to 90°. In $\pi/4$ QPSK, the set of constellation points is toggled to each symbol. So, transitions through zero cannot occur. This scheme produces the lowest envelope variations. All QPSK schemes require linear power amplifiers. Offset and $\pi/4$ QPSK are described afterwards in the chapter.

7.2.3 M-ary Phase Shift Keying (M-PSK)

In general, M-ary PSK waveform is represented by

$$s(t) = f(t)\cos\left[\omega_c t + \frac{2\pi}{M}(m-1)\right] \qquad 0 \le t \le T_s \tag{7.2}$$

where $m = 1,2,\ldots,M$ denotes the M possible phases of the carrier $f_c = \omega_c/2\pi$ corresponding to M possible data symbols represented by Eq. (7.1) and $f(t)$ is a real-valued pulse waveform (normally rectangular). All the M different M-PSK waveforms have the same energy

$$E_s = \frac{1}{2}E_{f(t)} \tag{7.3}$$

where $E_{f(t)}$ denotes the energy of the basic pulse $f(t)$.

This signal set is two dimensional. So, PSK waveforms may be represented by a linear combination of two basis functions $\psi_1(t)$ and $\psi_2(t)$.

In general, M-PSK schemes are more spectrally efficient. As we go on increasing the number M, the consumption of the spectrum reduces. M is related to the power of 2 values: 8, 16, 32, and so on. In 8PSK, 3 bits/symbol are read together to be mapped into phases. As there are eight combinations of 3 bits, eight points will be created on the constellation and so on. The constellation diagram for 16PSK is given in Fig. 7.12.

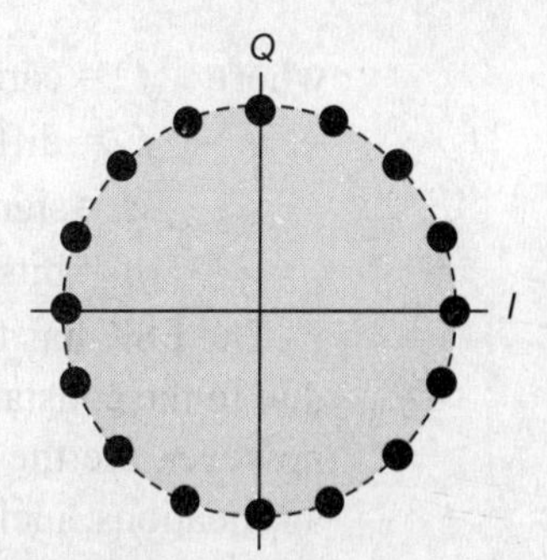

Fig. 7.12 Constellation of 16PSK

Equivalently popular is another version—M-QAM described in the next section. The M-PSK is a power-efficient technique. Wherever the large amplitude variations are there, M-PSK can be used, while if the possibility of phase errors is more, M-QAM schemes are utilized more. Thus, depending upon the wireless channel environment, the selection of modulation scheme is made. Higher M-ary scheme has less spacing between the constellation points and is, therefore, more affected by noise. Thus, M-ary schemes are more bandwidth efficient but more susceptible to noise. Hence, in case of BER analysis of such systems, for higher M-ary scheme, the bit error rate reduces only with higher SNR value.

7.2.4 Frequency Shift Keying (FSK)

The generic name for digital family of frequency modulation is *frequency shift keying* (FSK). In FSK, the frequency of the carrier is changed as a function of the modulating signal (data) being transmitted. The amplitude remains unchanged in FSK. In binary FSK (BFSK or 2FSK), a '1' is represented by one frequency and a '0' is represented by another frequency (Fig. 7.13).

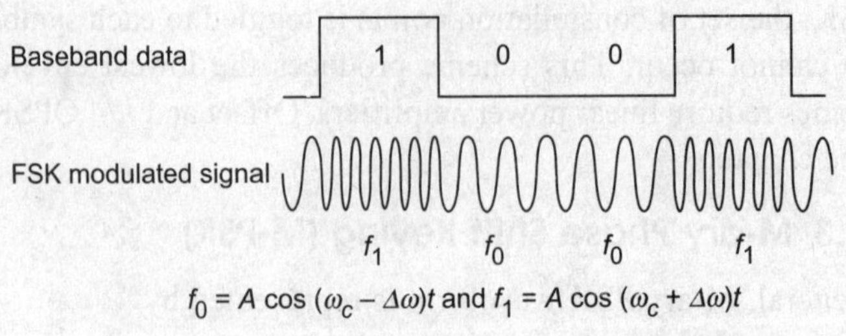

$f_0 = A\cos(\omega_c - \Delta\omega)t$ and $f_1 = A\cos(\omega_c + \Delta\omega)t$

Fig. 7.13 FSK modulated time domain waveform

Frequency modulation and phase modulation are closely related. A static frequency shift of +1 Hz means that the phase is constantly advancing at the rate of 360 degrees per second (2π rad/sec), relative to the phase of the unshifted signal. Bandwidth occupancy of FSK is dependant on the spacing of the two symbols. A frequency spacing of inverse of 0.5 times the symbol period is typically used. The FSK can be expanded to an M-ary scheme called MFSK, employing multiple frequencies as different states as given in Eq. (7.4).

$$s_i(t) = A\cos 2\pi[f_c + (2i - 1 - M)f_d]t, \qquad 1 \le i \le M \tag{7.4}$$

where f_c = carrier frequency
f_d = difference frequency
M = signalling elements = 2^n
n = bits per symbol

The FSK has the advantage of being very simple to generate and demodulate. Also, due to the constant amplitude, it can utilize a non-linear PA. Significant disadvantages, however, are the poor spectral efficiency and BER performance. FSK is used in many applications, including cordless and paging systems. Some of the cordless systems are CT2 (Cordless Telephone 2), DECT (Digital Enhanced Cordless Telephone), and Bluetooth, which uses GFSK.

Note M-ary FSK requires a considerably increased bandwidth in comparison with M-ary PSK.

7.2.5 Minimum Shift Keying (MSK)

Minimum shift keying is FSK with a modulation index of 0.5. Therefore, the carrier phase of an MSK signal will be advanced or retarded by 90° over the course of each bit period to represent either 1 or 0. Due to this exact phase relationship, MSK can be considered as either phase or frequency modulation. That is why it is also called *continuous phase frequency shift keying* (CPFSK). The result of this exact phase relationship is that MSK cannot practically be generated with a voltage controlled oscillator and a digital waveform. Instead, an *IQ* modulation technique, as for PSK, is usually implemented. Coherent demodulation is usually employed for MSK due to the superior BER performance. This is practically achievable and widely used in real systems due to the exact phase relationship between each bit.

If we have two orthogonal signals

$$s_0(t) = \cos(2\pi f_0 t + \theta) \tag{7.5a}$$

$$s_1(t) = \cos(2\pi f_1 t + \theta) \tag{7.5b}$$

where θ is some arbitrary phase offset. It can be shown that the minimum frequency separation will be according to the orthogonality constraints, that is $\Delta f = f_1 - f_0$ will be equal to $(1/2)T_b$, where T_b is the bit duration. This result is not dependent on the phase offset θ and hence we may design the system to have continuous phase (no discontinuity as in case of M-PSK). This will result in lower spectral side lobes. In other words, the value of θ in each symbol period is such that the phase continues smoothly from the previous symbol. If f_c is the centre frequency then f_1 and f_0 can be represented as

$$f_0 = f_c - (1/4)T_b \tag{7.6a}$$

$$f_1 = f_c + (1/4)T_b \tag{7.6b}$$

Since a frequency shift produces an advancing or retarding phase. Frequency shifts can be detected by sampling phase at each symbol period. Phase shifts of $(2N + 1)\ \pi/2$ radians (integer $N = 0, 1, 2, \ldots$) are easily detected with *I/Q* demodulator. At even-numbered symbols, the polarity of the *I* channel conveys the transmitted data, while at odd-numbered symbols, the polarity of the *Q* channel conveys the data. This orthogonality between *I* and *Q* simplifies detection algorithms and hence reduces power consumption in a mobile receiver. The minimum frequency shift that yields orthogonality of *I* and *Q* is that resulting in a phase shift of $\pm\pi/2$ radians per symbol (90 degrees per symbol). FSK with this deviation is called MSK. The deviation must be accurate in order to generate repeatable 90 degree phase shifts. A phase shift of +90 degrees represents a data bit equal to '1', while that of –90 degrees represents a '0'. The polar diagram is shown in Fig. 7.14. The peak-to-peak frequency shift of an MSK signal is equal to one-half of the bit rate.

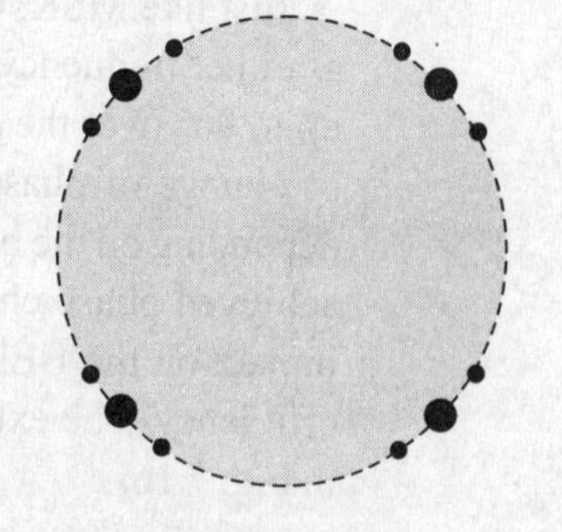

One bit per symbol

Fig. 7.14 Polar diagram of MSK

The FSK and MSK produce constant envelope carrier signals, which is a desirable characteristic for improving the power efficiency of transmitters. Amplitude variations can exercise non-linearities in an amplifier's amplitude transfer function, generating spectral growth, a component of adjacent channel power. Therefore, more efficient amplifiers (which tend to be less linear) can be used with constant envelope signals, reducing power consumption. MSK has a narrower spectrum than wider deviation forms of FSK. The width of the spectrum is also influenced by the waveforms causing the frequency shift. If those waveforms have fast transitions or a high slew rate, then the spectrum of the transmitter will be broad.

In practice, the waveforms are filtered with a Gaussian filter. Hence, instead of rectangular pulse, Gaussian pulses enter in the modulator, resulting in a narrow spectrum. In addition, the Gaussian filter has no time domain overshoot, which would broaden the spectrum by increasing the peak deviation. MSK with a Gaussian filter is termed GMSK (Gaussian MSK). GMSK is used in the GSM cellular standard (Chapter 13). Gaussian filter is a pulse shaping filter. It is described in Chapter 2.

7.2.6 Gaussian Minimum Shift Keying (GMSK)

The GMSK is a derivative of MSK where the bandwidth required is further reduced by passing the modulating waveform through a Gaussian filter as shown in Fig. 7.15. The Gaussian filter minimizes the instantaneous frequency variations over time. The major advantages of GMSK are that it has a constant envelope, spectral efficiency, and good BER performance and is self-synchronizing.

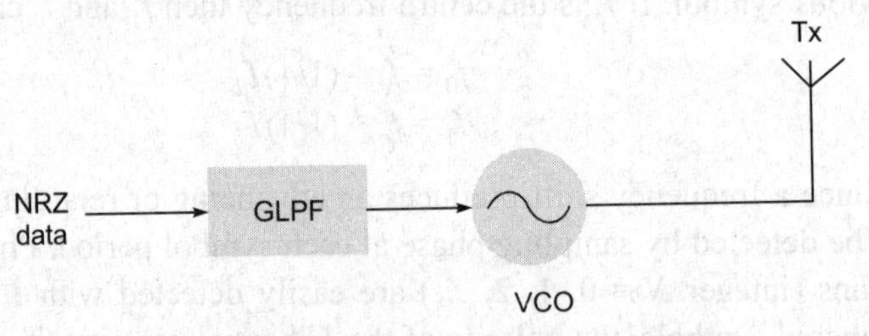

Fig. 7.15 Conceptual diagram of GMSK

Just like MSK, GMSK is a constant amplitude scheme. Again GMSK can be viewed as either frequency or phase modulation. The phase of the carrier is advanced or retarded up to 90° over the course of a bit period depending on the data pattern, although the rate of change of phase is limited with a Gaussian response. The net result of this is that, depending on the bandwidth time product (BT), effectively the severity of the shaping, the achieved phase change over the bit may fall short of 90°. This will obviously have an impact on the BER, although the advantage of this scheme is the improved bandwidth efficiency. The extent of this shaping can clearly be seen from the 'eye' diagrams given in Fig. 7.16.

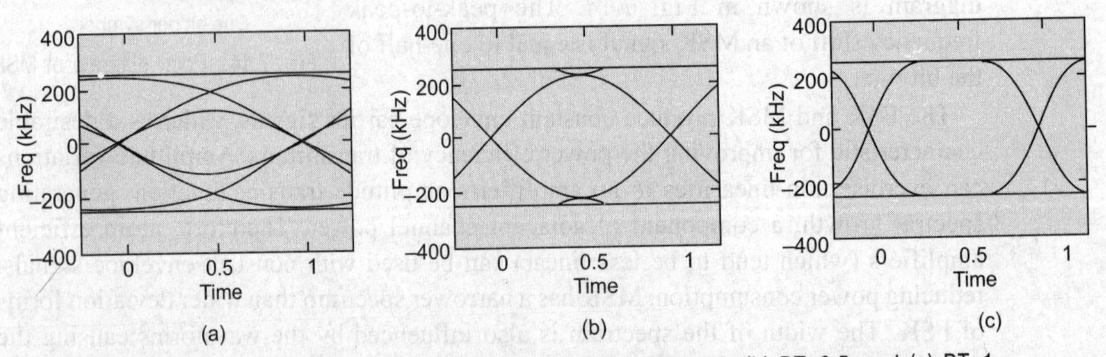

Fig. 7.16 Eye diagrams for GMSK schemes with (a) BT=0.3, (b) BT=0.5, and (c) BT=1

This resultant reduction in the phase change of the carrier for the shaped symbols (i.e., 101 and 010) will ultimately degrade the BER performance as less phase has been accrued or retarded, therefore less noise will be required to transform a zero to a one and vice versa. The spectral efficiency of this scheme can be proved with the help of the results given in Figures 7.17 and 7.18, which compare the spectra of GMSK with different BT values and MSK.

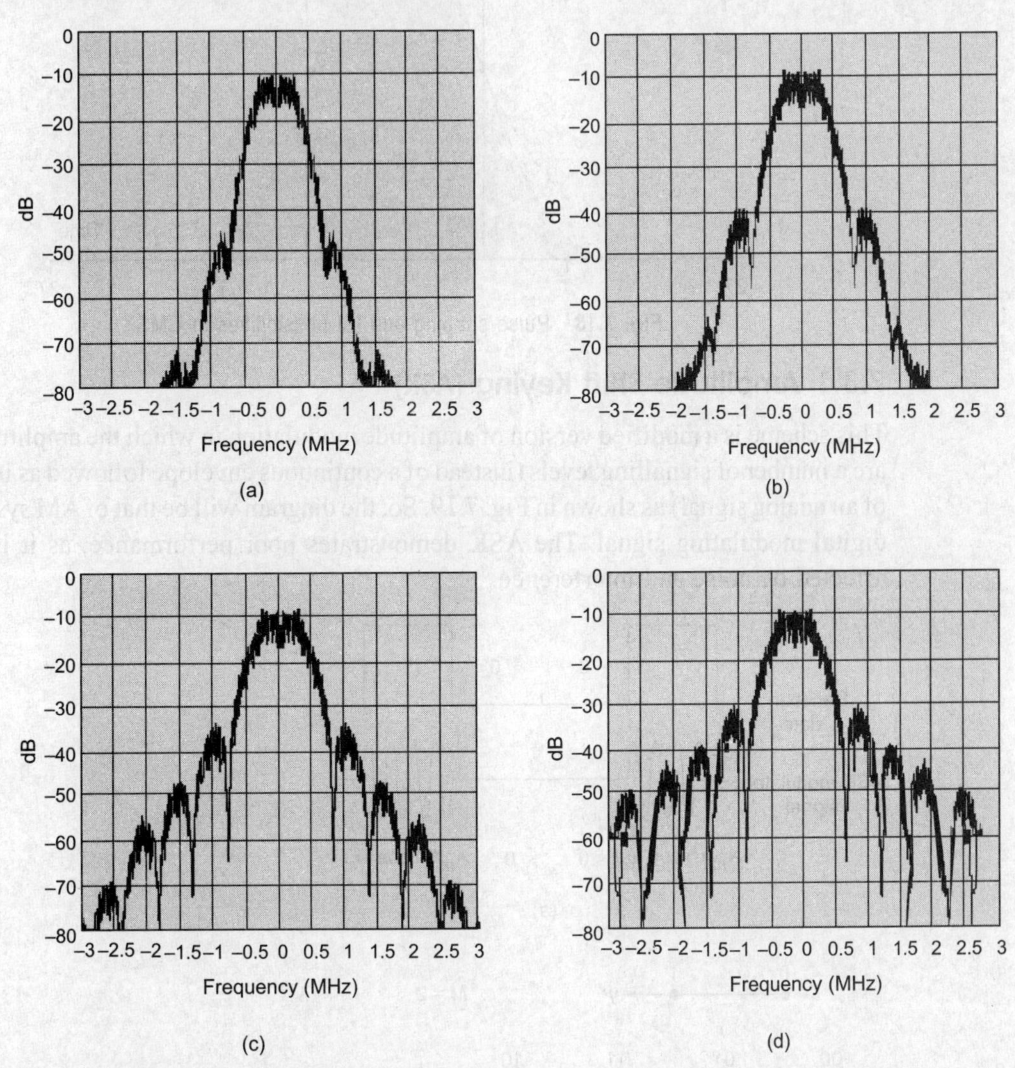

Fig. 7.17 Spectral saving in GMSK: (a) GMSK-BT=0.3, (b) GMSK-BT=0.5, (c) GMSK-BT=1, and (d) MSK

7.3 VARIABLE ENVELOPE MODULATION SCHEMES

Other types of modulation schemes are the variable envelope modulation schemes. Here both amplitude as well as phase information is required to detect the signal properly.

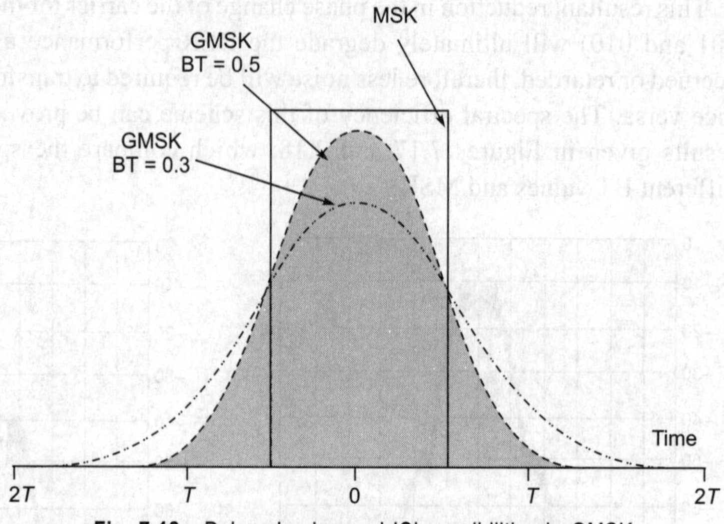

Fig. 7.18 Pulse shaping and ISI possibilities in GMSK

7.3.1 Amplitude Shift Keying (ASK)

This scheme is a modified version of amplitude modulation in which the amplitude levels are a number of signalling levels (instead of a continuous envelope followed as in the case of an analog signal) as shown in Fig. 7.19. So, the diagram will be that of AM system with digital modulating signal. The ASK demonstrates poor performance, as it is heavily affected by noise and interference.

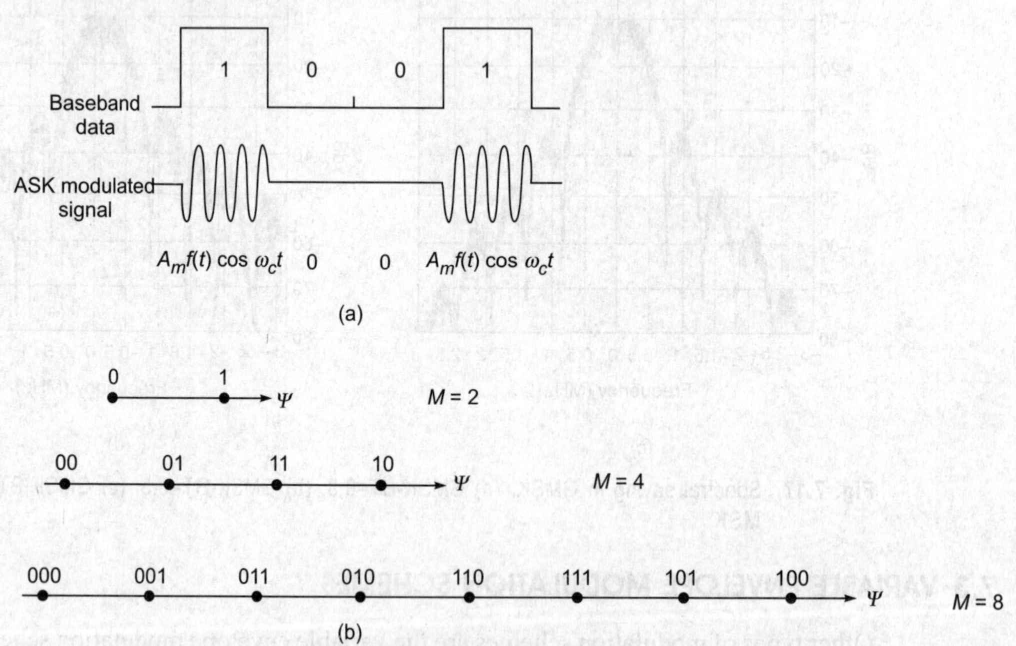

Fig. 7.19 (a) Waveform for ASK for a binary case and (b) one-dimensional signal space diagram for ASK signals in which mapping of the symbols to information bits is in the form of grey code and the adjacent signal levels differ in one bit position

An ASK waveform is represented by the following equation in general:

$$s(t) = [A_m f(t)] \cos \omega_c t \quad 0 \le t \le Ts \tag{7.7}$$

where $m = 1, 2,..., M$ denotes the M possible signalling (amplitude) levels corresponding to M possible data symbols represented by Eq. (7.1) and $f(t)$ is a real valued pulse waveform (normally rectangular).

Here A_m takes the following different values:

$$A_m = \frac{(2m-1-M)x}{2} \tag{7.8}$$

where x is the distance between the adjacent signal points in the signal space diagram represented by basis functions $\psi(t)$. Here A_m takes only the discrete set of values and does not change during the bit interval. If the symbol rate for the input is R_s, then the amplitude of the ASK signal changes every $1/R_s$ seconds. The energy of the ASK signal is [from Eq. (2.11)]

$$E_s = \frac{1}{2} A_m^2 E_{f(t)} \tag{7.9}$$

where $E_f(t)$ denotes the energy of the basic pulse $f(t)$.

7.3.2 Quadrature Amplitude Modulation (QAM) and M-ary QAM

Quadrature amplitude modulation (QAM) is the method of combining two amplitude modulated signals into one channel. It may be an analog QAM or a digital QAM. Analog QAM combines two amplitude modulated signal using the same carrier frequency with a 90 degree phase difference. In digital QAM, two ASK signals are combined in the same way. The QAM is used in applications, including microwave digital radio, DVB-C (digital video broadcasting-cable), and modems. Reader can compare QPSK schemes described under constant envelope modulation schemes with this scheme. A block diagram of QAM is shown in Fig. 7.20.

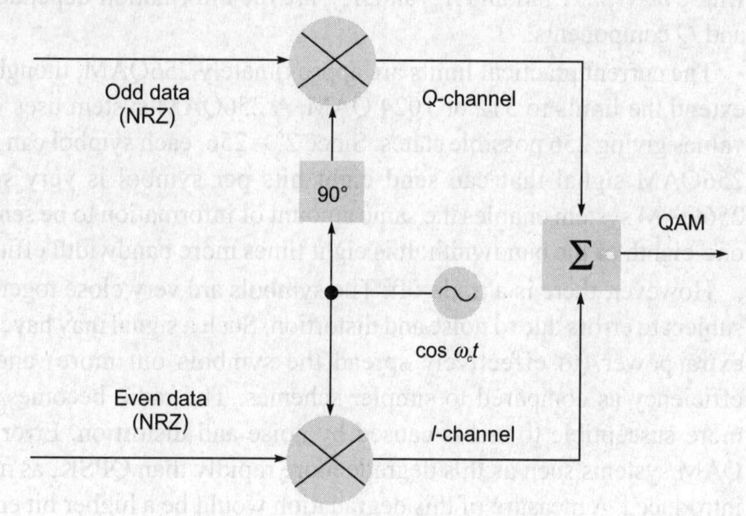

Fig. 7.20 Block diagram of QAM

Other variations in QAM are 8, 16, 32QAM, etc., i.e., M-ary QAM. In 16-state quadrature amplitude modulation (16QAM), there are a total of 16 possible states for the signal [Fig. 7.21(a)]. It can have transition from any state to any other state at every symbol time. Since $16 = 2^4$, four bits per symbol can be sent. This consists of two bits for *I* and two bits for *Q*. The symbol rate is one-fourth of the bit rate. So, this modulation format produces a more spectrally efficient transmission. In case of 32QAM, there are six *I* values and six *Q* values resulting in a total of 36 possible states ($6 \times 6 = 36$)—too many states for a power of two (the closest power of two is 32). So, the four corner symbol states, which take the most power to transmit, are omitted [Fig. 7.21(b)]. This reduces the amount of peak power the transmitter has to generate. Since $2^5 = 32$, there are five bits per symbol and the symbol rate is one-fifth of the bit rate.

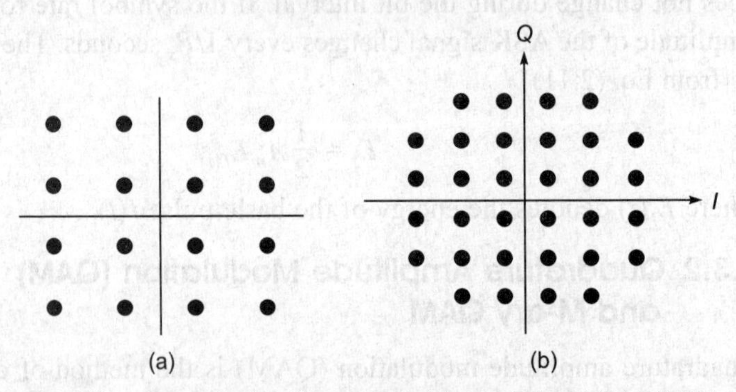

Fig. 7.21 Constellation diagrams of (a) 16QAM and (b) 32QAM

The equation can be represented as

$$s(t) = [A_{mI} f(t)] \cos \omega_c t - [A_{mQ} f(t)] \sin \omega_c t \qquad 0 \le t \le T_s \tag{7.10a}$$

or

$$s(t) = \sqrt{A_{mI}^2 + A_{mQ}^2}\, f(t) \cos(\omega_c t + \theta_m) \tag{7.10b}$$

where $m = 1, 2, \ldots, M$ and A_{mI} and A_{mQ} are the information-dependent amplitudes of the *I* and *Q* components.

The current practical limits are approximately 256QAM, though work is underway to extend the limits to 512 or 1024 QAM. A 256QAM system uses 16 *I*-values and 16 *Q*-values giving 256 possible states. Since $2^8 = 256$, each symbol can represent eight bits. A 256QAM signal that can send eight bits per symbol is very spectrally efficient. A 256QAM system enables the same amount of information to be sent as BPSK, using only one-eighth of the bandwidth. It is eight times more bandwidth efficient.

However, there is a trade-off. The symbols are very close together and are thus more subject to errors due to noise and distortion. Such a signal may have to be transmitted with extra power (to effectively spread the symbols out more) and this reduces power efficiency as compared to simpler schemes. The radio becomes more complex and is more susceptible to errors caused by noise and distortion. Error rates of higher-order QAM systems such as this degrade more rapidly than QPSK, as noise or interference is introduced. A measure of this degradation would be a higher bit error rate. In any digital modulation system, if the input signal is distorted or severely attenuated, the receiver will eventually lose symbol clock completely. If the receiver can no longer recover the symbol clock, it cannot demodulate the signal or recover any information. With less degradation, the symbol clock can be recovered, but it is noisy and the symbol locations themselves are noisy. In some cases, a symbol will fall far enough away from its intended position that it will cross over to an adjacent position. The I and Q level detectors used in the demodulator would misinterpret such a symbol as being in the wrong location, causing bit errors. The QPSK is not as spectral efficient, but the states are much farther apart and the system can tolerate a lot more noise before suffering symbol errors. As QPSK has no intermediate states between the four corner-symbol locations, there is less opportunity for the demodulator to misinterpret symbols. QPSK requires less transmitter power than QAM to achieve the same bit error rate.

7.4 DIFFERENTIAL MODULATION SCHEMES

The differential modulation can be seen in differential QPSK (DQPSK), differential 16QAM (D16QAM), etc. Differential means that the information is not carried by the absolute state; it is carried by the transition between states. In some cases there are also restrictions on allowable transitions. This occurs in $\pi/4$ DQPSK, where the carrier trajectory does not go through the origin. A DQPSK transmission system can have transition from any symbol position to any other symbol position. The $\pi/4$ DQPSK modulation format is widely used in many applications, including the following:

- Cellular: NADC- IS-54 (North American digital cellular), PDC (Pacific digital cellular)
- Cordless: PHS (personal handyphone system)
- Trunked radio: TETRA (Trans European trunked radio)

The $\pi/4$ DQPSK modulation format uses two QPSK constellations offset by 45 degrees ($\pi/4$ radians). Transitions must occur from one constellation to the other. This guarantees that there is always a change in phase at each symbol, making clock recovery easier. The data is encoded in the magnitude and direction of the phase shift, not in the

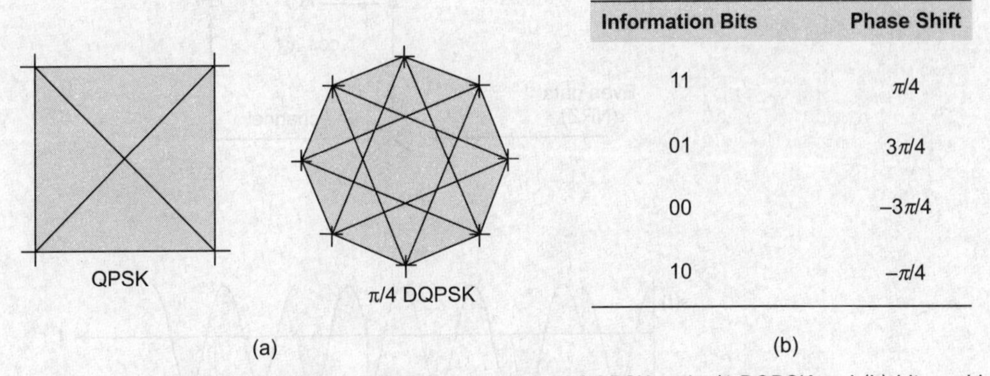

Information Bits	Phase Shift
11	$\pi/4$
01	$3\pi/4$
00	$-3\pi/4$
10	$-\pi/4$

Fig. 7.22 (a) Difference between constellation diagrams of QPSK and $\pi/4$ DQPSK and (b) bit combinations and corresponding phases in $\pi/4$ DQPSK

absolute position on the constellation. One advantage of $\pi/4$ DQPSK is that the signal trajectory does not pass through the origin, thus simplifying transmitter design. Another is that $\pi/4$ DQPSK, with root raised cosine filtering, has better spectral efficiency than GMSK, the other common cellular modulation type. Because of differential modulation schemes correlation between consecutive symbol is established and hence it becomes easier to detect the errors. One can eliminate equalization or channel estimation scheme at the receiver as differential modulation scheme but due to this slight degradation in the performance is observed.

7.5 I/Q OFFSET MODULATION SCHEMES

The final variation is offset modulation. One example of this is the offset QPSK (OQPSK). As previously discussed, the potential for 180° phase shift in QPSK results in the requirement for better linearity in the power amplifier and spectral re-growth. OQPSK reduces this tendency by adding a time delay of one bit period (half a symbol) in the Q arm of the modulator. The result is that the phase of the carrier is potentially modulated every bit (depending on the data), not every other bit as for QPSK, hence the phase trajectory never approaches the origin. In OQPSK, the phase transitions take place every T_b seconds. In QPSK, the transitions take place every $2T_b$ seconds.

As with the other phase modulation schemes considered, shaping of the phase trajectory between constellation points is typically implemented with a raised cosine filter to improve the spectral efficiency. Due to the similarities between QPSK and OQPSK similar signal spectra and probability of error are achieved. OQPSK is utilized in the North American IS-95 CDMA cellular system for the reverse link from the mobile to the base station.

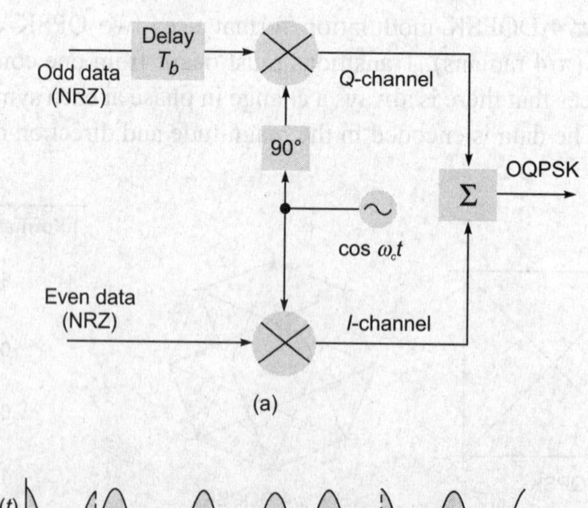

(a)

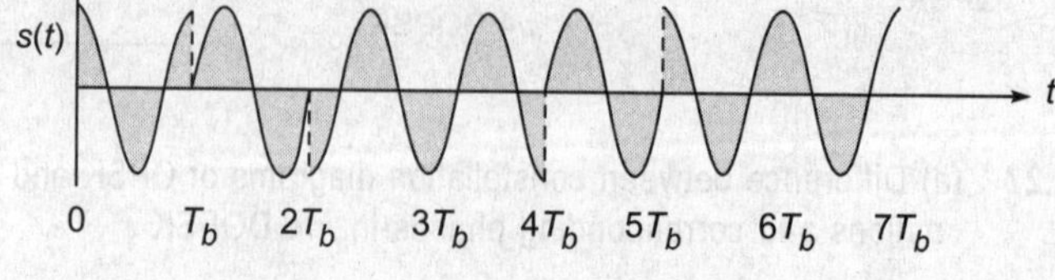

(b)

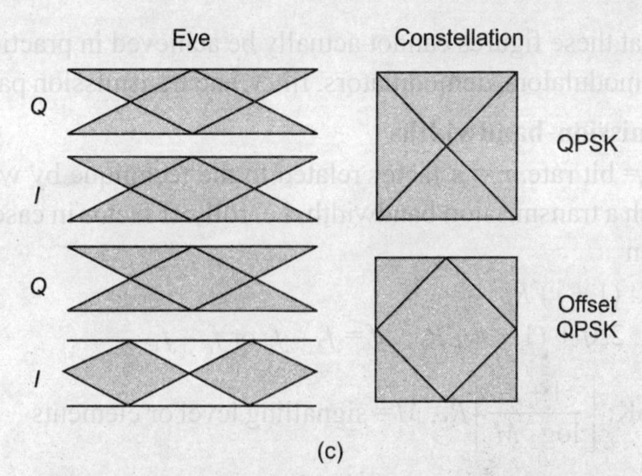

Fig. 7.23 (a) OQPSK diagram (only delay part being additional), (b) OQPSK waveforms on time axis, and (c) difference between QPSK and offset QPSK eye and constellation diagrams

In QPSK, *I* and *Q* bit streams are switched at the same time. The symbol clocks, or *I* and *Q* digital signal clocks, are synchronized. In offset QPSK, *I* and *Q* bit streams are offset in their relative alignment by one bit period (one-half of a symbol period). This is shown in the diagram. Since the transitions of *I* and *Q* are offset, at any given time, only one of the two bit streams can change values. This creates a dramatically different constellation, even though there are still just two *I*/*Q* values. This has power efficiency advantages. In OQPSK, the signal trajectories are modified by the symbol clock offset so that the carrier amplitude does not go through or near zero (the centre of the constellation). The spectral efficiency is the same with two *I* states and two *Q* states. The reduced amplitude variations (perhaps 3 dB for OQPSK versus 30 to 40 dB for QPSK) allow a more power-efficient, less linear RF power amplifier to be used.

7.6 THEORETICAL BANDWIDTH EFFICIENCY LIMITS

Bandwidth efficiency describes how efficiently the allocated bandwidth is utilized or the ability of a modulation scheme to accommodate data, within a limited bandwidth. Table 7.1 shows the theoretical bandwidth efficiency limits for the main modulation types.

Table 7.1 Spectrum efficiency limits for different modulation schemes

Modulation Format	Theoretical Spectrum Efficiency Limits
MSK	1 bit/sec/Hz
BPSK	1 bit/sec/Hz
QPSK	2 bits/sec/Hz
8PSK	3 bits/sec/Hz
16 QAM	4 bits/sec/Hz
32 QAM	5 bits/sec/Hz
64 QAM	6 bits/sec/Hz
256 QAM	8 bits/sec/Hz

Note that these figures cannot actually be achieved in practical radios since they require perfect modulators, demodulators, filter, and transmission paths.

Transmission bandwidths

If R_b= bit rate, r = a factor related to the technique by which the signal is filtered to establish a transmission bandwidth, i.e. roll-off factor in case of raised cosine filter, $0 \le r \le 1$, then

ASK: $(1 + r) R_b$

FSK: $2\Delta f + (1 + r) R_b$, $\Delta f = f_2 - f_c$ or $f_c - f_1$

MPSK: $\left[\dfrac{1+r}{\log_2 M}\right] R_b$, M= signalling level or elements

MFSK: $\left[\dfrac{(1+r)M}{\log_2 M}\right] R_b$

If the radio has a perfect (rectangular in the frequency domain) filter, then the occupied bandwidth can be made equal to the symbol rate.

An example of how symbol rate influences spectrum requirements in eight-state phase shift keying (8PSK) is shown in Fig. 7.24. It is a variation of PSK. There are eight possible states that the signal can transit to at any time. The phase of the signal can take any of eight values at any symbol time. Since $2^3 = 8$, there are three bits per symbol. This means that the symbol rate is one-third of the bit rate. This is relatively easy to decode.

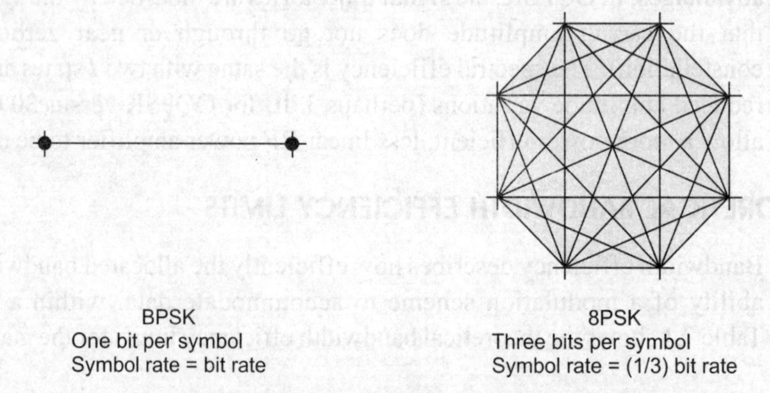

Fig. 7.24 Constellation showing that the number of bits/symbol is more in case of 8PSK and hence the symbol rate reduces compared to BPSK

7.7 INCREASING SPECTRUM EFFICIENCY

The ratio of bit rate to available bandwidth is called the spectral efficiency (Table 7.1). Practically, the bandwidth of the transmitting (modualted) signal can be controlled by applying external measures like premodulation or postmodulation filtering. As described in Section 2.6.5, pulse shaping is very much helpful in spectrum saving. The effect of variation of α on the polar diagram of QPSK while using the raised cosine filter is shown in Fig. 2.40. Filtering allows the transmitted bandwidth to be significantly reduced without losing the content of the digital data. This improves the spectral efficiency of the signal.

Different varieties of filters are described in Chapter 2, such as raised cosine, square root raised cosine, Gaussian, and Chebyshev.

Instead of post-modulation filtering, windowing may also be used as described in Section 2.6.6. This technique also saves the spectrum by eliminating out-of-band components.

Other techniques for maximizing spectral efficiency include the following:

- Relate the data rate to the frequency shift (as in GSM)
- Restrict the types of transitions

7.8 TRANSMISSION POWER RELATED ISSUES

As with any natural resource, it makes no sense to waste the RF spectrum by using channel bands that are too wide. Therefore, narrower filters are used to reduce the occupied bandwidth of the transmission. Narrower filters with sufficient accuracy and repeatability are more difficult to build. Smaller values of α increase ISI because more symbols can contribute. This tightens the requirements on the clock accuracy. These narrower filters also result in more overshoot and, therefore, more peak carrier power. The power amplifier must then accommodate the higher peak power without distortion. The bigger amplifier causes more heat and electrical interference to be produced since the RF current in the power amplifier will interfere with other circuits. Larger, heavier batteries will be required. The alternative is to have shorter talk time and smaller batteries. Constant envelope modulation, as used in GMSK, can use class-C amplifiers, which are the most efficient. In summary, spectral efficiency is highly desirable, but there are penalties in power, cost, size, weight, complexity, talk time, and reliability.

In radios, today there is a need for significant excess power beyond the power needed to transmit the symbol values themselves. A typical value of the excess power needed at an α of 0.2 for QPSK with Nyquist filtering would be approximately 5 dB. This is more than three times as much peak power because of the filter used to limit the occupied bandwidth. These principles apply to QPSK, offset QPSK, DQPSK, and the varieties of

Table 7.2 Modulation format and application

Modulation Format	Application
MSK, GMSK	GSM, CDPD
BPSK	Deep space telemetry, cable modems
QPSK, $\pi/4$ DQPSK	Satellite, CDMA, TETRA, PHS, PDC, LMDS, DVB-S, cable (return path), cable modems, TFTS
OQPSK	CDMA, satellite
FSK, GFSK	DECT, paging, AMPS, CT2, land mobile, public safety, Bluetooth
8PSK	Satellite, aircraft, telemetry pilots for monitoring broadband video systems
16 QAM	Microwave digital radio, modems, DVB-C, DVB-T
32 QAM	Terrestrial microwave, DVB-T
64 QAM	DVB-C, modems, broadband set top boxes, MMDS
256 QAM	Modems, DVB-C (Europe), Digital video (US)

QAM, such as 16QAM, 32QAM, 64QAM, and 256QAM. Not all signals will behave in exactly the same way, and exceptions include FSK, MSK, and others with constant-envelope modulation. The power of these signals is not affected by the filter shape.

Various modulation formats and their applications are listed out in Table 7.2

Let us consider an example. A radio has an 8 bit sampler, sampling at 10 kHz for voice. The bit rate, the basic bit stream rate in the radio, would be eight bits multiplied by 10 k samples/second, or 80 kbps. (For the moment, we will ignore the extra bits required for synchronization, error correction, etc.). The symbol rate is the bit rate divided by the number of bits that can be transmitted with each symbol. If one bit is transmitted per symbol, as with BPSK, then the symbol rate would be the same as the bit sent with each symbol and the same amount of data can be sent in a narrower spectrum. This is why rate of 80 kbps. If two bits are transmitted per symbol, as in QPSK, then the symbol rate would be half of the bit rate or 40 kbps. If more bits can be modulation formats that are more complex and use a higher number of states can send the same information over a narrower piece of the RF spectrum.

CASE STUDY MATHEMATICAL MODEL FOR CALCULATING PHASE ERRORS

This is a general model. The model developed is based on the transmission modulation technique being M-ary phase shift keying (M-PSK) and the channel noise being white Gaussian. As the constellation diagram is very important in the PSK, based on the phasor plot, the scheme can be developed to find phase errors.

If we assume that the transmission modulation method used is phase shift keyed, then any noise added to the transmitted signal will result in a phase error. If we look at the *IQ* diagram of the transmitted signal, then the transmitted signal will be a phasor of fixed magnitude and of the phase corresponding to the data to be transmitted. The noise can then be considered as the random vector added to the transmitted signal (Fig. 7.25). The magnitude of the phase error depends on two things, the relative phase angle of the noise vector and the magnitude of the noise vector.

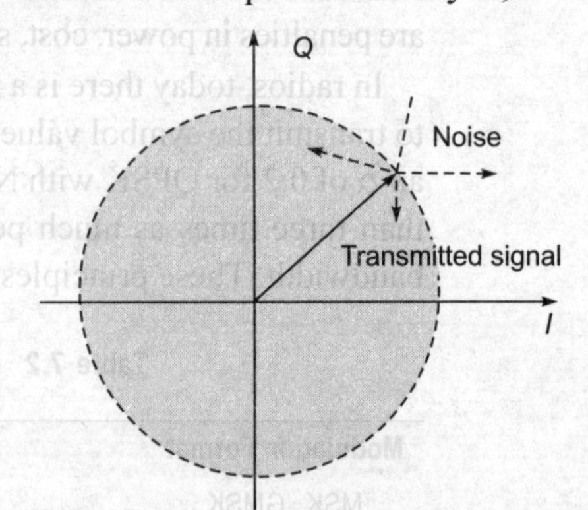

Fig. 7.25 Relative phase and magnitude of the received vector

The received vector will be the vector sum of the transmitted signal and the noise. If we assume that the noise is a constant magnitude vector equal to its RMS magnitude and that it has a random phase angle, then the problem of working out the received angle would be as follows.

Figure 7.26 shows the effect of noise on the received phase angle. If we let the amplitude of the transmitted signal be 1 and the length of the noise vector be ***a*** with angle φ, then the received phase error is θ_{err} as shown in Fig. 7.26.

$$\theta_{err} = \tan^{-1}\left(\frac{y}{x}\right) \tag{7.11}$$

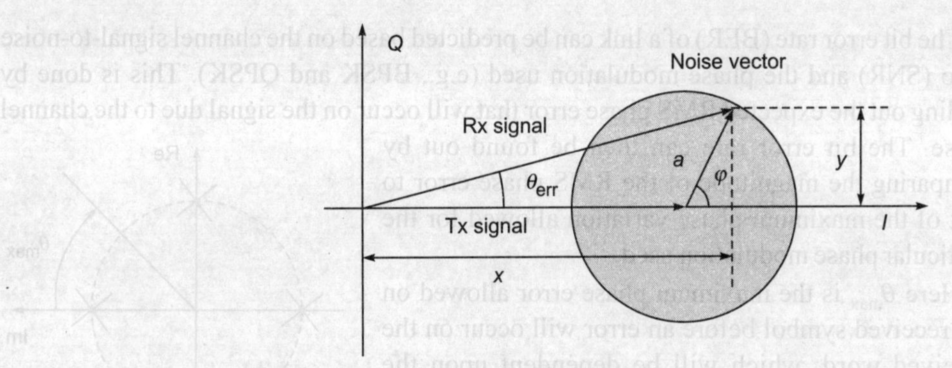

Fig. 7.26 Received phasor showing the effect of noise on the received phase angle

Using trigonometry,

$$x = 1 + a\cos\varphi \text{ and } y = a\sin\varphi$$

Therefore,

$$\theta_{err} = \tan^{-1}\left(\frac{a\sin\varphi}{1 + a\cos\varphi}\right) \tag{7.12}$$

The signal-to-noise ratio determines the relative amplitude of the received signal and the noise level. Since the signal is scaled to amplitude of 1, the amplitude of the noise is

$$a = \frac{1}{\text{SNR}} \tag{7.13}$$

Note The SNR is based on the amplitudes of the signals thus must be scaled correctly when converting it to dB.

If we substitute this in Eq. (7.12) and simplify, we get

$$\theta_{err} = \tan^{-1}\left(\frac{\sin\varphi}{\text{SNR} + \cos\varphi}\right) \tag{7.14}$$

The noise signal can be of any phase angle. What we need is to find out the RMS phase error. So, if we find the average phase error (assuming that the noise phase angle is always positive), then this can be scaled to find the RMS error. The average phase angle can be found out by integrating θ_{err} from 0 to φ over a half circle, i.e. π:

$$\text{Av } \theta_{err} = \frac{1}{\pi}\int_0^{\pi} \tan^{-1}\left(\frac{\sin\varphi}{\text{SNR} + \cos\varphi}\right) d\varphi \tag{7.15}$$

The RMS phase error will be greater by root 2. Thus,

$$\text{RMS } \theta_{err} = \frac{\sqrt{2}}{\pi}\int_0^{\pi} \tan^{-1}\left(\frac{\sin\varphi}{\text{SNR} + \cos\varphi}\right) d\varphi \tag{7.16}$$

This equation can be used to predict the RMS phase error for different channel SNRs.

As RMS phase error has been calculated, the BER can be easily calculated using simple statistics Bayes' theorem. The RMS phase error is the standard deviation of the phase error. An error will occur if the phase error gets bigger than the maximum allowable phase angle for the modulation method used.

The bit error rate (BER) of a link can be predicted based on the channel signal-to-noise ratio (SNR) and the phase modulation used (e.g., BPSK and QPSK). This is done by finding out the expected RMS phase error that will occur on the signal due to the channel noise. The bit error rate can then be found out by comparing the magnitude of the RMS phase error to that of the maximum phase variation allowed for the particular phase modulation used.

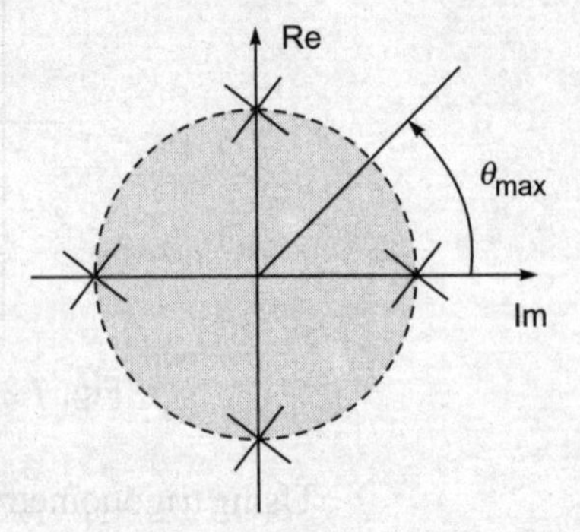

Fig. 7.27 *IQ* diagram for QPSK with θ_{max}

Here θ_{max} is the maximum phase error allowed on the received symbol before an error will occur on the received word, which will be dependent upon the maximum allowable phase angle (Table 7.3). For example, the *IQ* diagram for QPSK scheme shown in Fig. 7.27 has four phases. The figure shows the phase locations for data (crosses) and that θ_{max} is 45 degrees.

Table 7.3 Maximum allowable phase error

Modulation Technique	Maximum Phase Error Allowed (θ_{max} in degrees)
BPSK	90
QPSK	45
16PSK	11.25
256PSK	0.70313

Once θ_{max} and $\theta_{err(rms)}$ have been established, Z can be calculated as

$$Z = \frac{\theta_{max}}{\theta_{err(rms)}} = \text{number of standard deviations} \tag{7.17}$$

Summary

- Choice of digital modulation schemes depends upon the application and channel environment.
- The expected parameters for digital modulation schemes are spectrum efficiency, power efficiency, and less susceptibility to noise.
- There are two types of systems: coherent and non-coherent. Coherent systems are phase dependent.
- Diagrams like polar, constellation, eye, trellis, etc. are used for the performance analysis of system with different modulation schemes.
- Constant envelope schemes are M-PSK, FSK, MSK, GMSK, etc.
- M-ary PSK schemes are power-efficient techniques and become spectrally efficient with higher value of *M*.
- QPSK modulation is very robust but requires some form of linear amplification.
- ASK, QAM, etc., are variable amplitude type modulation schemes and hence have degraded power efficiency. They can be used when there are less amplitude variations over channel or more possibilities of phase errors.

- For fast fading environment, both amplitude as well as phase change occur very fast. So, selection of the scheme is a crucial issue. Adaptive schemes are utilized.
- Other types of modulation are differential and IQ offset.
- Constant envelope schemes, such as GMSK, can be employed since they provide efficient non-linear amplifier.
- Coherent reception provides better performance than differential but requires a complex receiver.
- By using a mathematical model, phase errors can be calculated in case of PSK schemes.

REVIEW QUESTIONS

1. Why does choice of modulation scheme depend upon the environment?
2. How can we categorize different modulation schemes?
3. Simulate QPSK scheme using MATLAB and observe the waveforms, polar diagram, constellation diagram, and its spectrum. Also extend this exercise for all other modulation schemes.
   (**Hint:** *pskmod* and *pskdemod* as well as *qammod* and *qamdemod* functions are given in MATLAB in which you can set the value of *M*.)
4. What is the difference between QAM and QPSK schemes?
5. Why is VCO used in the diagram of MSK many times? Is it possible practically?
6. What is the difference between MSK and GMSK?
7. Discuss the different types of QPSK schemes with difference in constellations.
8. How are four corner points eliminated from the constellation of 32QAM?
9. What is the major difference between the block diagrams of QPSK and OQPSK?
10. How can you justify that for higher M-ary schemes spectral consumption reduces?
11. Discuss the power constraints of a transmitter.
12. How can you increase the bandwidth/spectral efficiency in a particular modulation scheme?
13. Why does bit error rate decrease with increase in SNR in case of M-ary PSK-based systems while increasing *M*?

PROBLEMS

1. What is the bandwidth efficiency for FSK, ASK, BPSK, and QPSK for a bit error rate of 10^{-5} on a channel with SNR of 15 dB?
2. In a QPSK transmitter, the information bit rate is 100 kbps. If the available bandwidth is 200 kHz, find out the spectral efficiency of the scheme. Compare it with the case of 16PSK.
3. Calculate the minimum required bandwidth for a QPSK system so that 15% of the spectral power contributes to out-of-band components. The symbol duration is 0.1 ms.
4. Calculate the minimum bandwidth requirement for a non-coherent BFSK system if the symbol duration is 0.2 ms. What is the minimum bandwidth required for MSK system having the same symbol duration?
5. If it is required to send 1.024×10^6 binary digits (bits) per second with $P_e \leq 10^{-6}$ with BPSK and 16PSK, the channel noise power spectral density is $N_o = 10^{-8}$. Determine the transmission bandwidth and the

signal power required at the receiver input in both the cases.

6. In case of PSK, prove that the bit error probability is

$$P_e = Q\sqrt{\frac{2E_b}{N_o}}$$

7. The QPSK scheme is used to modulate the bit stream 001010011100. Sketch the transmitted waveform. Assume $f_s = f_b/2$. Also get the waveform for 8PSK for the same stream assuming $f_s = f_b/3$.
8. For the sequence in Problem 7, draw MSK signal waveform.
9. Find out the equations for determining the distance between the constellation points on the signal space diagram for ASK, FSK, and M-PSK modulation schemes.
10. Prove that the phase in MSK is continuous.
11. For a band-pass channel using BPSK signalling, the data rate is 9.6 kbps. Find out the transmission bandwidth at (a) the first null and (b) the second null. Also find out the peak value of the spectrum.
12. A telephone line is equalized to allow bandpass data transmission over a frequency range of 300–3100 kHz. Design a 16QAM signalling scheme that will allow a data rate of 9.6 kbps to be transferred over the channel so that the channel is ISI-free.

Multiple Choice Questions

1. Wireless MODEM is
   (a) a circuit that carries out the modulation and demodulation of a carrier frequency
   (b) an ARQ device for correcting errors
   (c) a system for transmitting high-speed burst
   (d) an anti-jamming technique invariably installed on all communication devices
2. Which of the following modulation schemes is the most spectral-efficient scheme?
   (a) 256PSK (b) 8PSK
   (c) QPSK (d) 16PSK
3. Which of the following modulation schemes is the most bandwidth-efficient scheme?
   (a) 256PSK (b) 8PSK
   (c) QPSK (d) 16PSK
4. Spectral efficiency can be improved by
   (a) increasing power
   (b) using shaping filters
   (c) reading more symbols simultaneously
   (d) all of these
5. In QPSK, the phase reversals occur at
   (a) T_b interval (b) $4T_b$ interval
   (c) $T_b/2$ interval (d) $2T_b$ interval
6. Which of the following is a power-efficient technique?
   (a) 16PSK (b) QAM
   (c) ASK (d) 16QAM
7. Linear amplifier is required in
   (a) QPSK
   (b) OQPSK
   (c) DQPSK
   (d) all of the above
8. The MSK is the special case of CPFSK with modulation index of
   (a) 1 (b) 0.5
   (c) 0.25 (d) none of these
9. On the constellation diagram, the allowable phase margin in 8PSK scheme is
   (a) 45° (b) 90°
   (c) 22.5° (d) 180°

10. In GMSK, of two cases with BT = 0.3 and BT = 0.5, the ISI effect will
   (a) be more in the second case
   (b) be equal in both the cases
   (c) be more in the first case
   (d) not be there

11. The amplitude vs time plot with multiple traces is represented on
   (a) polar plot
   (b) constellation diagram
   (c) trellis diagram
   (d) eye diagram

chapter 8

Wideband Modulation Techniques 1: Spread Spectrum Techniques (Single Carrier Modulation)

Key Topics

- Spread spectrum concept
- Concept of SS transmission bandwidth from Shannon's theorem and SNR
- Code properties: autocorrelation and cross correlation
- Sequence or PN codes and generation: ML, Walsh, Gold, etc.
- DSSS transmitter and receiver
- PN signal characteristics, spectral density, bandwidth, and processing gain
- Rake receiver
- Cyclic prefix in DS-CDMA
- Signal processing at the rake receiver
- Interference rejection and anti-jamming characteristics in DSSS
- Energy and bandwidth efficiency
- Near–far problem and power control
- Frequency hopping spread spectrum
- Time hopping spread spectrum
- Comparison of various hopping systems
- Hybrid systems
- Chirped spread spectrum

Chapter Outline

This is a different type of modulation scheme in which the spread spectrum effect is achieved with the help of a unique code. The code repeatedly becomes *pseudo-noise* (PN) sequences, which must be known at the receiver side to despread the signal again. The advantage of the system is the high level of secure communication, initially started for military applications. Because of very large spectrum, it rejects the interfering signals efficiently after despreading and also exhibits anti-jamming characteristics. Because of good features, this modulation scheme is utilized in multi-user environment in terms of *code division multiple access* (CDMA). Carefully designed rake receivers are required for the detection of the signal back. Rake receivers exploit multipath diversity. There are two main schemes for spread spectrum technology, direct sequence, and frequency hopping. There are others but are not so popular. The features of the spread spectrum modulation are described in detail in this chapter.

8.1 SPREAD SPECTRUM MODULATION (SSM) CONCEPT

Spread spectrum (SS) techniques were originated to meet the needs of military communications. They are based on signalling schemes (code or sequence) that greatly expand the transmitted spectrum relative to the data rate. Now, the techniques are used for commercial applications also. There is a growing interest in these techniques for use in mobile radio networks and for both communication and positioning applications in satellites.

A spread spectrum transmission offers three main advantages over a fixed frequency transmission:

1. Spread-spectrum signals are highly resistant to noise and interference. The process of re-collecting a spread signal spreads out noise and interference, causing them to recede into the background.
2. Spread-spectrum signals are difficult to intercept. A frequency-hop spread-spectrum signal simply sounds like an increase in the background noise to a narrowband receiver.
3. Spread-spectrum transmissions can share a frequency band with many types of conventional transmissions with minimal interference. The spread-spectrum signals add minimal noise to the narrow-frequency communications, and vice versa. As a result, bandwidth can be utilized more efficiently.

Spread spectrum techniques also meet the following objectives:

- Operation with a low-energy spectral density
- Multiple-access capability without external control
- Security (difficult for unauthorized receivers to observe the message)
- Anti-jamming capability
- Multipath protection
- Ranging

One way of classifying the spread spectrum system is given below.

Averaging system In this system, the interference reduction takes place because the interference can be averaged over a large time interval.

Avoidance system In this system, the reduction of interference occurs because the signal is made to avoid the interference over a large fraction of time.

Spread spectrum technique is a wideband modulation technique. Here bandwidth expansion factor is very large. In fact, bandwidth expansion does not combat white noise as it does in FM, PCM, and other wideband modulation methods, because bandwidth expansion is achieved by something that is independent of the message, rather than being uniquely related to the message. Though the spread spectrum system is not useful in combating white noise, it has advantages as mentioned above, which make the technique worth considering.

As far as classification by modulation technique is concerned, two major techniques used in spread spectrum systems are direct sequence (DS) and frequency hopping (FH), of which, perhaps, frequency hopping is easier to visualize. Other methods are also available, which are described afterwards.

In direct sequence spread spectrum (DSSS), the unique code (PN code), which appears as random sequence to unauthentic user) is used to spread and despread the signal using the logic as shown in Fig. 8.1(a). In that, for logic 1, inverted PN sequence appears in the output, while for logic 0 same sequence appears because of the XOR operation. As shown in Fig. 8.1(b) for the data voltage level, +1 is considered for logic 1 and −1 is considered for logic 0. Here multiplication operation rather than XOR is considered. Thus, wherever logic 0 occurs, the inverted PN sequence appears in the output.

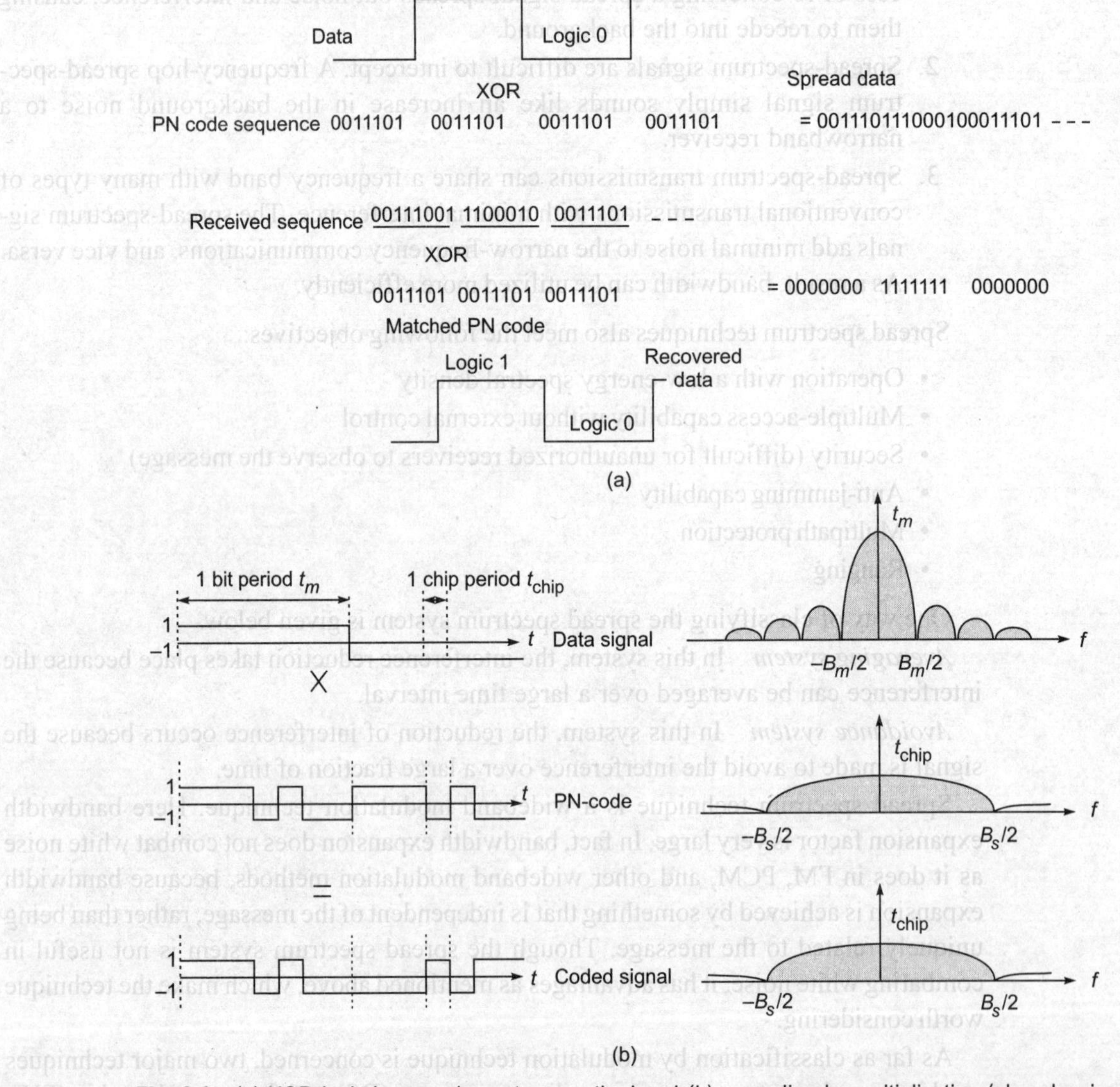

Fig. 8.1 (a) XOR logic in spread spectrum method and (b) spreading by multiplication (also showing relation between bit and chip durations)

Here assume that the clock rate is provided for generating each bit of a PN code (with one clock cycle one bit of PN code comes out). Bit rate of PN code is called the *chip rate*, which is 10 times or more than the data bit rate. Thus finally, we can say that in this technique PN code directly does phase shift keying of the data, increasing its bandwidth. In a typical direct sequence system a double balanced mixer is driven by the PN code to switch a carrier's phase between 0 and 180 degree. This is known as BPSK. Unlike a frequency hopping transmitter where the pseudo random sequence commands a synthesizer to change frequency, the direct sequence signal is directly generated by long duration PN code which repeats the same cycle.

In a frequency hopping system, the frequency or channel in use changes rapidly. The transmitter hops from channel to channel in a predetermined but pseudo-random sequence. The receiver has an identical list of channels to use (the hop set) and an identical pseudo-random sequence generator to that of the transmitter. A synchronizing circuit in the receiver ensures that the pseudo-random code generator in the receiver synchronizes to the one in the transmitter. When the transmitter and receiver are synchronized, the users are unaware that the transmitter and receiver are changing the frequencies rapidly, which provides anti-jamming characteristics. However, should the receiver not be synchronized to the transmitter or a conventional receiver be used, nothing will be heard unless the transmitter hops on to the receiver's tuned frequency. As a frequency hopping transmitter typically hops over tens to thousands of frequencies (fast frequency hopping/FFH) per second (the hop rate), the time it stays on a particular channel (the dwell time) is very short and as a result the signal would appear as a burst of interference.

The DS receiver despreads this wideband signal by using an identical synchronized pseudo-random sequence to that in the transmitter. As with the frequency hopper, the receiver must use a circuit to adjust its clock rate so that the receiver's PN code is at the same point in the code as the transmitter and hence the hopping pattern is also same. A tracking circuit is necessary to maintain the synchronism once it has been attained.

Other spread spectrum modulation techniques are as follows:

- Time hopping spread spectrum (THSS)
- Hybrid methods
- Chirped spread spectrum (CSS)

Direct sequence system is an averaging system, while frequency hopping, time hopping, and chirping systems are avoidance systems.

8.2 CONCEPT OF SSM BANDWIDTH FROM SHANNON'S THEOREM AND SNR

We can rewrite Eq. (2.7) as

$$C = W \log_2\left(1+\frac{S}{N}\right) \tag{8.1}$$

or

$$\frac{C}{W} = \log_2\left(1+\frac{S}{N}\right) \tag{8.2}$$

Changing bases, $\frac{C}{W} = \log_2 e \cdot \log_e\left(1+\frac{S}{N}\right)$ (8.3)

Now, $\log_a b = \frac{1}{\log_b a}$

So, $\frac{C}{W} = \frac{1}{\log_e 2} \cdot \log_e\left(1+\frac{S}{N}\right)$

$$= 1.44\log_e\left(1+\frac{S}{N}\right)$$

$$\cong 1.44\log_e\left(\frac{S}{N}\right) \quad (8.4)$$

By logarithmic expansion,

$$\log_e\left(\frac{S}{N}\right) = \frac{S}{N} - 1/2\left(\frac{S}{N}\right)^2 + 1/3\left(\frac{S}{N}\right)^3 - 1/4\left(\frac{S}{N}\right)^4 \ldots \quad (8.5)$$

In a spread spectrum system, the signal-to-noise ratio is typically small and much less than 0.1. Hence, higher order terms are neglected.

Hence, $\frac{C}{W} \approx \frac{1.44S}{N}$

Thus, $W \approx \frac{NC}{1.44S}$ (approximation for the wider bandwidths) (8.6)

From the derived relationship, it can be clearly seen that a desired signal-to-noise ratio for a fixed data rate C can be achieved by increasing the transmission bandwidth.

Two criteria for the spread spectrum system are as follows:

1. Transmission bandwidth is much larger than the bandwidth of the information being sent.
2. Some function other than the information being sent (here PN code) determines the resulting RF bandwidth.

Example 8.1 If data rate is 32 kbps and SNR is 0.001(–30 dB), then with the help of Shannon's theorem, find out the necessary bandwidth for SSM signal. Comment on your result.

Solution

We have $W \approx \frac{NC}{1.44S} \approx \frac{32 \times 10^3 \times 1000}{1.44} \approx 22 \text{ MHz}$

Comment For data rate of 32 kbps, operation with lower value of SNR of –30dB is achievable by spreading the signal over a bandwidth of 22 MHz. By using a much wider bandwidth than that of the original data is possible to maintain the channel capacity without increasing the transmitter output power. It is an extreme example of power-bandwidth trade-off.

8.3 OPERATIONS RELATED TO PN CODE OR SEQUENCE

In the spread spectrum based multiuser system (multiple access), unique codes are to be assigned to different users to differentiate them. For the identification of unique user, the code must undergo some operations so that the data may not be diverted to wrong user.

8.3.1 Autocorrelation

Autocorrelation is introduced in Chapter 2. It is the procedure by which matching of a signal is done with itself. The autocorrelation should be maximum for the DS or PN codes with itself so that correct PN signal can be identified from the number of co-existing signals. The autocorrelation function of a typical PN sequence is shown in Fig. 8.2. Note that on a normalized basis, it has a maximum value of 1 that repeats itself every period, but in between these peaks, the level is at a constant value of $(1/N_c)$. If N_c is a very large number, the autocorrelation function is very small in this region.

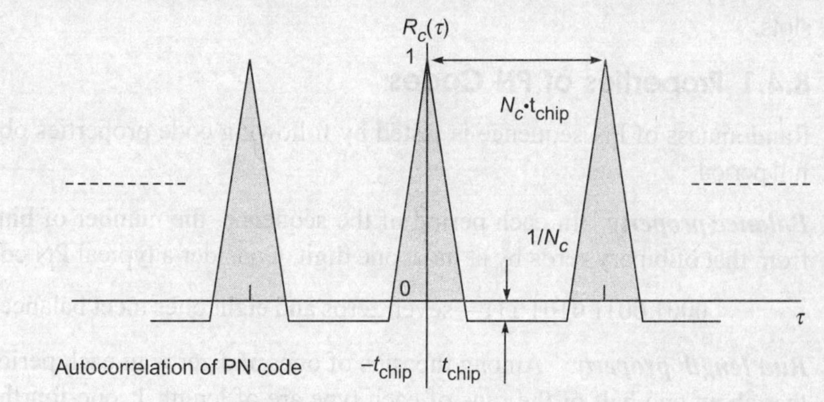

Fig. 8.2 Autocorrelation function (normalized) for the codes with respect to time

The autocorrelation of the spreading waveform (PN signal) $c(t)$ is represented mathematically as

$$R_a(\tau) = \frac{1}{T_{\text{code}}} \int_0^{\infty} c(t)c(t+\tau)\,dt \tag{8.7}$$

where $T_{\text{code}} = N_c t_{\text{chip}}$ is the code period and τ represents a time shift variable.

8.3.2 Partial Autocorrelation

The partial autocorrelation is similar to the formula given in Eq. (8.7) but integrated only over a portion of T_{code}, maybe over a message bit duration. This is done to avoid long processing time for matching purposes. If the partial sequence is autocorrelated, then obviously the full sequence will be autocorrelated because of uniqueness of the code. This is the concept behind the partial autocorrelation.

8.3.3 Cross Correlation

Cross correlation is also described in Chapter 2. Different signals have different spreading codes. The cross correlation between the signals of two codes i and j is

$$R_c(\tau) = \frac{1}{T_{\text{code}}} \int_0^{\infty} c_i(t) c_j(t + \tau)\, dt \tag{8.8}$$

which is equal to the autocorrelation if $i = j$. It is desirable to have poor cross correlation between two different codes so that unwanted code will be rejected easily by the receiver of the CDMA system.

8.4 VARIOUS PSEUDO-NOISE (PN) CODES OR DIRECT SEQUENCES (DS) FUNDAMENTALS

The DS is used to modulate the data. In DSSS, the user signal is multiplied by a pseudo-noise code sequence of high bandwidth. This code sequence is also called the chip sequence as it occurs with chip rate. The resulting modulated signal is transmitted over the radio channel. DS is responsible for spreading of bandwidth. In case of frequency or time hopping, the PN sequence is used to generate hopping frequencies or hopping time slots.

8.4.1 Properties of PN Codes

Randomness of PN sequence is tested by following code properties observed over one full period.

Balance property In each period of the sequence, the number of binary ones differs from that of binary zeros by at most one digit. Consider a typical PN code:

0001 0011 0101 111—seven zeros and eight ones meet balance condition

Run length property Among the runs of ones and zeros in each period, it is desirable that about one-half of the runs of each type are of length 1, one-fourth are of length 2, one-eighth are of length 3, and so on. Consider the same code again:

Number of runs = 8

$\underline{000}\ \underline{1}\ \underline{00}\ \underline{11}\ \underline{0}\ \underline{1}\ \underline{0}\ \underline{1111}$

3 1 2 2 1 1 1 4

Autocorrelation property Autocorrelation function of a maximal length sequence (explained in Section 8.4.3) is periodic and binary valued. We can state autocorrelation function as

$R_a(\tau) = \frac{1}{N_c}$ [No. of agreements (a) – No. of disagreements (d) in comparison of one full period]

0 **0 0** 1 0 **0** 1 **1** 0 1 0 1 **1 1 1**
1 **0 0** 0 1 **0** 0 **1** 1 0 1 0 **1 1 1**

---

d a a d d a d a d d d d a a a

$$R(\tau) = -\frac{1}{\ldots}$$

In short, a periodic sequence will repeat PN code of N_c chip durations (1 period) for infinity times. Such a sequence is said to be pseudo-random if it satisfies the following conditions:

1. In every period, the number of +1's differs from the number of –1's by exactly 1 (balance property). Hence, N_c is an odd number.
2. In every period, half of the runs of the same sign have length 1, one-fourth have length 2, one-eighth have length 3, and so forth. Also, the number of positive runs equals the number of negative runs (run property).
3. The autocorrelation of a periodic sequence is two valued, that is N_c for shifts 0, N_c, $2N_c$, $3N_c$, etc., and –1 otherwise (without normalization).

8.4.2 Aperiodic and Periodic Sequences

A PN sequence consists of a sequence of plus or minus ones (+1 or –1) that possesses certain specified properties. There are two general classes of PN sequences—aperiodic and periodic. An aperiodic sequence is one that does not repeat itself in a periodic fashion. It is usually assumed that the sequence has a value of zero outside its stated interval. The periodic sequence, however, is a sequence of plus or minus ones that repeats itself exactly with a specified period.

An ideal aperiodic sequence of N_c chips has an autocorrelation to be N_c for no shift and 0 or ±1 with a shift of one or more bits. Such sequences (also called Barker sequences) are known to exist for very few values of N_c. Specifically, they are N_c = 1, 2, 3, 4, 5, 7, 11, and 13. No larger sequences have been found and it has been hypothesized. Also such sequences are normally too short for the spread spectrum system. Sometimes, they may be used for the synchronization purposes under some conditions. Periodic sequences are much more important in a spread spectrum system because of the limitations of aperiodic sequences and for communicating continuously for required time.

Example 8.2 A seven digit aperiodic Barker sequence has the form +1,+1,+1,–1, –1,+1,–1. Determine the autocorrelation for this code. Does a cyclic shift of this sequence (e.g., +1,+1,–1,–1,+1,–1,+1) have the same correlation properties?

Solution

Considering the discrete signal for the limited duration, we can establish the autocorrelation as follows.

For Barker sequence:

Theoretically, autocorrelation function, $R_a(k) = \sum_{n=1}^{N_c - k} a_n a_{n+k}$

$= N_c$ for $k = 0$

$= 0$ or ± 1 for $k \neq 0$

where k is the number of shifts in bit positions. It can be verified as follows.

For this particular sequence, for $k = 0$:

$$\begin{array}{l} +1\ +1\ +1\ -1\ -1\ +1\ -1 \\ +1\ +1\ +1\ -1\ -1\ +1\ -1 \\ \hline +1\ +1\ +1\ +1\ +1\ +1\ +1 = N_c \text{ (summing up all the values)} \end{array}$$

For one shift, i.e. $k = 1$:

$$\begin{array}{l} +1\ +1\ +1\ -1\ -1\ +1\ -1 \\ \quad\ \ +1\ +1\ +1\ -1\ -1\ +1\ -1 \\ \hline \quad\ \ 0\ +1\ +1\ -1\ +1\ -1\ -1\ 0 = 0 \text{ (summing up all the values)} \end{array}$$

For two shifts, i.e. $k = 2$:

$$\begin{array}{l} +1\ +1\ +1\ -1\ -1\ +1\ -1 \\ \qquad\quad +1\ +1\ +1\ -1\ -1\ +1\ -1 \\ \hline 0\ \ 0\ +1\ -1\ -1\ -1\ +1\ \ 0\ \ 0 = -1 \text{ (summing up all the values)} \end{array}$$

Similarly, for more shifts, we can check the results.

Checking for the cyclic shift

For one cyclic shift, i.e. $k = 1$:

$$\begin{array}{l} +1\ +1\ +1\ -1\ -1\ +1\ -1 \\ -1\ +1\ +1\ +1\ -1\ -1\ +1 \\ \hline -1\ +1\ +1 -1\ +1\ -1\ -1 = -1 \text{ (summing up all the values)} \end{array}$$

With the cyclic shift, this sequence will have the same autocorrelation properties, i.e. it will be a two-valued function. With no shift it is N_C and for other cases -1 (think why).

Popular periodic code sequences used in spread-spectrum transmission are:

- Maximum length sequences
- Walsh–Hadamard sequences
- Gold codes/sequences
- Kasami codes, etc.

The names 'pseudo-noise' (PN) and 'maximal length' are used in some papers synonymously, while other authors use the name PN sequences to indicate a very broad class of sequences, including for instance ML and Gold sequences.

8.4.3 Maximum Length (ML) Sequences

The ML sequences and method of its generation is described first. PN code is generated in a maximal length shift register, which is nothing but a linear feedback shift register (LFSR). PN code generators are periodic in that the sequence that is produced repeats itself after a period of time. Such a typical periodic sequence generator is shown in Fig. 8.3(a). The smallest time increment in the sequences of duration is t_{chip} and is known as a *time chip*. The total period consists of N_c time chips, where the code is generated by a maximal linear PN code generator. The value of N_c is $2^p - 1$, where p is the number of stages in the code generator. The short length sequences are $N_c = 7, 15, 31, 63, 127, 255, \ldots$. The reason for using shift register codes is that the period of the PN sequence can easily be made very large by increasing the number of register stages, so that apparently it looks random to the users.

The LFSR can be implemented in two ways:

1. Fibonacci implementation [Fig. 8.3(b)]
2. Galois implementation [Fig. 8.3(c)]

The Fibonacci implementation consists of a simple shift register in which a binary-weighted modulo-2 sum of the taps is fed back to the input. The Galois implementation consists of a shift register, the contents of which are modified at every step by a binary-weighted value of the output stage.

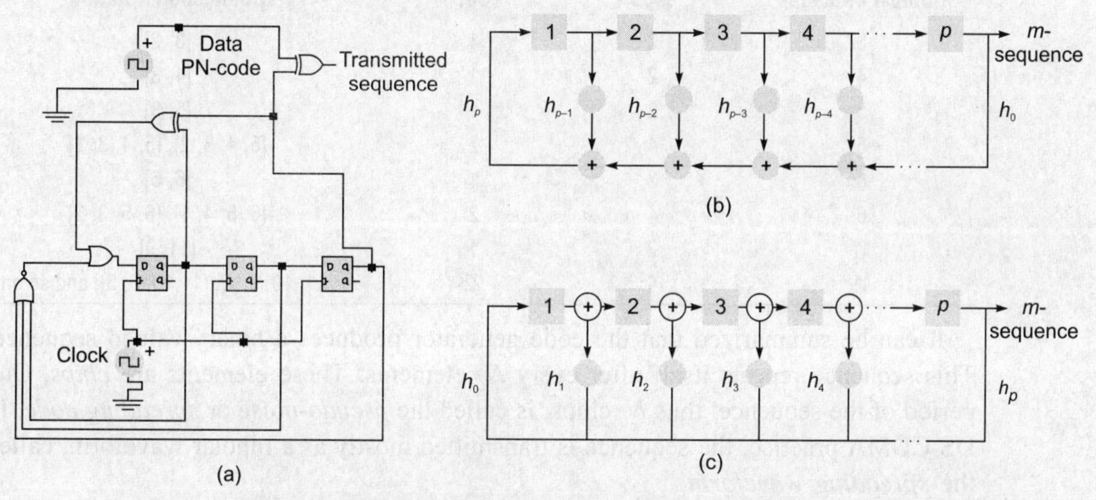

Fig. 8.3 (a) Typical PN code generator with three shift register stages and its application for data spreading, (b) Fabonacci implementation of LFSR, and (c) Galois implementation of LFSR

In general, shift register implements a primitive polynomial $h(x)$:

$$h(x) = x^p + h_{(p-1)}\, x^{(p-1)} + \cdots + h_1\, x + 1 \tag{8.9}$$

which is the generator polynomial for a code. In the polynomial $h(x)$, the coefficient h_i takes on the binary value 0 (no connection) or 1 (feedback or connection). Two exceptions are h_0 and h_p, which are always 1 and thus always connected. Note that h_p is not really a feedback connection, but rather is the input of the shift register. In short, the coefficients h_i determine the feedback connections in the code generator.

Thorough inspection reveals that the order of the Galois weights is opposite to that of the Fibonacci weights. Given identical feedback weights, the two LFSR implementations will produce the same sequence. However, the initial states of the two implementations must necessarily be different for the two sequences to have identical phase. The initial state of the Fibonacci form is called the *initial fill*, which comprises the first p bits output from the generator, while the initial state of the Galois generator must be adjusted appropriately to attain the equivalent initial fill. The Galois form is generally faster due to the reduced number of gates in the feedback loop.

For these, feedback connection tables exist. The feedback connections, or the corresponding realized polynomial, are represented by the notation

$$[p, h_{(p-1)}\,(p-1), ..., h_1] \tag{8.10}$$

where zero entries are not written explicitly. A few entries of feedback connection are given in Table 8.1. The table entries can be extended with increase in the number of stages, say, 12, 24, 32, and more.

Table 8.1 Feedback connection table

Number of Stages	Taps	Set	Connection Notation
3	2	1	[3, 2]
4	2	1	[4, 3]
5	2	1	[5, 3]
5	4	2	[5, 4, 3, 2], [5, 4, 3, 1]
6	2	1	[6, 5]
6	4	2	[6, 5, 4, 1], [6, 5, 3, 2]
11	2	1	[11, 9]
11	4	22	[11, 10, 9, 7], [11,10, 9, 5], and so on

It can be summarized that the code generator produces a binary-valued sequence. This sequence repeats itself after every N_c elements. These elements are *chips*. One period of the sequence, thus N_c chips, is called the *pseudo-noise* or *spreading code*. In DS-CDMA practice, the sequence is transmitted mostly as a bipolar waveform, called the *spreading waveform*.

Thus, with fairly modest amount of hardware, it becomes possible to make arbitrarily longer sequences. Today's VLSI circuits and ASICs have made possible the SSM-based system designs very much compact. The modulation of the spread spectrum carrier can be either biphase or quadriphase as described later on. The features of the ML/PN code can be summarized as follows.

- ML linear feedback shift-register sequences have desirable autocorrelation properties for spread spectrum applications. The autocorrelation function has a clear maximum if $\tau = 0$, but is small for any other offset. Due to this property,
  - a spread-spectrum receiver can 'easily' recognize code synchronization between the received code and the locally generated (identical) code by examining their correlation value and
  - a receiver locked to the dominant wave sees only little interference from the delayed waves, which may be present in a multipath channel.

- The ML sequences can be used in systems that need to operate in channels with large delay spreads in multipath channels.

Example 8.3 The clock frequency is 10 MHz and hence $t_{\text{chip}} = 10^{-7}$ seconds. If a shift register with 32 stages is used for generating the PN sequence, then compute the time to complete one cycle of the sequence.

Solution

Total number of chips in a period, $N_c = 2^{32}-1 = 4.29 \times 10^9$

So,
$$\begin{aligned} N_c \times t_{\text{chip}} &= 4.29 \times 10^9 \times 10^{-7} \\ &= 4.29 \times 10^2 \text{ seconds} \\ &= \mathbf{429} \text{ seconds} \end{aligned}$$

(If the stages are increased to 41, the total time to generate one sequence will be 2.545 days. The exercise is left to the readers.)

Example 8.4 Draw a [3,1] PN code generator that realizes the polynomial and generates [3,1] codes. Also draw the waveforms for the *m*-sequence for 3 periods.

Solution

Bracket [3,1] for ML code generation can be interpreted as follows.

The [3, 1] code has register length $n = 3$

So, code length, $N_c = 2^3 - 1 = 7$

The primitive polynomial $h(x)$ for this case is $h(x) = x^3 + 0 + x + 1$ (The polynomial suggests the stages from which the output is fed back.)

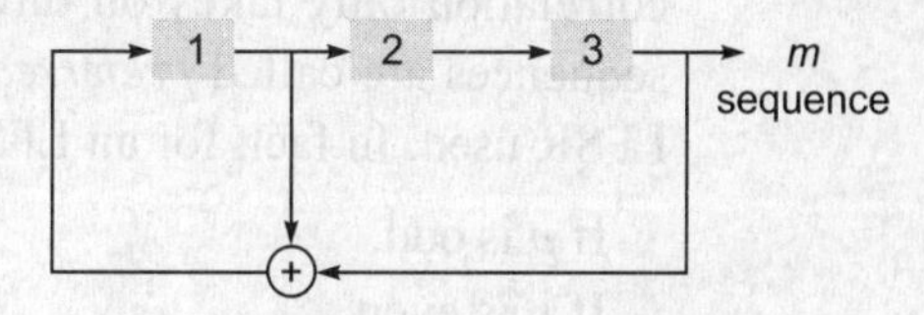

Fig. 8.4 [3,1] ML code generator

As can be seen $h_1 = 1$, $h_2 = 0$, and $h_3 = 1$. The generator is shown in Fig. 8.4 and the waveforms are shown in Fig. 8.5.

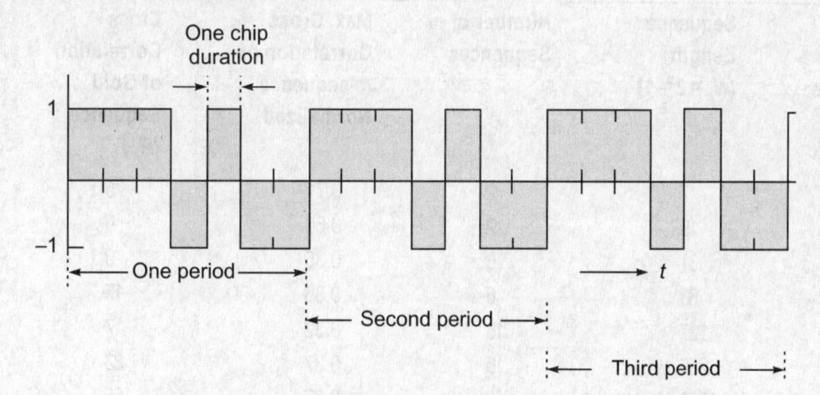

Fig. 8.5 Waveforms of the [3,1] code sequence

8.4.4 Walsh–Hadamard Sequences

If perfectly synchronized with respect to each other, W–H codes are perfectly orthogonal. That is, these are optimal codes to avoid interference among users in the link from base station to terminals.

The simplest matrix of two orthogonal Walsh–Hadamard codes is

$$C(_2^{\,1}) = \begin{bmatrix} 1 & 1 \\ 1 & -1 \end{bmatrix} \tag{8.11}$$

The code of user 1 is the first column, i.e. (1, 1), the code of user 2 is the second column, i.e. (1, – 1). Clearly, (1, 1) is orthogonal to (1, – 1). This matrix can be extended using a recursive technique. For $K = 2^n$ users, the matrix is found from the code matrix for $2^{(n-1)}$ users, according to

$$C(2^n) = \begin{bmatrix} C[2^{(n-1)}] & C[2^{(n-1)}] \\ C[2^{(n-1)}] & -C[2^{(n-1)}] \end{bmatrix} \tag{8.12}$$

8.4.5 Gold Codes/Gold Sequences

Gold sequences have been proposed by Gold in 1967 and 1968. These are constructed by EXOR-ing two or more *m*-sequences of the same length with each other. Thus, for a Gold sequence of length $N_c = 2^p - 1$, one uses two LFSRs, each of length $2^p - 1$. If the LFSRs are chosen appropriately, Gold sequences have better cross correlation properties than maximum length LSFR sequences as shown in Table 8.2.

Gold (and Kasami) showed that for certain well-chosen *m*-sequences, the cross correlation only takes on three possible values, namely –1, $-R_c$, or R_c–2. Two such sequences are called *preferred sequences*. Here R_c depends solely on the length of the LFSR used. In fact, for an LFSR with *p* memory elements,

If *p* is odd, $$R_c = 2^{(p+1)/2} + 1 \tag{8.13}$$

If *p* is even, $$R_c = 2^{(p+2)/2} + 1 \tag{8.14}$$

Table 8.2 Calculation of cross correlation of Gold sequences

Number of LFSR Elements (*p*)	Sequence Length ($N_c = 2^p - 1$)	Number of *m* Sequences	Max. Cross Correlation of *m*-sequence Normalized	Cross Correlation of Gold Sequence (R_c)	Cross Correlation of Gold Sequence, Normalized [$R_c/(2^p-1)$]
3	7	2	0.71	5	0.71
4	15	2	0.60	9	0.60
5	31	6	0.35	9	0.29
6	63	6	0.36	17	0.27
7	127	18	0.32	17	0.13
8	255	16	0.37	33	0.13
10	1023	60	0.37	65	0.06
12	4095	144	0.34	129	0.03

Thus, a Gold sequence formally is an arbitrary phase of a sequence in the set $G(x,y)$ defined by

$$G(x, y) = \{x, y, x \oplus y, x \oplus Ty, x \oplus T^2 y, x \oplus T^{(N_c-1)} y\} \tag{8.15}$$

where T^k denotes the operator that shifts vectors cyclically to the left by k places, $\oplus$ is the exclusive OR operator, and x, y are m-sequences of period generated by different primitive binary polynomials.

It is well known that the 'partial cross correlation' values can be altered by changing the phases of the code sequences. In theory, then, it is possible to find optimal phases that minimize the interference in the desired data signal. However, for K users, each employing a sequence of period N_c, there are a total of $N_c K$ different sets of sequence phases possible.

A few other sequences are described below.

8.4.6 Quadratic Residue Sequences (*q-r*) Sequences

These sequences have lengths that are prime numbers of the form $N_c = 4q-1$ = a prime number, where q is an integer. Since prime numbers are fairly distributed and since half of the prime numbers can form quadratic residue sequences, there are many more sequences of this type than m-sequences. Its implementation is somewhat complex as it is not generated directly by using shift registers. There are some values of m for which both *q-r* and m sequences may exist, $N_c = 3, 7, 31, 127, 8191, 131071, \ldots$. However, except $N_c = 3$ and 7, *q-r* and m sequences are distinct. The *q-r* sequences have the same correlation properties as m-sequences.

8.4.7 Hall Sequences

Hall sequences also have a length that is a prime number described by $N_c = 4q-1 = 4r^2 + 27$ = a prime number, where both q and r are integers. Since both the conditions are to be satisfied, there are very few Hall sequences. It has same properties as *q-r* sequences but there is no particular advantage of using such sequences.

8.4.8 Twin-Prime Sequences

The twin-prime sequences are those sequences in which the length of the sequence is defined by $N_c = p_p(p_p+2)$, where both p_p and p_p+2 are prime numbers. The only advantage of twin-prime sequences is that it is possible to achieve some lengths that are not readily achieved by any of other forms; otherwise the implementation is not simple.

8.4.9 Criteria to Select Code/Sequence

In synchronous DS-CDMA, the link performance is affected by

- Multi-user interference
- Asynchronous multipath interference arising from the delayed signals from the other users as well as the user himself/herself

As the user capacity of the channel is limited by the interference, we would like to make the partial cross correlations of codes small, ideally zero for all time shifts. The values that the partial cross correlation functions take depend upon the choice of codes. For example, Walsh codes have the desirable property of perfect orthogonality at zero phase offset but have poor autocorrelation properties for other phase offsets. Moreover,

Walsh codes do not spread the signal uniformly over the spectrum. To counteract this, one scheme is proposed, e.g. for IS-95, whereby each user's code consists of his unique Walsh code modulated by a ML sequence common to all users. Good autocorrelation properties are also crucial for timing recovery and coherent detection.

The user capacity of a synchronous CDMA system, in which the various user signals exhibit no time offsets when they arrive at the receiver, is limited by the number of different codes. With a spread factor N_c and K users, perfectly orthogonal codes are chosen if K is less or equal to N_c. A bound for the maximum normalized cross correlation $R_{c\max}$ (at zero time offset) between the user codes can be given by

$$R_{c\max}^2 = \frac{(K/N_c)-1}{K-1} \tag{8.16}$$

For Walsh–Hadamard codes, $N_c = K$, so $R_{c\max} = 0$, and for Gold codes, $N_c = K - 1$. So, $R_{c\max} = 1/N_c$.

Synchronous CDMA is used in the downlink of cellular systems, but multipath reception can destroy the time alignment of components traveling over different paths. In asynchronous CDMA, codes arrive in a time-shifted manner, and the cross correlation values typically are much larger. Asynchronous CDMA occurs in the uplink of cellular systems.

8.5 GENERAL BLOCK DIAGRAMS OF DSSS TRANSMITTER AND RECEIVER

In the transmitter (Fig. 8.6) side, for biphase modulation, an MOD 2 adder is used. While using quadriphase modulation two MOD 2 adders are used with two alternate chips available from PN code generator. Two balanced modulators are fed with 90 degree phase shifted carriers (just like QPSK generation). Adding both the signals, the SSM RF output is available. It may be noted that the message modulation is using binary data and code.

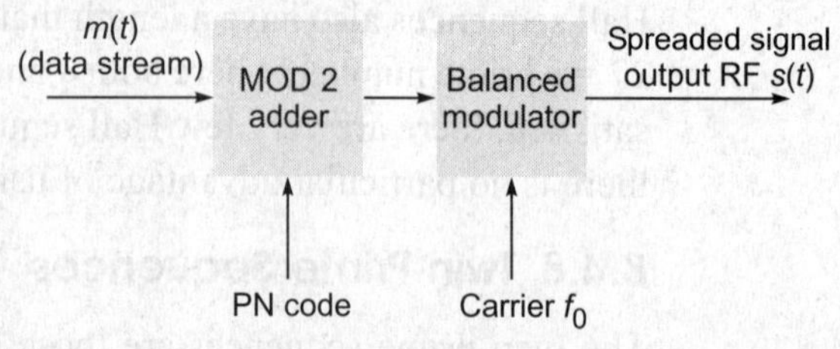

Fig. 8.6 Simplified diagram for biphase modulation

Receiver diagram is shown in Fig. 8.7. Receiver for a spread signal must perform three distinct functions: detection of presence of signal, carrier removal, and despreading or demodulation using PN sequence. Detection of signal and despreading operations can be either active or passive.

The active method involves searching for the signal's presence in both time and frequency domain and tracking the sequence after it has been acquired, despreading the signal with the correlator and demodulating the signal in a usual way described previously, i.e. using PN sequence again.

The passive methods only require that the signal be searched for in frequency, i.e. only carrier, since the passive system will respond whenever the signal occurs. The despreading is accomplished in a matched filter rather than a correlator. The demodulation is performed in a usual manner as mentioned previously.

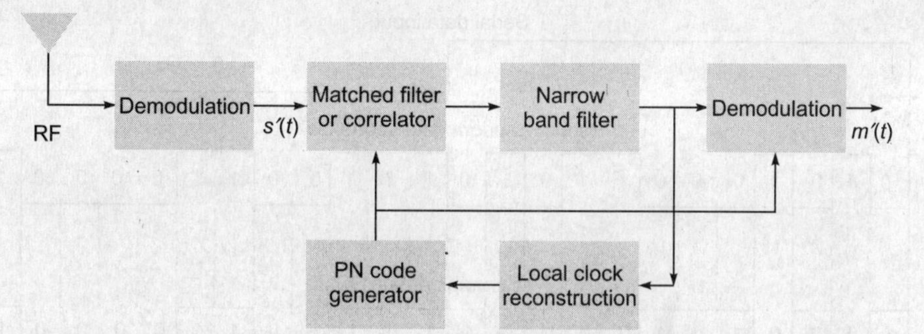

Fig. 8.7 Carrier demodulation and despreading of SSM signal to get original data

The choice of method, whether active or passive, depends upon the conditions. Active methods are preferred when the sequence is too long and the processing gain (define in Section 8.6) is very large. On the other hand, passive methods may be preferred when the sequence is short or when used as an aid to acquisition. It is possible to combine both the approaches in a single receiver. Presently, active method based rake receiver designs with antenna diversity is developed for the better performance.

The complete process of spread spectrum modulation and demodulation can be understood with the help of the following example.

Example 8.5 Draw the waveforms of the different stages of DSSS modulator and demodulator for input data bits 1 followed by 0. The PN sequence during these two bits is (0101000110101110011000001).

Solution

Considering XOR operation, we get the waveforms as shown in Fig. 8.8.

8.6 PN SIGNAL CHARACTERISTICS, SPECTRAL DENSITY, BANDWIDTH, AND PROCESSING GAIN

The PN signal means the waveform generated due to PN sequence (Even the spread spectrum output with carrier allocation also deserves the same nature and can be considered in the same manner) The frequency domain representation of a time domain pulse is a sinc (sin x/x type) form of envelope. The PN code as well as the data are both pulsed/square wave in nature. If the binary or quadriphase PN sequence is considered to be purely random (due to long sequence, though it may be repeated) rather than the periodic, it is straightforward to show that the spectral density of it is

$$s(f) = \frac{t_{\text{chip}}}{2}\left\{\left[\frac{\sin \pi (f - f_0) t_{\text{chip}}}{\pi (f - f_0) t_{\text{chip}}}\right]^2 + \left[\frac{\sin \pi (f + f_0) t_{\text{chip}}}{\pi (f + f_0) t_{\text{chip}}}\right]^2\right\} \tag{8.17}$$

The expression is normalized to represent a signal having unit average power. The spectral density for the positive and negative frequencies is shown in Fig. 8.9. f_o represents the centre frequency of the sync-shaped spectrum (which may be carrier after upconversion).

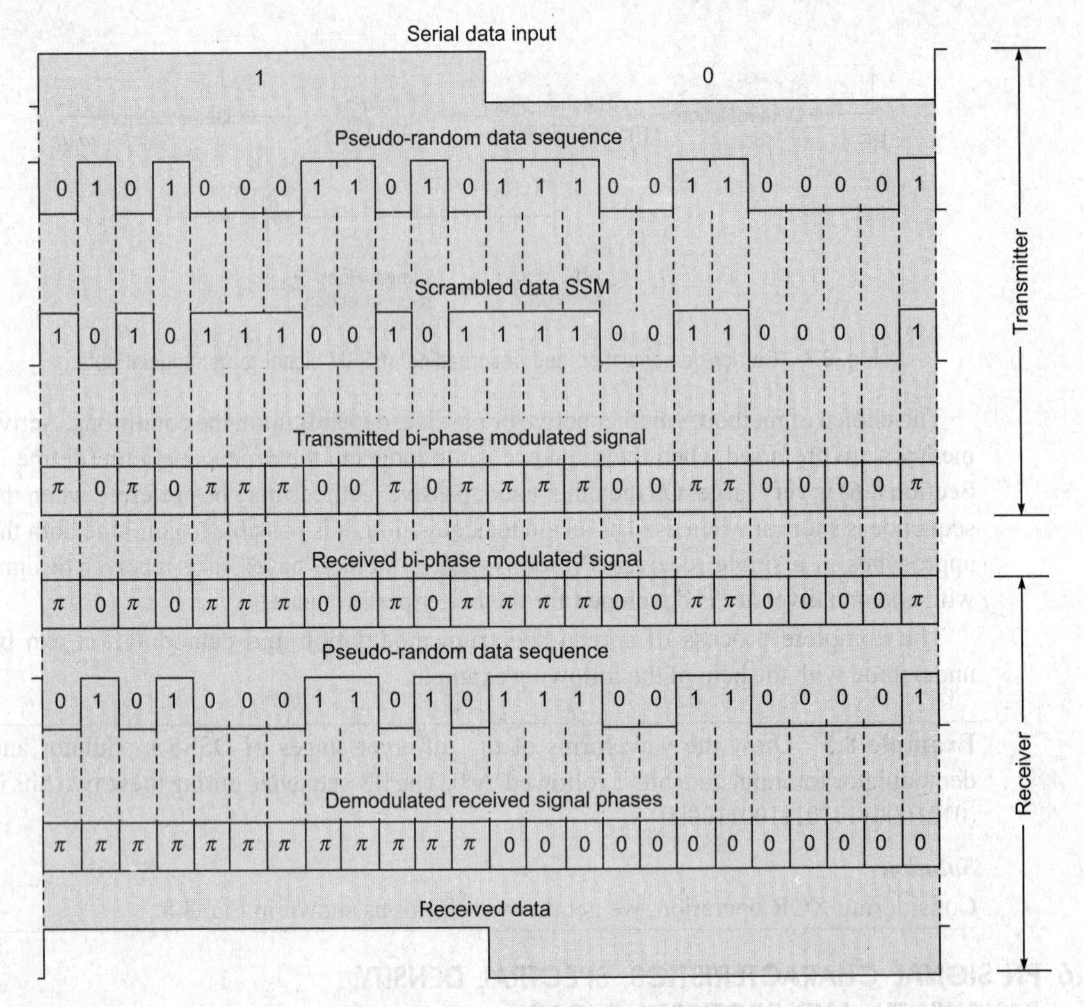

Fig. 8.8 Waveforms for Example 8.5

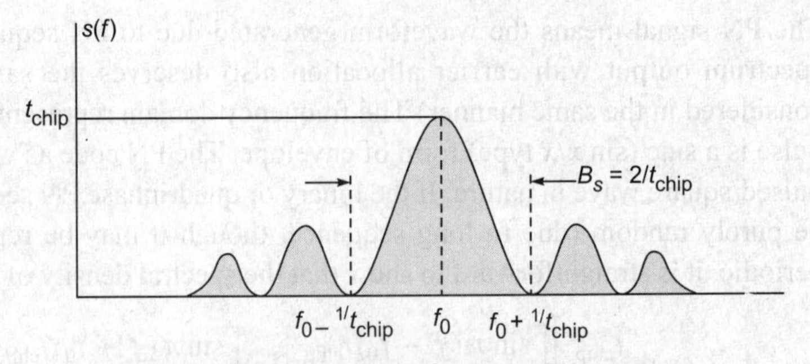

Fig. 8.9 Spectral density of a binary PN sequence

It is customary to define the bandwidth B_s of a PN signal as the frequency increment between the two zeros of the spectral density that are closest to the centre frequency. It is clear from Fig. 8.9 that the bandwidth of the biphase signal is $2/t_{chip}$.

Since the message is binary/square wave it will have similar spectral density but centred on zero. Thus the message signal spectral density is

$$s_m(f) = t_m \left[\frac{\sin \pi f\, t_m}{\pi f\, t_m} \right]^2 \tag{8.18}$$

The bandwidth of the message B_m with the similar considerations will be $1/t_m$, because it is customary to use only the positive frequency portion of the spectrum in defining the bandwidth. Let

B_s = Bandwidth of the SSM or PN signal

B_m = Bandwidth of the message signal

The processing gain is defined as

$$\text{PG} = \frac{B_s}{B_m} \tag{8.19}$$

It gives the amount of spreading of the signal.

If the biphase modulation is used, then the processing gain is given by

$$\text{PG} = 2\, t_m / t_{\text{chip}} \tag{8.20}$$

For quadriphase modulation, the processing gain

$$\text{PG} = t_m / t_{\text{chip}} \tag{8.21}$$

Example 8.6 If the chip rate of a DSSS transmitter is 20 Mcps, the message bit rate is 10 kbps. Find out the processing gain achieved finally if the biphase modulation is used.

Solution

$$t_{\text{chip}} = \frac{1}{20 \times 10^6} \Rightarrow \text{SSM bandwidth } B_s = 2/t_{\text{chip}} = 40 \times 10^6$$

$$\text{Message BW} = 1/t_m = 10^4$$

Hence, processing gain = $B_s/B_m = 2t_m/t_{\text{chip}} = 4000$

In dB, the processing gain can be represented as $10 \log_{10}(4000) = 36$ dB

Example 8.7 A recorded conversation is to be transmitted by a pseudo-noise spread spectrum system. Assume that the spectrum of the speech waveform is bandlimited to 3 kHz and use 128 quantization levels. (a) Find out the chip rate required to obtain a processing gain of 20 dB. (b) Given that the sequence length is to be greater than 5 hours, find out the number of shift register stages required.

Solution

(a) Processing gain = 20 dB = $10\log_{10}$ PG

$$\Rightarrow \quad \text{PG} = 10^2 = 100$$

Now, 128 quantization levels means that 7 bits per sample will be utilized.

Also 3 kHz speech signal means that the sampling rate will be 6 kHz (Nyquist criteria).

$$\begin{aligned} \text{Message bit rate} &= \text{sampling rate} \times \text{bits/sample} \\ &= 6 \times 7 \\ &= 42 \text{ kbps} \end{aligned}$$

Now, message signal bit duration $t_m = 1/42$ kbps $= 0.023$ ms

Processing gain $= 2\, t_m/t_{\text{chip}}$

$$\Rightarrow \quad 100 = 2 \times (0.023 \times 10^{-3})/t_{\text{chip}}$$

$$\Rightarrow \quad t_{\text{chip}} = (0.046 \times 10^{-3})/100 = 0.46\ \mu\text{s}$$

(b) Sequence length is 5 hours $= 5 \times 60 \times 60 = 18{,}000$ seconds

Now, $(N_c t_{\text{chip}})/t_{\text{chip}} = 18{,}000/0.46 \times 10^{-6} = 3.9 \times 10^{10} = 2^p$

where p is the number of stages.

We have $\quad p\log_{10}2 = \log_{10}(3.9 \times 10^{10})$

$\Rightarrow \quad p = 35.23$ rounded to higher value $\Rightarrow$ 36 stages will be required.

8.7 MATHEMATICS ASSOCIATED WITH THE SPREAD SPECTRUM MODULATION/DEMODULATION

Spreading

If binary data is $m(t)$, PN signal (at the transmitter end) is $c_t(t)$, and transmitted baseband is $s(t)$, then the binary data is directly multiplied with the PN sequence, which is independent of binary data, to produce the transmitted baseband signal.

$$s(t) = m(t)c_t(t) \tag{8.22}$$

The effect of multiplication of $m(t)$ with PN sequence is to spread the baseband bandwidth B_m of $m(t)$ to a baseband bandwidth of B_s.

Despreading

In the receiver, the received baseband signal $s'(t)$ is multiplied with the PN sequence (at the receiver end) $c_r(t)$.

- If $c_t(t) = c_r(t)$, and synchronized to the PN sequence in the received data, then the recovered binary data $m'(t)$ is produced. The bandwidth is also despread.

  We have
  $$m'(t) = s'(t)\, c_r(t) \tag{8.23a}$$
  $$m'(t) = (m(t)c_t(t))\, c_r(t) \tag{8.23b}$$
  when PN sequence is multiplied with itself (with perfect synchronization), –1 will be multiplied with –1, and 1 will be multiplied with 1. Hence, alterations will be destroyed and $c_t(t)\, c_r(t) = 1$ for all t. Thus, $m'(t)$ will be equal to $m(t)$, same as the original data.
- If $c_t(t) \neq c_r(t)$, then there is no despreading action. A receiver not knowing the PN sequence of the transmitter cannot reproduce the transmitted data.

In presence of interference:

$$s'(t) = s(t) + i(t) \tag{8.24a}$$
$$s'(t) = m(t)c_t(t) + i(t) \tag{8.24b}$$

To recover the original data, the received signal is multiplied with the locally generated PN sequence $c_r(t)$ with $c_t(t) = c_r(t)$ condition and perfect synchronization.

$$m'(t) = s'(t)\, c_r(t)$$

$$= [s(t) + i(t)]\, c_r(t) \quad (8.25a)$$

$$= [m(t)c_t(t) + i(t)]\, c_r(t) \quad (8.25b)$$

Finally, the multiplier output becomes

$$m'(t) = m(t) + i(t)\, c_r(t) \quad (8.26)$$

Here, the term $i(t)c_r(t)$ means that the spreading code will affect the interference just as it did with the information bearing signal at the transmitter. Noise and interference being uncorrelated with the PN sequence, become noise, like increase in bandwidth and decrease in power density, after the multiplier. After despreading, the data component is narrowband whereas the interference component is wideband. By applying the $m'(t)$ signal to a low-pass filter with a bandwidth just large enough to accommodate the recovery of the data signal, most of the interference component is filtered out. The effect of the interference is reduced by the processing gain.

8.8 DIRECT SEQUENCE SPREAD SPECTRUM RECEIVER CONSIDERATIONS (RAKE RECEIVER)

The reception can be explained with the help of an example. We assume that a logical '1' is represented by voltage level 1 and a logical '0' is represented by –1. In direct sequence spread spectrum, the user data symbols are multiplied with a fast code sequence. For instance, the maximum length LSFR code {1,1,1,–1,1,–1,–1} (Fig. 8.10) has autocorrelation value 7 for zero time-offset and –1 for time-offsets 1, 2, …, or 6.

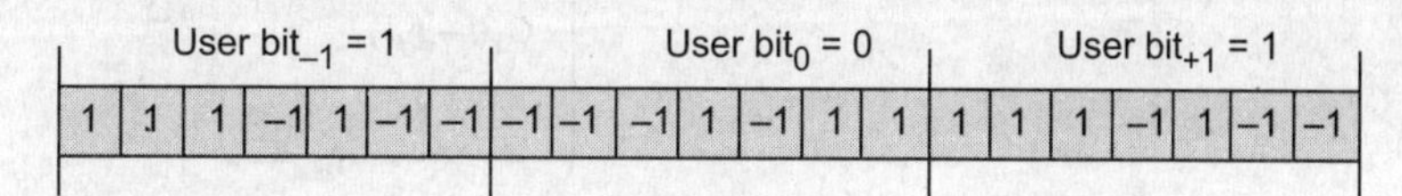

Fig. 8.10 Pattern after SSM (shown for three bit intervals)

After transmission over a multipath channel, the received signal consists of multiple delayed copies of the transmit signal. How the signals can be dealt with will be clear once the rake receiver is introduced.

The receiver for DSSS system is a specially designed rake receiver, which exploits multipath and its hardware is mainly based on digital signal processing. Rake means a long-handled tool with a row of prongs at the end for smoothing the soil. If we see the configuration of rake receiver, it does almost the same task by extracting and smoothing the available multipath signals. The front end looks like a row with prongs (parallel channels as shown in Fig. 8.11) each known as fingers of the rake receiver.

The rake receiver consists of multiple correlators, each synchronized to one of the time offsets of the received signal. The rake receiver correlates the received signal with different time offsets of the spreading and scrambling codes and performs a channel correction to compensate for the differing channel characteristics of the individual multipaths. These steps are performed for all multipaths in the environment. The results of these operations are combined to drive the decision-making process for the value of the received symbol.

Two configurations with five-finger rake receiver are shown in Fig. 8.11.

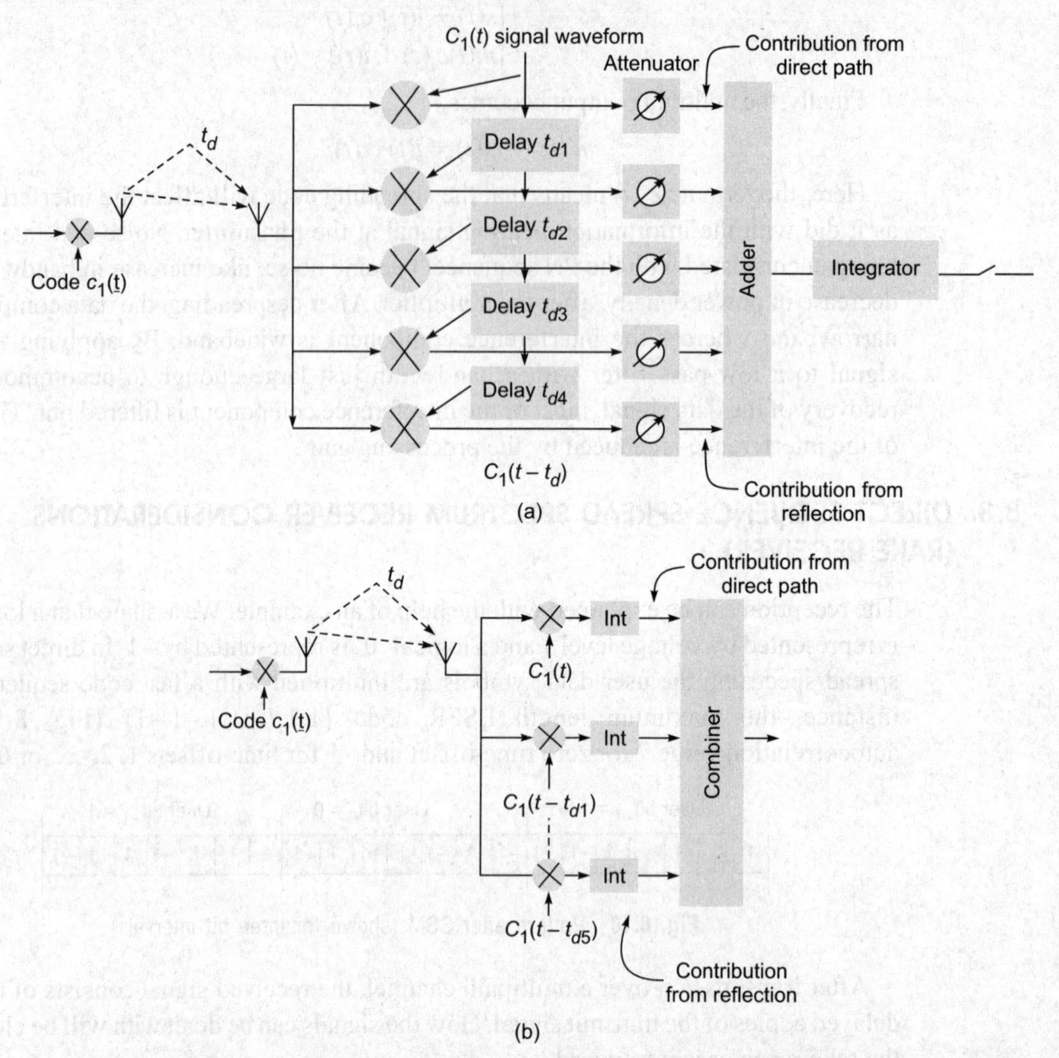

Fig. 8.11 (a) Rake receiver with five fingers, one combiner, and one integrator and (b) scenario of another way of reception with multiple integrators at rake receiver and processing

If in a mobile radio channel, reflected waves arrive with small relative time delays, self-interference occurs. The DSSS is often claimed to have particular properties that makes it less vulnerable to multipath reception. In particular, the rake receiver architecture allows an optimal combining of energy received over different paths with. It avoids wave cancellation (fades) if delayed paths arrive with phase differences and appropriately weighs the signals coming in with different signal-to-noise ratios. In short, the rake receiver is designed to optimally detect a DS-CDMA signal transmitted over a dispersive multipath channel. It is an extension of the concept of the matched filter as we mentioned in passive method.

In the matched filter receiver, the signal is correlated with a locally generated copy of the signal waveform. If, however, the signal is distorted by the channel, the receiver should correlate the incoming signal by a copy of the expected received signal, rather than by a copy of transmitted waveform. Thus, the receiver should estimate the delay profile of the channel and adapt to its locally generated copy according to this estimate.

The spreading code is chosen to have a very small autocorrelation value for any non-zero time offset. This avoids crosstalk between fingers. In practice, the situation is less ideal. It is not the full periodic autocorrelation that determines the crosstalk between signals in different fingers, but rather two partial correlations, with contributions from two consecutive bits or symbols. It has been attempted to find sequences that have satisfactory partial correlation values, but the crosstalk due to partial (non-periodic) correlations remains substantially more difficult to reduce than the effects of periodic correlations.

In contrast to the second-generation mobile telecommunications standards, the UMTS/W-CDMA infrastructure is capable of handling a 'soft' handover, whereby a mobile terminal is in contact with multiple base stations at the same time. This can occur, for example, when a mobile terminal is at a cell border and the signals from all surrounding base stations have near equal strength. The signals from multiple base stations differentiate themselves in their scrambling code. Thus, to handle the 'soft' handover scenario, rake receiver in the mobile terminal must be capable of correlating the received signal with the individual scrambling codes of all the base stations involved. Furthermore, the rake receiver is also capable of collecting and using the energy from multipath components of a signal.

In addition to the actual signal reception tasks, the rake receiver must also perform a set of timing and synchronization tasks. A path searcher performs a correlation of a fixed set of pilot signals over a sliding window to detect the paths with the strongest signal values in a multipath environment. The offsets of these paths are stored within a control context and are used to generate the required offsets for the individual rake fingers that descramble and despread the chip rate signals. The path searcher divides itself into a coarse and a fine searcher, with differing repetition intervals and accuracies. A path tracker is responsible for the tracking and the resynchronization of the paths that are currently being received.

In short, like a garden rake, the rake receiver gathers the energy received over the various delayed propagation paths. According to the maximum ratio combining principle, the SNR at the output is the sum of the SNRs in the individual branches, provided that we assume that only AWGN is present (no interference) and codes with a time offset are truly orthogonal.

The order of diversity achieved by the rake receiver depends on the number of resolvable paths L with

$$L = \text{round}(\tau_{\text{max}}/t_{\text{chip}}) + 1 \tag{8.27}$$

where τ_{max} is the maximum delay spread of the multipath channel as described in Chapter 5.

Now coming to the example signal shown in Fig. 8.12 again, in ith finger of the rake, the incoming signal is multiplied with the code sequence synchronized to the signal arriving over path i. Let a_i be the autocorrelation of the signal received over path i.

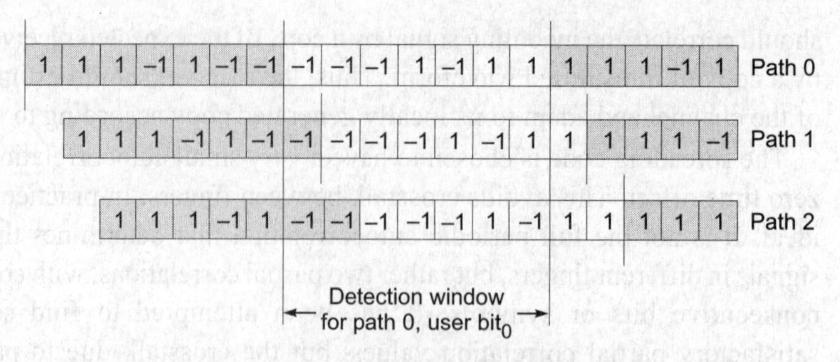

Fig. 8.12 Reception of three delayed waves over paths 0, 1, and 2 and matching with the detection window contributing towards F_0

Assume total three paths. If the autocorrelation is good, crosstalk into other fingers is weak, typically *m* times weaker than the wanted signal, with *m* being the length of the sequence. Because of modulation, the crosstalk is much more severe in a practical situation. The signal F_i seen in the *i*th finger consists of three terms, corresponding to contribution from the signal over all three paths. In this case arbitrarily,

$$F_0 = -7\,a_0 - a_1 - 3\,a_2$$
$$F_1 = -7\,a_1 - a_0 + a_2$$
$$F_2 = -7\,a_2 - 3\,a_0 - a_1$$

Note that the interference from the signal over paths 0 and 2 is substantially larger (three times) than in case of unmodulated carriers. Hence, ISI occurs.

In the same manner, we can draw the delection windows for path 1 and path 2 also to find out the amount of interference from other paths.

8.8.1 Partial Correlation of PN Sequences at the Rake Receiver

The rake receiver for DSSS can separate energy received over multiple propagation paths with different time delays. Each finger in the rake sees signals with a specific time delay, but it also sees some unwanted residual crosstalk from signals over differently delayed paths. If bit transitions occur in delayed reflection, the crosstalk between two fingers in the rake detector is determined by two partial correlation functions, which exhibit substantially inferior properties than can be achieved for the full (or periodic) autocorrelation function.

The rake receiver consists of multiple correlators as shown in Fig. 8.11, in which the received signal is multiplied by time-shifted versions of a locally generated code sequence. The intention is to separate signals such that each finger sees only the signals coming in over a single (resolvable) path. The spreading code is chosen to have a very small autocorrelation value for any non-zero time offset. This avoids crosstalk between fingers. In practice, the situation is less ideal. It is not the full periodic autocorrelation that determines the crosstalk between signals in different fingers, but rather two *partial* correlations, with contributions from two consecutive bits or symbols. It has been attempted to find sequences that have satisfactory partial correlation values, but the crosstalk due to partial (non-periodic) correlations remains substantially more difficult to reduce than the effects of periodic correlations.

It can be useful to use a correlation window in the receiver that is equal to the full period of the spreading code. However, the spreading sequence used by the transmitter is longer than the period used by the correlator for detection. The extension consists of a cyclic prefix for DS-CDMA.

8.8.2 Cyclic Prefix in DS-CDMA

The use of a correlation window in the receiver is proposed, which is equal to the full period of the spreading code. However, the spreading sequence used by the transmitter is longer than the period used by the correlator for detection. The extension consists of a cyclic prefix (Fig. 8.13). In the event of maximum length (m) LFSR codes, the total code used to spread one user symbol then becomes the concatenation of one full period of the code plus the prefix. Cyclic prefixes are well established for OFDM (Chapter 9), and also used in equalizer training sequences to estimate the channel (Chapter 6). Their use for direct sequence CDMA, in particular in conjunction with a rake receiver, has been explored to a much lesser extent.

In the prefix example of an m-sequence $\{1, 1, 1, -1, 1, -1, -1\}$, the transmitter creates the spreading code $\{-1, -1, 1, 1, 1, -1, 1, -1, -1\}$. The receiver correlates during the same window for every branch of the rake but uses different, cyclically shifted sequences.

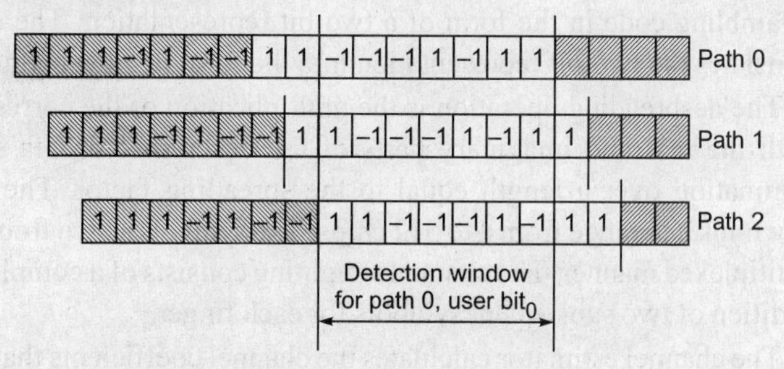

Fig. 8.13 Delayed signals (A cyclic prefix allows the choice of a detection window that avoids ISI.)

Now, the ISI is reduced to levels determined by the periodic autocorrelation function and strength in each finger becomes

$$F_0 = -7\, a_0 + a_1 + a_2$$
$$F_1 = -7\, a_1 + a_0 + a_2$$
$$F_2 = -7\, a_2 + a_0 + a_1$$

Due to cyclic prefix, all ISI terms disappear influencing the BER performance but at the same time it has the following disadvantages:

- Cyclic prefixes waste some transmit bandwidth and energy.
- Correlated fading of the signal amplitudes in the various diversity branches of the rake degrades the performance. The best performance is achieved with uncorrelated fading.

8.9 SIGNAL PROCESSING AT THE RAKE RECEIVER

The basic partitioning of tasks between DSP, dedicated hardware, and reconfigurable hardware is shown in Fig. 8.14. Dataflow-oriented tasks that operate on a word-level granular data stream are executed using the reconfigurable hardware. A DSP is used to execute the control-flow and synchronization tasks. Bit-level data processing tasks that execute continuously are mapped onto dedicated hardware resources. The individual components of the rake receiver finger are illustrated in Fig. 8.14.

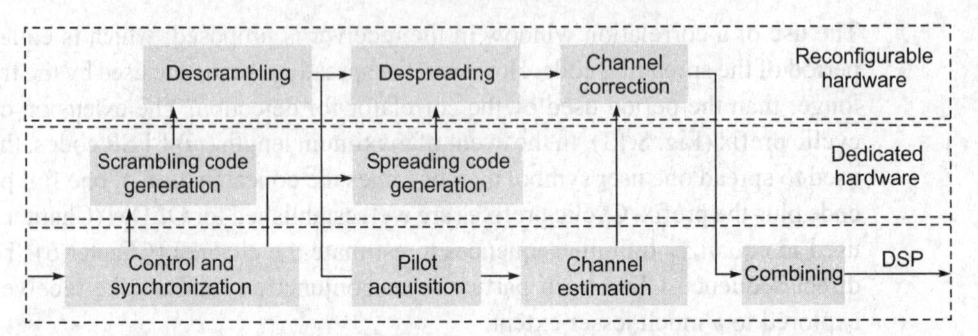

Fig. 8.14 Partitioning the rake receiver showing one rake finger typically

The descrambling operation involves the complex multiplication of the aligned incoming data with then scrambling codes. The scrambling code generator provides the scrambling code in the form of a two-bit representation. The reconfigurable hardware translates the two-bit representation into the form of $\pm1\pm j$ by the use of multiplexers.

The despreading operation is the multiplication of the corresponding spreading code with the real and imaginary parts of the descrambled data sequence followed by a summation over a length equal to the spreading factor. The spreading factor in the downlink can range from 4 to 512 chips. The symbols arrive from the despreader in a time multiplexed manner. The channel weighting consists of a complex multiplication and the addition of two subsequent symbols for each finger.

The channel estimator calculates the channel coefficients that are used for the channel correction, which are then transferred to the reconfigurable hardware. The channel coefficients are calculated on the basis of a specific sequence of pilot signals. The FIFOs store the channel coefficients for finger. Channel correction unit, in addition to the actual channel correction, takes two symbols from the despreader at half the symbol rate and also performs the space time transmit diversity (STTD) decoding of the symbols. In STTD encoding, the symbol stream is divided into two streams each with half the transmit frequency. Each stream is transmitted over a locally separate antenna. The first symbol stream remains unchanged. The second symbol stream is reordered and the conjugate complex of the symbol is transmitted. The antennas are located far away from each other such that each stream has its own channel coefficient, but close enough so that both symbols arrive at the receiver at the same time.

8.10 CHARACTERISTICS OF DSSS SYSTEM

The main advantageous characteristics of DSSS systems are described in the following sections.

8.10.1 Interference Rejection

Interference rejection is introduced in Section 8.7 along with mathematics, which can be confirmed with Fig. 8.15. The SNR is an important parameter for a spread spectrum system because it decides the detectability of the signal as well as the output signal quality. Maximum SNR will occur when there is no delay spread. If $s'(t)$ is the received signal at the rake receiver, the maximum SNR is simply the ratio of the square of the mean value of $s'(t)$ to the variance of $s'(t)$. Finally, it can be shown that output SNR

$$\text{SNR}_{\text{out}} = \frac{P_r}{N_o/2t_m} \tag{8.28}$$

where P_r is the received power, N_o is the noise power spectral density, and t_m is the message bit duration. It may be noted that SNR depends upon the ratio of received power to noise spectral density and bit duration. By increasing the message bit time and P_r/N_o, we can increase the SNR.

Now, the mean square value of the output interference signal is given by

$$\overline{j^2} = \frac{J}{\text{PG}} \tag{8.29}$$

where J is the total interference power and PG is the processing gain. This is the case when interfering signal bandwidth is less than transmission bandwidth. We can note that the interference power can be reduced by a factor equal to processing gain. This is an average result, however the averaging takes place over all phases of interference. For some fixed phase relationships the mean square value of the output interference may be twice the value given by Eq. (8.29).

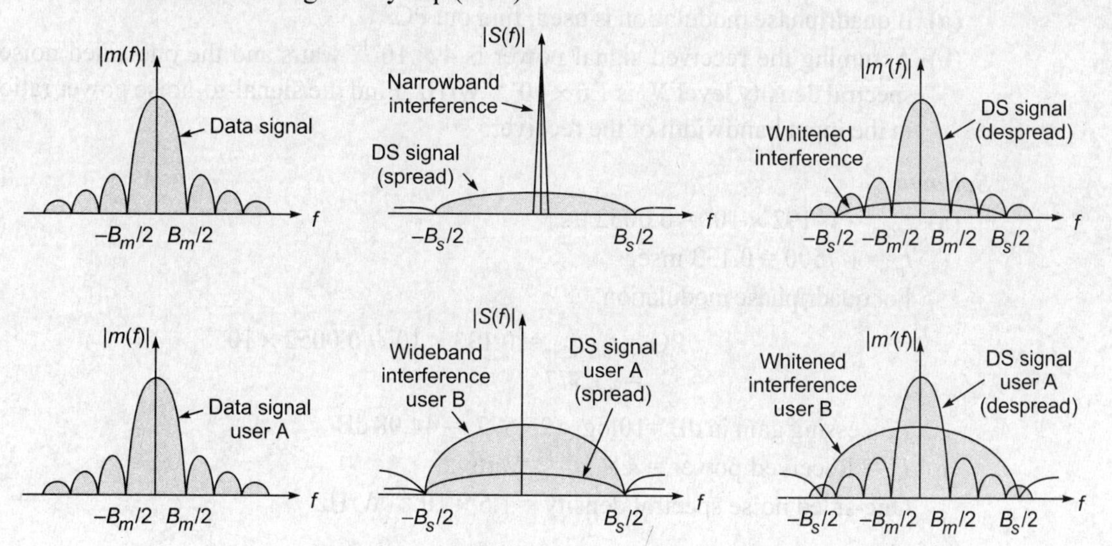

Fig. 8.15 Narrowband and wideband interference rejection

It is also interesting to consider the situation in which the interference bandwidth B_j is greater than spread spectrum bandwidth B_s. In this case

$$\overline{j^2} = \frac{B_s}{2B_j} \cdot \frac{J}{PG} \tag{8.30}$$

Since J is the interference power in its total bandwidth, the factor $(B_s/2\ B_j) \cdot J$ represents simply that portion of the interference power that exists in the receiver bandwidth. This power is again reduced by a factor equal to the processing gain. Considering the interference signal as a noise, the SNR is again to be calculated as follows:

$$\text{SNR}_{\text{out}} = \frac{P_r}{(N_o/2t_m) + (J/\text{PG})} \tag{8.31}$$

But
$$\frac{\text{PG}}{t_m} = \frac{B_s}{B_m} \cdot B_m = B_s$$

Hence, the output SNR becomes

$$\text{SNR}_{\text{out}} = \frac{\text{PG} \cdot P_r}{(N_o\,B_s/2) + J} \tag{8.32}$$

Now,

$$\text{SNR}_{\text{in}} = \frac{P_r}{(N_o\,B_s/2) + J} \tag{8.33}$$

The new relationship can be written as $\text{SNR}_{\text{out}} = (\text{PG}).\ \text{SNR}_{\text{in}}$ (8.34)

Here the losses in the system are not considered.

Example 8.8 A direct sequence system has a PN code rate of 192×10^6 chips per second and a binary message bit rate at 7500 bps.

(a) If quadriphase modulation is used, find out PG.

(b) Assuming the received signal power is 4×10^{-14} watts and the one-sided noise spectral density level N_o is 1.6×10^{-20} W/Hz. Find the signal-to-noise power ratio in the input bandwidth of the receiver.

Solution

(a) $t_{\text{chip}} = 1/\ 192 \times 10^6 = 0.0052\ \mu\text{s}$

$t_m = 1/7500 = 0.133$ msec

For quadriphase modulation,

$$\text{PG} = t_m/t_{\text{chip}} = 0.133 \times 10^{-3}/\ 0.0052 \times 10^{-6} = 25{,}577$$

Processing gain in dB $= 10\log_{10}(25{,}577) = 44.08$ dB

(b) P_r = Received power $= 4 \times 10^{-14}$ watts

One-sided noise spectral density $= 1.6 \times 10^{-20}$ W/Hz

For biphase modulation:

Signal bandwidth = $2/t_{chip}$ = $2/\,0.0052 \times 10^{-6} = 384.6$ MHz

Noise power = $N_o \times$ B.W. = $1.6 \times 10^{-20} \times 384.6 \times 10^{6} = 615 \times 10^{-14}$ W

Signal-to-noise ratio = $P_r/N_o = 4 \times 10^{-14}/615 \times 10^{-14} = 0.0065$

In dB scale (SNR) dB = $10\log_{10}(P_r/N_o) = -21.86$ dB

For quadriphase modulation:

Signal bandwidth = $1/t_{chip}$ = $1/0.0052 \times 10^{-6} = 192.3$ MHz

Noise power = $N_o \times$ B.W. = $1.6 \times 10^{-20} \times 192.3 \times 10^{6} = 307.7 \times 10^{-14}$ W

Signal-to-noise ratio = $P_r/N_o = 4 \times 10^{-14}/307.7 \times 10^{-14} = 0.013$

In dB scale (SNR) dB = $10\log_{10}(P_r/N_o) = -18.86$ dB

8.10.2 Anti-jam Characteristics

One of the important reasons for being interested in the interference rejection capabilities of a spread spectrum system is to be able to evaluate the degree to which such a system can reduce the effects of intentional jamming. This ability can be expressed in terms of jamming margin or sometimes called anti-jam (AJ) margin and it is usually expressed in decibels. If we write the SNRs in terms of dB, Eq. (8.34) can be rewritten as

$$\text{SNR}_{in}\ (\text{dB}) = \text{SNR}_{out}\ (\text{dB}) - \text{PG}\ (\text{dB}) \tag{8.35}$$

Anti-jam margin can be defined as

$$\text{Margin (AJ)} = -\text{SNR}_{in}\ (\text{dB}) - \text{L}\ (\text{dB}) \tag{8.36}$$

where L represents the losses in the system expressed in dB. Also the anti-jam margin can be written as

$$\text{Margin (AJ)} = \text{PG}\ (\text{dB}) - \text{L}\ (\text{dB}) - \text{SNR}_{out}\ (\text{dB}) \tag{8.37}$$

Example 8.9 Consider the spread spectrum system in which the chip rate is 10^7 chips per second and the message bit rate is 100 bps. If it is desired to obtain an output SNR of 25, which is about 15 dB and if losses are determined to be 2 dB, then find out anti-jam margin.

Solution

Considering biphase modulation:

Processing gain PG = $2\ t_m/t_{chip} = 2 \times 10^7/\,100 = 2 \times 10^5 = 53$ dB

Margin (AJ) = 53 – 2 – 15 = 36 dB

It means that the desired output SNR can be obtained if the jamming signal is less than 36 dB. If we consider some practical circumstances at the receiver, the actual anti-jam margin may be something less than this value.

8.10.3 Energy and Bandwidth Efficiency

The output SNR of the spread spectrum receiver in presence of noise and interference is given as

$$\text{SNR}_{\text{out}} = \frac{P_r}{(N_o/2t_m) + (J/\text{PG})} \tag{8.38}$$

However, the energy associated with a message bit is

$$E_b = P_r\, t_m = \text{Energy per bit} \tag{8.39}$$

The output SNR can be represented again with the following expression:

$$\text{SNR}_{\text{out}} = \frac{P_r \cdot t_m}{(N_o/2) + (Jt_m/\text{PG})} = \frac{E_b}{(N_o/2) + (J/B_s)}$$

$$= \frac{E_b/N_o}{(1/2) + (J/N_o B_s)} \tag{8.40}$$

Rewriting the above equation in terms of E_b/N_o,

$$\frac{E_b}{N_o} = \left[\frac{1}{2} + \frac{J}{N_o B_s}\right]\text{SNR}_{\text{out}} \tag{8.41}$$

Two important comments can be made from this result:

1. When the interference is narrowband, increasing the spreading improves the energy efficiency.
2. When the interference is wideband, such as white noise, increasing the spreading does not improve the energy efficiency because the interference power J increases directly with the signal bandwidth B_s.

To obtain the bandwidth utilization efficiency B/R,

$$\frac{B}{R} = \frac{B_s}{2R_m} \tag{8.42}$$

where R_m is the message bit rate

The term $B_s/2$ is used since this is the equivalent energy bandwidth of the signal. It is important to note that B/R increases linearly with spreading, thus the bandwidth utilization efficiency becomes poorer with the increase in bandwidth of the spread spectrum. If the interference is narrowband compared to the bandwidth of the spread spectrum signal, then combining the results of Equations (8.41) and (8.42),

$$\frac{E_b}{N_o} = \left[\frac{1}{2} + \frac{J}{2N_o R_m (B/R)}\right]\text{SNR}_{\text{out}} = \frac{1}{2}\left[1 + \frac{J/N_o R_m}{B/R}\right]\text{SNR}_{\text{out}} \tag{8.43}$$

If the interference is wideband compared to the bandwidth of the spread spectrum signal, then Eq. (8.43) can be written as

$$\frac{E_b}{N_o} = \frac{1}{2}\left[1 + \frac{J_o}{N_o}\right]\text{SNR}_{\text{out}} \tag{8.44}$$

The quantity J_o is the spectral density of the interfering signal.

8.10.4 Near–Far Problem and Power Control

Near–far problem in spread spectrum system relates to the problem of very strong signals at the receiver swamping out the effects of weaker signals. This problem is particularly serious in the case of DSSS systems. The problem can be considered with the help of a receiver and two transmitters (Fig. 8.16). One transmitter is close to the receiver; the other is far away from the receiver. If both transmitters transmit simultaneously with equal powers, then the receiver will receive more power from the nearer transmitter. This creates the problem for the farther transmitter. Since one transmission is the other's noise, the signal-to-noise ratio for the farther transmitter must be much higher. If the nearer transmitter transmits a signal in magnitudes of the order over the farther transmitter, then the SNR for the farther transmitter may be below the required value making the signal undetectable and the farther transmitter may just as well not transmit. This effectively jams the communication channel. To achieve the successful communication, the far transmitter would have to drastically increase transmission power that simply may not be possible. In short, the near–far problem is one of detecting and receiving a weaker signal amongst the stronger signals. Moreover, the received level fluctuates quickly due to fading. In order to maintain the strength of received signal level at BS, power control technique must be employed in CDMA systems.

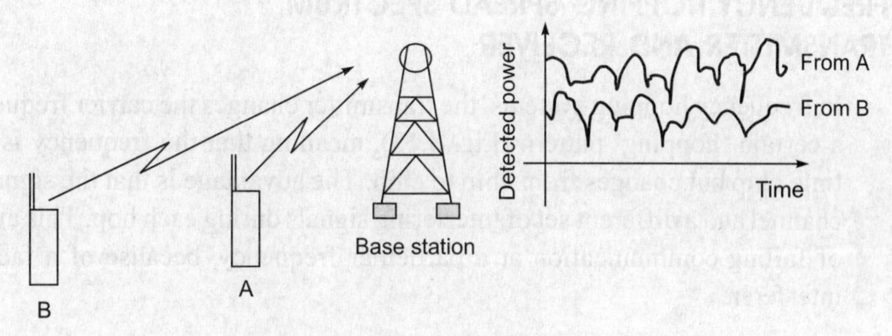

Fig. 8.16 Near–far problem scenario

In CDMA systems or other cellular phone-like networks, this is commonly solved by dynamic output power adjustment of the transmitters. That is the closer transmitters use less power so that the SNR for all transmitters at the receiver is roughly the same. Since all mobiles transmit at the same carrier frequency, internal interference generated within the system plays a critical role in determining the system capacity and voice quality. The two conditions must be satisfied simultaneously:

1. The transmit power from each mobile must be controlled to limit interference.
2. The power level should be adequate for a satisfactory voice quality.

The objective of the power control is to limit transmitted power on the forward and reverse links while maintaining link quality under all conditions (Fig. 8.17).

Power control is capable of compensating the fading fluctuation. Received power from all MS are controlled to be equal. Near–far problem is mitigated by the power control. Now, the detected power strengths of users A and B shown in Fig. 8.16 will be near equal. A frequency hop system, which is described next, is much less susceptible to the near–far problem because it is an avoidance system rather than averaging system.

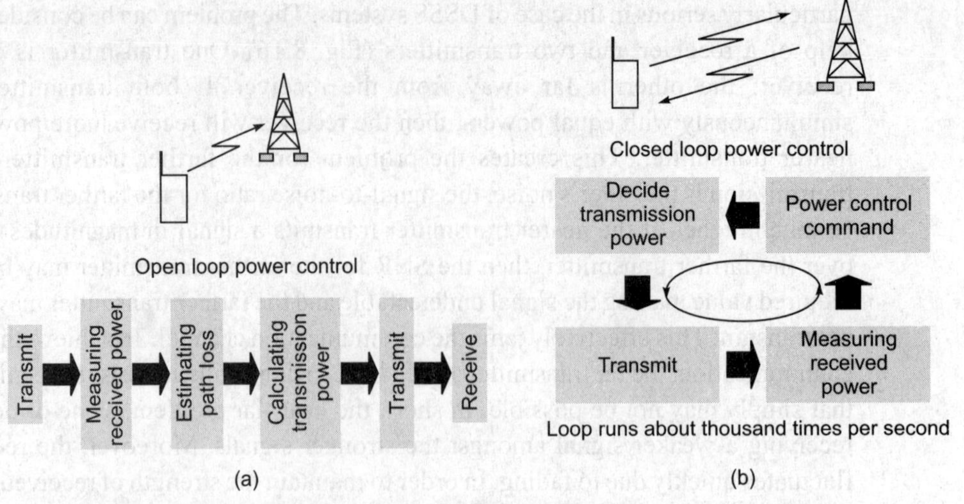

Fig. 8.17 (a) Open loop power control and (b) closed loop power control

8.11 FREQUENCY HOPPING SPREAD SPECTRUM TRANSMITTER AND RECEIVER

In frequency hopping systems, the transmitter changes the carrier frequency according to a certain 'hopping' pattern (Fig. 8.18), meaning that the frequency is constant in each time chip but changes from chip to chip. The advantage is that the signal sees a different channel and a different set of interfering signals during each hop. This avoids the problem of failing communication at a particular frequency, because of a fade or a particular interferer.

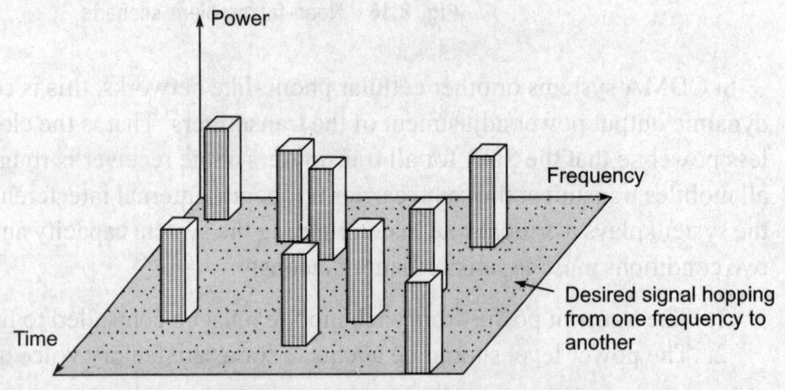

Fig. 8.18 Concept of frequency hopping

There are two kinds of frequency hopping and there is a considerable difference in performance for both the systems—slow frequency hopping (SFH) and fast frequency hopping (FFH).

- In *SFH*, one or more data bits are transmitted within one hop, i.e. hopping rate is less than the message bit rate. An advantage is that coherent data detection is possible. Often, systems using slow hopping also employ (burst) error-control coding to restore loss of (multiple) bits in one hop.

- In *FFH*, one data bit is divided over multiple hop, i.e. frequency hopping rate is greater than message bit rate. In fast hopping, coherent signal detection is difficult, and seldom used. Mostly, FSK or MFSK modulation is used.

There is, of course, an intermediate situation in which the hop rate and message bit rate are of the same order. Slow frequency hopping is a popular technique for wireless LANs. In GSM telephony, slow frequency hopping can be used, at the discretion of the network control software. It avoids the problem of the stationary terminal that happens to be located in a fade losing its link to the base station. Fast frequency hopping is adopted in Bluetooth.

As nearby hopping interferers are unlikely to continuously transmit in the same frequency slot as the reference user, the near–far problem is less severe than in DS-CDMA. Particularly, for wireless LANs, where terminals can be located anywhere, this advantage has made SFH popular.

For purposes of illustration, a fast hop system is considered here in which there are k frequency hops in every message bit duration t_m. Thus, the chip duration is

$$t_{\text{chip}} = \frac{t_m}{k}, \; k = 1, 2, 3, \ldots \tag{8.45}$$

The number of frequencies over which the signal may hop is usually a power of 2, although all these frequencies are not necessarily used in a given system. The reason for making number of frequencies power of 2 is obvious because it is generated by PN sequence generator control and PN sequences are related to power of 2. The frequency hopping is accomplished by means of a digital frequency synthesizer, which in turn is driven by PN code generator. The ML-sequence, i.e. c chips will produce $M = 2^c$ frequencies for each distinct combination of these digits.

As shown in Fig. 8.19, 1 bit from message and c–1 bits come from PN code generator. If a bit from message produces the smallest frequency change, then by itself it will

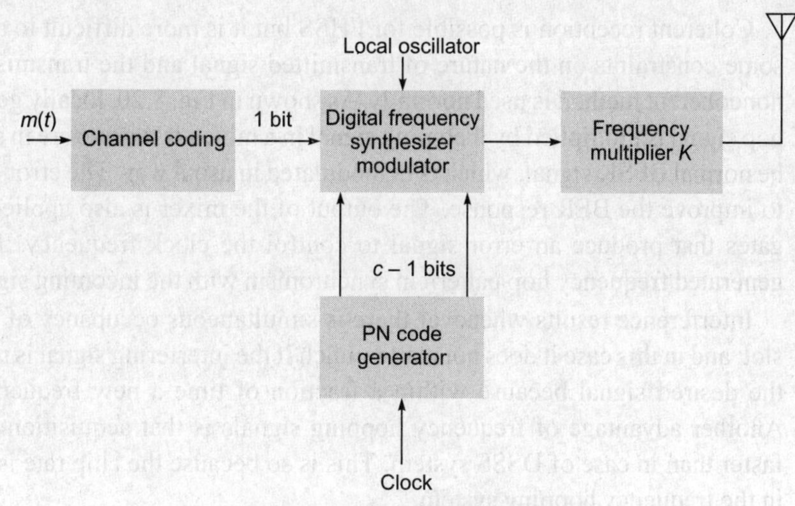

Fig. 8.19 FHSS generator diagram

produce a binary FSK signal. The c–1 bits from PN code generator then hop this FSK signal over the range of possible frequencies. The data is channel coded to combat the bit errors. It may also be noted that there is a frequency multiplier K at the output of the system. It is to increase the bandwidth and thereby increase the processing gain. It also changes the shape of the spectrum.

Considering again fast hopping case, if M frequencies are separated by $f_1 = 1/t_{chip} = k/t_m$, then the signal bandwidth is given by

$$B_s = KMf_1 = KM/t_{chip} \tag{8.46}$$

So the processing gain $PG = B_s/B_m = \dfrac{KM/t_{chip}}{1/t_m} = \dfrac{kKM/t_m}{1/t_m} = kKM$ (8.47)

The FHSS receiver is shown in Fig. 8.20.

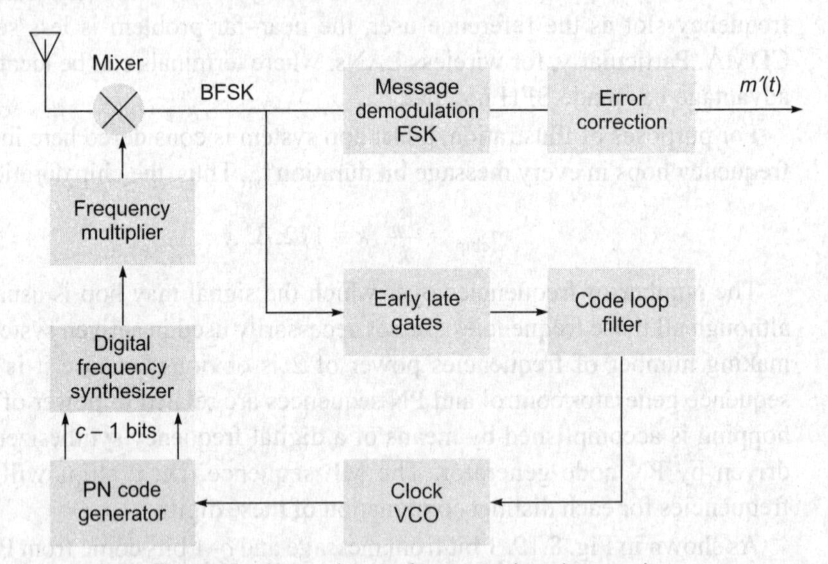

Fig. 8.20 Non-coherent frequency hopping receiver

Coherent reception is possible for FHSS but it is more difficult to achieve and places some constraints on the nature of transmitted signal and the transmission medium. So, noncoherent method is used normally. As shown in Fig. 8.20, locally generated frequency hop signal is multiplied by incoming signal in a mixer. If the two are in step, the result will be normal BFSK signal, which is demodulated in usual way. The error correction is made to improve the BER response. The output of the mixer is also applied to early and late gates that produce an error signal to control the clock frequency. This keeps locally generated frequency hop pattern in synchronism with the incoming signal.

Interference results whenever there is simultaneous occupancy of a given frequency slot, and in this case it does not matter much if the interfering signal is much stronger than the desired signal because within a fraction of time a new frequency will come up. Another advantage of frequency hopping signals is that acquisition is normally much faster than in case of DSSS system. This is so because the chip rate is considerably less in the frequency hopping system.

A disadvantage of FHSS system is that one can not easily use coherent demodulation techniques. This results in poor performance against thermal noise.

Example 8.10 In a frequency hopping spread spectrum system for the generation of hopping frequencies, a 3 bit PN code generator is used. The carrier frequency is 8 kHz and the frequency spacing is 0.5 kHz. Corresponding to the code pattern 000, the frequency is 9.75 kHz. Find out all the hopping frequencies. Show all these frequencies on the frequency time plane.

Solution

Control Codes	Carrier Frequency (kHz)	Derivation of Output Frequency (kHz)	Output Frequency (kHz)
000	8	8+1.75	9.75
001	8	8+1.25	9.25
010	8	8+0.75	8.75
011	8	8+0.25	8.25
100	8	8–0.25	7.75
101	8	8–0.75	7.25
110	8	8–1.25	6.75
111	8	8–1.75	6.25

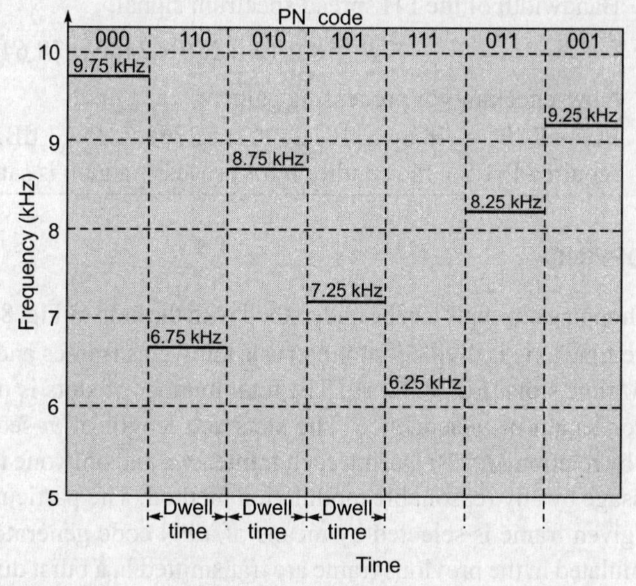

Fig. 8.21 Time frequency plane showing the frequency hopping for Example 8.10

Example 8.11 A frequency hopping spread spectrum system is to have the following parameters:

Message bit rate = 2400 bps

Hops per message bit =16
Frequency multiplication = 8
Processing gain ≥ 45 dB

(a) Find out the smallest number of frequencies required if this number is to be a power of 2.
(b) Find out the bandwidth of the FH spread spectrum signal.

Solution

(a) $B_m = 2400$ bps
$k = 16$, $K = 8$, $M = ?$
Now, processing gain 45 dB = 10 $\log_{10}$ PG

$$\text{PG} = B_s/B_m = 31623$$
$$B_s = 31623 \times B_m = 31623 \times 2400 = 75.9 \text{ Mbps}$$

For frequency hopping system:

$$B_s = KMf_1 = KM/t_{chip} = kKM/t_m$$
$$B_s = 8 \times 16 \times M \times 2400/240 = 75.9 \text{ Mbps}$$
$$\Rightarrow \quad M = 75.9 \times 10^6/8 \times 16 \times 2400 = 247$$

But to make M power of 2, the next selected value = 256 = the number of hopping frequencies

(b) Bandwidth of the FH spread spectrum signal,

$$B_s = 8 \times 16 \times 256 \times 2400 = 78.64 \text{ Mbps}$$

Now, checking for processing gain:
PG = B_s/B_m = 78.64 × 10^6/2400 = 32768 = 45.13 dB, which is higher than the required PG. So, the condition for processing gain is satisfied.

8.12 TIME HOPPING

Time hopping concepts can be understood with the help of Fig. 8.22 with different methods.

The time axis is divided into intervals known as frames and each frame is subdivided into M time slots [Fig. 8.22(a)]. The total number of slots is decided on the basis of PN code or length of m-sequence. The slots and length of m-sequence are related to each other by relation $M = 2^c$. During each frame, one and only one time slot is modulated with a message by any reasonable modulation method. The particular time slot that is chosen for a given frame is selected by means of a PN code generator. All of the message bits accumulated in the previous frame are transmitted in a burst during the selected time slot. The frame duration T_f, the number of message bits k, and the message bit duration t_m are related to each other by

$$T_f = kt_m \tag{8.48}$$

The width of each time slot in a frame is T_f/M and width of each bit in the time slot is T_f/kM or simply t_m/M.

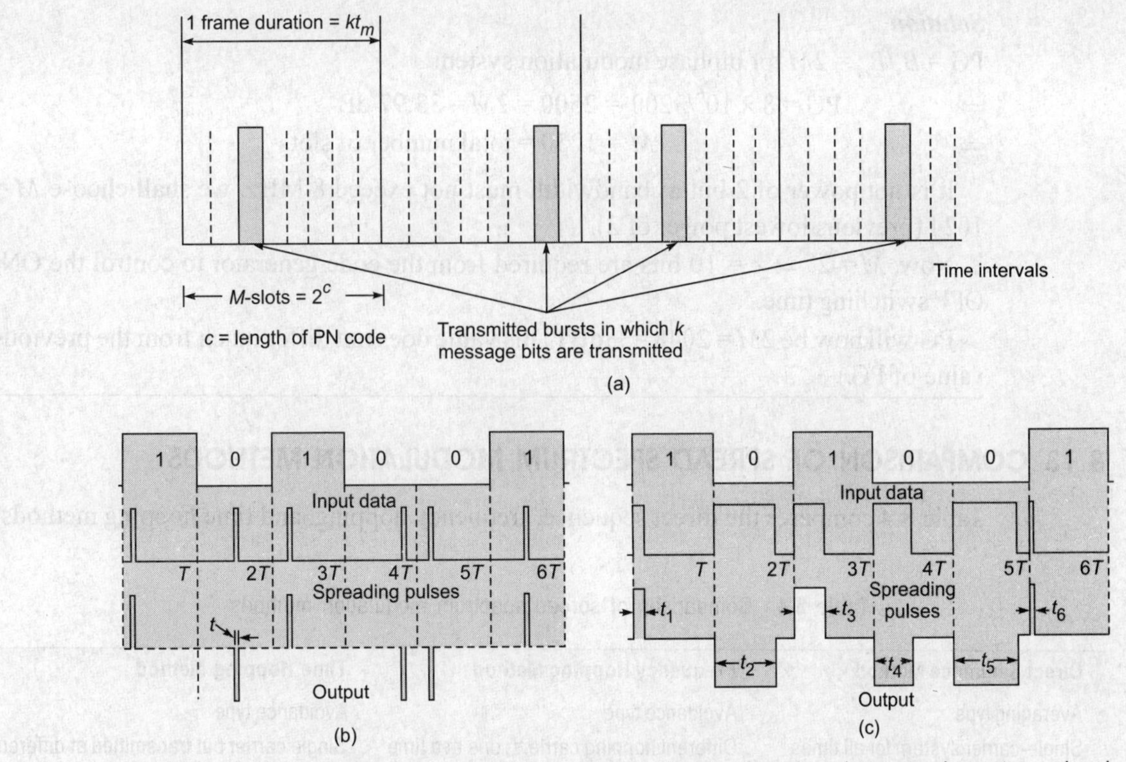

Fig. 8.22 (a) Time hopping concept, (b) waveforms showing time hopping spread spectrum signal formation on bit-by-bit basis, and (c) time hopping with variable time slots (bit by bit)

$$\text{Processing gain} = \frac{B_s}{B_m} = 2\frac{t_m}{(t_m/M)} = 2M \quad \text{for biphase modulation} \tag{8.49a}$$

$$= \frac{t_m}{(t_m/M)} = M \quad \text{for quadriphase modulation} \tag{8.49b}$$

This indicates that the transmitted signal bandwidth is $2M$ times the message bandwidth and hence the processing gain of the time hopping system is simply twice the number of time slots in each frame when biphase modulation is used and half this when quadriphase modulation is used. Figures 8.22(b) and (c) show other possible variations.

Interference among simultaneous users in time hopping system can be minimized by coordinating the times at which each user can transmit a signal. This also avoids the near–far problem. In a non-coordinated system, overlapping transmission bursts will result in message errors and this will normally require the use of error correction coding to restore the message bits. The acquisition time is similar to that of direct sequence systems for a given bandwidth. Implementation is simpler than FH system.

Example 8.12 A certain time hopping spread spectrum system is allocated a maximum signal bandwidth of 8 MHz. Assuming that the message bit rate after channel coding is 3200 bps, find the number of bits necessary from the code generator to control the ON/OFF switching time. If biphase modulation is used, what is the processing gain?

Solution

PG = $B_s/B_m = 2M$ for biphase modulation system

$\Rightarrow$ PG = $8 \times 10^6/3200 = 2500 = 2M = 33.97$ dB

$\Rightarrow$ $M = 1250$ = total number of slots

It is not power of 2 but as bandwidth must not exceed 8 MHz, we shall choose M = 1024 (previous lowest power of 2).

Now, $M = 2^c \Rightarrow c = 10$ bits are required from the code generator to control the ON/OFF switching time.

PG will now be $2M$ = 2048 =33dB (This value does not differ much from the previous value of PG.)

8.13 COMPARISON OF SPREAD SPECTRUM MODULATION METHODS

Table 8.4 compares the direct sequence, frequency hopping, and time hopping methods.

Table 8.4 Comparison of spread spectrum modulation methods

Direct Sequence Method	Frequency Hopping Method	Time Hopping Method
Averaging type	Avoidance type	Avoidance type
Single-carrier system for all times	Different hopping carriers, one at a time	Single carrier but transmitted at different times
Simplest implementation	Complex implementation because the synthesizers are required	Implementation is simpler than FH
Long acquisition time	Short acquisition time	Long acquisition time
Near–far problem	No near–far problem	Near–far problem is avoided by proper time synchronization
Advantages	**Advantages**	**Advantages**
• Best noise and anti-jamming performance	• Greatest amount of spreading	• Highest bandwidth efficiency of all three
• Most difficult to detect	• Can be programmed to avoid portion of the spectrum	• Useful when transmitter is average power limited but not peak power limited
• Best discrimination against multipath		
Limitations	**Limitations**	**Limitations**
• Requires wideband channel with minimum phase distortion	• Not useful for range and range rate measurement	• Error correction needed
• Fast code generator required	• Error correction is required	• Synchronization is very critically required
• Long chip sequence is required		• Can be jammed easily hence is not used generally

8.14 HYBRID SPREAD SPECTRUM SYSTEMS

The use of hybrid techniques attempts to capitalize upon the advantage of a particular method while avoiding the disadvantages. For example, evaluation of direct sequence and frequency hopping shows that both of those techniques have their specific advantages and disadvantages. Direct sequence, on one hand, suffers heavily from the near–far effect, which makes this technique hard to apply to systems without the ability of power control. On the other hand, its implementation is inexpensive. The *PN code* generators are easy to implement and the spreading operation itself can be simply performed by EXOR-ports. Frequency hopping effectively suppresses the near–far effect and reduces the need for power control. However, implementation of the (fast) hopping frequency synthesizer required for a reasonable spreading gain is more problematic in terms of higher silicon cost and increased power consumption.

Applying both techniques allows for combining their advantages while disadvantages can be reduced. This results in a reasonable near–far resistance at an acceptable hardware cost.

Many different hybrid combinations are possible. Some of them are PN/FH, PN/TH, FH/TH, and PN/FH/TH.

While designing a hybrid system, the designer should make the choice of applying either fast or slow frequency hopping. The fast frequency hopping increases the cost of the frequency synthesizer but provides more protection against the near–far effect. The slow frequency hopping combines a less expensive synthesizer with a worse near–far rejection and the need for a more powerful error correction scheme (several symbols are lost during a 'hit' jamming). Such considerations can be applied to other types of systems also.

Example 8.13 A PN/FH hybrid spread spectrum system has the following parameters:

Message bit rate = 9600 bps after error correction coding

PN code rate = 1,53,600 chips per second

Chips per hop = 16

Hops per message bit =1

Number of hopping frequencies used = 4096

(a) Find out the processing gain.

(b) Find out the bandwidth of the spread spectrum signal.

Solution

(a) PN code rate = 1,53,600 bps, which is the spacing between the two frequencies.

Number of hopping frequencies, $M = 4096$

Hopping rate =153600/16 = 9600 hops per second

$B_s = Mf_1 = 4096 \times 153600 = 629$ MHz

Now, PG = $629 \times 10^6/9600 = 65536$

PG in dB = $10 \log_{10}$ PG = $10\log_{10}(65536)$

$= 48.16$ dB

(b) Bandwidth of the SS signal is 629 MHz without RF multiplication factor.

8.15 CHIRP SPREAD SPECTRUM

The standard IEEE-802.15.4a uses chirp spread spectrum. The system diagram is shown in Fig. 8.23. A chirp spread spectrum system utilizes linear frequency modulation of the carrier to spread the bandwidth. Though it is a common technique in radar system, it is also used in communication systems. The relationship between the frequency and the time is shown in Fig. 8.24, in which T is the duration of a given signal waveform and B is the bandwidth over which the frequency is varied. In this case, the processing gain is simply BT. It is also possible to use non-linear frequency modulation and in some cases this may be desirable. Self-intersymbol interference reduction is achieved in chirp spread spectrum communication systems.

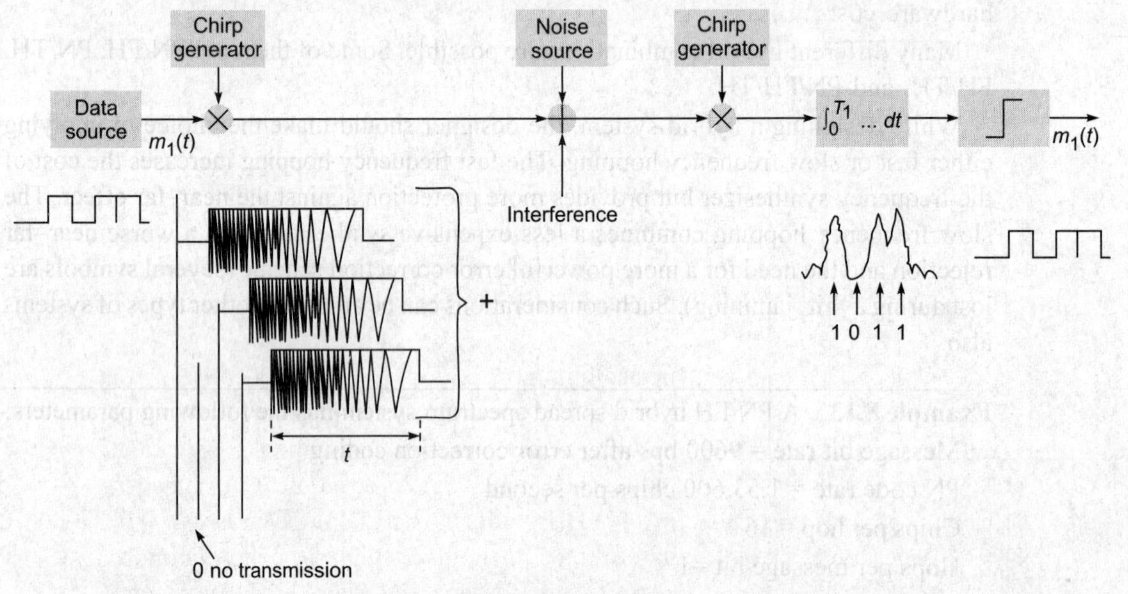

Fig. 8.23 Chirp spread spectrum transceiver

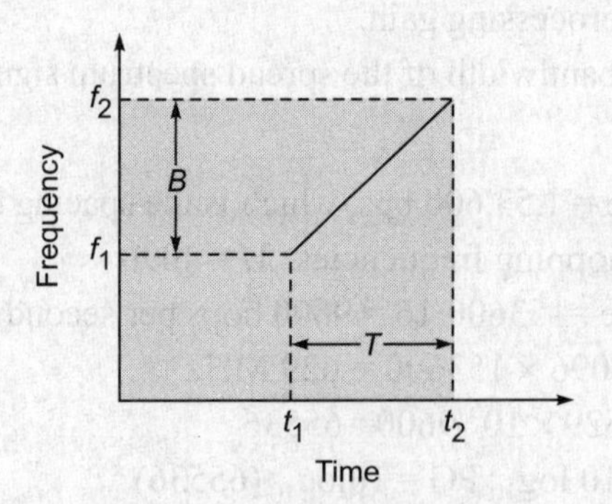

Fig. 8.24 A linear chirp signal

Summary

- Spread spectrum modulation scheme is different from conventional single carrier methods because it uses the code of higher rate for the data modulation purpose and, due to this, achieves the spectrum spreading.
- Two major SSM methods are direct sequence method and frequency hopping method. Other SSM methods are time hopping method, hybrid method, and chirping.
- The code used for spreading purpose is nothing but random sequence whose smallest element is known as chip. Chip duration is decided by clock rate.
- Chip rate is much higher than the data bit rate. This is a required condition for spreading.
- The amount of spreading is decided by its processing gain.
- The receiver designed for DSSS reception is a rake receiver with a number of correlators.
- In a rake receiver, the multipath effect is exploited.
- Near–far problem is a serious problem in DSSS system and power control is required to eliminate this.
- The DSSS exhibits interference rejection and anti-jamming characteristics.
- In the frequency hopping method, the PN code is used for hopping pattern generation.
- Complex synthesizers are required for hopping frequency generation, which makes the system complex.
- In time hopping, the slots are decided on the basis of PN code.
- The hybrid systems can be designed using the combination of PN, FH, or TH systems. This is done to take the advantages of the scheme and to eliminate disadvantages.
- The chirping spread spectrum is nothing but linear frequency modulation of pulses over required range of frequencies.

Review Questions

1. What do you understand by spreading of spectrum?
2. How can you say that spread spectrum system becomes spectrally efficient and in which case?
3. What should be the requirement of selection of chip rate and message bit rate so that the spreading can be achieved?
4. Design any suitable five-stage PN code generator. Observe that the length of the sequence is an odd number. Why?
5. Give the required autocorrelation properties for DSSS in general.
6. Discuss the suitability of ML sequence, Walsh code, and Gold sequence for the DSSS-based multi-user system.
7. Demonstrate that the codes in an 8×8 Walsh matrix are orthogonal to each other by showing that multiplying any code by any other code and adding them produces a result of zero.
8. Discuss in detail the quadriphase modulation based direct sequence spread spectrum transmitter with all its aspects.
9. Find the theoretical spectrum of PN signal and the information signal and from that derive the equation for processing gain. Why should the processing gain be large?
10. How does the rake receiver exploit multipath?
11. What is the importance of the correlator in a rake receiver?
12. How can you say that FHSS is an avoidance type of system?
13. What is the near–far problem? Find out the methods of power control to eliminate this problem in DSSS.

14. What are the steps of receiving the spread spectrum signal at the receiver?
15. What do you mean by partial correlation? Why is partial correlation important for the rake receiver rather than full correlation?
16. Comment on the SNR of DSSS system. Can you say that anti-jamming characteristics of DSSS depend upon the available SNR? How can the SNR be improved in that case?
17. Compare the DSSS and FHSS types of systems with their pros and cons. Find out the existing systems/standards in which these schemes are utilized.
18. Draw the block diagram of a PN/FH hybrid system and compare it with the block diagram of FHSS system. Why are hybrid methods suitable many times in spite of their complexity?
19. In FHSS systems, what are the parameters upon which the processing gain depends?
20. Why is time hopping not much used practically?
21. Explain that FH/TH type systems are nothing but chirped spread spectrum.

PROBLEMS

1. If 48 kbps data stream is to be transmitted using spread spectrum modulation method, find out the channel bandwidth required when SNR = 0.1, 0.01, and 0.001.
2. Show that the periodic PN signal normalized autocorrelation function corresponding to the binary case with equal probability of 1's and 0's (+1 and −1) and with N= 7 is of the form as shown in Fig. 8.25.

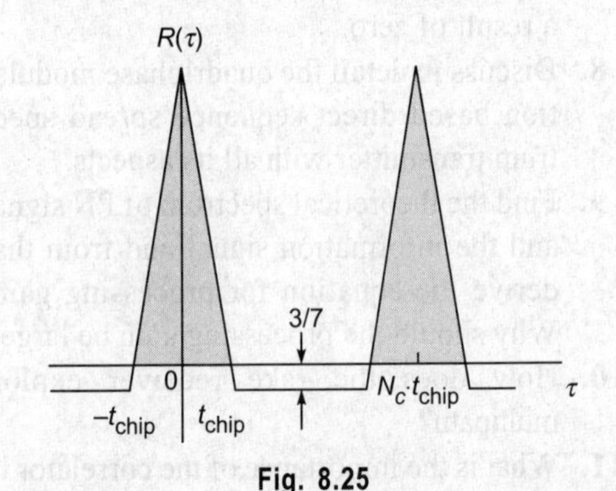

Fig. 8.25

3. An aperiodic PN waveform is defined for the interval N_c = 7 with a sequence +1, −1, −1, +1, −1, +1, +1. Find the autocorrelation function of this process. Is this an ideal aperiodic sequence?

Hint: Different shifts will give different values of the correlation functions.

4. Evaluate the 15 bit spreading code '100110101111000' for balance, run property, and correlation.
5. A DSSS system is operating under the conditions of two independent jammers, which have very close centre frequencies. The chip rate is 125 Mbps and the message bit rate is 2500 bps. If the combined bandwidth of the two jammers is 60 kHz and their respective received signal powers are 0.2×10^{-7} watts and 10^{-6} watts, what is the approximate interference power at the output of the correlator?
6. In a DSSS system, the message bit rate is 4800 bps and the PN clock rate is 220 Mbps.
   (a) Find out the processing gain.
   (b) Find out the output signal-to-noise power ratio if the received signal power is 6×10^{-10} watts and one-sided noise spectral density is 10^{-20} watts/Hz.
   (c) If the coherent reference signal is off by 10 degrees and the synchroniza-

tion error is 30%, will the code noise term be a determining factor in the output SNR? How?

7. A speech signal bandlimited to 4 kHz and 256 quantization levels is to be transmitted by a DSSS transmitter. Find out the required chip rate to obtain a processing gain of 6400.

8. A frequency hopping spread spectrum system utilizes a fast hop system in which 10 hops per message bit and 1024 hopping frequencies are there. The message bit rate is 2400 bps and the final RF multiplication factor is 10. Find out (a) RF signal bandwidth, (b) processing gain in dB, (c) PN code generator clock rate, and (d) frequency separation in kHz.
   For the found value of processing gain, what must be an equivalent PN code rate when $t_m = 1/42{,}000$?
   Hint: PN code generator clock rate $= k(c-1) \times$ message bit rate.

9. In a frequency hopping system, the switching speed of the synthesizer is 5 μs and the message bit rate is 5 kbps after error-correction coding. There are 5 hops per message bit and the final frequency multiplication is 8. What is the maximum processing gain obtainable?

10. Consider a hybrid system PN/FHSS. The PN code rate is 250 kchips per second. For the frequency hopping, total 8192 frequencies are used with the spacing of 250 kHz. For error correction, 1/3 rate convolution coding is used. Find out the processing gain if the final message rate is (a) 75 bps and (b) 2400 bps.

11. For the polynomials $h(x) = x^4 + x^3 + x^2 + 1$ and $h(x) = x^4 + x^2 + x + 1$, assuming the initial condition (1111), (a) calculate and plot the autocorrelation functions for each, (b) calculate and plot the cross correlation functions, and (c) draw the block diagrams using both the Fabonacci and Galois methods.

12. Using the polynomials given in Problem 11, find out the Gold sequence.

13. In a PN sequence with the length $2^8 - 1$, how many runs of 1111 would be expected?

14. In an FHSS system, the total bandwidth of $B_s = 400$ MHz and the individual channel bandwidth $B_m = 100$ Hz. What is the minimum number of PN bits required for each frequency hop?

Multiple Choice Questions

1. Rake receiver's front end detects the required signal by
   (a) correlation
   (b) superheterodyning
   (c) convolution
   (d) demodulation

2. Which of the following is an averaging type of system?
   (a) FHSS (b) PN/FH
   (c) THSS (d) DSSS

3. The chip duration of PN code
   (a) can be anything
   (b) must be much less than the message bit duration
   (c) must be equal to the message bit duration
   (d) must be much larger than the message bit duration

4. If two different PN codes are correlated, then
   (a) their cross correlation should be high
   (b) their cross correlation should be less
   (c) their autocorrelation should be high
   (d) none of the above is true

5. For good interference rejection, the preferable value of PG should be
   (a) 0.5 (b) 10
   (c) 1 (d) 6500
6. In every period, the number of +1's differs from the number of –1's by exactly 1. This represents
   (a) the autocorrelation of PN code
   (b) the balance property of PN code
   (c) the run property of PN code
   (d) all of the above
7. N_c must be
   (a) a very large number
   (b) a very large odd number
   (c) an odd number
   (d) an even number
8. Near–far problem can occur in
   (a) FHSS (b) PN/FH
   (c) THSS (d) DSSS
9. B/R represents the
   (a) spectral efficiency
   (b) signal-to-noise ratio
   (c) bandwidth utilization efficiency
   (d) processing gain
10. Greatest amount of spreading can be obtained in
   (a) FHSS (b) PN/FH
   (c) THSS (d) DSSS
11. Which of the following relationships represents interference rejection?
   (a) $\text{SNR}_{\text{out}} = (\text{PG}) \cdot \text{SNR}_{\text{in}}$
   (b) $\dfrac{E_b}{N_o} = \left[\dfrac{1}{2} + \dfrac{J}{N_o B_s}\right] \text{SNR}_{\text{out}}$
   (c) $\overline{j^2} = \dfrac{J}{\text{PG}}$
   (d) $\text{SNR}_{\text{out}} = \dfrac{P_r}{(N_o/2t_m) + (J/\text{PG}}$

chapter 9

Wideband Modulation Techniques 2: OFDM (Multicarrier Modulation)

Key Topics

- OFDM multiple subcarrier allocation to OFDM frame
- OFDM spectrum setting and spectrum efficiency issues
- OFDM digital signal processing
- OFDM block diagram
- OFDM RF upconversion
- Mathematics of OFDM
- Pulse shaping, windowing, and synchronization issues in OFDM
- OFDM amplitude limitations
- OFDM implementation constraints due to FFT points
- Comparison CDMA and hybrid OFDM

Chapter Outline

The orthogonal frequency division multiplexing (OFDM) is a wideband wireless digital communication technique that is based on block modulation. With the wireless multimedia applications becoming more and more popular, the required bit rates are achieved due to OFDM multicarrier transmissions. For video communication, very high bit rate/high-speed communication is required. To satisfy this, we must have the modulation scheme that can read more number of bits at a time and send it with considerably low bit errors. The quality of the reception must be good enough. The OFDM is a digital modulation scheme that can support high-speed video communication along with audio with elimination of ISI and ICI. At the same time, it can accommodate more number of users showing the spectral efficiency. It is a multiplexing/multiple access scheme that has many favourable features required for the fourth-generation wireless communication systems. The OFDM scheme is mainly based on DSP techniques. The chapter describes all the basic concepts and systems related to the OFDM technique.

9.1 BASIC PRINCIPLES OF ORTHOGONALITY

Orthogonality between two signals means that the two coexisting signals are independent of each other in a specified time interval and do not interact with each other. The concept of orthogonal signals is essential for the understanding of orthogonal frequency division multiplexing (OFDM) system. In the normal sense, it may look like a miracle that one can separately demodulate overlapping carriers. Orthogonality is a property that allows multiple information signals to be transmitted perfectly over a common channel and detected without interference. Loss of orthogonality results in blurring between these information signals and degradations in communications.

Many common multiplexing schemes (described in the next chapter) are inherently orthogonal. Time division multiplexing (TDM) allows transmission of the multiple signals over a single channel by assigning unique time slots for each of the information signals. During each time slot, one and only one information signal is transmitted, thus preventing any interference between the multiple interference sources. Because of these, TDM is orthogonal in nature. Of course, time synchronization problem is the limitation.

In the frequency domain, most FDM systems are orthogonal in the sense that each of the separate transmission signals is well spaced out in frequency, preventing the interference and hence it consumes a lot of spectrum.

Although the above two methods preserve orthogonality with a compromise, the term *OFDM* reserves a special feature. It is orthogonal FDM in which the subcarriers are spaced as close as is theoretically possible maintaining orthogonality between them. Here, saving of spectrum is achieved. The OFDM arranges the subcarriers in the frequency domain by allocating partly the information signals onto different subcarriers. Single information stream or frame is split into multiple symbols and each symbol or a group of symbols will be assigned a separate carrier. All the split information is then transmitted in parallel through multiple carriers.

There are two ways in which the subcarriers are assigned. Consider one OFDM frame and a set of subcarriers assigned to it.

1. Read the bits/symbol and assign one subcarrier to it. Then read another group of bits per symbol and assign another carrier that is orthogonal to the previous one and continue. T_s will be the duration of symbol in this case. Do this in parallel.
2. Read the bits/word of the whole OFDM frame and then form a matrix in which each element of it represents one word. The whole frame is represented in terms of time vs frequency plot (a 2 D lattice) in which each point of lattice represents a word. (Sometimes, word is also referred to as symbols and there is a difference between symbol and OFDM symbol—symbols are the elements of OFDM symbol.) Frequency axis will support the total number of carriers to be assigned to the OFDM symbols and time axis will represent OFDM symbol duration that is formed by breaking the OFDM frame into blocks and each block prolonged to OFDM symbol duration T_s. These blocks contain one or more symbols. Thus, the statement that 'OFDM is a block modulation scheme' will be true. Figure 9.1 represents a typical case of carrier assignment according to this method. The block duration will be T_s after the serial-to-parallel conversion procedure. Thus, the effective bit duration and hence the duration between two consecutive pulses on all parallel lines is prolonged. (Bits/symbol are as per the mapping scheme used.)

The OFDM signals are made up of a sum of sinusoids, each representing a modulated subcarrier. Figure 9.1 shows a typical case of 4 subcarriers and 3 symbols/carrier (a block of 3 symbols will be assigned one subcarrier), i.e. 3 symbols modulating a subcarrier. Hence, the bits/symbol and symbols/carrier is necessary to pre-decide or plan,

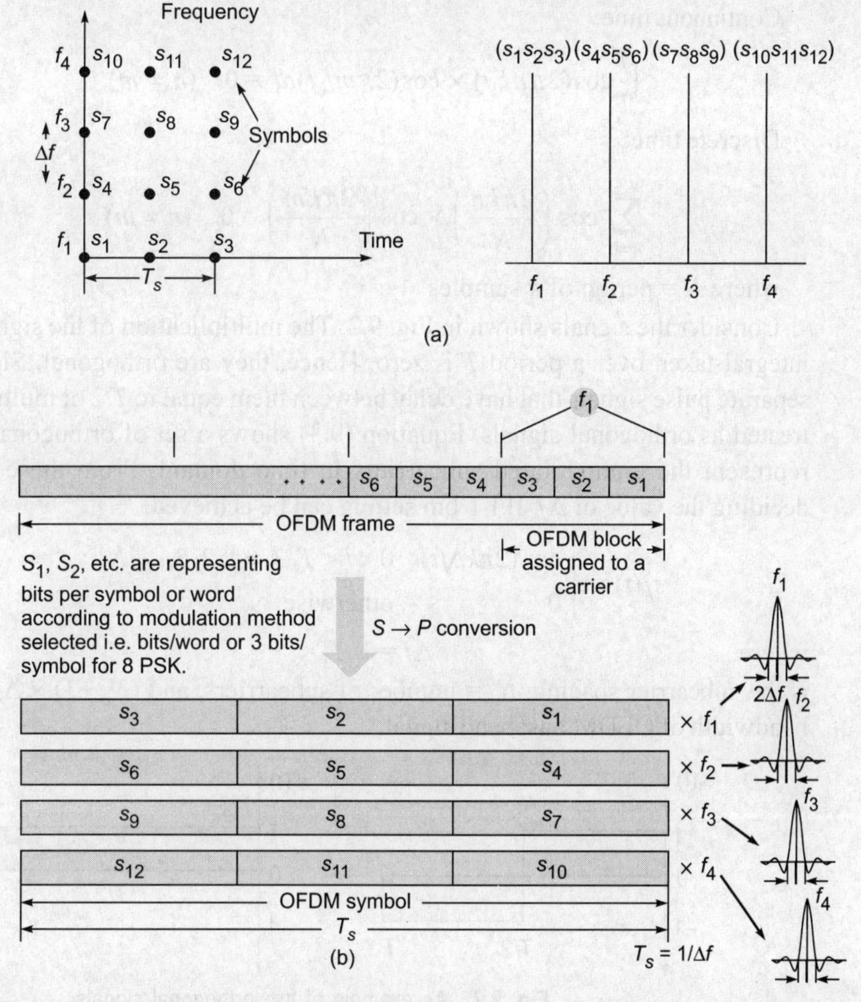

Fig. 9.1 (a) Lattice representation and carrier assignment planning and (b) concept of carrier frequency allocation to the symbols (a typical case in which three symbols are assigned to a carrier), serial-to-parallel conversion, and mapping into frequency domain components after modulation

such that the total OFDM frame can be broken up appropriately matching the total number of subcarriers to be assigned to it.

9.1.1 Orthogonality and Subcarrier Setting in the Spectrum

Two periodic signals are orthogonal when the integral of their product, over one period, is equal to zero and they have integral number of cycles in the fundamental period, i.e. the peak of the next carrier must occur at the null of the previous carrier. The mathematical representation of the orthogonal continuous time and discrete time signals is shown in Equations (9.1) and (9.2); these are the conditions for orthogonality.

Continuous time:

$$\int_0^T \cos(2\pi n f_0 t) \times \cos(2\pi m f_0 t)dt = 0 \quad (n \neq m) \tag{9.1}$$

Discrete time:

$$\sum_{k=0}^{N-1} \cos\left(\frac{2\pi kn}{N}\right) \times \cos\left(\frac{2\pi km}{N}\right) = 0 \quad (n \neq m) \tag{9.2}$$

where N = period of k samples

Consider the signals shown in Fig. 9.2. The multiplication of the signals and then the integral taken over a period T is zero. Hence, they are orthogonal. Similarly, the two separate pulse signals that have delay between them equal to $T/2$ or multiple of it are also treated as orthogonal signals. Equation (9.3) shows a set of orthogonal signals, which represent the unmodulated subcarriers in time domain. From these subcarriers, by deciding the value of Δf, IFFT bin setting can be achieved.

$$s_k(t) = \begin{cases} \cos(2\pi k \Delta f t) & 0 < t < T_s, k = 1, 2, 3, \ldots, N_c \\ 0 & \text{otherwise} \end{cases} \tag{9.3}$$

where $$\Delta f = 1/T_s \tag{9.4}$$

is the subcarrier spacing, N_c = number of subcarriers, and $(N_c + 1) \times \Delta f$ = transmission bandwidth of OFDM baseband signal.

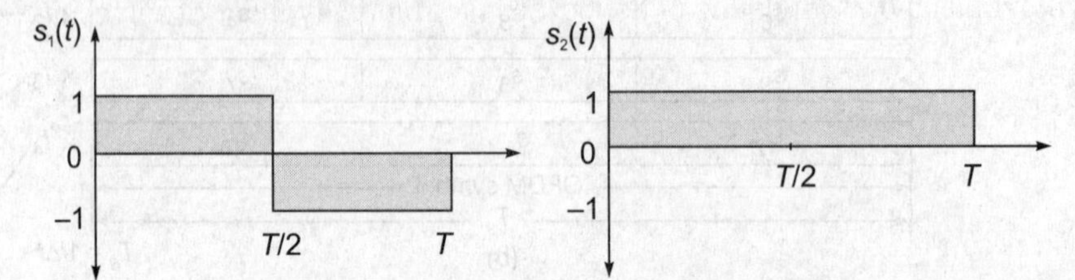

Fig. 9.2 An example of two orthogonal signals

Figure 9.3 illustrates how the orthogonal subcarriers are considered in frequency and time domain and how they are added up to get OFDM baseband signal.

Once the bin setting is over, each OFDM block with OFDM symbol duration is assigned to those subcarriers according to the predecided pattern, i.e. the carriers are modulated now. If a pulse (in time domain) is assigned to a carrier, then it will take sinc (sin x/x) shape in the frequency domain while centring the carrier. Instead of a narrow pulse, if long symbol is assigned to that carrier, its sinc shape [Fig. 9.1(b)] will shrink because time expansion will consume less spectrum. The same concept is applied to OFDM while modulating each subcarrier by OFDM block assigned to it and the same is followed for all the subcarriers assigned to whole OFDM frame. The carriers can be placed as near as possible maintaining the orthogonality, i.e. peak of one carrier should coincide with the null of the nearest subcarrier. As the width of the sinc depends upon T_s, it puts the limit on the subcarrier spacing. Each narrow bandwidth corresponding to each subcarrier is decided by the symbol duration T_s. Hence, the subcarrier spacing will be

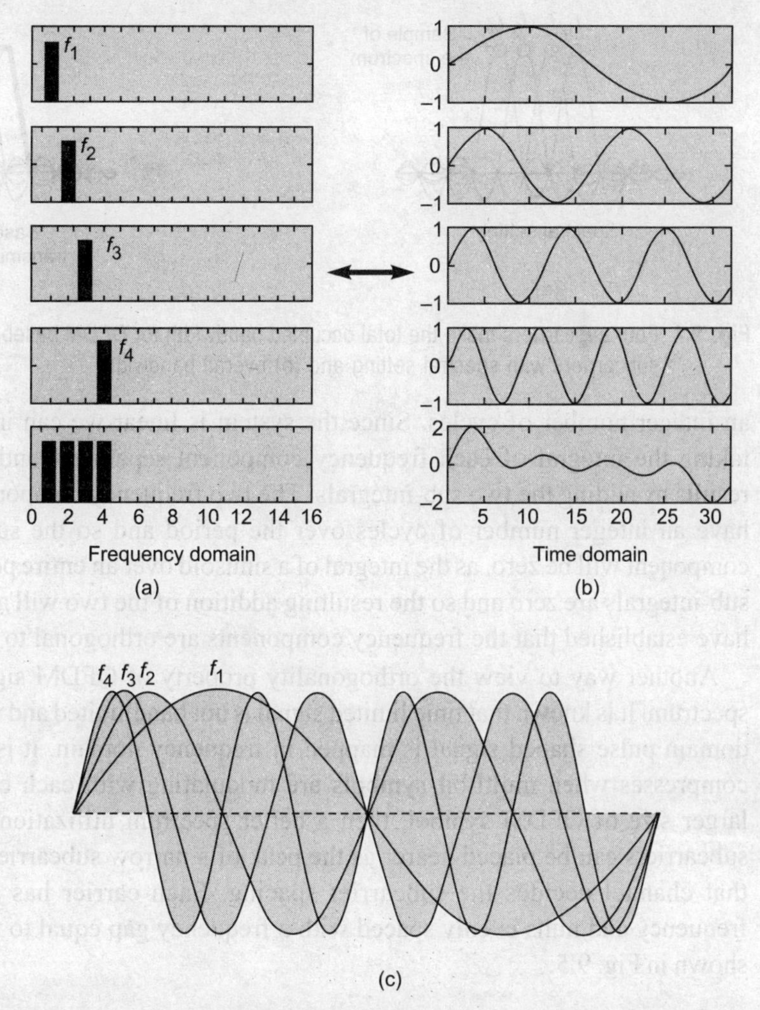

Fig. 9.3 Frequency to time domain conversion: (a) orthogonal subcarriers setting in the frequency domain with 32 point IFFT bin (the IFFT bin is symmetrical about centre with real and imaginary parts), (b) corresponding time domain interpretations for interval $N = 32$ samples, and (c) plot of four subcarriers on the same time axis for addition of subcarriers in time domain to get OFDM baseband (conceptual representation without symbols assigned to subcarriers—not to scale)

inversely proportional to the symbol duration. Figure 9.4(a) represents four subcarriers modulated by assigned symbols to it and hence narrow channels exhibit sinc shape. Subcarriers are also maintaining the orthogonality. Figure 9.4(b) represents the overall transmission bandwidth.

The orthogonality of the subcarriers can be proved by multiplying the time waveforms of any two subcarriers and integrating over the symbol period. The result will be zero. Multiplying the two orthogonal sine waves together is the same as mixing these subcarriers. This results in sum and difference frequency components, which will always be the integer subcarrier frequency, as the frequency of the two mixing subcarriers has

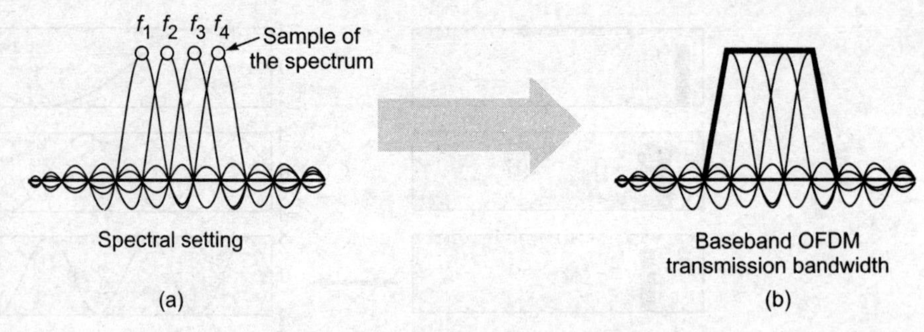

Fig. 9.4 Four subcarriers make the total occupied bandwidth for OFDM baseband signal: (a) modulated subcarriers with spectral setting and (b) overall bandwidth

an integer number of cycles. Since the system is linear we can integrate the result by taking the integral of each frequency component separately and then combining the results by adding the two sub-integrals. The two frequency components after the mixing have an integer number of cycles over the period and so the sub-integral over each component will be zero, as the integral of a sinusoid over an entire period is zero. Both the sub-integrals are zero and so the resulting addition of the two will also be zero. Thus, we have established that the frequency components are orthogonal to each other.

Another way to view the orthogonality property of OFDM signals is to look at its spectrum. It is known that time limited signal is not bandlimited and vice versa. When time domain pulse shaped signal is mapped in frequency domain, it is a sinc shape, which compresses when multi-bit symbols are modulating with each carrier. If we select a larger size of OFDM symbol, then a better spectrum utilization can be made as the subcarriers can be placed nearer as the peak of a narrow subcarrier channel and null of that channel decides the subcarrier spacing. Each carrier has a peak at the centre frequency and nulls evenly spaced with a frequency gap equal to the carrier spacing as shown in Fig. 9.5.

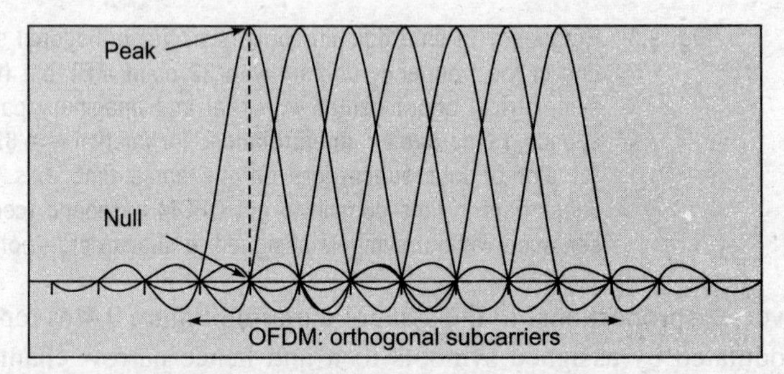

Fig. 9.5 Orthogonal signals in the frequency domain: peak of one signal occurs at the null of nearer subcarrier

The sinc shape has a narrow main lobe that decays fast and many side lobes. Because the subchannels are very narrow, they act as samples of the spectrum. This will be shown in the mathematical part. When the spectrum setting is done for IDFT stage, the spectrum is not treated as continuous as shown in Figures 9.4 and 9.5 but is considered in terms of discrete samples. The sampled spectrum is shown with 'o's in Fig. 9.4(a). If the DFT is time synchronized, the frequency samples of the DFT correspond to just the peaks of the subcarriers, Thus, the overlapping frequency region between subcarriers does not affect the receiver.

In short, in OFDM modulation, the available channel is divided into several independent subcarriers and all the subcarriers are transmitted at a time. This is achieved by making all the subcarriers orthogonal to each other preventing ICI. The received signals are retrieved in the reverse way. The orthogonal frequency difference is not selected arbitrarily but is selected on the basis of the data rate or symbol time.

To understand the concept of orthogonality, it is very helpful to interpret subcarrier signals as vectors in the signal space. Vector signals can be added, multiplied by a scalar, expanded into a base, and many more properties can be considered.

Example 9.1 A 64 kbps voice frame is to be modulated by OFDM scheme. The duration of OFDM symbol is 1000 μs. Total of 32 subcarriers are to be assigned to this frame. Find out

(a) the null-to-null subchannel bandwidth assuming square signal,
(b) the total bandwidth occupied, and
(c) the number of bits in OFDM frame.

Solution

(a) T_s = symbol duration = 1000 μs

So, Δf = subcarrier spacing

$= 1/T_s = 1/1000\ \mu s = 0.001$ MHz = 100 kHz

⇒ Null-to-null subchannel bandwidth = 200 kHz (sinc shape)

(b) N_c = number of subcarriers = 32

⇒ Transmission bandwidth $= (N_c + 1) \times \Delta f \approx N_c \times \Delta f$

$= 32 \times 100$

$= 3200$ kHz

(c) Bit duration $= 1/(64 \times 10^3) = 15.625\ \mu s$

⇒ No. of bits per frame $= 1000/15.625$

$= 64$ bits

9.1.2 FDM vs Orthogonal FDM

Some differences between FDM and orthogonal FDM are as follows:

- The OFDM signals must be time and frequency synchronized, which is not necessary in case of FDM.
- FDM is a single-carrier oriented system while OFDM is a multicarrier system.
- FDM transmission signals need to have a large frequency guard-band between channels in comparison to OFDM to prevent ICI interference.

Each carrier in an OFDM system is a sinusoid with a frequency such that the spacing between two consecutive subcarriers depends upon the symbol rate and hence on the bit rate, which again can be derived from the condition of orthogonality. If the OFDM symbol period is too long, the spacing between the carriers can be reduced to be as close as theoretically possible.

In FDM, there is no relationship between the carriers as they carry unique user's information. The separation between two channels depends upon the basic bandwidth and the modulation scheme adopted. In OFDM, multiple carriers are used for the transmission of the multiplexed multi-user information in a frame and hence the carriers are related in one channel bandwidth. However, shaping the spectrum by applying windowing or pulse shaping along with maintaining the orthogonality is the matter of separate study for the better achievement of spectral efficiency in OFDM. Windowing and pulse shaping issues are involved in FDM but on individual user channel basis. Figure 9.6 shows how OFDM is a spectrally efficient technique.

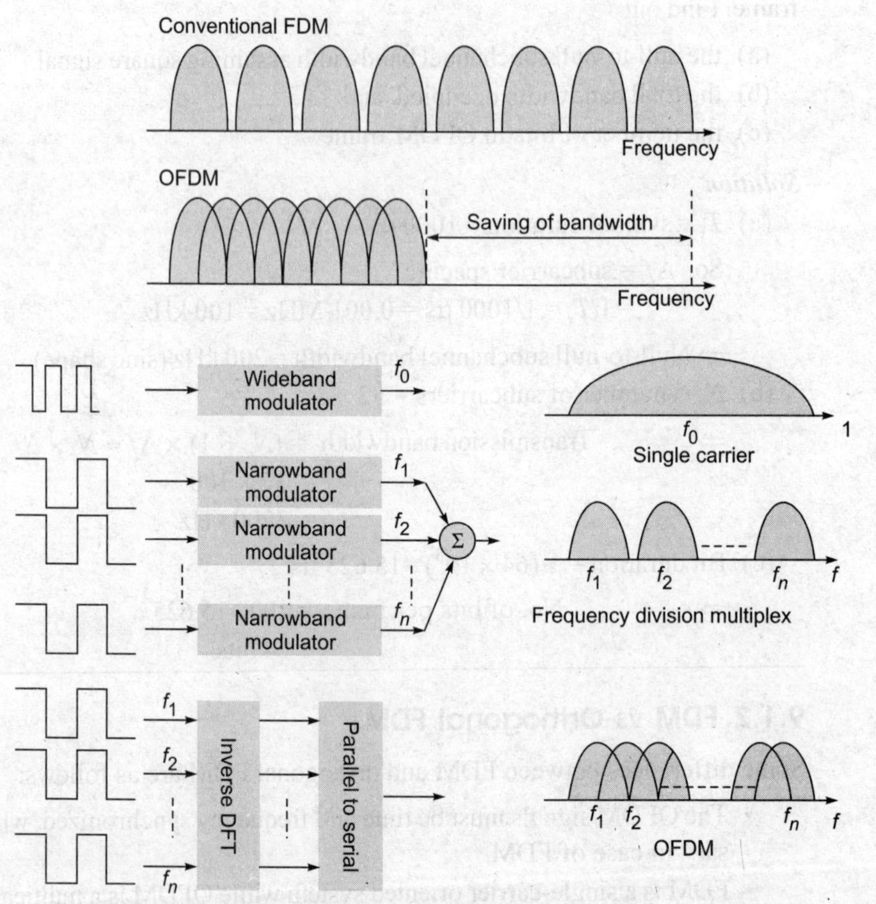

Fig. 9.6 FDM vs OFDM–comparing the spectral efficiency: (a) spectrum saving due to multicarrier modulation and (b) OFDM spectrally efficient compared to other techniques

Example 9.2 Considering the typical case, five 100 kHz channels are placed nearer to each other modulated by single-carrier scheme with five different carriers. If the same five carriers are placed orthogonally to each other, compare the occupied bandwidth in both the cases and find out the saving in bandwidth due to multicarrier scheme.

Solution

(Note: Here specific modulation schemes are not considered; otherwise the results may differ. It is just an arbitrary example with approximate comparison.)

For single-carrier system:

The total occupied bandwidth will be $100 \times 5 = 500$ kHz without guard interval

For orthogonal carriers:

Considering the same bandwidth of 100 kHz as subcarrier bandwidth, the spacing between the orthogonal carriers will be approximately 50 kHz. So, the total occupied bandwidth will be $5 \times 50 + 50 = 250 + 50 = 300$ kHz.

The saving of bandwidth is 200 kHz.

9.2 SINGLE VS MULTICARRIER SYSTEMS

All conventional communication systems are of single-carrier type. Although the multicarrier concept is very old, it was introduced in OFDM just few years back. The principle of OFDM was already around in the 50's and 60's as an efficient MCM technique. But the system implementation was delayed due to technological difficulties like digital implementation of FFT/IFFT, which were not possible to solve then. In 1965, Cooley and Tukey presented the algorithm for FFT calculation and later its efficient implementation on chip made the OFDM into application. Multiple carriers together can withstand the channel effects as they exploit the diversity. Figure 9.7 shows the conventional single-carrier communication links.

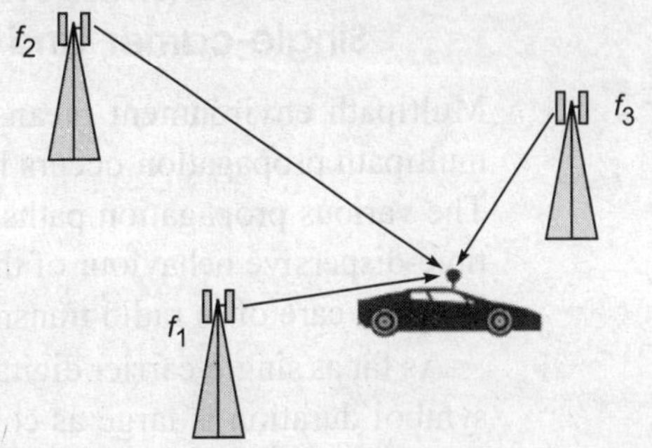

Fig. 9.7 Single-carrier frequency allocation to users in a network

9.2.1 Single-Carrier Systems

Efficient use of radio spectrum in single-carrier systems includes placing modulated carriers as close as possible without causing ICI with frequency division multiplexing. Ideally, the bandwidth associated with each carrier should be adjacent to its neighbours for efficient utilization of spectrum. In practice, a guard band must be placed between each carrier bandwidth to provide a space where a filter can attenuate an adjacent carrier's signal. These guard bands are wasted bandwidth. To transmit information at high data rates, symbol period will be relatively shorter for the same information. The symbol period is the inverse of the baseband data rate ($T_{sc} = 1/R_{sc}$). So, when R_{sc}

increases, T_{sc} must decrease and vice versa. In a multipath environment, a shorter symbol period leads to a greater chance of intersymbol interference. This occurs when a delayed version of symbol n arrives during the processing period of symbol $n+1$. M-ary schemes with higher value of M provides better spectral efficiency and data rate at the cost of SNR.

9.2.2 Multicarrier Systems

The OFDM, a multicarrier system, addresses both the problems of single-carrier system mentioned above. The basic idea of OFDM is to divide the available spectrum into several subchannels (or subcarriers). By making all subchannels narrowband, they experience almost flat fading, which makes equalization very simple. The OFDM provides a DSP-based technique allowing the bandwidths of modulated carriers to overlap without interference. Bandwidth and spectral efficiency is achieved compared to single-carrier system. To obtain a high spectral efficiency, the frequency responses of the subchannels are overlapping and orthogonal. This orthogonality can be completely maintained, even though the signal passes through the time-dispersive channel, by introducing cyclic prefix as described later on. It also supports a high data rate due to serial-to-parallel conversion of symbols acquiring long symbol duration, thus helping eliminate ISI. Time and frequency synchronization is the main limitation of multicarrier systems.

9.2.3 Data Transmission in a Multipath Environment by Single-carrier and Multicarrier Systems

Multipath environment means a time-dispersive channel. In typical radio channels, multipath propagation occurs because of multiple reflections of the transmitted signal. The various propagation paths are characterized by different delays and this leads to a time-dispersive behaviour of the channel. Intersymbol interference is caused and has to be taken care of in radio transmission systems (refer to Chapter 5).

As far as single-carrier digital transmission is concerned, if the data rate is low and the symbol duration is large as compared to the maximum delay of the channel, it can be possible to cope up with the resulting ISI without any equalization. As the distance range or the data rate of the system increases, ISI becomes more severe and channel equalization has to be provided. To have equalization in this situation, complicated tasks like channel estimation, filter coefficient calculation, etc., are required with computational complexity. Adaptive equalizers were the better solution till now. The alternate transmission technique using multicarrier modulation (MCM) with orthogonal frequency division multiplexing is better than the above approach. The idea of OFDM is to convert the high-rate data stream to many low-rate data streams that are transmitted in a parallel way over many subchannels. Thus, in each subchannel, the symbol duration is large as compared to the maximum delay of the channel and ISI can be handled.

The common representation of the multipath channel is the channel impulse response (CIR), which is the response at the receiver if a single narrow pulse is transmitted (Fig. 9.8). Indirectly it is the power delay profile, which we have described in Chapter 5.

According to Figures 9.8 to 9.10, $T_{sc} = 1/R_{sc}$ (subscript 'sc' with T and R is used purposely to represent single carrier and 'mc' for multicarrier), R_{sc} = symbol rate, and τ_{max} = delay of the longest path with respect to the earliest path.

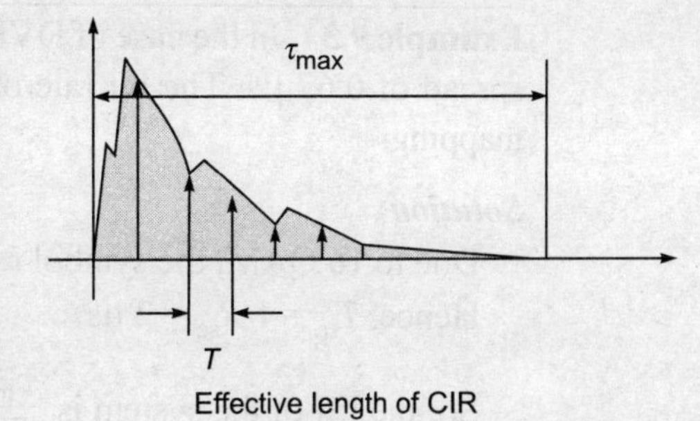

Fig. 9.8 CIR when single pulse is transmitted

A received symbol can theoretically be influenced by τ_{max}/T_{sc} previous symbols. This influence has to be estimated and compensated for in the receiver. For single-carrier system, this results in an ISI of $\tau_{max}/T_{sc} \approx$ very large value. The complexity involved in removing this interference in the receiver is tremendous.

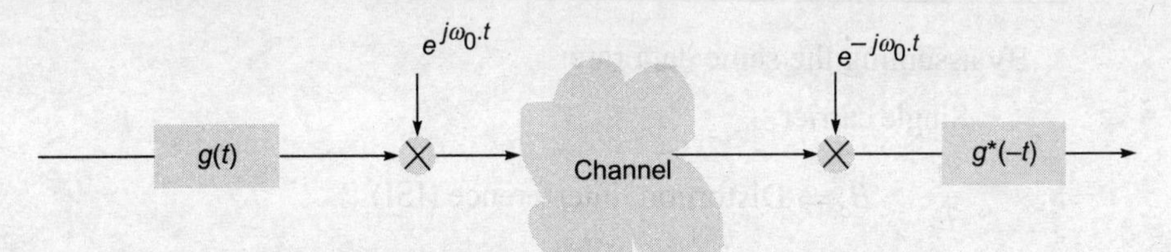

Fig. 9.9 Block diagram of a single-carrier system

As shown in the multicarrier system in Fig. 9.10, the original stream of rate R_{sc} is multiplexed into N_c parallel data streams of rate

$$R_{mc} = \frac{1}{T_{mc}} = \frac{R_{sc}}{N_c}$$

Each of the data streams is modulated with a different frequency and the resulting signals are transmitted together in the same band. Correspondingly, the receiver consists of N_c parallel receiver paths. Due to the prolonged distance in between the transmitted symbols, the ISI for each subsystem reduces to

$$\frac{\tau_{max}}{T_{mc}} = \frac{\tau_{max}}{N_c \cdot T_{sc}}$$ (Note that here ISI represents the efficiency of the system.)

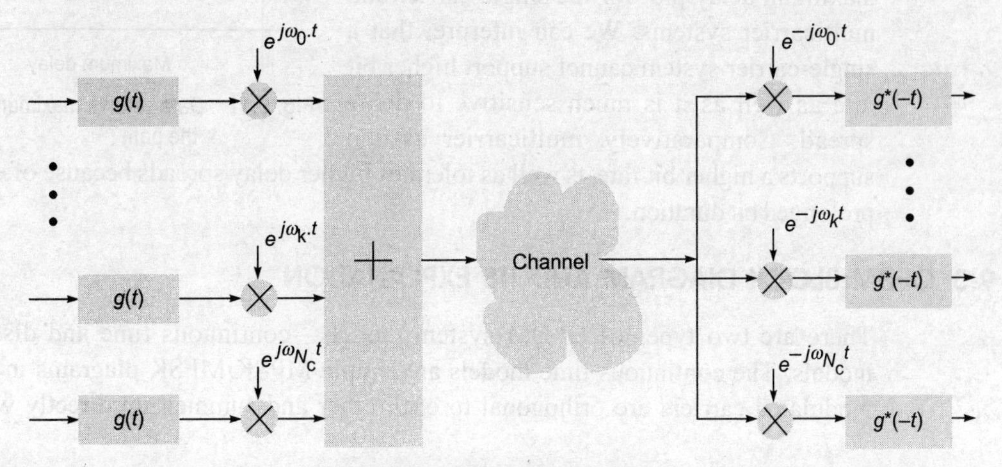

Fig. 9.10 A multicarrier system

Example 9.3 In the case of DVB, if $N_c = 6817$, find out the ISI for the maximum delay spread of 0.67 μs. The bit rate of the system is 2 Mbps. Assume 16QAM modulation mapping.

Solution

Due to 16 QAM, the symbol rate will be $R_{sc} = 2/4 = 0.5$ Mbps.

Hence, $T_{sc} = 1/R_{sc} = 2$ μs

So, the ISI of the system is $\frac{\tau_{max}}{T_{mc}} = \frac{0.67\,\mu s}{6817 \times 2\,\mu s} = 0.00004$

Such a little ISI can often be tolerated and less extra countermeasure, such as an equalizer, is needed. (In case of time-varying channel, the situation differs from the simple multipath and in vehicular mobility channel, the scenario becomes worst.)

By assuming the same data rate:

- Single carrier

  $\frac{1}{T_{sc}} > B_c \Rightarrow$ Distortion, interference (ISI)

  Thus, large amount of signal processing is required in the equalizer.
- Multicarrier

  $\frac{1}{N_c T_{sc}} < B_c$ (B_c = coherence bandwidth) $\Rightarrow$ No interference

  Thus, the date rate can be increased by using a large number of subcarriers and there is less equalization effort (as ISI is reduced by a factor of N_C)

As far as complexity of a receiver is concerned in a system with 6817 parallel paths, the simple approach of system design is not feasible. This asks for a slight modification of the approach that leads us to the concept of OFDM. Figure 9.11 gives the data rate vs maximum delay plot for the single-carrier and multicarrier systems. We can interpret that a single-carrier system cannot support higher bit rate as well as it is much sensitive to delay spread. Comparatively, multicarrier system supports a higher bit rate as well as tolerates higher delay spreads because of effectively prolonged bit duration.

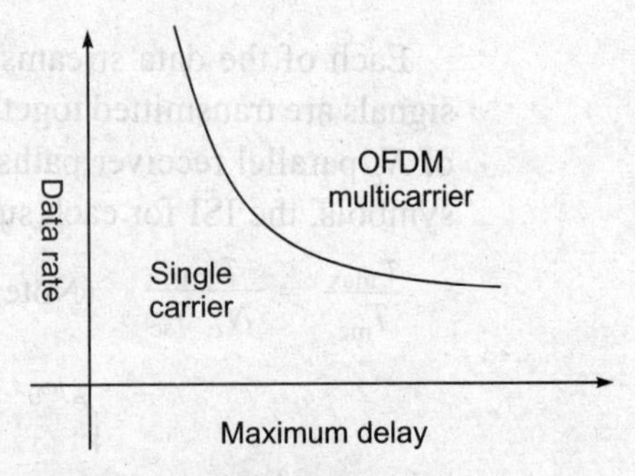

Fig. 9.11 Data rate vs maximum delay over the path

9.3 OFDM BLOCK DIAGRAM AND ITS EXPLANATION

There are two types of OFDM system models—continuous time and discrete time models. The continuous time models are simple MPSK/MFSK diagrams in which the modulated carriers are orthogonal to each other and summed up directly without the

IDFT stage. Phases are generated according to symbols read by modulator stage input. We shall consider the discrete model throughout the chapter.

Assumptions in Discrete Time Model

- The digital implementation of OFDM system is achieved through the mathematical operations called *discrete Fourier transform* (DFT) and its counterpart *inverse discrete Fourier transform* (IDFT). These two operations are extensively used for transforming data between the time domain and frequency domain. In case of OFDM, these transforms can be seen as summing or splitting of orthogonal modulated subcarriers. The resultant OFDM baseband signal will be the output of IDFT operation.
- A cyclic prefix is added in two different ways as discussed below:
  1. The IFFT output available in parallel in the form of samples is converted into a serial output. Few samples are padded at the end in the form of circular shift. The additional block is without any useful information and acts as a guard interval in time domain to eliminate the effect of multipath delay spread. Cyclic prefix will be removed immediately after RF down-conversion stage. So, CP is a purely overload and normally used method.
  2. Sometimes, the guard interval is added at the end of each OFDM symbol in terms of copy of part of OFDM symbol. The spectrum due to this addition will not be much affected and guard interval T_g will be removed immediately after demodulation stage. Prior calculations are required.
- The OFDM symbol time is very large compared to the guard interval.
- Cyclic prefix or guard interval is larger than the channel impulse response or delay over the channel.
- Transmitter and receiver are perfectly synchronized.
- Flat fading over each narrow subcarrier channel is considered, while all the subcarriers are received with different strengths. So, overall transmission bandwidth is said to undergo the frequency-selective fading as shown in Fig. 9.12(b).

According to simplified diagram shown in Fig. 9.12(a), the transmitter of OFDM link starts from channel coding stage if the source coded data is available in the required form. By scrambling, convolutional coding, code puncturing, and interleaving operations, the redundancy will be added in the data to combat the fading channel. Thereafter, the OFDM modulation stage starts.

To generate OFDM successfully, the relationship between all the subcarriers must be carefully controlled to maintain the orthogonality of the carriers. For this reason, OFDM is generated by firstly choosing the spectrum required based on the input data and the modulation scheme used. Each subcarrier to be produced is assigned some data to transmit. The required amplitude and phase of the subcarrier is calculated based on the modulation scheme (typically differential BPSK, QPSK, or QAM). Some additional subcarriers, called pilot subcarriers, are added to create the reference at the receiver end, which carries out the channel estimation procedure to remove channel impairments.

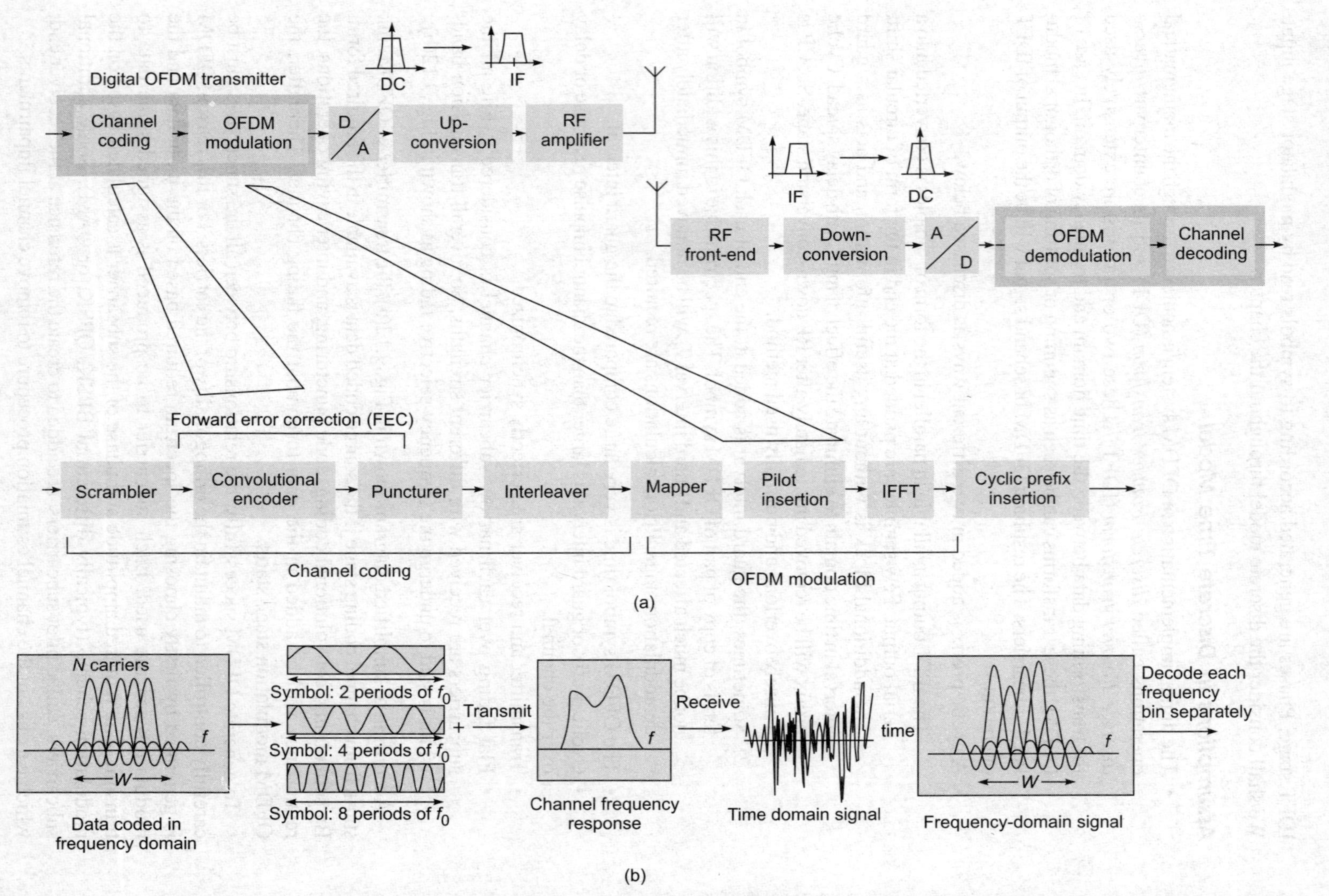

Fig. 9.12 (a) Conceptual block diagram for discrete OFDM system (Channel coding and OFDM modulation blocks are highlighted better. Exactly opposite task is followed at the receiver along with equalization and decision-making blocks in addition.) and (b) effect of frequency-selective fading on the received spectrum

The modulated subcarriers spectrum setting acts as sampled IDFT bin setting. In order to perform frequency domain data into time domain data, IDFT correlates the frequency domain input data with its orthogonal basis functions, which are sinusoids at certain frequencies (subcarriers here). In other ways, this correlation is equivalent to mapping the input data onto the sinusoidal basis functions. It provides complete spectrum setting.

The spectrum setting is then converted into its equivalent time domain signal using an inverse discrete Fourier transform. In practice, an inverse fast Fourier transform (IFFT) is used to perform IDFT. The IFFT performs the transformation very fast. After IFFT stage, the time domain signal must be received in serial form through P/S. The signal thus generated is called OFDM baseband and to generate an upconverted RF signal, the signal must be filtered and mixed to the desired transmission frequency.

Before RF upconversion, the cyclic prefix is added to remove the ISI effect. Also the output of the IFFT stage is in the form of discrete samples. So, D-to-A conversion is required before RF conversion stage.

The IFFT performs the function of transforming a spectrum (amplitude and phase of each component) into a time domain signal. An IFFT converts a number of complex data points, of length that is a power of 2, into the time domain signal of the same number of points. Each data point in frequency spectrum used for an FFT or IFFT combined together makes a bin. It is not necessary that the number of subcarriers and the IFFT bin size should be equal. Zero padding can be applied to IFFT bin to increase the IFFT points and improve the time domain resolution or guard carriers can be used.

At the receiver after A-to-D conversion, the FFT transforms a time domain signal into its equivalent frequency spectrum. The amplitude and phase of the sinusoidal components represent the frequency spectrum of the time domain signal. Since each bin of an IFFT corresponds to the amplitude and phase of a set of orthogonal sinusoids, the reverse process guarantees that the subcarriers generated are orthogonal. Of course, Doppler effect may exhibit some shifts in the received carrier frequency that must be corrected by perfect synchronization. Figure 9.13 gives some additional blocks representing some other important OFDM operations at the receiving end. These bocks are in addition to the inverse blocks required for OFDM demodulation and channel decoding.

For subcarrier offset removal, pilot signals are considered which may have slightly higher power than the data subcarriers. Some other methods may also be used. Channel estimation is the process by which the channel impulse response is estimated by adaptive feedback system and hence phase correction is applied. This is done by extracting the pilot subcarriers. Pilot information can be sent in time domain or in frequency domain and accordingly time or frequency domain methods are utilized to estimate the channel.

Each of the important stages are explained in detail in the next few sections. Figure 9.14 represents all those stages.

9.3.1 Serial-to-Parallel Conversion and Symbol Mapping

The input serial data stream (OFDM block) is read into the word size, e.g., 2 bits/word for QPSK, 8 bits/word for 256-PSK, etc. and converted into a parallel format just like conventional M-PSK method. For lattice style of subcarrier allocation, after reading

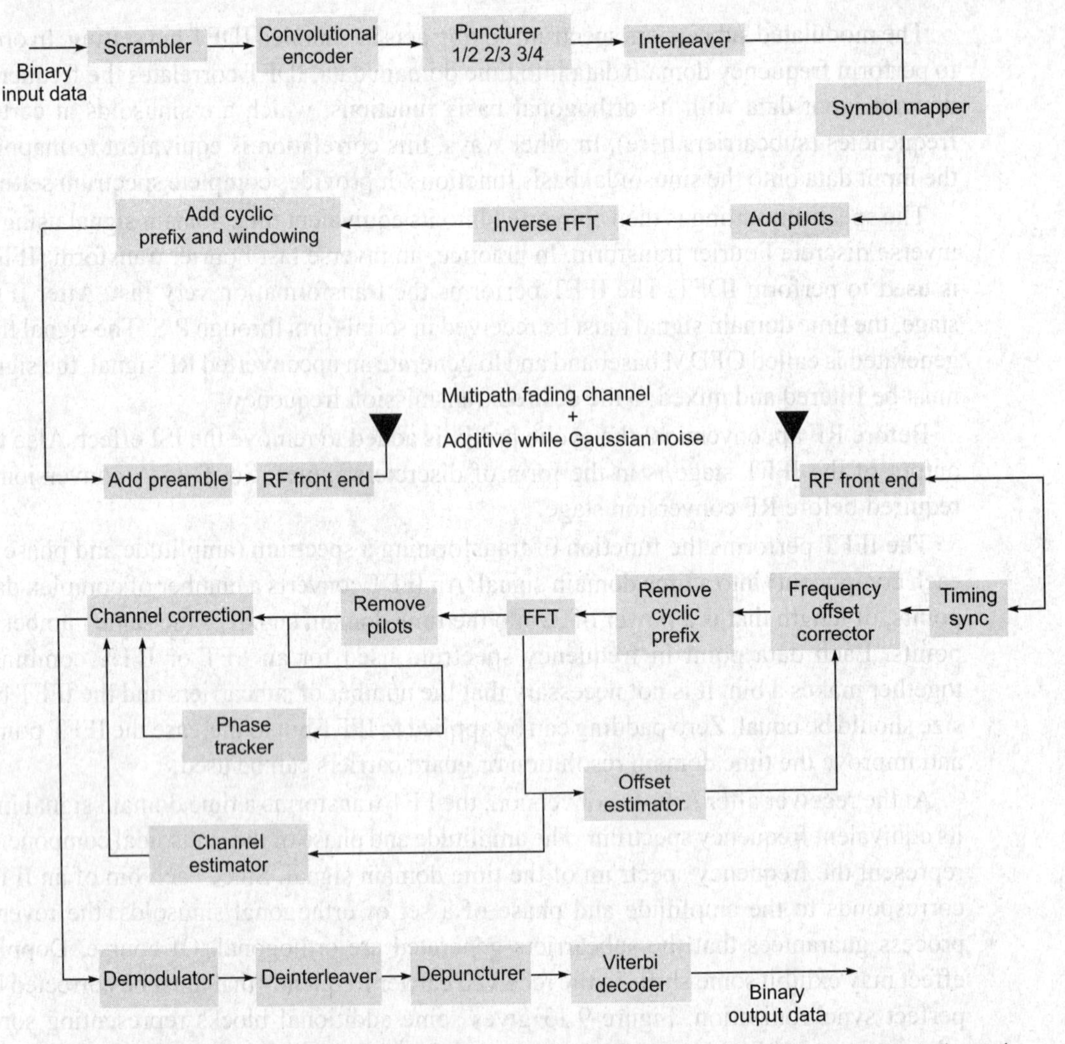

Fig. 9.13 Typical OFDM block diagram in detail showing channel coding stages and channel estimation and corrections at the receiver

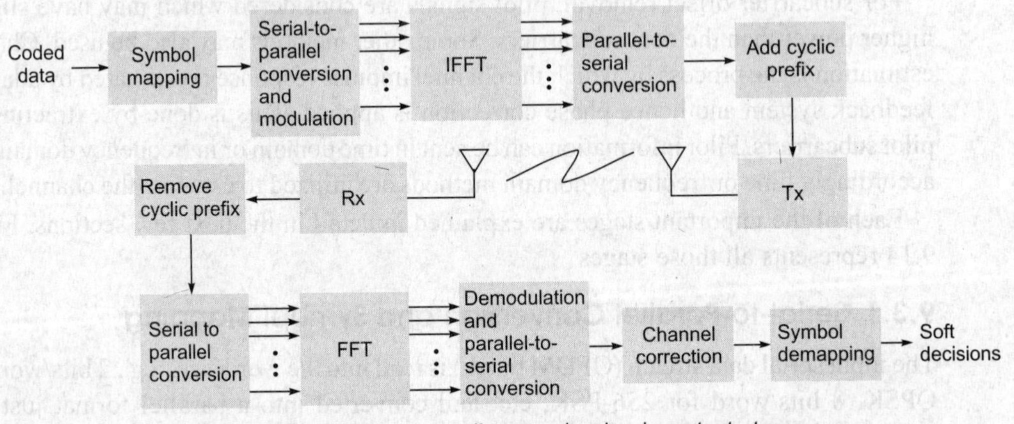

Fig. 9.14 OFDM system diagram showing important stages

symbols, all symbols are reorganized into N_c parallel lines, which is equal to the number of subcarriers. The data is then transmitted in parallel by assigning different blocks to different carriers in the transmission, f_1, f_2, etc. as shown in Fig. 9.1. The OFDM is a high data rate communication. So, a high probability of ISI is expected as the symbol time duration may become less than the maximum delay of the RF channel; but here reading the data in the form of word is advantageous (Fig. 9.15). Indirectly, the time duration between two consecutive pulses is prolonged and hence ISI can be minimized.

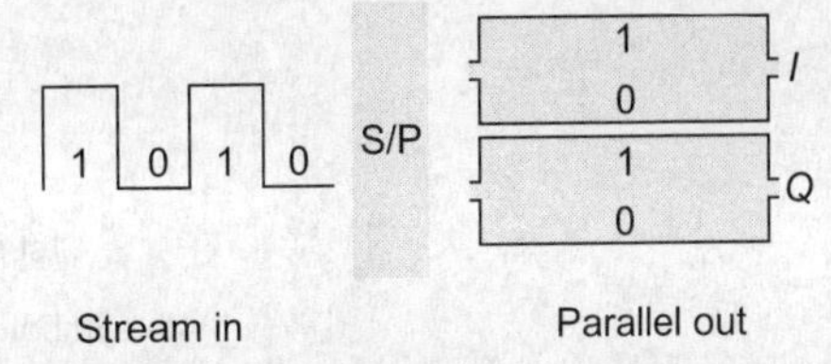

Fig. 9.15 Serial-to-parallel conversion of the data

Suppose a symbol transmission takes four seconds. Then, as per Fig. 9.15, each piece of data on the left has duration of one second. On the other hand, OFDM would send the four pieces simultaneously as shown on the right. In this case, each piece of data has duration of four seconds. The M-ary shift keying schemes or QAM can be used for symbol mapping in OFDM. Symbols are mapped onto the constellation diagram. Thus, this method of parallel transmission can support a very high bit rate.

Sometimes, the data to be transmitted on each subcarrier can be differential encoded with previous symbols (this is not a compulsory process but improves the correlation between consecutive symbols and helps, while detecting the errors), then mapped into PSK format. Since differential encoding requires an initial phase reference, an extra (null) symbol is added at the start for this purpose. The data on each symbol is then mapped to a phase angle based on the modulation method. For example, for QPSK, the phase angles used are 0, 90, 180, and 270 degrees. The use of phase shift keying produces a constant amplitude signal and is normally chosen in comparison with QAM, for its simplicity and to reduce problems with amplitude fluctuations due to fading.

9.3.2 Modulation of Data

The following steps may be carried out in order to apply modulation to the carriers in case of long frames and more symbols per carrier.

- After reading data in terms of bits/symbol, apply differential coding to each symbol sequence (optional) and convert the serial symbol stream into parallel segments according to the number of carriers and symbols per carrier.
- Assign OFDM block to the appropriate subcarrier in IFFT bin.
- Taking the IFFT of the result will give the discrete time domain signal.

There is another approach by which the modulation of data is achieved (especially in short frames). In that, the incoming frame is read symbol-wise as shown in Fig. 9.16 and then the serial-to-parallel conversion splits the symbols in the parallel lines equal to the number of carriers (1 symbol/carrier). Multiply the symbols with the carriers and perform IFFT to have summing operation. This will output the OFDM baseband signal after P/S conversion that can be modulated by RF carrier thereafter.

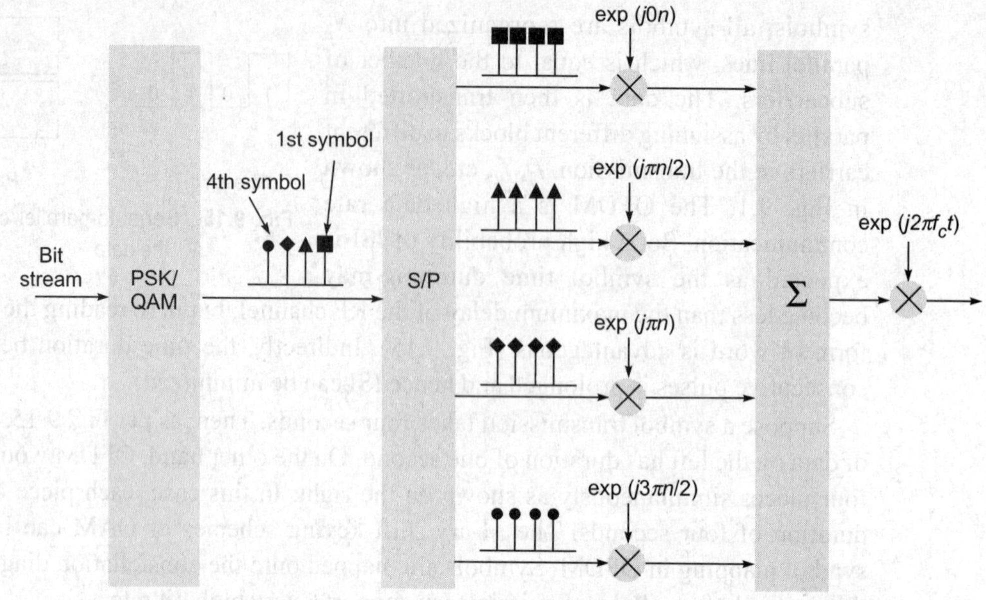

Fig. 9.16 Frequency to time domain conversion and IFFT acting as a summer

The condition for the selection of the number of subcarriers is as follows:

Number of subcarriers < (IFFTsize/2)–2 (for real-valued time signal)

Number of subcarriers < (IFFTsize)–1 (for complex-valued time signal)

One set of subcarriers makes transmission bandwidth of OFDM.

Note that the modulated OFDM signal is nothing more than a group of delta (impulse) functions, each with a phase determined by the modulating symbol. In addition, also note that the frequency separation between each pair of delta is proportional to inverse of the symbol duration.

9.3.3 Guard Period

The ISI is a common problem found in high data rate communication. It occurs when the transmission interferes with itself and the receiver cannot decode the transmission correctly. This is because as the data rate increases, the time duration between the consecutive pulses decreases. For avoiding ISI, the pulse time duration should be greater than the maximum delay of the channel. The ISI is one of the major drawbacks of multipath single-carrier transmission as discussed earlier. To avoid this, guard interval is provided along with the data period so that the ISI effect observed in the guard interval can be removed afterwards and the data can be retrieved.

One of the most important properties of OFDM transmissions is its high level of robustness against multipath delay spread. This is a result of the very long symbol period used compared to path delay, which minimizes the intersymbol interference. The level of multipath robustness can be further increased by addition of a guard period between the transmitted symbols. The guard period allows time for multipath signals from the pervious symbol to die away before the information from the current symbol is gathered. The most effective guard period to use is a cyclic extension of the symbol. If a mirror in time at the end of the symbol waveform is put at the start of the symbol as the guard period, this effectively extends the length of the symbol, while maintaining the orthogonality of the waveform. Using cyclic extended symbol, multipath immunity as well as symbol time synchronization tolerance can be achieved.

Transmitting a cyclic prefix of the data during guard interval transforms the linearly convolutive channel into circularly convolutive channel. Hence, the channel equalization problem is simplified. Specifically, channel equalization in the frequency domain can be done using one tap filters. This is because cyclic prefixing makes the channel matrix circulant, which is diagonalized by the IDFT and DFT operations.

The added guard interval and its effect in reducing ISI are shown in Figures 9.17 and 9.18.

As long as the multipath delay echoes stay within the guard period duration, there is strictly no limitation regarding the signal level of the echoes: they may even exceed the signal level of the shorter path. The signal energy from all paths add up at the input to the receiver, and since the FFT is energy conservative, the whole available power feeds the decoder. If the delay spread is longer then the guard interval, then they begin to cause inter symbol interference. However, provided the echoes are sufficiently small they do not cause significant problems. This is true most of the time, as multipath echoes delayed longer than the guard period they will have been reflected from very distant objects.

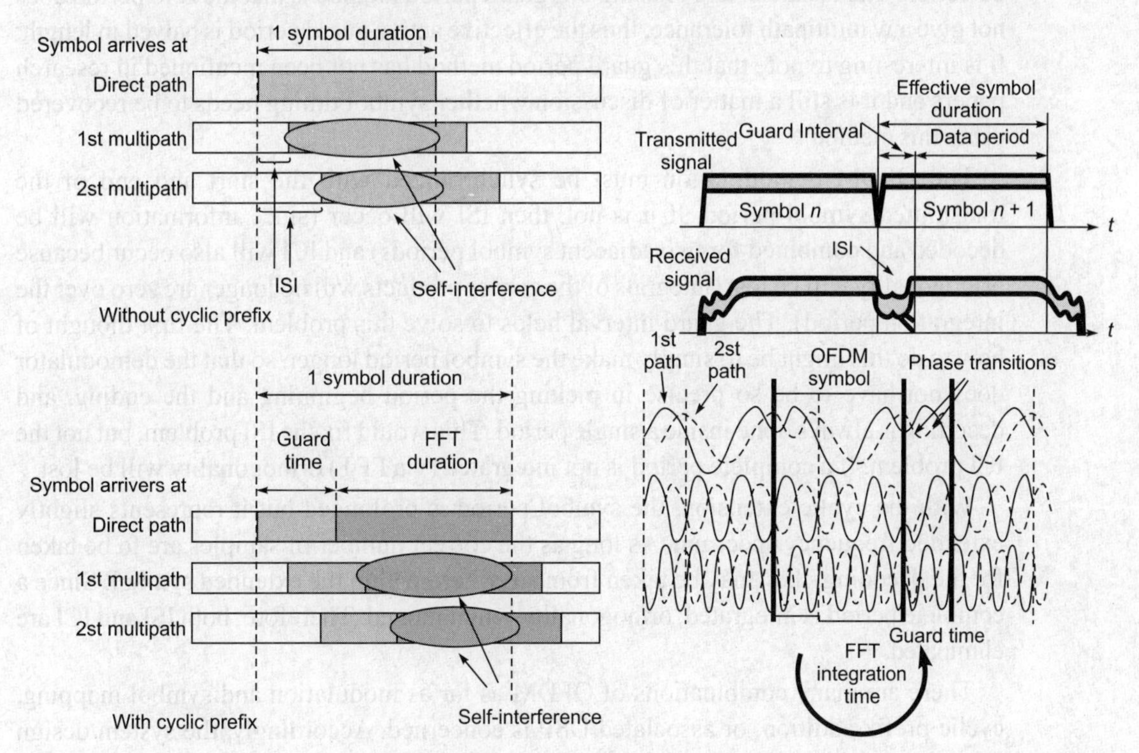

Fig. 9.17 Concept of removal of multipath effect due to guard interval addition in each symbol block by padding copy of part of the symbol (Phase errors are mainly in guard interval.)

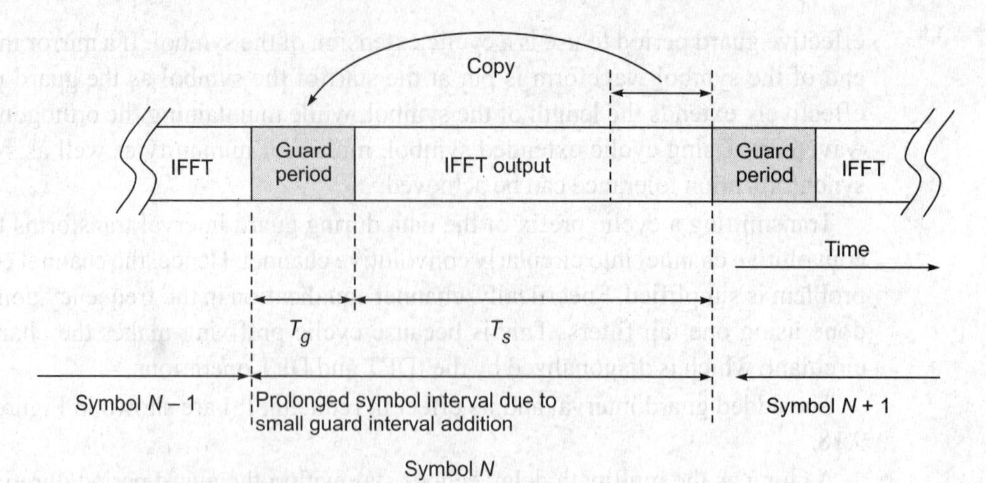

Fig. 9.18 Addition of guard period in terms of cyclic prefix after IFFT output increases effective symbol duration making circularly convolutive channel

Other variations of guard periods are possible. One possible variation is to have half the guard period a cyclic extension of the symbol, as above, and the other half a zero amplitude signal. Using this method the symbols can be easily identified. This possibly allows for symbol timing to be recovered from the signal, simply by applying the envelop detection. The disadvantage of using this guard period method is that the zero period does not give any multipath tolerance, thus the effective active guard period is halved in length. It is interesting to note that this guard period method has not been mentioned in research papers and it is still a matter of discussion whether symbol timing needs to be recovered using this method.

The OFDM demodulation must be synchronized with the start and end of the transmitted symbol period. If it is not, then ISI will occur (since information will be decoded and combined for two adjacent symbol periods) and ICI will also occur because orthogonality will be lost (integrals of the carrier products will no longer are zero over the integration period). The guard interval helps to solve this problem. The first thought of how to do this might be to simply make the symbol period longer, so that the demodulator does not have to be so precise in picking the period beginning and the ending, and decoding is always done inside a single period. This would fix the ISI problem, but not the ICI problem. If a complete period is not integrated (via FFT) orthogonality will be lost.

With the cyclic extension, the symbol period is prolonged but it represents slightly extended frequency spectrum. As long as the correct number of samples are to be taken for the decoding, they may be taken from anywhere within the extended symbol. Since a complete period is integrated, orthogonality is maintained. Therefore, both ISI and ICI are eliminated.

There are many combinations of OFDM as far as modulation and symbol mapping, cyclic prefix addition, or associated DSP is concerned. Accordingly, the system design may vary.

Although the cyclic prefix introduces a loss in SNR, This may be considered as a small price to pay to mitigate interference. Note that some bandwidth efficiency is lost with the addition of the guard period, but the advantages that we get are many, and hence compromising solution is required.

Example 9.4 An OFDM symbol duration is of 1280 μs. If such a frame is modulated by OFDM scheme, transmitted over the channel and received after the delay of 450 ns at the receiver. What should be the minimum duration of the cyclic prefix? Show that cyclic prefix does not affect the occupied spectrum much, if the number of subcarriers is 64 and 64 point FFT is used. Will ISI occur?

Solution

Obviously, the cyclic prefix minimum duration must be 450 ns.

If cyclic prefix of 0.45 μs is appended to OFDM symbol duration, 1280.45 μs is the final symbol duration (expected).

Spacing between two consecutive subcarriers without cyclic prefix will be

$$1/T_s = 781.25 \text{ Hz}$$

So, the spectrum occupied will be $(64 + 1) \times 781.25 \approx 50.781$ kHz

The 64 point IFFT reflects in time domain as 64 samples of the OFDM symbol with the sampling duration of 20 μs ideally. If at least 1 sample is cyclically prefixed, 1 point added in IFFT bin, the additional spectrum consumption will be 781.25 Hz, total spectrum ≈ 50.781 kHz + 781.25 Hz ≈ 51.562 kHz. One sample CP is sufficient to combat 450 ns path delay. So, no ISI will occur and the spectrum is also not much affected.

9.3.4 Transmission of OFDM Signal over the Channel

It is not currently practical to generate the OFDM signal directly at RF. So, OFDM baseband signal must be upconverted for transmission. To remain in the discrete form, the OFDM baseband can be up-sampled and added to a discrete carrier frequency. This carrier can be an intermediate frequency whose sample rate is handled by current technology. It can then be converted to analog and increased to the final transmit frequency using analog frequency conversion methods. Alternatively, the OFDM

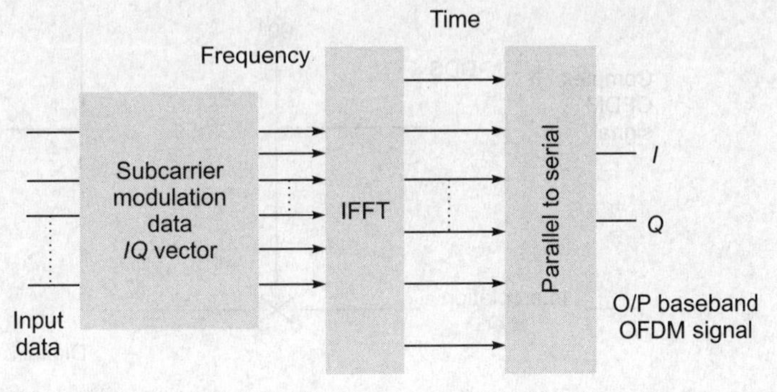

Fig. 9.19 IFFT stage before upconversion stage giving the signal in *IQ* form

modulation can be immediately converted to analog and directly increased to the desired RF transmit frequency. Either way, the selected technique would have to involve some form of linear AM or QAM (possibly implemented with a mixer).

RF Modulation

The output of the OFDM modulator generates a baseband signal, which must be mixed up to the required transmission frequency. This can be implemented using an analog technique or a digital up-converter as shown in Figures 9.20 and 9.21. Both techniques perform the same operation. However, the performance of the digital modulation will tend to be more accurate due to improved matching between the processing of the I and Q channels and the phase accuracy of the digital IQ modulator. In both the figures, the levels are shown mentioning whether the signal is in analog form or digital form.

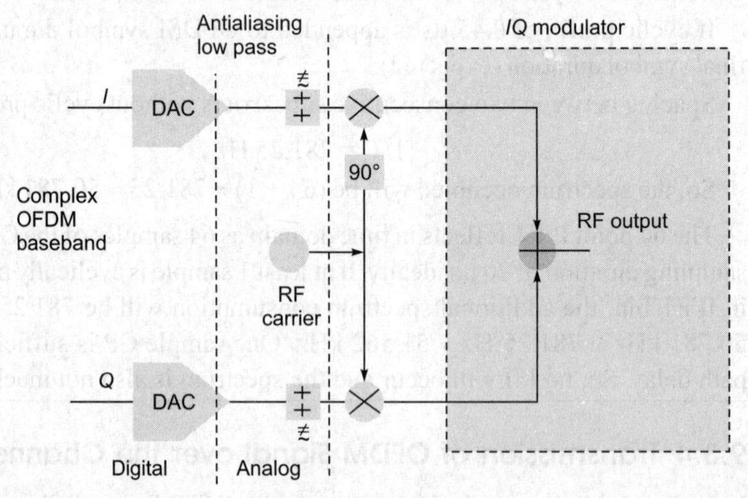

Fig. 9.20 RF modulation of complex baseband OFDM signal using analog techniques like M-ary schemes

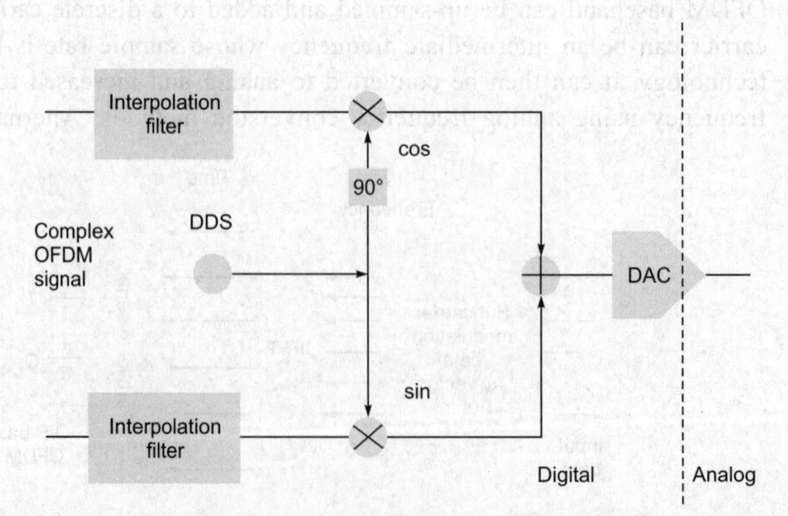

Fig. 9.21 RF modulation of complex OFDM signal using digital techniques

The role of interpolation filter is to increase the sampling rate for better analog conversion. In digital techniques direct digital synthesizer (DDS) is used that can generate the required discrete RF carrier by controlling the phase of samples and then converts into an analog form. Most of the latest ASIC designs use DDS for RF carrier generation that can generate very high carrier frequencies.

After RF conversion and transmission, the transmitted signal will get corrupted by the various channel effects and then it will be received by the receiver.

9.3.5 OFDM Signal Reception and Demodulation

The received RF signal is down-converted and also converted into digital form. The guard period is removed in case of cyclic prefix. Demodulation is done to get back the required setting in the frequency domain (just as transmitter). For that the offset removal is required. The subcarrier offset condition is shown in Fig. 9.22. The receiver basically does the reverse operation to the transmitter. The FFT of each OFDM symbol is performed to find the original transmitted spectrum. The phase angle of each transmission carrier is then evaluated and converted back to the data word by demodulating the received phase (demapping). The data words are then split back to the same pattern as the original bits to have the serial data again by parallel-to-serial conversion.

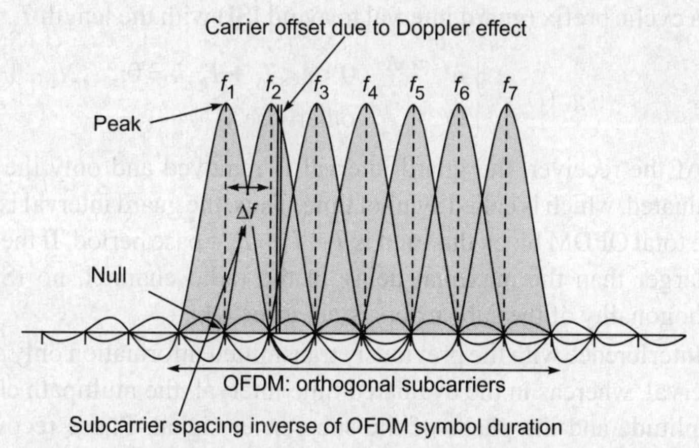

Fig. 9.22 Carrier offset condition at the receiver end (loss of orthogonality)

The following steps may be taken to demodulate the OFDM:

- Partition the input stream into vectors representing each symbol period.
- Take the FFT of each symbol period vector.
- Extract the carrier FFT bins and calculate the phase of each.
- Calculate the phase difference, from one symbol period to the next, for each carrier.
- Decode each phase into binary data.
- Sort the data into the appropriate order.

9.4 OFDM SIGNAL MATHEMATICAL REPRESENTATION

The following mathematics is for lattice-type OFDM structure in which pilot insertion for channel estimation is easier. The OFDM is a hybrid multiple access multi carrier modulation scheme. An OFDM signal consists of N_c subcarriers spaced by the frequency distance Δf. Thus, the total system bandwidth W is divided into N_c equidistant subchannels. All subcarriers are mutually orthogonal with a time interval of length $T_s = 1/\Delta f$. This serves two purposes, efficient use of spectrum and N_c streams transmitted with different carriers orthogonal to each other reducing effect of ICI and ISI simultaneously.

The kth subcarrier signal is described analytically by the function $\hat{g}_k(t)$:

$$\hat{g}_k(t) = \begin{cases} e^{j2\pi k\Delta ft} & 0 < t < T_s,\ k = 0,\ldots,N_c - 1 \\ 0 & \text{otherwise} \end{cases} \tag{9.5}$$

Since the OFDM system bandwidth W is subdivided into N_c narrowband subchannels (subcarrier bandwidth), the OFDM symbol duration T_s is N_c times large compared to a single-carrier transmission system covering the same bandwidth W. For a given system bandwidth, the number of subcarriers is chosen such that the symbol duration is large compared to the maximum delay of the channel. This subcarrier signal $\hat{g}_k(t)$ is extended by a cyclic prefix (guard interval to avoid ISI) with the length T_g yielding the signal

$$\hat{g}_k(t) = \begin{cases} e^{j2\pi k\Delta ft} & 0 < t < T_s + T_g, k = 0,\ldots,N_c - 1 \\ 0 & \text{otherwise} \end{cases} \tag{9.6}$$

At the receiver, the guard interval is removed and only the time interval $[0, T_s]$ is evaluated, which is called symbol time. Thus, the guard interval is a pure system overload. The total OFDM block duration is $T = T_s + T_g$ = base period. If the guard interval length T_g is larger than the maximal delay in the radio channel, no ISI occurs at all and the orthogonality of the subcarriers is not affected.

Interference with the previously transmitted information only appears within the guard interval, whereas in the evaluated time interval, the multipath channel only changes the amplitude and the phase of the sub-carrier signal. Phase recovery can be done using decision making on the basis of particular range of phases. Better representation of OFDM signal in frequency as well as time domain is given as follows.

OFDM Frequency Domain Representation

Single real OFDM subcarrier:

$$S(k) = e^{j\theta_m}\delta\left(k - m - \frac{N}{2}\right) + e^{-j\theta_m}\delta\left(k + m - \frac{N}{2}\right) \tag{9.7}$$

Composite (real) OFDM subcarriers (corresponding bin setting is shown in Fig. 9.23):

$$S(k)_{\text{ofdm}} = \sum_{m=c\text{ first}}^{c\text{ last}} \left[e^{j\theta_m}\delta\left(k - m - \frac{N}{2}\right) + e^{-j\theta_m}\delta\left(k + m - \frac{N}{2}\right)\right] \tag{9.8}$$

where

k = frequency (0 to N–1) point/sample
m = mth OFDM carrier frequency
N = IFFT bin size (N_c may or may not be equal to N)
c = first to last OFDM carriers

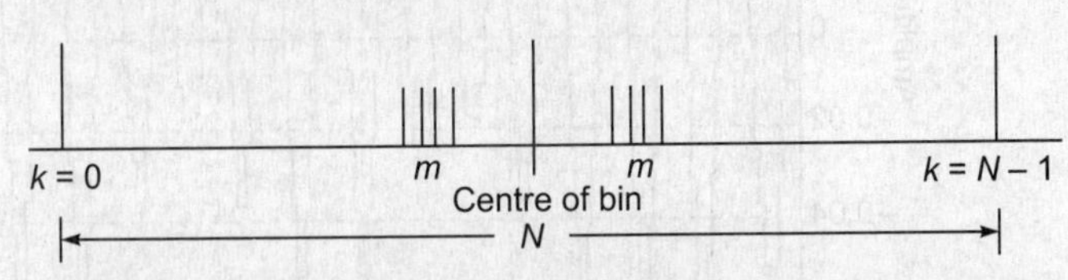

Fig. 9.23 IFFT bin setting correlated with given mathematics

(Here the two variables k and m are defined separately. Think why.)

After the symbol mapping is applied, an IFFT is performed to generate one OFDM symbol period in the time domain by adding all the subchannels. It is clear that the OFDM signal has varying amplitude. After addition of all the subchannels the amplitude of time domain OFDM baseband signal may cross the limit of linear power amplifier. It is very important that the amplitude variations be kept intact as they define the content of the signal. If the amplitude is clipped or modified, then an FFT of the signal would no longer result in the original frequency characteristics, and the modulation may be lost.

Note: One of the drawbacks of OFDM is the fact that it requires linear amplification. In addition, very large amplitude peaks may occur depending on how the sinusoids line up, so the peak-to-average power ratio is high. This means that the linear amplifier has to have a large dynamic range to avoid distorting the peaks. The result is a linear amplifier with a constant, high bias current resulting in very poor power efficiency. Peak power clipping is an important issue of OFDM system and is described afterwards.

OFDM Time Domain Representation

We have

$$s(n) = \sum_{m=c_{\text{first}}}^{c_{\text{last}}} \sum_{n=0}^{N-1} \cos\left(\frac{2\pi mn}{N} + \theta_m\right) \tag{9.9}$$

where n = time sample
m = OFDM carrier
N = IFFT bin size
θ_m = pahse modulation for OFDM carrier (m)
$c_{\text{first}}, c_{\text{last}}$ = OFDM carriers (first and last)

OFDM System Spectral Settings According to Lattice

The data symbols are transmitted as per lattice points at allocated subcarrier and the spacing in the frequency domain is then $\Delta f = \dfrac{1}{T_S + T_g}$, where T_g is the length of the guard interval.

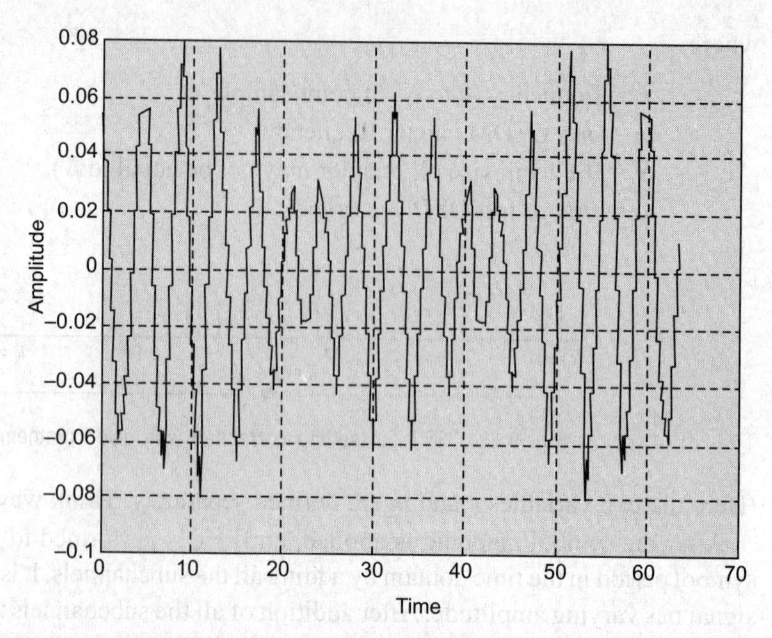

Fig. 9.24 OFDM time signal after addition of all the carriers (OFDM baseband signal) 64 points

Each transmitted data symbol in the lattice experiences flat fading, which simplifies equalization and channel estimation. The channel attenuations at the lattice points are correlated and by transmitting known symbols at some positions, the channel attenuations can be estimated with an interpolation filter. The lattice is a 2D version for pilot-symbol assisted modulation, which has been proposed for several wireless OFDM systems and described afterwards.

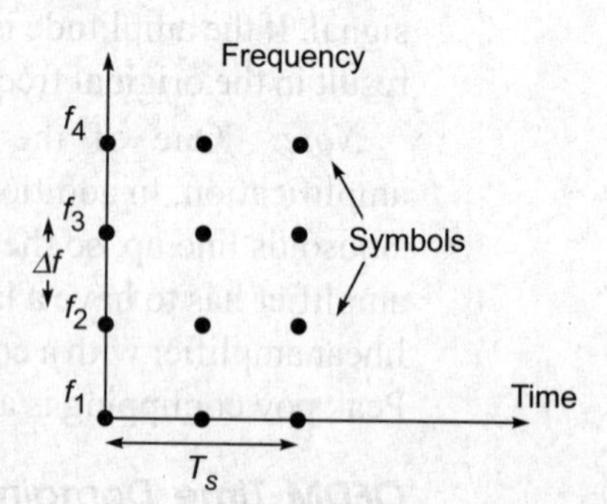

Fig. 9.25 Lattice in the time frequency plane (as given in Fig. 9.1)

Individual subchannel sinc shape in the frequency domain can be adjusted according to the Nyquist criteria to optimize the setting of spectrum.

Each subcarrier can be modulated independently with the complex modulation symbol $S_{n,k}$, where the subscript n refers to the nth time interval and k to the kth subcarrier in the considered OFDM block (lattice-wise). Thus, within the symbol duration $T = T_S + T_g$, the following signal of the nth OFDM symbol block is formed:

$$s_n(t) = \frac{1}{\sqrt{N_c}} \sum_{k=0}^{N_C-1} S_{n,k}\, g_k(t - nT) \tag{9.10}$$

The total continuous-time signal consisting of all OFDM blocks is

$$s(t) = \frac{1}{\sqrt{N_c}} \sum_{n=0}^{\infty} \sum_{k=0}^{N_C-1} S_{n,k}\, g_k(t - nT) \tag{9.11}$$

Thus, by modulation, a rectangular symbol blocks are applied to each OFDM carrier. The shaping is applied to each symbol block. All shaping functions differ by one symbol period. Due to the rectangular pulses applied to subcarriers, the spectra of the subcarriers are sinc-functions, e.g. for the *k*th subcarrier:

$$G_k(f) = T \operatorname{sinc}\,[\pi T(f - k\Delta f)]$$

where $$\operatorname{sinc} x = \frac{\sin x}{x} \tag{9.12)}$$

The spectra of the subcarrier overlap, as shown in Figure 9.5 OFDM spectrum, but the subcarrier signals are mutually orthogonal and modulation symbols $S_{n,k}$ can be recovered by a correlation technique as shown in Eq. (9.13).

Correlation between the two consecutive shaping functions $g_k(t)$ and $g_l(t)$

$$= \langle g_k, g_l \rangle = \int_0^{T_s} g_k(t)\,\overline{g_l}(t)\,dt = T_s \delta_{k,l} \tag{9.13}$$

or $$S_{n,k} = \frac{\sqrt{N_c}}{T_s} \overline{\langle s_n(t), g_k(t - nT) \rangle} \tag{9.14}$$

where $\overline{g_k(t)}$ is the conjugate of $g_k(t)$.

In practical applications, the OFDM signal $s_n(t)$ is generated in a first step as a discrete time signal in the digital signal processing part of the transmitter. Since the bandwidth of an OFDM system is $W \approx N_c \Delta f$, the signal must be sampled with the sampling time $\Delta t = 1/W = 1/N_c \Delta f$. Let us introduce one more term *i* for *i*th sample. The samples of the signal are written as $S_{n,i}$, $i = 0, 1, \ldots, N_c - 1$ (N is equal to N_c in case of IFFT bin of N-point without zero padding) and can be calculated as

$$s_{n,i} = \frac{1}{\sqrt{N_c}} \sum_{k=0}^{N_c - 1} S_{n,k} e^{j2\pi ik/N_c} \tag{9.15}$$

This equation describes exactly the inverse discrete Fourier transform (IDFT), which is typically implemented as an IFFT. After IFFT, further signal processing can be applied to avoid out-of-band radiation. Out-of-band radiation can occur if amplitude peaks of the OFDM signal are limited by the power amplifier and can occur to the side lobes of the subcarrier spectra. Finally, the signal is D/A converted and transmitted. The efficiency η of the system depends upon delay spread and symbol duration.

$$\eta = \frac{\text{delay spread}}{\text{symbol duration}} \tag{9.16}$$

The effect of the number of subcarriers and the guard time duration on the system performance is summarized as follows.

- For a given number of subcarriers, increasing guard time duration reduces ISI due to the decrease in delay spread relative to the symbol time, but reduces the power efficiency and bandwidth efficiency.
- For a given signal bandwidth, increasing the number of subcarriers increases the power efficiency but also increases the symbol duration and results in a system more sensitive to Doppler spread.

Interference with the previously transmitted information only appears within the guard interval, whereas in the evaluated time interval, the multipath channel only changes the amplitude and the phase of the subcarrier signal. Phase recovery can be done using decision making on the basis of particular margin of phases.

As far as the receiver part is concerned $r_n(t)$ can be separated into the orthogonal subcarrier signals by a correlation technique according to Eq. (9.13),

$$R_{n,k} = \frac{\sqrt{N_C}}{T_S}\left\langle r_n(t), \overline{g_k(t-nT)}\right\rangle \tag{9.17}$$

Equivalently the correlation at the receiver can be implemented as a DFT or an FFT, respectively:

$$r_{n,i} = \sum_{k=0}^{N_C-1} R_{n,k} e^{-j2\pi ik/N_C} \tag{9.18}$$

where $r_{n,i}$ is the ith sample of the received signal $r_n(t)$ and $R_{n.k}$ is the received complex modulation symbol of the kth subcarrier. The FFT and IFFT algorithms must be implemented efficiently as the number of FFT points will introduce processing delay.

If the symbol duration T_s is chosen to be much smaller than the coherent time of the channel, then the transfer function of the channel $H(f,t)$ can be considered constant within the duration of each modulation symbol $S_{n,k}$. In this case, the effect of the radio channel is only a multiplication of each subcarrier channel by a complex transfer function $H_{n,k}$. As a result, the received modulation symbol after FFT is

$$R_{n,k} = H_{n,k} S_{n,k} + N_{n,k} \tag{9.19}$$

where $N_{n,k}$ is the additive noise of the channel. As the CIR for each subchannel is different, the above equation represents mentioned subscripts.

The mathematics here gives the main modulation and demodulation part, i.e. symbol processing. Reading the data in the form of symbols, converting them into parallel, channel coding, etc. are the additional supporting tasks for the OFDM modulation.

9.5 SELECTION PARAMETERS FOR MODULATION

The OFDM system depends on the following four requirements:

- *Available bandwidth* The bandwidth limit will play a significant role in the selection of number of subcarriers. Large amount of bandwidth will allow obtaining a large number of subcariers with reasonable CP length.
- *Required bit rate* The system should be able to provide the data rate required for the specific purpose.
- *Tolerable delay spread* An user environment specific maximum tolerable delay spread should be known beforehand in determining the CP length.
- *Doppler values* The effect of Doppler shift due to user movement should be taken into account.

Let W be the total available bandwidth. Let the maximum delay spread of the channel be T_d seconds. To prevent ISI, choose guard interval T_g for OFDM symbol much greater than the maximum delay spread T_d, say, $T_g = 4 \times T_d$.

To reduce the overhead introduced by cyclic prefix, choose OFDM symbol time T_s much greater than the guard time T_g, say, $T_s = 8 \times T_g$, where T_s = symbol time without guard interval. $T = T_S + T_g$, then subcarrier spacing $\Delta f = 1/T$. The number of subcarriers, $N_c = W/\Delta f$ (Nearest power of 2).

To minimize the signal-to-noise ratio (SNR) loss due to the guard time, the symbol duration should be much larger than the guard time. However, symbols with long duration are susceptible to Doppler spread, phase noise, and frequency offset.

Two observations are made from the above calculations:

1. Increasing the symbol duration decreases the frequency spacing between subcarriers. Thus, for a given signal bandwidth, more subcarriers can be accommodated. On the other hand, for a given number of subcarriers, increasing the symbol duration decreases the signal bandwidth.
2. Increasing the number of subcarriers increases the number of samples per OFDM symbol. However, it does not necessarily imply that the symbol duration increases. If the OFDM symbol duration remains the same, the duration between two samples decreases as a result. This implies the increase in the OFDM signal bandwidth. On the other hand, if the OFDM signal bandwidth is fixed, then increasing the number of subcarriers decreases the frequency spacing between two subcarriers, which in turn increases the symbol duration.

CASE STUDY BANDWIDTH REQUIREMENTS

If a bandpass modulation scheme is used with single-carrier selection, minimum bandwidth required will be $BW_{RF} = R(1 + \alpha)$ when raised cosine filtering is applied, where R = bit rate and α = raised cosine filter pulse shaping factor. If we have BPSK or any M-ary digital data analog transmission methods, different bandwidth per information may be consumed. So, we restrict the comparison with respect to the above equation only. Calculation of bandwidth for OFDM can be done as follows with W being the total available bandwidth:

$W = N_c \times \Delta f$, where $\Delta f = 1/T_s$

As T_s is proportional to the bit time, frequency spacing of the carriers is indirectly proportional to bit rate or data rate. So, the spectral efficiency of OFDM can be proved with the calculations given in Table 9.1.

9.6 PULSE SHAPING IN OFDM SIGNAL AND SPECTRAL EFFICIENCY

To obtain high spectral efficiency, the subchannels are overlapping and orthogonal. Hence the name OFDM. For improving the spectrum efficiency, each individual subcarrier channel width as well as overall transmission bandwidth, both must be shaped so that out-of-band components can be minimized. For narrow subcarrier channels, Nyquist pulse shaping is applied to input signal at the input stages. Final OFDM spectrum setting is achieved thereafter as shown in Fig. 9.26.

The Nyquist pulse shaping of the transmitted pulses results in a desired sinc-shaped frequency response for each channel. That is why this shape is considered in previous

Table 9.1 Calculations for single- and multiple-carrier systems

For Single-carrier System	For Multiple-carrier System (OFDM)
τ_{max} = 10 μsec = delay spread	τ_{max} = 10 μsec = delay spread
Data rate (R) = 56 kbps (arbitraily)	Data rate (R) = 56 kbps (arbitraily)
Bit time T_b = 1/R = 17.86 μsec	N_c = 100 carriers
Raised cosine filter pulse shaping α = 0.3	T_S = OFDM symbol time without guard period
BW_{RF} = $R(1 + \alpha)$ = 56 kbps (1+0.3) = 72.8 kHz	ΔT_S = Sampling duration
Also, τ_{max} > 17.86 μsec/10 ⇒ Frequency selective fading and ISI	IFFT size = IFFT bin count
	$R_{carrier}$ = 560 bps [(56000/100) bps] due to parallel conversion
	$T_{carrier}$ = 0.0018 sec = 1800 μsec = IFFTsize × ΔT_s
	BW_{RF} = (N_c + 1)/IFFT size × T_s
	= (100 + 1)/1800 μsec = 56.1 kHz
	Also, τ_{max} < 1800 μsec/10 ⇒ flat fading and reduced ISI.
	(adding guard period completely removes ISI)
	Thus, OFDM is 29% more bandwidth efficient for this example.

Note: If $\tau_{max} > T_s/10$, then channel has frequency-selective fading. $B_c = 1/5\,\tau_{max}$ = 20 kHz = coherence bandwidth for both the cases.

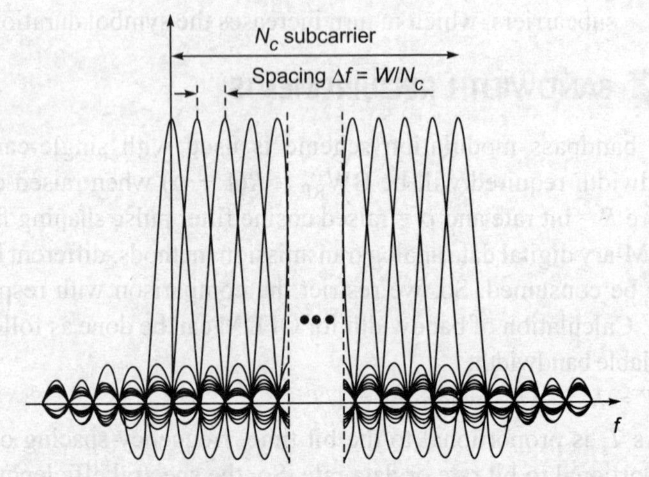

Fig. 9.26 Overlapping subcarriers with rectangular pulse shaping and orthogonality

description also. Thus, the power spectrum of the OFDM system decays as f^{-2}. In some cases, this is not sufficient and methods have been proposed to shape the spectrum. If a raised cosine pulse is used, the roll-off region also acts as a guard space. If the flat part is the OFDM symbol, including cyclic prefix, both ICI and ISI are avoided. The signal with this kind of pulse shaping is shown in Fig. 9.27, when it is compared with the rectangular pulse, spectrum response is as shown in Fig. 9.28. The overhead introduced by an extra guard band with a graceful roll-off can be a good investment for ISI elimination, since the spectrum falls much more quickly and reduces the interference to adjacent frequency bands. Choice of roll-off plays an important roll as far as spectral efficiency is concerned.

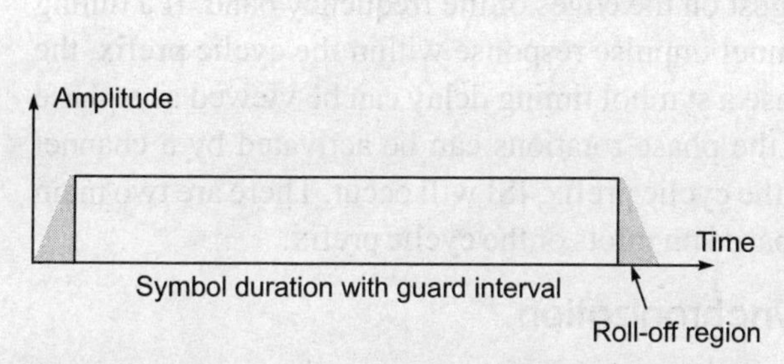

Fig. 9.27 Pulse shaping using raised-cosine function (The grey part of the signal shows the roll off.)

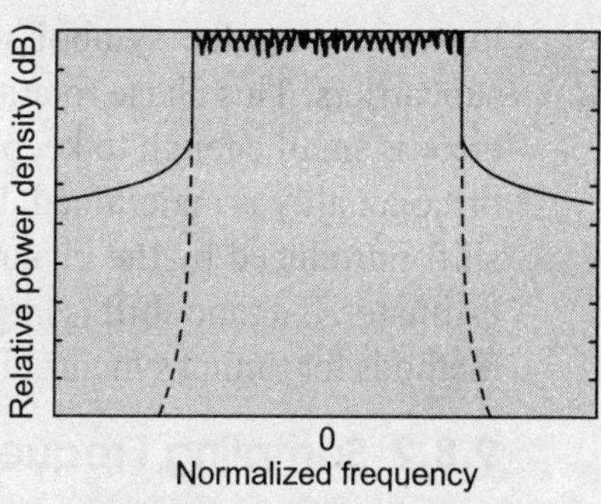

Fig. 9.28 Spectrum shape with rectangular pulse (solid lines) and raised cosine pulse (dashed lines)

9.7 WINDOWING IN OFDM SIGNAL AND SPECTRAL EFFICIENCY

If the overall spectrum of OFDM is concerned, because of the sinc shape of each narrowband channel, the side lobes with main lobe will appear. These are unwanted components and are to be eliminated from the final transmission bandwidth. This can be achieved by applying the windowing function to final OFDM baseband signal in which effect of all the subcarriers is present. For example, Hamming window, Hanning window, Bartlet window, Keiser window, etc. are defined by specific mathematical function. Passing OFDM baseband signal through the windowing block will perform the task of removing out of band components. Hence it can be considered as a shaping filter also. Dark lines in Fig. 9.29 represents the concept of windowing. This will allow designing next transmission bandwidth closer to the previous one and maintaining the orthogonality.

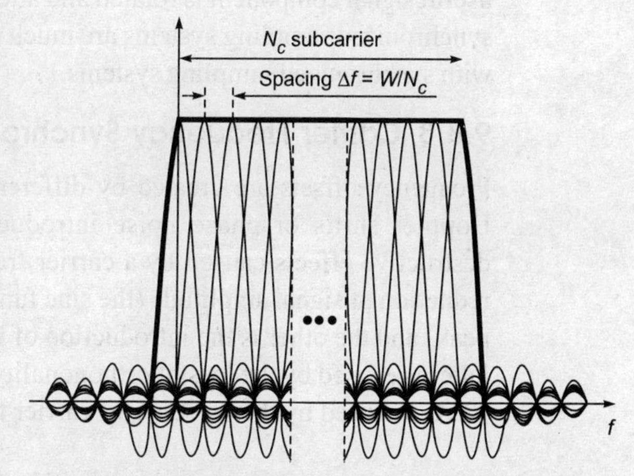

Fig. 9.29 Applying the windowing function to the OFDM baseband signal

9.8 SYNCHRONIZATION IN OFDM

One of the drawbacks in OFDM is that it is highly sensitive to synchronization errors, in particular, to frequency error. Hence, three synchronization problems arise—symbol, sampling frequency, and carrier frequency synchronization.

9.8.1 Timing Errors and Symbol Synchronization

A great deal of attention is given to symbol synchronization in OFDM system. However, by using a cyclic prefix, the timing requirements are relaxed somewhat. The objective is to know when the symbol starts. The timing offset gives rise to a phase rotation of subcarriers. This phase rotation is largest on the edges of the frequency band. If a timing error is small enough to keep the channel impulse response within the cyclic prefix, the orthogonality is maintained. In this case a symbol timing delay can be viewed as a phase shift introduced by the channel, and the phase rotations can be activated by a channel estimator. If a time shift is larger than the cyclic prefix, ISI will occur. There are two main methods for timing synchronization: based on pilots or the cyclic prefix.

9.8.2 Sampling Frequency Synchronization

The received continuous-time signal is sampled at the instants determined by the receiver clock. There are two types of methods of dealing with the mismatch in sampling frequency. In synchronized-sampling system, a timing algorithm controls a VCO in order to align the receiver clock with the transmitter clock. The other method is non-synchronized sampling, where the sampling rate remains fixed, which requires post processing in the digital domain. The effect of the clock frequency offset is twofold: the useful signal component is rotated and attenuated and in addition ICI is introduced. Non-synchronized sampling systems are much more sensitive to a frequency offset compared with synchronized sampling systems.

9.8.3 Carrier Frequency Synchronization

Frequency offsets are created by differences in oscillator in transmitter and receiver, Doppler shifts or phase noise introduced by non-linear channels. There are two destructive effects caused by a carrier frequency offset in OFDM systems. One is the reduction of signal amplitude (the sinc functions are shifted and no longer sampled at the peak) and the other is the introduction of ICI from the other carriers (see Fig. 9.30). The latter is caused by the loss of orthogonality between the subchannels. Degradation of the BER is caused by the presence of carrier frequency offset and carrier phase noise for an

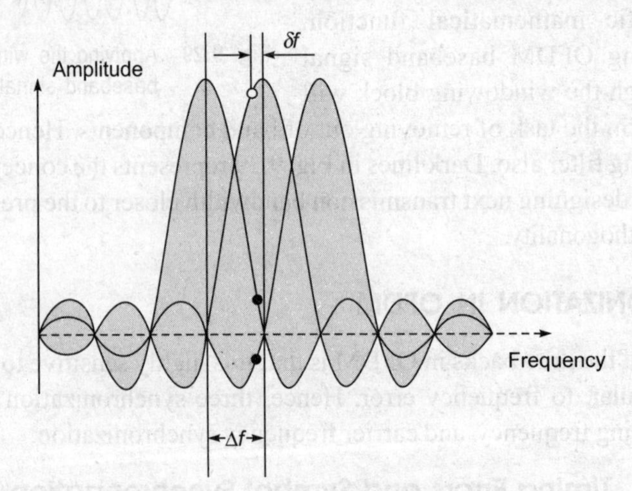

Fig. 9.30 Effects of a frequency offset δf: reduction in amplitude (o) and intercarrier interference (•)

AWGN channel. It is found that multi carrier system is much more sensitive than the single-carrier system. Denote the frequency offset by δf.

Several carrier synchronization schemes have been suggested, which can be divided into two categories—based on pilots and on the cyclic prefix.

It is interesting to note the relationship between time and frequency synchronization. If the frequency synchronization is a problem, it can be reduced by lowering the number of subcarriers, which will increase the subcarrier spacing. This will, however, increase the demands on the time synchronization, since the symbol length get shorter, that is, a larger relative timing error will occur. Thus, the synchronization in time and frequency are closely related to each other.

9.9 PILOT INSERTION IN OFDM TRANSMISSION AND CHANNEL ESTIMATION

Introduction to channel estimation technique is given in Chapter 6. Channel estimators will help in equalization at the receiver end. These usually need some kind of information as a reference and may be transmitted purposely or based on channel statistics as given in Table 6.3.

- A channel estimate is only a mathematical estimation of what is truly happening in the natural environment.
- Channel estimation
  - allows the receiver to approximate the effect of the channel on the signal to eliminate it,
  - is essential for removing inter symbol interference and noise, and
  - is used in diversity combining, maximum likelihood (ML) detection, angle of arrival estimation, etc.
- Pilot is basically a reference carrier/tone or reference signal/symbol that is known at the receiver end in terms of position or sequence/pattern and used for the channel estimation because it provides channel state information as it has undergone the most recent channel behaviour.

There are two main problems in designing channel estimators for wireless OFDM systems:

1. The arrangement of pilot information, where pilot means the reference signal used by both transmitters and receivers
2. The design of an estimator with both low complexity and good channel tracking ability

Coherent modulation allows arbitrary signal constellations, but efficient channel estimation strategies are required for coherent detection and decoding. Time variant and frequency-selective fading channels present a severe challenge to the designer of a wireless communication system. An OFDM receiver plays dual role to tackle this problem: phase correction by channel estimation or equalization and demodulation of the signal with channel decoding. Several choices are possible for the implementation of a receiver depending on the modeling of the channel and complexity invested in each task. The nature of the OFDM enables powerful estimation and equalization techniques.

In a communication system, the estimation is generally performed using the received signal and applying some knowledge about the transmitted sequence/known symbols or channel statistics, e.g. block-based estimation methods (explained later on) use a batch of received symbols to estimate the average channel in an interval where the channel is treated as time invariant. We can say that a snapshot of the channel is obtained from each estimation interval. The continuous impulse response of the channel is thus sampled, not only in delay to form a discrete FIR channel but also in time, resulting in snapshots of the time-varying channel. The estimated samples may have an estimation error that depends on the measurement noise, the transmitted symbols, the properties of the channel, the estimation algorithm and the deviation from a time invariant channel in the estimation interval.

It is well understood that the phase noise effect on OFDM signal reception consists of two components: an ICI term that can be modelled as additional Gaussian noise, and a common phase error (CPE) that rotates all the sub carriers equally. The pilot signal will also undergo the same effects. While the ICI is difficult to remove due to its noise like characteristics, the CPE can be easily corrected by estimating such rotation through continuous pilot tones embedded in OFDM symbols.

However, some CPE compensation schemes assume *known channel states* or at least static channels, allowing the CPE to be separated from actual channel effects. Such an assumption does not hold true for time varying channels. Reversely, the Wiener filtering channel estimation approaches may not be directly applicable in the presence of severe phase noise (use of Kalman filter may be more suitable). Only joint consideration of CPE and channel estimation can address these problems, although in practice the two effects are often dealt with separately. The effect of phase noise on channel estimation should be first analysed, based on which the existing time domain LSE or MMSE channel estimators can be modified to accommodate the CPE. A new CPE estimator suitable for fast fading channels that requires no explicit CSI is then developed based on continuous pilot tones.

9.9.1 Basics of Channel Impulse Response Estimation

In general, channel processes are WSSU.

Conceptually, on stationary channel, if an impulse is transmitted, then at the receiver multiple delayed versions of impulses will be received at different instants of time that are non correlated and with reducing amplitudes with time. They are just like delayed samples and hence a differential equation with coefficient values can be correlated with this concept.

In this work, the radio channel is described by a discrete time transfer function, that is, a discrete time impulse response. This impulse response $h_e(n)$ is time varying and the goal of the identification procedure is to estimate the time-dependent parameters in $\{h_e(n)\}$ as accurately as possible. For the linear time-varying discrete systems (see Appendix A), the received time domain OFDM signal $y(n)$ is a function of the transmitted signal $x(n)$, the channel transfer function and AWGN $w(n)$. It can be expressed as

$$y(n) = h_e(n) * x(n) + w(n), \quad 0 \le n \le N-1 \tag{9.20a}$$

or

$$y(n) = \sum_{k=0}^{\infty} h_e(n)x(n-k) + w(n) \tag{9.20b}$$

where '*' in Eq. (9.20a) denotes convolution process, N is the FFT bin size/period of N samples, and $h_e(n)$ is the estimated channel impulse response (maybe with interpolation to make N sample points). This equation represents one channel snapshot. The whole time domain scenario can be mapped into frequency domain. The convolution in time domain represents multiplication in frequency domain.

Over short time intervals (a batch of data), the time-varying channel can be approximately described by a time invariant impulse response. A further simplification is to assume that the transfer function, $\{h_e(n)\}$, can be described by a time invariant FIR model of length M in each time interval. The model is then modified to

$$y(n) = \sum_{k=0}^{M-1} h_e(n)\, x(n-k) + w(n) \tag{9.20c}$$

where M has to be chosen large enough to encompass all significant contributing paths. By expressing using in the unit delay operator q^{-1}, $(q^{-1}x(n) = x(n-1))$, we obtain

$$y(n) = H(q^{-1})\, x(n) + w(n) \tag{9.20d}$$

where $H(q^{-1}) = \sum_k h_e(q^{-1})$. This model is a valid approximation for time segments that are short related to the channel variation.

For the time-varying channels, this whole scenario will become dynamic and in each snapshot/state, the CIR will be different. The next state of the channel can be estimated by previous impulse responses. The channel model for time-varying channel can be approximated by autoregressive (AR) model also (explained in Chapter 6). For the next state estimation based on the present state condition, one can write

$$h(n+1) = a\, h(n) + w(n) \tag{9.21}$$

where a is the coefficient value to estimate the next channel state. Similarly, many delayed components will make the differential equation with a's coefficients.

One method of sending additional information for channel estimation purpose is sending the pilots, may be inserted in time domain or in frequency domain. A fading channel requires constant tracking, so pilot information is to be transmitted more or less continuously.

As far as multicarrier system like OFDM is concerned, carriers are transmitted maintaining the orthogonality among them. Over the channel, there may be a little shift in the frequency that is called offset as mentioned previously and it is to be corrected at the receiver end. Pilot carriers help in this. Hence, the pilot information is unavoidable part in the OFDM system as it serves dual purpose.

There are two main problems in the design of pilot-based channel estimators for OFDM system. The first problem concerns with the choice of how pilot information should be transmitted. The second problem is the design of an estimator with both low complexity and good channel-tracking ability. These two problems are interconnected, since the performance of the estimator depends on how pilot information is transmitted.

There are many different ways suggested for pilot information transmission. Especially in OFDM, when the lattice type of carrier assignment technique is used, an efficient way of allowing a continuously updated channel estimate is achieved by transmitting pilot symbols or pilot carriers along with data at certain locations of the OFDM time frequency lattice. It is shown in Fig. 9.31, in which two methods are shown, comb-type and block-type pilot arrangements. In the comb-type method, special pilot symbols are assigned to few pilot carriers dedicatedly where pilot carriers are spaced equally along with the data subcarriers, while in block type method each OFDM symbol block assigned to a subcarrier is having some time duration for sending training sequences.

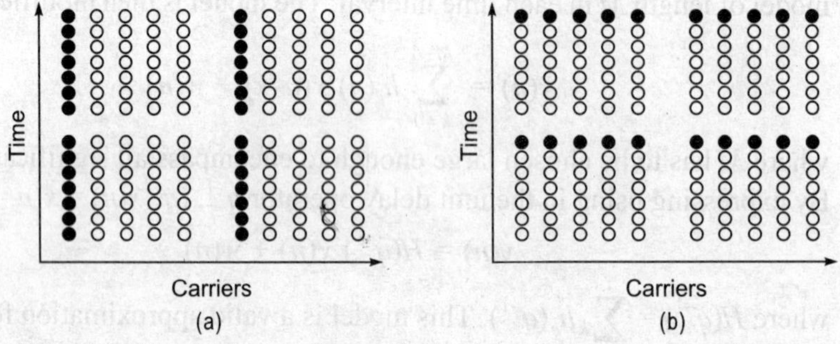

Fig. 9.31 Pilot carrier insertion methods: (a) comb-type pilot insertion and (b) block-type pilot insertion

In the case of time varying channels, the pilot signal should be repeated frequently. The spacing between pilot signals in time and frequency domain depends on coherence time and coherence bandwidth of the channel. One can reduce the pilot signal overhead by using a pilot signal with a maximum distance of less than the coherence time and coherence bandwidths. Then, by using time and frequency interpolation, the CIR and frequency response of the channel can be calculated.

If f_d is Doppler spread and τ_{max} is delay spread, a suitable choice for pilot spacing in time N_{pt} and in frequency N_{pf} is as follows:

$$N_{pt} \approx 1/f_d \,.\, T_s \qquad (9.22)$$

$$N_{pf} \approx 1/\Delta f \,.\, \tau_{max} \qquad (9.23)$$

where Δf is subcarrier bandwidth and T_s is symbol time.

The method of channel estimation implied by the frame structure and scattered pilots is, channel estimation via interpolation. The basic principle is depicted in Figures 9.32(a) and (b).

Embedded into the data stream are training symbols [depicted as arrows in Fig. 9.32(b)] that can be used to obtain snapshots of the channel transfer function $\hat{H}_{n,l}$ (*n*th column and *l*th row of the lattice).

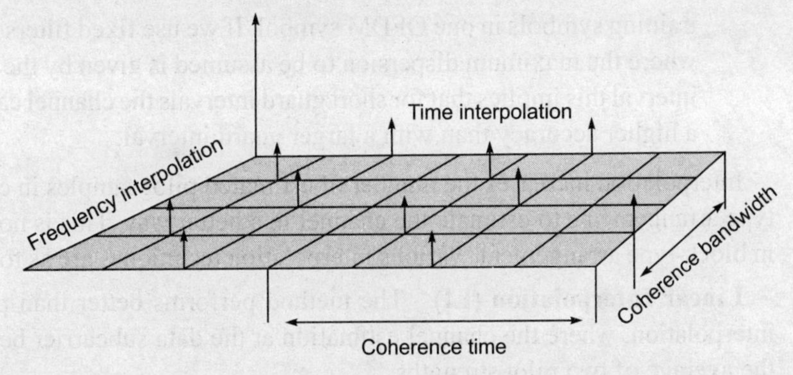

(a) Scattered pilot positioning in time and frequency domain implies interpolation requirement

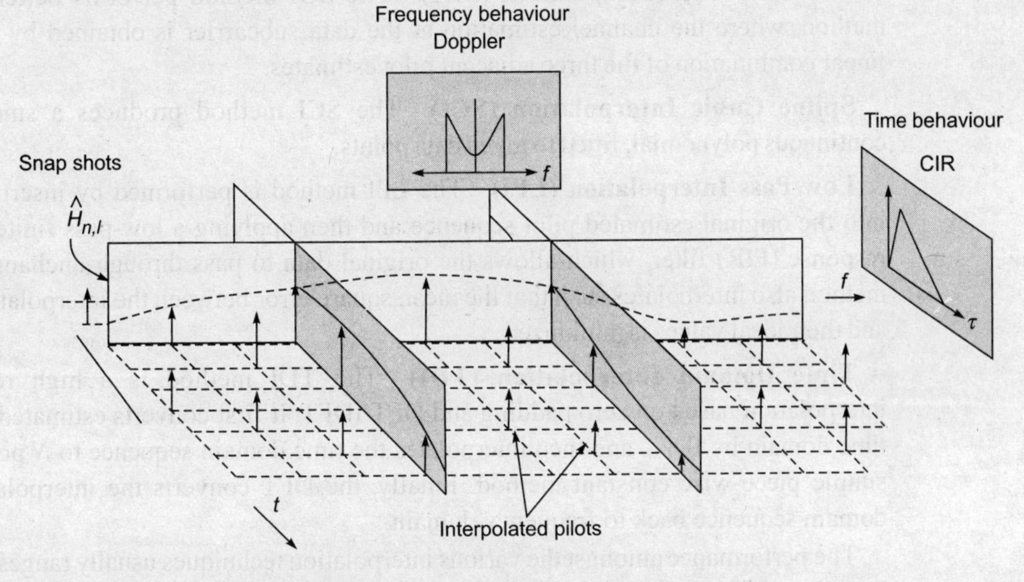

(b) Interpolation estimates better channel impulse response

Fig. 9.32 Interpolation technique in channel estimation

The values of the CIR coefficient in between the samples can then be obtained via interpolation procedure. Generally, we have a two-dimensional interpolation problem. Fortunately, the problem can be separated into interpolation in time and in frequency. The most critical task is the design of the interpolation filters used. Both interpolations must agree with the sampling theorem:

- The interpolation in time is bandlimited by the time-variant behaviour of the channel. This is caused by a movement of the receiver and by uncompensated synchronization errors. The maximum allowable bandwidth of these disturbances is determined by the number of training symbols in one subcarrier.
- Due to the duality of time and frequency, the interpolation in frequency is bandlimited by the length of the CIR. The maximum allowable CIR length thus is not only determined by the length of the guard interval but also by the number of

training symbols in one OFDM symbol. If we use fixed filters for implementation where the maximum dispersion to be assumed is given by the length of the guard interval this implies that for short guard intervals the channel can be estimated with a higher accuracy than with a larger guard interval.

Interpolation increases the number of estimated pilot samples in comb and scattered-type arrangements to estimate the channel in a better way. This is not very critical issue in block-type arrangement. Various interpolation techniques are as follows.

Linear Interpolation (LI) The method performs better than piece-wise constant interpolation, where the channel estimation at the data subcarrier between two pilots is the average of two pilot strengths.

Second Order Interpolation (SOI) The SOI method performs better than LI method, where the channel estimation at the data subcarrier is obtained by weighted linear combination of the three adjacent pilot estimates.

Spline Cubic Interpolation (SCI) The SCI method produces a smooth and continuous polynomial, fitted to given data points.

Low Pass Interpolation (LPI) The LPI method is performed by inserting zeros into the original estimated pilot sequence and then applying a low-pass finite impulse response (FIR) filter, which allows the original data to pass through unchanged. This method also interpolates such that the mean square error between the interpolated points and their ideal values is minimized.

Time Domain Interpolation (TDI) The TDI method is a high resolution interpolation based on zero padding and DFT/IDFT. It first converts estimated pilots to time domain by IDFT and then interpolates the time domain sequence to N points with simple piece-wise constant method. Finally, the DFT converts the interpolated time domain sequence back to frequency domain.

The performance amongst the various interpolation techniques usually ranges from the best to worst as follows: LPI, SCI, TDI, SOI, and LI. The LPI and SCI best performs almost same in the low and middle SNR scenario, while LPI outperforms SCI at the higher SNR scenario. In terms of complexity, TDI, LPI and SCI have roughly the same computational complexity, while SOI and LI have less complexity.

The use of pilot signals for the purpose of channel estimation in multi carrier systems does not only consume bandwidth but also signal power that could otherwise be invested in the information symbols to be transmitted. There must be optimal transmitter power distribution between the pilot and information signals maintaining the good performance of the system. There will be some impact of pilot power to signal power ratio on the channel estimation. Indirectly, speaking power invested in the pilot symbols is power deducted from the information symbols. The optimal pilot to signal power ratio (PSR) varies with respect to the ratio of pilots to subcarriers. It is also evident that the optimal performance can be obtained irrespective of the ratio of pilots to subcarriers so long as the Nyquist criterion is met and the optimal PSR is selected.

9.10 AMPLITUDE LIMITATIONS IN OFDM

An OFDM signal is the sum of many subcarrier signals that are modulated independently by different modulation symbols. As the amplitude of the OFDM signal is a stochastic process, according to the central limiting theorem, it obeys a complex Gaussian distribution if the number of subcarriers is large. Thus, OFDM signals have a very large peak to average power ratio. In the transmitter, the maximal output power of the amplifier therefore limits the peak amplitude of the signal. This effect produces interference both within the OFDM band and in adjacent frequency bands. The amplitude threshold is shown in Fig. 9.33; few peaks of OFDM signal cross that limit.

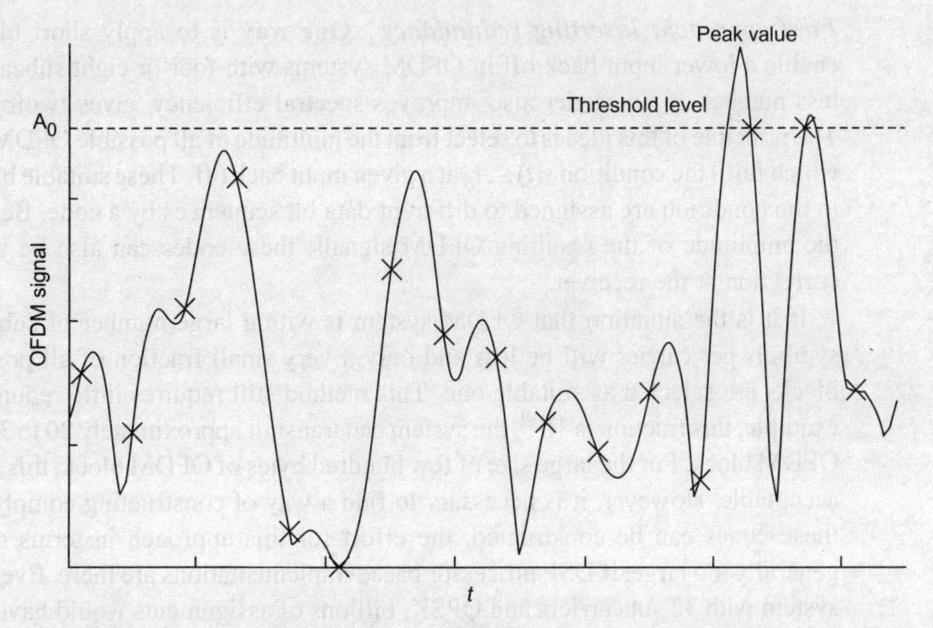

Fig. 9.33 Continuous time OFDM baseband signal exceeds the amplitude threshold results in clipping by amplifier

Hence, we propose a modification of the OFDM signal to remove the signal peaks that exceed a given amplitude threshold. The out-of-band interference produced by these corrections can be kept within clearly defined limits while minimum interference within the OFDM band is obtained.

Approaches to Remove the Amplitude Limitations

In most of the literatures about amplitude limitation of OFDM signals, it is assumed that the limitation can be achieved by pre distortion of the signal, that is, the amplifier behaves like an ideal limiter. This means that the signal is amplified linearly up to a maximum input amplitude A_0, and larger amplitudes are limited to A_0. The input power of the amplifier as compared to the threshold is described by the input back off $IBO = 10 \log (A_0^2/P_s)$, where P_s is the average power of the baseband OFDM signal.

Two approaches can be investigated, which ensure that the transmitted OFDM signal $s(t)$ does not exceed the amplitude A_0 if a given input back off is used. The first method makes use of redundancy in such a way that any data sequence leads to an OFDM signal with $s(t) \leq A_0$ or that at least the probability of higher amplitude peaks is greatly reduced. This approach does not result in interference of OFDM signal. [Note: There exist genetic algorithms by which low-energy pilot symbols are generated with special features.] In the second approach, the OFDM signal is manipulated by a correcting function that eliminates the amplitude peaks. The out of band interference caused by the correcting function is zero or negligible. However in any case the BER performance degrades and hence interference of the OFDM signal itself is tolerated to a certain extent.

First approach: inserting redundancy **One way** is to apply short block codes to enable a lower input back off in OFDM systems with four or eight subcarriers. Use of less number of subcarrier also improves spectral efficiency, gives twofold advantage. The principle of this idea is to select from the multitude of all possible OFDM blocks those which fulfil the condition $s(t) \leq A_0$ at a given input back off. These suitable blocks meeting to the condition are assigned to different data bit sequences by a code. Beyond limiting the amplitude of the resulting OFDM signals, these codes can also be used for error correction at the receiver.

If it is the situation that OFDM system is with a large number of subcarriers, then symbols per carrier will be less and only a very small fraction of all possible OFDM blocks are selected as suitable one. This method still requires little redundancy. If, for example, this fraction is 10^{-9}, the system can transmit approximately 20 to 30 bits less per OFDM block. For the large size of few hundred bytes of OFDM block, this figure may be acceptable. However, it is necessary to find a way of constructing complying codes. If these codes can be constructed, the effort for this approach in terms of memory is generally too large, if DSP processor based implementations are there. Even for a simple system with 32 subcarriers and QPSK, billions of assignments would have to be stored. For this reason, practical application of this scheme is limited to systems with very few subcarriers.

Second way is the same data sequence can be represented by several different OFDM blocks. The transmitter generates all possible signals corresponding to a data sequence and chooses the most suitable one for transmission. The receiver must additionally be told which of the signals has been chosen. This can be achieved with a little redundancy. If differential modulation is applied between adjacent subcarriers, the receiver does not even need any side information. However, in this case, on several subcarriers, reference symbols are transmitted for the differential demodulation. This scheme allows us, for example to decrease the input backoff from 12 dB to 10 dB at the same level of out-of-band interference.

Third way is suggested in some literature that realizes an OFDM transmission with a constant envelope using 50% redundancy. In this scheme, instead of one OFDM block, two blocks are transmitted, calculated from $s(t)$. However, this calculation is non-linear and causes out-of-band interference. The objective of this approach is not to avoid out-of-band interference but to avoid interference of the OFDM signal.

Second approach: correcting the OFDM signal In this approach, the OFDM signal is corrected with suitable functions to avoid out-of-band interference but it tolerates interference of the OFDM signal itself. In the simplest case, the sampled signal is limited to the amplitude A_0. This method is termed as clipping or *clip compress*. Clipping does not cause out-of-band interference if $s(t)$ is not over sampled. However, without over-sampling, the analog signal after the D/A conversion will exceed the amplitude threshold as shown in Fig. 9.32 in spite of the clipping applied. This effect is to be considered. Furthermore, the OFDM signal must be filtered because of the rectangular modulation pulse. For both reasons, over sampling of the signal is necessary. The proposal is made to apply clipping to the over sampled signal that causes out of band interference. This interference is taken care of by a FIR filter that also removes the side lobes of the modulation pulse. However, the filter leads to new amplitude peaks in the signal, but after all, the peak to average power ratio of the signal is reduced by this method.

Different corrective functions can be chosen. They may be multiplicative corrections or additive corrections, for example, $k(t)$ is some correcting function. If the signal exceeds the amplitude threshold A_0 at the times t_n, then the corrected signal $c(t)$ with multiplicative correction can be written as

$$c(t) = s(t)^* \, k(t) \tag{9.24}$$

$$k(t) = 1 - \sum_n A_n g(t - t_n) \tag{9.25}$$

$$g(t) = \exp(-t^2/2\sigma^2) \tag{9.26}$$

$$A_n = \frac{|s(t_n)| - A_0}{|s(t_n)|} = \text{Amplitude of } n\text{th sample} \tag{9.27}$$

Thus, the signal is attenuated by a Gaussian function at all positions where it has high amplitude peaks. The spectrum $C(f)$ of the corrected signal is

$$C(f) = S(f) \times K(f) \tag{9.28}$$

$$= S(f) - S(f) \times k \tag{9.29}$$

where

$$k = \left[\exp\left(-f^2/2\sigma_f^2 \sum_n B_n \times \exp(j2\pi t_n f \right) \right] \tag{9.30}$$

where B_n represents the coefficient values of the corrective function in frequency domain and σ^2 is the variance of the Gaussian function $g(t)$.

The correction broadens the signal spectrum by the width of a Gaussian function with the variance $\sigma_f^2 = 1/2\pi\sigma^2$. The Gaussian function used for the correction should be narrow in the time domain so that only small interval of the signal is attenuated. It should also be narrow in the frequency domain so that the signal spectrum is broadened as little as possible.

With this scheme, we can limit the amplitude of the over sampled signal without causing out of band interference, except in a narrow frequency band adjacent to the OFDM band. However, if many amplitude peaks have to be corrected, the entire signal is attenuated and the peak to average power ratio of the signal cannot be improved beyond a certain limit. This scheme of correcting the OFDM signal can be realized for any number of sub carriers and it does not need any redundancy. It causes interference of the OFDM signal, but this is of secondary importance in a fading environment in which OFDM is typically applied. The polar plot shown in Fig. 9.34 gives the typical results of applying clip compress. Here QPSK mapping is used for modulation.

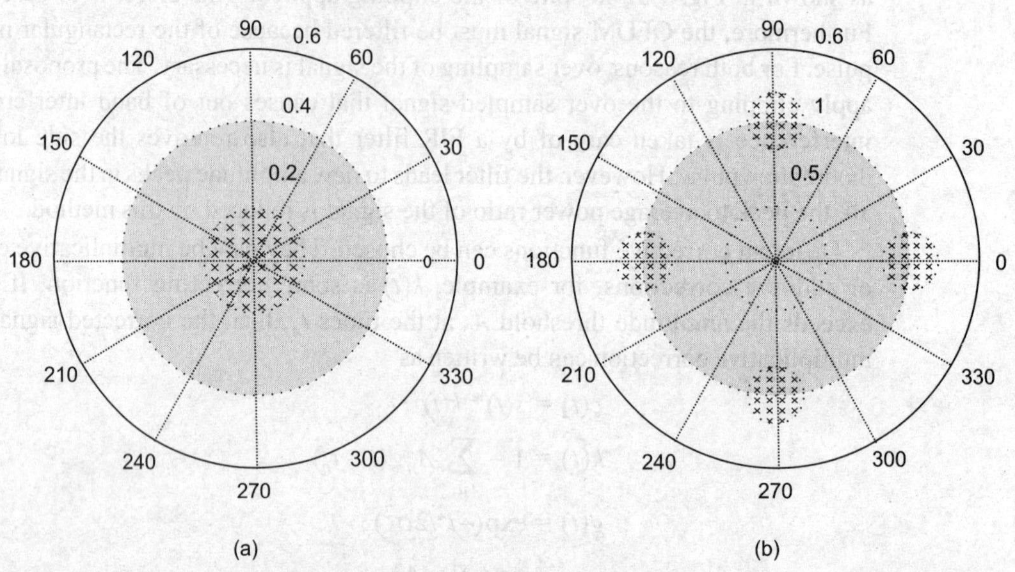

Fig. 9.34 Polar plots showing received phases for OFDM signal after clip compress is applied: (a) with maximum clip compression—maximum phase error and maximum bit error and (b) with tolerable clip compress applied—minimum phase error and minimum bit error

9.11 FFT POINTS SELECTION CONSTRAINTS IN OFDM

As far as hardware implementation is concerned, the more the FFT points the more the processing power consumed that can be taken care of by the fast processors but in general more number of FFT points increases the resolution of the OFDM signal that gives better results, especially if the SNR is of the order of 12 dB.

General-purpose DSP-based implementation

There are several processing stages required to generate and receive an OFDM signal. However, most of the processing is required in performing the FFT and IFFT.

The complexity of performing an FFT is dependent on the size of the FFT. The larger the FFT, the greater the number of calculations required. However, since the symbol period is longer, the increased processing required is less than the straight increase in processing to perform a single FFT. Table 9.2 shows the number of calculations required for an FFT (radix–2) of size N, and also the relative processing for various FFT sizes. It can be seen that because the symbol period increases with a larger FFT that the extra processing required is minimal. The equations for the number of calculations for an N-point FFT according to Fig. 9.35 are as follows:

$$\text{Number of complex multiplications} = M = \frac{N}{2}\log_2 N = \frac{N\log_{10} N}{2\log_{10} 2} \tag{9.31}$$

$$\text{Number of complex additions} = A = N\frac{\log_{10} N}{\log_{10} 2} \tag{9.32}$$

$$\text{Total number of calculations} = M + A \tag{9.33}$$

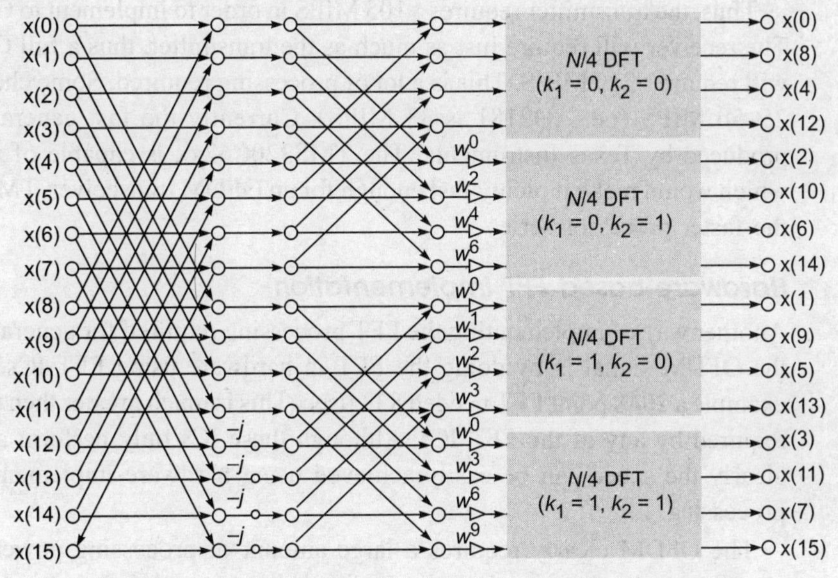

Fig. 9.35 Butterfly diagram for FFT implementation for 16-point

Table 9.2 FFT calculations

Size FFT(N)	Total Number of Complex Calculations
32	240
64	576
128	1344
256	3072
512	6912
1024	15360
2048	33792
4096	73728

The processing efficiency of a DSP processor depends on the architecture of the processor. However, for most single instruction DSPs, the number of cycles required to calculate an FFT is twice the total number of calculations shown in Table 9.1. This is due to complex calculations requiring two operations per calculation.

Required Processing Power

From Table 9.2, the number of complex calculations required for a 2048 point FFT is 33,792. The maximum time that can be taken in performing the calculation is once every symbol, thus once every 833 μs. If we assume that the processor used requires two instructions to perform a single complex calculation, and that there is an overhead of 30% for scheduling of tasks and other processing, the minimum processing power required for this is then

$$\text{MIPS} = \frac{33{,}792 \times 2}{833 \times 10^{-6}} \times 1.3 \times 10^{-6} = 105$$

Thus, the transmitter requires >105 MIPS in order to implement to OFDM transmitter. The receiver will require just as much as the transmitter, thus a full OFDM transceiver will require >210 MIPS. This is a lot of processing required. Some cheap DSPs are only 25–50 MIPS (i.e. AD2181 is 33 MIPS). Currently the fast general purpose DSP is produced by Texas Instruments. The TMS320C62xx is capable of up to 1600 MIPS which would make it plenty fast enough for an OFDM transceiver. TMS 32067xx is also the faster processor series.

Hardware-based FFT Implementation

Another way of implementing the FFT processing required for generating and receiving the OFDM signal is by doing the FFT in hardware using FFT ICs. For the previous example a 2048 point FFT needed 833 μsec. This is much greater then the execution time required by any of the FFT ICs. Although these ICs only perform a 1024 point FFT, clearly the speed can be easily achieved using hardware implementation of the FFT processing.

The OFDM clearly requires a large amount of processing power. However, since computer technology is advancing so fast, this may not become a problem in future.

9.12 CDMA VS OFDM

This topic investigates the effectiveness of OFDM as a modulation technique for wireless radio application. Most third generation mobile phone systems are proposing to use DS-CDMA as their modulation technique. For this reason, CDMA is also investigated so that the performance of CDMA could be compared with OFDM. At the end, it is concluded that the good features of both the modulation schemes can be combined in an intelligent way to get the best modulation schemes as a solution for wireless communication high speed requirement, channel problems and increased number of users. The CDMA and OFDM both are wideband wireless digital communication systems in general.

As explained in Chapter 8, CDMA is based on spread spectrum technique that uses neither frequency channels nor time slots. With CDMA, the narrowband messages (typically digitized voice data) are multiplied by a large bandwidth signal that is a unique pseudo random noise code (PN code). All users in a CDMA system use the same frequency band and transmit simultaneously with different codes. The transmitted signal is recovered by correlating the received signal with the PN code used by the transmitter.

Orthogonal frequency division multiplexing is a multicarrier modulation transmission technique, which divides the available spectrum into many carriers, each one being modulated by a low rate data stream. The OFDM can be similar to FDMA in that the multiple user access is achieved by subdividing the available bandwidth into multiple channels, allocated to users. However, OFDM uses the spectrum much more efficiently by spacing the channels much closer together. This is achieved by making all the carriers orthogonal to one another, preventing ICI. (Multiple access schemes are covered in Chapter 10.)

It is found that OFDM performs extremely well compared with CDMA, providing a very high tolerance to multipath delay spread, peak power clipping and channel noise. OFDM is found to have total immunity to multipath delay spread provided the reflection time is less than the guard period used in the OFDM signal. In a typical system a delay spread of up to 100 msec could be tolerated, corresponding to multipath reflections of 30 km. Also OFDM is spectral efficient and supports high bit rates with appropriate selection of M-PSK mapping.

Considering cellular base, CDMA capacity is limited by multiple access interference (MAI), which results from the imperfection of autocorrelation and cross correlation characteristics of spreading codes. Although, zero cross correlated orthogonal codes could result in no MAI in flat fading channels, the orthogonality will not be guaranteed in frequency-selective fading channels because of interchip interference, which wili cause MAI and degrade the system performance. Though CDMA is very good as far as security aspects are concerned, it is found to perform poorly in a single cellular system. Typically, with each cell, only 10–20 simultaneous users are allowed in a cell compared to 128 users for OFDM.

Some other performance comparison is discussed below.

Multipath Delay Spread Performance

To eliminate ISI, the guard period insertion is done in OFDM. For a delay spread that is longer than the effective guard period, BER rises rapidly due to ISI. The maximum BER will occur when the delay spread is greater than the symbol time (see Fig. 9.36).

As shown in Fig. 9.37, CDMA is inherently tolerant to multipath delay spread signals as any signal delayed by more than one chip time becomes uncorrelated to the PN code used to decode the signal. This results in the multipath simply appearing as noise. This noise lead to an increase in the amount of interference seen by each user subjected to the multipath and thus increases the received BER.

Peak Power Clipping Performance

If a transmission technique is tolerant to peak power clipping (allows signal power to be clipped), reduces the peak to RMS signal power ratio thus allowing the signal power to be increased for the same sized transmitter. The transmitted OFDM signal could be heavily clipped with little effect on the received BER. In fact, the signal could be clipped by up to 9 dB without a significant increase in the BER. This means that the signal is highly resistant to clipping distortions caused by the power amplifier used in transmitting the signal. It also means that the signal can be purposely clipped by up to 6 dB so that the

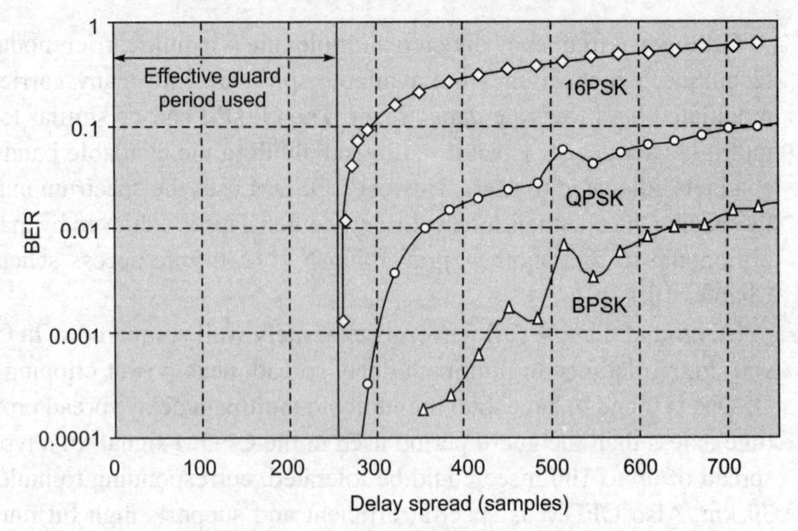

Fig. 9.36 Delay spread tolerance of OFDM

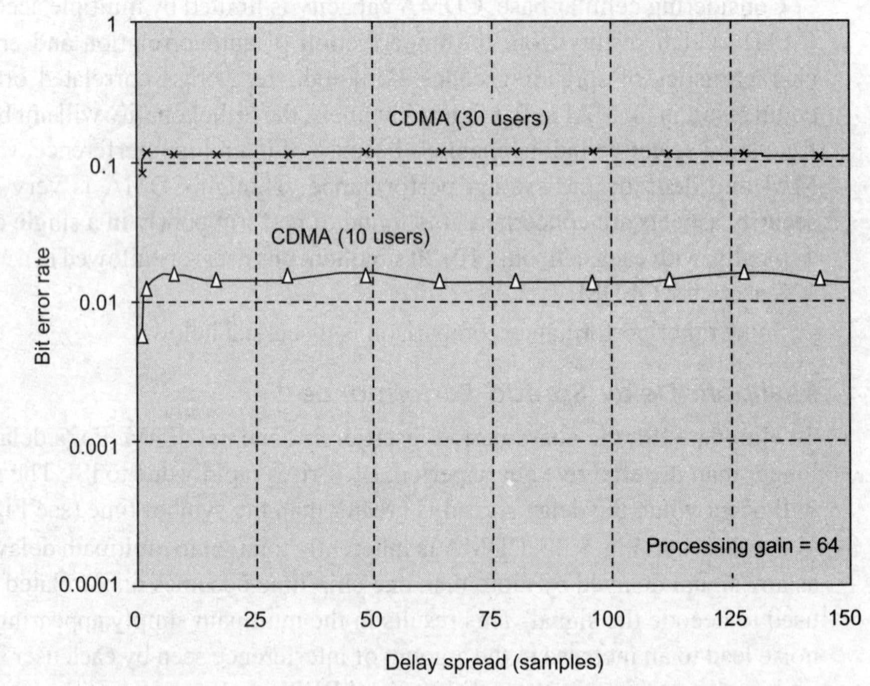

Fig. 9.37 Effect of multipath delay spread on the reverse link of a CDMA system.

peak to RMS ratio can be reduced allowing an increased transmitted power. The performance is shown in Fig. 9.38.

In CDMA, according to Fig. 9.39, for the reverse link the BER starts high initially due to inter user interference. The peak power clipping of the signal has little effect on the reverse link because the extra noise due to the distortion is not very high with the inter-user interference. Moreover, any added noise is reduced by the processing gain of the system.

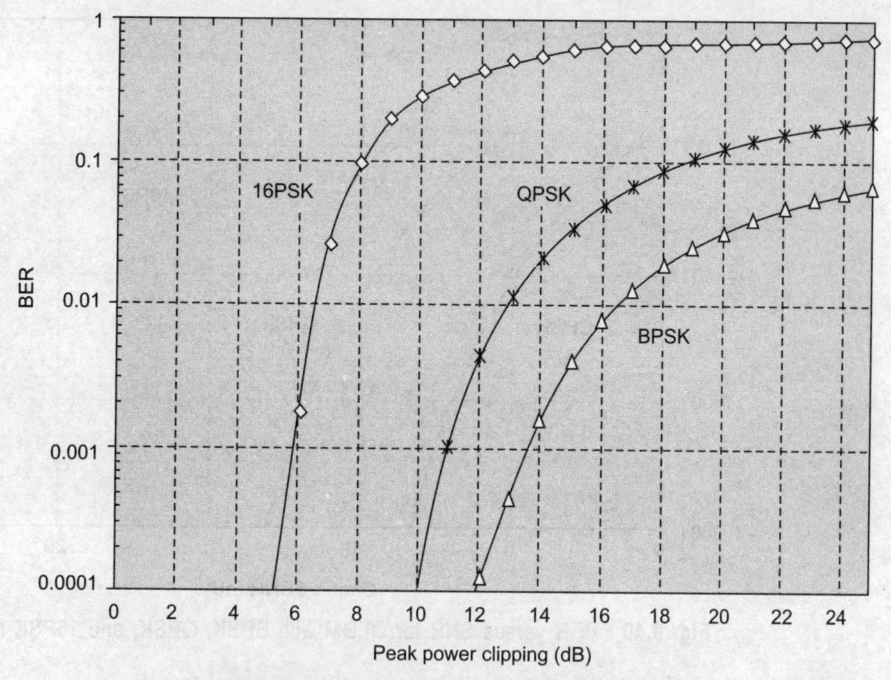

Fig. 9.38 Effect of peak power clipping for OFDM

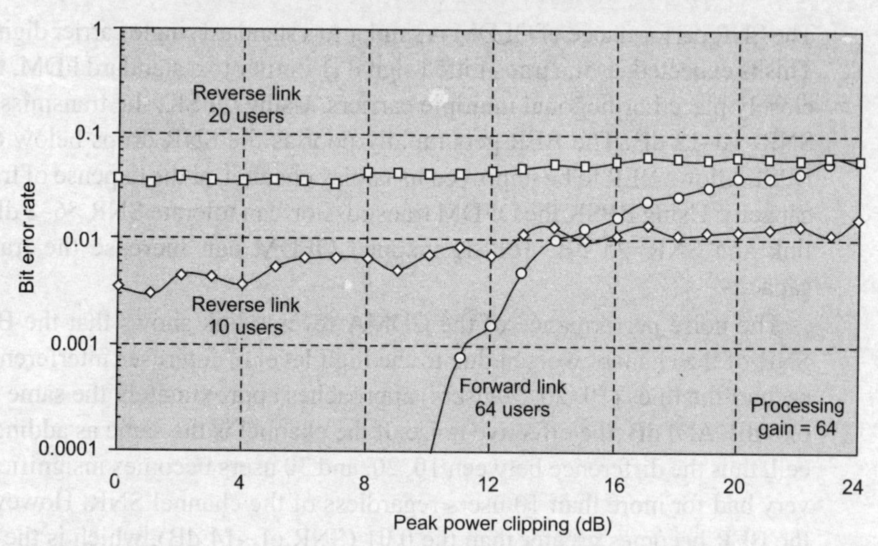

Fig. 9.39 Effect of peak power clipping on the BER for the forward and reverse links of CDMA

Peak power clipping for the reverse link is also likely to be small, as clipping would only ever occur due to distortion in the base station receiver, as this is the only point where all the signals are combined. A well-designed receiver is unlikely to cause significant clipping of the signal. The forward link result is more important as significant clipping of the transmitted signal could occur at the base station transmitter. The result for the forward link is completely different to the reverse link. The peak power clipping tolerance of the forward link is very similar to OFDM. The BER is low for clipping of less than 10 dB.

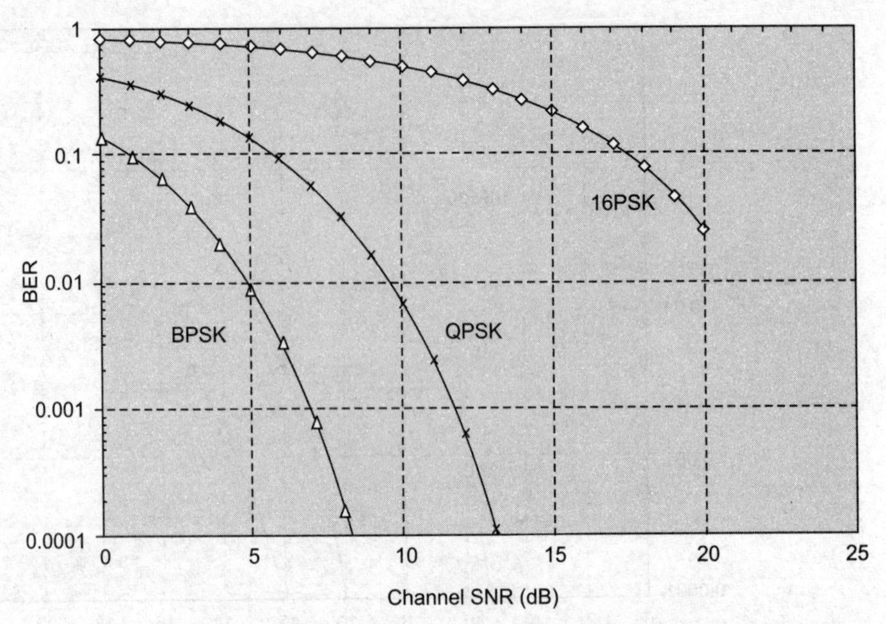

Fig. 9.40 BER versus SNR for OFDM with BPSK, QPSK, and 16PSK mapping

Gaussian Noise Tolerance

The SNR performance of OFDM is similar to a standard single carrier digital transmission. This is expected, as the transmitted signal is similar to a standard FDM, though it is with closely placed orthogonal multiple carriers. Using QPSK, the transmission can tolerate SNR>10–12 dB. The BER gets rapidly poor as the SNR drops below 6 dB. However, BPSK allows BER to be improved in a noisy channel, at the expense of transmission data capacity. Using BPSK the OFDM transmission can tolerate SNR >6–8 dB. If a low noise link and SNR>25 dB, 16PSK mapped OFDM can increase the transmission data capacity.

The noise performance of the CDMA reverse link shows that the BER rises as the SNR of the channel worsen due to the high level of interuser interference. The BER of each of the lines (10, 20, 30 users) approaches approximately the same BER at an SNR of 0 dB. At 0 dB, the effective noise of the channel is the same as adding 60 users to the cell, thus the difference between 10, 20, and 30 users becomes insignificant. The BER is very bad for more than 10 users regardless of the channel SNR. However, for 10 users the BER becomes greater than the 0.01 (SNR of ~14 dB), which is the maximum BER that can be normally tolerated for voice communications (refer to Fig. 9.41).

The problem associated with OFDM is frequency-selective fading, which results in carriers being heavily attenuated due to destructive interference at the receiver. This may result in the carriers being lost in the noise. Another weak point is that it is very sensitive to frequency and phase errors between the transmitter and the receiver. The main sources of these errors are frequency stability problems, phase noise of the transmitter, and any frequency offset errors between the transmitter and the receiver due to moving mobiles and Doppler effects. This may lead to error in finding the start time of the FFT

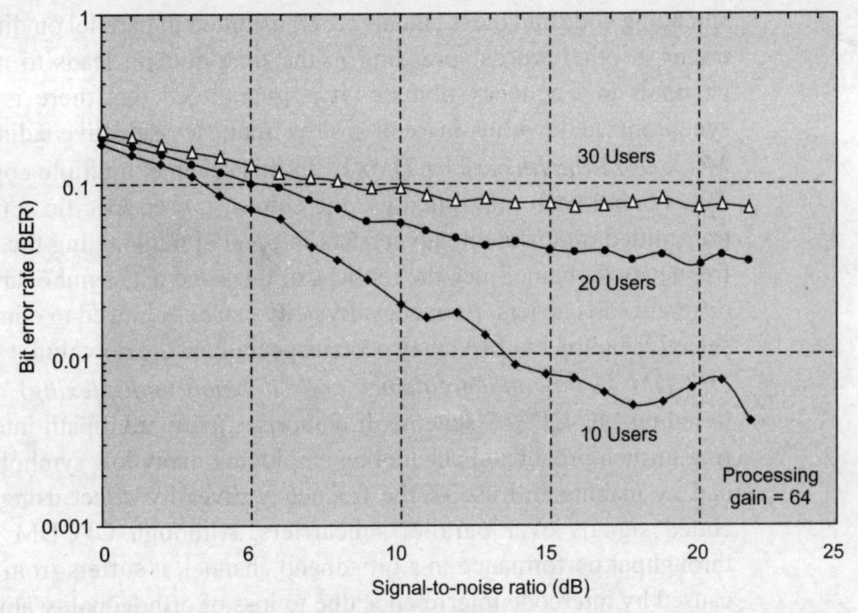

Fig. 9.41 BER versus the Gaussian channel SNR for the reverse link of a CDMA system

symbol. This problem can be overcome by synchronizing the clocks between the transmitter and the receiver, by designing the system appropriately or by reducing the number of carriers used. The noise performance of OFDM is found dependent on the modulation technique used for mapping each carrier of the signal. The performance of the OFDM signal is found to be the same as for a single carrier system, using the same modulation mapping. The minimum SNR required for BPSK is ~8 dB, where for QPSK ~12 dB and for 16PSK ~25 dB are the SNR requirements for better performance same as for OFDM. The OFDM signal is contaminated by non-linear distortion of transmitter power amplifier, because it is combined amplitude frequency modulation and it is necessary to maintain the linearity.

The main problem associated with CDMA is near–far problem, limited users in a cell and complex rake receiver design.

9.13 HYBRID OFDM

After studying the various advantages and disadvantages of both CDMA and OFDM systems, the good issues of both the systems can be combined as the OFDM-CDM transmission technique, where the information to be transmitted is spread over several OFDM subcarriers. Arbitrary orthogonal codes can be used for spreading. The effect of this additional measure is that narrowband fading is avoided. This can be seen as frequency diversity effect. Thus, at high code rates, an OFDM-CDM system performs better than a pure OFDM system.

The combined CDMA systems with OFDM can be mainly categorized into three types.

1. *MC/DS–CDMA (multicarrier direct sequence CDMA)* In this scheme, the serial-to-parallel converted data symbols are DSSS modulated using a specific

spreading code and these signals are transmitted in parallel on different sub carriers or in other words spreading in the time domain leads to narrowband sub-channels in frequency domain. It is guaranteed that there is no MAI in the synchronized downlink in a cell in slow frequency-selective fading channels.

2. *MC-CDMA (multicarrier CDMA)* In this scheme, multiple copies of the same data symbol each multiplied by one chip of a user specific spreading code are transmitted on different sub carriers in parallel maintaining the orthogonality in frequency domain. Since the replicas of the same data symbol are transmitted on different sub carriers, frequency diversity can be achieved to eliminate frequency selective fading but MAI may occur especially in the downlink.
3. *OFCDM (orthogonal frequency code division multiplexing)* It is originally based on MC-CDMA scheme. It comprises many multipath interference occurring in such broadband channel by employing many low symbol rate subcarriers and by making full use of the frequency diversity effect using the spread and coded signals over parallel subcarriers. Although OFCDM achieves better throughput performance in a broadband channel, it suffers from the degradation caused by intercode interference due to loss of orthogonality among code multiplexed channels. So, despreading the signal in frequency domain is a key technique in order to compensate for the destruction of orthogonality.

However, if multipath is not eliminated properly, equalization and despreading in the receiver lead to noise amplification. For this reason, the pure OFDM system performs better at low code rates where sufficient frequency diversity is provided by the code. However, the performance of OFDM-CDM systems can be improved with more sophisticated detection methods like iterative despreading and decoding. A major disadvantage of OFDM-CDM systems is the fact that coherent detection is required. So, the channel estimation and equalization cannot be avoided.

If a Fourier matrix is used for spreading in an OFDM-CDM system, then spreading and the IFFT at the transmitter cancel each other, resulting in the single-carrier transmission system with a guard interval results. At the receiver, the signal processing can be interpreted as frequency domain equalization, with the additional advantage that the transmitted signal does not have the typical amplitude peaks of an OFDM signal. Thus, it seems to be reasonable to use the Fourier matrix for spreading if spreading is to be applied at all.

Finally, OFDM is found to perform very well compared with CDMA, with its out-performing CDMA in many areas for a single-cell and multicell environment. The OFDM is found to allow up to 2 – 10 times more users than CDMA in a single-cell environment and from 0.7 – 4 times more users in a multi-cellular environment. The difference in user capacity between OFDM and CDMA is dependent on whether cell sectorization and voice activity detection is used.

It was found that CDMA performs well only in a multicellular environment where a single frequency is used in all cells. This increases the comparative performance against other systems, which require a cellular pattern of frequencies to reduce intercellular interference.

One important thing that needs to be mentioned here is the problems that may be encountered when OFDM is used in a multiuser environment. One possible problem is that the receiver may require a very large dynamic range in order to handle the large signal strength variation between the users. However, COFDM with forward error correction may solve many problems improving the BER.

9.14 OTHER VARIANTS OF OFDM

Flash OFDM This is a variant that was developed by Flarion and is a fast hopped form of OFDM. It uses multiple tones and fast hopping to spread signals over a given spectrum band.

VOFDM (vector OFDM) This form of OFDM uses the concept of MIMO technology. It is being developed by CISCO systems. The MIMO stands for multiple input-multiple output (described in Chapter 15) and it uses multiple antennas to transmit and receive the signals so that multipath effects can be utilized to enhance the signal reception and improve the transmission speeds that can be supported.

WOFDM (wideband OFDM) The concept of this form of OFDM is that it uses a degree of spacing between the channels that is large enough that any frequency errors between the transmitter and the receiver do not affect the performance. It is particularly applicable to Wi-Fi systems.

Adaptive OFDM While communicating, the wireless channel effects on the OFDM received signal differ due to selection of different parameter values as well as existing SNR condition at that time. Different allowable bandwidths with same number of subcarriers, different number of subcarriers with same bandwidth, different modulation mapping schemes, different FFT points, variation in pilot power and pilot positions, variations in cyclic prefix interval, etc., are considered as important OFDM parameters requiring critical selection, whose effect is directly reflected in the performance due to various reasons. These parameters can be made adaptive with SNR conditions of the channel. The SNR variations may arise due to distance variations between the transmitter and the receiver and fast or slow mobility, exhibiting time varying channel. Indirectly, by making the whole system adaptive in terms of such parameters, effective and efficient transmission control can be achieved. Such control can be adopted in frame-based systems with at least half duplexity between the transmitter and the receiver. This results in quality reception.

Summary

- The OFDM is a wideband scheme for wireless digital communication.
- It is a block modulation technique.
- The OFDM combats ISI by inserting the guard interval or cyclic prefix in the OFDM symbols.
- The OFDM combats ICI by maintaining the orthogonality between consecutive carriers.
- All the subcarriers may be assigned multiple users' information and hence orthogonal FDM is spectrally efficient method compared to FDM.

- The set of orthogonal carriers makes one transmission bandwidth.
- In OFDM system, appropriate source coding, channel coding, and encryption methods can be incorporated to have good error handling.
- The OFDM method can use M-PSK or M-QAM mapping for assigning amplitude and phase to the subcarriers.
- The OFDM technique can have continuous time or discrete time model. Discrete time model use IFFT for generating OFDM baseband signal while FFT is used at the receiver side for the reverse process.
- The OFDM block size, the number of subcarriers, and the number of IFFT point selections are independent of each other. However, some conditions must be satisfied.
- Each narrow subchannel will have sinc shape due to Nyquist pulse shaping.
- Narrow subchannels will undergo flat fading while overall transmission bandwidth will be considered under frequency-selective fading.
- Pulse shaping and windowing techniques can improve the OFDM spectral efficiency.
- Time and frequency synchronization is must in OFDM and that may be considered as a limitation of the system.
- Amplitude limitations in OFDM must be corrected and that is considered as another drawback of the system.
- Pilot carriers are transmitted for channel estimation.
- Advantages of OFDM are high spectral efficiency, combat to channel, simple implementation through FFT, high data rate, flexibility and adaptation features, and medium complexity multiple access scheme.
- Disadvantages of OFDM are high sensitivity to time and carrier offset and synchronization, high peak-to-average power ratio (PAPR) compared to single-carrier systems.
- Good features of OFDM and CDMA can be combined to get better hybrid systems.

Review Questions

1. Differentiate between single-carrier and multicarrier systems. List out all the single-carrier modulation schemes and compare them with OFDM.
2. Draw the conceptual diagrams of FDM and TDM and justify that these schemes indirectly maintain orthogonality.
3. What is the role of M-PSK mapping in OFDM? Is it exactly same as the independent M-PSK scheme? When do we use M-PSK mapping and when M-QAM?
4. Compare FDM and OFDM in all respects. Prove that OFDM is spectrally more efficient than FDM.
5. How are the carriers assigned to OFDM input signal? Write the necessary steps giving the final spectral setting.
6. What is the role of IFFT stage in OFDM? Explain the effect of IFFT points selection over the OFDM baseband signal and over the implementation of the system.
7. Explain the methods used in OFDM system to combat the ISI effect over wireless channel.
8. Describe the methods of RF upconversion for OFDM baseband signal.
9. Why is it necessary to apply pulse shaping and windowing to OFDM system?
10. List out the parameters that are to be decided in advance when the OFDM system design starts. What may be the criteria in deciding those parameters?
11. Explain the need to transmit pilot carriers. How will they help the system?
12. How is the timing and synchronization important in OFDM systems?
13. Explain the amplitude limitations in OFDM system. How can these be eliminated?

14. Prove that an OFDM system is more bandwidth-efficient technique compared to a single-carrier system.

15. Describe the performance of OFDM system in presence of Gaussian noise.

16. Why are hybrid OFDM systems designed? What are the alternate possible configurations for such hybrid systems?

PROBLEMS

1. If a total of 52 subcarriers, spaced at 312.5 kHz, are defined, find out the total occupied bandwidth excluding the secondary lobes.

2. If the OFDM bandwidth is 19.2 MHz and the total of 48 subcarriers are modulated with convolution coded 2/3 rate data using (a) BPSK and (b) QPSK. In each case, what must be the input data rate?

3. In an OFDM system, 52 subcarriers are defined with spacing of 300 kHz. If for the channel estimation purpose, 8 pilot carriers are added at equal distance maintaining the same spacing, what will be the percentage rise in occupied spectrum?

4. In OFDM system, frequency domain setting is such that the spacing between two subcarriers is 312.5 kHz. The frequency spacing includes cyclic prefix of 800 ns added to OFDM symbol.

(a) What will be the final symbol duration with cyclic prefix?

(b) What will be the OFDM symbol duration without cyclic prefix?

(c) Will there be any loss of spectrum efficiency due to addition of cyclic prefix?

5. Find out the additions and multiplications required for 256 point FFT and 1024 point FFT. By selecting 1024 point IFFT instead of 256, will you be able to perform OFDM modulation? If yes, at what cost?

6. A data rate of 5 Mbps is targeted in a multipath radio environment by using BPSK modulation. The maximum delay spread is 25 μs. The number of subcarriers is 128 for multicarrier transmission. Compare the ISI effect if the system is (a) single carrier and (b) multicarrier.

MULTIPLE CHOICE QUESTIONS

1. The OFDM is a method of

(a) multiplying

(b) orthogonal coding

(c) parallel transmissions

(d) domain transformations

2. Due to orthogonality,

(a) subcarriers can be demodulated independently

(b) subcarriers cannot be overlapped

(c) same subcarrier can be used for parallel transmissions

(d) subcarriers interfere

3. Orthogonality means

(a) $\int v_1(t)/v_2(t) = 0$

(b) $v_1(t)v_2(t) = 0$

(c) $\int_T v_1(t)v_2(t) = 0$

(d) none

4. If there are 256 symbols in a frame to be modulated by 16 subcarriers, each subcarrier will modulate

(a) 16 symbols (b) 256 symbols
(c) 4 symbols (d) 8 symbols

5. For which of the following modulation mappings, SNR must be highest to maintain the required BER?

(a) QPSK (b) QAM
(c) 16PSK (d) 16QAM

6. Cyclic prefix is added to

(a) eliminate ICI
(b) eliminate bit errors
(c) increase the spectral efficiency
(d) eliminate ISI

7. Addition of pilots will be helpful in

(a) frequency offset removal
(b) synchronization
(c) channel estimation
(d) all of the above

8. In which of the following types of pilot arrangements, all the subcarriers will act as pilots?

(a) Comb (b) Block
(c) Superimposed (d) Scattered

chapter 10

Multiplexing and Multiple User Access Techniques

Key Topics

- Multiplexing and multiple access
- FDM, TDM, CDM, and SDM
- FDMA, TDMA, CDMA, and SDMA: fixed allocation of resource
- Random access—ALOHA, slotted ALOHA, CSMA/CD, ISMA, and DAMA
- Reservation-based access: PRMA, polling, token passing, etc.

Chapter Outline

Any communication system should be designed efficiently when it is shared by many users because all have to play within the limited allocated spectrum and also the wireless channel is commonly shared by many applications, where interference is likely to occur. The efficiency may be in terms of efficient utilization of available spectrum and all the users must be treated equally. Multiplexing and multiple access schemes provide efficient ways to design communication systems. The spectrum can be shared on time-sharing or frequency-sharing basis or any other basis as described in this chapter; of which any one or combination can be selected. For example, GSM system uses TDMA/FDMA as a combination of two multiple access schemes over SDMA.

10.1 INTRODUCTION TO MULTIPLEXING AND MULTIPLE ACCESS

Multiplexing schemes are used when multiple users' signals are to be combined and to be sent on a single channel as a single input stream. The multiplexing describes how several users can share a medium with a minimum or no interference in the same communication link. All the incoming signals are upconverted or downconverted simultaneously afterwards being a single stream and transmitted together as one signal. For wireless communication, multiplexing can be carried out in four dimensions: frequency, time, code, and space. There are mainly four schemes for multiplexing based on four possibilities:

1. Frequency division multiplexing (FDM)
2. Time division multiplexing (TDM)
3. Code division multiplexing (CDM)
4. Space division multiplexing (SDM)

The task of multiplexing is to assign frequency, time, code, or space to each user with a minimum of interference and maximum of medium utilization. But here the association of senders and receivers over the communication channel is the key feature. The concept of multiplexing indirectly gives birth to the concept of individual's independent medium access as described afterwards.

In FDM, individual user is provided an individual channel, which will in combination make the whole transmission bandwidth. Figure 10.1 shows that seven user bandwidths make one transmission bandwidth. In TDM, each individual user is preassigned a time slot in which he or she can send the information and once that slot is over, the slot for the next user will start. The result of FDM is a large bandwidth while the result of TDM is nothing but a frame of certain duration covering all the users in a cycle. The scenario for two users is given in Fig. 10.2. All the users are scanned in order in a cyclic manner to collect the data. For n users, the bit rate of the TDM stream will increase n fold. Time synchronization is a very important issue for TDM. However, there are two methods for TDM: synchronous and asynchronous. In a multiplexing technique, allocation of frequency and time is fixed and if the user does not use, it will be considered as a waste.

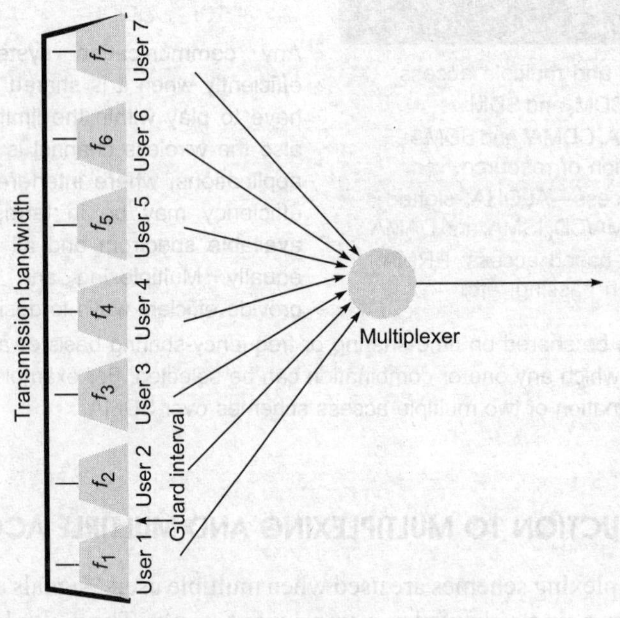

Fig. 10.1 Frequency division multiplexing

Another scheme, CDM, is shown in Fig. 10.3. Such a scheme can be combined with OFDM to create the hybrid configurations as discussed in Chapter 9. In CDM, separation is achieved by assigning each user channel its own code. Guard spaces are realized by using codes with necessary distance in code space, orthogonal codes. Good protection against unauthorized reception is the main advantage of CDM. In SDM, signal can be transmitted by different directional antennas or the signals received by the multidimensional antenna can be combined to get all of them back. Three directional antennas along with their lobes are shown in Fig. 10.4. Sometimes, differently polarized

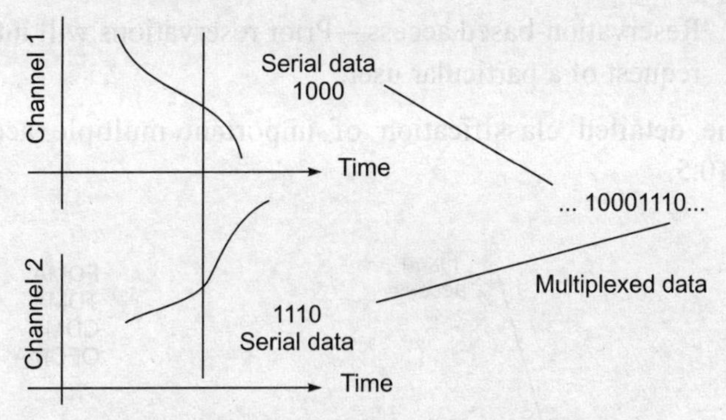

Fig. 10.2 Time division multiplexing of two channels

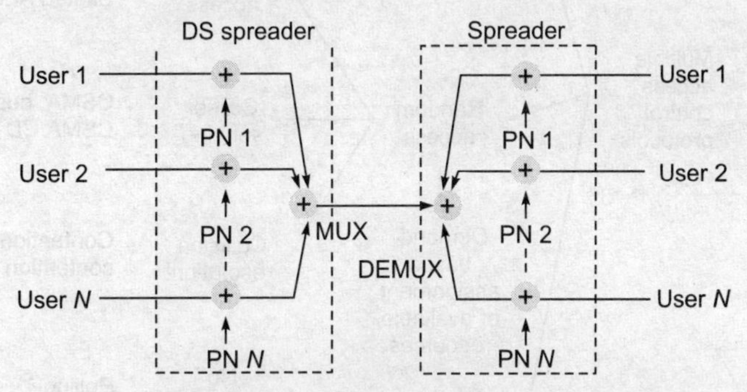

Fig. 10.3 Code division multiplexing for *N* users

signals can be transmitted simultaneously. Space division multiplexing is just like a highway with provision of different lanes.

Multiple access schemes allow many simultaneous users to share the same available channel bandwidth or radio spectrum on an individual basis on desire of the user. Whenever channel-sharing or channel-access issues arise, multiple access schemes are to be used. In any radio system, the available bandwidth that is allocated to the users is always limited. For example, in mobile phone systems, the total bandwidth is typically 50 MHz, which is split in half to provide the forward and reverse links of the system. Further, the forward and reverse links must be established for thousands of subscribers. The sharing of the spectrum schemes is required in order to increase the user capacity of any wireless network.

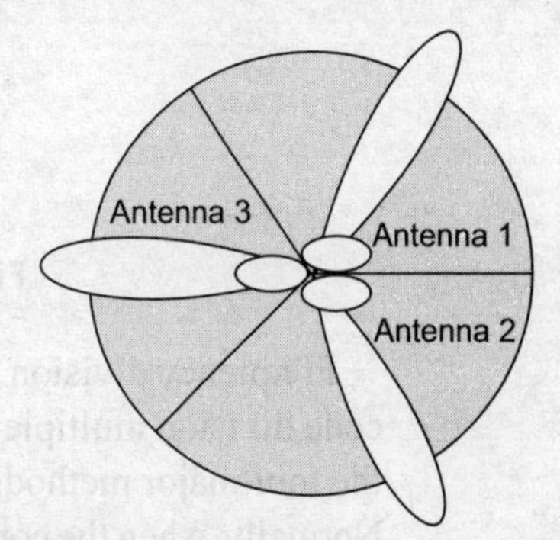

Fig. 10.4 Space division multiplexing

Multiple access may be achieved by three different manners as follows:

1. Multiple access by a fixed assignment of resources in terms of carrier allotment, time slot allocation, code allocation, or area allocation to specific users.
2. Random access, i.e. a dynamic assignment of spectrum resources in time or bandwidth to the users, according to their needs or maybe on the basis of demand.

3. Reservation-based access—Prior reservations will intimate other users about the request of a particular user.

The detailed classification of important multiple access schemes is given in Fig. 10.5.

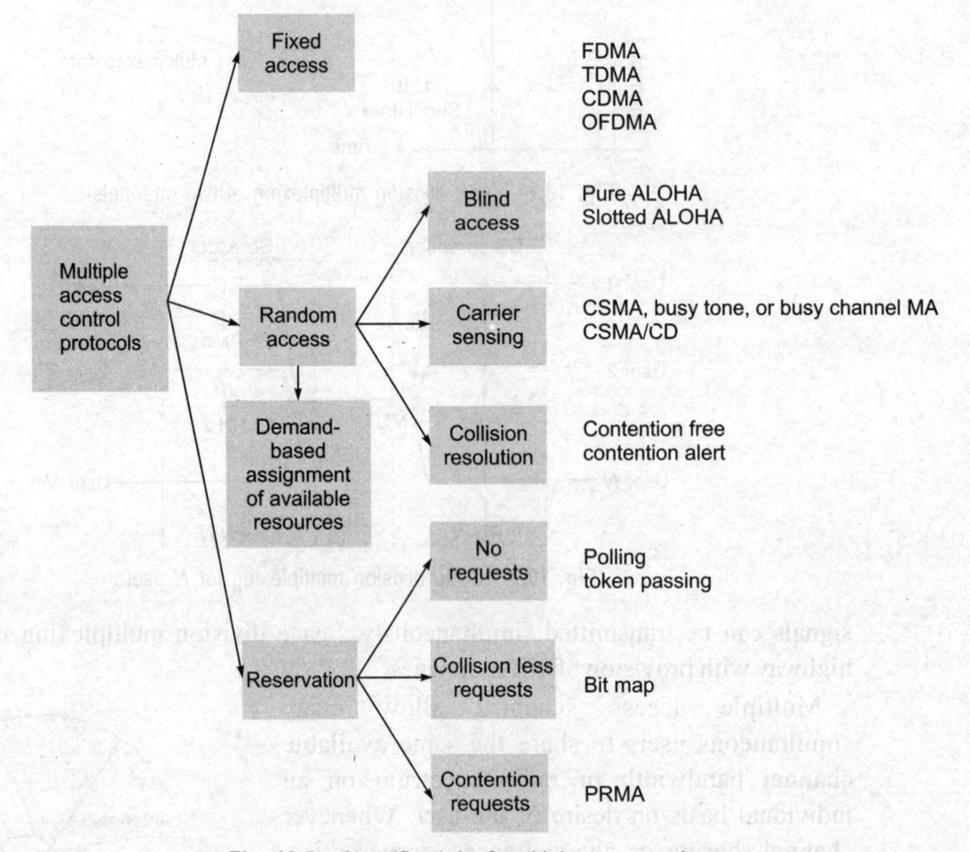

Fig. 10.5 Classification of multiple access schemes

Frequency division multiple access (FDMA), time division multiple access (TDMA), code division multiple access (CDMA), and space division multiple access (SDMA) are the four major methods of multiple access by fixed assignment of resources to the users. Normally, when the continuous transmission or streaming is required just like the real-time transmissions, fixed assignment types of schemes are better. These methods follow the same concepts of FDM, TDM, CDM, and SDM, correspondingly. The only way they differ in is that multiplexed information is transmitted on combined basis while multiple access is done on individual basis. In Figures 10.6, 10.7, and 10.8, FDMA, TDMA, and CDMA are compared on the basis of bandwidth sharing considering three users

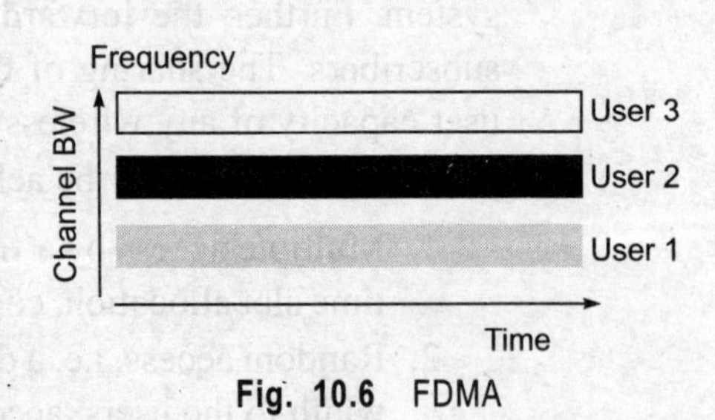

Fig. 10.6 FDMA

sharing the available channel bandwidth. This concept can be extended for n users. The SDMA is a special case described later on.

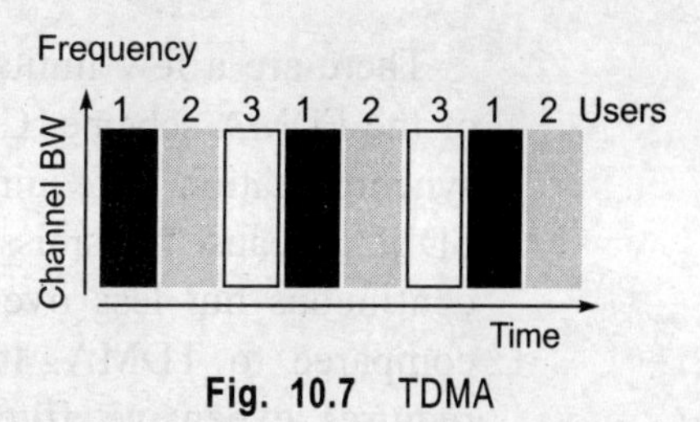

Fig. 10.7 TDMA

- In FDMA, for each user, it is not possible to use the entire bandwidth but only limited one is allocated.
- In TDMA, the user can use the total bandwidth but for the limited slot duration only.
- In CDMA, the total available bandwidth can be shared by all the users at a time.

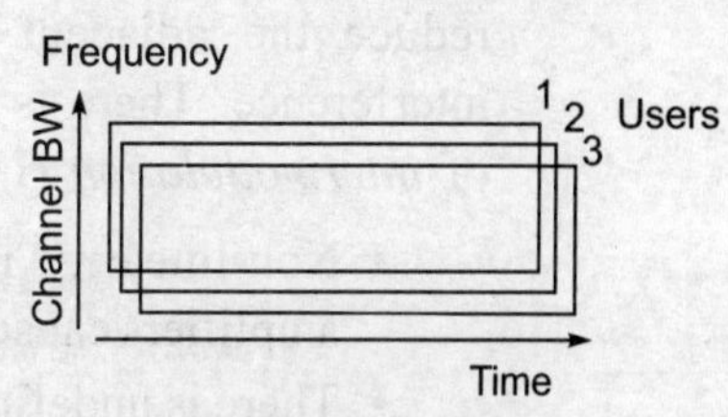

Fig. 10.8 CDMA

There are many extensions and hybrid techniques for fixed assigned multiple access, such as OFDMA and hybrid TDMA and FDMA systems. However, an understanding of the three major methods is required for understanding of any extension to these methods. Basically, in OFDMA, the information of the various users may be collected using FDM or TDM. Thus, a frame of the order of thousands of bytes is generated. Digital signal processing is applied to this block and converted into multicarrier modulated time domain signal, subchannels can be allocated to users on demand basis.

Random access mainly deals with packet radio and mostly storage data is utilized. Reservation-based access deals with channel reservation in advance whenever data transmission is needed. Demand-based channel assignment deals with allocation of free channel at the time of request. All are described in this chapter.

10.2 FIXED ASSIGNMENT TYPE OF MULTIPLE ACCESS SCHEMES

10.2.1 Frequency Division Multiple Access (FDMA)

The FDMA is the most common multiple access scheme that deals with the RF carriers. It is a technique whereby spectrum is equally divided up into frequencies and then assigned to different users as shown in Fig. 10.9. Thus, the available bandwidth is subdivided into a number of narrower band channels. With FDMA, only one subscriber at any time is assigned to a channel. The channel, therefore, is closed for other conversations (or until it is handed off to a different channel in case of demand assignment-based FDMA). A 'full-duplex' or frequency division duplex (FDD) FDMA has been used for first-generation analog systems. Each user is allocated a unique frequency band in which to transmit and receive. During communication, no other user can use the same frequency band. The FDMA transmission requires two channels, each being one way, one for transmitting and the other for receiving to avoid interference. The channel bandwidth used in most FDMA systems is typically low as each channel only needs to support a single user and associated baseband. Direct analog or digital information transmissions are possible as no buffering is required, like in AM, FM, FSK, etc.

There are a few limitations of the FDMA scheme. Carrier synchronization is required in FDMA because transmission is continuous but less overhead compared to TDMA. It also requires expensive filters to reduce the adjacent channel interference. There is problem of *intermodulation* (IM).

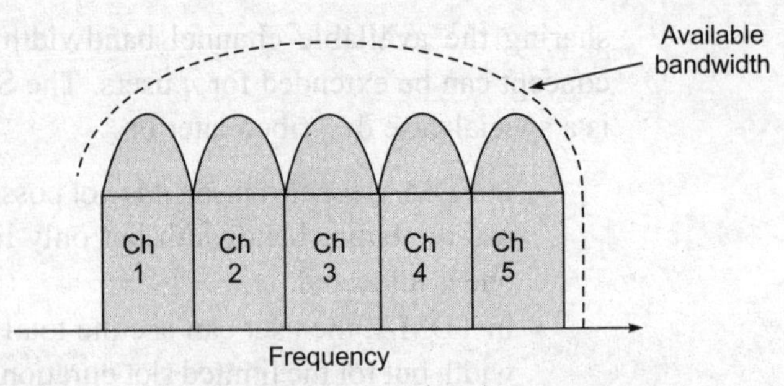

Fig. 10.9 Another way to represent FDMA compared to Fig. 10.6 (Guard interval is implied between two channels.)

- Non-linearity in power amplifiers cause signal spreading in frequency domain.
- There is undefined RF radiation that leaks into other channels in FDMA systems.
- There is generation of undesirable harmonics that cause interference to other users in mobile system or other systems in adjacent spectrum bands.

FDMA capacity:

It is the number of channels N_c. If total bandwidth $=W_{\text{channel}}$, guard band $= W_{\text{guard}}$, and each channel has a bandwidth W_{signal}, then

$$N_c = \frac{W_{\text{channel}} - (N_c - 1)\, W_{\text{guard}}}{W_{\text{signal}}} \tag{10.1}$$

More specifically

$$N_u = N_c - N_{\text{cch}} \tag{10.2}$$

where N_u is the number of users supported and N_{cch} is the number of control channels.

Example 10.1 Consider an AMPS uplink (mobile to base station). The total available bandwidth is 12.5 MHz, per channel bandwidth is 30 kHz, and the guard band is 10 kHz, of which 21 are used for control (setting up voice calls). Find the number of user channels.

Solution

$W_{\text{channel}} = 12.5$ MHz, $W_{\text{guard}} = 10$ kHz, $W_{\text{signal}} = 30$ kHz

So, $$N_c = \frac{12.5 \times 10^6 - (N_c - 1) \times 10^4}{30{,}000} \Rightarrow N_c \approx 312 \text{ channels}$$

$$N_u = 312 - 21 = 291 \text{ channels can be assigned to the users.}$$

10.2.2 Time Division Multiple Access (TDMA)

The TDMA improves spectrum capacity by splitting its use into time slots. It allows each user to access the entire radio frequency channel for the short period of a call. Other users share this same frequency channel at different time slots. Thus, TDMA divides the available spectrum into multiple time slots as shown in Fig. 10.10 by giving each user (or

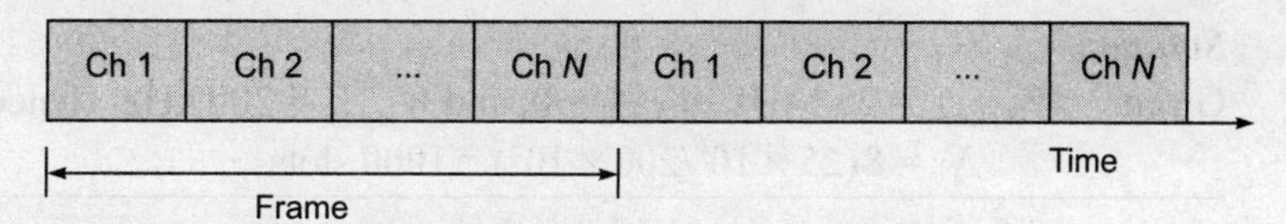

Fig. 10.10 TDMA frame generation

channel) a time slot in which they can transmit or receive. The figure shows how the time slots are provided to users in a round robin fashion making time frames, with each user being allotted one time slot per frame. The base station continually switches users on the channel. TDMA may be asynchronous or synchronous just like TDM. It is the dominant technology for the second-generation mobile cellular networks.

The TDMA systems transmit data in a buffer and hence it is a bursty communication method. Thus, the transmission of each channel is non-continuous. The input data to be transmitted is buffered over the previous frame and burst the transmitted data at a higher rate during the time slot for the channel. As TDMA cannot send an analog signal directly due to the buffering required, it is used for transmitting digital data. The TDMA can suffer from multipath effects, as the transmission rate is generally very high, resulting in significant intersymbol interference.

Advantages of TDMA are as follows:

- Narrowband filters are not required and hence it is cheaper than FDMA.
- Mobile devices can save battery power by turning off transmitter and receiver during slots when not transmitting or receiving data.
- It does not require a duplexer or multiple antennas for transmitting and receiving even when using FDD.
- It can allocate multiple time slots to a user to provide increased data rate.

The TDMA has the following disadvantages:

- It requires guard time between the time slots to separate users and accommodate time inaccuracies due to clock instability, delay spread of transmitted symbols, and transmission time delay.
- It requires signal processing techniques and high overhead for synchronization due to burst transmissions.

TDMA capacity:

Suppose the total number of slots is N_s and the number of users per channel is n. Then the TDMA capacity is given by

$$N_s = \frac{n[W_{\text{channel}} - (N_c - 1)\, W_{\text{guard}}]}{W_{\text{signal}}} \tag{10.3}$$

This is rather FDMA-TDMA case in general.

Example 10.2 Find out the total number of slots for a GSM system having 25 MHz forward link. Per channel bandwidth is 200 kHz and 8 speech channels are supported per radio channel. No guard band is assumed.

Solution

Given $W_{\text{channel}} = 25$ MHz, $W_{\text{guard}} = 0$, and $W_{\text{signal}} = 200$ kHz. Hence,

$$N_s = 8(25 \times 10^6/200 \times 10^3) = 1000 \text{ slots}$$

The TDMA is used in conjunction with FDMA to subdivide the total available bandwidth into several channels. This is done to reduce the number of users per channel allowing a lower data rate to be used. This helps reduce the effect of delay spread on the transmission. Figure 10.11 shows the use of TDMA with FDMA. Each channel based on FDMA is further subdivided using TDMA so that several users can transmit of the one channel. This type of transmission technique is used by most digital second-generation mobile phone systems. For GSM, the total allocated bandwidth of 25 MHz is divided into 125, 200 kHz channels using FDMA. These channels are then subdivided further by using TDMA so that each 200 kHz channel allows 8–16 users.

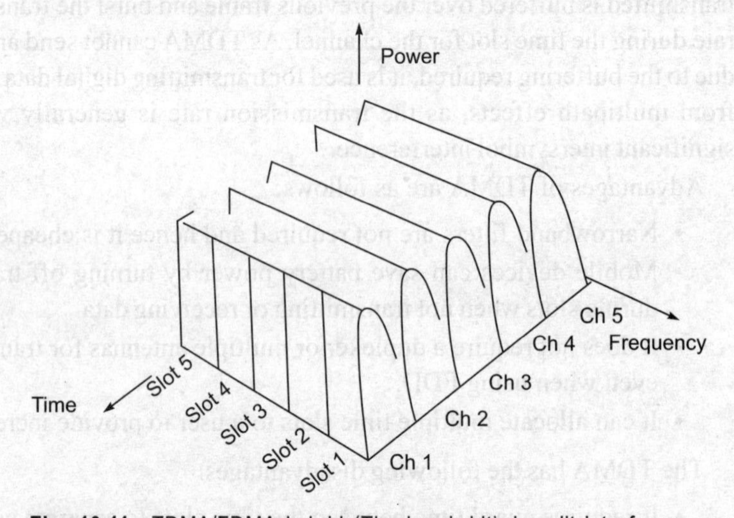

Fig. 10.11 TDMA/FDMA hybrid (The bandwidth is split into frequency channels and time slot.)

So far we have described the fixed TDMA pattern. If the bidirectional link is to be handled or if a duplex channel between a base station and a mobile station is to be established, another approach is utilized. Assigning different slots for uplink and downlink using the same frequency is called *time division duplex* (TDD). The DECT and UMTS systems use this method.

Spectral Efficiency Calculation for FDMA and TDMA

The FDMA requires guard bands in frequency to avoid adjacent channel interference and reduces channel bandwidth for data transmission. The TDMA requires guard bands and synchronization sequences reduce the time for data transmission. Spectral efficiency measures the total time frequency domain dedicated for voice/data transmission to the total time frequency domain available to the system. As discussed previously, the spectral efficiency for FDMA system is defined by

$$\eta_{\text{fdma}} = \frac{W_{\text{signal}}\, N_u}{W_{\text{channel}}} \leq 1 \tag{10.4}$$

For TDMA system if

N_r = number of reference bursts per frame
b_r = number of overhead bits per reference burst
N_t = number of data slots per frame
b_p = number of bits in each slot preamble
b_g = number of bits in each guard interval
T_f = frame time
R = channel bit rate

and if overhead bits $b_{\text{OH}} = N_r b_r + N_t b_p + N_t b_g + N_r b_g$ and total bits $b_T = T_f R$, then the spectral efficiency is given by

$$\eta_{tdma} = \left(1 - \frac{b_{\text{OH}}}{b_T}\right) \leq 1 \tag{10.5}$$

Total available bandwidth is broken into smaller channels. N_u channels are used for data transfer by FDMA. Each of the N_u channels are broken into time frames. b_{OH} of the total bits b_T is the overhead. Then the combined spectral efficiency of FDMA-TDMA becomes

$$\eta_{fdma\text{-}tdma} = \eta_{fdma} \times \eta_{tdma} = \left(1 - \frac{b_{\text{OH}}}{b_T}\right) \frac{W_{\text{signal}}\, N_u}{W_{\text{channel}}} \tag{10.6}$$

10.2.3 Spread Spectrum Multiple Access (SSMA)

There are many different methods of spread spectrum communication as described in Chapter 8. The SSM is a unique modulation method and for such uniqueness, the unique multiple access methods are also required. Depending upon the method of modulation selected, different multiple access methods can also be derived. There are mainly two multiple access schemes that are more popular for SSM type of systems as given below.

1. *Direct sequence spread spectrum modulation technique* enables code division multiple access with sufficient fading rejection and security. The near–far effect, a notorious problem in CDMA, can be reduced by adjusting the transmitting power of the mobiles.
2. *Frequency hopping spread spectrum modulation technique* enables frequency hopped multiple access (FHMA). With the Bluetooth technology, FHMA has become popular. These two schemes are described in the following sections.

Code Division Multiple Access (CDMA)

The CDMA is used along with spread spectrum modulation technique, which uses neither frequency channels nor time slots. In SSM, spreading is achieved by PN code. If such unique code is assigned to each individual user, the demodulation will be possible only if the code matches at the receiver end. This enables the multiple access. The users in a

CDMA system use the same frequency band and transmit simultaneously. They can use the whole available bandwidth for all the time. The transmitted signal is recovered by correlating the received signal with the PN code used by the transmitter. Figure 10.12 shows the general use of the spectrum using CDMA. At one frequency, four users with different codes are shown sharing the same bandwidth.

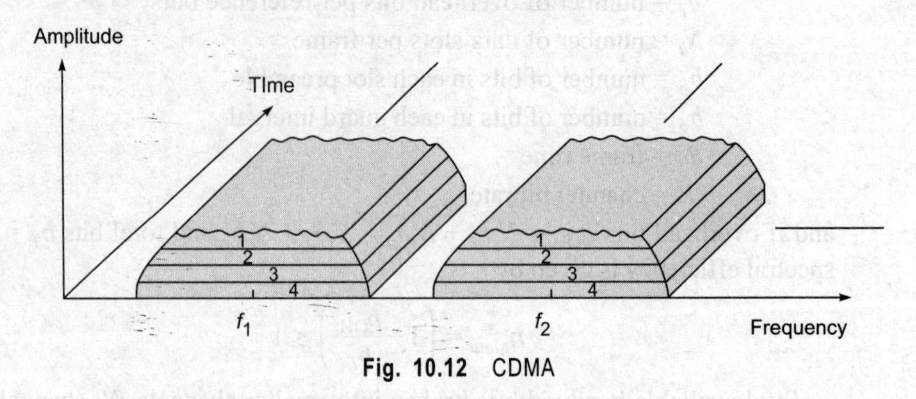

Fig. 10.12 CDMA

Some of the properties that have made CDMA useful are as follows:

- Signal hiding and non-interference with the existing systems
- Anti-jam and interference rejection
- Information security
- Accurate ranging
- Multipath tolerance

For many years, spread spectrum technology was considered solely for military applications. However, with rapid developments in LSI and VLSI designs, commercial systems have also developed. CDMA increases spectrum capacity by allowing all the users to occupy all channels at the same time. Transmissions are spread over the whole radio band and each voice or data call is assigned a unique code to differentiate it from the other calls carried over the same spectrum. CDMA allows for a 'soft hand off', which means that terminals can communicate with several base stations at the same time. The dominant radio interface for third-generation mobile, or IMT-2000, will be a wideband version of CDMA with three modes (IMT-DS, IMT-MC, and IMT-TC), which are described in Chapter 13.

DS-CDMA capacity calculation If all the users receive power P_r and there are $N-1$ interferers and no channel noise, the signal-to-interference ratio is

$$\text{SIR} = \frac{P_r}{(N-1)P_r} = 1/(N-1) \tag{10.7}$$

$$\text{SIR}_{\text{bit}} = \frac{E_b}{I_o} = \frac{P_r/R}{(N-1)(P_r/B)} = \frac{B/R}{(N-1)} \tag{10.8}$$

Including the channel noise, we get

$$\mathrm{SIR_{bit}} = \frac{E_b}{I_o} = \frac{P_r/R}{(N-1)(P_r/B)+N_o} = \frac{B/R}{(N-1)+N_oB/P_r} \quad (10.9)$$

So, the number of users in a cell, $N = 1 + \dfrac{B/R}{E_b/I_o} - \dfrac{N_oB}{P_r}$ (10.10)

We can increase the capacity of DS-CDMA system as follows:

1. *Cell sectorization* If cell is split in to s sectors, then the capacity increases by a factor s.(This is applied to other multiple access schemes too.)
2. *Voice activity factor* Users do not speak continuously, so turn off the transmitter when silence is detected. Percentage of time user speaking is equal to the voice activity factor v_f. (FDMA and TDMA cannot take advantage of voice activity factor easily.)

   The multiuser interference term becomes $(N_{us} - 1)\, v_f(Pr/B)$,

   where N_{us} = number of users per sector (sector is defined in Chapter 11)

   $\Rightarrow N_{us} = N/s$

   The signal-to-interference ratio for bit then becomes

$$\mathrm{SIR_{bit}} = \frac{B/R}{(N_{us}-1)v_f + N_oB/P_r} \quad (10.11)$$

So, the number of users in a sector,

$$N_{us} = 1 + \frac{1}{v_f}\frac{B/R}{E_b/I_o} - \frac{1}{v_f}\frac{N_oB}{P_r} \quad (10.12)$$

Example 10.3 The parameters for an IS-95 system are: the total available bandwidth is 1.25 MHz, bit rate is 9600 bps, SIR_{bit} = 10 dB, s = 3 sector, and v_f = 3/8. Find out the number of users supported avoiding the interference from each other. Avoid channel noise.

Solution

The total number of users = Number of sectors × Number of users per sector

So, $$N = 3\left(1 + \frac{1}{3/8}\,\frac{1250000/9600}{10}\right) = 107 \text{ users}$$

Multi-user Detection in CDMA

The CDMA implemented with DSSS signalling is among the most promising multiplexing technologies for cellular telecommunications. In the DS-CDMA framework, all the users transmit data at the same time and frequency but use distinct signature sequences to allow signal separation at the receiver. Demodulation of CDMA signals by the conventional matched filter suffers from multiple access interference and the near–far problem as described in Chapter 8. So, the efficient user detection techniques are required. Over the past decade, considerable attention focuses on multi-user detection technique, which refers to optimum or near-optimum demodulation in the presence of MAI.

Recently, there has been considerable interest in linear multi-user detection based on *minimum mean square error* (MMSE) criterion, which was proposed and developed in the early years. It is shown that MMSE detector, relative to other previously proposed detection schemes, has the advantage that explicit knowledge of interference parameters is not required, since filter parameters can be adapted to achieve the MMSE solution. Although it does not achieve minimum bit error rate, MMSE detector has been proved to achieve the optimal near–far resistance. More recently, a detailed error probability analysis of MMSE multi-user detection was presented. It suggested that the MAI plus noise at the output of MMSE detector could be accurately approximated by Gaussian noise in many cases. The environment considered in all the above work focuses on AWGN channel. However, in practical CDMA communication systems, channels are always demonstrated as multipath fading channels.

Frequency Hopped Multiple Access (FHMA)

The FHSS is a method of transmitting radio signals by rapidly switching carriers among many frequency channels, using a pseudo-random sequence known to both the transmitter and the receiver. Such spread spectrum signals are difficult to intercept. A frequency hop spread spectrum signal simply sounds like an increase in the background noise to a narrowband receiver. In FHSS, the frequency is constant in each time chip but changes from chip to chip. There may be two types of FHSS systems: with slow hopping and with fast hopping.

The different users can be allocated the different codes and hence the carrier sequence generated by different codes will also be different (Fig. 10.13). There will be less possibility that the same carrier transmitted from different users interferes with each other. Thus, multiple access on the basis of different hopping pattern will be possible.

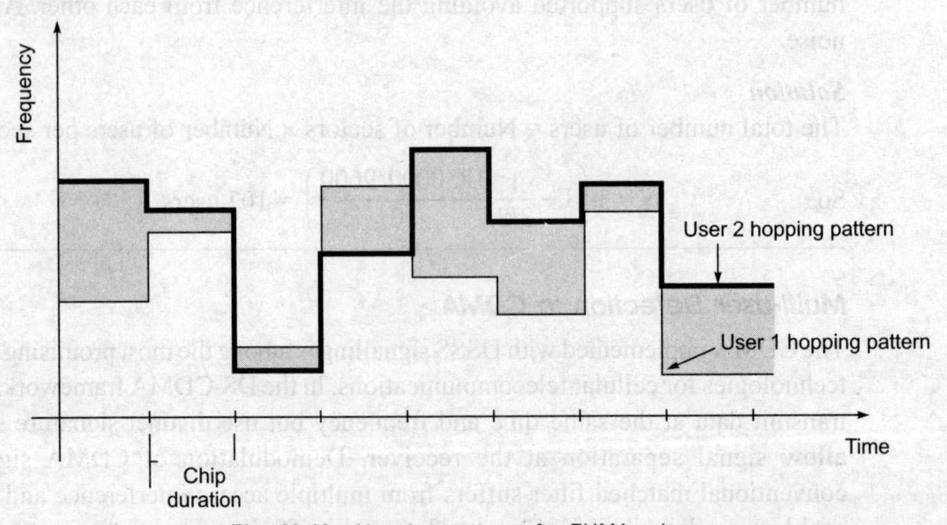

Fig. 10.13 Hopping patterns for FHMA scheme

Similar to frequency hopping, there may be the method in which time hopping is utilized for multiple access.

10.2.4 Space Division Multiple Access (SDMA)

Space division multiple access is geographical or cellular as shown in Fig. 10.14. The idea behind the concept is that if two transmitter/receiver pairs are far enough, such that they do not interfere, they can operate on the same frequency (by reusing the carrier) without interfering with each other. Next chapter covers the fundamentals of cellular technology, which utilizes the geographical division to accommodate all the users belonging to different areas. Cell site design is a very important issue for mobile communication networks design as SDMA.

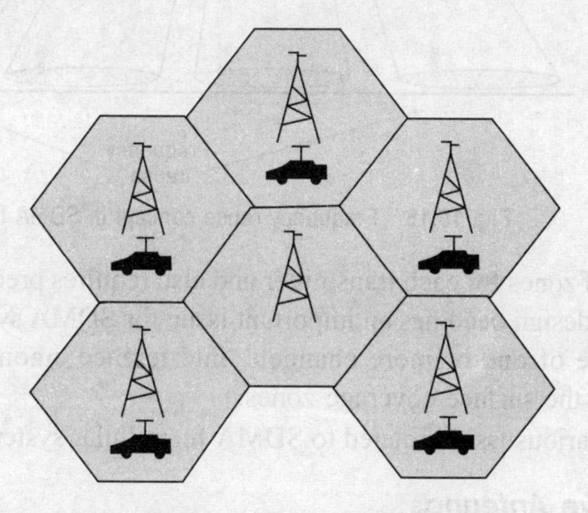

Fig. 10.14 Space division multiple access scenario

The SDMA uses the directional antenna for splitting the coverage, as explained in Section 11.9 on Sectorization in the next chapter. It is also a satellite communication mode that optimizes the use of radio spectrum and minimizes the system cost by taking advantage of the directional properties of dish antennas. In SDMA, satellite dish antennas transmit signals to numerous zones on the earth's surface. The antennas are highly directional, allowing duplicate frequencies to be used for multiple surface zones.

Consider a scenario in which signals must be transmitted simultaneously by one satellite to mobile or portable wireless receivers in 20 different surface zones. In a conventional system, 20 channels and 20 antennas would be necessary to maintain channel separation. In SDMA, there can be far fewer channels than zones. If duplicate channel zones are sufficiently separated, the 20 signals can be transmitted to the earth using 3–4 channels. The narrow signal beams from the satellite antennas ensure that interference will not occur between zones using the same frequency. In Fig. 10.15, it is shown that the fourth zone is using the frequency of the first zone as they are far enough not to interfere. Thus, fewer channels can support a large number of zones.

In case of cellular structure, it will be necessary to restrict the power to avoid the co-channel and interchannel interference while satellite-based SDMA requires careful

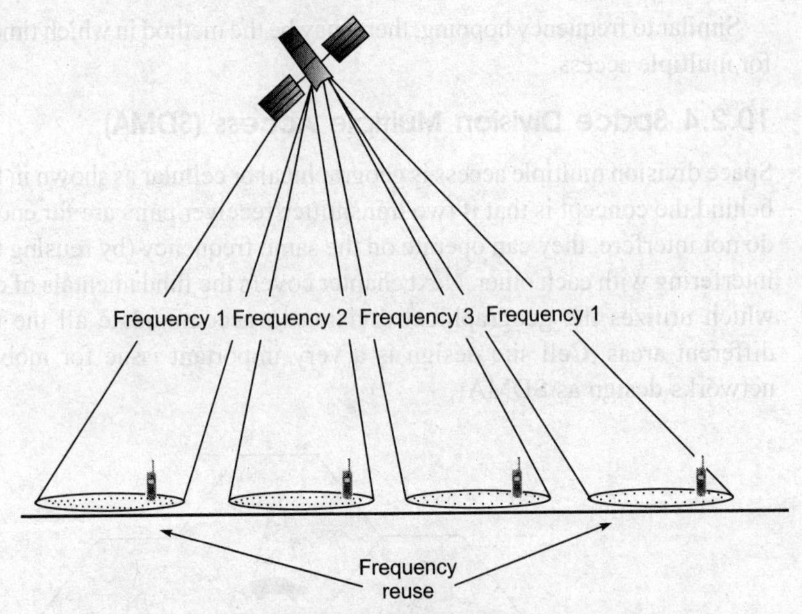

Fig. 10.15 Frequency reuse concept in SDMA for satellite system

choice of zones for each transmitter and also requires precise antenna alignment. Hence, antenna design becomes an important issue for SDMA systems. A small error can result in failure of one or more channels, interference among channels, and/or confusion between the surface coverage zones.

The various issues related to SDMA for cellular systems are described as follows.

Adaptive Antennas

The growing spread of wireless communication necessitates technical measures to enhance spectral efficiency and subscriber capacity. Adaptive antennas at the base station estimate the location of the mobile terminal and receive and transmit only from and to this desired direction. This

- improves range,
- improves the carrier-to-interference ratio, and
- allows denser frequency reuse, thus higher capacity.

Research is being conducted into such systems; meanwhile, the first products have entered in the market. Ericsson and Mannesmann conducted tests in Germany in 1997. Their system uses a 6 by 8 matrix of patch antennas. Typically, a horizontal element with spacing of $\lambda/2$ is used. The technology of using such 'smart antennas' is often called space division multiple access. The SDMA is mainly being researched for (the future extensions of) GSM and IS-95.

Adaptive power control appears necessary to achieve the best performance gains.

Combining Signals from Different Antenna Branches

In a noise-limited environment, the best diversity combining scheme is maximum ratio combining. If interference is present, interference rejection combining optimizes the C/I at the output. Improvements of more than 10 dB appear feasible in the uplink. For the downlink, such improvements are more difficult to achieve. Typically, a 6 dB gain appears practical, except in bad urban multipath environments with small cell sizes.

SDMA systems with dense frequency reuse may require many intracell handovers, to keep the angles of arrival well balanced. In particular, signals arriving from almost the same direction and signals coming from side lobes of the main beams should be avoided. The complexity of the base station software increases to accommodate SDMA. The software of existing base stations needs modifications.

10.2.5 Orthogonal Frequency Division Multiple Access (OFDMA)

The OFDM or OFDMA is a hybrid multiplexing/multiple access technique with multicarrier modulation, which divides the available spectrum into many carriers, each one being modulated by a low rate data stream as described in Chapter 9. In OFDM, different users' information is processed in combination and then allocated to multiple carriers, while in OFDMA, out of the total available bandwidth, each narrow channel can be used by an individual user for having access. The OFDMA is similar to the FDMA. However, OFDMA uses the spectrum much more efficiently by spacing the channels much closer together. This is achieved by making all the carriers orthogonal to one another, preventing interference between the closely spaced carriers.

How is OFDMA Superior to FDMA and TDMA

In FDMA, each user is typically allocated a single channel, which is used to transmit all the user information. The bandwidth of each channel is typically 10 kHz–30 kHz for voice communications. However, the minimum required bandwidth for speech is only 3.1 kHz. The allocated bandwidth is made wider than the minimum amount required, preventing the channels from interfering with one another. This extra bandwidth is used to allow for signals from neighbouring channels to be filtered out and for any drift in the centre frequency of the transmitter or receiver. In a typical system, up to 50% of the total spectrum is wasted due to the extra spacing between the channels. This problem becomes worse as the channel bandwidth becomes narrower and the number of frequency bands increases.

Most digital phone systems use vocoders to compress the digitized speech. This allows for an increased system capacity due to a reduction in the bandwidth required for each user. Current vocoders require a data rate somewhere within 4–13 kbps, depending on the quality of the sound and the type used. Thus, each user only requires a minimum bandwidth of somewhere between 2–7 kHz, using QPSK modulation. However, simple FDMA does not handle such narrow bandwidths very efficiently.

The TDMA partly overcomes this problem by using wider bandwidth channels, which are used by several users. Multiple users access the same channel by transmitting in their data in time slots Thus, many low data rate users can be combined together to transmit in a single channel that has sufficient bandwidth so that the spectrum can be used efficiently.

There are, however, two main problems with TDMA. There is an overhead associated with the changeover between the users due to time slotting on the channel. A changeover time must be allocated to allow for any tolerance in the start time of each user due to the propagation delay variations and synchronization errors. This limits the number of users that can be sent efficiently in each channel. In addition, the symbol rate of each channel is high (as the channel handles the information from multiple users) resulting in problems with multipath delay spread.

The OFDMA overcomes most of the problems with both FDMA and TDMA. It splits the available bandwidth into many narrowband channels (typically, 100–8000, depending upon the application). The carriers for each channel are made orthogonal to one another, allowing them to be spaced very close together, with almost no overhead as in the FDMA example. Because of this, there is no great need for the users to be time multiplex as in TDMA. Thus, there is no overhead associated with switching between users. Each carrier in an OFDMA signal has a very narrow bandwidth and thus the resulting symbol rate is low. This results in the signal having a high tolerance to multipath delay spread, as the delay spread must be very long to cause significant intersymbol interference (e.g., > 100 μsec).

In short, multiple users' information can be jointly or on individual basis transmitted using multiple orthogonal carriers in OFDM/OFDMA. Figure 10.16 shows that with more number of carriers, we get better spectral efficiency for more number of users in case of OFDMA where all the carriers are orthogonal to each other. Guard interval is omitted without loss of generality.

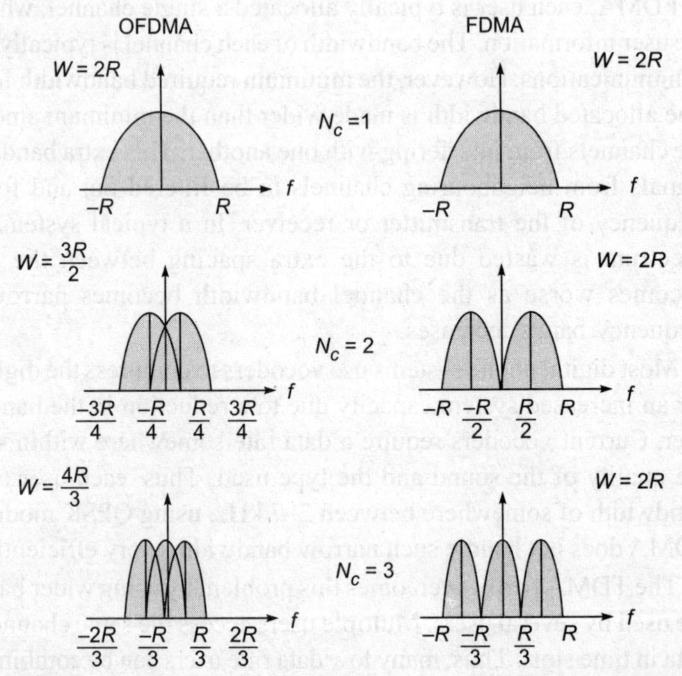

Fig. 10.16 Comparison between OFDMA and FDMA in terms of spectral consumption for equal number of users (W = available bandwidth, R = maximum bit rate, and N_c = number of carriers)

10.2.6 Hybrid Methods of Multiple Access

A few combinations of multiple access techniques are as follows.

1. FDMA+CDMA Spectrum is divided into channels and each channel is a narrowband CDMA system with processing gain lower than original CDMA system.
2. DSSS+FHSS The direct sequence modulates signal and hops centre frequency using pseudo-random hopping pattern. The method avoids near–far effect.
3. TDMA+CDMA Different spreading codes are assigned to different cells. One user per cell is allotted particular time slot. Only one CDMA user transmits in each cell at any given time. The method avoids near–far effect.
4. TDMA+FHSS It involves hop involves to new frequency at start of new TDMA frame. The method avoids severe fades on particular channel. Hopping sequences are predefined and unique per cell. It avoids cochannel interference if other base stations transmit on different frequencies at different times.

The following sections will describe another form of multiple access—dynamic allocation of resources.

10.3 MULTIPLE ACCESS FOR PACKET RADIO SYSTEM (RANDOM ACCESS)

When wireless networks are established, they especially follow the layered protocol stack. The lowermost layer is the physical layer, which takes care of actual physical transmission of the packets by choosing an appropriate modulation scheme, bit rate, synchronization methods, etc. As far as the packet access control (or user access control) is concerned, these issues are handled by data link layer according to ISO/OSI reference model. Data link control layer is subdivided into two layers. The lower sublayer is the medium access control (MAC) layer while the upper sublayer is logical link control (LLC) layer. In networks, whenever the request, acknowledgement, or data packets are to be transmitted, the channel will be utilized. This will not be a continuous process but will be bursty in nature. At the same time, for the datagram approach, all the packets will be routed independently towards the destination. In this situation, the task of data link layer is to establish a reliable point-to-point or point-to-multipoint connection between different devices. The same is true for wired as well as wireless medium. Here we shall discuss only MAC sublayer.

First of all, it is necessary to explain why special MACs are needed in wireless domain and why standard MAC schemes known from wired networks often fail. Actually, in contrast to a wired network, the wireless network faces serious problems like hidden and exposed terminals, near and far terminals, etc. Fixed assignment type of schemes can accommodate lesser number of users and also most of the time active terminals will release the packets while idle terminals may release packets with a long gap of time, which will result in waste of allocated fixed resources. Hence, some sort of dynamism is required for sharing the channel. Some random access schemes are described here, which may be considered for wired as well as wireless system. However, the problems of wireless system will be described separately, which will decide the suitability of methods.

The three types of methods are described here. Pure ALOHA and slotted ALOHA are the blind access schemes while CSMA or CSMA/CD are the carrier sensing based random access schemes. Two more schemes are collision-free methods, which are also called collision avoidance type of schemes or contention-free protocols.

10.3.1 Pure ALOHA

The ALOHA protocol provides the fundamental solution for wireless access to computer systems. The main advantage of the ALOHA random access scheme is simplicity. According to ALOHA, terminals can transmit their data regardless of the activity of other terminals. Any terminal is allowed to transmit without considering whether the channel is idle or busy. If the packet is received correctly, the base station transmits an acknowledgement. If no acknowledgement is received by the mobile computer,

1. it assumes the packet to be lost and
2. it retransmits the packet after waiting for a random time.

The delay or the random time is mainly determined by the probability that a packet is not received (because of interference from another transmission, called a 'collision') and the average value of the random waiting time before a retransmission is made. The basic flow of ALOHA is illustrated in Fig. 10.17.

The ALOHA needs some adaptive control of the retransmission scheme. Otherwise, the system will become unstable. A popular method is to increase the mean waiting time if too many collisions occur. Shorter delay and higher throughput can be achieved if collision resolution schemes are used, which exploit feedback about the collisions. One example is the stack algorithm, which can be studied from suitable books.

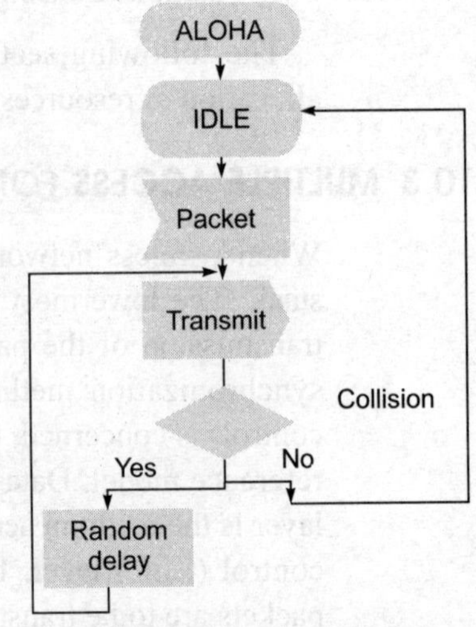

Fig. 10.17 Terminal behaviour in ALOHA random access network

ALOHA in Mobile Radio Sets

The ALOHA concept is very commonly used in modern wireless communication systems. The call set-up procedure of almost any (analog or digital) cellular telephone system uses ALOHA-type random access. But the performance differs from what one would expect in a wireline network.

In a radio channel, packets may be lost because of signal fading even if no other contending signal is present. On the other hand, packets may be received successfully despite interference from competing terminals. This is called *receiver capture*. This has a significant influence on the throughput.

The optimum frequency reuse for ALOHA random access networks differs from the frequency reuse for telephony, because the performance criteria differ (throughput/delay versus outage probability). The best reuse pattern for an ALOHA system is to use the same frequency in all cells.

Collision Resolution

Later studies revealed that for an infinite population of users and under certain channel conditions, the ALOHA system is unstable. Packets lost in a collision are retransmitted but the retransmission again experiences a collision. This may set off an avalanche of retransmission attempts. The number of previously unsuccessful packets that need to be retransmitted grows beyond any finite bound. One method to mitigate instability is to dynamically adapt the random waiting times of all terminals, if the base station notices that many collisions occur. There are two solutions to ensure stability (the description of which is beyond the scope of this book) as follows:

1. *Dynamic frame length* (DFL) ALOHA, by Frits Schoute It uses a centralized control mastered by the base station.
2. *The stack algorithm* by Boris Tsybakov, et al. The stack algorithm is a decentralized method.

10.3.2 Slotted ALOHA

In unslotted or pure ALOHA, a transmission may start at any time. In the slotted ALOHA, the time axis is divided to slots. All terminals are assumed to know the times at which a new slot begins, that is time synchronization is needed in the network. Packets may only be transmitted at the beginning of a new slot. The slotted ALOHA has significantly better throughput than unslotted ALOHA.

GSM Call Set-Up: An Example of Slotted ALOHA

If the transmitter-to-receiver propagation time is large and unknown, the slot time must be equal to the sum of the packet length and a sufficiently large guard time. In the call set-up of GSM, random access packets are substantially shorter than normal telephone speech blocks. During call set-up, the propagation time is still unknown because the subscriber can be anywhere in the cell. During the call, the propagation times are measured and the terminal transmitter will compensate for it, by sending all blocks a bit in advance. A closed loop control circuit, sending adaptive timing advance/delay feedback information, is used to ensure that the timing remains correct even if the subscribers move in the cell.

Performance analysis is done by measuring the throughput, drift, stability, and delay for both ALOHA and slotted ALOHA.

10.3.3 Carrier Sense Multiple Access (CSMA)

For high offered traffic loads, collisions occur frequently in the most basic random access system, the ALOHA system. This reduces the throughput and may lead to instability if the collided packets collide again during their retransmission. A number of other protocols have been proposed to mitigate this problem.

Busy channel multiple access (BCMA) is the class of multiple access schemes in which no new packet transmissions are allowed when the inbound channel is busy.

Various strategies have been proposed to acquire information on the channel state. One of them, *carrier sense multiple access* (CSMA), is the scheme in which the carrier sensing activity is supported to identify the busy channel condition before transmission. CSMA in its basic form does not employ feedback other than the positive acknowledgements of correct reception of a data packet. However, all terminals listen to the inbound (terminal-to-base) channel. No new packet transmission is initiated when the inbound channel is sensed busy by the mobile terminal. This requires that all mobile terminals can receive each other's signals on the inbound frequency.

However, in mobile radio networks with fading channels, a mobile terminal might not be able to sense a transmission by another (remote) terminal. This effect, known as the 'hidden terminal' problem, is avoided in inhibit sense multiple access (ISMA), where the base station transmits a busy signal on an outbound channel to inhibit all other mobile terminals from transmitting as soon as an inbound packet is being received. A disadvantage of ISMA is the necessity of a real-time (continuous) feedback channel.

Similar to ISMA, *busy tone multiple access* (BTMA) has been proposed. If the feedback channel contains only a narrowband 'busy tone', the mobile terminal may erroneously miss the presence of this tone if the outbound signal happens to be in a deep fade. This may be mitigated by casting an (active) idle tone, rather than an (active) busy tone, or by transmitting busy reports with error control coding. Even if signalling messages on the feedback channel are always received correctly by all mobile terminals, collisions can nonetheless occur in ISMA for two reasons:

- New packet transmissions can start during the delay in reception of the inhibit signal.
- Packets from two or more persistent terminals, awaiting the channel to become idle, can collide immediately after the termination of the previous packet transmission.

In CSMA-CD with collision detection, the receiver continuously informs all terminals about the incoming signal. If a collision is detected, the transmissions are aborted by all terminals immediately. Otherwise, the channel would remain occupied without any useful signal being recoverable. This enhances the throughput, compared to a system that only acknowledges reception after transmission of the full message or packet.

The CSMA is used in

- Military packet radio
- Amateur (Ham) packet radio
- Proprietary wireless local area networks
- Wireline local area networks

10.3.4 Inhibit Sense Multiple Access (ISMA)

The ISMA is the modified CSMA scheme. Compared to ALOHA or CSMA, the inhibit sense multiple access radio system is supplemented by an outbound signalling of the status of the channel: either 'busy' or 'idle'. When the base station receives an inbound packet, a 'busy' signal is broadcast to all mobiles to inhibit them from transmission. In a practical system, this only occurs after a short processing delay d_1. The effect of this delay depends on its magnitude relative to the duration of the data packet. After termination of (all contending) transmissions, the base station starts transmitting an 'idle' signal after a delay of duration d_2. The whole scenario is represented in Fig. 10.18.

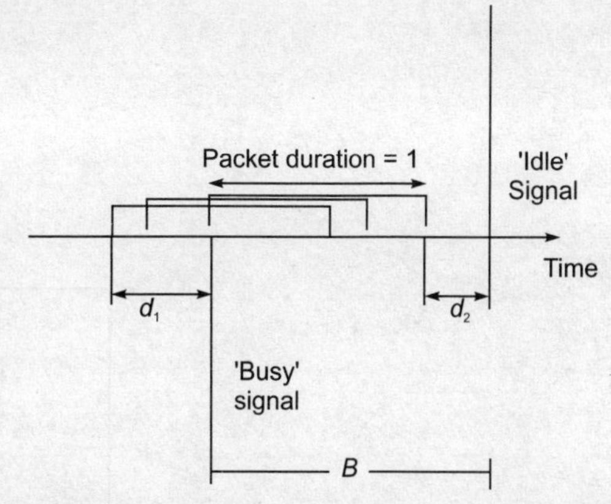

Fig. 10.18 Scenario of inhibit sense multiple access

In CSMA, the delay is mainly caused by the time a mobile terminal takes to switch from reception to transmission mode (power up), after sensing the radio channel for carriers from other active terminals.

The *busy period* is defined as the sum of the period during which the base station broadcasts a busy signal and the preceding signalling delay d_1. For memoryless Poisson arrivals, the duration of the idle period, that is the time interval starting at the release of the channel until the first packet arrival, is exponentially distributed with mean value $I = 1/G$, where G is the traffic.

The busy period is the time interval between the first arrival of a packet until the moment that the channel becomes idle. During the initial period d_1 of the busy period, the outbound channel still reports an idle inbound channel. If packet duration is of 1 unit time, the duration of the busy period is at least $1 + d_2$, but may be longer if a collision is caused by a packet arrival during the inhibit delay. Moreover, persistent terminals, which sense a busy signal, may start to transmit immediately after the channel becomes idle. In such cases, the busy period has a duration longer than two (or more) units of time.

Hence, the average duration of the busy period B depends on the signalling delay d_1 and d_2 and on the persistency p in (re-) scheduling inhibited packets.

Non-persistent ISMA

For non-persistent CSMA and ISMA, rescheduling (with random back-off time) always occurs if the channel is busy at the instant of sensing. So, if a packet arrives at a non-persistent terminal when the base station transmits a 'busy' signal, the attempt is considered to have failed. If the feedback channel (or, in CSMA, the channel sensing mechanism) is imperfect, a transmission may erroneously be started in the period. The packet is rescheduled for later transmission.

1-Persistent ISMA

Terminal behaviour in 1-persistent ISMA/CSMA random access networks is shown in Fig. 10.19.

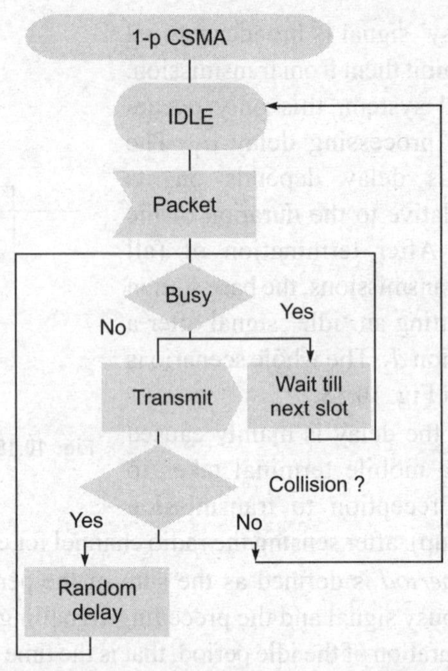

Fig. 10.19 Description of terminal behaviour in 1-persistent ISMA/CSMA random access network

p-Persistent ISMA

Terminal behaviour in *p*-persistent ISMA/CSMA random access network in shown in Fig. 10.20.

With probability $I/(I + B)$, a test packet starts at an instant when the channel is idle. A collision can occur if one or more other terminals start transmitting during the time delay d_1 of the inhibit signal. This allows to compute the conditional probability of n transmissions overlapping with the test packet that initiated the busy period.

Alternatively, the test packet itself starts during a period of duration d_1 when the channel is busy because of a transmission by another terminal, but seems idle since the inhibit signal is not yet being broadcast. This event occurs with probability $d_1/(B + I)$. The test packet thus experiences interference from the packet that initiated the busy period, but possibly also from other arriving packets. The additional contending signals occur with a Poisson arrival rate during the interval d_1.

Taking account of the above three possible events, the unconditional probability of successful transmission can be derived. The derivation differs from techniques typically used for wireline LANs, because in radio systems, we mostly want to be able to consider expressions for capture probabilities, depending on the location of one particular terminal.

10.3.5 Throughput of Random Access Schemes

Throughput is an important measure of performance.

- The *offered* or *attempted* traffic, denoted as G, is the number of transmission attempts that is made on the channel, per unit of time. In non-persistent CSMA,

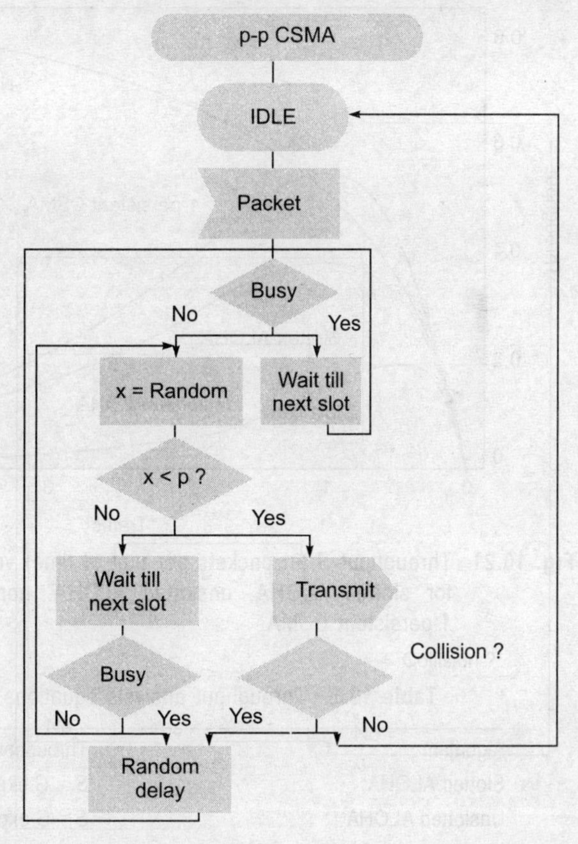

Fig. 10.20 Description of terminal behaviour in *p*-persistent ISMA/CSMA random access network

attempts made when the channel busy are inhibited. These attempts contribute to the offered traffic.

- Throughput, denoted as *S*, is the average fraction of time that the radio channel takes to successfully carry packets.
- In slotted systems, the slot length is taken as the unit of time.

In virtually all protocols (including ALOHA, CSMA, and ISMA), throughput S is approximately equal to G for a small G, that is if the channel is lightly loaded. At larger traffic loads (larger G), conflicts, called *collisions*, occur. So, $S < G$.

Throughput versus attempted traffic graph is shown in Fig. 10.21.

Table 10.1 shows throughput analysis equations.

Non-persistent CSMA/ISMA is the only curve for which S does not reduce to zero for large G. If terminals can indeed infinitely fast sense the channel, as is assumed here, collisions can be avoided completely.

Note that the average number of (re) transmission attempts only equals G/S in a stable system. Stability is problematic if retransmission waiting times are short in a network with many terminals.

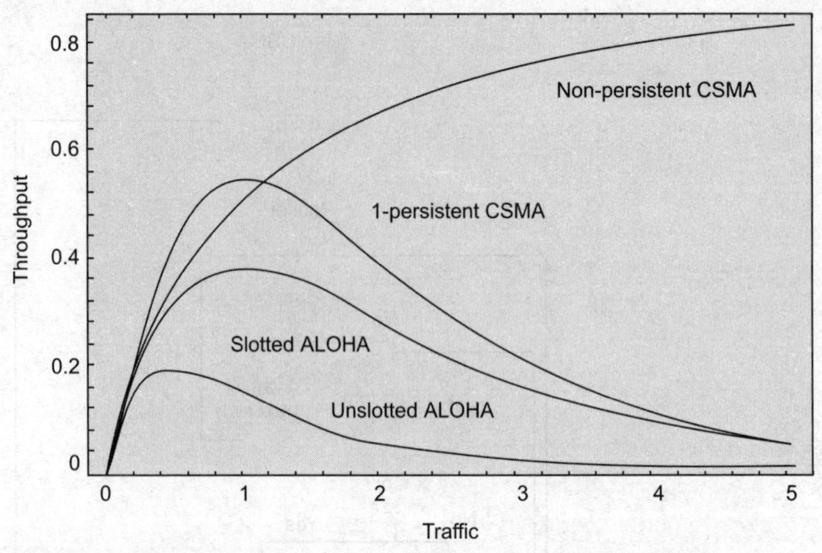

Fig. 10.21 Throughput S (in packets per unit of time) versus attempted traffic G for slotted ALOHA, unslotted ALOHA, non-persistent CSMA, and 1-persistent CSMA

Table 10.1 Throughput analysis equations

Algorithm	Throughput S
Slotted ALOHA	$S = G \exp[-G]$
Unslotted ALOHA	$S = G \exp[-2G]$
Nonpersistent CSMA and ISMA	$S = G/(1+G)$
1–persistent CSMA and ISMA	$S = (G + G^2) / (1 + G \times \exp[G])$

10.3.6 Demand Assigned Multiple (Random) Access (DAMA)

Demand assigned multiple access/random access schemes dynamically assign radio resources as per availability of the resources to users. This scheme is more suitable in a mobile scenario and in a satellite system. The request for communication from the user will be transferred to the central authority that can assign the available/free carrier or time slot to the user on the basis of current situation. SPADE system for satellite communication, GSM system, etc., use such a method of multiple access in support.

The DAMA is also called reservation ALOHA, a scheme typical for satellite systems. In this scheme, DAMA has two modes. During a contention phase, following the slotted ALOHA scheme, all stations can try to reserve future slots. Thus, collisions during the reservation phase do not destroy data transmission but only the short requests for data transmission. If successful, a time slot in the future is reserved and no other station is allowed to transmit in this slot. Therefore, the satellite collects all successful requests destroying the others and sends back a reservation list indicating access rights for future slots. All ground stations have to obey this list. For such continuous processes, the stations have to be synchronized from time to time. In DAMA, each transmission slot has to be reserved explicitly.

10.3.7 Why All Random Access Schemes are not Suitable in Wireless Media

Let us consider a typical case of carrier sense multiple access with collision detection (CSMA/CD) working over the wired line as follows. A sender senses the medium to see if it is free. If the medium is busy, the sender waits until it is free. Once the medium is free, the sender starts transmitting data and continues to listen into the medium. If the sender now detects a collision while sending, it stops at once and sends a jamming signal.

Why does this scheme fail in wireless networks? The CSMA/CD is not really interested in collisions at the sender, but rather in those at the receiver. The signal should reach the receiver without collisions. But the sender is the one detecting collisions. Now the collision may occur at a far distance also. This is not a problem in the wired medium as more or less same signal strength can be assumed all over the wire, especially in case of LAN. If a collision occurs anywhere in the wire, everybody will notice it.

The situation is different in wireless networks. The strength of a signal decreases proportionally to the square of the distance to the sender. The sender may now apply carrier sense and detect an idle medium. Thus, the sender starts sending but the collision occurs at the receiver due to a second sender. This may be called *hidden terminal problem*. The second sender terminal may be valid but does not exist within the detection range of first sender, that is if the receiver is in the middle of the first and second senders, the second sender will be at a future distance from the first one. This will not detect any collision though it occurs. The sender detects no collision and assumes that the data has been transmitted without errors, but actually a collision might have destroyed the data at the receiver. This is very common in an MAC scheme of wired medium and sometimes fails in wireless scenario.

In a wireless channel, performance differs substantially from the performance over cabled local area network as follows:

- The packets may be lost due to fading even if no collision occurs.
- If collisions occur, not all packet involved are lost. Some may nonetheless be detected successfully, particularly if these are received with strong power. This effect is called *capture*.

The mobile channel gives a better performance than the wired channel with capture.

10.4 RESERVATION-BASED MULTIPLE ACCESS SCHEMES

In such schemes, there is the provision that the users can reserve their slot or resources in advance and then the transmission will be followed in that order. Such reservation times are separately provided.

10.4.1 Packet Reservation Multiple Access (PRMA)

Packet reservation multiple access (PRMA) is an implicit reservation scheme. Here slots can be reserved implicitly as follows. A certain number of slots form a frame. The frame is repeated in time just like a TDM pattern. A base station now broadcasts the status of each slot to all mobile stations. All stations receiving this vector will then know which slot is occupied and which slot is currently free. If in a vector, all the slots are reserved, except one, then more than one station attempts to access this free slot. Thus, a collision occurs and the new status will be generated with the same free slot indicating that still one slot is free for reservation. Again stations can compete for this slot and the procedure continues. The final reservation status will be followed in that order in the network for actual transmission of packets.

10.4.2 Polling and Token Passing

Polling is a strictly centralized scheme with one master station and several slave stations. The master can poll the slaves according to many schemes: round robin (only efficient if traffic patterns are similar over all the stations) or randomly according to the reservations, etc. The master can also establish a list of stations wishing to transmit during a contention phase. After this phase, the station polls each station on the list. Similar schemes are used, for example in the Bluetooth wireless LAN and as one possible access function in IEEE 802.11 systems.

Token passing is mainly suitable for the wired networks. IEEE 802.4 is the token bus protocol on the LAN while IEEE 802.5 is a token ring protocol. Here the key part is the token, which is a small bit pattern. The station that is transmitting captures the token during the transmission and after reception of the complete data, it releases the token again circulating it among the various users. Thus, a logical or physical ring configuration is necessary for the token passing type of scheme.

Summary

- Multiple access schemes are necessary in multiple user environment to share the available bandwidth efficiently and to accommodate more number of users in the system.
- Multiple access schemes are of three types: fixed assignment-based access, random access, and reservation-based access.
- Fixed assignment of the resources can be done by choosing one of the four parameters: frequency, time, code, and space.
- Wireless channel is more susceptible to noise and interference and hence multiple access schemes may not give the exact performance as in the case of wired lines.
- FDMA does not use the spectrum efficiently while TDMA requires synchronization critically.
- CDMA schemes use both time and frequency efficiently but there is a limit on the number of users due to interference noise.
- SDMA is mainly suitable in cellular systems.
- ALOHA, slotted ALOHA, CSMA, CSMA/CD, ISMA, etc., are the random access MAC schemes.
- DAMA is a special type of random multiple access scheme, which is the combination of random and reservation-based access.
- Performance of non-persistant CSMA gives better throughput with rise in offered traffic.
- PRMA is a method based on packet transmission slot reservation.
- Polling, token passing, etc., are the types of reservation-based medium access schemes.

REVIEW QUESTIONS

1. What is the main reason for the failure of the MAC schemes sometimes in wireless networks that is there in wired networks?
2. Give the detailed classification of various medium access control schemes.
3. Explain the problem of hidden and exposed terminals.
4. How can you use the combinations of multiple access schemes? Explain with typical examples.
5. Justify that for real-time transmission, fixed assignment type of multiple access schemes are more suitable.
6. Why is near-far problem more serious in CDMA-based schemes?
7. Assume all the stations can hear all other stations. One station wants to transmit and senses that the carrier is idle. Why does collision still occur after the start of transmission?
8. Explain the term 'interference' in the frequency, time, code, and space domains. What are the countermeasures to avoid such problems in FDMA, TDMA, CDMA, and SDMA?
9. What is the difference between random access and reservation-based access?
10. What are the advantages of reservation schemes? How are collisions avoided during data transmission? Why is the probability of collisions lower compared to classical ALOHA? What are the disadvantages of reservation schemes?
11. Compare the performance of the various random access schemes and find out the best scheme suitable in the wireless environment.
12. Prove that the OFDMA is more spectrally efficient than FDMA.

PROBLEMS

1. Eight PCM channels, each bandlimited to 4 kHz, are time division multiplexed. Each sample is coded into a 8 bit word. Find the output rate and the required bandwidth.
2. How many broadcast stations can be accommodated in a 100 MHz bandwidth if the highest modulation bandwidth is 2 MHz? The guard band is 50 kHz.
3. If the bandwidth is 1.75 MHz and the bit rate is 9.6 kbps along with minimum acceptable E_b/I_o due to interference signals from other users found to be 9 dB. Avoid channel noise. Determine the maximum number of users that can be supported in a single-cell CDMA system, if (a) omnidirectional BS antenna without vioce activity detection and (b) three sectors at the base station along with voice activity detection with $v_f = 3/8$.

MULTIPLE CHOICE QUESTIONS

1. TDM
   (a) can be used with PCM only
   (b) interleaves pulses belonging to different transmissions
   (c) combines fine groups into a supergroup
   (d) stacks channels in adjacent frequency slots

2. Synchronous TDM is called synchronous because
   (a) synchronous transmission is used
   (b) time slots are preassigned to sources and fixed
   (c) preamble is transmitted
   (d) of none of the above
3. DAMA stands for
   (a) data accessibility master aerial
   (b) demand assignment multiple access
   (c) digital attenuator microwave antenna
   (d) dual accessibility mode antenna
4. A communication satellite is a repeater between
   (a) a transmitting station and a receiving station
   (b) a transmitting station and many receiving stations
   (c) many transmitting stations and one receiving station
   (d) many transmitting stations and many receiving stations
5. Each channel in the TDM input has bit rate of 16 kbps. If there are 4 channels, then the output bit rate will be
   (a) 16 kbps (b) 4 kbps
   (c) 32 kbps (d) 64 kbps
6. Which of the following multiple access schemes is used by the SPADE system in combination?
   (a) SDMA, TDMA, FDMA
   (b) DAMA, FDMA
   (c) SDMA, DAMA
   (d) DAMA, FDMA, TDMA
7. GSM system uses TDMA-FDMA together
   (a) to accommodate many low bit rate voice channels
   (b) to accommodate more users
   (c) to eliminate lower bandwidth problems
   (d) for all of the above
8. Which of the following is a random access method?
   (a) PRMA
   (b) FDMA
   (c) ISMA
   (d) Token passing
9. In which of the following multiple access methods, the whole bandwidth is utilized simultaneously by all the users?
   (a) TDMA (b) FDMA
   (c) CDMA (d) SDMA
10. In which of the following multiple access schemes, a user can use only a part of the total bandwidth?
    (a) TDMA (b) FDMA
    (c) CDMA (d) SDMA
11. In which of the following multiple access schemes, the frequency reuse concept is utilized?
    (a) DAMA (b) FDMA
    (c) CDMA (d) SDMA
12. In which of the following random access methods, the best performance is obtained?
    (a) ALOHA
    (b) Slotted ALOHA
    (c) 1-persistent CSMA
    (d) Non-persistent CSMA
13. For real-time voice transmissions, the suitable multiple access type is
    (a) random access
    (b) fixed assignment based access
    (c) reservation access
    (d) none of the above
14. In which of the following techniques, fast and accurate power control is required because each cell must be tightly synchronized?
    (a) TDMA (b) CDMA
    (c) OFDMA (d) FDMA

15. Which of the following statements is true?
   (a) The OFDMA provides benefits of frequency diversity and out-of-cell interference.
   (b) The OFDMA provides the code sequences, which are orthogonal.
   (c) The OFDMA is wideband and hence least spectrally efficient.
   (d) The OFDMA is nothing but FDMA.

16. Which of the following statements is true?
   (a) The CDMA uses frequency reuse concept in cellular infrastructure.
   (b) The CDMA system has code sequence that are truly orthogonal in the presence of multipath delay.
   (c) Due to voice activity factor, inclusion of the number of users in a CDMA system decreases.
   (d) None of the above is true.

chapter 11

Infrastructure to Develop Mobile Communication Systems: Cellular Theory

Key Topics

- Why cellular technology
- Cellular communication infrastructure: cells, clusters, and cell splitting
- Frequency reuse concept and reuse distance calculation
- Cellular system components
- Operations of cellular systems and handoff
- Channel assignment: fixed and dynamic
- Cellular interferences: co-channel and adjacent channel
- Antennas for the base station
- Sectorization
- Mobile traffic calculation
- Attributes of CDMA in cellular systems

Chapter Outline

Mobile telephone networks require the infrastructure in which the total area is divided geographically. Each small area is known as a cell and one tower in each cell is required for communication. *Mobile telecommunication switching office* (MTSO) or *mobile switching centre* (MSC) handles various calls and towers. The SDMA concept can be applied to cell-based systems. The cellular concept allows cells to be sized according to the subscriber density and demand of a particular area. As the population grows, cells can be split to accommodate that growth. Frequencies used in one cell can be reused in other cells. This makes the whole technique spectrally efficient. Services can be handed over from cell to cell to maintain uninterrupted conversation as the user moves from one cell to another. For understanding the use of cellular infrastructure, the complete cellular theory must be studied, which is given in this chapter. This infrastructure is mainly used for GSM, WLL, UMTS, and WiMAX systems.

11.1 WHY CELLULAR TECHNOLOGY

If mobility is to be provided to the subscriber, it is obvious that wireless links and radio trunking are required in the telephone system and these may be provided to the users systematically by developing a kind of infrastructure. The concept of cellular technology describes nothing but the various aspects of such infrastructure, like division of area, frequency management, call handling, etc. On the basis of such infrastructure, various mobile telephone systems are developed, which are described in Chapter 13. Particularly, the term *cell phone* is often used by the public when a wireless mobile phone is meant. The cellular approach was proposed and developed predominantly by the Bell System, in the US in the early 70's. The initial requirements were thought of as follows:

- A large subscriber capacity
- Efficient use of spectrum resources
- Nationwide coverage
- Adaptability to traffic density
- Telephone service to both vehicle and portable user terminals
- Toll quality
- Affordability, which can eventually make it a mass-market service

To satisfy these requirements, a simple hypothetical system should have the following components:

1. *Central station* This must contain the switching equipment and an RF transmitter and receiver.
2. *Mobile telephone* (one for each subscriber) This must contain microphone, speaker, dialing facility, a radio transmitter, and a receiver.

Apart from these, some infrastructure is required to support the mobility of the user as well as coverage of different areas with seamless connectivity.

Present use of cellular technology fulfils all the above-mentioned requirements. Cellular technology covers the whole country and international calls also and uses the available frequency spectrum efficiently as it is limited. All mobile operators use available radio spectrum to provide their services to the customers. Spectrum is generally considered as a scarce resource and has been allocated systematically. It has traditionally been shared by a number of industries, including broadcasting and mobile communications, as well as the military.

Before the advent of cellular technology, capacity was enhanced through a division of frequencies and the result was addition of available channels. However, this reduced the total bandwidth available to each user, affecting the quality of service. For example, if division of frequency-based scheme is used for cellular telephone system and 1,00,000 subscribers are to be accommodated, then 2,00,000 radio channels are required, as one channel for uplink and one channel for downlink are necessary. Let us now calculate the bandwidth. This will depend on the nature of modulation. Let us assume frequency modulation (one of the early mobile systems was thought of with FM). For speech, f_{max} is about 4 kHz. If we assume frequency deviation of 12 kHz, we need bandwidth of about 2(12+4) = 32 kHz. For simplicity, let us say 30 kHz bandwidth is required for each channel, then the total bandwidth required is 2,00,000 × 30 kHz = 6000 MHz. This is a ridiculously large requirement and practically impossible while approximately a total bandwidth of 50 MHz is allocated for the cellular telephony application. Division of time was also used by some systems, but only division of frequency or division of time would not be sufficient to accommodated the large number of subscribers and something extra was needed for the system. Cellular technology allowed for the division of geographical areas, rather than only frequencies or time, and then assigning the channel frequencies reusing them geographically. This geographical reuse of radio channels is known as *frequency reuse* and leads to a more efficient use of the radio spectrum. Presently, GSM system uses a combination of FDMA/TDMA along with frequency reuse concept, which will be discussed in detail in Chapter 13.

To describe cellular systems in general, it is necessary to include a discussion of the basics of the cellular systems, their performance criteria, the uniqueness of the mobile radio environment, the operation of the cellular systems, reduction of channel interferences, handoffs, and so forth. There are basically two types of cellular systems; one is analog or digital circuit switched and the other is the packet switched system. The description of these two types of systems is given later on in this chapter.

11.2 CELLULAR RADIO COMMUNICATION INFRASTRUCTURE

In modern cellular telephony, rural and urban regions are divided into areas according to specific provisioning guidelines. Deployment parameters, such as amount of cell splitting and cell sizes, are determined by engineers, experimented in cellular system architecture. Provisioning for each region is planned according to an engineering plan, which includes cells, clusters, frequency reuse, and handovers. Presently, software tools are also available for the cellular designing.

11.2.1 Cells, Clusters, and Cell Splitting

A cell is the basic geographic unit of a cellular system. When cluster maps are to be drawn, the question is how the cell should be depicted. Circle seems to be natural choice since an omnidirectional antenna with circular radiation pattern can be placed at the centre of the cell. However, if the large area is to be filled with circles, either overlaps or gaps may be there. Hence, the selection of the hexagonal shape is most suitable. The term *cellular* comes from the honeycomb (hexagonal) shape of the areas into which a coverage region is divided theoretically. Cell-wise, one base station is providing transmission over small geographic areas. Each cell size varies depending on the landscape. Because of constraints imposed by natural terrain and human-made structures, the true shape of cells is not a perfect hexagon.

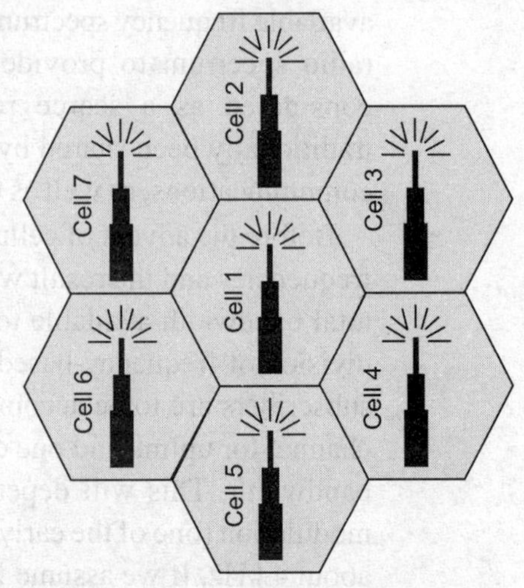

Fig. 11.1 Cluster of seven cells ($N = 7$)

A cluster is a group of cells and is denoted as N. No channels are reused within a cluster. Figure 11.1 illustrates a seven-cell cluster. In different clusters, same group of frequencies can be reused as mentioned in frequency reuse concept.

Restrictions on the value of N

If we choose any cluster size arbitrarily, we may not be able to cover the entire area using clusters without leaving gaps. To ensure that there are no gaps, N is restricted by the following equation:

$$N = i^2 + ij + j^2 \tag{11.1}$$

This can be proved by the use of geometry of hexagonal cells. Possible cluster sizes are obtained only with Eq. (11.1) with non-negative integers i and j. The integers i and j determine the relative location of co-channel cells as shown in Fig. 11.3.

$$i = 1, j = 1 :: N = 3$$
$$i = 2, j = 1 :: N = 7$$
$$i = 2, j = 2 :: N = 12$$
$$i = 3, j = 2 :: N = 19$$

Example 11.1 Show that the area of a hexagonal cell is 2.598 R^2, where R is the radius of the hexagon.

Solution

A hexagon is made up of 6 triangles as shown in Fig. 11.2.

From the geometry of the figure, $a = R \cos \theta$

$$\Rightarrow \quad \frac{a}{R} = \cos 60° = 1/2$$

$$\Rightarrow \quad a = \frac{R}{2}$$

Also $\quad b^2 = R^2 - a^2 = R^2 - R^2/4 = 3\,R^2/4$

$$\Rightarrow \quad b = \sqrt{\frac{3}{4}}\,R$$

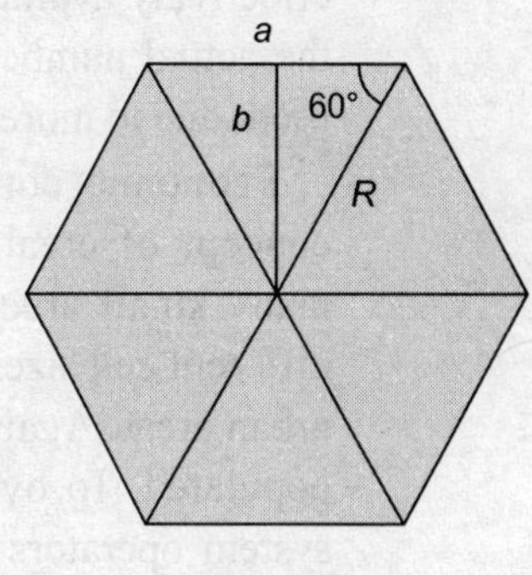

Fig. 11.2

Now, area = 6 × area of triangles

$$= 6 \times \frac{1}{2}\,(2ab)$$

$$= 6ab$$

$$= 6 \times \frac{R}{2} \times \sqrt{\frac{3}{4}}\,R$$

$$= 2.598\,R^2$$

There is a relationship between cluster design equation and frequency reuse as well as reuse distance. The frequency reuse is described in Section 11.2.2. However, according to Fig. 11.3, to find the nearest co-channel cell, one must do the following:

1. Move i (here $i = 2$) cells along any chain of hexagons (here vertically up).
2. Turn 60° clockwise and move j (here j=1) cells.

The approached cell is the co-channel cell of the host cell from which moves are started.

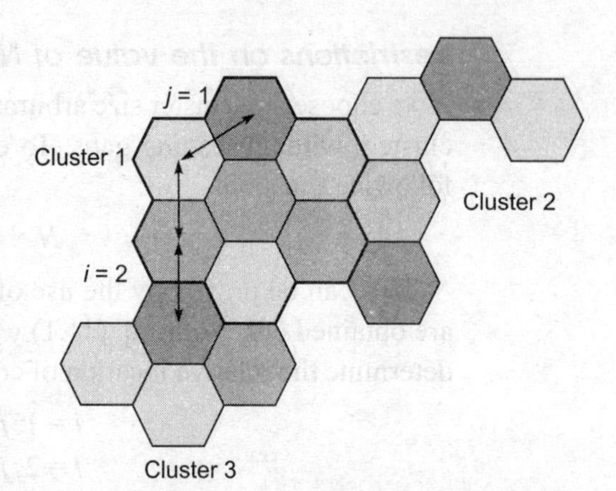

Fig. 11.3 Co-channel cell search in seven-cell cluster with i = 2 and j =1 (The cluster is represented with thick outlines.)

The size of a cell depends on the density of subscribers in an area. For instance, in a densely populated area, the capacity of the network can be improved by reducing the size of a cell or by splitting into more cells along with low power base stations. According to the size of the cell, cells may be categorized as macrocells, microcells, and picocells. Smaller cells use low-power transceivers to cover a smaller region; so, the frequency reuse can also be increased without interferences. Smaller cells increase the number of channels effectively available without increasing the actual number of frequencies being used, due to more divisions of area.

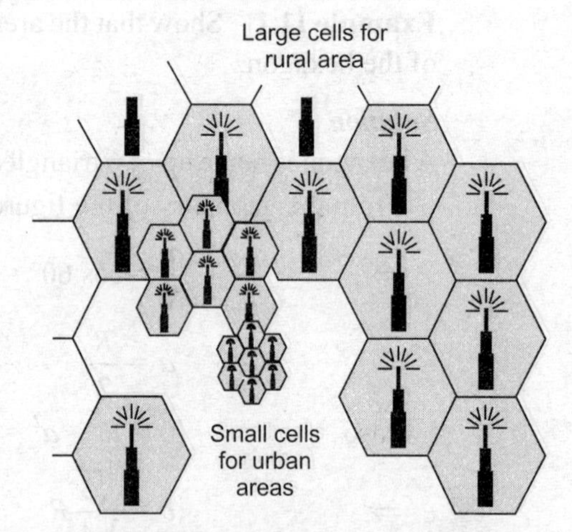

Fig. 11.4 Cell splitting concept

Economic considerations made the concept of creating full systems with many small areas impractical. Hence, different cell sizes are used for rural and urban areas. Again urban area is densely populated. To overcome this difficulty, system operators developed the idea of cell splitting. As a service areas becomes full of users, this approach is used to split a single area into smaller ones. In this way, urban centres can be split into as many areas as necessary to provide acceptable service levels in heavy traffic regions, while less expensive bigger cells can be used to cover remote rural regions. Figure 11.4 shows the division of an area in small cells.

11.2.2 Principles of Cellular Frequency Reuse

As limited radio channel frequencies were available for mobile systems, engineers had to find a way to carry number of circuit switching-based conversations at a time. The solution adopted is called *frequency planning* or *frequency reuse*. A radio channel consists of a pair of frequencies for full duplex operation. Each frequency has its coverage within the cell radius R. Another cell with same coverage, but at distance D can use the same frequency. If the frequency reuse is not properly designed, serious interference may occur. This type of interference is called *co-channel interference*.

The concept of frequency reuse is based on assigning to each cell, a group of radio channels used within a small geographic area, a set of N different frequency groups $\{f_1, ..., f_N\}$ are used for each cluster of N adjacent cells. Cells are assigned a group of channels that are completely different from the neighbouring cells. The coverage area of cells is called the *footprint*. The footprint is limited by a boundary so that the same group of channels can be used in different cells that are far enough away from each other so that their frequencies do not interfere. Figure 11.5 represents the frequency reuse concept without interference.

According to Fig. 11.5(b), cells with the same number have the same set of frequencies. Here if the number of available frequency set is 7, the frequency reuse factor is 1/7. That is, each cell is using 1/7 of the available cellular channel for reuse. Also from Figures 11.5(a) and (b), it is clear that with increase in cluster size, the co-channel cell distance also increases.

With the help of the following example, the clustering, frequency reuse concept, and effectiveness of frequency reuse will be understood in a better way.

Let L = total number of duplex channels available for reuse (i.e. frequencies per cluster)

k = number of duplex channels allocated to each cell of a cluster (obviously, $k < L$)

N = cluster size (in which N cells are there)

M = number of times the cluster is repeated

C = total effective number of duplex channels available in the area

Thus,

$$L = k \times N \tag{11.2}$$

$$C = M \times L \tag{11.3}$$

For example, if the service provider is allocated 21 duplex radio channels and if he or she uses 1000 clusters, then the effective number of channels will be 21,000. Using

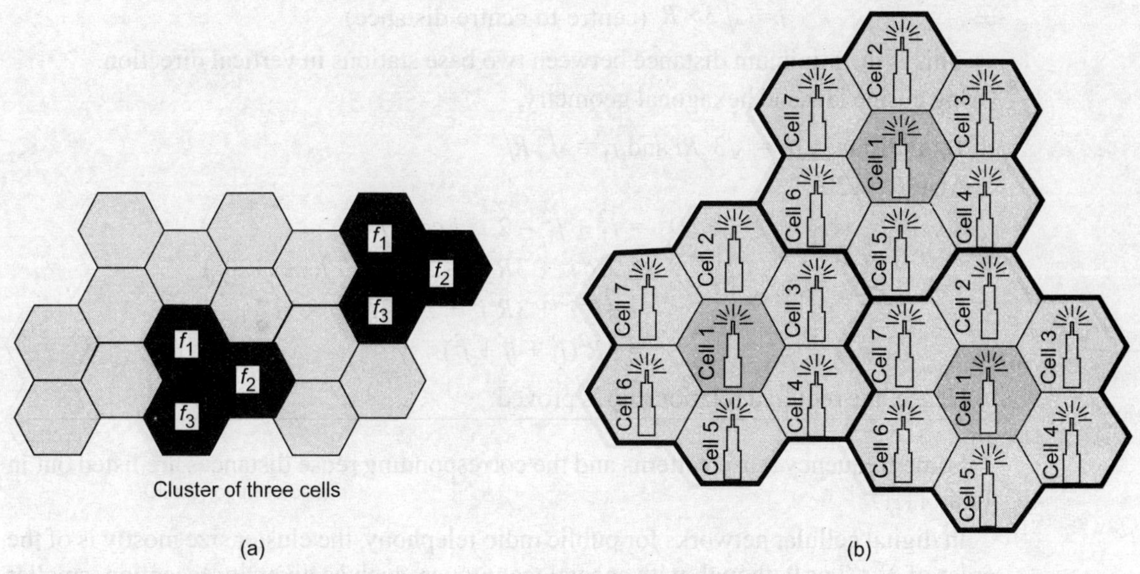

Fig. 11.5 (a) Frequency reuse plan for N = 3 (black cells show clusters) and (b) frequency reuse plan for N = 7 in three clusters without interference

demand assignment and assuming statistical results, as many as 2 lakh customers could be served. If TDMA is also accommodated and single slot per user is assigned, then arbitrarily for 8 slots per frame per channel, almost 16 lakh subscribers can be served. Thus, the effective number of channels can be increased by using techniques efficiently.

It should be remembered that this customer base is distributed over a wide area (for example, an entire state), it is obvious that larger the value of M, larger is the capacity. This means that the cluster size has to be small. This will reduce the distance between the cells having the same set of frequencies and hence the interference increases. Therefore, there has to be a compromise between the requirement of larger capacity and smaller interference. The total bandwidth for the system is N times the bandwidth occupied by a single cell.

Reuse Distance Calculations

The closest distance D between the centres of two cells using the same frequency (in different clusters) is determined by the choice of the cluster size N and the layout of the cell cluster. This distance is called *frequency reuse distance*. It can be shown that the reuse distance D, normalized to the size of each hexagon, is

$$D = \sqrt{3N} \times R \tag{11.4}$$

Example 11.2 Show that $D = \sqrt{\{3(i^2 + ij + j^2)\}} \times R$.

Solution

Looking at the geometry of Fig. 11.2, one hexagon means six triangles with equal sides. Radius of the hexagon is R. Hence, all sides of the triangles are also R.

So, $$(i/2)^2 = R^2 - (R/2)^2$$

$\Rightarrow$ $$i = \sqrt{3} \times R \text{ (centre-to-centre distance)}$$

$\Rightarrow$ This is the minimum distance between two base stations in vertical direction.

Using cosine law and hexagonal geometry,

Total distance, $i_1 = \sqrt{3}\ Ri$ and $j_1 = \sqrt{3}\,Rj$

Also $\theta = 120°$

So,
$$\begin{aligned} D^2 &= i_1^2 + j_1^2 - 2\, i_1 j_1 \cos\theta \\ &= 3R^2i^2 + 3R^2j^2 - 2 \times 3R^2\, i\, j \cos(120°) \\ &= 3R^2i^2 + 3R^2j^2 + 2 \times 3R^2 i\, j \times ½ \\ &= 3R^2(i^2 + ij + j^2) \end{aligned}$$

Hence, the required relationship is proved.

Some frequency reuse patterns and the corresponding reuse distances are listed out in Table 11.1.

In digital cellular networks for public radio telephony, the cluster size mostly is of the order of $N = 7$ or 9, though with special techniques, such as diversity reception, smaller

Table 11.1 Some reuse distances

Cluster Size (N)	Reuse Distance (D)
4	$3.46R$
7	$4.6R$
12	$6R$
19	$7.55R$

reuse distances can be used. The GSM can also work with $N = 3$ or 4. Cellular CDMA systems can use $N = 1$ and the same frequency is used in all cells.

11.3 REAL-WORLD CELLS

In the practice of cell planning, cells are not hexagonal as in the theoretical studies. Computer methods are being used for optimized planning of base station location and cell frequencies. Path loss and link budgets are computed from the terrain features and antenna data. This determines the coverage of each base station and interference to other cells. An example of cellular planning tool is provided in Fig. 11.6.

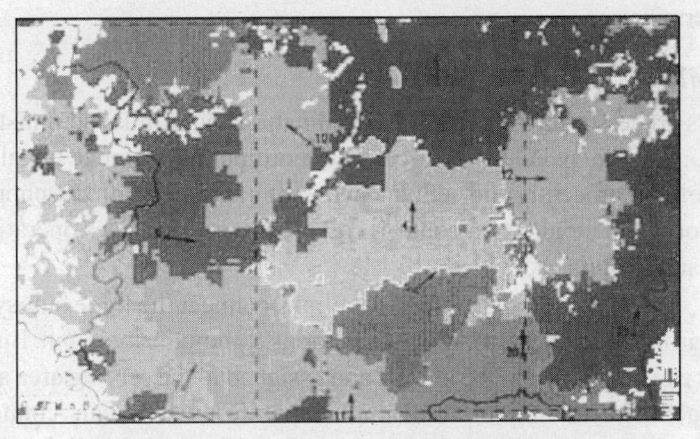

Fig. 11.6 An example of cellular planning tool

11.4 CELLULAR SYSTEM COMPONENTS

Along with cell structure, some system components are required to make the whole cellular system work. The cellular system offers mobile and portable telephone stations with the same service provided by the fixed stations over conventional wired loops. It has the capacity to serve tens of thousands of subscribers in a major metropolitan area. The cellular communications system consists of the following major components that work together to provide mobile service to subscribers.

11.4.1 Analog Circuit Switched Cellular System Components

Basically, there are three subsystems of analog circuit switched cellular system—mobile unit, cell site, and MTSO e.g. in AMPS.

Mobile Unit

It is also called *mobile subscriber unit* (MSU). A mobile telephone unit contains a control unit with a battery, a transceiver, and an antenna system that transmits and receives radio transmissions to and from a cell site. The following three types of MSUs are available:

- Mobile telephone (typical transmit power being 4.0 watts)
- Portable (typical transmit power being 0.6 watts)
- Transportable (typical transmit power being 1.6 watts)

The mobile telephone is installed in the trunk of a car, and the handset is installed in a convenient location to the driver. Portable and transportable telephones are hand-held and can be used anywhere. The use of portable and transportable telephones is limited to the charge life of the internal battery.

Cell Site or Base Station (BS)

The term is used to refer to the physical location of the radio equipment that provides coverage within a cell. The cell site provides interface between the MTSO and the mobile units. A list of hardware located at a cell site or base station includes control unit, interface equipment, radio frequency transmitters and receivers, antennas, a power plant, and data terminals.

MTSO

Its processor provides central coordination and cellular administration. The central office for mobile switching, the central coordinating element for all cell sites, contains the cellular processor and cellular switch. It interfaces with telephone company zone offices, does field monitoring, controls call processing, provides operation and maintenance, and handles billing activities.

The radio and high-speed data links connect the three sub-systems. Each mobile unit can only use one channel at a time for its communication link. But the channel is not fixed; it can be any one in the entire band assigned to the serving area as mentioned earlier, with each site having multichannel capabilities that can connect many mobile units simultaneously. The MTSO is the heart of the analog cellular mobile system.

The cellular switch, which can be either analog or digital, calls to connect mobile subscribers to other mobile subscribers and to the nationwide telephone network. It uses voice trunks similar to telephone company interoffice voice trunks. It also contains data links providing supervision links between the processor, the switch, and between the cell sites and the processor. The radio link carries the voice and signalling between the mobile unit and the cell site. The high-speed data links cannot be transmitted over standard telephone trunks and, therefore, must use either microwave links or T-carriers (wire lines). Microwave radio links or T-carriers carry both voice and data between cell site and the MTSO.

11.4.2 Digital Circuit Switched Cellular System Components

The digital circuit switched cellular system basically has four subsystems as discussed below e.g. GSM-900.

Mobile Station (MS)

It consists of two parts, mobile equipment with battery and subscriber identity module (SIM). The SIM contains all the subscriber specific data stored on the MS side.

Base Station or Base Transceiver Station (BTS)

Besides having the same function as the analog BS/BTS, it has the *transcoder/rate adapter unit* (TRAU), which carries out coding and decoding as well as rate adaptation in case the data rate varies.

Base Station Controller (BSC)

It is a new element in digital systems that performs the radio resource (RR) management for the cells under its control. It also handles handovers, power management, time and frequency synchronization, and frequency reallocation among BTSs.

Switching Subsystems

Mobile switching centre (MSC) is the main element that coordinates the set-up of calls between MS and PSTN users. Visitor location register (VLR), home location register (HLR), authentication centre (AUC), equipment identity register (EIR), and operation and maintenance centre (OMC) are the other elements in the switching subsystem. These elements will be dealt with in detail in Chapter 13 while studying GSM system.

11.4.3 Packet Switched Cellular System Components

An example of the cellular packet switched system is the GPRS, which has six elements: MS (user equipment), node B (the name for base station in GSM), radio network controller (RNC), service GPRS support node (SGSN), gateway GPRS support node (GGSN), and charging gateway function (CGF). All these elements and their functions will be described in Chapter 13 along with GPRS system.

11.5 OPERATIONS OF CELLULAR SYSTEMS

The operations of cellular systems can be divided into five different tasks, namely mobile originated call, mobile unit initialization to receive a call, network originated or landline phone originated call, call termination, and handoff procedure.

11.5.1 Mobile Originated Call

There are two possibilities: mobile-to-mobile call and mobile-to-landline call.

The user places the called number into an originating register in the mobile unit, and pushes the *send* button. A request for service is sent on a selected set-up channel obtained from a self-location scheme. The cell site receives it and in directional cell sites (or sectors in case of sectorization), selects the best directive antenna for the voice channel to use. At the same time, the cell site sends a request to MTSO/MSC via a high-speed data link. The MTSO/MSC selects an appropriate voice channel for the call, and the cell site acts on it through the best directive antenna to link the mobile unit. The

MTSO/MSC also connects to landline telephones through the telephone company zone office (local exchange).

11.5.2 Mobile Unit Initialization to Receive a Call

There are two possibilities: mobile-to-mobile unit call and landline-to-mobile unit call.

When a user activates the receiver of the mobile unit, the receiver scans the set-up channels. It then selects the strongest channel frequency and locks on for a certain time. Because each site is assigned a different set-up channel, locking onto the strongest set-up channel usually means selecting the nearest cell site. This self-location scheme is used in the idle stage and is user independent. It has a great advantage because it eliminates the load on the transmission at the cell site for locating the mobile unit. The disadvantage of the self-location scheme is that no location information of idle mobile units appears at each cell site. Therefore, when a call initiates from the landline to a mobile unit, the paging process is longer. For a large percentage of calls originating at the mobile unit, the use of self-location schemes is justified. After a given period, the self-location procedure is repeated. A feature called 'registration' is used while landline originated calls occur.

11.5.3 Network Originated or Landline Phone Originated call

A landline party dials a mobile unit number. The telephone company zone office recognizes that the number is mobile and forwards that call to the MTSO/MSC. The MTSO/MSC sends a paging message to certain cell sites based on the mobile unit number and the search algorithm. Each cell site transmits the page on its own set-up channel. If the mobile unit is registered, the registered site pages the mobile. The mobile unit recognizes its own identification on a strong set-up channel, locks onto it, and responds to the cell site. The mobile unit also follows the instruction to tune to an assigned voice channel and initiates user alert.

11.5.4 Call Termination

When the mobile user turns off the transmitter, a particular signalling tone transmits to the cell site and both sides free the voice channel. The mobile unit resumes monitoring pages through the strongest set-up channel.

11.5.5 Handoff or Handover Procedure

Handover basically may be hard, soft, softer, etc. Here first the hard handoff is described and rest are described later on. The hard handover occurs when the change in assigned channel frequency occurs during transition from one cell to another. The obstacle in the development of the cellular network involved the problem created when a mobile subscriber travelled from one cell to another during a call. As adjacent areas do not use the same radio channels, a call must either be dropped or transferred from one radio channel to another when a user crosses the line between adjacent cells. Because dropping the call is unacceptable, the process of handoff is created. The handoff occurs when the mobile telephone network automatically transfers a call from radio channel to radio channel as the mobile crosses adjacent cells.

A few terms associated with handoff procedure are defined below.

Handoff probability It is the probability that a handoff is executed before call termination.

Rate of handover It is the number of handovers per unit time.

Interruption duration It is the duration of time during the handover procedure in which a mobile is not connected to either base station.

Handoff delay It is the delay over the distance the mobile moves from the point at which the handoff should occur to the point at which it does occur.

Probability of unsuccessful handoff It is the probability that a handoff is executed while the reception conditions are inadequate.

Handoff blocking probability It is the probability that a handover cannot be completed successfully.

The scenario of handover is shown in Fig. 11.7(a). Here the handover procedure is between base station A and base station B and the mobile is moving towards base station B from A crossing the cell area of A.

Each mobile uses a separate temporary radio channel to talk to the cell site. The cell site talks to many mobiles at once, using one channel per mobile. As radio energy dissipates over distance, mobiles must stay near the base station to maintain communications. When the mobile unit moves out of the coverage area of a given cell site, the reception becomes weak. At this point, the cell site in use requests a handoff.

When a mobile terminal moves outside the coverage area of its base station, the network management is assumed to take appropriate measures. Immediate connection to another base station is required to ensure sufficient quality of reception, including acceptable interference power levels. Cellular technology allows the 'handoff' of subscribers from one cell to another as they travel around. A computer constantly tracks mobile subscribers of units within a cell and when a user reaches the border of a cell, the computer automatically hands off the call and the call is assigned a new channel in a different cell. International roaming arrangements govern the subscriber's ability to make calls to and receive calls from the home network's coverage area. A mobile user experiences the worst link quality if the terminal is located at the boundary of two cells where the distances to base stations are longest.

For calculating handoff threshold, some considerations are taken into account. The system designer must specify an optimum signal level at which to initiate the handoff. Once a particular signal level is specified as the minimum usable signal for acceptable voice quality at the base station receiver, a slightly stronger signal level is used as a threshold at which a handoff is made as shown in Figure 11.7(b). This margin is given by

$$\delta = P_{r\,\text{handoff}} - P_{r\,\text{min usable}} \tag{11.5}$$

where δ should not be too large or too small. If δ is too large, unnecessary handoffs may occur, which may burden the MSC. If it is small, there may be insufficient time to complete a handoff before a call is lost due to weak signal conditions. So, it must be chosen very carefully.

Referring to Fig. 11.7(c), the handoff occurs only if the new base station is sufficiently stronger than the current one by a margin δ. It occurs at the mentioned level HO. This

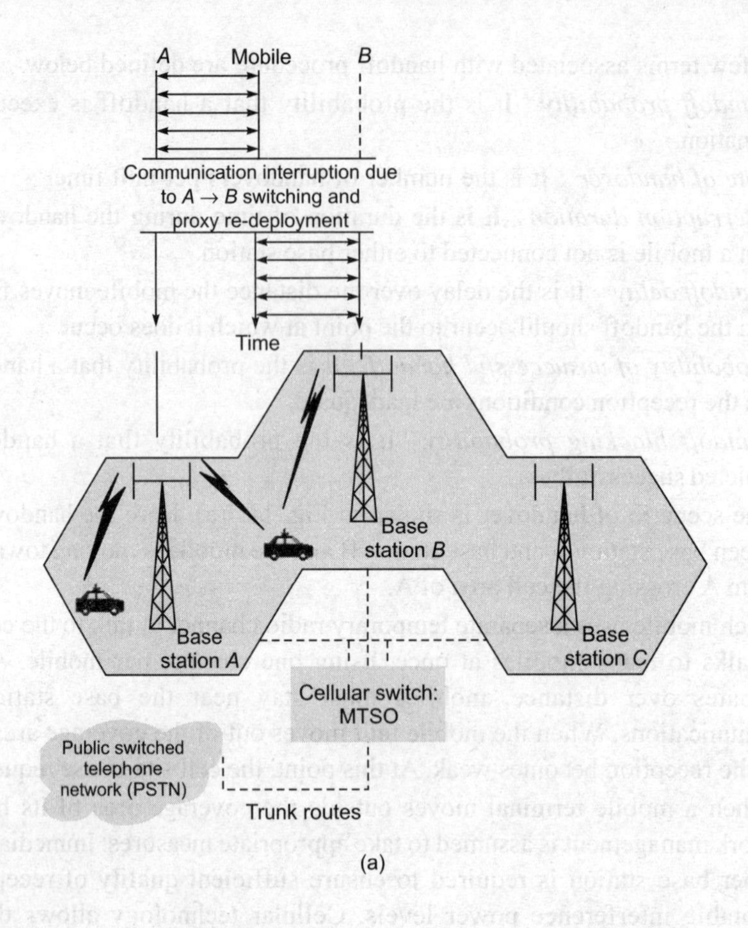

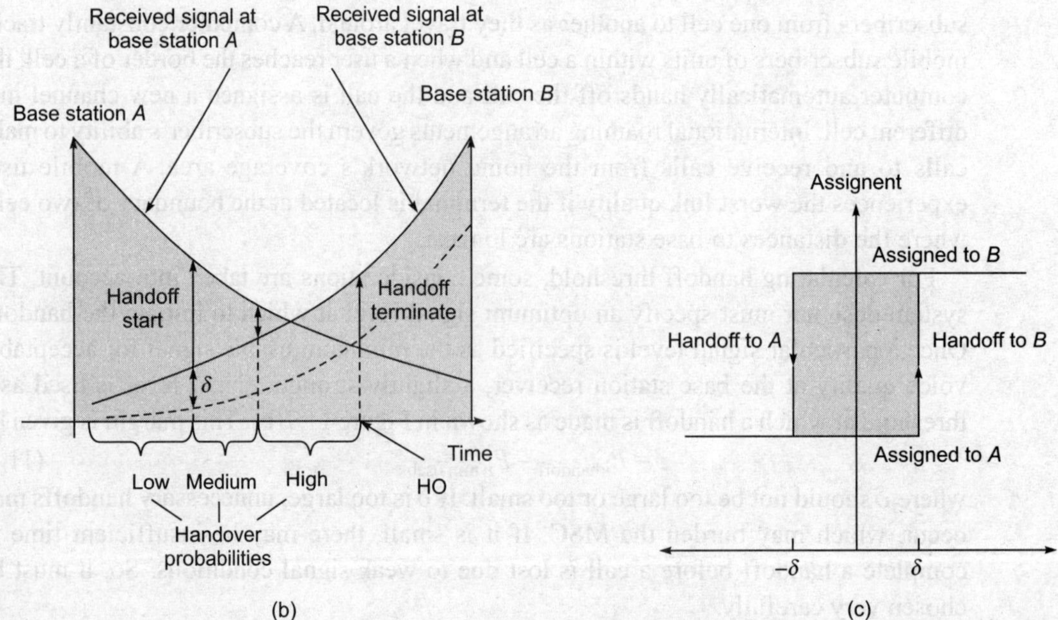

Fig. 11.7 (a) Cellular infrastructure with base stations and handover concept, (b) received signal strength variations and timing during handoff, and (c) hysteresis mechanism for handoff.

scheme prevents the ping pong effect because once the handover occurs, the effect of the margin δ is reversed. It exhibits hysteresis. We can think of the handoff mechanism as having two states. While the mobile is assigned to base station A, the mechanism will generate a handoff when the relative signal strength reaches or exceeds δ. Once the mobile is assigned to B, it remains so until the relative signal strength falls below δ, at which point it is handed back to A.

Different strategies for handover exist as given below:

- Centralized methods [for instance, used in GSM—Chapter 13]
- Decentralized methods [for instance, used in DECT (WLL)—Chapter 13]

Handovers are mostly made for uninterrupted services along with improvement of the received signal power or the C/I ratio of a particular link. It can also be made to free spectrum resources in one cell if that cell becomes fully loaded with traffic. In such a case, handovers become a part of the dynamic frequency assignment (DFA) scheme.

The monitoring of handoff can be done either by measurement done at the base stations or measurements done by the mobile. Thus there are two methods of handoff,

Mobile Assisted Handoff

Every mobile measures the received power from the surrounding base stations and continuously reports the results of the measurements to the serving base station. A handoff is initiated when the power received from the base station of a neighboring cell begins to exceed the power received from the current base station by a certain level or a certain period of time. This method is used in the current mobile systems.

Base Station Assisted Handoff

In the first-generation systems, the strength measurements are made by the base stations and supervised by the MTSO. Here the base station measures the signals from mobiles served by it and also from the mobiles in the neighbouring cells and reports to MTSO. MTSO decides whether handoff is necessary or not, and for whom. Here MTSO remains loaded more compared to previous method and hence handoffs are slower.

Many times, prediction techniques are used in handover. The handoff decision is based on the expected future value of the received signal strength. There is another important issue during the handoffs. The changes in the signal level occur not only because of moving far from the base station (i.e. increased distance) but also because of the relative movement of the nearby objects. This is because the reflections from these objects also determine the signal level. This variation in the signal level from time to time is called *fading*, which is described in Chapter 5. Hence, handoffs may occur even when they are not needed. To avoid this situation, handoffs are based on running average rather than the instantaneous values.

The queuing of handoff request is a method to decrease the probability of forced termination of a mobile call due to lack of available channels. This is possible because of the fact that there is a finite time interval between the time the received signal level drops below the handoff threshold and the time when the call may be terminated due to insufficient signal level. The queue size and the delay time depend upon the traffic condition. Thus, the priority-based handover may be possible. The queuing does not guarantee a zero probability of forced termination.

We have so far described the hard handoff. However, the spread spectrum (CDMA) based mobiles share the same channel in every cell. Here, the term *handoff* does not mean a physical change in the assigned channel, but rather that a different base station handles the radio communication task. By simultaneously evaluating the received signals of neighbouring base stations, the MSC may decide which version of the user's signal is best at any moment in time. This technique exploits space diversity provided by the different physical locations of the base stations and allows the MSC to make a soft decision as to which version of the user's signal to pass. This ability to select between instantaneous received signals from a variety of base stations is called *soft handover*. The technique is described in detail in Chapter 13.

Example 11.3 A mobile is moving along a straight line from base station BS1 to base station BS2 as shown in Fig. 11.8. The distance between the base stations is 2200 metre. For simplicity, small-scale fading is neglected. The received power $P_{r,i}(d_i)$ in dBm at base station i (here $i = 1, 2$ for two base stations) from the mobile station is modelled as a function of distance d_o on the reverse link. All distances are in metre.

$$P_{r,i}(d_i) = P_o - 10n \log_{10}\left(\frac{d_i}{d_o}\right) \text{ dBm}$$

Given that P_o = received power at base station $BS2$ at distance d_o from mobile antenna = 0 dBm at the time of handoff, $d_o = 1$ metre, the path loss exponent is 3, and the minimum usable power $P_{r\text{ min usable}} = -87$ dBm. The threshold level used by the switch for handoff initialization is $P_{r\text{ handoff}}$. The mobile is currently connected to BS1 and moving towards BS2 allowing handoff procedure and there is no call loss during the handoff. Also the antenna heights are negligible compared to the distance between the mobile and the base station. The handoff time is 4.5 seconds and the velocity of the mobile station is 100 km/hr. Determine the minimum required margin for handoff δ and the effect of the margin on the performance of the cellular systems.

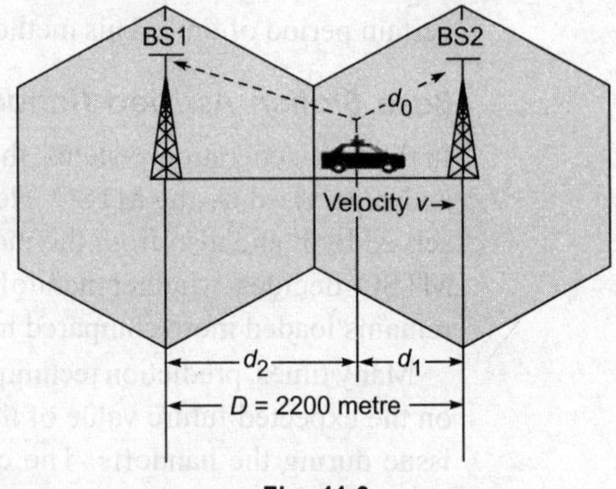

Fig. 11.8

Solution

We have $v = 100$ km/hr $= 100 \times 1000/3600 = 1000/36$ m/s

Distance at which the handoff starts,

$$d_1 = \text{velocity/ handoff time}$$

$$= (1000 \times 4.5)/36$$

$$= 125 \text{ metre}$$

We have $P_{r,i}(d_i) = P_o - 10n\log_{10}\left(\frac{d_i}{d_o}\right)$ dBm

Here, $P_o = 0$ dBm, $\frac{d_i}{d_o} = 125/1 = 125$, and path loss exponent $n = 3$

So, the second term in the received power equation will be 62.9 dBm.

$$P_{r,2}(d_2) = P_{r\,\text{handoff}} = 0 - 62.9 = -62.9 \text{ dBm} \quad \text{and} \quad P_{r\,\text{min usable}} = -87 \text{ dBm}$$

Now, $\delta = P_{r\,\text{handoff}} - P_{r\,\text{min usable}}$

$$= -62.9 \text{ dBm} - (-87 \text{ dBm})$$

$$= 24.1 \text{ dBm}$$

For minimum usable power, the distance from BS2 is 794.3 metre because

$$87/3 \times 10 = \log_{10}\left(\frac{d_i}{d_o}\right) \Rightarrow \frac{d_i}{d_o} = 794.3$$

The handover procedure will have a margin of 669.3 metre, which is large compared to the total distance of 2200 metre.

The received power when the handoff starts is stronger than the minimum usable power, thus successful handover can be completed. The margin should not be too large or too small. If it is too large, unnecessary handoff burdens the MSC, and if it is too less, there may be insufficient time to complete a handoff before a call is lost due to weak signal condition. Larger margin is beneficial when the queuing is used due to unavailability of the channel at BS2.

11.6 CHANNEL ASSIGNMENT

The aim of channel assignment is to use the available radio spectrum efficiently. There are different ways of channel assignment, such as fixed channel assignment, borrowing strategy, and dynamic channel assignment.

Fixed Channel Assignment

In this method, each cell is allocated a predetermined set of voice channels. Any call attempt within the cell can only be served by the unused channels in that particular cell. If all the channels are occupied, then the call will be blocked and the subscribers will not get any service.

Example 11.4 If a total of 36 MHz of bandwidth is allocated to a particular frequency division duplex cellular telephone system, which uses two 25 kHz simplex channels to provide full duplex voice and control channels, compute the number of channels available per cell if a system uses (a) 7 cell reuse and (b) 12 cell reuse.

Solution

The total bandwidth given is 36 MHz.

Channel bandwidth $= 25 \text{ kHz} \times 2$ simplex channels

$= 50$ kHz per duplex channel

The total available channels within 36 MHz range = 36 MHz/50 kHz
= 720 channels

(a) If $N = 7$, then the total number of available channels per cell is 720/7 = 102.85 (rounded to 102 channels)

(b) If $N = 12$, then the total number of available channels per cell is 720/12 = 60

Borrowing Strategy

It is the variation in the fixed assignment type scheme. A cell is allowed to borrow a channel from a neighbouring cell if all of its own channels are occupied. The MSC supervises such a borrowing procedure such that borrowing of a channel does not disrupt or interfere with any of the calls in progress in the donor cell.

Dynamic Channel Assignment

Here voice channels are not allocated permanently in any of the cell. Each time a call request is made, the serving base station requests a channel from the MSC. The switch then allocates a channel to the requested cell. While allocating the channel, it will follow an algorithm that will take into account the traffic statistics and likelihood of future blocking within the cell, the frequencies already used, reuse distance of the channel, cost functions, etc. The MSC only allocates a given frequency if that frequency is not presently used in the cell or any other cell within the minimum reuse distance to avoid cochannel interference.

There are a few advantages of the dynamic channel assignment strategy. The likelihood of call blocking reduces, which increases the trunking capacity of the system because all the available channels to the service provider are accessible to all of the cells. Thus, it increases the channel utilization.

The disadvantage is that MSC must collect real-time data on channel occupancy, traffic distribution, and radio signal strength indications of all channels on a continuous basis. This requires faster processor. Also this increases the storage and computational load on the system.

11.7 CELLULAR INTERFERENCES

Co-channel interference and adjacent channel interference are two major types of cellular interferences. We have already described the co-channel interference. It cannot be combated by simply increasing the carrier power of a transmitter. This is because an increase in the carrier transmit power increases the interference to the neighbouring co-channel cells. Frequency reuse distance calculation becomes a critical issue here.

Signal-to-Interference Ratio Calculation for Co-channel Interference

If we observe the co-channel cells surrounding a particular cell, they exist in a circular (or rather hexagonal) pattern. The nearest circle of cells will interfere maximum. They are also called first-tier cells and they are always 6 in number. Depending upon the cluster size selection, the size of the tier will also change. However, for a larger cluster size, there is very less or negligible co-channel interference from the cells of second and higher number tiers. If we consider in general and if i_{int} is the total number of co-channel cells that create the interference to the cell in the centre, then the signal-to-interference ratio (SIR) for a mobile receiver, monitors a forward channel, can be expressed by

$$\frac{S}{I} = \frac{S}{\sum_{i=1}^{i_{\text{int}}} I_i} \tag{11.6}$$

where S is the desired signal power from the desired base station and I_i is the interference power of the base station of interfering co-channel cell (neglecting noise). If the signal levels of the co-channel cells are known, the SIR for the forward link can be found. For the received power calculation due to propagation loss, Eq. (5.14) can be applied, where instead of square of the distance ratio, nth power will be used and n is the path loss exponent. The value of n is 2 in free space. However, in mobile systems, rarely a signal will reach the receiver by line-of-sight path. There will be a scattering and absorption loss by an obstacle. Hence, the empirical value of n ranges from 2 to 5. A typical value assumed for cellular systems is $n = 4$ (mostly assumed). If the first layer of the six interfering cells is considered, if all the interfering base stations are equidistant from the desired base station and this distance is equal to the distance D between the cell centres (this assumption is valid because $R << D$), and if the path loss exponent is same throughout the coverage area, then Eq. (11.6) can be written in terms of the radius of the cell R, the distances of the interfering base stations to the desired mobile receiver D_i's, D/R ratio, and the cluster size N.

$$\frac{S}{I} = \frac{R^{-n}}{\sum_{i=1}^{6} (D_i)^{-n}} \tag{11.7}$$

In other words, we can write

$$S = k \times R^{-n}$$

where k is the proportionality constant, which depends on the power radiated by other base stations.

Now, $$I = 6\,k \times D^{-n}$$

So, $$\frac{S}{I} = \frac{(D/R)^n}{6} = \frac{(\sqrt{3N})^n}{6} \text{ or } \frac{1}{6}(3N)^{n/2} \tag{11.8}$$

From Eq. (11.8), it is clear that N should be designed critically for the desired SIR. It can be found that to have the SIR at least 18 dB or more, the cluster size must be 7, assuming the path loss exponent $n = 4$. Obviously, these equations are derived assuming hexagonal geometry. Practically, these results may differ. The D/R ratio is also called co-channel reuse ratio.

A well-maintained S/I level helps to reduce or eliminate co-channel interference. The target levels for the various wireless technologies are as follows:

- AMPS: 17 dB

- GSM: 13 dB
- CDMA: No need for well-maintained S/I (due to N=1 basis of system and the fact that it is a noise-based system)

Example 11.5 The desired SIR for a cell-based system is 15 dB. Considering only the first tier of the interfering cell, find out the frequency reuse factor for maximum capacity. Assume path loss exponent $n = 3$.

Solution

We have to find out the minimum value of N that satisfies the required SIR. We shall do it by trial and error. Let us first assume $N = 7$.

Thus, $$\frac{S}{I} = \frac{(\sqrt{3N})^n}{6} = (1/6) \times (3 \times 7)^{3/2} = 16.04$$

or $$\frac{S}{I}\ \text{dB} = 10 \log (16.04) = 12.05\ \text{dB}$$

Since this value is less than the desired value, we will next calculate it with $N = 12$. We can show that for this value of N, SIR becomes 15.56 dB. Hence, $N = 12$ can be used.

Signal-to-Interference Ratio Calculation for Adjacent Channel Interference

The interference resulting from the signals that are adjacent in frequency to the desired signal is called *adjacent channel interference*. It results from an imperfect receiver filter that allows nearby frequencies to enter into the passband. This problem is serious when the adjacent channel user transmits in a very close range compared to a subscriber's receiver, while the receiver attempts to receive a base station on the desired channel. The scenario is shown in Fig. 11.9. The base station in the diagram is in communication with mobile B and mobile A is the interfering mobile. Mobile B is at the distance that is 20 times the distance of mobile A from the base station. Assuming that both mobiles transmit same power, the signal-to-interference ratio picked up by the base station receiver will be approximately given by

$$(S/I) = (20)^{-n} \tag{11.9}$$

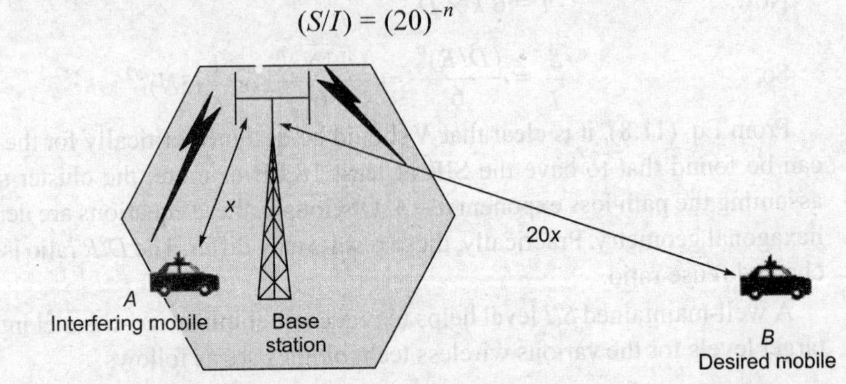

Fig. 11.9 Near–far problem creates adjacent channel interference

If the path loss exponent is assumed to be 4, the SIR will become –52 dB.

Alternatively, the near–far effect occurs when a mobile close to a base station transmits on a channel close to one being used by a weak mobile, at a far distance mobile. The base station may have difficulty in discriminating the desired mobile user and the close adjacent channel mobile. If the frequency reuse factor is large or N is small, the separation between the adjacent channels may not be sufficient to keep the adjacent channel interference level within tolerable limits.

The adjacent channel interference can be minimized through careful filtering at IF stages and careful channel assignments. Since each cell is given only a fraction of the available channels, a cell need not be assigned channels that are all adjacent in frequency. Keeping large frequency separation between each channel in a given cell, adjacent channel interference can be reduced considerably. There should not be sequential allocation of the channels in a cell; instead distant channels can form a group of channels per cell.

11.8 ANTENNAS FOR THE BASE STATION

There are two main types of antennas used in the wireless industry. All antennas fall under one of these two categories—*omnidirectional* or *directional* antennas (Fig. 11.10). There are a multitude of omnidirectional and directional antennas available for deployment in a wireless system. An omnidirectional antenna radiates equally in 360 degrees.

The antenna elements that comprise the collinear array antenna will be longer for the antennas that are designed and used for lower frequencies and shorter for the antennas that are designed and used for higher frequencies. This length in the antenna elements

Fig. 11.10 Omnidirectional and directional antennas for a base station

correlates directly to the fact that lower frequencies emit longer radio wavelengths and higher frequencies emit shorter wavelengths. These differences are directly attributed to the properties of radio physics. Thus, an antenna used for lower frequencies will be physically longer than that used for higher frequencies, if the gain assigned to both is the same. If a wireless carrier deploys a 9 dB antenna that operates at 850 MHz and a 9 dB antenna that operates at 1900 MHz, the 850 MHz antenna will be physically longer due to the longer wavelength inherent with RF emitted at 850 MHz.

Today, 0 dB (unity antennas) or 3 dB gain antennas are frequently used in urban areas. These antennas are mounted on the sides of buildings or at street level and are used to cover very small areas to support enhancer or microcell deployments. A directional antenna is an antenna that shapes and projects a beam of radio energy in a specific direction and receives radio energy only from a specific direction, employing various horizontal beamwidths. The directional antennas are effectively omniantennas that use a reflecting element, which directs or focuses the RF signal (energy) over a specified beamwidth. They produce more gain than a typical omni base station antenna produces. The most popular beamwidth is a 120-degree beamwidth, which supports three-sectored base stations. Other beamwidths are used though, such as 90-degree antennas and even 60-degree antennas. There are many types of directional antennas used by wireless carriers: log periodic, Yagi, phased array, and panel antennas.

At cell sites with a very high tower and a high gain antenna, coverage shadows may be created near the tower. To compensate for coverage shadows, electrical downtilt antennas and mechanical downtilt kits were developed specifically for the wireless industry by antenna manufacturers. A downtilted antenna is an antenna whose radiation pattern is electrically or mechanically tilted at a specified number of degrees downward. Downtilting of antennas decreases distance coverage horizontally but increases signal coverage closer to the cell site. Omniantennas can only be downtilted electrically, which is accomplished in the manufacturing process of the antenna by adjusting the phasing of the RF signal that is fed to the collinear antenna elements. Electrical downtilting is the way in which a specific antenna is manufactured, similar to how given antennas are manufactured with specific gains assigned to them. Electrical downtilt antennas are manufactured to downtilt to a preset amount of degrees. Directional antennas can be downtilted either electrically or mechanically. Mechanical downtilting is accomplished by actually manipulating the antennas so that they tilt towards the ground. Although the mechanical downtilting is less expensive than the electrical downtilting, it distorts (expands) the side lobes of the radiation pattern and might lead to interference with adjacent sectors.

A common place to install a downtilt antenna is at a cell site that is on a very tall tower or a hill, or near a large body of water. Downtilt antennas are also used to reduce the impact of what is known as the *far-field effect* in wireless networks. The far-field effect occurs when the radio coverage projected from site A may completely and unintentionally overwhelm the intended coverage area of site B or other nearby sites. Site A may transmit and receive into site B or other sites, theoretically leaving these sites unused. This would not only be terribly inefficient but would also be a terrible waste of equipment and frequency resources at cell site B and the other nearby base stations. The deployment of a downtilt antenna at cell site A would ensure that the intended radio coverage from site A stays within its designated coverage boundary.

The far-field effect can occur for any of the following reasons:

- RF power level is too high at the base station.
- Downtilt antennas are not being used at the base station.
- The tower is too high at the base station or the base station transmit antenna is too high on the tower.
- The antenna gain is too high at the base station, exceeding its intended coverage area.

The base station antennas are much more sophisticated and utilize a much wider variety of designs than the mobile phone antennas. One reason for this difference is that the base station antennas are required to have a higher degree of gain, ordinarily between 6 and 12 dB for omnidirectional antennas and between 4 and 18 dB for directional antennas. In some cases, 0 dB gain omniantennas are used to support microcell deployments.

The type of base station antenna that is chosen in any situation depends on many factors such as given below:

- The size of the area to be covered
- Neighbouring cell sites' configurations
- Whether the antenna is omnidirectional or directional
- The antenna's beamwidth in case of directional antenna
- The allotted RF spectrum the antenna can utilize

11.9 SECTORIZATION

As more co-channel cells were added to a wireless system over time (1983 to 1993), it became necessary to develop a means to increase system capacity without constantly having to split cells. Cell splits could be very costly undertakings. The industry developed a way to migrate from an omniantenna configuration at cell base stations and begin to sectorize base stations in order to obtain more capacity from each base station deployment. Sectorized base stations are created by subdividing an omnicell into sectors that are covered using directional antennas mounted in the same base station location (i.e. the same tower or rooftop). Operationally, each sector is treated as a different cell, the range of which is greater than an omnicell. The directional antennas always produce more gain than the omniantennas. All subcell directional antennas supporting each sector are co-located at the same base station. All radio equipment for each subcell, or sector, is housed in the same base station. To sectorize a cell, a horizontal, equilateral platform that resembles a triangle is deployed on a tower. Each side of the platform is called a *face* and has three-, four-, or six-directional antennas installed depending on the number of sectors. The directional antennas propagate the different frequencies/channels assigned within each respective face.

The sectorization facilitates wireless engineering and operations in the following ways:

- It minimizes or eliminates co-channel interference.
- It optimizes the frequency-reuse plan. This is facilitated through another concept known as the *front-to-back ratio*, which is described at the end of this section.
- At a minimum, it triples the capacity of any given coverage area when compared to the capacity that would be offered by deploying omniantennas. Most wireless carriers in the United States usually deploy three sectors per cell site and, in some cases, four. In some parts of the world, six-sectored base stations are used. In cases where three-sectored base stations are deployed, the directional antennas mounted in each sector will have 120-degree beamwidths. From a graphical viewpoint in this scenario, sectorization takes a circle (representing an omni base station) and converts it into a three-section pie chart.
- In cases where four-sectored sites are deployed, 90-degree beamwidth directional antennas are used.
- In cases where six-sectored base stations are deployed, 60-degree beamwidth directional antennas are used.

Obviously, the objective when implementing sectorized base stations is to support 360-degree coverage from a single location. The amount of sectors will dictate the beamwidth of the directional antennas used within each sector. Each sector has its own assignment of radio channels and its own channel set or sets. Each sector also has its own control channels and will hand off calls to its adjacent sectors that are housed on the same tower, rooftop, water tank, and so forth. A typical illustration of the sectorization concept in a cell is given in Fig. 11.11.

The front-to-back (F-B) ratio is defined as the ratio of the forward gain of a given cell's sector (based on the placement of the directional antenna) to the gain 180 degrees to the rear of co-channel sectors. As directional antennas are used in sectorized cells, RF

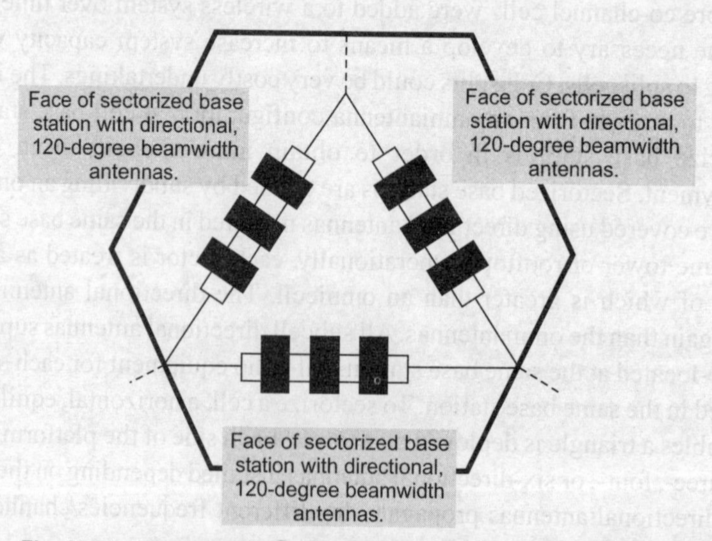

Fig. 11.11 Sectorization in a cell with 120-degree beamwidth directional antenna

engineers must take the front-to-back ratio into account to reduce the potential of picking up signals from co-channel sectors that are 180 degrees to the rear. This reduces the possibility of co-channel interference caused by the back lobe of the antenna propagation. The back lobe is RF that is projected 180 degrees to the rear of the antenna's front. Antennas with high F-B ratio ratings are sought by engineers and purchasing managers when trying to reduce co-channel interference in a sectorized environment. Directional antennas are assigned an F-B ratio by manufacturers, based on the electrical, frequency, and gain characteristics assigned to any particular antenna.

Example 11.6 A cellular service provider decides to use a TDMA scheme that can tolerate a signal-to-interference ratio of 16 dB in worst case. Find the optimum value of cluster size N in case of (a) omnidirectional antenna, (b) 120° sectoring, and (c) 60° sectoring. What will be the advantage of sectoring? Which sectoring will be better, 60° or 120° ? Assume path loss exponent $n = 4$.

Solution

We have $(S/I)\ \text{dB} = 16\ \text{dB} = 10 \log (S/I)$

So, $S/I = 39.81,\ n = 4$

(a) For omnidirectional antenna:

$$\frac{S}{I} = \frac{(\sqrt{3N})^n}{6}$$

or $$39.81 \times 6 = (\sqrt{3N})^4$$

or $$3.931 = \sqrt{3N}$$

or $$3N = 15.45$$

or $$N = 5.15$$

We have to select next possible integer value as the worst-case S/I is given. By selecting the higher value of N, the interference reduces and again N should be the value that satisfies Eq. (11.1). Hence, $N = 7$.

(b) For 120° sectoring:

$$\frac{S}{I} = 3 \times \frac{(\sqrt{3N})^n}{6}$$

or $$39.81 \times 2 = (\sqrt{3N})^4$$

or $$2.98 = \sqrt{3N}$$

or $$3\,N = 8.9$$

or $$N = 2.967$$

We have to select the next possible integer value as the worst case S/I is given. By selecting the higher value of N, the interference reduces, and again N should be the value that satisfies Eq. (11.1). Hence, $N = 3$.

(c) For 60° sectoring:

$$\frac{S}{I} = 6 \times \frac{(\sqrt{3N})^n}{6}$$

or $$39.81 = (\sqrt{3N})^4$$
or $$2.51 = \sqrt{3N}$$
or $$3\,N = 6.3$$
or $$N = 2.1$$

We have to select next possible integer value as the worst case *S/I* is given. By selecting the higher value of *N*, the interference reduces and again *N* should be the value that satisfies Eq. (11.1). Hence, $N = 3$.

Due to sectoring, the interferences will reduce because the antennas will be more directional and have more coverage angle and hence the area reduces. For this example, whether we go for 60° sectoring or 120° sectoring, the cluster size requirement is same but 120° sectoring saves three antennas. So, it is a better choice.

11.10 MOBILE TRAFFIC CALCULATION

The traffic calculation is applied to both mobile as well as landline telephone systems. On the basis of traffic condition, the telephone system must be designed such that the minimum number of calls may be blocked and the service provided to the subscriber remains excellent. The traffic varies considerably throughout a day, but most of the systems are designed to handle the traffic of peak busy hour in a day. In teletraffic engineering, the term *trunk* is used to describe any entity that will carry one call. In cellular systems, radio trunking is used.

The traffic intensity, more often called simply the *traffic*, is defined as the average number of calls in progress. Although this is a dimensionless quantity, a name has been given to the unit of traffic—erlang (abbreviated as E). On a group of channels, the average number of calls in progress depends on both the number of calls that arrive and their duration. The duration of a call is often called its *holding time*. One erlang of traffic can result from one trunk being busy all of the time, from each of the two trunks being busy for half of the time, or from each of the three trunks being busy for one-third of the time and so on. Averaged over time, one erlang of telephone traffic occupies exactly one channel. However, the arrival and closing of telephone calls are random processes. As time elapses, one erlang of traffic may occupy zero, one, or multiple channels. The definition of the unit erlang does not say anything about how the traffic behaves statistically about this average. Thus, from definition of the erlang, the traffic carried by a group of trunks is given by

$$A = \frac{ch}{T} \tag{11.10}$$

where A is the traffic in erlangs, c is the average number of call arrivals during time T, and h is the average call holding time. From Eq. (11.10), if $T = h$, then $A = c$. Thus, the traffic in erlangs is equal to the mean number of calls arriving during a period equal to the mean duration of the call. In short, one erlang of traffic can be generated, for instance, by

- one call of infinite duration or

- a random process of many calls arriving and closing, such that the average number of active calls is one.

From the above discussion, we can say that $A \leq 1$ for single trunk or channel. The unit of telephone traffic is named after A.K. Erlang, a Danish mathematician, who published in 1914 and 1917 the first basic results on the number of subscribers that can be served with a given number of channels at a required quality of service (QoS), known as blocking probability.

Example 11.7 A user has the average duration of call as 2 minutes. If the average number of requests for call per hour is 6, find out the traffic intensity in erlangs. If the total number of users is 100, find out the total traffic.

Solution

Let T = average duration of call

R = average number of call requests per unit time

Then the traffic intensity of single user, $A = T \times R$

$= 2 \times 6/60$

$= 0.2$ erlangs

For the total number of users, the total traffic intensity $= 0.2 \times 100 = 20$ erlangs

It can be interpreted as follows: Twenty channels can handle the total traffic in the worst case, 1 call per hour in a fully occupied condition. If more than 20 channels are used, the call will never be blocked.

11.10.1 Call Handling and Grade of Service

In a telephone system, a finite number of N channels are available. New calls are assigned a channel until all the channels are full. When all the channels become occupied, the situation that the system cannot accept further calls can arise. This state is known as *congestion*. The new call arrival is then

- Blocked (for lost call system based on circuit switching)
- Queued (for delayed system based on message or packet switching)

The result of congestion is that the traffic actually carried is less than the traffic offered to the system. We may, therefore, write

$$\text{Traffic carried} = \text{Traffic offered} - \text{Traffic lost} \tag{11.11}$$

The proportion of the calls that are lost or delayed due to congestion is utilized to measure the service quality. It is called the *grade of service* (GOS). The grade of service B for the lost call system may be defined as

$$B = \frac{\text{Number of calls lost}}{\text{Number of calls offered}} \tag{11.12a}$$

or

$$B = \frac{\text{Traffic lost}}{\text{Traffic offered}} \tag{11.12b}$$

$$= \text{Probability that a call will be lost due to congestion}$$

It is nothing but the proportion of the time for which the congestion exists or the probability of congestion. Thus, if traffic of A erlangs is offered to a group of trunks or channels having a grade of service B, the traffic lost is AB and the traffic carried is $A(1-B)$ erlangs. The larger is the GOS value, the worse is the service provided. The GOS is normally specified for the traffic at the busy hour as it is the worst case condition and checks the maximum capacity. At other times, it is always better.

Trunking efficiency is a measure of the number of users that can be offered a particular GOS with a particular configuration of fixed channels. The way in which channels are grouped can substantially alter the number of users handled by a trunked system. Sectoring changes the trunking efficiency.

Example 11.8 During the busy hour, 1300 calls were offered to a group of trunks and six calls were lost. The average call duration is 2.5 minutes. Find out the (a) traffic offered, (b) traffic carried, (c) traffic lost, (d) grade of service, and (e) total duration of period of congestion.

Solution

(a) $A = \dfrac{ch}{T} = 1300 \times 2.5/60 = 54.17$ E = offered traffic

(b) Six calls are lost.

$\Rightarrow 1300 - 6 = 1294$ calls are successful

$\Rightarrow 1294 \times 2.5/60 = 53.917$ E = carried traffic

(c) $6 \times 2.5/60 = 0.25$ E = lost traffic

(d) $B = 6/1300 = 0.0046$

(e) The total duration of period of congestion can be found out from GOS. All the traffic is measured on one-hour basis. So, fraction of 1 hour under congestion is found as follows:

$0.0046 \times 3600 = 16.56$ seconds

11.10.2 Poisson Arrival Process and Mathematical Modelling

A simple mathematical model is based on the following assumptions:

- Pure chance traffic
- Statistical equilibrium

Pure Chance Traffic

It means that call arrivals and call terminations are independent random events. Actually, calls are made by an individual user, but not made at random. However, the total traffic generated by a large number of users is observed to behave as if calls were generated at random. If call arrivals are independent random events, their occurrence is not affected by previous calls. The traffic is, therefore, sometimes called *memoryless traffic*. It also implies that the number of sources generating calls is very large. If the number of sources is small and several are already busy, then the rate at which new calls can be generated is less than what it would be if all the sources were free.

A commonly used model for random, mutually independent message arrivals is the Poisson process. The Poisson distribution can be obtained by evaluating the following assumptions for arrivals during an infinitesimal short period of time δt:

- The probability that one arrival occurs between t and $t + \delta t$ is λt, where λ is a constant, independent of the time t and independent of arrivals in earlier intervals. Here λ is called the arrival rate, which is expressed in the average number of arrivals during a unit of time.
- The number of arrivals in non-overlapping intervals is statistically independent.
- The probability of two or more arrivals happening during δt is negligible compared to the probability of zero or one arrival.
- The distribution of the number of arrivals in a time interval of t to $t + T$ is independent of the starting time t. The interval T is the interval between the call arrivals or the interval between two random events.

The probability of the number of call arrivals in a given time has a Poisson distribution given by

$$P(x) = \frac{\mu^x}{x!} e^{-\mu} \tag{11.13}$$

where x is the number of call arrivals in time T and μ is the mean number of call arrivals in time T. Here μ is nothing but λt. It can be shown that these intervals have a negative exponential distribution

$$P(T \geq t) = e^{-t/T_{\text{mean}}} \tag{11.14}$$

where T_{mean} is the mean interval between call arrivals, which is equal to h, the holding time. Now, the call arrival and the call termination are also the independent random events. Then T can also be the interval between these two random events that also has the negative exponential distribution as given in Eq. (11.14), because most of calls are short and few are long. The probability of number arrivals during period of duration T is

$$P(T) = e^{-\lambda T} \tag{11.15}$$

where $P(T)$ is the probability density function of the duration between two arrivals.

Statistical Equilibrium

It means that the generation of traffic is a stationary random process, that is the probabilities do not change during the period being considered. Consequently, the mean number of calls in progress remains constant. This condition is satisfied during the busy hour and it is, of course, the busy hour GOS that one wishes to determine. Statistical equilibrium is not obtained immediately before the busy hour, when the calling rate is increasing, or at the end of the busy hour, when the calling rate is falling.

Example 11.9 In a company, one call arrives every 5 minutes, on average. During a period of 10 minutes, what is the probability that (a) no calls arrive, (b) one call arrives, (c) two calls arrive, and (d) more than two calls arrive. If the average call duration is 2 minutes and a call has lasted for 4 minutes, what is the probability that the call will last for at least another 4 minutes?

Solution

We have $$P(x) = \frac{\mu^x}{x!} e^{-\mu}$$

where x = number of call arrivals in time T

μ = mean number of call arrivals in time $T = 2$ because during the period of observation the maximum possibility of calls is 2 only as it arises every 5 minutes.

So, the probability of no call arrival in duration of 10 minutes is

$$P(0) = \frac{2^0}{0!} e^{-2} = e^{-2} = 0.135$$

The probability of one call arrival in duration of 10 minutes,

$$P(1) = \frac{2^1}{1!} e^{-2} = 0.27$$

The probability of two calls arrival,

$$P(2) = \frac{2^2}{2!} e^{-2} = 0.27$$

The probability of arrival of more than two calls,

$$P(>2) = 1 - P(0) - P(1) - P(2) - 1 - 0.135 - 0.27 - 0.27 = 0.325$$

Now, the probability of the call lasting for another 4 minutes, for 8 minutes, when the average call duration is 2 minutes,

$$P(T \geq t) = e^{-t/T_{mean}}$$

Here, T_{mean} is average call duration = 2 minutes

So, $$P(8 \geq 4) = e^{-4/2} = e^{-2} = 0.135$$

Markov model New calls arrive according to the *Poisson* process with rate λ calls per unit of time. Calls have a (memoryless) exponential duration with mean $1/\mu$. The number of active calls is a Markov process. Such a process is called a simple Markov chain. It is represented in Fig. 11.12. Here $P(1)$, $P(2)$,..., $P(N)$ are the *state probabilities* and $P_{0,1}$, $P_{1,2}$, etc. are the probabilities of a state increment. Similarly, the decrement probabilities are also there. They are all the *conditional probabilities*, also

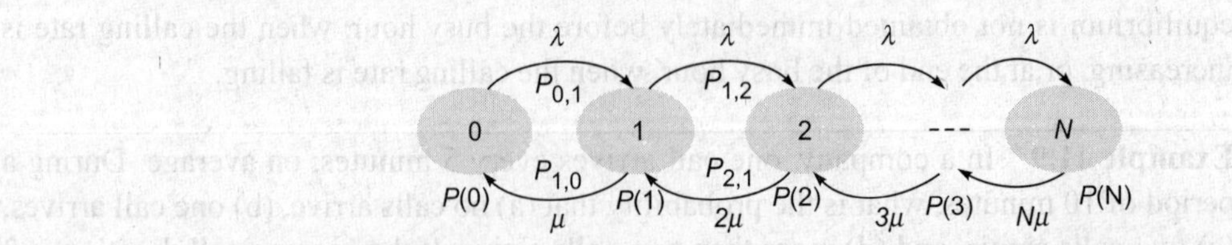

Fig. 11.12 Markov model for a number of occupied channels in a network with N channels

called the *transition probabilities*. If there is statistical equilibrium, these probabilities do not change and the process is said to be the regular Markov chain.

11.10.3 Erlang B Formula for Lost Call System

We can find from the Poisson distribution that

$$P(x) = \frac{A^x}{x!} e^{-A} \tag{11.16}$$

where $P(0) = e^{-A}$

In Eq. (11.16), $P(x)$ is the probability of x calls in progress. Here x can have any value between zero and infinity and sum of their probabilities must be unity. Thus, if call arrivals have a Poisson's distribution, so does the number of calls in progress. This requires an infinite number of trunks to carry the calls. If the number of trunks available is finite, then some calls can be lost or delayed and the distribution is no longer Poissonian. The distribution that then occurs is derived in Erlang's formula.

Erlang determined the grade of service (i.e. the loss probability) of a lost call system having N trunks, when the offered traffic is A. His solution depends on the following assumptions:

- Pure chance traffic.
- Statistical equilibrium.
- Full availability.
- The call that encounters congestion is lost.

The probability of congestion, of a lost call, or the grade of service B for the full availability group of N trunks and the offered traffic A erlangs is given by

$$B = \frac{A^N / N!}{\sum_{k=0}^{N} A^k / k!} \tag{11.17}$$

This is also called Erlang's lost call formula or Erlang B formula.

Example 11.10 How many users can be supported for 1% blocking probability if the number of channels is 20? Assume that each user generates 0.1 erlangs of traffic. Find out peak hour traffic with average duration of call being 2 minutes. The offered traffic intensities for a number of channels is given below:

Number of channels	A
2	0.153
4	0.869
5	1.36
20	12

Solution

When the number of available channels is 20, $A = 12$.

Each user generates 0.1 erlangs of traffic.
So, the number of users supported = 12/0.1 = 120
Peak hour traffic in terms of calls = number of calls in peak hour
When the traffic load is 1 erlang, the number of calls in 1 hour is T/h.
Thus, the number of total calls = $A \times T/h$

$$= 12 \times 60/2 = 360 \text{ calls in 1 hour}$$

11.10.4 Erlang C Formula for Delayed System

In an Erlang C telephone system, N channels are available. New calls are assigned a channel until all channels are full. When all the channels become occupied, a new call is queued until it can be served. This is in contrast to an Erlang B system, in which new calls are blocked.

Assumptions here are the same as in Erlang B system, except the fourth one, where the calls that encounter congestion enter a queue and are stored there until a server becomes free. Here the second assumption, that is statistical equilibrium, implies that $A \leq N$. If $A \geq N$, the calls are entering the system at a greater rate than they leave. As a result, the length of the queue will continually increase towards infinity. This is not statistical equilibrium.

Probability of delay formula or Erlang's delay formula can be given as

$$P_D = \frac{A^N}{A^N + N!\left(1 - \frac{A}{N}\right)\sum_{k=0}^{N} \frac{A^k}{k!}} \tag{11.18}$$

This is also called Erlang C formula.

11.11 SPECTRUM EFFICIENCY OF CELLULAR SYSTEM

In most cellular systems, each base station can carry more than one telephone call in its cell.

Let us have

k = the number of channels per cell
N = cluster size
W_{channel} = total bandwidth for the cellular net
W_{signal} = occupied bandwidth per channel

Then

$$W_{\text{channel}} = k\,N\,W_{\text{signal}} \tag{11.19}$$

The spectrum efficiency S_E of a cellular net can be defined as the carried traffic per cell A_c, expressed in erlang, divided by the bandwidth of the total system W_{channel} and divided by the area of a cell S_u. So,

$$S_E = \frac{A_c}{W_{\text{signal}}\,Nk\,S_u} \tag{11.20}$$

Here A_c is mostly computed from Erlang B formulas, with A_c equal to the attempted traffic multiplied by the probability of success (= 1 – blocking probability). Mostly, the spectrum efficiency is expressed in erlang/MHz/km^2.

Thus, we observe that the spectrum efficiency decreases with the cluster size N. The system performance, for instance, expressed in terms of the outage probability or the bit error rate experienced by the user, improves with increasing reuse distance, and so it improves with the cluster size. Hence, achieving high system performance and efficient use of the radio spectrum are conflicting objectives for a network designer.

11.12 NUMBER OF CUSTOMERS IN THE CELLULAR SYSTEM

When we design a system, the traffic conditions in the area during a busy hour are some of the parameters that will help to determine both the sizes of different cells and the number of channels therein. This is because the channels are divided into the cells of a cluster. The channels will be shared by the users on the demand assignment basis. More the traffic, more reuse of the channels.

The maximum number of calls per hour per cell is driven by the traffic conditions at each particular cell. After the maximum number of frequency channels per cell has been allocated in each cell, then the maximum number of calls per hour can be taken care of in each cell. If maximum number of calls per hour in each cell is Q_i, sum them up over all cells of a cluster to have the total number of calls per hour using the set of frequencies. If we assume that 65% of the phones were active during the busy hour, then the number of customers in a cluster can be estimated by dividing the total number of calls per hour per cluster by 0.65.

11.13 ATTRIBUTES OF CDMA IN CELLULAR SYSTEMS

In CDMA-based cellular systems, though more than one channel (a set of frequencies) may be allocated per cell for handling more calls at a time, strictly speaking, other cells may repeat the same set of frequencies. The conventional frequency reuse concept in a cluster is not required here because the users are differentiated on the basis of their codes even though they may be transmitting at the same frequency. There are many attributes of CDMA that are of great benefit to the cellular system.

Soft Handoff

Since every cell uses the same radio frequency band, the only difference between the user channels is the spreading code sequence. Therefore, there is no jump from one frequency to another when a user moves between the cells. The mobile terminal receives the same signal in one cell as it does in the next and thus there is no harsh transition from one receiving mode to another. Two or more neighbouring base stations can receive the signal of a particular user, because they all use the same channel. Moreover, two base stations can simultaneously transmit to the same user terminals. The mobile (rake) receiver can resolve the two signals separately and combine them. This feature is called *soft handoff*.

Soft Capacity or Graceful Degradation

In FDMA and TDMA, N channels can be used virtually without interference from other users in the same cell but potential users $N+1$, $N+2$, ..., are blocked until a channel is released. The capacity of FDMA and TDMA is, therefore, fixed at N users and the link quality is determined by the frequency reuse pattern. In theory, it does not matter whether the spectrum is divided into frequencies, time slots, or codes, the capacity provided from these three multiple access schemes is the same. However, in CDMA, all the users in all cells share one radio channel and are separated by codes. Therefore, an additional user may be added by sacrificing somewhat the link quality, with the effect that voice quality is just slightly degraded compared to that of the normal N-channel cell. Thus, degradation of performance with an increasing number of simultaneous users is 'graceful' in CDMA systems compared with the hard limits placed on FDMA and TDMA systems.

Multipath Tolerance

Spread spectrum techniques are effective in combating the frequency selective fading that occurs in multipath channels. The underlying principle is that when a signal is spread over a wide bandwidth, a frequency selective fade will corrupt only a small portion of the signal's power spectrum, while passing the remaining spectrum unblemished. As a result, upon despreading, there is a better probability that the signal can be recovered correctly. For an unspread signal whose spectral density happens to be misplaced in a deep fade, an unrecoverable signal at the receiver is virtually assured. To optimally combine signals received over various delayed paths, a rake receiver can be used.

No-Channel Equalization Needed

When the transmission rate is much higher than 10 kbps in both FDMA and TDMA, an equalizer is needed for reducing the intersymbol interference caused by time delay spread. This is because when the bit period becomes smaller than about ten times the time delay spread, the intersymbol interference becomes significant. However, in CDMA, a correlator is needed at the minimum. To achieve good performance, a rake receiver, is needed to combat delay spread.

Privacy

An important requirement of spreading signals is that they should be 'noise-like' or pseudo-random. Despreading the signal requires knowledge of the user's code, and for a binary code with spreading factor N, there exist 2^N possible random sequences. In military systems, these codes are kept secret. So, it is very difficult for an unauthorized attacker to tap into or transmit on another user's channel. Often, it is even difficult to detect the presence of a spread spectrum signal because it is below the noise that is present in the transmit bandwidth. Note that in cellular systems, the codes are fully described in publicly available standards. In digital systems, security against eavesdropping (confidentiality) is obtained through encryption. This is a highly desirable alternative to the analog FDMA cellular phone system widely used today, where with an inexpensive scanner, one can tune into the private conversations of unwary neighbours.

There are, of course, a number of disadvantages associated with CDMA. Two of the most severe are the problem of 'interference' and the related problem of the 'near–far' effect. These have been described in Chapter 8.

Summary

- Cellular technology is required to have the mobile telephony with seamless connectivity.
- The smallest area into which the total area is divided is called the cell and number of cells makes the clusters. Clusters are designed with specific mathematical calculation.
- Due to cellular concept, the total area is divided into cells and cell-wise base stations are provided with transceiver facility to communicate with the mobiles in that area.
- Theoretically, a cell has a hexagonal shape.
- Cell splitting is the further division of a cell area to accommodate more traffic.
- A group of channels are assigned to a cell, which is not repeated in the same cluster but repeated in the different clusters in the same manner.
- Frequency reuse concept is used in the cellular systems so that the limited available spectrum can be used efficiently. The reuse distance can be calculated by using some mathematical relationships so that co-channel interference can be minimized.
- Along with the frequency reuse, FDMA or TDMA can be used in combination.
- Cellular systems may be of two types—analog or digital circuit switched and packet switched. In both these types, the components of the system differ slightly but the concept remains same.
- MTSO/MSC is the central switch to handle the calls and the channel assignments.
- Handoff procedure is necessary during transition from one cell area to another for continuous conversation.
- Channel assignment may be of two types—fixed and dynamic; Borrowing strategy can be used sometimes.
- Cellular interferences are of two types—co-channel and adjacent channel.
- Antennas for a base station may be omnidirectional or directional. The directional antennas are used for sectorization.
- Sectorization increases the traffic handling capacity.
- The unit of traffic is erlang.
- Mobile traffic calculation can be done by considering the traffic with Poisonian distribution. Markov chain model can be used for calculation of Erlang B formula for lost call system.
- Erlang C formula can be used for delayed system.
- Attributes of CDMA make the technology possible in cellular systems.

Review Questions

1. What do you mean by a cell? Why is the hexagonal shape assumed theoretically for the cell?
2. Comment on the size of a cluster and frequency reuse distance relationship.
3. How is the reuse distance decided theoretically?
4. Justify that cellular communication increases the spectral efficiency but spectral efficiency also depends upon the cluster size.
5. What are the parameters that will be affected by the size of a cluster?
6. What will happen if the power of a cell site transmission is detected in the adjacent cell? What will be the situation of the mobile user at the boundary of the cell?

7. Why is the value of path loss exponent more than 2?
8. How are the mobile-originated and landline-originated calls handled?
9. What do you mean by channel assignment? Why is this process required?
10. Discuss the constraints of the dynamic channel assignment.
11. 'Directional antennas produce more gain than omnidirectional antennas.' Justify. How is this feature utilized in cellular communication?
12. Why is the sectorization a critical issue, though it looks simple due to directional antenna?
13. What are the differences between outdoor and indoor propagation issues of a cellular system?
14. Explain why operators of cellular networks in densely populated areas prefer to put their base stations in the valleys, rather than on top of mountains. How does this differ from the situation in rural areas?
15. Compare the traffic handling in landline and mobile telephone systems. What are the aspects of these systems that differ except in the transmission media?
16. What is the difference between Erlang B and C formulas?
17. Define Ergodic, Poisson, and Markov processes from the basic theory of random processes and try to correlate them with telephone traffic.
18. Analyse the trade-off between the sectoring and the trunking efficiency loss. Compare the same for 120° and 60° sectoring.
19. List out the advantages of digital cellular technology in terms of user capacity and performance.

PROBLEMS

1. Explain the relative merits and demerits of large and small cluster sizes in cellular systems. Prove that

$$D/R = (3N)^{0.5}$$

2. Find out the appropriate cluster size N for a cellular system if S/I requirement is 15 dB. Here I corresponds to co-channel interference. Assume path loss factor $n = 3$. Find your answer for omnidirectional and 120-degree sectored antennas.
3. For the different cluster sizes N=1, 3, 4, 7, 12, etc., find out the S/I ratio in dB for all the cases. Assume that all cells have equal radii and the base stations have equal power and are located at the centre of each cell. Comment on your result in the form of some conclusions. What will be the effect of first tier, second tier, and so on?
4. Show that the frequency reuse factor for a cellular system is given by k/S, where k is the average number of channels per cell and S is the total number of channels available to the service provider.
5. If 24 MHz of total spectrum is allocated to a duplex wireless cellular system and each simplex channel has 25 kHz RF bandwidth, find out the number of duplex channels and the total number of channels per cell site if N= 4 cell reuse is used.
6. For a cell, the number of available channels is 30. Assume average call length of 2 minutes. Also the average number of calls per hour per user is 1. Find out the capacity loss when going from omnidirectional to 60-degree sectored antenna. The blocking probability desired is 1%. A section of blocking table is given below. (The entries show the offered load.)

Number of channels	A
5	1.361
10	4.462
20	12.03
30	20.34
40	27.38
51	44.2

7. On an average, during busy hour, a business firm makes 100 outgoing calls of average duration of 3 minutes. It receives 120 calls, which are incoming of average duration of 2.5 minutes. Find out the outgoing traffic, incoming traffic, and total traffic in erlangs.

8. Show that the probability of delaying a call is

$$P = \frac{A^N}{N!}\frac{1}{1-(A/N)}$$

where A is the offered traffic expressed in erlang (λ/μ).

9. A total of 25 MHz of bandwidth is allocated to a particular FDD cellular telephone system that uses two 30 kHz simplex channels to provide full duplex voice and control channels. Assume that each cell phone user generates 0.2 erlangs of traffic. Find out the number of channels in each cell for a four-cell frequency reuse system. If each cell is to offer capacity, that is 90% of perfect scheduling, find out the maximum number of users that can be supported per cell for an omnidirectional antenna case. What is the blocking probability of the system using Erlang B formula, when the maximum users are available in the user pool?

(**Hint** Assume that each channel can carry 1 erlang of traffic.)

10. The cluster size is of seven cells with blocking probability $P_r = 1\%$ and average call holding time is 2 minutes. Find out the traffic capacity loss in percentage due to trunking for 51 channels when going from omnidirectional antennas to 120° and 60° sectoring. Assume that the blocked calls are cleared and the average per user call rate $\lambda = 1$ per hour. Use the table provided in Problem 6 for total traffic intensity A. Comment on your results.

11. Assume that there are six co-channel cells in the first tier and all of them are at the same distance from the mobile. If an SIR of 12 dB is needed for satisfactory forward channel performance in a cellular system, what should be the optimum frequency reuse factor and cluster size if the path loss exponent is (a) $n = 3$ and (b) $n = 4$.

MULTIPLE CHOICE QUESTIONS

1. Cellular theory is applicable to
(a) WLL
(b) GSM
(c) mobile satellite
(d) all of the above

2. Cluster is nothing but a
(a) set of reuse frequencies
(b) group of channels in a cell
(c) group of cells using the same frequencies
(d) group of cells using different frequencies without reuse

3. A group of channels assigned to a cell
(a) is repeated in the same cluster
(b) is repeated in the different clusters in a different manner
(c) is repeated in the different clusters in the same manner
(d) varies cell by cell

4. Which of the following multiple access schemes can be accommodated by cellular technology based networks?
(a) FDMA
(b) TDMA
(c) SDMA
(d) All of the above

5. If the cluster size is increased,
(a) the interchannel interference reduces
(b) the transmission power is to be increased
(c) the cochannel interference reduces
(d) the number of reuse frequency increases

6. Hard handover means
(a) handing over the call to another mobile
(b) changing over the communication channel
(c) increase in power while moving to another base station
(d) handing over the call to mobile switching centre

7. Cell splitting is done to
(a) accommodate more traffic
(b) accommodate more area
(c) save the power
(d) increase the frequency reuse

8. In CDMA-based cellular networks, the near–far effect may appear due to
(a) distant users
(b) imperfect orthogonality between codes
(c) interfering signals
(d) orthogonal codes

9. Sectorization means
(a) dividing clusters into sectors
(b) dividing the channels
(c) replacing omnidirectional antenna by directional antennas
(d) cell splitting

10. Which of the following angle is suitable for sectorization?
(a) 45°
(b) 60°
(c) 100°
(d) 80°

11. The number of handoffs will increase in case of
(a) micro-cellular structure
(b) cell splitting
(c) sectorization
(d) (b) and (c)

12. Interference effects in cellular systems are a result of
(a) the distance between areas
(b) the height of the antennas
(c) the ratio of the distance between areas to the transmitted power of the areas
(d) power of the transmitters

13. Larger cells are more useful in
(a) lightly populated urban area
(b) rural areas
(c) densely populated urban areas
(d) hilly areas

14. Rayleigh fading occurs when there is
(a) an LOS component present
(b) no LOS component present
(c) intercell interference
(d) none of the above

15. The frequency reuse can be maximized by
(a) increasing the size of cells
(b) decreasing the size of cells
(c) increasing the size of cluster
(d) increasing the number of users

16. Which of the following statements is true?
(a) In TDMA, a digital air interface standard has twice the capacity of analog.
(b) In practice, cells are always hexagonal in shape.
(c) The more the hanoff margin, more the benefit always.
(d) By increasing the sectors, the cochannel interference can be reduced.

chapter 12

Wireless Communication Systems and Standards 1: Broadcast Networks

Key Topics

- DAB standard
- DRM and HD radio systems
- DVB standard
- DTH (direct-to-home) television system

Chapter Outline

Wireless communication systems fall in two broad categories—broadcast networks and mobile telephone/ data networks. This chapter describes a few of the broadcast systems with their key features and standards. In broadcast systems, there is one transmitter and a number of receivers, which receive the information as and when desired. These systems are one-way systems. There is no provision for an acknowledgement. Proper multiple-access schemes are a must in these systems to accommodate the number of channels. However, most of the recent systems are based on OFDM multicarrier modulation, which forms a single-frequency network concept. So, high-speed, high-quality communication has become possible. Most of the applications are audio- and video-based entertainment; however, some data services are also supported.

12.1 INTRODUCTION TO BROADCAST SYSTEMS

Broadcasting systems require a common transmitter at the broadcast point and a number of portable radio receivers. These systems have existed for long in terms of AM/ FM radio and analog television broadcasting. However, the scenario has changed. Because of digital modulation schemes like OFDM, revolutionary changes in the broadcast systems have been made. High quality audio and video transmissions have now become possible. Concentrated work has been carried out in this direction and as a result some standards have been developed, like DAB and DVB standards. Based on these standards, many systems are developed, which has made the conventional systems obsolete. In this chapter, details of these systems are provided with considerations of various hardware stages such that the importance of the initial chapters in the design of overall system can be established.

12.2 DIGITAL AUDIO BROADCASTING (DAB)

Digital audio broadcasting (DAB) is a digital radio broadcasting standard that is designed to replace the analog FM/AM radio transmissions. The development of T-DAB was carried out in the EUREKA 147 consortium formed by the broadcasting companies, network operators, consumer electronic industries, and research institutes. The development started officially in 1987, and in 1995, DAB was standardized by the European Telecommunication Standard Institute (ETSI). The European Telecommunication Standard (ETS) 300 401 became the first standard to include OFDM. In 1997, a second edition of ETS 300 401 was released and the commercial employment of DAB started in 1998. Later on, the DAB included satellite as well as hybrid satellite/terrestrial broadcasting options.

The DAB has several benefits over the analog radios. With DAB, significantly higher spectral efficiency can be achieved. Due to adoption of the OFDM scheme, DAB supports *single-frequency networks* (SFNs) enhancing the spectral efficiency of networks even further. In single-frequency network, transmitters use the same set of carrier frequencies. Digital transmission is more robust against interference and also enables the employment of efficient source and channel coding methods. Therefore, the quality of received sound is improved, approaching the sound quality of the compact disc. Efficient multiplexing of different kinds of data streams is feasible with digital modulation and hence new radio services (like text, pictures, and video clips) can be offered to the consumer.

There are two main areas of the system that are of interest in digital radio, namely the audio digital *encoding and compression system* and the *modulation system.*

The encoding and compression of the data is of great importance. For the system to be viable, the data rate has to be considerably reduced but the quality should be that of a standard CD. Also the audible characteristics of human ear must be taken into account. The digital radio system adopted a reduced data rate, down to 128 kbps, one-sixth of the bit rate for a similar quality linearly encoded signal. To achieve these reductions, the incoming audio signal is carefully analysed. It is found that the ear has a certain threshold

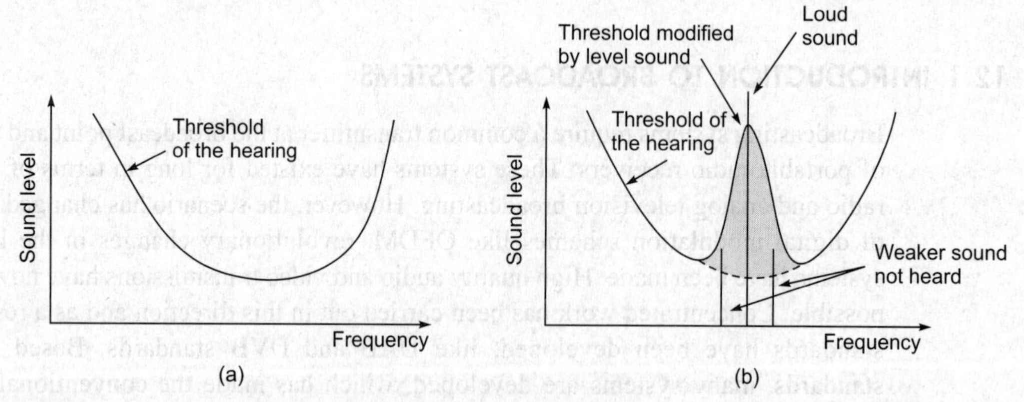

Fig. 12.1 Human ear audio perception characteristics: (a) threshold of hearing variation with no sound present and (b) threshold modified by level sound

of hearing ability. Below this, the signals are not heard. Additionally, if a strong sound is present on one frequency, then weaker sounds close to it may not be heard because the threshold of hearing is modified. By analysing the incoming audio and encoding only those constituents that the ear can hear, significant reductions can be made. Further reductions in data rate can be achieved by reducing the audio bandwidth. This is implemented on some channels such as those used only for speech.

The other key to the operation of digital radio is the modulation system. It is coded orthogonal frequency division multiplex (COFDM). It is a form of multicarrier modulation that provides the robustness required to prevent multipath and other forms of interference from disrupting reception as described in Chapter 9. Here coded means incorporating channel-coded data for modulating with OFDM scheme.

The system uses about 1500 individual subcarriers that fill around 1.5 MHz of spectrum as shown in Fig. 12.2. The carriers are spaced very close to one another. The interference between the carriers is prevented by making the individual signals orthogonal to each other. This is done by spacing each one by a frequency equal to the data rate being carried. In this way, the nulls in the modulation sidebands fall at the position where the next carrier is located. The audio data is then split among the carriers so that each carrier takes only a small proportion of the data rate. This has the advantage that if interference is encountered in one area, then sufficient data is received to reconstitute the required signal. Guard bands are also introduced at the beginning of each symbol and the combined effect is such that the system is immune to delays.

The advantages of SFN topology are as follows:

1. There is only one frequency for national deployment.
2. No frequency planning is required.
3. There is reduced number of transmitters compared to multifrequency network.

However, the SFN topology also has some disadvantages as given below:

1. Frequency has to be free all over the country.
2. There is a need to synchronize all transmitters in both timing and content.

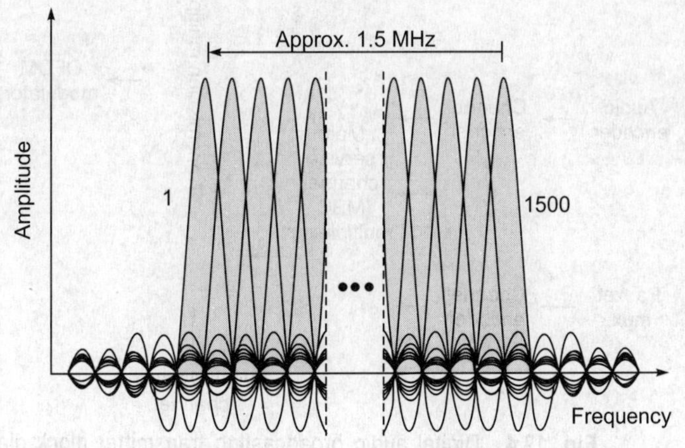

Fig. 12.2 Spectrum of a digital radio signal constituted from multiple subcarriers

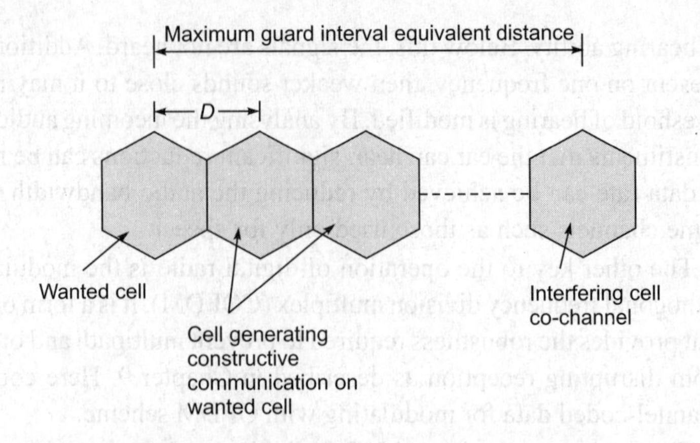

Fig. 12.3 SFN size limited by guard interval

3. It is suitable for national network only.
4. The SFN size is limited by guard interval because if the signal falls into guard interval, there is no interference.

It is possible to set up a system where all the transmitters for a network operate on the same frequency. This means that it is possible to set up single-frequency networks throughout an area where a common 'multiplex' is used. It is found that the out-of-area signals tend to augment the required signal. It also means that small areas of poor coverage can have a small transmitter on exactly the same frequency filling in the hole and further improving reception in adjacent areas. A further advantage of this digital radio system is that it requires less power than the more traditional transmitters. Figure 12.4

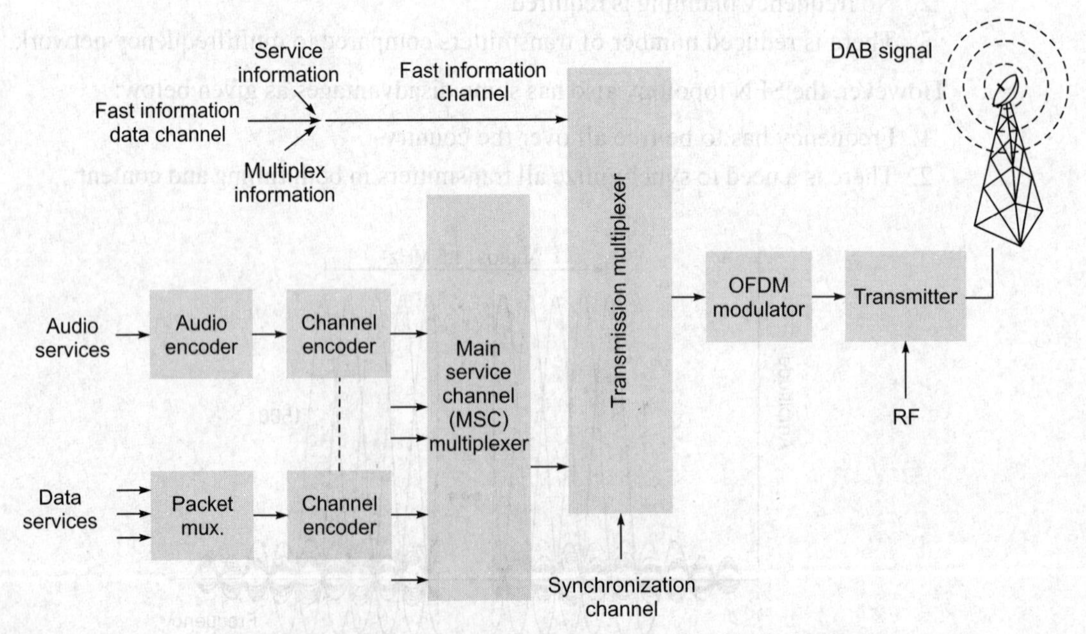

Fig. 12.4 Digital audio broadcasting transmitter block diagram

gives the simplified diagram of the DAB transmitter while Fig. 12.5 is the detailed one. The functional blocks of the system will be clear with the details given further.

12.2.1 Overview of DAB Standard

In the UK, a spectrum allocation between 217.5 and 230 MHz has been reserved for digital radio transmissions. This gives a total of seven blocks of 1.5 MHz, each able to carry a multiplex of services. In other countries as well, the spectrum is being made available. Within Europe, it is being made available either in Band III as in the UK or in L band between 1452 and 1467 MHz. The upper part of the band between 1467 and 1492 MHz will be reserved for satellite delivery of digital radio.

The DAB digital radio can be broadcast on a wide number of frequencies. There are both terrestrial and satellite allocations for digital audio broadcasting. The terrestrial broadcasting is just like conventional FM broadcast using a transmitter within a range, another transmitter for another area and so on. This creates the feel of cells along with the base stations. Thus, a network of transmitters is created. Currently, the main frequencies where it is being deployed are within the Band III frequencies as given in Table 12.1. Here a number of channels have been allocated. A complete list of the channels is given in Table 12.1, although in many countries the full number of channels is not available.

Although it may appear that comparatively few channels are available, each multiplex is able to carry many stations. If high-quality audio is required, then fewer stations can be accommodated. However, it is often possible to accommodate around four or five high-quality broadcasts along with several low-quality ones. In addition to this, the data can also be carried.

The DAB standards support SFN broadcasting. It means that the same DAB ensemble is broadcast on the same frequency through the whole network of transmitters, whereas in the conventional broadcasting systems, neighbouring transmitters use

Table 12.1 Band III channel frequencies for DAB

Channel	Frequency MHz	Channel	Frequency MHz	Channel	Frequency MHz
5A	174.928	8B	197.648	11C	220.352
5B	176.640	8C	199.360	11D	222.064
5C	178.352	8D	201.072	12A	223.936
5D	180.064	9A	202.928	12B	225.648
6A	181.936	9B	204.640	12C	227.360
6B	183.648	9C	206.352	12D	229.072
6C	185.360	9D	208.064	13A	230.748
6D	187.072	10A	209.936	13B	232.496
7A	188.928	10B	211.648	13C	234.208
7B	190.640	10C	213.360	13D	235.776
7C	192.352	10D	215.072	13E	237.448
7D	194.064	11A	216.928	13F	239.200
8A	195.936	11B	218.640		

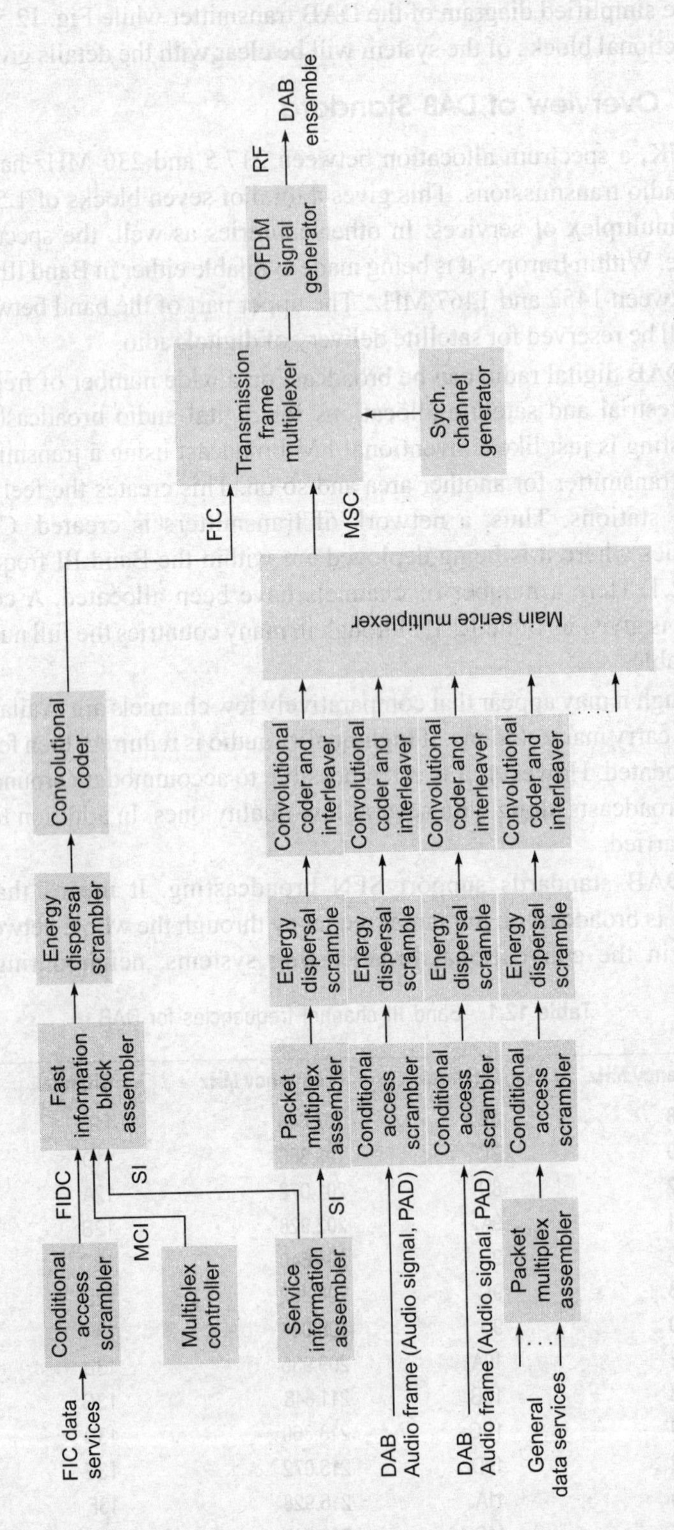

Fig. 12.5 Detailed diagram showing the multiplexing and coding in a DAB system

different frequencies. Clearly, SFN achieves higher spectral efficiency due to better frequency reuse. Also the reception is better on the fringes of the cell (coverage area), since the signals from all nearby transmitters are utilized in the receiver. Thus, SFN allows 'extra' transmitters in the network. However, SFN requires that the transmitters are synchronized and that the signals from all nearby transmitters are received within the guard period of OFDM symbol.

One of the main problems with the initial launch of digital radio was the availability of the equipment. A large investment was required from the equipment manufacturers. The heavy reliance on digital signal processing techniques meant that large development programmes were needed to develop the equipment. There were also problems with the fact that the early implementations required high current levels. These solutions would not have been suitable for portable receivers and for car and home applications, heat dissipation was a problem. Furthermore, the multichip solutions made the equipment large and bulky as well as the manufacturing costs high. Manufacturers solved the problem. Specific chip sets for DAB were developed, which enabled costs to be reduced dramatically such that DAB is nowhere near as high as it was when compared to FM receivers.

Accordingly, DAB digital radio is now the broadcasting medium for the twentieth century. The detailed diagram is shown in Fig. 12.5 and explained as follows.

The DAB standard supports not only audio services, normal radio programmes, but also a variety of new data services. The services can be categorized as audio services, data services associated with an audio programme, and data services independent of audio programmes. Due to the nature of OFDM, several services are collected into an ensemble that is transmitted and received as a one entity. Naturally, the user must be informed about the services included in the ensemble and the service information must be provided.

MPEG-1 and MPEG-2 audio layer II coding is used for the source coding of PCM audio signal with a sampling rate of either 24 kHz or 48 kHz. The spectrum of audio signal is divided into 32 sub-bands, which are coded utilizing a psycho-acoustic model of human perception. The rate of resulting signal varies between 8 kbps and 384 kbps depending on the required quality of signal. Four audio modes are possible—mono-channel mode, dual-channel mode (i.e. two mono-channels), stereo mode, and joint-stereo mode. The joint-stereo mode utilizes the redundancy of the two channels in a stereophonic programme to enhance the quality of received audio signal, and it is particularly suitable for lower bit rates. The DAB also supports the extension to multichannel audio coding (up to 5 channels) defined in MPEG-2 audio standard.

There are three channels in the DAB ensemble signal as shown in Figures 12.4 and 12.5:

1. Synchronization channel for receiver synchronization
2. Fast information channel (FIC)
3. Main service channel (MSC)

As far as the generation of such channels is concerned, the input blocks are to be concentrated. Each audio service channel also contains the programme associated data (PAD) channel for conveying information that is closely linked to the audio programme. This kind of data can be for dynamic range control information for sound decoder, dynamic labelling for programme titles or lyrics, and speech/music indicator. The data is included at the end of DAB audio frame and the capacity of PAD channel varies between 667 bps and 65 kbps. (Minimum is 333 bps in case of half sampling frequency of audio signal.)

In the DAB system, text, picture, and even video clips can be conveyed as a separate service, independent of audio programmes. Such separate data channels can be used for providing traffic messages warning about traffic problems and giving route suggestions (traffic navigation, roadmaps, travel information about local hotels, events, car parks, etc.), electronic newspapers, paging services, and emergency warning system for authorities among others.

In every DAB ensemble, there is service information (SI) describing the contents of the ensemble, thus helping the user in the selection of the programme. For the selection of the programme, service information includes basic programme service label (name of radio station), type of programme service (news, sports,...), programme language, dynamic programme label (names of artists, songs, lyrics, etc.), as well as time and date. For the control of receiver, information about frequency, transmitter identification, and regional identification is included in the service information as well as cross-references to the same programme being transmitted on another DAB, FM, or AM signal.

Fast information channel contains multiplex configuration information (MCI), fast information data channel (FIDC), and some parts of service information. The MCI describes the multiplexing of the main service channel and it is required before the MSC can be successfully demultiplexed. The FIDC includes traffic messages, paging messages, and emergency warnings. The FIC carries information that the receiver should have easy and rapid access to, say, MCI. Therefore, it is not time-interleaved but is heavily protected with 1/3 rate channel coding. The MSC carries most of the audio and data services and contains several sub-channels. For the transmission, the logical channels are time multiplexed into frames. The frame duration is of 96 ms, 48 ms, or 24 ms, depending on the transmission mode. Frame structures are given in Figures 12.6, 12.7, and 12.8.

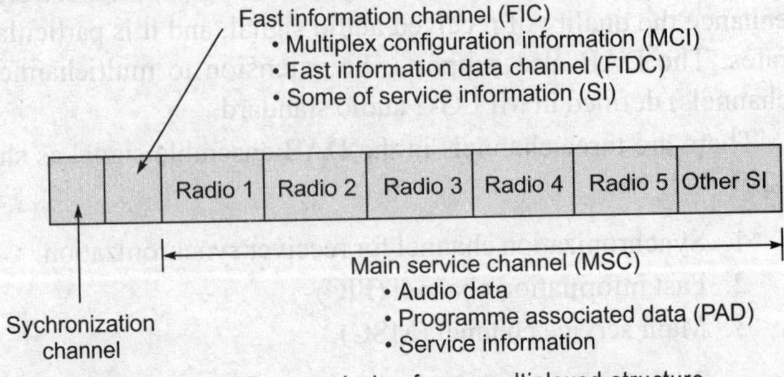

Fig. 12.6 DAB transmission frame multiplexed structure

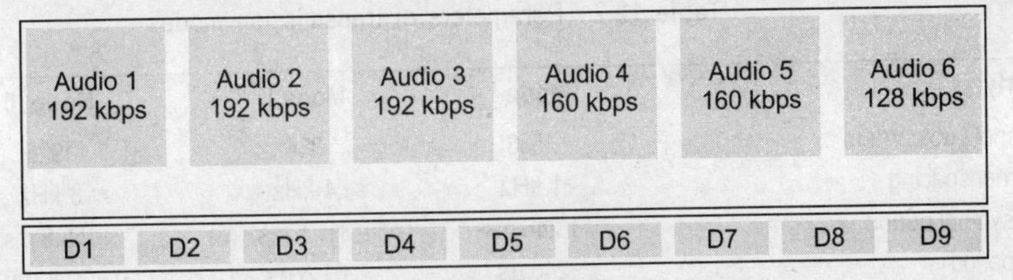

Fig. 12.7 DAB MSC multiplexed structure

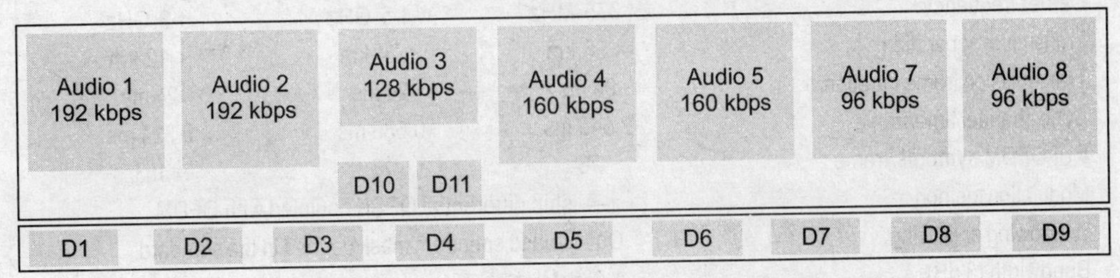

Fig. 12.8 DAB MSC multiplexed reconfigured

Conditional access can be created to any of services by scrambling the data stream before channel coding and multiplexing operations. This means that a scrambled service is available only to the users with an appropriate scrambling code, in a similar fashion as in cable/satellite TV broadcasting. Each of the subchannels in MSC and FIC is individually coded and subchannels of MSC are also individually time-interleaved. Both unequal error protection and equal error protection schemes are supported. Only equal error protection is available for audio services below 144 kbps rate, but unequal error protection is recommended for audio service above 144 kbps rate. Varying error protection level is achieved by employing different channel coding rates. Punctured convolutional coding is used as channel coding. The mother code is 1/4 rate convolutional code with a constraint length of 7 and the final rates of punctured codes range from 1/3 to 3/4. Afterwards, the MSC and FIC are multiplexed together by a transmission frame multiplexer, the whole signal of DAB ensemble is frequency interleaved prior to OFDM. This means that consecutive symbols are not sent on adjacent OFDM subcarriers.

The DAB supports four transmission modes designed to different frequency ranges. Main parameters of the transmission modes are given in Table 12.2. Special attention should be paid to mode IV, which is designed for the same frequency range as mode II. However, mode IV provides wider coverage at the expense of increased sensitivity to the terminal velocity. Due to SFN operation, an increase in the transmitter separation dictates a corresponding increase in guard period and to maintain an efficient transmission also in OFDM symbol period. Longer symbol period makes the signal more sensitive to the changes in the channel, i.e. to the velocity of terminal.

The total capacity of a DAB ensemble occupying 1.5 MHz band is roughly 2.4 Mbps. From this, MSC takes 2.3 Mbps and FIC 96 kbps (or 128 kbps in transmission mode III).

The net capacity of MSC, channel coding excluded, varies between 0.6 Mbps and 1.7 Mbps depending on the employed coding rates. The net capacity of FIC is 32 kbps (or

Table 12.2 Parameters of transmission modes

Property	Mode I	Mode II	Mode III	Mode IV
Number of sub carriers	1536	384	1925	768
Subcarrier spacing	1 kHz	4 kHz	8 kHz	2 kHz
OFDMsymbol period	1.246 ms	311.5 ms	155.8 ms	623 ms
Guard period	246 µs	61.5 µs	30.8 µs	123 µs
Useful symbol time	1 ms	250 µs	125 µs	500 ms
Carrier frequency	< 375 MHz	< 1.5 GHz	< 3 GHz	< 1.5 GHz
Transmitter separation	< 96 km	< 24 km	< 12 km	< 48 km
Transmission frame duration	96 ms	24 ms	24 ms	48 ms
Sync channel time/frame	2.543 ms	0.636 ms	0.324 ms	1.271 ms
# of OFDM symbols/frame	76	76	153	76
Modulation method	π-4–shift differential QPSK coulpled with OFDM			
Windowing or filtering	Out-of-band spectrum mask defined in the standard			
Bandwidth (3 dB)	1.536 MHz			

42.67 kbps in mode III), and multiplex configuration information occupies roughly 30% of FIC capacity.

All India Radio (AIR), the public broadcaster, started an experiment in regular DAB broadcasts in New Delhi in late 1997. The area covered was only 1% of the country's population. More regular services were started at the end of 2005 in Delhi, extending to Kolkata, Mumbai, and Chennai in the second phase. Six channels of AIR programming in stereo were provided.

12.3 DIGITAL RADIO MONDIALE (DRM)

Digital radio mondiale (DRM) is set to revolutionize broadcasting on the long, medium, and short wavebands. The technology is supported by All India Radio in India. Since the earliest days of broadcasting, these wavebands have been filled with signals that are amplitude modulated. These transmissions are of low audio quality and, particularly in recent years, there has been a move away from these bands to find higher quality transmissions. Afterwards, broadcasts in the VHF FM band also have received far more listeners. Now, DAB digital radio is available in many countries and this has set new standards in broadcasting. The next stage is to improve the transmissions on the long, medium, and short wavebands. As the requirements are very different from those experienced on the higher frequencies, the DAB standard is not directly applicable but similar concepts are utilized and as a result a new system has been developed, known as DRM. It provides many of the improvements that are needed along with the flexibility to allow for future developments.

The DRM is a consortium of broadcasters, network operators, equipment manufacturers, broadcasting unions, regulatory bodies, and other organizations representing 29 countries. Now, with 82 members, the wide base of its membership has been partly the reason behind its success. It has gained the experience by the EUREKA project, set up to develop the DAB digital radio. As a result, the DRM system is very similar to DAB and, therefore, has come to fruition remarkably swiftly. A preliminary system was designed and tested within a laboratory and this was later extended to include field trials on air to ensure that the new system would successfully meet all the requirements.

When the specification for DRM was being drawn up, there were a number of key requirements that needed to meet. The main thrust of the development was to ensure that greater audio quality could be achieved, but this needed to be achieved while keeping the transmissions in a form where they could operate alongside the existing AM transmissions. In the US, 10 kHz channel spacing is used on the medium waveband while in Europe there is 9 kHz spacing. On the short wavebands, 5 kHz channel spacing has been adopted. It is necessary for the new standard to be compatible with these, while offering the possibility of other bandwidth options for the future.

There are many advantages of the DRM system along with the good-quality audio transmissions. Data can also be transmitted using DRM system. One particularly useful feature for the short wavebands is that a list of alternative frequencies is transmitted so that the listeners can be transferred to better channels very easily as conditions change.

Another advantage of the DRM system is that it can support SFN. This allows a single frequency to be reused even within the coverage area of the first transmitter, without mutual interference. Currently, the frequencies can only be reused outside the coverage area of the first transmitter to avoid interference problems. By using an SFN, far more efficient use can be made of the available channels. With spectrum always in shortage, this is another important feature.

The transmitted signal uses COFDM. This form of modulation is being used more frequently and is very resilient to many common forms of interference and fading. Its main drawback has been that it requires a significant level of signal processing to extract the data from the carriers and reassemble it in the correct fashion. However, signal processing ICs are now sufficiently powerful and have a reasonable cost to make the use of this form of modulation viable. Interestingly, COFDM is also used by DAB digital radio. This makes it similar to DRM.

There are two main elements or rather special features of the DRM transmission system. These are the audio coding and the RF modulation used.

The main audio encoding system employs two main techniques. The first is called advanced audio coding (AAC). It is found that the human ear does not perceive all the sounds that are heard. A strong sound on one frequency will mask out others close in frequency that may be weaker. AAC, therefore, analyses each section of the spectrum and only encodes those sounds that will be perceived (as we have studied in case of DAB).

However, AAC on its own does not provide sufficient compression of the data to enable the transmissions to be contained within the narrow transmission bandwidths required. To provide the additional data compression required, a scheme known as *spectral band replication* (SBR) is employed. This analyses the sounds in the highest octave that are normally from sounds, like percussion instruments, of those that are harmonically related to other sounds lower in frequency. It analyses them and sends the data to the receiver, which will enable them to be reconstituted later.

As in DAB, the data to provide the different functions on the transmission is organized into a number of channels that are then applied to the overall modulating signal. The main payload for the signal is known as the *main service channel* (MSC) and this includes the audio signal data. Two subsidiary channels are also available. These are known as the *fast access channel* (FAC) providing the essential data required to fully decode the signal and the *service description channel* (SDC).

The DRM has been trialled for some time with experimental transmissions from broadcasters, including the BBC. Now, scheduled transmissions are taking place from a number of broadcasters, with more commencing their transmissions all the time.

12.4 HD RADIO TECHNOLOGY

The DAB digital radio, established in some areas of the globe, is known as high-definition (HD) radio. Using HD radio will enable high-quality audio to be received along with the ability to incorporate many new features and facilities.

The HD radio system has been developed by iBiquity and has now been selected by the Foderal Communications Commission (FCC) in the US. It will take the place of both the existing AM and FM transmissions and offers many advantages for both listeners and broadcasters alike as discussed below.

Improved audio quality It is claimed that HD radio broadcasts on the AM bands will be as good as current FM services and those on the FM band will offer CD quality audio.

Reduced levels of interference AM transmissions in particular are prone to static pops and bangs as well as high levels of background noise. HD radio will almost eliminate this.

Opportunity to use additional data services By using digital technology, HD radio provides the opportunity to add data services such as scrolling programme information, song titles, artist names, and much more. There is also the possibility of adding more advanced services, such as surround sound, multiple audio sources, and on-demand audio services.

Easy transition for broadcasters and listeners Although new HD radio receivers are required to receive the new transmissions in their digital format, there is considerable reuse of infrastructure and spectrum.

The HD radio uses a variety of technologies to enable it to carry digital audio in an acceptable bandwidth with the new high quality that is required. The transmission uses COFDM combined with specialized codec to compress the audio. One of the requirements for HD radio was that it would maintain compatibility with existing stations. To achieve this, there are two versions; one HD radio system for AM and the other for FM.

In what is termed as a hybrid mode, the AM version has a data rate of 36 kbps for the main audio channel and the version of HD radio for the FM bands carries 96 kbps. In addition to this, HD radio can also be used to carry multiple audio channels, and in addition to this, secondary channels for services, such as weather, traffic, and the like, may be added. However, adding additional channels will reduce the available bandwidth for the primary channel and the audio quality may be impaired.

In the hybrid mode, a radio receiver will first lock onto an analog signal. If this is possible, then it will try to find a stereo component (FM only) and finally it will endeavour to decode a digital signal. If the digital signal is lost, then it will fall back to the analog signal. The success of this process depends upon the transmitting station being able to synchronize the digital and analog signals. Often, the digitization process takes a noticeable amount of time and the digital and analog signals may not be transmitted in time with each other.

Once the HD radio is fully established, the hybrid mode may be removed and then no analog information will be transmitted. However, it is envisaged that this will take some time as this can only be viable when only very few analog radios are in use.

12.5 DIGITAL VIDEO BROADCASTING (DVB)

Technically, there are two main ways of delivering mobile TV in recent scenario—via a two-way cellular network and a one way dedicated broadcast network. Few examples of mobile TV technologies are DVB-H, satellite digital multimedia broadcast (S-DMB)(mainly used in South Korea, Japan), T-DMB (South Korea, Germany) TDtv (based on TD-CDMA technology from IPWireless), China mobile multimedia broadcasting (CMMB), 1seg (one segment) (based on Japan's integrated service digital broadcasting (ISDB-T)), MediaFLO, GPRS, and 3G.

Digital video broadcasting is a suite of internationally accepted specifications that became open standards for digital television. The DVB is a set of standards that define digital broadcasting using existing satellite, cable and terrestrial infrastructures. The scenario of DVB is given in Figures 12.9 and 12.10. The DVB project consists of over 220 organizations in more than 29 countries worldwide and its different standards are published by a Joint Technical Committee (JTC) of European Telecommunications Standards Institute (ETSI), European committee for electrotechnical standardization (CENELEC) and European Broadcasting Union (EBU). DVB uses mostly MPEG standards for the compression of audio and video signals. According to the modes of distribution, there are four different standards—DVB-S, DVB-C, DVB-T, and DVB-H. The last letter indicates the type of system.

The DVB-S is based on satellites, DVB-C is based on the domestic cable network, and DVB-T is based on terrestrial transmission and for audio/video streaming. All the technologies supersede the existing analog TV via satellite, cable, and terrestrial antenna. But the DVB standard is broader than just—S/-C/-T. DVB-H (H stands for hand-held) is an upcoming standard to broadcast TV content to mobile devices like PDAs or mobile phones. DVB-S2 is an extension of DVB-S standard, which is more efficient and offers more services. It is also geared towards the transmission of HDTV content. These distribution systems differ mainly in the modulation schemes due to different technical constraints. DVB-S (SHF range) uses QPSK, DVB-T (VHF/UHF range) (Fig. 12.11) uses COFDM and DVB-H also uses COFDM. We shall study DVB-H standard in detail as it is of great interest for the next generation.

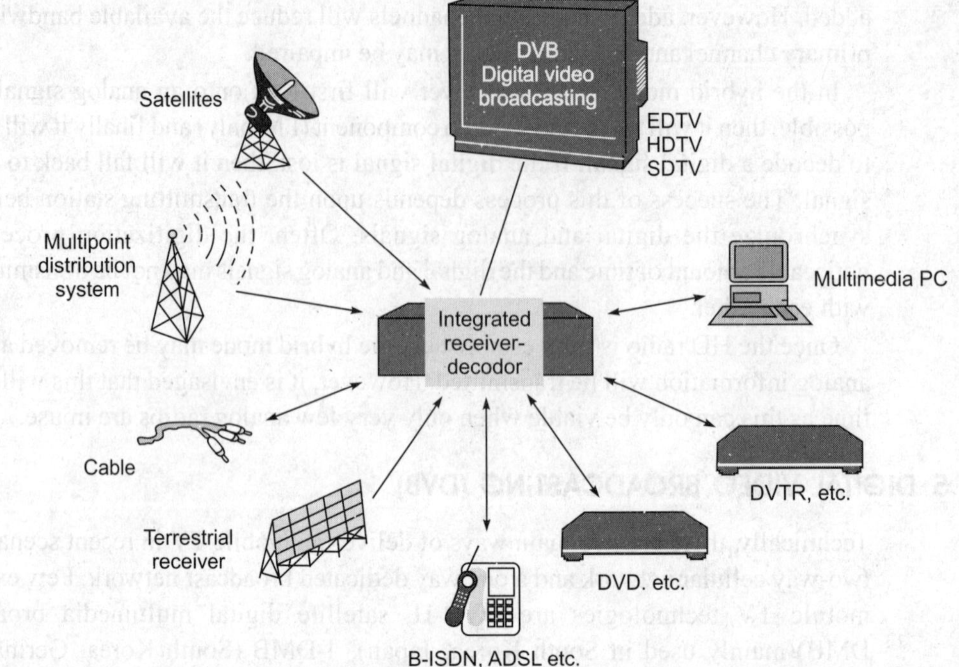

Fig. 12.9 DVB scenario

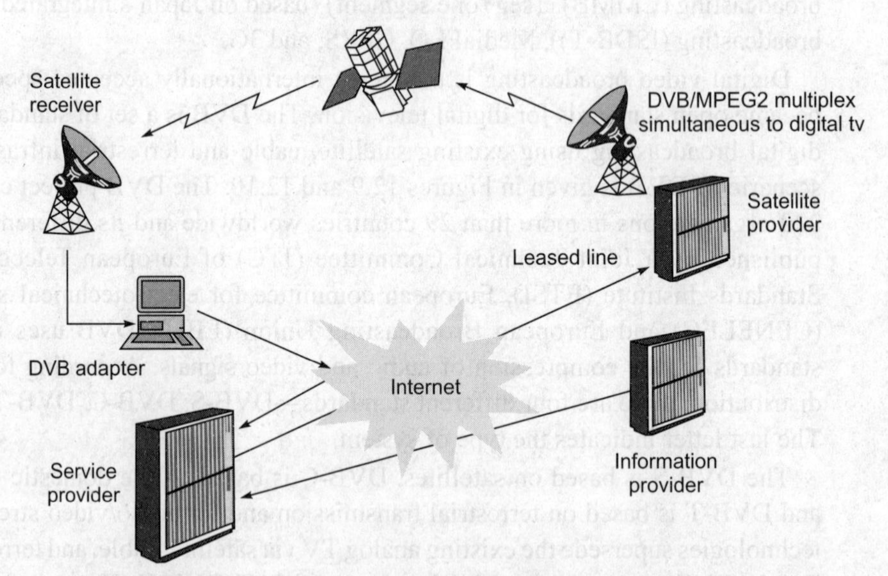

Fig. 12.10 High bandwidth Internet access using digital video broadcasting

12.5.1 DVB-H

Digital video broadcast-handheld (DVB-H) is one of the major systems to be used for mobile video and television for cellular phones and handsets. It is of much importance today. The DVB-H has been developed from the DVB-T (terrestrial) television standard,

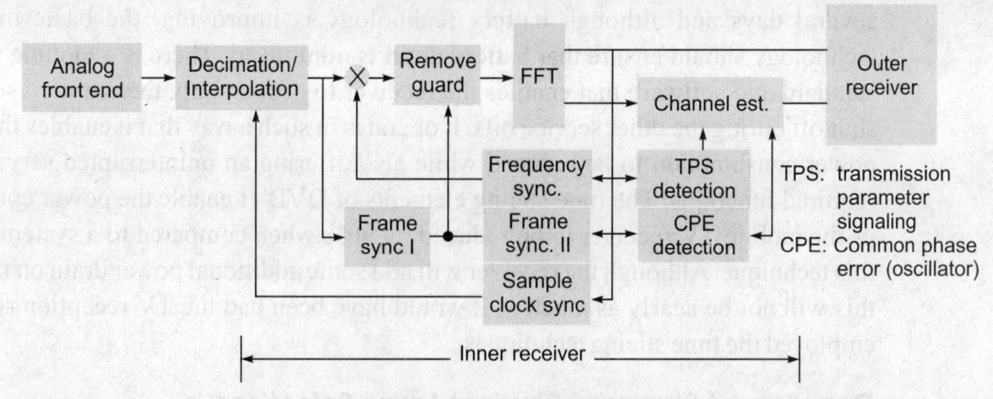

Fig. 12.11 Typical DVB-T receiver

which is used in many countries around the globe, much of Europe, including the UK, and also other countries, including the US. The DVB-T standard has been shown to be very robust and in view of its widespread acceptance, it forms a good platform for further development for hand-held applications. The DVB-H has taken up this basic standard and adapted so that it is suitable for use in a mobile environment, particularly with the electronics incorporated into a mobile phone. A typical DVB-T receiver diagram is given in Fig. 12.11, which is also very similar to DVB-H. The DVB-H standard has been adopted by ETSI and in this way the system can be truly international, and this will prevent compatibility problems caused by different countries and operators using different variants of the same system. The documents for the physical layer were ratified in 2004, with the upper layers defined in 2005.

The environment for hand-held devices is considerably different to that experienced by most televisions. Normally, domestic televisions have good directional antenna systems and, in addition to this, the reception conditions are fairly constant. Additionally, most televisions receiving DVB-T will be powered by mains supplies. As a result, current consumption is not a major issue.

Further, in the first instance, the antennas will be particularly poor because they will need to be small and integrated into the handset in such a way that they either appear fashionable or are not visible. Additionally, they will obviously be mobile, and this will entail receiving signals in a variety of locations, many of which will not be particularly suitable for video reception. Not only will the signal be subject to considerable signal variations and multipath effects, but it may also experience high levels of interference. Also some difficulties are presented by the fact that the handset could be in a vehicle and actually on the move. The operation of DVB-H has to be sufficiently robust to accommodate all these requirements.

While DVB-H proved to be remarkably robust under many circumstances, one of the major problems was that of current consumption. Battery life for handsets is a major concern where users anticipate the life between charges to be several days.

One of the key requirements for any mobile TV system is that it should not give rise to undue battery drain. Mobile handset users demand that the battery life extends over several days and although battery technology is improving, the basic mobile TV technology should ensure that battery drain is minimized. There is a module within the standard and software that enables the receiver to decode only the required service and shut off during the other service bits. It operates in such a way that it enables the receiver power consumption to be reduced while also offering an uninterrupted service for the required functions. The time slicing elements of DVB-H enable the power consumption of the mobile TV receiver to be reduced by 90% when compared to a system not using this technique. Although the receiver will add some additional power drain on the battery, this will not be nearly as much as it would have been had the TV reception scheme not employed the time slicing techniques.

Overview of Standard Physical Layer Selections

The DVB-H standard like DVB-T uses OFDM technology. This has been adopted because of its high data capacity and suitability for applications such as broadcasting. It also offers a high resilience to interference and can tolerate multipath effects and is able to offer the possibility of a single frequency network (SFN).

There are a variety of modes in which the DVB-H signal can be configured. These conform to the same concepts as those used by DVB-T. These are 2K, 4K, and 8K modes, each having a different number of carriers as defined in Table 12.3. The 4K mode is a further introduction beyond what is available for DVB-T. The different modes balance the different requirements for network design, trading mobility for single frequency network size, with the 4K mode being that which is expected to be most widely used.

Table 12.3 DVB-H signal modes

Parameter	2K Mode	4K Mode	8K Mode
Number of active carriers	1705	3409	6817
Number of data carriers	1512	3024	6048
Individual carrier spacing	4464 Hz	2232 Hz	1116 Hz
Channel width	7.61 MHz	7.61 MHz	7.61 MHz

The standard will support a variety of different types of modulation within the OFDM signal. The QPSK, 16QAM, and 64QAM will all be supported, chipsets being able to detect the modulation and receive the incoming signal. The choice of modulation is again a balance, QPSK offering the best reception under low signal and high noise conditions, but offering the lowest data rate. The 64QAM offers the highest data rate but requires the highest signal level to provide sufficiently error-free reception.

Interleaving is a technique where sequential data words or packets are spread across several transmitted data bursts. In this way, if one transmitted burst or group is lost as a result of noise or some other dropout, then only a small proportion of the data in each original word or packet is lost and it can be reconstructed using the error detection and correction techniques employed. The basic mode of interleaving used on DVB-T and which is also available for DVB-H is a native interleaver that interleaves bits over one OFDM symbol. However, DVB-H provides a more in-depth interleaver that interleaves bits over two OFDM symbols (for the 4K mode) and four bits (for the 2K mode).

Using the in-depth interleaver enables the noise resilience performance of the 2K and 4K modes to be brought up to the performance of the 8K mode and it also improves the robustness of the reception of the transmissions in a mobile environment.

In view of the particularly difficult reception conditions that may occur in the mobile environment, further error correction schemes are included. A scheme known as MPE-FEC provides additional error correction to that applied in the physical layer by the interleaving. This is a forward error correction scheme that is applied to the transmitted data and after reception and demodulation, allows the errors to be detected and corrected.

The DVB-H is a development of DVB-T and as a result it shares many common components. It has also been designed such that it can be used in 6, 7, and 8 MHz channel schemes although the 8MHz scheme will be the one most widely used. There is also a 5 MHz option that may be used for non-broadcast applications.

In view of the similarities between DVB-H and DVB-T it is possible for both forms of transmission to exist together on the same multiplex. In this way a broadcaster may choose to run two DVB-T services and one DVB-H service on the same multiplex. This feature may be particularly attractive in the early days of DVB-H when separate spectrum is not available.

The DVB-H has been used in a number of trials and appears to perform well. It has support from a number of the major industry players and is likely to achieve a considerable degree of acceptance worldwide. Accordingly, it is likely to be one of the major standards used for mobile video.

12.6 DIRECT TO HOME (DTH)

In direct-to-home (DTH) telecast, TV channels/programmes are directly distributed via satellite to the subscribers' homes without the intervention of a cable operator. The signals are transmitted in Ku band (10.7 GHz to 18 GHz) and are received by the subscribers through a small dish antenna (about 45cm in dia.) and a set top box (or an integrated receiver decoder).

The DTH system can also provide many value-added services, such as the Internet, e-mail, data casting, e-commerce, and interactive multimedia. It has the provision for a subscriber management system similar to the one for *conditional access system* (CAS). The current means of broadcasting in India do not provide quality reception in shadow areas, particularly in the North-Eastern region. The DTH can fill this void easily. In short, DTH offers immense opportunities to both the broadcasters and the viewers. Detailed guidelines for starting DTH service in India were issued by the government on March 15, 2001, followed by guidelines in March 2003 for uplinking of foreign-owned news channels. Doordarshan has also planned to launch DTH television satellite with Prasar Bharati.

There are five major components involved in a direct-to-home (DTH) satellite system:

- Programming source
- Broadcast centre
- Satellite
- Satellite dish
- Receiver

Programming sources are simply the channels that provide programming for broadcast. The provider (the DTH platform) does not create original programming itself. It pays other companies for the right to broadcast their content via satellite. Satellite TV providers get programming from two major sources: international turnaround channels (such as HBO, ESPN and CNN, STAR TV, SET, and B4U) and various local channels (Suchas Sahara TV and Doordarshan). Most of the turnaround channels also provide programming for cable television. So, sometimes some of the DTH platforms will add some special channels exclusive to it to attract more subscriptions. The turnaround channels usually have a distribution centre that beams their programming to a geostationary satellite. The broadcast centre uses antennas or large satellite dishes to pick up these analog and digital signals from several sources. The complete scenario for direct to home is given in Fig. 12.12.

In this way, the DTH service provider is like a broker between the viewer and the actual programming sources. (Cable television networks also work on the same principle.) The broadcast centre is the hub of the system. At the broadcast centre, the DTH television service provider receives signals from various programming sources, compresses it using digital compression, if necessary scrambles it and beams a broadcast signal to satellites being used by it. The satellites receive the signals from the broadcast

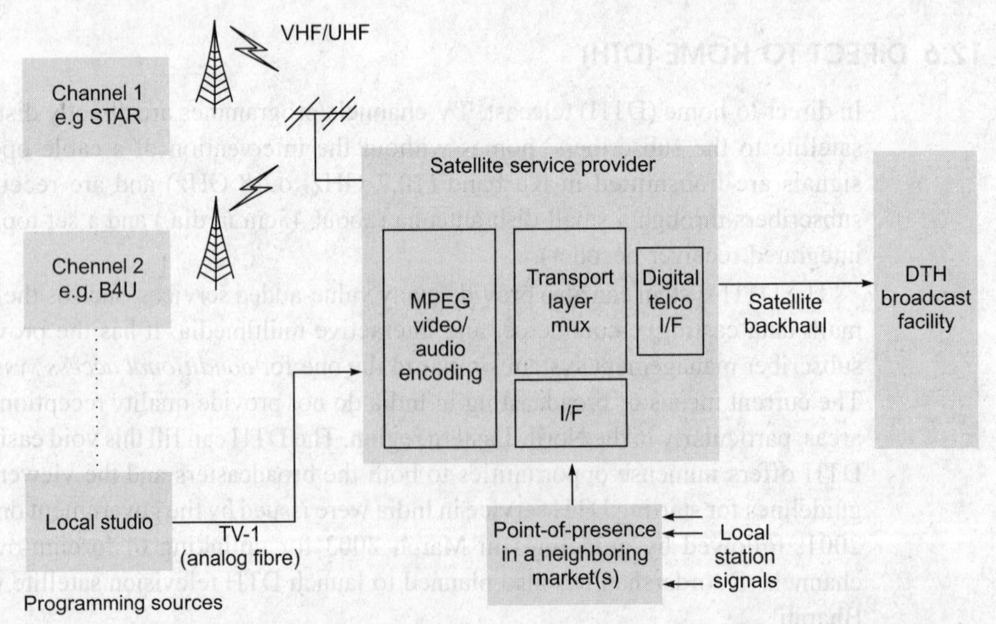

Fig. 12.12 Scenario for the programming sources connected to DTH service provider

station and rebroadcasts them to the ground. The viewer's dish picks up the signal from the satellite (or multiple satellites in the same part of the sky) and passes it onto the receiver in the viewer's house. The receiver processes the signal and satellite and cable TV passes it onto a standard television.

The broadcast centre converts all of this programming into a high-quality, uncompressed digital stream. At this point, the stream contains a vast quantity of data — about 270 Mbps for each channel. In order to transmit the signal from there, the broadcast centre has to compress it. Otherwise, it would be too big for the satellite to handle. The providers use the MPEG-2 compressed video format—the same format used to store movies on DVDs. With MPEG-2 compression, the provider can reduce the 270-Mbps stream to about 3 or 10 Mbps (depending on the type of programming). This is the crucial step that has made DTH service a success. With digital compression, a typical satellite can transmit about 200 channels. Without digital compression, it can transmit about 30 channels. At the broadcast centre, the high-quality digital stream of video goes through an MPEG-2 encoder, which converts the programming to MPEG-2 video of the correct size and format for the satellite receiver in your house.

After the video is compressed, the provider needs to encrypt it in order to keep people away from accessing it for free. Encryption scrambles the digital data in such a way that it can only be decrypted (converted back into usable data), if the receiver has the correct decoding satellite receiver with decryption algorithm and security keys. Once the signal is compressed and encrypted, DVB-S based modulation stages will be followed and after converting to RF the broadcast centre will beam it directly to one of its satellites. The transmitter blocks are shown in Fig. 12.13. Reverse are the receiver stages.

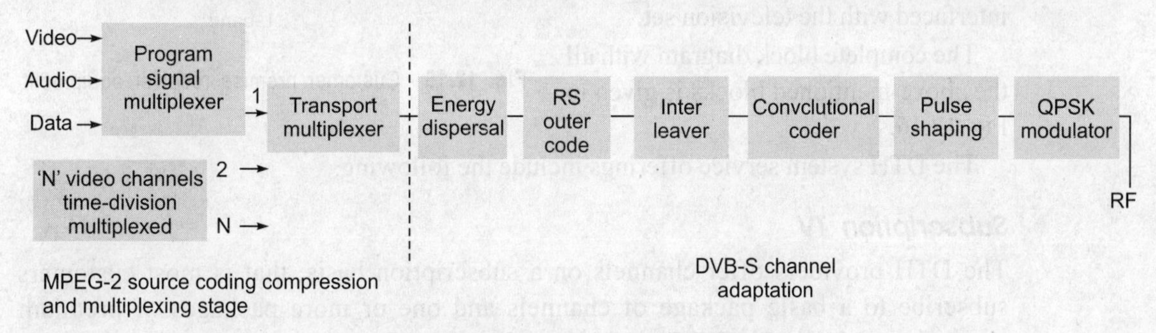

Fig. 12.13 DTH transmitter blocks

The *satellite* picks up the signal, amplifies it, and beams it back to the earth, where viewers can pick it up.

A *satellite dish* is just a special kind of antenna designed to focus on a specific broadcast source as shown in Fig. 12.14. This makes the system spectrally efficient. The standard dish consists of a parabolic (bowl-shaped) surface and a central feed horn. To transmit a signal, a controller sends it through the horn and the dish focuses the signal into a relatively narrow beam. The dish is connected with the receiver set.

A *receiver* does an exactly opposite task to that of the transmitter and gets compatible signals for various conventional display devices but before that it must be down

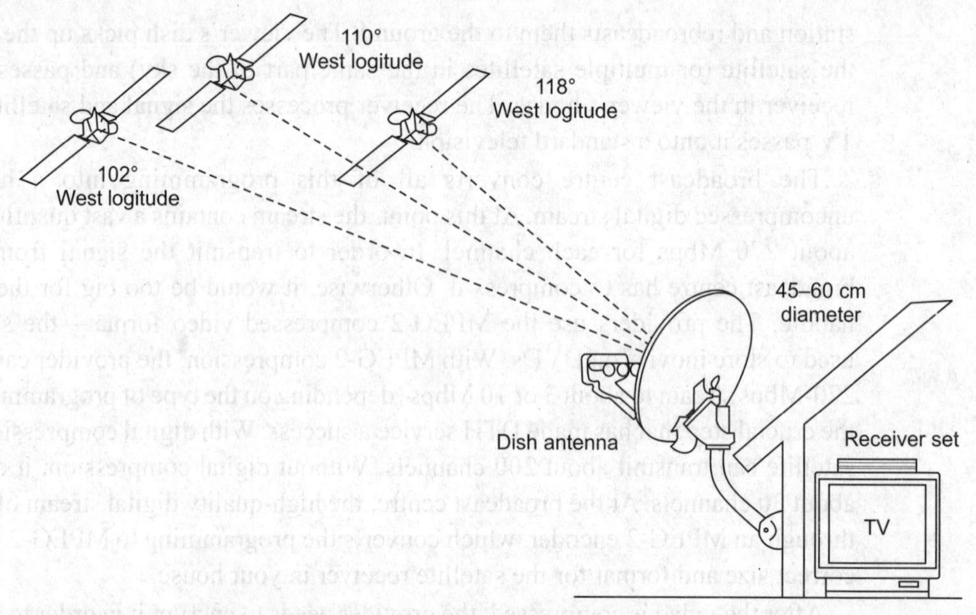

Fig. 12.14 Use of multiple orbital locations allow spectral reuse making it spectrally efficient

converetd. The simplified diagram of customer premise receiver equipment with RF down conversion part by superheterodyning is given in Fig. 12.15. The receiver device is normally interfaced with the television set.

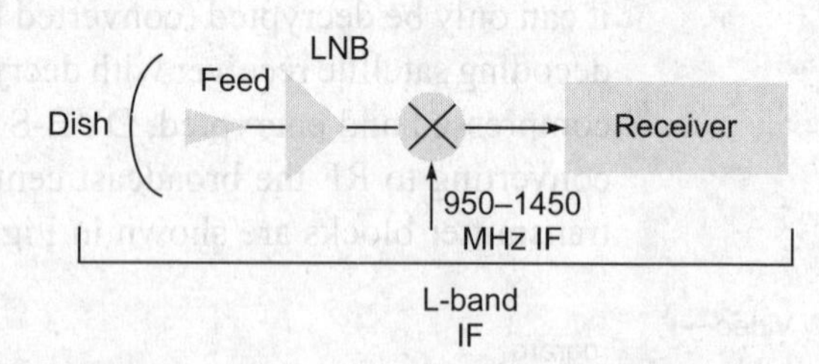

Fig. 12.15 Customer premise receiver equipment

The complete block diagram with all the above-mentioned blocks is given in Fig. 12.16.

The DTH system service offerings include the following.

Subscription TV

The DTH providers offer channels on a subscription basis, that is most customers subscribe to a basic package of channels and one or more packages of premium channels, such as the HBO multiplex. Special television subscriptions, such as international channels, are available. Typically, the basic packages include access to an on-screen electronic programme guide and a number of audio-only music channels. Subscriptions are also offered for series of events, such as games.

Pay per View (PPV)

The PPV services give customers the option to pick a specific programme or series of programmes and pay for the selected content as a one-time transaction. As indicated on their web sites, the DTH systems provide PPV offerings, including films, concerts, and sports events. In certain cases, the PPV concept has been extended to selling viewing rights for a movie for an entire day rather than for a single showing.

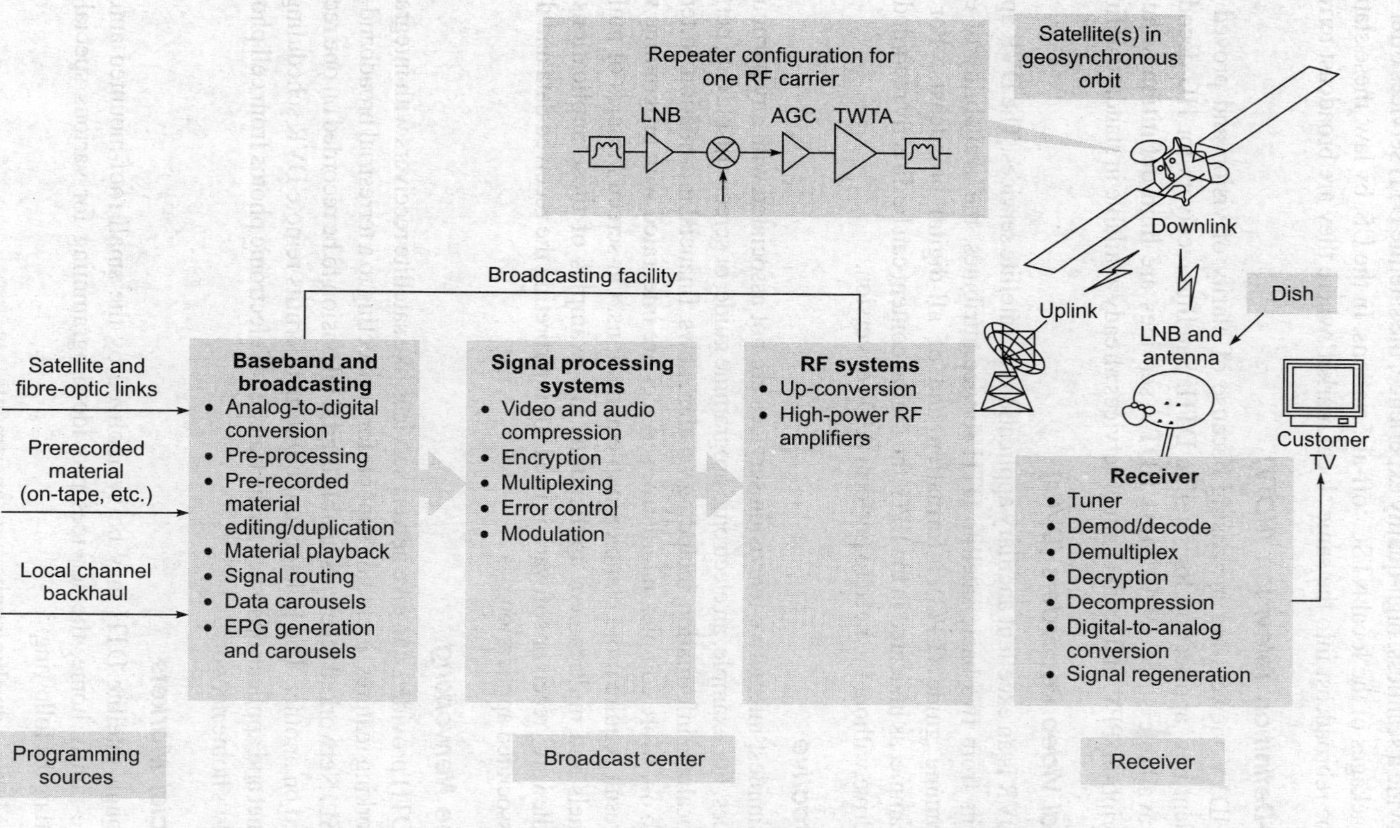

Fig. 12.16 Complete DTH system block diagram

Local Channel Rebroadcasts

To provide a seamless, high-quality experience, the satellite DTH services offer subscription packages of the local NTSC 'off-air' stations. In the US, by law, these stations may only be rebroadcast into the same 'local market' where they are broadcast terrestrially.

High-Definition Television (HDTV)

The HDTV viewers are increasing because of clarity of vision and proved to be an excellent new application for satellite DTH. Satellite receivers for HD decoding have been available since 1999. Most HDTV services are high-definition simulcasts of subscription, PPV, and local channel services already available in standard definition.

Digital Video Recorders (DVRs)

The DVR is an excellent ancillary application for satellite services. The DVR application benefits from two basic satellite DTH service attributes, the availability of electronic programme guide (EPG) information and of all-digital broadcasts. For a given programme, as indicated in the EPG, the digital content can be directly recorded to a hard disk drive without the need to perform A/D conversion.

Interactive

The simplest interactive television services are not associated with any particular video services. For example, an electronic programme guide, or screens displaying personalized and localized information, including weather, news, financial information, lottery results, and so on. More complex interactive services are integrated with programme video and as a result require more complex implementations. On-screen mosaics of multiple live channels and multicamera applications are examples of these applications. Special 'middleware' receiver software is used for interpreting the received data and displaying the associated application.

Home Networking

The DTH providers can give newest services like satellite receivers with integrated home networking features, including support for connecting to a terrestrial broadband path such as DSL. Networked receivers enable digital television to be recorded on one receiver and played on another. The linkage to the Internet permits remote DVR scheduling over the Internet and applications, such as the transfer of electronic photos from cell phones to the family's home network.

Special Markets

Although satellite DTH may be symbolized by the small roof-mounted antenna on a single-family home, the services provide programming for various special markets, including the following:

1. Multiple dwelling units, e.g., apartments
2. Hospitality market of hotels, bars, and restaurants
3. Mobile vehicles
4. Commercial aircraft

Summary

- Most of the future broadcast systems will be digital as the analog systems cannot give quality of reception.
- Broadcast systems mainly deal with audio and video transmission and data.
- The DAB standard for audio is based on COFDM modulation scheme that can form a single-frequency network. All the channels will form the DAB frames that can be broadcast at the same time using SFN.
- The SFN concept is spectrally efficient.
- For digital broadcast systems, efficient source and channel coding schemes are required. MPEG standards are utilized.
- The Digital radio modiale and HD radio are the specially designed systems for digital audio broadcasting. Both are based on COFDM and SFN concept.
- The DVB mostly uses MPEG standards for the compression of audio and video signals. According to distribution, there are four different standards: DVB-S, DVB-C, DVB-T, and DVB-H. The last letter indicates the way the signal is transmitted.
- DVB-S is based on satellites, DVB-C is based on the cable network in house, and DVB-T is based on terrestrial transmission and for audio/video streaming.
- DVB-H (H stands for hand-held) is an upcoming standard to broadcast TV content to mobile devices like PDAs or mobile phones.
- DVB-S2 is an extension of DVB-S standard, which is more efficient and offers more services.
- DTH is a satellite-based television broadcast system.
- In DTH, dish antenna and customer premise equipment is required.
- Apart from audio and TV, latest broadcast systems can provide many other services like forecasting, news, etc.

Review Questions

1. What is the difference between analog and digital broadcasting? Can you list out the suitable modulation schemes for both types of systems?
2. Why is it necessary to form DAB frames?
3. Discuss the suitability of selection of COFDM for the broadcasting applications.
4. Describe the concept of single-frequency network.
5. Find out the similarities and differences between DRM and HD radio systems? How similar are these systems to the DAB standards?
6. Why is it necessary to have sophisticated source and channel coding schemes for digital broadcasting applications? List out the important methods and point out the suitability and advantages of each.
7. What may be the simple receiver diagram for digital radios?
8. What are the differences and similarities between DAB and DVB standards?
9. Describe the various DVB standards and associated applications. Why is it necessary to define different DVB standards? Can we have a common standard?
10. What are the tools required for DTH reception?
11. List out the possible applications on DTH. Can you suggest some more applications at your end with justification?
12. Write down the constraints of any satellite-based communication system? How are these constraints eliminated from DTH?

Multiple Choice Questions

1. DAB and DVB work on
   (a) spread spectrum modulation
   (b) multicarrier COFDM
   (c) QPSK
   (d) QAM
2. Spectrum efficiency in DAB and DVB is enhanced because of
   (a) multiple carriers
   (b) single-frequency network concept
   (c) assigning less number of channels
   (d) narrow channels
3. For any audio or video transmission system, which of the following stage(s) is/are of great importance for the best-quality reception?
   (a) Encoding and compression
   (b) Modulation
   (c) RF
   (d) Both (a) and (b)
4. The spectrum of DAB signal is
   (a) 1.2 MHz (b) 4 kHz
   (c) 1.5 MHz (d) 150 kHz
5. Which of the following is the requirement of single-frequency network?
   (a) All transmitters in the network must be synchronized.
   (b) Signals from all nearby transmitters must be received within guard period of OFDM.
   (c) Carriers are multiple and orthogonal.
   (d) All of the above.
6. The maximum number of subcarriers forming the channel of DAB system is approximately
   (a) 1200 (b) 8000
   (c) 1500 (d) 256
7. Which of the following source encoding techniques is used for DAB system?
   (a) MPEG-4
   (b) Only PCM
   (c) MPEG-4 audio layer
   (d) MPEG-2 audio layer
8. Which of the following channels is contained by each audio service channel for conveying information that is closely linked to the audio program and provide dynamism?
   (a) PAD (b) FIC
   (c) MSC (d) SI
9. The total capacity of a DAB channel is approximately
   (a) 2.4 kbps (b) 10 Mbps
   (c) 2.4 Mbps (d) 1.5 Mbps
10. In DRM system, which of the following methods is used for additional data compression along with advanced audio coding?
    (a) Spectral band splitting
    (b) Spectral band replication
    (c) Subband coding
    (d) None
11. Which of the following broadcasting systems can work in the hybrid mode with AM or FM versions (bands)?
    (a) DAB
    (b) DRM
    (c) HD radio
    (d) All of the above
12. Which of the following DVB standards is designed for the satellite transmissions?
    (a) DVB-C (b) DVB-T
    (c) DVB-S
    (d) DVB-H
13. Which of the following standards is suitable for the mobile cellular phones?
    (a) DVB-C (b) DVB-T
    (c) DVB-S (d) DVB-H

14. In DVB-H system, the maximum number of active carriers will be
(a) 6817 (b) 8048
(c) 1500 (d) 3614

15. By which of the following modulation schemes supported OFDM, the maximum data rate can be achieved in DVB-H system?
(a) QPSK (b) 64QAM
(c) 64QPSK (d) 16QAM

16. Which of the following is the band of transmission for the DTH system?
(a) S band (b) C band
(c) Ka band (d) Ku band

chapter 13

Wireless Communication Systems and Standards 2: Infrastructure-Based/ Cellular Networks

Key Topics

- GSM system
- GSM upgradations
- GPRS addition to GSM system
- EDGE technology
- IS-95 CDMA system and evolution up to CDMA2000
- WLL and CorDECT WLL system
- UMTS and IMT2000 system
- Mobile satellite communication
- Convergence in the networks and beyond 3G system

Chapter Outline

Mobile networks are divided into two types: permanent infrastructure-based networks and ad hoc or personal area networks. These provide both voice and data services. For both types, different protocols are designed. In the protocol stacks, provisions for voice and data transfer with reliability are made. Only infrastructure-based networks are discussed in this chapter. Again, the infrastructure is used in such a way that mobility is also maintained. Complete details of all the systems are not possible to be given here, but a brief overview is provided. GSM, GPRS, and EDGE systems, WLL, CDMA-based IS-95, and 3G UMTS based on WCDMA, all these provide services for mobile phones, with voice and data applications using the infrastructure based on cellular theory described in Chapter 11. Mobile networks can be combined with satellites called mobile satellite communication. The scenario for 3G and beyond is also given here, which again gives rise to the concept of converged networks.

13.1 INTRODUCTION TO MOBILE NETWORKS

There are two types of mobile networks that can be formed: infrastructure networks and ad hoc networks or infrastructure independent networks. Both types can be designed for voice/data services.

The infrastructure application is aimed at office areas or to provide a 'hotspot'. It can be installed instead of a wired system and can provide considerable cost savings, especially when used in established offices. A backbone wired network may be still required and is connected to a server. The wireless network is then split up into a number of cells, each serviced by a base station or *access point* (AP), which acts as a controller for the cell. Each access point may have a range dependent upon the environment and the location of the access point.

The other type of network that may be used is termed as an ad hoc network. These are formed when a number of computers and peripherals are brought together to establish their own network in a small domain. They may be needed when several people come together and need to share data, or if they need to access a printer without the need for having to use wire connections. In such a situation, the users only communicate with each other and not with a larger wired network. As a result, there is no access point. Special algorithms within the protocols are used to enable one of the peripherals to take over the role of the master to control the network with the others acting as slaves. These networks are described in the next chapter.

13.2 GSM SYSTEM

The GSM system is the most widely used mobile telecommunications system in the world today. The GSM originally stood for groupe speciale mobile, but afterwards it was changed to global system for mobile communications. Since it was first deployed in 1991, the use of GSM has grown steadily, and it is now the most widely spread out cell phone system in the world. The GSM reached the one billion subscriber point in February 2004 and continues to grow in popularity.

Global system for mobile telecommunication (GSM) standards were developed for mobile telephony in digital form using a cellular structure. Though the development scenario is introduced in Chapter 1, it is repeated here for the continuity. Wireless technology growth for mobile telephony and among all position of GSM is as follows.

- Firstly, there was the analog mobile phone and a limited area of coverage of 1G.

This type of system offered limited mobility and voice communication within a group only. No data communication was possible.

- Secondly, the analog cellular phone with a wide area coverage came.

This system offered complete mobility within the service area and connectivity to PSTN and other networks, however, still without data communication facility.

- Thirdly, the digital cellular phone systems (GSM) of 2G came.

This system offers complete mobility within the service area and connectivity to PSTN and other networks. They provided data communication facility also along with other advanced features. The limitation was low spectral efficiency and low data rates.

- Lastly, came the digital cellular communication systems (3G).

This system offers complete mobility within the service area and connectivity to PSTN and Internet/intranet. They have a very high data rate communication facilities and other advanced features.

The GSM system was designed as a second-generation (2G) cellular telecommunication system. It is a fully digitized technology for better speech quality. One of the basic aims was to provide a system that would enable a greater capacity to be achieved than the previous first-generation analog systems. The GSM achieved this by using a digital

TDMA along with FDMA. By adopting this technique, more users can be accommodated within the available bandwidth. In addition to this, ciphering of the digitally encoded speech was adopted to retain privacy. Because of the encryption of user information, GSM calls cannot be captured by unauthorized persons. It is fully compatible with the existing fixed line network. Operation with a single number with worldwide roaming is possible in GSM. Very well-defined interfaces make GSM a truly open system. Available versions of GSM are GSM 900, 1800, and 1900. The GSM is a successful technology and well established in India also.

GSM-900 System Specifications

- The frequency range is 890 MHz to 915 MHz for uplinking and 935 MHz to 960 MHz for downlinking.
- It uses FDMA/TDMA multiple access technology for downlinking / uplinking.
- A duplex technique FDD is incorporated.
- It has a modulation GMSK (BT = 0.3).
- There are 124 reusable spot frequencies (channels) of 200 kHz bandwidth each.
- Each spot frequency carries eight time slots for traffic/signalling, i.e. eight speech channels per RF channel.
- It has separate logical signalling and traffic channels.
- The transmit/receive frequency spacing is 45 MHz while there is FDD-based FDMA.
- The transmit/receive time slot spacing is of three time slots.
- The typical channel data rate is 270.833 kbps.
- The time for frame duration is 4.615 ms.
- Voice encoding by various methods is possible, originally RPE-LTP, the voice coder bit rate is 13.4 kbps.
- It is compatible to ISDN and public switched packet data network (PSPDN).
- It uses the SS7 method for signalling.

In addition to the voice services, GSM supports a variety of other data services. Although their performance is nowhere near the level of those provided by 3G, they are nevertheless still important and useful. Various data services are supported with user data rates up to 9.6 kbps. The services, including group 3 facsimile, videotext, and teletext, can be supported. Many enhanced one like voicemail, RMCA, etc., can also be supported.

One service that has grown enormously is the short message service. Developed as a part of the GSM specification, it has also been incorporated into other cellular systems. It can be thought of as being similar to the paging service but is far more comprehensive allowing bidirectional messaging, store, and forward delivery. It also allows alphanumeric messages of a reasonable length. This service has become popular due to its simplicity and low fixed cost.

13.2.1 Components of GSM System

In general, the components of the cellular systems are introduced in Chapter 11. However, the architecture with its hardware for the various parts of the GSM system are

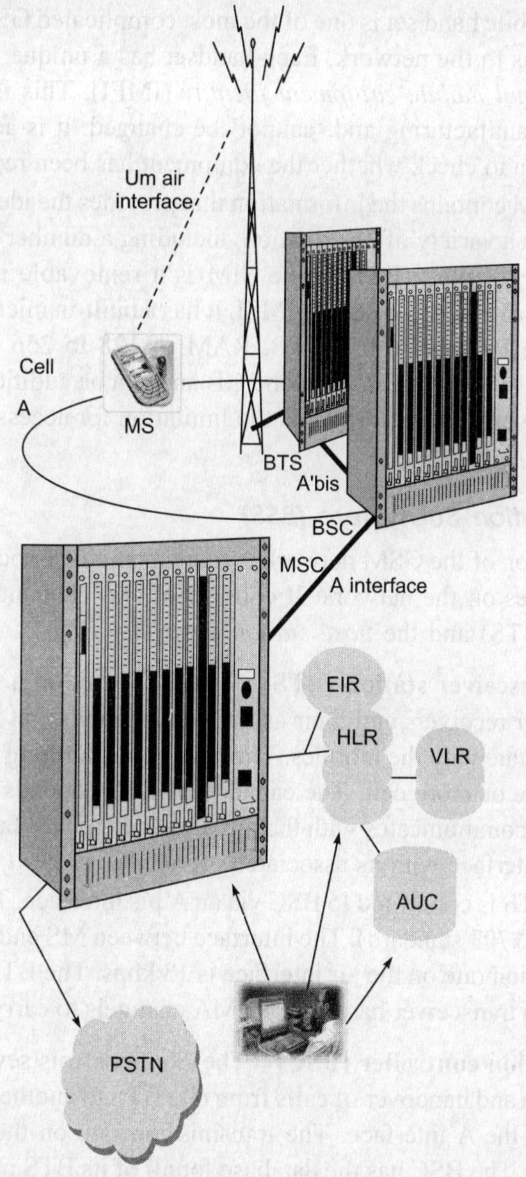

Fig. 13.1 A complete GSM system scenario with interconnection between subsystems

described here in brief. The GSM system diagram is shown in Fig. 13.1. The architecture of the GSM system with its hardware can broadly be grouped into three main areas: the mobile station, the base station subsystem, and the network subsystem. Each area performs its own functions and when used together, they enable the fully operational capability of the system to be realized.

Mobile Station (MS)

The mobile station consists of two units—mobile handset with battery and subscriber identity module (SIM).

The mobile hand set is one of the most complicated GSM devices. It provides the user with access to the network. Each handset has a unique identity number known as the. *international mobile equipment identity* (IMEI). This is installed in the phone at the time of manufacturing and 'cannot' be changed. It is accessed by the network during registration to check whether the equipment has been reported as stolen.

The SIM contains the information that provides the identity of the user to the network. It contains a variety of information, including a number known as *international mobile subscriber identity* (IMSI). The SIM is a removable module that fits in the mobile handset. Each SIM has unique IMSI. It has a built-in microcomputer and memory in it. It contains a ROM of 6 to 16 kB, RAM of 128 to 256 kB (latest phones have phone memories in Mbytes and even one GB and can be additionally provided) and EEPROM of three to eight kB. Memory is the limitation for accessing large web pages on mobile phones.

Base Station Subsystem (BSS)

This section of the GSM network is fundamentally associated with communication with the mobiles on the network. It consists of two elements, namely the *base transceiver station* (BTS) and the *base station controller* (BSC).

Base transceiver station (BTS) The BTS used in a GSM network comprises radio transmitter receivers and their associated antennas that transmit and receive to directly communicate with the mobiles. The BTS is the defining element for each cell. One BTS covers one or more cell. The capacity of BTS depends on the number of transceivers. The BTS communicates with the mobiles and the interface between the two is known as the *Um* interface with its associated protocols.

The BTS is connected to BSC via an A'bis interface. The transmission rate on A'bis is 2 Mbps (G.703 standard). The interface between MS and BTS is called air interface. The transmission rate on the air interface is 13 kbps. The BTS controls the RF parameters of MS. Each transceiver has eight TDMA channels to carry voice and signalling.

Base station controller (BSC) The BSC controls several BTSs. It manages channel allocation and handover of calls from one BTS to another BTS. The BSC is connected to MSC via the A interface. The transmission rate on the A interface is 2 Mbps (G.703 standard). The BSC has the database for all of its BTS parameters. It provides path from MS to MSC.

Network Subsystem

The network subsystem contains a variety of different elements and is often termed the core network. It provides the main control and interfacing for the whole mobile network. It includes the elements, such as MSC, HLR, VLR, AUC, and more, as described here (Fig. 13.1).

Mobile switching centre (MSC) The MSC is heart of the entire network connecting the fixed line network to the mobile network. It manages all call-related functions and billing information. It is connected to the HLR and VLR for subscriber identification and for routing incoming calls. The MSC capacity is in terms of the number of subscribers. The MSC is connected to BSC at one end and to the fixed line network at other end. *Call detail record* (CDR) is generated for each and every call in the MSC.

Home location register (HLR) All the subscriber's data is stored in HLR. It has a permanent database of all the registered subscribers. The HLR has a series of numbers for all subscribers. When a user switches on the phone, the phone registers with the network and from there, it is possible to determine which BTS it communicates with so that the incoming calls can be routed appropriately. Even when the phone is not active (but switched on), it re-registers periodically to ensure that the network (HLR) is aware of its latest position. There is one HLR per network, although it may be distributed across various subcentres too for operational reasons.

Visiting location register (VLR) An active subscriber is registered in VLR. It has a temporary database of all the active subscribers used for their call routing. The HLR validates the subscriber before registration. The MSC asks VLR before routing incoming call.

Authentication centre (AUC) The AUC is a protected database that contains the secret key also contained in the user's SIM card. It is used for authentication and for ciphering on the radio channel. Authentication is a process to verify the subscriber SIM. Secret data and verification algorithm are stored in the AUC. The AUC and HLR combine to authenticate the subscribers. Subscriber authentication can be done on every call, if required.

Equipment identity register (EIR) All subscribers' mobile handset data is stored in EIR. The EIR is the entity that decides whether a given mobile equipment may be allowed onto the network. Each mobile equipment has a unique IMEI. This number, as mentioned above, is installed in the equipment and is checked by the network during registration. The MSC asks the mobile to send its IMEI and then checks it with the data available in EIR.The EIR has different classifications for mobile handsets like white list, grey list, and black list. Depending upon the information held in the EIR, the mobile may be allocated one of three states—allowed onto the network, barred access, or monitored in case of problems. According to the categorization, the MS can make calls or can be stopped from making calls.

Operation and maintenance centre (OMC) All the network elements are connected to the OMC, which monitors the health of all the network elements and carries out any maintenance operation, if required. The OMC links to BTS are via the parent BSC. The OMC keeps records of all the faults occurred. It can also do traffic analysis and prepares the MIS report for the network.

Apart from these, there is the gateway mobile switching centre (GMSC). The GMSC is the point to which a mobile terminating call is initially routed, without any knowledge of the MS' location. The GMSC is thus incharge for obtaining the *mobile station roaming number* (MSRN) from the HLR based on the MSISDN (mobile station ISDN number, the 'directory number' of an MS) and routing the call to the correct visited MSC. The 'MSC' part of the term 'GMSC' is misleading since the gateway operation does not require any linking to an MSC.

The *SMS-G* or *SMS gateway* (Fig. 13.2) is the term used to collectively describe the two *short message services gateways* defined in the GSM standards. The two gateways handle messages directed in different directions: (a) The SMS-GMSC (short message service gateway mobile switching centre) is for short messages being sent to a mobile. (b) The SMS-IWMSC (short message service interworking mobile switching centre) is used for short messages originating with a mobile on that network. The SMS-GMSC role is similar to that of the GMSC, whereas the SMS-IWMSC provides a fixed access point to the short message service centre. Similarly, voicemail services are also provided by VMS-G (Fig. 13.2).

The short message service gateway (SMS-G) provides text message service and sends short messages from mobile to another mobile subscriber. Messages can also be sent by manual terminal connected to SMS-G. The voice mail service gateway (VMS-G) provides voice mail service. It has database for all the VMS subscribers and also stores voice messages for them.

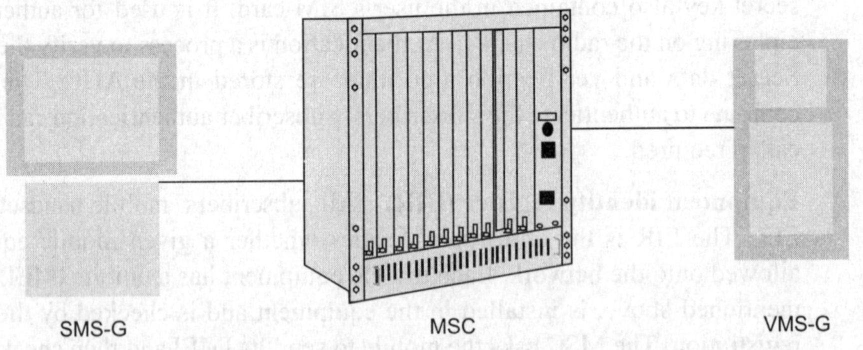

Fig. 13.2 Short message and voice mail services

13.2.2 Call Handling in GSM

We will discuss here how mobile-originated and mobile-terminated calls are handled in GSM.

Mobile-Originated Calls

Following are the steps that describe the minimum procedure to handle the mobile-originated calls (Fig.13.3):

- Mobile-originated calls go to BTS first and then to BSC.
- The BSC forwards this call to MSC.
- The MSC does the authentication and call routing as per the dialled digits.
- If the call is to another mobile subscriber, then for another mobile the process for that call is the same as for mobile-terminated call as discussed next.

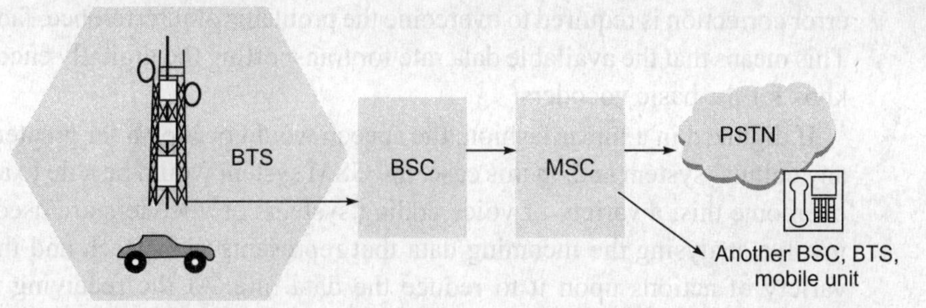

Fig. 13.3 Blocks followed towards PSTN (or to another mobile) for mobile-originated calls

Mobile-Terminated Calls

Following are the steps that describe the minimum procedure to handle the mobile-terminated calls (Fig. 13.4):

- Mobile-terminated calls come to MSC first, where HLR/VLR inquiry is carried out and, as per the information, the MS is paged in suitable BSC.
- The BSC forwards this page to all children BTS where the actual paging is done.
- After BTS gets a response from the mobile, it allocates a channel for this call.
- On ending of the call, BTS informs BSC and MSC.

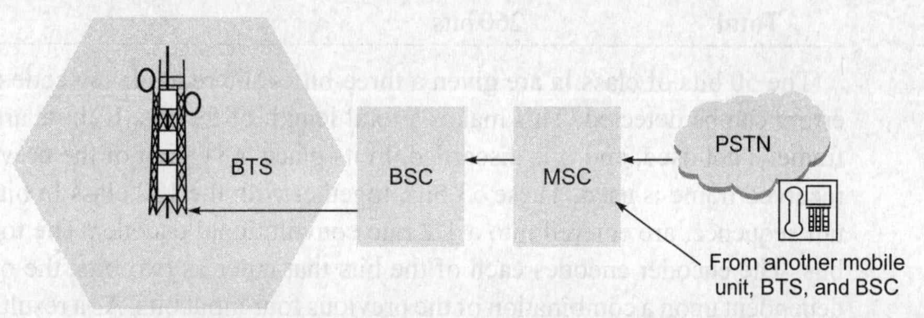

Fig. 13.4 Blocks followed towards BTS (with which the mobile unit is logged) for mobile-terminated calls

Handovers are managed as discussed in Chapter 11. There are a number of elements to the GSM radio or air interface. There are the aspects of the physical power levels, channels and the like. Additionally, there are the different data channels that are employed to carry the data and exchange the protocol messages, which enables the radio subsystem to operate correctly.

13.2.3 Key Selections at GSM Physical Layer

The carrier is modulated using GMSK as described in Chapter 7. It is resilient to noise when compared to some other forms of modulation, occupies a relatively narrow bandwidth, and has a constant power level. The data transported by the carrier serves up to eight different users under the basic system. Even though the full data rate on the carrier is approximately 270 kbps, some of this supports the management overhead and, therefore, the data rate allotted to each time slot is only 24.8 kbps. In addition to this, the error correction is required to overcome the problems of interference, fading, and the like. This means that the available data rate for transporting the digitally encoded speech is 13 kbps for the basic vocoders.

If digitized in a linear fashion, the speech would occupy a far greater bandwidth than any cellular system and, in this case, the GSM system would be able to accommodate. To overcome this, a variety of voice coding systems or vocoders are used. These systems involve analysing the incoming data that represents the speech and then performing a variety of actions upon it to reduce the data rate. At the receiving end, the reverse process is undertaken to reconstitute the speech data such that it can be understood. In GSM, a variety of vocoders are used, including LPC-RPE, EFR, etc., as described below.

The vocoder that was originally used in the GSM system was the LPC-RPE (linear prediction coding with regular pulse excitation) vocoder. This vocoder took each 20 ms block of speech and then represented it using just 260 bits. This actually equates to a data rate of 13 kbps. In GSM, it is recognized that some bits are more important than others. If some bits are missed or corrupted, it is more important to the voice quality than others. Accordingly, the different bits are classified as follows:

Class Ia	50 bits	Most important and sensitive to bit errors
Class Ib	132 bits	Moderately sensitive to bit errors
Class II	78 bits	Least sensitive to bit errors
Total	260 bits	

The 50 bits of class Ia are given a three-bit cyclic redundancy code (CRC) such that errors can be detected. This makes a total length of 53 bits. If there are any errors, the frame is not used, and it is discarded. In its place, a version of the previously corrected received frame is used. These 53 bits, together with the 132 class Ib bits with a four-bit tail sequence, are entered into a 1/2 rate convolutional encoder. The total length is 189 bits. The encoder encodes each of the bits that enter as two bits, the output also being dependent upon a combination of the previous four input bits. As a result, the output from the convolutional encoder consists of 378 bits. The remaining 78 class II bits are considered the least sensitive to errors and they are not protected and simply added to the data. In this way, every 20 ms speech sample generates a total of 456 bits. Accordingly, the overall bit rate is 22.8 kbps. Once in this format, the data is interleaved to add further protection against interference and noise. The 456 bits output by the convolutional encoder are divided into eight blocks of 57 bits and these blocks are transmitted in eight consecutive time-slots, i.e. a total of four bursts as each burst takes two sets of data.

Later, another vocoder, called the enhanced full rate (EFR) vocoder, was added in response to the poor quality perceived by the users. This new vocoder gave much better sound quality and was adopted by GSM. Using the algebraic code excitation linear prediction (ACELP) compression technology, it gave a significant improvement in quality over the original LPC-RPE encoder. It became possible as the processing power that was available increased in mobile phones as a result of higher levels of processing power combined with their low-current consumption.

There is also a half-rate vocoder. Although this gives much inferior voice quality, it does allow for an increase in network capacity. It is used in some instances when the network loading is very high to accommodate all the calls.

A variety of power levels are allowed by the GSM standard, the lowest being only 800 mW (29 dBm). As mobiles may only transmit for one-eighth of the time, i.e. for their allocated slot, which is one of eight, the average power is one-eighth of the maximum. Additionally, to reduce the levels of transmitted power and hence the levels of interference, mobiles are able to step the power down in increments of two dB from the maximum to a minimum 13 dBm (20 milliwatts). The mobile station measures the signal strength or signal quality (based on the bit error rate) and passes the information to the BTS and hence to the BSC, which ultimately decides if, and when, the power level should be changed.

A further power saving and interference reducing facility is the *discontinuous transmission* (DTX) capability that is incorporated within the specification. It is particularly useful because there are long pauses in speech, for example, when the person using the mobile is listening, and during these periods there is no need to transmit a signal. In fact, it is found that a person speaks for less than 40% of the time during normal telephone conversations. The most important element of DTX is the voice activity detector. It must correctly distinguish between voice and noise inputs, a task that is not trivial. If a voice signal is misinterpreted as noise, the transmitter is turned off, an effect known as clipping results, and this is particularly annoying to the person listening to the speech. However, if noise is misinterpreted as a voice signal too often, the efficiency of DTX is dramatically decreased.

13.2.4 Multiple Access in GSM and GSM Channels

The GSM uses a combination of both TDMA and FDMA techniques as shown in Fig. 13.5(a) along with cellular structure (SDMA). Combination of time slot number along with absolute radio frequency channel numbers (ARFCN) constitutes the physical channel for both forward and reverse link.

GSM-900 calculations

Forward link (BS to mobile) $\rightarrow$ 960 MHz – 935 MHz = 25 MHz

Reverse link (Mobile to BS) $\rightarrow$ 915 MHz – 890 MHz = 25 MHz

$$\text{Total number of channels} = \frac{\text{Total available bandwidth}}{\text{Per channel bandwidth}} = \frac{25\,\text{MHz}}{200\,\text{kHz}} = 125$$

ARFCN = 0 to 124

Per channel eight time slots are assigned and the channel data rate is 270.833 kbps.

So, Effective channel transmission rate = 270.833/8 = 33.854 kbps

In GSM with overhead user data is actually sent at 24.7 kbps instead of 33.854 kbps,

Signalling bit duration = 1/270.833 = 3.692 μs

The carriers are then divided in time using a TDMA scheme. Starting from bottom to top, according to Fig. 13.5(b), the fundamental unit of time (one slot) is called a burst period and it lasts for approximately 0.577 ms (15/26 ms). Eight of these burst periods are

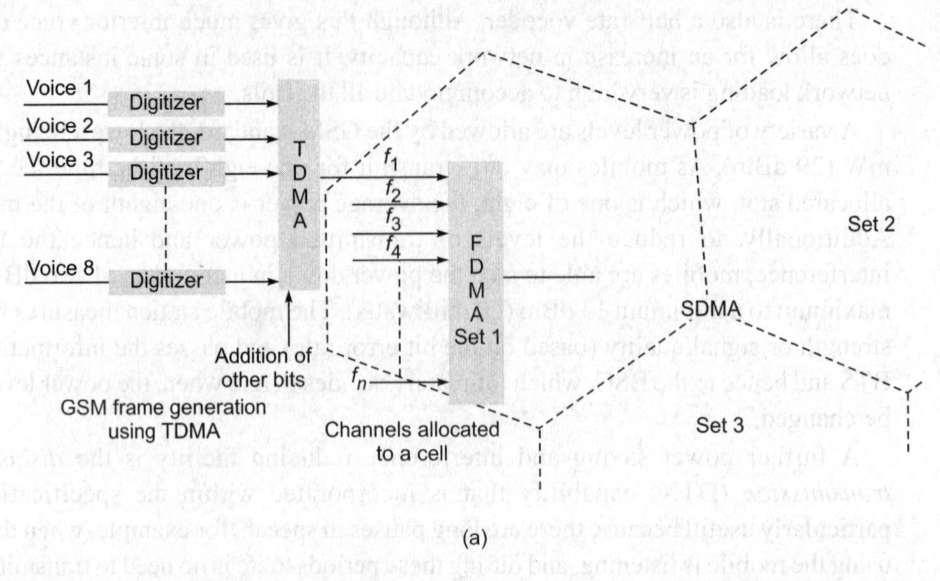

(a)

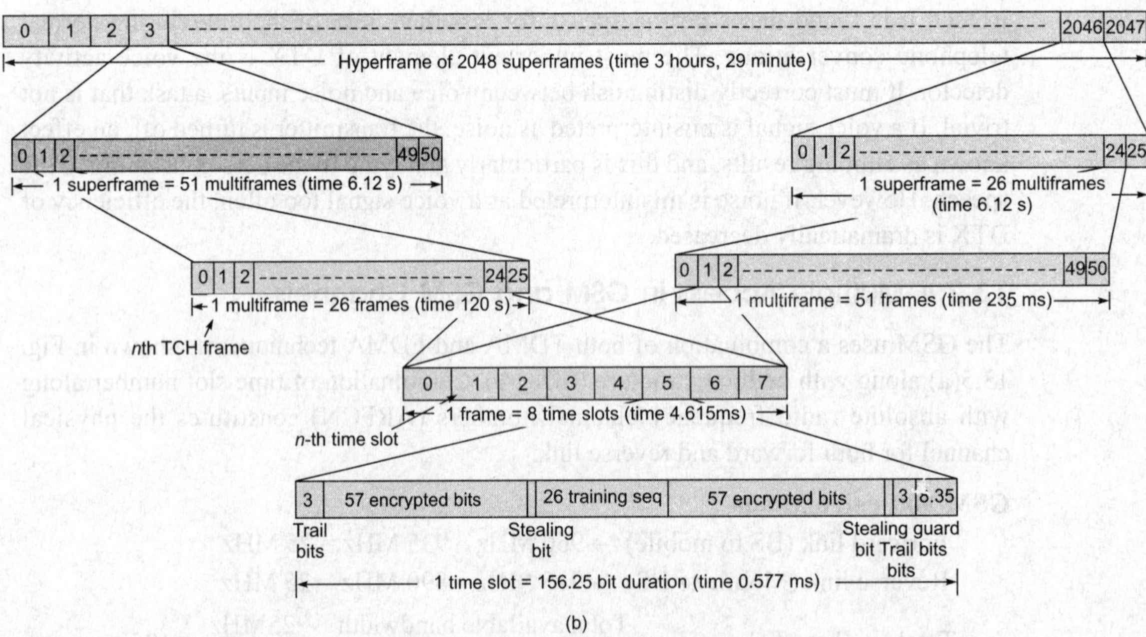

(b)

Fig. 13.5 (a) Concept of TDMA and FDMA used in GSM along with SDMA using cellular structure and (b) TDMA frame formats in GSM

grouped into what is known as a TDMA frame. This lasts for approximately 4.615 ms (i.e.120/26 ms) and it forms the basic unit for the definition of logical channels. One physical channel is one burst period allocated in each TDMA frame. (Channels are described later on in this section.) There are different types of frames that are transmitted to carry different data and the frames are organized into what are termed multiframes and superframes also to provide overall synchronization.

Due to digital TDMA technology combined with a channel bandwidth of 200 kHz, the system is able to offer a higher level of spectrum efficiency. As there are many carrier frequencies that are available, one or more can be allocated to each base station. The system also operates using frequency division duplex (FDD) and, as a result, paired bands are needed for the uplink and downlink transmissions. The frequency separation is dependent upon the band in use.

Due to the cellular architecture, nearby cells use a different frequency set and the whole set of frequencies in a cluster is repeated in another cluster due to frequency reuse concept. Thus, TDMA-FDMA along with SDMA support can accommodate billions of users in the network.

The GSM uses a variety of channels in which the data is carried. In GSM, these channels are separated into *physical channels* and *logical channels*. The physical channels are determined by the timeslot, whereas the logical channels are determined by the information carried within the physical channel. It can be further summarized by saying that several recurring time slots on a carrier constitute a physical channel. These are then used by different logical channels to transfer information. These channels may either be used for user data (payload) or signalling to enable the system to operate correctly (Fig. 13.6).

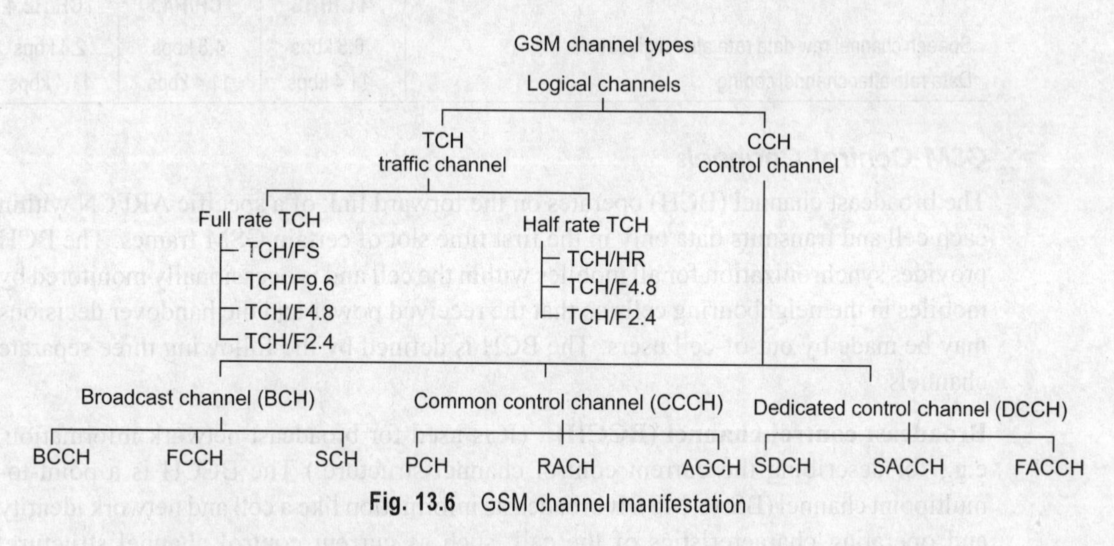

Fig. 13.6 GSM channel manifestation

The following logical channels and their functions are defined in GSM:

1. Traffic channels (TCHs)
2. Control channels (CCHs)

Traffic channels carry digitally encoded user speech or user data and have identical functions and formats on both the forward and reverse links. Control channels carry signalling and synchronizing commands between the mobile and BS.

GSM Traffic Channels

TCH-f (full-rate traffic channel) When transmitted at full rate, the user data is contained within one time slot per frame (Table 13.1).

Table 13.1 TCH-f channel rates

	TCH/FS	TCH/F9.6	TCH/F4.8	TCH/F2.4
Speech channel raw data rate after digitization	13 kbps	9.6 kbps	4.8 kbps	2.4kbps
Data rate after channel coding	22.8 kbps	22.8 kbps	22.8 kbps	22.8 kbps

TCH-h (half-rate traffic channel) When transmitted as half rate, the user data is mapped onto the same slot but is sent in alternate frames. Thus, two half-rate channel users can use the same time slot but alternately transmit during every other frame (Table 13.2).

For every 26 frames of TCH, data is broken up every at the thirteenth frame by either slow associated control channel (SACCH) or idle frames. The twenty-sixth frame contains idle bits for the case when full-rate TCHs are used and contains SACCH data when half-rate TCHs are used.

Table 13.2 TCH-h channel rates

	TCH/HS	TCH/H4.8	TCH/H2.4
Speech channel raw data rate after digitization	6.5 kbps	4.8 kbps	2.4 kbps
Data rate after channel coding	11.4 kbps	11.4 kbps	11.4 kbps

GSM Control Channels

The broadcast channel (BCH) operates on the forward link of a specific ARFCN within each cell and transmits data only in the first time slot of certain GSM frames. The BCH provides synchronization for all mobiles within the cell and is occasionally monitored by mobiles in the neighbouring cells so that the received power and the handover decisions may be made by out-of-cell users. The BCH is defined by the following three separate channels.

Broadcast control channel (BCCH) (It is used for broadcast network information, e.g., for describing the current control channel structure.) The BCCH is a point-to-multipoint channel (BS-to-MS). It broadcasts information like a cell and network identity and operating characteristics of the cell, such as current control channel structure, channel availability, congestion, etc. It also broadcasts a list of channels that are currently in use within the cell.

Synchronization channel (SCH) It is used for synchronization of the MSes. It is used to identify the serving base station while allowing each mobile to frame synchronize with the base station.

Frequency correction channel (FCCH-MS) It is used for frequency correction and allows each subscriber unit to synchronize its internal frequency standard or local oscillator to the exact frequency of the base station.

The common control channels (CCCHs) are of two main types—forward and return channels. The *forward* common channels are used for paging to inform a mobile of an incoming call, responding to channel requests, and broadcasting bulletin board information. The *return* common channel is a random access channel used by the mobile to request channel resources before the timing information is conveyed by the BSes.

Access grant channel (AGCH) The BS acknowledge channel requests from MS and allocates an SDCCH. The AGCH is used by the base station to provide the forward link communication to the mobile and carries data, which instructs the mobile to operate in a particular physical channel (with a time slot and ARFCN) and with a particular dedicated control channel. It is the final CCCH message sent by the base station before a subscriber is moved off the control channel and is also used by BS to respond to a RACH sent by MS in a previous CCCH frame.

Paging channel (PCH) It is used for terminating call announcement by providing paging signals from the base stations to all mobiles in the cell and notifies a specific mobile of an incoming call, which originates from PSTN. The PCH transmits IMSI of the target subscriber that is with a request for acknowledgement from the mobile on RACH. The PCH may be used for another function to provide 'cell broadcast' ASCII text messages to all MSes in form of SMS.

Random access channel (RACH-MS) Its main functions are access requests, response to call announcement, location update, etc. It is used by the subscriber unit to acknowledge a page from the PCH and also to originate a call by the mobile. Slotted ALOHA is the suitable access method. All mobiles must request access or respond to the PCH alert within the 0th time slot of a GSM frame. At the BTS, every frame, including an idle one, will accept RACH transmissions from mobiles during the 0th slot. To establish the service, the BS must respond to the RACH transmission by allocating a channel and assigning an SDCCH for signalling during a call. This connection is confirmed by the BS over AGCH.

The *dedicated* channels are of two main types: those used for signalling and those used for traffic. The signalling channels are used for maintenance of the call and for enabling call set-up, providing facilities such as handover when the call is in progress, and finally, terminating the call. The traffic channels handle the actual payload.

Fast associated control channel (FACCH) It is used for supervisory data transmissions between MS and BS during a call. It is associated with traffic channel and SDCCH. It is used for time critical signalling over the TCH (e.g., for handover signalling). Traffic burst is stolen for a full signalling burst. FACCH carries urgent messages and essentially the same type of information as the SDCCH. FACCH is assigned when SDCCH has not been dedicated for a particular user and an urgent message comes (handover).

Slow associated control channel (SACCH) It is associated with traffic channel TCH and SDCCH in-band signalling, e.g., for link monitoring. In both the cases, SACCH maps them onto some physical channel. Thus, each ARFCN systematically carries

SACCH data for all of its current users. On the forward link, SACCH is used to send a slow but regularly changing control information to the mobile, e.g., transmit power level instructions, specific timing advance instructions for each user on the ARFCN. On the reverse link, SACCH carries information about the received signal strength and quality of the TCH and also the BCH measurement resulting from the neighbouring cells. The SACCH is transmitted in the thirteenth frame (also in twenty-sixth frame when half-rate traffic is used) of every speech /dedicated control channel multiframe. Within the SACCH frame, eight time slots are allocated to provide the SACCH data to each of the eight full rate (or 16 half rate) users on the ARFCN.

Stand-alone dedicated control channel (SDCCH) It is used for signalling exchanges, e.g., during call set-up, registration / location updates. The SDCCH carries the signalling data following the connection of MS with the BS and just before a TCH assignment is issued by the BS. It ensures that the mobile and base station remain connected while the base station and MSC verify the subscriber unit and then it allocates resources for the mobile. It can be considered as an intermediate temporary channel accepting a newly completed call from the BCH. It holds the traffic while waiting for the base station to allocate a TCH channel. The SDCCH is used to send authentication and alert messages.

13.2.5 GSM Enhancements

The GSM mobile service started in India with second-generation service capabilities, which supported voice and low-speed circuit switched data. To support an ever-increasing demand of the customers, many more advanced technologies are being developed on the GSM platform, like GPRS and EDGE. This brought a revolution in the services provided by GSM operators. A few of them are listed below:

- High-speed packet switched data connectivity
- Multimedia applications
- High-speed remote corporate LAN access
- Web browsing, e-mail, fax, and wireless imaging
- Video telephony and digital TV reception on mobile phones

Existing GSM services are with enhanced features called GSM phase 2 and 2+.

HSCSD (GSM Phase 2+) Services

GSM phase 2 services are as follows:

- Voice services
- FAX and data services up to 9.6 kbps and short message services

The limitations of these services are that only circuit switching is possible and the spectral efficiency is low.

A straightforward improvement of GSM's data transmission capabilities is *high-speed circuit switched data* (HSCSD), which is already available with some service providers. In this system, higher data rates are achieved by bundling several TCHs. Here MS requests one or more TCHs from the GSM network, i.e. it allocates several TDMA slots within a TDMA frame. This allocation may be asymmetrical, i.e. more slots can be allocated on the downlink than on the uplink, which fits the typical user behaviour of downloading more data compared to uploading. Basically, HSCSD requires only the software upgrades in MS and MSC.

- Normal data rate enhanced up to 57.6 kbps
- Typical applications : Internet on mobile, mobile LAN, wireless real-time applications, file transfer

Limitations Only circuit switching is possible.

The HSCSD, though appearing advantageous, exhibits some major disadvantages. It still uses the connection-oriented mechanisms of GSM, which are not at all efficient for bursty computer data traffic. While downloading, a large file may require all channels to be reserved and typical web browsing will keep channels idle for most of the time. Allocating channels is directly reflected in the service costs. Furthermore, for n channels, HSCSD requires n times the signalling during handover, connection set-up, and release. Each channel is treated separately. The probability of blocking or service degradation increases during handovers because BSC has to check resources for n channels and not only for one.

After all, HSCSD may be an attractive interim solution for higher bandwidth and rather constant traffic, like file download. However, it does not make much sense for bursty Internet traffic as long as the user is charged for each channel allocated for communication. GPRS is the next step towards more flexible and powerful data transmission avoiding the problems of HSCSD, as it is fully packet oriented.

13.3 GENERAL PACKET RADIO SERVICE (GPRS)

The GSM system was developed for voice services but it was not having the capabilities of data services that are based on packet switching. To develop a higher data rate capability in the mobile phones, GPRS protocol is developed on the GSM platform to enhance the services. The GPRS and GSM are able to operate alongside one another on the same network, using the same base stations. The scenario and GPRS architecture is shown in Fig. 13.7. The GPRS was developed in 1999. GPRS became the first stepping stone on the path between the second-generation GSM cell phone system and the W-CDMA / UMTS system. With GPRS offering data services with data rates up to 115 kbps, facilities such as web browsing and other services requiring data transfer became possible. Although some data could be transferred using GSM, the rate was too slow for real data applications.

The key element of GPRS is that it uses packet switched data rather than circuit switched data, this technique makes a much more efficient use of the available capacity. This is because most data transfer occurs in what is often termed a 'bursty' fashion. The transfer occurs in short peaks, followed by breaks when there is little or no activity, while in circuit switching dedicated connections are provided to the user till the communication is over. In packet switching, the overall capacity can be shared between several users. To

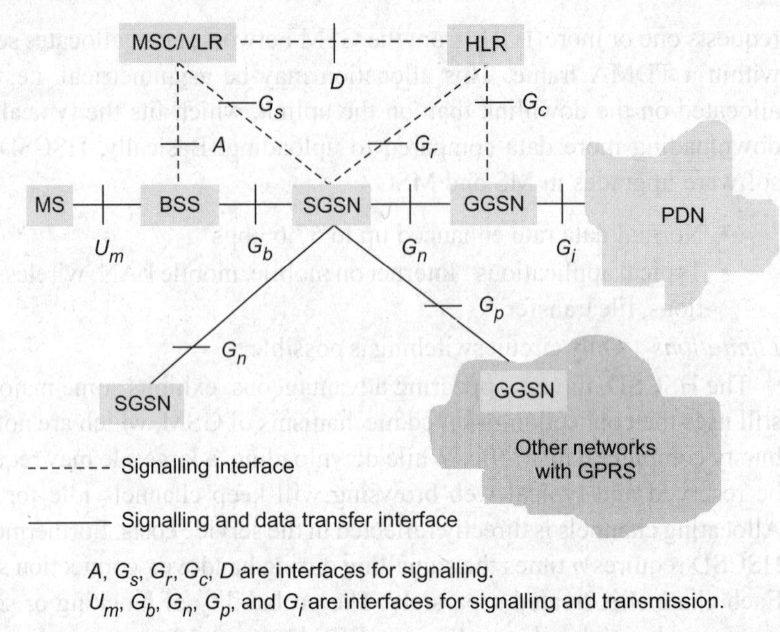

Fig. 13.7 GPRS architecture with modifications in the GSM, including additional GPRS components (GGSN: gateway GPRS support node; SGSN: serving GPRS support node; PDN: packet data network)

achieve this, the data is split into packets and tags inserted into the packet to provide the destination address. Packets from several sources can then be transmitted over the link. As it is unlikely that the data burst for different users will occur all at the same time, by sharing the overall resource in this fashion, the channel, or combined channels can be used far more efficiently. This approach is known as packet switching, and it is at the core of many cellular data systems, and in this case GPRS.

General requirements of any packet based system is a complete layered approach that should be strictly followed including the network layer as the packet routing procedures are required. This may not be the case in circuit switched systems, where signalling and information transfer is done over temporary but a dedicatedly established link. Packet formation, addition of address of source, and destination in the packet, checksum calculation, packet reassembling, quality of service, etc., are the general requirements of packet based systems.

GSM frames cannot be called packets as they are synchronous TDM frames, but GPRS packet format is defined. GPRS is TDMA based and follows the IS-136 protocol.

Generally for identification of the destination in computer networks (Internet), the physical address of NIC and IP address are required and for application identification the port address is required. For GPRS based system, the SIM and IMSI numbers can be used as physical address. In GPRS, IP address and *domain name system* (DNS), which are developed for computer networks are adopted directly. In order to accommodate the packet data within GPRS it has been necessary to develop coding schemes. Additionally, the layers based on the OSI system have become more important. There are few points that make the GPRS different than the computer networks. Some of then are listed below:

- Wireless channel for data transfer
- Mobility management is necessary
- GSM infrastructure, which incorporates HLR, VLR, AUC, MS, BSS, BTS, and MSC, etc. (Fig. 13.1) is used
- For routing of packets, GPRS requires supporting hardware and software, which are different from the computer networks

Although designed to run alongside the GSM system, the core network structure updated for GPRS has several new elements added to enable it to carry the packet data. The network between the BSC and BTS is similar but behind this, there is new infrastructure to support the packet data.

For GPRS, the data from the BSC is routed through what is termed as *serving GPRS support node* (SGSN). This forms the gateway to the services within the network and then a *gateway GPRS support node* (GGSN), which forms the gateway to the outside world.

The SGSN serves a number of functions for GPRS mobiles. It enables authentication to occur and then tracks the location of the mobile within the network and ensures that the quality of service is to the required level. For the network protocols, there are two layers that are used and supported by GPRS, namely X.25 and IP. In operation, the protocols assign addresses (Packet Data Protocol or PDP addresses) to the devices in the network for the purpose of routing the data through the system. Thus, the GGSN appears as a data gateway to the public packet network, and thus the fact that the users are mobiles cannot be seen. In operation, the mobile must attach itself to the SGSN and activate its PDP address. This address is supplied by the GGSN, which is associated with the SGSN. As a result, a mobile can only attach to one SGSN, although once assigned its address, it can receive data from multiple GGSNs using multiple PDP addresses.

13.3.1 GPRS Functional Groups

The GRPS functional groups (defined according to required functions for GPRS) are as follows:

- Network access
- Packet routing and transfer
- Mobility management
- Logical link management
- Radio resource management
- Network management

GPRS Network Access

It supports the standard point-to-point data transfer and anonymous access (without authentication and ciphering). The functions include the following:

- Registration, which associates the mobile station identity with the packet data protocols
- Authentication and authorization.
- Packet terminal adaptation, which adapts data transmission across the GPRS network
- Admission control, which determines the radio and network resources to be used for communication of an MS
- Message screening, which filters out unsolicited messages
- Charging information collection for packet transmission in GPRS and external networks

Packet Routing and Transfer Functions

Route the data between an MS and the destination through the SGSNs and GGSNs. Figure 13.7 shows the GPRS architecture in which SGSN and GGSN nodes are shown within the highlighted boxes. The functions include the following:

- Relay function, which is used by the BSS to forward packets between an MS and a serving GSN; it is also used by a serving GSN to forward packets between BSS and serving or gateway GSN.
- Routing, which determines the destinations of packets.
- Address translation and mapping, which converts a GPRS network address to an external data network address and vice versa.
- Encapsulation and tunnelling, which encapsulate packets at the source of a tunnel, deliver the packets through the tunnel, and decapsulate them at the destination.
- Domain name service functions, which resolve logical GSN names to their IP addresses.
- Compression and ciphering.

Mobility Management

It keeps track of the current location of an MS. Three different scenarios can exist when the MS enters a new cell and, possibly, a new routing area:

- Cell update
- Routing area update
- Combined routing area and location area update

Logical Link Management

It maintains the communication channel between the MS and the GSM network across the radio interface, which includes the following:

- Logical link establishment
- Logical link maintenance
- Logical link release

Radio Resource Management

It allocates and maintains radio communication paths, which includes the following:

- Um interface management, which determines the amount of radio resources to be allocated for the GPRS usage.
- Cell selection, which enables the MS to select the optimal cell for radio communication.
- Um-tranx, which provides packet data transfer capability, such as medium access control, packet multiplexing, packet discrimination, error detection and correction, and flow control across the radio interface between the MS and the BSS.
- Path management, which maintains the communication paths between the BSSs and serving GSNs.

Network Management Functions

These provide mechanisms to support operational, authentication, and maintenance (OA&M) related functions related to GPRS.

13.3.2 Coding Schemes in GPRS

The GPRS offers a number of coding schemes with different levels of error detection and correction. These are used depending upon the radio frequency signal conditions and the requirements for the data being sent. These are given labels from CS-1 to CS-4.

CS-1

This applies the highest level of error detection and correction. It is used in scenarios when interference levels are high or signal levels are low. By applying high levels of detection and correction, this prevents the data having to be re sent too often. Although it is acceptable for many types of data to be delayed, for others there is a more critical time element. This level of detection and coding results in a half code rate, i.e., for every 12 bits that enter the coder, 24 bits result. It results in an actual throughput of 9.05 kbps data rate.

CS-2

This error detection and coding scheme is for better channels. It effectively uses a 2/3 encoder and results in a real data throughput of 13.4 kbps, which includes the RLC/MAC header, etc.

CS-3

This effectively uses a 3/4 coder and results in a data throughput of 15.6 kbps.

CS-4

This scheme is used when the signal is high and interference levels are low. No correction is applied to the signal, allowing for a maximum throughput of 21.4 kbps. If all eight slots were used then this would enable a data throughput of 171.2 kbps to be achieved.

In addition to the error detection and coding schemes, GPRS also employs interleaving techniques to ensure the effects of interference and spurious noise are reduced to a minimum. As blocks of 20 ms data are carried over four bursts, with a total of 456 bits of information, a total of 181, 268, 312, or 428 bits of payload data are carried depending upon the error detection and coding scheme chosen, i.e. from CS-1 to CS-4, respectively.

13.3.3 GPRS Layers and Functions

Software plays a very large part in the current cellular communications systems. To enable it to be sectioned into the areas that can be addressed separately, the concept of layers has been developed. In GPRS, the main functions are divided into layers, one, two, and three.

Layer one concerns the physical link between the mobile and the base station. This is often subdivided into two sub layers, namely the *physical RF layer*, which includes the modulation and demodulation and the *physical link layer*, which manages the responses and controls required for the operation of the RF link. These include elements such as error correction, interleaving and the correct assembly of the data, power control, and the like. Above this are the *radio link control* (RLC) and the *medium access control* (MAC) layers. These organize the logical links between the mobile and the base station. They control the radio link access and they organize the logical channels that route the data to and from the mobile. There is also the *logical link layer* (LLC) that formats the data frames and is used to link the elements of the core network to the mobile.

The MAC layer is central to this and there are three MAC modes that are used to control the transmissions. These are named as fixed allocation, dynamic allocation, and extended dynamic allocation.

The *fixed allocation mode* is required when a mobile requires a data to be sent at a consistent data rate. To achieve this, a set of PDCHs (see channels in GPRS) are allocated for a given amount of time. When this mode is used, there is no requirement to monitor for availability, and the mobile can send and receive data freely. This mode is used for applications such as video conferencing.

The *dynamic allocation mode* is required for the network to allocate time slots as they are required. A mobile is allowed to transmit in the uplink when it sees an identifier flag, known as the uplink status flag (USF), that matches its own. The mobile then transmits its data in the allocated slot. This is required because up to eight mobiles can have potential access to a slot, but obviously only one can transmit at any given time.

The *extended dynamic allocation* mode allows for much higher data rates to be achieved because it enables mobiles to transmit in more than one slot. When the USF indicates that a mobile can use this mode, it can transmit in the number allowed, thereby increasing the rate at which it can send data.

Like other cellular systems, GPRS uses a variety of physical and logical channels to carry the data payload as well as the signalling required to control the calls.

13.3.4 Channels in GPRS

The GPRS builds on the basic GSM structure. It uses the same modulation and frame structure that is employed by GSM and in this way it is an evolution of the GSM standard. Slots can be assigned dynamically by the BSC to GPRS calls depending upon the demand, the remaining ones being used for GSM traffic.

There is a new data channel that is used for GPRS called the *packet data channel* (PDCH). The overall slot structure for this channel is the same as that used within GSM, having the same power profile, and timing advance attributes to overcome the different signal travel times to the base station depending upon the distance the mobile is from the base station. This enables the burst to fit in seamlessly with the existing GSM structure. Each burst of information for GPRS is 0.577 msec in length and is the same as that used in GSM. It also carries two blocks of 57 bits of information, giving a total of 114 bits per burst. It, therefore, requires four bursts to carry each 20 msec block of data, i.e. 456 bits of encoded data. The BSC assigns PDCHs to particular time slots and there will be times when the PDCH is inactive, allowing the mobile to check for other base stations and monitor their signal strengths to enable the network to judge when a handover is required. The GPRS slot may also be used by the base station to judge the time delay using a logical channel known as the *packet timing advance control channel* (PTCCH).

There are a variety of channels used within GPRS and they can be set into groups, depending upon whether they are for common or dedicated use. The system uses the GSM control and broadcast channels for the initial set-up, but all the GPRS actions are carried out within the GPRS logical channels carried within the PDCH.

Broadcast Channels

Packet broadcast central channel (PBCCH) This is a downlink only channel that is used to broadcast information to mobiles and informs them of incoming calls, etc. It is very similar in operation to the BCCH used for GSM. In fact, the BCCH is still required in the initial stage to provide a time slot number for the PBCCH. In operation, the PBCCH broadcasts general information, such as power control parameters, access methods and operational modes, network parameters, etc., required to set up calls.

Common Control Channels

Packet paging channel (PPCH) This is a downlink only channel and is used to alert the mobile to an incoming call and to alert it to be ready for receiving data. It is used for control signalling prior to the call set up. Once the call is in progress, a dedicated channel referred to as the PACCH takes over.

Packet random access channel (PRACH) This is an uplink channel that enables the mobile to initiate a burst of data in the uplink. There are two types of PRACH burst, one is an eight bit standard burst and a second one using an 11 bit burst has added data to allow for a priority setting. Both types of bursts allow for timing advance setting.

Packet access grant channel (PAGCH) This is also a downlink channel and it sends information telling the mobile the traffic channel assigned to it. It occurs after the PPCH has informed the mobile that there is an incoming call.

Packet notification channel (PNCH) This is another downlink only channel used to alert mobiles that there is broadcast traffic intended for a large number of mobiles. It is typically used in what is termed point-to-point multicasting.

Dedicated Control Channels

Packet associated control channel (PACCH) This channel is present in both the uplink and downlink directions and it is used for control signalling while a call is in progress. It takes over from the PPCH once the call is set up and it carries information, such as channel assignments, power control messages, and acknowledgements of received data.

Packet timing (advance) common control channel (PTCCH) This channel is present in both the uplink and downlink directions and is used to adjust the timing advance. This is required to ensure that messages arrive at the correct time at the base station regardless of the distance of the mobile from the base station. As timing is critical in a TDMA system and signals take a small but finite time to travel, this aspect is very important if long guard bands are not to be left.

Dedicated Traffic Channel

Packet data traffic channel (PDTCH) This channel is used to send the traffic and it is present in both the uplink and downlink directions. Up to eight PDTCHs can be allocated to a mobile to provide high-speed data.

13.3.5 GPRS Functioning and Modes in Mobiles

Not only does the network need to be upgraded for GPRS, new GPRS mobiles are also required. It is not possible to upgrade an existing GSM mobile for use as a GPRS mobile, although GSM mobiles can be used for GSM speech on a network that also carries GPRS. To utilize GPRS, new modes are required to enable it to transmit the data in the required format. All GPRS mobiles are not designed to offer the same levels of service. As a result, they are split into three basic categories according to their capabilities in terms of the ability to connect to GSM and GPRS facilities:

Class A This class describes mobile phones that can be connected to both GPRS and GSM services at the same time.

Class B These mobiles can be attached to both GPRS and GSM services but they can be used on only one service at a time. A class B mobile can make or receive a voice call, or send and or receive a SMS message during a GPRS connection. During voice calls or text, the GPRS service is suspended but it is re-established when the voice call or SMS session is complete.

Class C This classification covers phones that can be attached to either the GPRS or GSM services, but the user needs to switch manually between the two different types.

The GPRS mobiles are also categorized by the data rates they can support. Within GSM, there are eight time slots that can be used to provide TDMA, allowing multiple mobiles onto a single RF signal carrier. Within GPRS, it is possible to use more than one slot to enable much higher data rates to be achieved when these are available. The different speed classes of the mobiles are dependent upon the number of slots that can be used in either direction. There are a total of 29 speed classes. Class one mobiles are able to send and receive in one slot in either direction, i.e., uplink and downlink, and class 29 mobiles are able to send and receive in all eight slots. The class within these two limits is able to support sending and receiving in different combinations of uplink and downlink slots. There are two possibilities: SIM of GSM system may be GPRS aware or non-aware. If non-aware, a GPRS service pack must be installed in the mobile phone.

When looking at the way in which GPRS operates, it can be seen that there are three basic modes in which it operates. These are initialization/idle, standby, and ready.

Initialization/idle

When the mobile is turned on, it must register with the network and update the location register. This is very similar to that performed with a GSM mobile, but it is referred to as a location update. It first locates a suitable cell and transmits a radio burst on the RACH using a shortened burst because it does not know what timing advance is required. The data contained within this burst temporarily identifies the mobile and indicates that the reason for the update is to perform a location update.

When the mobile performs its location update, the network also performs an authentication to ensure that it is allowed to access the network. As for GSM, it accesses the HLR and VLR as necessary for the location update and the AUC for authentication. It is at registration that the network detects that the mobile has a GPRS capability. The SGSN also maintains a record of the location of the mobile so that the data can be sent there if required.

Standby

The mobile then enters a standby mode, periodically updating its position as required. It monitors to ensure that it has not changed base stations and also looks for a stronger base station control channels. The mobile will also monitor the PPCH in case of an incoming alert indicating that the data is ready to be sent. As for GSM, most base stations set up a schedule for paging alerts based on the last figures of the mobile number. In this way, it does not have to monitor all the available alert slots and can instead only monitor a reduced number where it knows alerts can be sent for it. Also, the receiver can be turned off for a longer period and hence the battery life can be extended.

Ready

In the ready mode, the mobile is attached to the system and a virtual connection is made with the SGSN and GGSN. By making this connection, the network knows where to route the packets when they are sent and received. In addition to this, the mobile is likely to use the PTCCH to ensure that its timing is correctly set so that it is ready for a data transfer should one be needed.

With the mobile attached to the network, it is prepared for a call or data transfer. To transmit data, the mobile attempts a packet channel request using the PRACH uplink channel. As this may be busy, the mobile monitors the PCCCH, which contains a status bit indicating the status of the base station receiver, whether it is busy or idle and capable of receiving data. When the mobile sees that this status bit indicates the receiver is idle, it sends its packet channel request message. If accepted, the base station will respond by sending an assignment message on the PAGCH on the downlink. This will indicate which channel the mobile is to use for its packet data transfer as well as other details required for the data transfer.

This only sets up the packet data transfers for the uplink. If data needs to be transferred in the downlink direction, then a separate assignment is performed for the downlink channel. When data is transferred, this is controlled by the action of the MAC layer. In most instances, it will operate in an acknowledge mode whereby the base station acknowledges each block of data. The acknowledgement may be contained within the data packets being sent in the downlink, or the base station may send data packets down purely to acknowledge the data.

While disconnecting, the mobile will send a packet temporary block flow message, and this is acknowledged. Once this has taken place, the USF assigned to the mobile becomes redundant and can be assigned to another mobile wanting access. With this, the mobile effectively becomes disconnected and although still attached to the network, no more data transfer takes place unless it is re-initiated. Separate messages are needed to detach the mobile from the network.

Packet reservation protocol is used in GPRS for multiple accesses. Nine kbps to 150 kbps bit rate per user can be achieved in GPRS. All securities of GSM system and ciphering are utilized in GPRS. The GPRS is relatively inexpensive mobile data service compared to SMS and circuit switched data.

13.4 EDGE TECHNOLOGY

The EDGE is an enhancement to the GSM mobile cellular phone system. The name EDGE stands for *enhanced data for GSM evolution* and it enables data to be sent over a GSM TDMA system at speeds up to 384 kbps. In some instances, EDGE systems may also be known as EGPRS or enhanced general packet radio service systems. Although strictly speaking a '2.5G' system, it is anticipated that it will be used to provide data services by operators who have not been able to secure the full 3G licences. It is highly spectrally efficient and a high-speed system.

It is generally expected that EDGE will be applied to networks where the enhancements provided by GPRS have already been added. Under the original GSM system, a circuit would be allocated to a given user whether the data was being transmitted or not. This was fine for voice communications because there would normally be some data present for most of the time. However, this is not the case for the data transmissions where high levels of data are transmitted in short bursts. To make more efficient use of the available capability, packet switching is used. Here individual packets of data are routed to the user, enabling the channel or channels to be shared by several users.

To achieve this, it requires the addition of two additional nodes to the network, namely GGSN and SGSN. Here the GGSN connects to packet-switched networks such as the Internet and other GPRS networks. The SGSN provides the packet-switched link to mobile stations.

In terms of implementation, EDGE systems require an EDGE transceiver unit to be added to each cell along with software upgrades to allow its use. This software upgrades may be implemented remotely. This change means that the inclusion of EDGE onto a network requires a significant investment in the infrastructure and, as a result, it is these upgrades that will normally be implemented over a period of time. However, GSM, GPRS, and EDGE can all coexist on the same network. As both GPRS and EDGE represent significant upgrades to handsets and they are not just software upgrades, new mobile handsets are required.

The EDGE is intended to build on the enhancements provided by the addition of GPRS where packet switching is applied to a network. It then enables a three-fold increase in the speed at which data can be transferred by adopting a new form of modulation. The GSM uses a form of modulation, known as GMSK, but EDGE changes the modulation to 8 PSK, which is a form of phase shift keying where eight phase states are used. The advantage is that it can transmit high data rates, although it is not as immune to interference and noise. The network, therefore, switches to 8 PSK to allow the high data transfer rates when signal strengths are sufficient to permit the data transfer with a sufficiently low bit error rate. By using 8 PSK, it is possible to transfer data at 48 kbps per channel rather than 9.6 kbps that is possible using GMSK. By allowing the use of multiple channels, the technology allows the transfer of data at rates up to 384 kbps. However, it should be remembered that these data transfer rates are possible only when the network is not highly loaded as access to all the channels would not be allowed.

The EDGE services have the following specifications:

- They offer data services up to 400 kbps.
- They are used on voice over packet switched network.
- They support IP-based application.
- Typical applications are Internet on mobile, mobile LAN, video phone, wireless real-time applications, and file transfer.

Table 13.3 gives the comparison between GSM and EDGE technologies.

Table 13.3 Comparison between GSM and EDGE

Parameter	GSM	EDGE
Modulation	Gaussian MSK	8PSK
Bit rate	270.833 kbps	812.499 kbps
Channel bandwidth	200 kHz	200 kHz
Pulse shaping	Gaussian prefilter BT = 0.3	Linearizes GMSK pulse
Modulation type	Non-linear, constant envelope	Linear

13.5 CDMA-BASED DIGITAL CELLULAR STANDARDS IS-95 TO CDMA-2000

The IS-95 is a standard for the cellular telephone system based on DS-CDMA multiple access. Thus, multiple users simultaneously share the same (wideband) channel. It was the first CDMA mobile phone system to gain widespread use and is found widely in North America. Gradually, evolution took place and different versions of CDMA systems came up, like CDMA2000. The major difference between GSM and CDMA-based systems initially was that CDMA mobiles did not have SIM cards, although recently this has changed. Instead, the subscriber data has simply been stored in memory of the mobile with a method of over-the-air programming of this data being available. The first offerings of CDMA under the standard IS-95 were voice as well as data up to a speed of 14.4 kbps.

Before CDMA2000, the scenario was like this. The GSM system used 200 kHz channels, whilst the CDMA, IS-95A, used a 1.25 MHz bandwidth and this was much wider than anything that had been used before. CDMA operates well with a wide bandwidth, but it was limited to 1.25 MHz to remain compatible with the spectrum allocations that were available. The maximum (speech coded) user data rate was 9.6 kbps. However, with the market moving towards data applications, the IS-95 specification was upgraded to IS-95B to cater to the needs of operators. This new specification allowed packet switched data transmission up to a speed of 64 kbps. IS-95B was first deployed in September 1999 in Korea and has since been adopted by operators in Japan and Peru. Often IS-95 A and B versions are marketed under the brand name cdmaOne. This is a registered trademark of the *CDMA development group* (CDG). Apart from voice, the mobile phone system is also able to carry data at rates up to 14.4 kbps for IS-95A and 115 kbps for IS-95B.The IS-95 system was introduced by Qualcomm.

13.5.1 The IS-95 System

The CDMA used for IS-95 is very different from the conventional systems. The CDMA is described in Chapter 10 and partly in Chapter 8. Although a complete summary of CDMA will not be included here, the basic principle of CDMA is that different codes are used to distinguish between the different users. The CDMA uses a form of modulation known as direct sequence spread spectrum. Here a signal is generated, which spreads out over a wide bandwidth. A code known as a spreading code is used to perform this action. By using orthogonal codes for spreading, it is possible to pick out a signal with a given code in the presence of many other signals with different orthogonal codes. In fact, many different baseband 'signals' with different spreading codes can be modulated onto the same carrier to enable many different users to be supported. By using different orthogonal codes, interference between the signals is minimal. Conversely, when signals are received from several mobile stations, the base station is able to isolate each one.

The advantage of using CDMA over FDMA and TDMA is that it enables a greater number of users to be supported. The improvement in efficiency is hard to define as it depends on many factors, including the size of the cells and the level of interference between cells and several other factors.

Unlike the more traditional cellular systems, where neighbouring cells use different sets of channels, a CDMA system reuses the same channels. Signals from other cells will appear as interference, but the system is able to extract the required signal by using the correct code in the demodulation and signal extraction process. Often more than one channel is used in each cell, and this provides additional capacity because there is a limit to the amount of traffic that can be supported on each channel.

Forward Link: Base to Mobile

The IS-95 specifies the 869–894 MHz band for forward link. The downlink transmission (i.e. base station to the mobile) within IS-95 consists of a number of elements. There are logical channels for pilot, paging, sync, and traffic. The pilot channel corresponds to the control channel in GSM and enables the mobile to estimate the path loss and as a result of this to set its power level accordingly. In addition to this, there are other channels for paging, speech, data, etc.

The forward link supports four types of channels within a total of 64 channels (bracket represents channel position):

Pilot channel (channel 0) It is a continuous signal on a single channel, allowing the mobile unit to acquire timing information, provide phase reference for the demodulation, and provide a means for signal strength comparison for the handover determination etc. The pilot channel transmits all zeros.

Synchronization channel (channel 32) It is to obtain identification information about the cellular system (e.g., system time, long code state, and protocol revision). It is a 1200 bps channel used by mobile station.

Traffic channel (channels 8 to 31 and 33 to 63) Originally, supported data rates of upto 9600bps. A subsequent revision added a second set of rates upto 14,400 bps.

Paging channel (channels 1 to 7) It contains messages for one or more mobile stations.

All of these channels use the same bandwidth. The chip code is used to distinguish them derived from 64×64 Walsh matrix. The speech is encoded using a voice encoder. Error correction is then applied to this data to enable it to be carried even under poor conditions. This brings the data rate up to 19.2 kbps. Data is then interleaved in blocks to reduce the effect of errors by spreading them out.

In the IS-95 system, all signals originate at the same transmitter. Thus, it is fairly simple to reduce mutual interference from users within the same cell, by assigning orthogonal Walsh–Hadamard codes to all the channels. As this is a 64 bit Walsh code, this multiplies the data rate by 64 to bring the overall data rate to 1.228 Mbps. This signal is then transmitted. One of the Walsh codes is the all 'one' word (1,1,1,1,...), which would result in a narrowband signal. Thus, a maximum length PN sequence is superimposed, which is the same for all users and has the same time phase for all users. The long PN code provides a measure of voice privacy and improves time synchronization. The short PN code in the forward link has a limited resolution but makes synchronization easier. A summary of the forward link parameters is given below:

- Chip rate: 1.2288 Mchip/s = 128 times 9600 bit/sec (128 spreading factor)
- Codes: Combines 64 Walsh-Hadamard (for orthogonality among users) and a maximum length code sequence (for effective spreading and multipath resistance)
- Transmit bandwidth: 1.25 MHz
- Convolutional coding with rate 1/2 for error correction
- Final carrier modulation scheme: QPSK

Reverse Link: Mobile to Base

The IS-95 specifies 824–849 MHz band for the reverse channel. On the reverse link, every user uses the same set of short sequences for modulation. The length of these sequences is 2^{15}, i.e. it is a modified 15 bit linear feedback shift register maximum length sequence ($2^{15}+1 - 1$). The reverse link consists of 94 logical CDMA channels each occupying the same 1228 kHz bandwidth. The link supports up to 32 access channels and up to 62 traffic channels.

The uplink signal for IS-95 is generated in a different way. The reverse link uses rate of the 1/3 convolutional coding. Although the same encoder is used, the resulting data has a greater degree of error correction or protection applied. Accordingly, the resulting data rate is brought up to 28.8 kbps. The data are then block interleaved. The next step is a spreading of the data using the Walsh matrix. The way in which the matrix used and its purposes are different than that of the forward channel. In the reverse channel, the data coming out of the block interleaver are grouped in units of six bits. Each six-bit unit serves as an index to select the row of the Walsh matrix. Walsh code ($2^6 = 64$ and 64×64 matrix) is used for spreading. However, this results in a 307.2 kbps data stream. The purpose of this encoding is to improve reception at the base station.

As the 64 possible codings are orthogonal, the block coding enhances the decision making algorithm at the receiver and is also computationally efficient. We can view this Walsh modulation as a form of block error correcting code with $(n, k) = (64, 6)$ and $dmin = 32$. The data burst randomizer is implemented to help reduce interference from other mobile stations. The operation involves using the long code mask to smooth the data out over each 20 ms frame.

The next step is the DSSS function. Further spreading is also required. This is provided by using a different form of orthogonal spreading code known as PN code. Each access channel and each traffic channel get a different long PN sequence. Access channels use distinct long code while traffic channel uses a user-specific long code. The long sequences are used to separate the signals from different users on the reverse link. This is multiplied with the signal to increase its data rate by four to bring it up to the final data rate of 1.228 Mbps, the same as the downlink signal. The bit stream is then modulated onto the carrier using an orthogonal QPSK modulation scheme. This differs from the forward channel as it uses the delay element in the modulator to produce orthogonality. The reason the modulators are different in than forward channel is that in the forward channel spreading, codes are orthogonal, all coming from the Walsh matrix, whereas in the reverse channel, orthogonality of the spreading codes is not guaranteed.

The reason that the uplink and downlink transmissions for IS-95 are generated in a different way results from the fact that it is difficult to synchronize the mobile handsets. Each one is a different distance away from the base station and the time delays will be different. As a result, synchronization is not possible. For the Walsh codes to maintain their orthogonality and to operate correctly, they must be properly synchronized. The PN codes do not require synchronization and can be used more successfully under these circumstances.

In IS-95, all base stations use the same channel (frequency reuse factor = 1). Thus, there are very few critical issues related to frequency reuse. The following factors enumerate the interference between the cells:

- Highly depends on path loss law
- Would theoretically diverge to infinity for free space loss with '20 log *d*'
- According to Qualcomm, the surrounding cells contribute to the total interference as follows:

  (i) 1st tier: six cells 6% per cell,
  (ii) 2nd tier: 12 cells 0.2% per cell
  (iii) 3rd tier: 18 cells 0.03% per cell
  (iv) 4th tier: 24 cells 0.01% per cell

  where percentages are relative to the power from own cell.

The Power control in IS-95 is achieved as follows:

- The CDMA performance is optimized if all signals are received with the same power.
- Update is needed every 1 ms.
- Performance is sensitive to imperfections of only 1 dB.
- For flat (frequency non-selective) Rayleigh and Rician fading, perfect power control is impossible. Fades are so deep that the average gain needed to compensate is unbounded.

13.5.2 Soft Handover in IS-95

Figure 13.8 illustrates the scenario for the soft handover.

The rake receiver is designed to optimally detect a DS-CDMA signal transmitted over a dispersive multipath channel. When a mobile is about to cross the cell boundary, the received CDMA signal comes from two different base stations, with slightly different delays. This can be used to support a soft handover from one base station to another. In the transition range, both base stations participate in maintaining the link with the user. This principle is used in the IS-95 cellular CDMA system.

One of the advantages of CDMA is the fact that handover can be made easier and more reliable. Normally, when handing over from a base station in one cell to the base

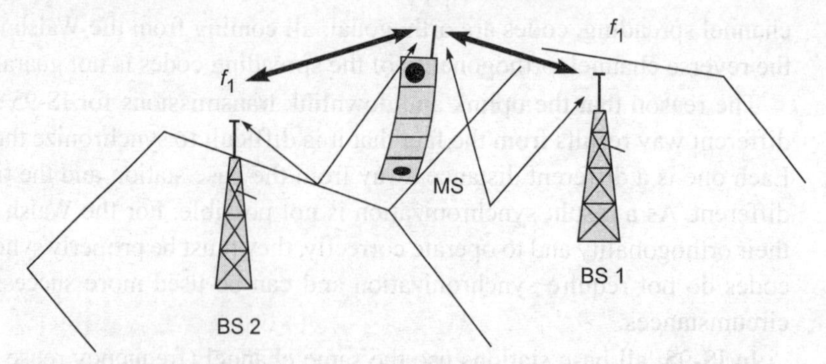

Fig. 13.8 DS-CDMA system with a soft handover (Two base stations transmit at the same frequency.)

station in the next, it is necessary for the system to arrange for a new channel to be used. The mobile then changes channel and hopes to be able to receive the signal on the new one satisfactorily. Obviously, there is a degree of risk and occasionally a handover does not proceed smoothly. This is achieved by soft handover.

As transmissions from the base stations in adjacent cells may be made on the same frequency, it is possible for a mobile to receive signals from two or more base stations at once. Thus, in soft handover, a mobile station is temporarily connected to more than one base station simultaneously. Considering the case of two base stations shown in Fig. 13.8, a mobile unit may start out assigned to a single cell. If the unit enters a region in which transmissions from two base stations are comparable within some threshold of each other, the mobile enters the soft handover state in which it is connected to the two base stations. During the period of the handover, the two base stations transmit the same signal enabling the mobile to receive the signal via two routes at the same time. This means that during this handover phase, the mobile should not loose signal. The mobile unit remains in this state until one base station clearly predominates and at that time it is assigned exclusively to that cell. Then as the mobile moves further into the second cell and the signal is firm, it can rely on one station only and the handover is complete. (The mobile would reject the signal from the second base station if it moves towards the first one.)

In the soft handover state, the transmissions from the mobile unit reaching the two base stations are both sent on to the MSC, which estimates the quality of both the signals and selects one, the switch sends data or digitized speech signals to both the base stations, which transmit them to mobile unit. The mobile unit then combines the two incoming signals to recover the information.

This approach considerably reduces the risk of loss of the connection during handover. It also minimizes the risk of a short break in the speech during this period. However it is not free and there is an associated cost. The mobile needs two decoders to monitor and decode the two signals and this increases the complexity of the mobile. On the network side, it means that two channels are used instead of one and this reduces the overall capacity. Some estimate that this could be as high as 40%. This is dependent upon the speed of handover and the degree of overlap in the cells. The figure given is obviously a worst case scenario, but despite this, the advantages are deemed to outweigh the reduction in capacity and increased mobile complexity.

13.5.3 IS-95 Upgradations upto CDMA-2000

The IS-95 has been successfully installed in many areas of the world, chiefly in North America. It also has the advantage of having an evolutionary migration path to 3G with CDMA2000 to give the higher data rates that are needed for video streaming and data transfer while retaining compatibility with the existing networks. The CDMA2000 cell phone system is designed to be used in existing cellular telecommunications frequency allocations in addition to those cell phone bands assigned to IMT2000 (3G). The actual band allocations used for cell phones vary from one country to another, dependent upon the frequency allocations available. Major points related to cdmaOne and CDMA2000 systems are given below.

The cdmaOne and CDMA2000 systems, being an evolutionary technology moving from the standards such as IS-95 (IS-95A and IS-95B) for cdmaOne through to standards including IS-2000 and IS-856 for CDMA2000 1X, 1 x Ev, 1 x EV-DO and 1 x EV-DV. CDMA2000 1X is the basic 3G standard, in fact, some people only consider it as a 2.75G system, and it is being developed beyond this. In what is termed CDMA2000 1x Ev, there are further developments to bring it in line with the UMTS or wideband CDMA system that is being deployed in Western Europe and many other areas. The first of these is known as CDMA2000 1 x EV-DO (EVolution Data Only) is something of a sideline from the main evolutionary development of the standard. It is defined under IS-856 rather than IS-2000 and is, as the name indicates, carries only data, but at speeds of up to 2.4 Mbps in the forward direction and the same as 1X in the reverse direction.

The forward channel forms a dedicated variable rate, packet data channel with signalling and control time multiplexed into it. The channel is itself time-divided and allocated to each user on a demand and opportunity driven basis. A data only format was adopted such that the system could be optimized for data applications, and if voice is required then a dual mode phone using a separate 1X channel for the voice call is required. In fact the 'phones' used for data only applications are referred to as access terminals or ATs. The first commercial CDMA2000 1 x EV-DO network was deployed by SK Telecom (Korea) in January 2002.Currently the standard uses one standard channel under a system known as 1X RTT, although for the future, three channels (3X RTT) may be used.

In view of the fact that the CDMA2000 system has been designed to be an evolutionary standard, it is relatively easy to introduce upgrades to the system. This has made it particularly popular with operators because the cost of upgrading to the new standards is much less and they can have users with a variety of types of phone on the same network. Thus, users may operate cdmaOne phones on the same network as CDMA2000 1X or CDMA2000 1X EV-DV phones. CDMA2000 1X can double the voice capacity of cdmaOne networks and delivers peak packet data speeds of 307.2 kbps in mobile environments, although today's commercial CDMA2000 1X networks (phase one) support a peak data rate of 153.6 kbps. CDMA2000 1X has been designated a 3G standard and it is now widely deployed.

13.6 WIRELESS LOCAL LOOP (WLL)

Fixed wireless access equipment is very much suited for rapid deployment of broadband connections in many instances. This approach has gradually become more popular for providing 'Last Mile' broadband local loop access, which merged the redundant point-to-point or point-to-multipoint private networks. The nature of the channel is time invariant in this case. Local loops can be thought of as the last mile of the telecommunication network that resides between the central office and the individual homes and business sectors in close proximity to the central office. The last mile technologies are:

- Wireline: PSTN, DSL, cable modem
- Wireless: Fixed wireless [local multipoint distribution service (LMDS), multichannel multipoint distribution service (MMDS), wireless LAN, and PCS

The systems based on optical fibres as well as microwave links (LOS) (with high frequencies of the order of 28 GHz and more, which are greater than ten times the carrier frequency of terrestrial cellular network) and satellite links (or VSAT) are categorized under long-haul communication while maximum coverage with respect to central unit is concerned.

When long-haul and last-mile systems are combined, the complete WLL architecture can be formed covering urban and suburban or rural area as shown in Fig. 13.9. Wireless equipment can usually be deployed in just a couple of hours and also once the wireless equipment is purchased, there is no additional cost except monthly bills. Many new services and applications have been proposed and commercialized. These services include the concept of LMDS. The unused spectrum of 27–31 GHz band was auctioned by the US government to support LMDS and similar other bands in other countries. It

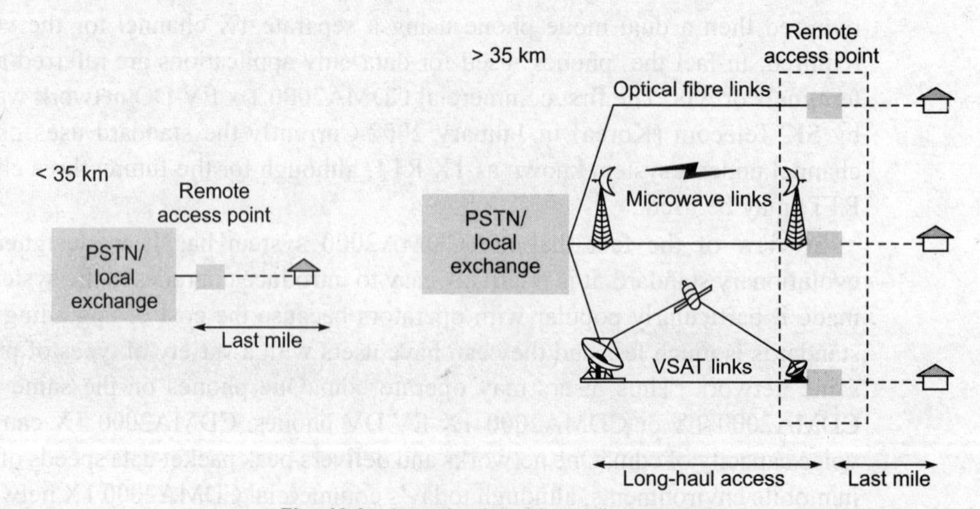

Fig. 13.9 Long-haul and last-mile systems

should be noted that most LMDS allocations share frequencies with the teledesic band, which was approved by the ITU World Radio Conference for broadband satellite systems. The teledesic band was originally established for the Motorola Iridium system, whose spectrum was later on merged into the teledesic system. The LMDS has the vast bandwidth capabilities for WLL applications. It exhibits LOS fixed wireless applications. One of the most promising applications for LMDS is in a local exchange carrier (LEC) network. The LEC may own wide bandwidth ATM or synchronous optical network (SONET) backbone switch capable of connecting huge traffic with the Internet, the PSTN, or to its own private network. Due to LOS, LMDS allows LECs to install wireless equipment on the customers' premises for rapid broadband connectivity without having to lease or install its own cable to the customers. The LMDS architecture is shown in Fig. 13.10. Remote places, large or small business houses, or residences may be connected is the Ethernet through LOS by antennas.

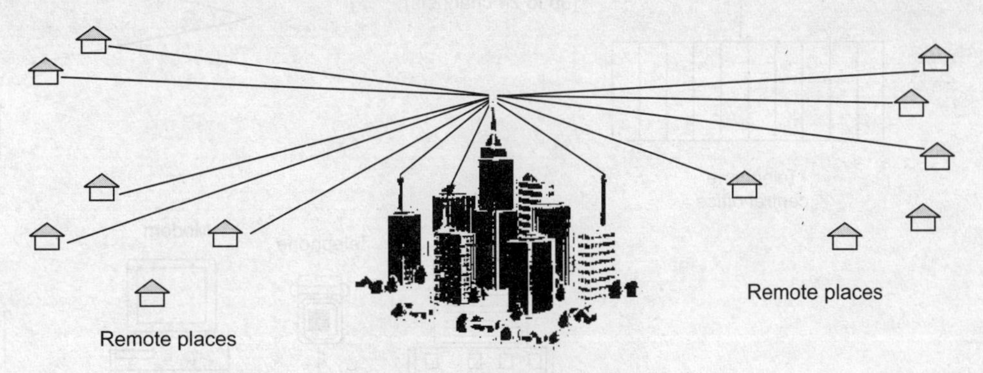

Fig. 13.10 Structure of LMDS

Wireless local loop (WLL) is defined in the last-mile system category. The WLL services may be defined as fixed wireless services intended to provide access to the telephone network. In general, local loop means exchange-to-home-to-exchange closed loop. Conventionally, telephone local loops are UTP or STP cables based. If we remove wires for the communication, i.e. establish the RF link, it becomes a wireless local loop, of course, with supporting hardware. Wireless local loop systems, as envisioned, will generally divide a geographic region into a number of similar sized cells (just like cellular telephony) or maybe like PCS. Each cell will be serviced by a base station [wireless access network unit (WANU)], which will communicate with all of the wireless local loop customers [wireless access subscriber units (WASUs)] within the cell as shown in Figures 13.11 and 13.12. The base station may be as simple as a small omnidirectional antenna and control box hanging from the overhead electrical lines. Each customer will be equipped with a transceiver and a small patch antenna. The transceiver may have several outputs, one for a telephone, one for a MODEM, and maybe even one for a television. The small antenna, which may be inside or outside, will be positioned to communicate with the base station. The WLL may be based on either analog (AMPS, etc.) or digital (GSM, DECT, CDMA, etc.) standards. Architecture of WLL is very important in which appropriate standard may be adopted.

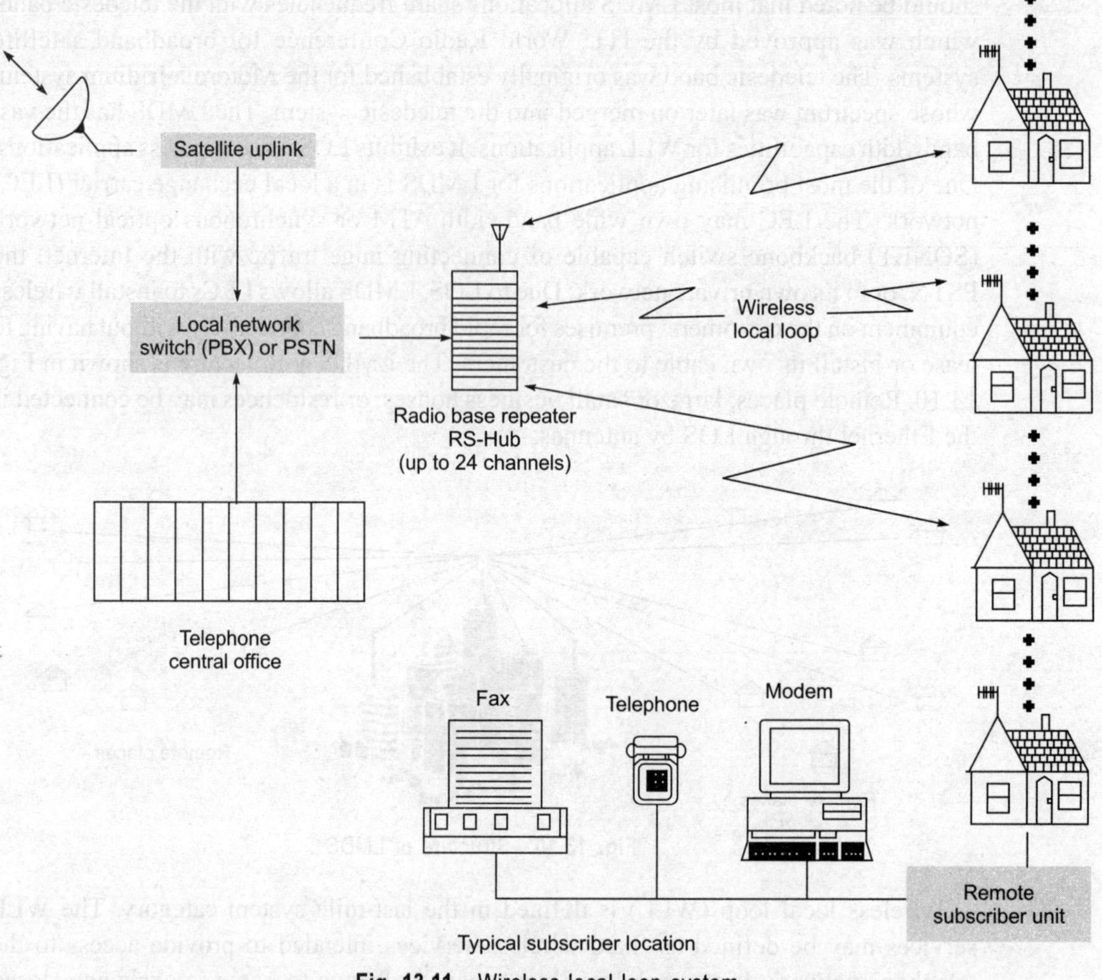

Fig. 13.11 Wireless local loop system

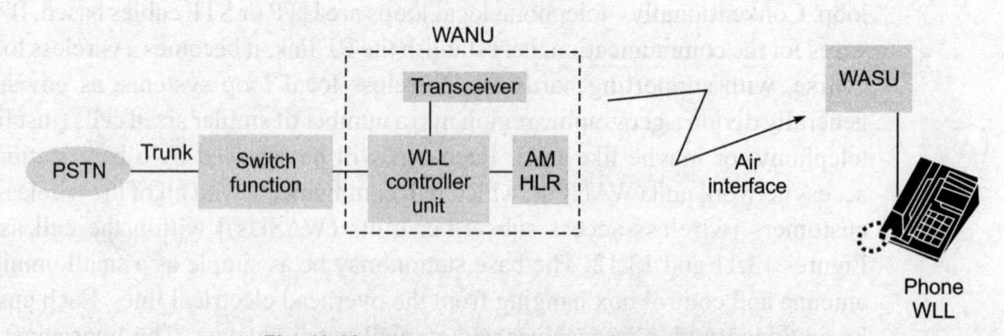

Fig. 13.12 WLL access units and their functions

The WANU, the interface between the underlying telephone network and the wireless link, consists of the following:

BTS: Base transceivers station

RPCU: Radio controller unit

AM: Access manager
HLR: Home location register
WASU: Located at the subscriber; translates the wireless link into a traditional telephone connection

What equipment and frequencies will WLL use? It depends. If the system is to provide basic local telephone service to a developing nation, then the bandwidth requirement will be modest and almost any infrastructure at any frequency will do. The major wireless infrastructure manufacturers, who provide the systems for mobile telephony, are naturally trying to use that same equipment for WLL to avoid any new developments (just like Reliance mobile network). Most of the WLL in underdeveloped countries will ultimately be at the same frequencies as the mobile telephony in developed countries and will utilize the same basic equipment, only slimmed down to provide fixed service only. Figure 13.13

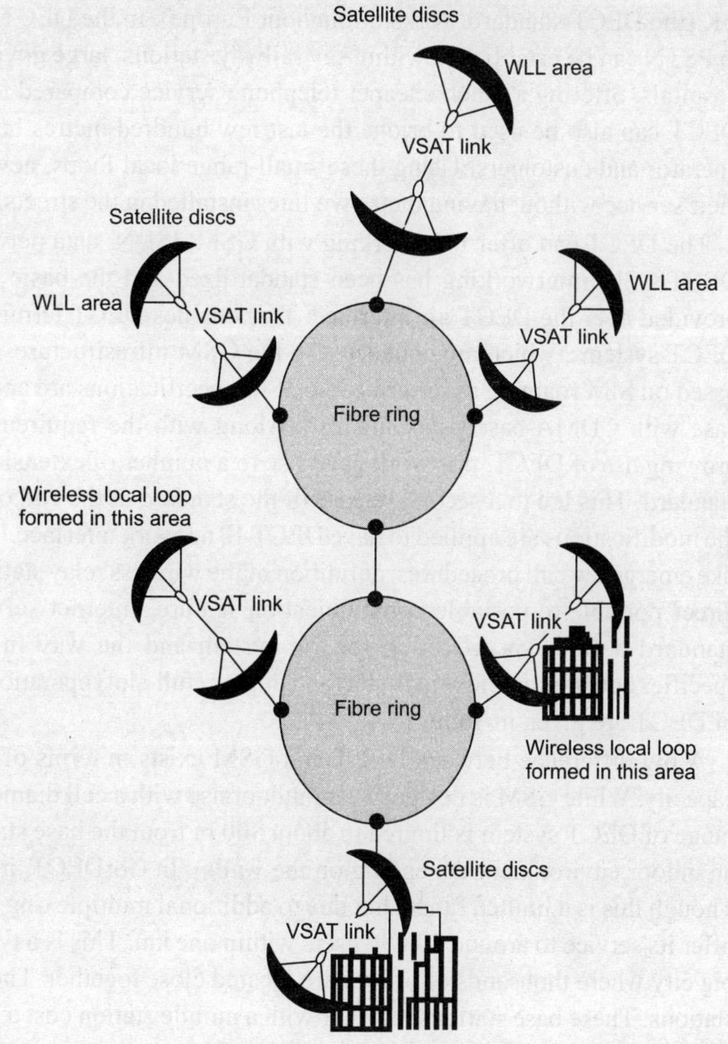

Fig. 13.13 WLL in combination of fibre as well as satellite links, exploiting the fibre ring

shows a typical scenario in which the combination of fibre ring along with VSAT links creates the infrastructure for WLL covering a very large area. The VSAT links may be replaced by microwave LOS links. In the WLL area as usual, base station as well as subscriber units will communicate. Base stations are linked with satellite disks or microwave antennas to complete the scenario.

13.6.1 DECT System

Digitally enhanced cordless telecommunications (DECT) system is specified by the European Telecommunications Standards Institute (ETSI) (DECT forum 1999). The DECT phone replaces the older analog cordless phone systems, such as CT1 and CT1+. These analog systems only ensured security to a limited extent as they did not used encryption for data transmission and only offered a relatively low capacity. But DECT is also a more powerful alternative to the digital system CT2, which is mainly used in the UK (the DECT standard works throughout Europe). In the DECT system, access points to PSTN can be established within say railway stations, large government buildings and hospitals, offering a much cheaper telephone service compared to a GSM system. The DECT can also be used to bridge the last few hundred metres between a new network operator and customers. Using these small-range local loops, new companies can offer their service without having their own lines installed in the streets.

The DECT can offer interworking with GSM, ISDN, data networks, etc. As a result, DECT/GSM interworking has been standardized and the basic GSM services can be provided over the DECT air interface. This enables DECT terminals to interwork with DECT systems, which are connected to the GSM infrastructure. All roaming scenarios based on SIM roaming as described in GSM specifications are applicable. Similar is the case with CDMA-based systems also. Along with the requirements arising from the growing use of DECT, this work gave rise to a number of extensions to the basic DECT standard. This led to a second release of the standard at the end of 1995 and up to 2005 the modifications are applied to have DECT-IP network interface. This included facilities like emergency call procedures, definition of the wireless relay station (WRS), an optional direct portable-to-portable communication feature, Internet services, etc. The DECT standard defines specification for the system and the way in which it operates. It specifies both the simplex (half-slot) and duplex (full-slot) operations. The salient features of DECT are given in Table 13.4.

A big difference between DECT and GSM exists in terms of cell diameter and cell capacity. While GSM is designed for outdoor use with a cell diameter of up to 70 km, the range of DECT system is limited to about 300 m from the base station (and even less for an indoor environment depending on the walls). In CorDECT, it is extended to 10 km. Though this is a limited range, but due to additional multiplexing techniques, DECT can offer its service to around 10,000 users within one km. This is a typical scenario within a big city where thousands of offices are located close together. The DECT also uses base stations. These base stations together with a mobile station cost a fraction of the amount,

Table 13.4 Specifications of DECT system

Parameter	Specification
Frequency band	1880 MHz to 1980 MHz and 2010 MHz to 2025 MHz (new)
Access technique	MC/TDMA/TDD
Channel bit rate	1152 kbps
Carrier spacing	1728 kHz
Frame duration	10 ms
Slots in a frame	24
Access channels/RF carrier	120 duplex 32 k bit/s channels
Traffic channel assignment	Instant dynamic
Control carriers	Not required
Modulation	GFSK/GMSK (BT = 0.5) and optional higher level modulation schemes possible
Portable average RF power	10 mW
Portable peak RF power	240 mW 24 dBm
Speech codec	32 kbit/s ADPCM
Base station sensitivity at 0.1% BER	–86 dBm (for GAP) (typically –90 dBm to –94 dBm)
Basic link budget	110 dBm (typically .114 dBm to 118 dBm)
Protected 64 kbit/s bearer service	Yes
Base station antenna	Diversity switched. Post detection selection optional. Dual antennas in handset optional
Tolerance to time dispersion with selection antenna diversity	200 ns (500 ns possible with low-cost non-coherent equalizer)

compared to which GSM set-up costs 100 times more. The DECT also can handle handover, but it is not designed to work at a higher speed.

The basic DECT system has a total of ten possible carrier frequencies between 1880 and 1900 MHz. In addition to this, the time dimensions for each carrier is divided to provide timeframes repeating every 10 ms. Each frame consists of 24 time slots (Fig. 13.14), each of which is individually accessible and may be used for either transmission or reception. The first 12 timeslots are used for the downlink transmissions and the remaining 12 are used for the uplink. This reduces the level of complexity and as this is not needed for basic implementations, it can provide some cost savings. The DECT TDMA structure enables up to 12 simultaneous basic voice connections per transceiver.

The system uses dynamic channel allocation and is thereby able to reduce the levels of interference, and ensure that links are set up on channels with the least interference. All DECT equipment scans the frequency allocation at least every 30 seconds as a background activity. This produces a list of free and occupied channels along with the available timeslots to be used for the channel selection, should this be required.

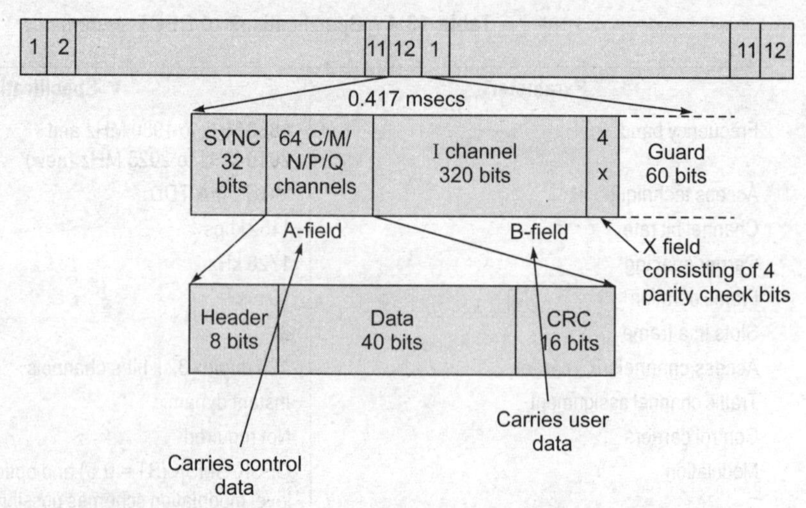

Fig. 13.14 DECT: a hybrid TDMA/FDMA system that uses TDD

Thus, DECT is a hybrid TDMA/FDMA system that uses TDD. For the basic DECT speech service, two time slots, with 5 ms separation, are paired to provide bearer capacity for typically 32 kbps (ADPCM G.726 coded speech) full duplex connections. It consists of 10 carriers. The carrier bit rate is 1.152 MHz. As mentioned earlier, each carrier is time slotted. The slot consists of 420 bits plus 60 bits guard time. The structure is shown in Figures 13.14 and 13.15 in detail.

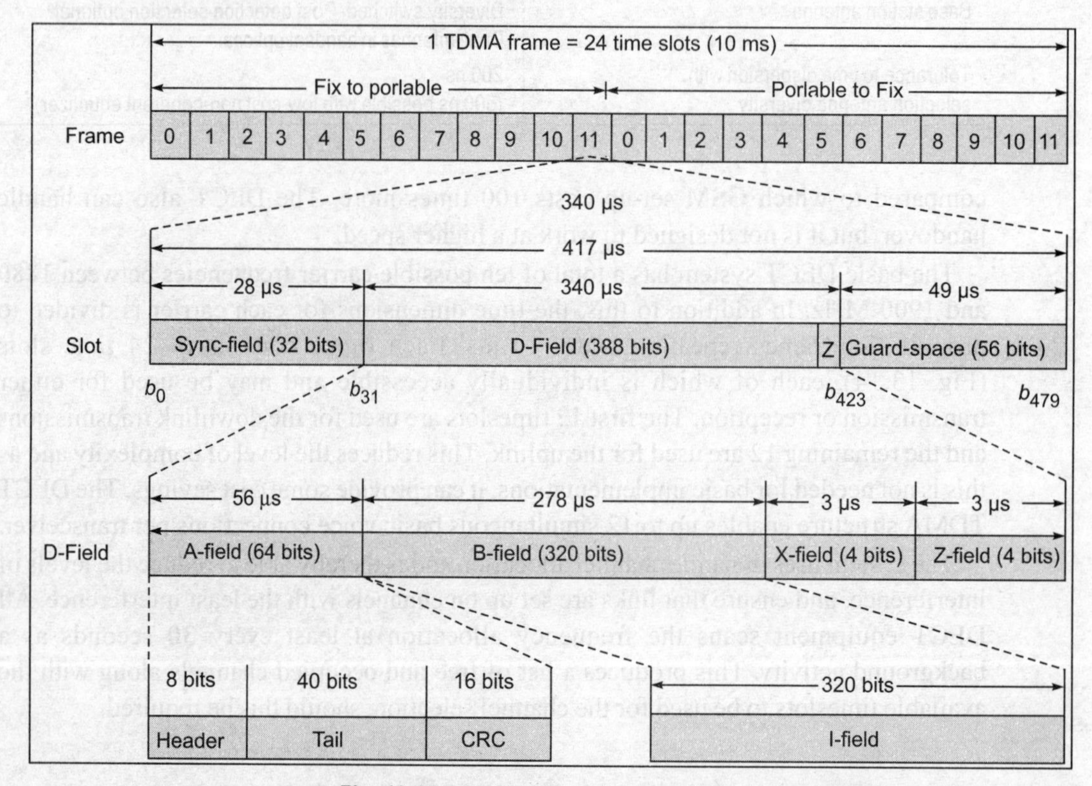

Fig. 13.15 DECT frame structure of Fig.13.14 in detail

A DECT system has the following features:

- The DECT specifies two logical channels for *user data* transmission and five logical channels for *network control*.
- The data channels (B-field):
  - (i) *Unprotected I_N data*, which uses 320 bits in every slot and rate is 32 kbps.
  - (ii) *Protected I_P data*, which uses 256 bits. The error control uses 64 bits. The rate is 25.6 kbps.
- The control channels (A-field):
  - (i) Five logical channels are muxed over this field in a slot.
  - (ii) An eight-bit header is used to identify the type of channel and nature of message.
  - (iii) Fourty bits are devoted to transmit the control channels:

    C: call management

    M: physical layer control, has priority

    N: hand shaking, carries an identification code to identify BS

    P: paging to locate a mobile terminal

    Q: system information
- The terminals are required to do most of the work in DECT. If a terminal wants to originate a call, it searches for an *idle channel* and starts to communicate with the BS.
- The DECT supports *soft handoff*. The mobile terminal can communicate with two BSs when necessary.
- The DECT supports authentication and encryption.

Additionally, the DECT portable continuously analyses the signals to ensure that the signals originate from the base station to which it is connected and has access rights. The portable locks onto the strongest base station and checks that it can access the base station, as detailed in the DECT standard, and the channels with the best signal strength receive signal strength indication (RSSI)) are used for the radio link as required. This dynamic channel selection and allocation mechanism guarantees that radio links are always setup on the channel available with the least interference and hence the best performance is obtained.

All DECT systems are based on a main standard, the *common interface* (CI), which is often used in association with the *generic access profile* (GAP). The GAP profile ensures interoperability of equipment from different providers for voice applications. The GAP defines the minimum interoperability requirements, including mobility management and security features. It has different requirements for the public and private systems. This means that the GAP is effectively the industry standard for a basic fallback speech service with mobility management. This basic service is not always used, but instead it forms the fallback, which is always be available, especially when requested by a roaming phone, etc.

13.6.2 CorDECT WLL

This section is dedicated to the architecture and the deployments of WLL system based on the DECT standard identified as CorDECT WLL. It is the technology adopted by India. CorDECT technology allows telephone and Internet services together on the same line. It is a European standard. This cost-effective technology can be a great boon to the millions of Indians who have no access to such facilities today, especially in the rural areas.

The visual diagram of PSTN and dial-up network access over the telephone line is shown in Fig. 13.16 (wired scenario).

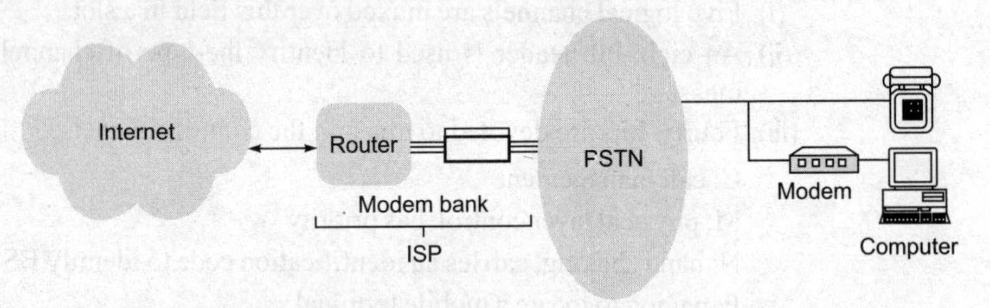

Fig. 13.16 PSTN dial-up wired network scenario

A simplified version of the CorDECT WLL system is shown in Fig. 13.17. In this figure, CorDECT WLL system is composed of the DECT interface unit (DIU), remote access switch (RAS), compact base station (CBS), and a subscriber unit at the receiving end. In contrast to PSTN, where individual copper lines must be routed and wired to each subscriber from the backbone network (and linked to the PSTN central office), WLL connects end-users to the backbone network wirelessly through an access unit. Dynamic frequency allocation along with a picture of DIU and BS antenna is shown in Fig. 13.18.

In Fig. 13.17, the subscriber terminal is a wall-set with Internet Port (WS-IP) as a standard interface (RJ-11) and a serial port interface (RS-232). It supports a standard

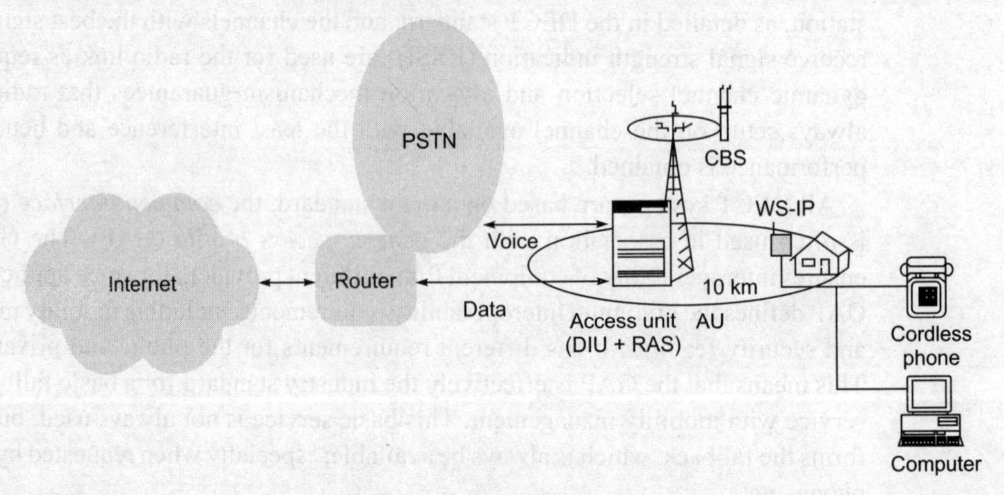

Fig. 13.17 CorDECT WLL scenario

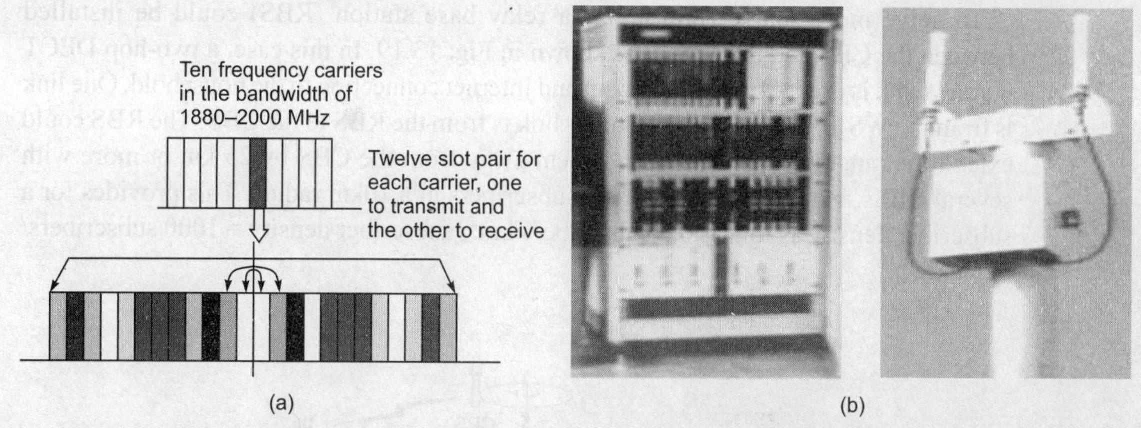

Fig. 13.18 (a) Frequency allocation on dynamic basis in CorDECT and (b) DECT interface unit and BS

telephone, modem, and fax through the RJ-11 port and provides a direct connection to a PC through the RS-232 port. The WS-IP provides high-speed Internet access at 40-70 kbps as well as normal telephone service for subscribers in the CorDECT system. The WS-IP is connected to the CBS using a wireless link. The CBS provides the radio interface between subscribers and an access unit (AU), which consists of a DIU and RAS. The DIU separates the voice traffic and directs it to the telecom network, as well as switches the Internet calls to a built-in RAS. The RAS then routes the traffic to the Internet network. Since Internet traffic does not have to pass through the telecom network, CorDECT WLL gets rid of the Internet tangle associated with POTS. Moreover, it offers a number of key advantages, such as faster deployment, immediate realization of revenues, lower construction costs, lower network maintenance, management and operating costs, and greater flexibility to meet uncertain levels of penetration and rates of growth.

Let us understand Fig. 13.17 with an example. In the CorDECT system, a DIU can be co-located with the PSTN through the backbone network at, say, any big company premises along with its township by utilizing the optical fibre network. Unlike the PSTN system, where copper wire is needed to connect remote subscribers to the backbone, CorDECT WLL system offers wireless telecom and Internet connectivity to the backbone network. Thus, the coverage range of CorDECT system can be extended from the central location in the company up to a distance of 35 km or more if required without the need for copper wires. Note that line of sight is required in every hop in CorDECT WLL system. The CorDECT system gives a good deployment opportunity for company premise. As shown in Fig. 13.17, the DIU and CBS are located at a tower, which should be installed at the central location, near the centre of the company premises. The CBS wireless link distance could be as long as 10 km by using line-of-sight connection. Thus, the demand can be concentrated at the tower or within a 10 km radius. In other words, this system enables more coverage area away from the company without using any copper wire. This provides a subscriber density of three subscribers/km^2 (subscriber density = 1000 subscribers/100π km^2). The subscriber density, however, can be increased by using more DIUs at the central premise.

To serve more sparse rural areas, a relay base station (RBS) could be installed between the CBS and the WS-IP as shown in Fig. 13.19. In this case, a two-hop DECT wireless link is used to provide telecom and Internet connection to the household. One link is from the WS-IP to the RBS. The other link is from the RBS to the CBS. The RBS could extend the range of the CorDECT system away from the CBS by 25 km or more with several RBSs. In turn, the RBS serves subscribers in a 10km radius. This provides for a subscriber density as low as 0.5 subscribers/km^2 (subscriber density = 1000 subscribers/ 625π km^2).

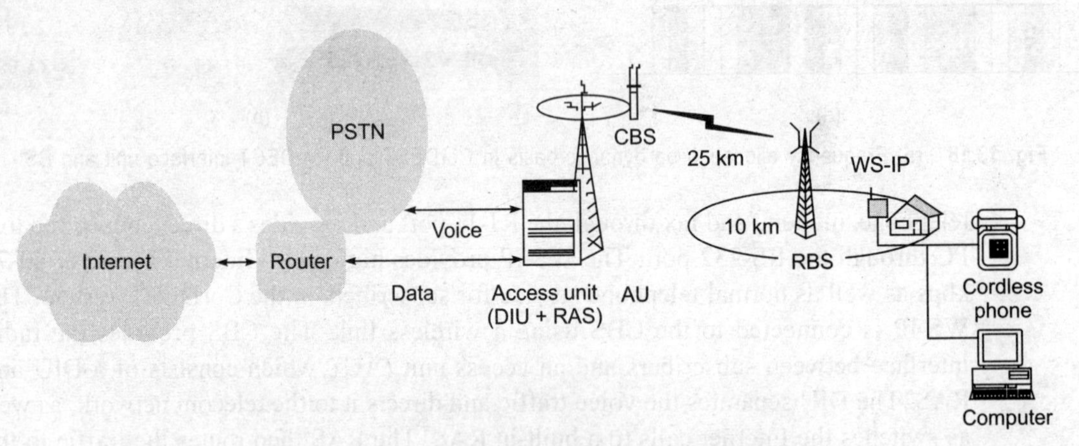

Fig. 13.19 Deployment with relay base station (RBS)

As shown in Fig. 13.20, the CorDECT WLL with a *base station distributor* (BSD) can be used for coverage distances beyond 35 km. The BSD is a remote unit connected to the DIU using a standard E1 interface. The maximum distance between the DIU and BSD depends upon the E1 link (radio, fibre, or copper). At the BSD site, a cluster of CBSs is mounted on a rooftop tower to serve an area of 10 km. The system is suitable for serving a load pocket in a remote midsized town or city.

As a result of becoming progressively less expensive than the wired alternatives and getting rid of Internet tangle, CorDECT WLL technologies offer advantages of rapid and

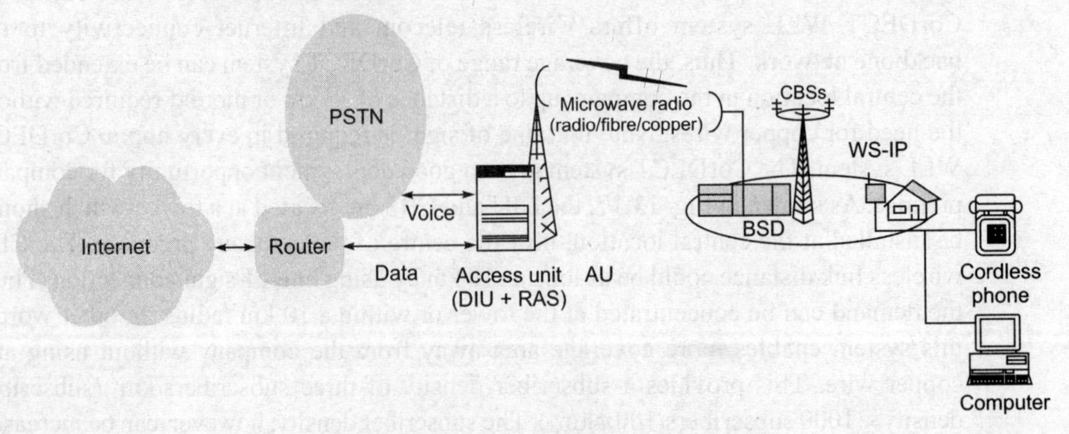

Fig. 13.20 Deployment with the BSD

flexible deployment in remote areas. In India, for example, the CorDECT technology has been installed in Kuppam (Andhra Pradesh), providing connections to about 65 villages. Moreover, the system is being installed by basic services operators in Punjab and Rajasthan and by several ISPs. CorDECT is specifically developed for rural area upliftment but has now been commercialized in the urban area also.

13.7 IMT-2000

International Telecommunication Union (ITU) asked for proposals for radio transmission technology (RTT) for the International Mobile Telecommunication (IMT)-2000 programme (ITU 1999). This was to establish the worldwide communication system that allows for terminal and user mobility supporting the idea of universal personal telecommunication (UPT). The ITU has created several recommendations for it. Here the number 2000 indicates the start of the system and also the spectrum used (around 2000 MHz).

International Mobile Telecommunications-2000 (IMT-2000) is the global standard for third-generation (3G) wireless communications, defined by a set of interdependent, ITU recommendations. The IMT-2000 provides a framework for worldwide wireless access by linking the diverse systems of terrestrial and/or satellite-based networks. It will exploit the potential synergy between the digital mobile telecommunications technologies and the systems for fixed and mobile wireless access systems.

The ITU activities on IMT-2000 comprise international standardization, including frequency spectrum and technical specifications for radio and network components, tariffs and billing, technical assistance, and studies on regulatory and policy aspects. The main areas of ITU activities on IMT-2000 are discussed below.

Telecommunication Development Group (ITU-D)

It is responsible for studies, activities, and direct assistance relating to the implementation of IMT-2000 in developing countries.

Radio Communications Group (ITU-R)

It is responsible for the overall radio frequency spectrum and radio system aspects of IMT-2000 and systems beyond. It has the prime responsibility for issues related to the terrestrial component of IMT-2000 and systems beyond. It works on the issues related to the satellite component.

Telecommunication Standardization Group (ITU-T)

The ITU-T study group 'mobile telecommunication networks' is responsible for studies relating to the network aspects of mobile telecommunications networks, including International Mobile Telecommunications 2000 (IMT-2000) and beyond, wireless Internet, convergence of mobile and fixed networks, mobility management, mobile multimedia functions, Internetworking, interoperability, and enhancements to existing ITU-T recommendations on IMT-2000.

Research and Analysis (SPU), General Secretariat

The Strategy and Policy Unit (SPU) of the general secretariat provides research, analysis, and statistics on the telecommunications sector as a whole. One of its main areas of activity is mobile communications, including IMT-2000 (3G). In this context, the unit produces publications that track changing trends in the global telecommunication environment. In the area of mobile communications, reports include 'Internet for a Mobile Generation' and 'the Portable Internet'.

The IMT-2000 includes different environments, such as given below:

- Indoor use
- Vehicles
- Satellite
- Pedestrians

The IMT-2000 offers the capability of providing value-added services and applications on the basis of a single standard. The system envisages a platform for the distribution of converged, fixed, mobile, voice, data, Internet, and multimedia services. One of its key visions is to provide seamless global roaming, enabling the users to move across borders while using the same number and handset. The IMT-2000 also aims to provide seamless delivery of services, over a number of media (satellite, fixed, etc.). It is expected that IMT-2000 will provide higher transmission rates: a minimum speed of 2 Mbps for stationary or walking users and 348 kbps in a moving vehicle. The second-generation systems only provide speeds ranging from 9.6 kbps to 28.8 kbps.

In addition, IMT-2000 has the following key characteristics.

Flexibility

With the large number of mergers and consolidations occurring in the mobile industry, and the move into foreign markets, operators wanted to avoid having to support a wide range of different interfaces and technologies. The IMT-2000 standard addresses this problem, by providing a highly flexible system, capable of supporting a wide range of services and applications. The IMT-2000 standard accommodates five possible radio interfaces based on three different access technologies (FDMA, TDMA, and CDMA).

Affordability

There was agreement among industry that 3G systems had to be affordable, in order to encourage their adoption by consumers and operators.

Compatibility with Existing Systems

The IMT-2000 services have to be compatible with existing systems. For example, 2G systems, such as the GSM standard, will continue to exist for some time and compatibility with these systems must be assured through effective and seamless migration paths.

Modular Design

The vision for IMT-2000 systems is that they must be easily expandable, in order to allow for growth in users, coverage areas, and new services, with minimum initial investment.

For new IMT-2000 systems, recommended frequency bands 1885–2025 and 2110–2200 MHz (Reco-ITU-R M-1036). For different proposals for RTT, the European proposal for IMT-2000 prepared by ETSI is called universal mobile telecommunication systems (UMTS). Specific proposal for RTT under which is called *UMTS* terrestrial radio access (UTRA) (ETSI, 1998 and UMTS Forum, 1999).

The UMTS as proposed by ETSI represents an evolution from second-generation GSM system to the third generation that is completely new. One enhancement of GSM towards UMTS is enhanced data rates for GSM evolution (EDGE), which uses enhanced modulation schemes and other techniques for data rates of up to 384 kbps using the same 200 kHz wide carrier and the same frequencies as GSM (i.e. a data rate of 48 kbps per time slot is available). The EDGE can be introduced incrementally offering some channels with EDGE enhancement that can change between EDGE and GSM/GPRS. Besides enhancing the data rates, new additions to GSM like customized application for mobile enhanced logic (CAMEL) introduced an intelligent network support. This system supports, for example, the creation of a virtual home environment (VHE) for visiting subscribers. It enables worldwide access to operator specific IN applications, such as prepaid, call screening, and supervision.

The GSM MoU (1999) provides many proposals covering QoS aspects, roaming, services, billing, accounting, radio aspects, core networks, access networks, terminal requirements, security, application domains, operation and maintenance, and several migration aspects.

The UMTS fits into a bigger framework developed by ETSI, called global multimedia mobility (GMM):

- The GMM provides architecture to integrate mobile and fixed terminals.
- The GMM provides many different access networks, such as GSM, BSS, DECT, ISDN, UMTS, LAN, MAN, CATV and NSS+IN, ISDN+IN, BISDN+TINA, and TCP/IP (ETSI 1998).

13.7.1 UMTS

The UMTS (universal mobile telecommunications system) is the third-generation (3G) successor to the second-generation GSM-based technologies, including GPRS and EDGE. Although UMTS uses a totally different air interface, the core network elements of GSM have been migrating towards the UMTS requirements with the introduction of GPRS and then EDGE. In this way, the transition from GSM to the 3G UMTS architecture does not require a large instantaneous investment.

The UMTS uses wideband CDMA (WCDMA or W-CDMA) to carry the radio transmissions and often the system is referred to by the name WCDMA. It is also gaining a third name. Some call it 3GSM because it is a 3G migration from GSM. In order to create and manage a system as complicated as UMTS or WCDMA, it is necessary to develop and maintain a large number of documents and specifications. For UMTS or WCDMA, these are now managed by a group known as 3GPP—the third-generation partnership programme. This is a global cooperation between six organizational partners—ARIB, CCSA, ETSI, ATIS, TTA, and TTC. The scope of 3GPP was to produce globally applicable technical specifications and technical reports for a third-generation mobile telecommunications system. This would be based upon the GSM core networks and the radio access technologies that they support. Figure 13.21 shows integration of the GSM, GPRS, UMTS, and other networks in one architecture.

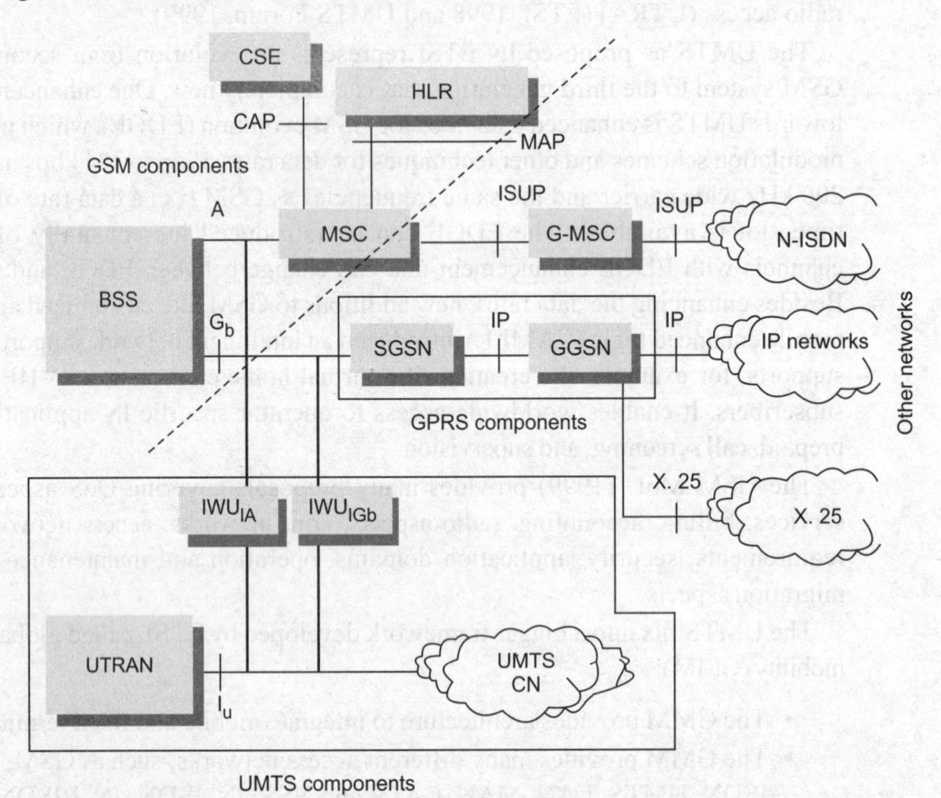

Fig. 13.21 Evolution of GSM towards UMTS, introduction to UMTS core network (2005) (Various standard interfaces are shown over the connecting lines.)

Many of the ideas that were incorporated into GSM have been carried over and enhanced for UMTS. Elements such as the SIM have been transformed into a far more powerful USIM (Universal SIM). In addition to this, the network has been designed so that the enhancements employed for GPRS and EDGE can be used for UMTS. In this way, the investment required is kept to a minimum.

The aim of the UMTS systems is to provide an 'anywhere, anytime' service. Thus, the operating environment will vary depending on the user location. The environment in which the wireless system must operate affects the system capacity and type of services that can be provided. Table 13.5 lists out some of the environments in which UMTS will be required to provide coverage. The maximum supported data rate for each environment is related to the cell size required to provide adequate coverage for the environment.

A cellular network is required to ensure that the UMTS can provide a high capacity network. As with any cellular system, the total capacity of the network is dependent on the size of the cells used. However, the cell size is limited by the amount of infrastructure. The cell size determines the maximum channel capacity for each cell, such as

Table 13.5 Maximum supported data rates for UMTS, for various environments

Environment	Maximum Supported Data Rate
Business (indoor)	384kbps
Suburban (indoor/outdoor)	144kbps
Urban vehicular (outdoor)	144kbps
Urban pedestrian (outdoor)	144kbps
Fixed (outdoor)	144kbps / 384kbps
Local high bit rate (indoor)	2Mbps

propagation effects, as multipath delay spread, and high path loss, force large cells to have a lower data rate. Large cells also have to service a large number of users, and since the cell capacity is approximately fixed, each user can only have a reduced data rate, with respect to a smaller cell. In order to optimize the cellular network three cell types are used. These are the pico-cell, micro-cell, and macro-cell. The three different cell type trade-offs of cell size will determine total capacity and services. The size and type of coverage of each cell type affects the radio propagation problems that will be encountered. This will determine the most suitable radio transmission technique to use. Table 13.6 shows the three cell types used in the UMTS system and some of the cell characteristics.

Table 13.6 Cell types and cell characteristics for UMTS

	Pico-cell	Micro-cell	Macro-cell
Cell radius	<100 m	<1000 m	<20 km
Antenna	Ceiling/wall mounted	Below roof top height	Roof top mounting
Maximum multipath delay spread	1 µs	5 µs	20 µs
Applications and environments	Indoor/outdoor within buildings, city centres, local high bit rate	Hight density outdoor business (indoor) fixed (outdoor) inner city areas	Low density areas, suburban areas, urban areas, fixed (outdoor)
Services and data rate supported	All services (up to 2 Mbps)	Up to 384 kbps	Limited sub-set (up to 144 kbps)

One of the aims identified for UMTS is to provide a wireless interface comparable to wired connections. The radio interface is currently undergoing substantial research, with the relative performance of CDMA and TDMA being investigated. Currently, CDMA appears to be the most likely candidate for supporting the high data rate required. However, other techniques, such as COFDM and CDMA-OFDM, hybrid solutions are also being investigated for UMTS.

The UMTS' aim is 'anywhere, anytime' and cellular networks can only cover a limited area due to the high infrastructure costs. For this reason, satellite systems will form an integral part of the UMTS network. Satellites will be able to provide an extended wireless coverage to remote areas and to aeronautical and maritime mobiles. The level of integration of the satellite systems with the terrestrial cellular networks is under investigation. A fully integrated solution will require mobiles to be dual mode terminals that would allow communications with orbiting satellites and terrestrial cellular networks. Low earth orbit (LEO) satellites are the most likely candidates for providing worldwide coverage.

Currently, several low earth orbit satellite systems are being deployed for providing global telecommunications. These include the Teledesic system, which was scheduled to begin operation by the end of 2002 with 288 satellites, to provide high bandwidth two-way communications to virtually anywhere in the world. However, the Teledesic system will not be able to meet even 20% of the demand. Hence, the need for broadband wireless networks.

The third-generation systems are expected to be working somewhere from 2000 to 2010. Manufacturers are creating several standards to meet requirements in each sector of the world. To date, the majority of telecom systems are based on CDMA standards. Before the infrastructure rolls out, third generations will be developed on a regional basis. This process is being guided by the effort of the International Telecommunications Union (ITU) to create the IMT 2000 standard. The ITU has produced the IMT 2000 with the aim of combining the regional systems into a unified standard.

Architecture of UMTS/WCDMA Elements

The network for UMTS can be split into three main constituents (Fig. 13.22):

1. Mobile station, called the *user equipment* (UE)

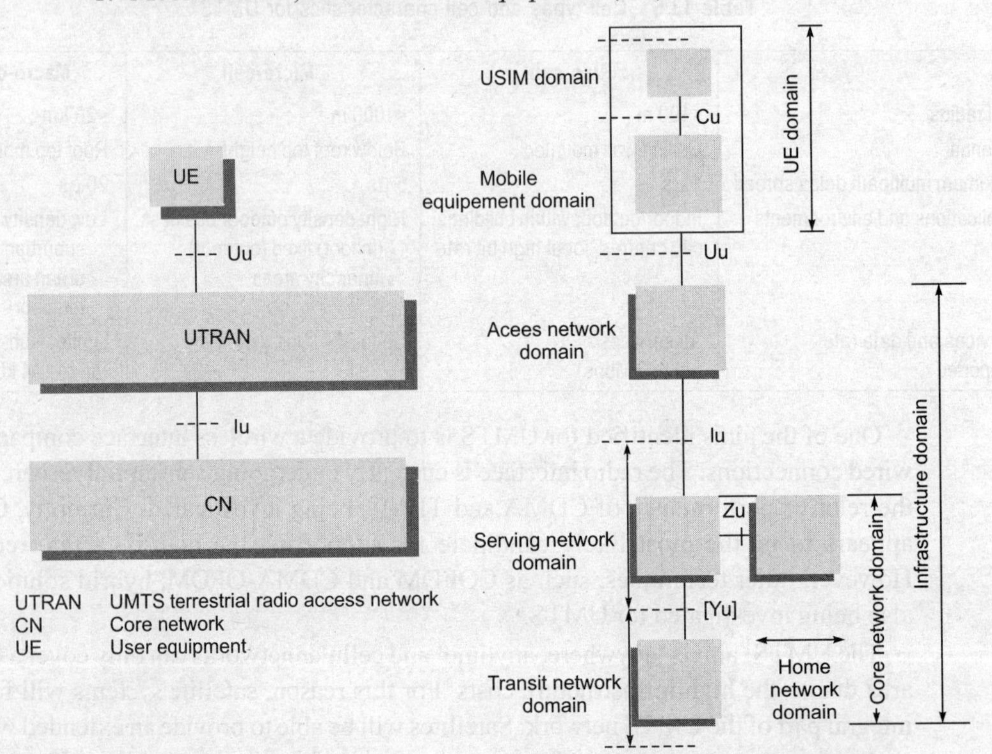

Fig. 13.22 UMTS architecture and domains along with defined interfaces

2. The base station subsystem, known as the radio (access) network subsystem (RNS)
3. The core network (CN)

UE The UE for UMTS / WCDMA is equivalent to the mobile equipment used on GSM networks. Essentially it is the handset, although having access to much higher speed data communications, it can be much more versatile, containing many more applications. It consists of a variety of different elements, including RF circuitry, processing, antenna, and battery.

For UMTS/WCDMA mobiles, as for any system, the circuitry used within the UE can be broadly split into the RF and baseband processing areas. The RF areas handle all elements of the signal, both for the receiver and for the transmitter. One of the major challenges for the RF power amplifier was to reduce the power consumption. The form of modulation used for W-CDMA requires the use of a linear amplifier. These inherently take more current than non-linear amplifiers, which can be used for the form of modulation used on GSM. Accordingly, to maintain battery life, measures were introduced into many of the designs to ensure the optimum efficiency.

The baseband signal processing consists mainly of digital circuitry. This is considerably more complicated than that used in phones for previous generations. Again this has been optimized to reduce the current consumption as far as possible.While current consumption has been minimized as far as possible within the circuitry of the phone, there has been an increase in current drain on the battery. With the users expecting the same lifetime between the charging batteries as experienced on the previous-generation phones, this has necessitated the use of new and improved battery technology. Now Lithium Ion (Li-ion) batteries are used. These phones are to remain small and relatively light while still retaining or even improving the overall life between charges.

The UE also contains a SIM card, although in the case of UMTS, it is termed as universal subscriber identity module (USIM). This is a more advanced version of the SIM card used in GSM and other systems, but embodies the same types of information. It contains the international mobile subscriber identity number (IMSI) as well as the mobile station international ISDN number (MSISDN). Other information that the USIM holds includes the preferred language to enable the correct language information to be displayed, especially when roaming, and a list of preferred and prohibited public land mobile networks (PLMN). The USIM also contains a short message storage area that allows messages to stay with the user even when the phone is changed. Similarly 'phone book' numbers and call information of the numbers of incoming and outgoing calls are stored.

UMTS radio access network subsystem This is the section of the UMTS/ WCDMA network that interfaces to both the UE and the core network. It contains what are roughly equivalent to the BTS and the BSC. Under UMTS terminology, the radio transceiver or the BTS equivalent (in GSM) is known as the Node B. This communicates with the various UEs. It also communicates with the radio network controller (RNC). This is undertaken over an interface known as the Iub. The overall radio access network is known as the *UMTS radio access network* (UTRAN). The RNC component of the radio access network (RAN) connects to the core network.

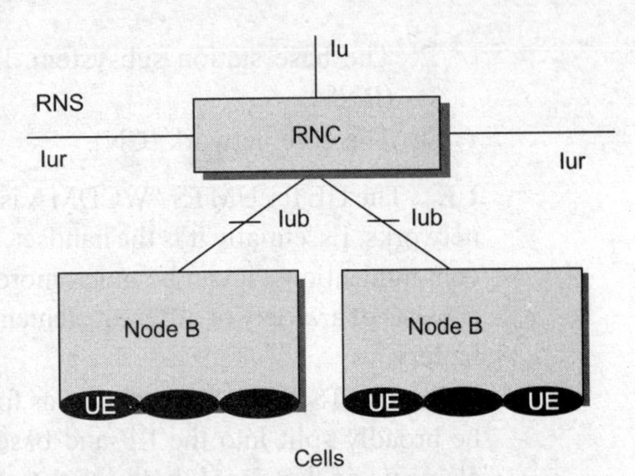

Fig. 13.23 UMTS radio network controller

Core network The core network used for UMTS is based upon the combination of the circuit switched elements used for GSM and the packet switched elements that are used for GPRS and EDGE. Thus, the core network is divided into circuit switched and packet switched domains. Some of the circuit switched elements are mobile switching centre (MSC), visitor location register (VLR) and gateway MSC. Packet switched elements are serving GPRS support node (SGSN), and gateway GPRS support node (GGSN). Some network elements, like EIR, HLR, VLR, and AUC mentioned earlier, are shared by both domains and operate in the same manner that they did with GSM.

The asynchronous transfer mode (ATM) is specified for UMTS core transmission. The architecture of the core network may change when new services and features are introduced. *Number portability database* (NPDB) will be used to enable the subscriber to change network provider while keeping their old phone number. Gateway location register (GLR) may be used to optimize the subscriber handling between network boundaries. The MSC, VLR, and SGSN can merge to become a UMTS MSC.

UMTS/WCDMA Radio or Air Interface

Physical layer within UMTS / WCDMA is totally different to that employed by GSM. It employs a spread spectrum transmission in the form of CDMA. Additionally, it currently uses different frequencies to those allocated for GSM.

There are currently *six bands*, which are specified for use for UMTS/WCDMA although operation on other frequencies is not precluded. However, much of the focus for UMTS is currently on frequency allocations around 2 GHz. At the World Administrative Radio Conference in 1992, the bands 1885–2025 and 2110–2200 MHz were set aside for use on a worldwide basis by administrations wishing to implement International Mobile Telecommunications-2000 (IMT-2000). The aim was that allocating spectrum on a worldwide basis would facilitate easy roaming for UMTS/WCDMA users.

Within these bands, the portions have been reserved for different uses:

- 1920–1980 and 2110–2170 MHz (FDD, WCDMA) Paired uplink and downlink, channel spacing is 5 MHz and raster is 200 kHz. An operator needs three to four channels (2×15 MHz or 2×20 MHz) to be able to build a high-speed, high-capacity network.

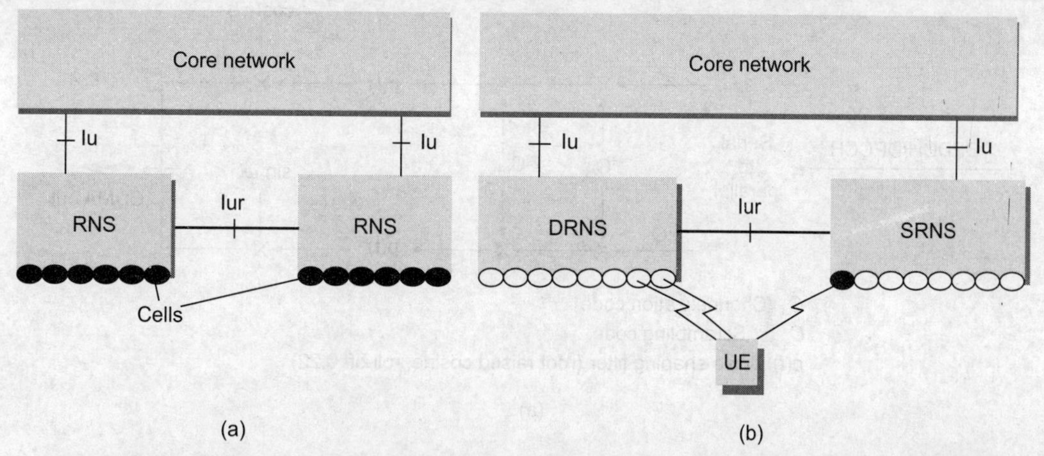

Fig. 13.24 (a) Structure of UTRAN and (b) service and drift radio access networks in UTRAN

- 1900–1920 and 2010–2025 MHz (TDD, TD/CDMA) Unpaired, channel spacing is 5 MHz and raster is 200 kHz. Transmit and receive transmissions are not separated in frequency.
- 1980–2010 and 2170–2200 MHz Satellite uplink and downlink.

Carrier frequencies are designated by UTRA absolute radio frequency channel number (UARFCN). This can be calculated as follows:

$$\text{UARFCN} = 5 \times (\text{frequency in MHz})$$

The channel coding may be convolution, turbo, or service-specific coding. The modulation that is used is different on the uplink and downlink. The downlink uses QPSK for all transport channels [Fig. 13.25(a)]. Data and control streams are de-interleaved by a serial-to-parallel conversion. Each output stream is applied to the I and Q paths. The I and Q paths are spread by a channel-specific code (*channelization code*). These paths are subsequently spread by a cell-specific code (*scrambling code*). Pulse shaping is applied to reduce spectrum occupancy and inter symbol Interference (ISI).

However, the uplink uses two separate channels as shown in Fig. 13.25(b), such that the cycling of the transmitter on and off does not cause interference on the audio lines, a problem that was experienced on GSM. The dual channels (dual channel phase shift keying) are achieved by applying the coded user data [data information (DPDCH)] to the I or In-phase input to the DQPSK modulator, and control data [control information (DPCCH)], which has been encoded using a different (*channelization)* spreading code to the Q or quadrature input to the modulator. A UE-specific spreading code (*scrambling code*) is subsequently applied. The data to be transmitted is encoded using a spreading code particular to a given user, this way only the desired recipient is able to correlate and decode the signal, all other signals appearing as noise. This allows the physical RF channel to be used by several users simultaneously.

The codes required to spread the signal must be orthogonal, if they are to enable multiple users and channels to operate without mutual interference. The codes used in W-CDMA are orthogonal variable spreading factor (OVSF) codes, and they must remain

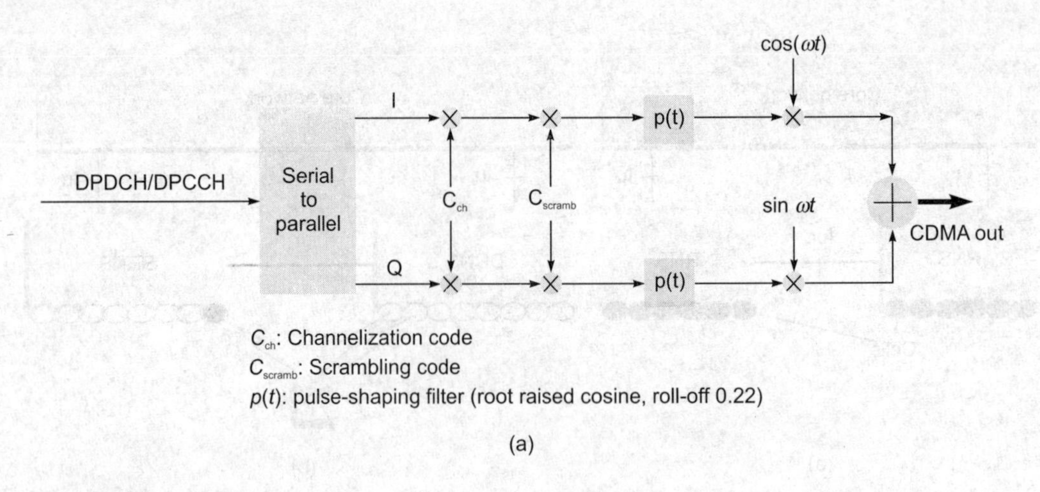

(a)

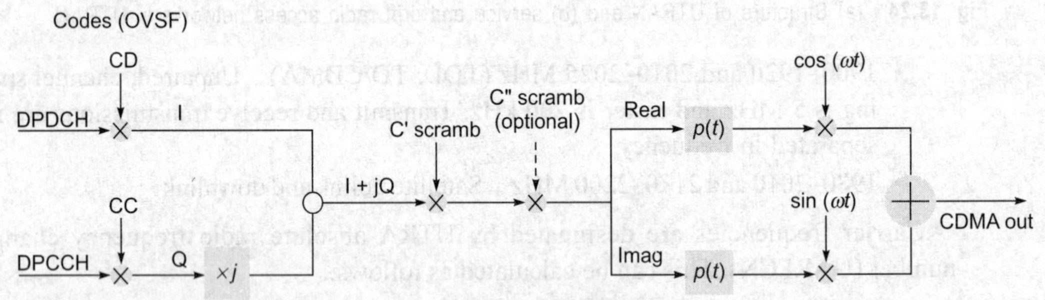

(b)

Fig. 13.25 (a) Downlink spreading and modulation of dedicated channels and (b) uplink spreading and modulation of dedicated channels

synchronous to operate. As it is not possible to retain exact synchronization for this, a second set of scrambling codes is used to ensure that interference does not result. This scrambling code is a PN code. Thus, there are two stages of spreading. The first using the OSVF code and the second using a scrambling PN code. These codes are used to provide different levels of separation. The OVSF spreading codes are used to identify the user services in the uplink and user channels in the downlink whereas the PN code is used to identify the individual Node B or UE.

On the uplink, there is a choice of millions of different PN codes. These are processed to include a masked individual code to identify the UE. As a result, there are more than sufficient codes to accommodate the number of different UEs likely to access a network. For the downlink, a short code is used. There are a total of 512 different codes that can be used, one of which will be assigned to each Node B.

The data of an SSM signal is multiplied with a chip or spreading code to increase the bandwidth of the signal. For WCDMA, each physical channel is spread with a unique and variable spreading sequence. The overall degree of spreading varies to enable the final signal to fill the required channel bandwidth. As the input data rate may vary from one application to the next, the degree of spreading needs to be varied accordingly.

For the downlink, the transmitted symbol rate is 3.84 Msps. As the form of modulation used is QPSK, this enables two bits of information to be transmitted for every symbol, thereby enabling a maximum data rate of twice the symbol rate or 7.68 Mbps. Therefore, if the actual rate of the data to be transmitted is 15 kbps, then a spreading factor of 512 is required to bring the signal up to the required chip rate for transmission in the required bandwidth. If the data to be carried has a higher data rate, then a lower spreading rate is required to balance this out. It is worth remembering that altering the chip rate does alter the processing gain of the overall system and this needs to be accommodated in the signal processing as well. Higher spreading factors are more easily correlated by the receiver and therefore a lower transmit power can be used for the same symbol error rate.

The level of *synchronization* required for the WCDMA system to operate is provided from the primary synchronization channel (PSCH) and the secondary synchronization channel (SSCH) (described later on in the chapter). These channels are treated in a different manner to the normal channels and as a result they are not spread using the OVSFs and PN codes. Instead they are spread using synchronization codes. There are two types of codes that are used: primary code used on the PSCH and secondary code used on the SSCH.

The primary code is the same for all cells and is a 256 chip sequence that is transmitted during the first 256 chips of each time slot. This allows the UE to synchronize with the base station for the time slot. Once the UE has gained time slot synchronization, it only knows the start and stop of the time slot, but it does not know any information about the particular time slot, or the frame. This information is gained using the secondary synchronization codes.

There are a total of sixteen different secondary synchronization codes. One code is sent at the beginning of the time slot, i.e. the first 256 chips. It consists of 15 synchronization codes and there are 64 different scrambling code groups. When received, the UE is able to determine before which synchronization code the overall frame begins. In this way the UE is able to gain complete synchronization.

The scrambling codes in the SSCH also enable the UE to identify, which scrambling code is being used and hence, it can identify the base station. The scrambling codes are divided into 64 code groups, each having eight codes. This means, that after achieving frame synchronization, the UE only has a choice of one in eight codes and it can therefore try to decode the CPICH channel. Once it has achieved this it is able to read the BCH information and achieve better timing and it is able to monitor the PCCPCH.

As with any CDMA system, it is essential that the base station receives all the UEs at approximately the same power level. If not, the UEs that are farther away will be lower in strength than those closer to the Node B and they will not be heard. This effect is referred to as the near–far effect. To overcome this, the Node B instructs those stations closer in to reduce their transmitted power, and those farther away to increase theirs. In this way, all stations will be received at approximately the same strength.

It is also important for the Node Bs to control their power levels effectively. As the signals transmitted by the different Node Bs are not orthogonal to one another, it is possible that signals from different nodes will interfere. Accordingly, their power is also kept to the minimum required by the UEs being served.

As described in Chapter 8, to achieve power control, there are three techniques that are employed—open loop, closed loop, and outer loop (Fig.13.26).

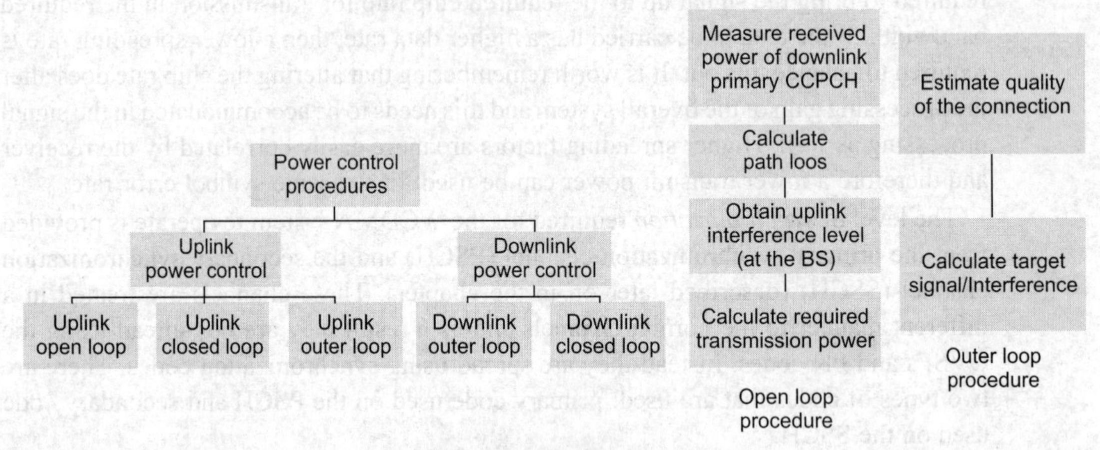

Fig. 13.26 Power control

Open loop techniques are used during the initial access before communication between the UE and Node B has been fully established. It simply operates by making a measurement of the received signal strength and thereby estimating the transmitter power required. As transmit and receive frequencies are different, the path losses in either direction will be different and, therefore, this method cannot be any more than a good estimate.

Once the UE has accessed the system and is in communication with the Node B, closed loop techniques are used. A measurement of the signal strength is taken in each time slot. As a result of this, a power control bit is sent requesting the power to be stepped up or down. This process is undertaken on both the uplink and downlink. The fact that only one bit is assigned to power control means that the power will be continually changing. Once it has reached approximately the right level, it would step up and then down by one level. In practice, the position of the mobile would change, or the path would change as a result of other movements and this would cause the signal level to move. So, the continual change is not a problem.

UMTS TDD and FDD

A communications system requires that communication is possible in both directions: to and from the base station to the remote station. There are a number of ways in which this can be achieved. The most obvious is to transmit on one frequency and receive on another. The frequency difference between the two transmissions being such that the two signals do not interfere. This is known as frequency division duplex (FDD). It is one of the most commonly used schemes and is used by most cellular schemes.

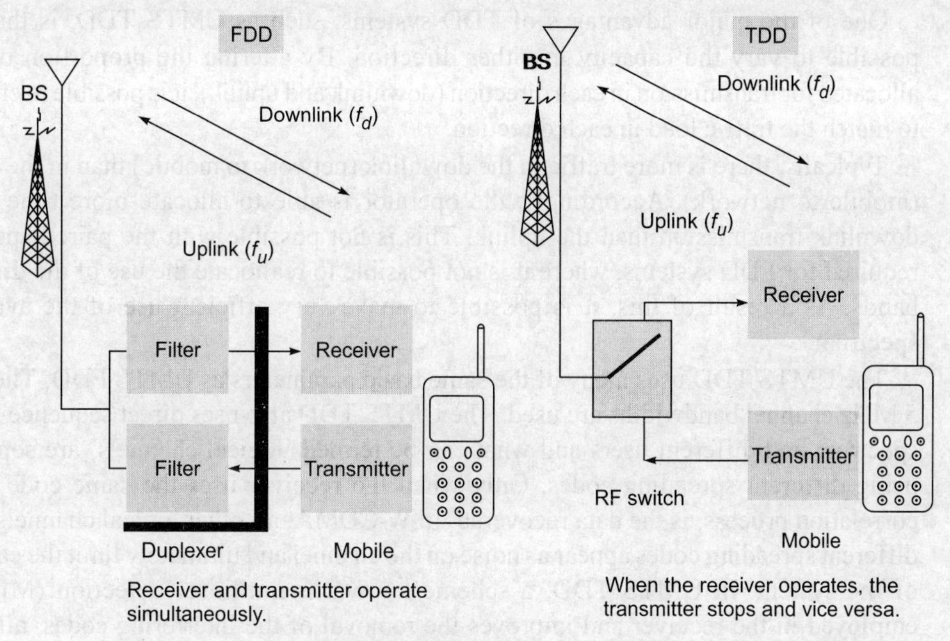

Fig. 13.27 FDD and TDD concept

It is also possible to use a single frequency and rather than using different frequency allocations, use different time allocations. If the transmission times are split into slots, then transmissions in one direction take place in one time slot, and those in the other direction take place in another. This scheme is known as time division duplex (TDD).

A new introduction for UMTS is that there are specifications that allow both FDD and TDD modes. The first modes to be employed are FDD modes where the uplink and downlink are on different frequencies. The spacing between them is 190 MHz for band 1 networks being currently used and rolled out. However, in the TDD mode, the uplink and downlink are split in time with the base stations and then the mobiles transmit alternately on the same frequency. It is particularly suited to a variety of applications. Obviously, the situation where spectrum is limited and paired bands are suitably spaced is not available. It also performs well where small cells are to be used. As a guard time is required between transmit and receive, this will be smaller when transit times are smaller as a result of the shorter distances being covered. A further advantage arises from the fact that it is found that far more data is carried in the downlink as a result of Internet surfing, video downloads, and the like. This means that it is often better to allocate more capacity to the downlink. Where paired spectrum is used, this is not possible. However, when a TDD system is used, it is possible to alter the balance between downlink and uplink transmissions to accommodate this imbalance and thereby improve the efficiency. In this way, TDD systems can be highly efficient when used in picocells for carrying Internet data. The TDD systems have not been widely deployed, but this may occur more in the future. In view of its character, it is often referred to as TD-CDMA (time division CDMA).

One of the major advantages of TDD systems, such as UMTS TDD, is that it is possible to vary the capacity in either direction. By altering the proportion of time allocated for transmission in each direction (downlink and uplink), it is possible to enable it to match the traffic load in each direction.

Typically, there is more traffic in the downlink (network to mobile) than in the uplink (mobile to network). Accordingly, the operator is able to allocate more time to the downlink transmission than the uplink. This is not possible with the paired spectrum required for FDD systems, where it is not possible to reallocate the use of the different bands. As a result of this, it is possible to make very efficient use of the available spectrum.

The UMTS TDD uses many of the same basic parameters as UMTS FDD. The same 5 MHz channel bandwidths are used. The UMTS TDD also uses direct sequence spread spectrum and different users and what can be termed 'logical channels' are separated using different spreading codes. Only when the receiver uses the same code in the correlation process, is the data recovered. In W-CDMA, all other logical channels using different spreading codes appear as noise on the channel and ultimately limit the capacity of the system. In UMTS TDD, a scheme known as multiuser detection (MUD) is employed in the receiver and improves the removal of the interfering codes, allowing higher data rates and capacity.

In addition to the separation of users by using different logical channels as a result of the different spreading codes, further separation between users may be provided by allocating different time slots. There are 15 time slots in UMTS TDD. Of these, three are used for overhead, such as signalling, etc., and this leaves twelve time slots for user traffic. In each time slot, there can be 16 codes. The capacity is allocated to the users on demand, using a two-dimensional matrix of time slots and codes.

As there fifteen time slots (TS) per frame of 10 ms, time slot = 2560 chips

In order for UMTS TDD to achieve the best overall performance, the transport format, i.e. the modulation and forward error correction, can be altered for each user. The schemes are chosen by the network and will depend on the signal characteristics in both directions. Higher order forms of modulation enable higher data speeds to be accommodated, but they are less resilient to noise and interference, and this means that the higher data rate modulation schemes are only used when signal strengths are high. Additionally the levels of forward error correction can be changed when the errors are likely, i.e. when signal strengths are low or interference levels are high, Similarly, higher levels of forward error correction are needed under single levels and lower levels require that additional data is to be sent. This slows the payload transfer rate. Thus, it is possible to achieve much higher data transfer rates when signals are strong and interference levels are low.

Standard allocations of radio spectrum have been made for 3G telecommunications systems in most countries around the globe. The allocations are mentioned along with UMTS radio or air interface previously. In addition to this, there are some other allocations around 3 GHz.

The UMTS TDD is able to support high peak data rates. Release 5 of the UMTS standard provides high-speed downlink packet access (HSDPA). The scheme allows the use of a higher order modulation scheme called 16-QAM, which enables peak rates of 10 Mbps per sector in commercial deployments. The next release increases the modulation to 64QAM and introduces intercell interference cancellation (called generalized MUD) and MIMO (Chapter 15). In combination, these increase the peak rate to 31 Mbps per sector.

The UMTS TDD, while not as widely deployed as UMTS FDD, nevertheless offers significant advantages for a number of applications. While currently being used for mobile broadband, it appears as if it can also serve to provide mobile TV and other data in a field where new methods of transport are being sought.

UMTS/WCDMA Channels

The data carried by the UMTS/WCDMA transmissions is organized into channels, frames, and slots. In this way, all the payload data as well as the control data can be carried in an efficient manner. UMTS uses CDMA techniques (such as WCDMA) as its multiple access technology, but it additionally uses time division techniques with a slot and frame structure to provide the full channel structure.

The channels carried are classified into three categories: logical, transport, and physical channels. The logical channels define the way in which the data will be transferred, the transport channel along with the logical channel again defines the way in which the data is transferred, and the physical channel carries the payload data and governs the physical characteristics of the signal.

The channels are organized such that the logical channels are related to what is transported, whereas the physical layer transport channels deal with how is transported and with what characteristics. The MAC layer provides data transfer services on logical channels. A set of logical channel types is defined for different kinds of data transfer services.

FDD mode

Downlink physical channels (Fig. 13.28)–FDD mode A channel is divided into 10 ms frames, each of which has sixteen time slots, each of 625 μs length. On the downlink,

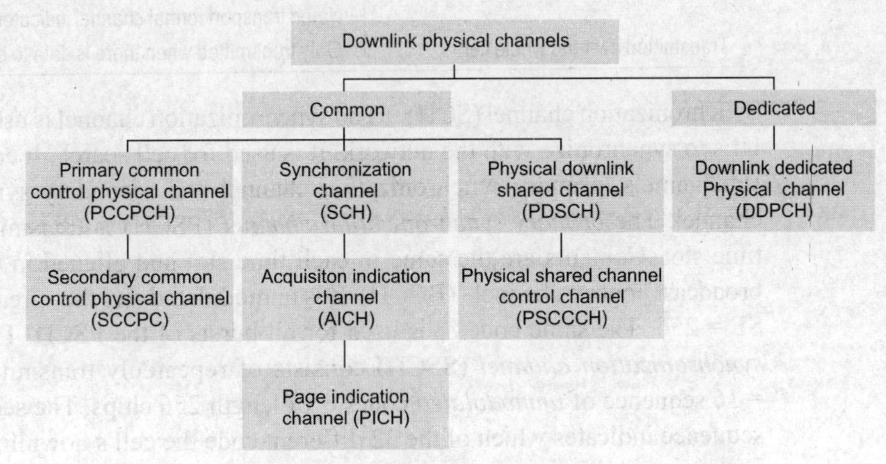

Fig. 13.28 UMTS downlink physical channels manifestation

the time is further subdivided so that the time slots contain fields that contain either user the data or control messages.

$$\text{Reference parameter: } k = 0, 1, 2, 3, 4, 5, 6$$

$$\text{Total quantity of bits} = 20 \times 2^k,$$

Here k is related to the spreading factor (SF) of the physical channel by means of the expression $\frac{512}{2^k}$.

1. Primary common control physical channel (PCCPCH) [Fig. 13.29(a)] This channel continuously broadcasts system identification and access control information. It is used to carry the broadcast control channel (BCH), fixed rate = 32 kbps, fixed spreading factor = 256. It is not transmitted during the first 256 chips of each time slot. This period is left to the primary and secondary synchronization channels.
2. Secondary common control physical channel (SCCPCH) [Fig. 13.29(a)] This channel carries the forward access channel (FACH) providing control information and paging channel (PCH) with messages for UEs that are registered on the network. The number of SCCPCHs depends on the cell traffic. There are two types of SCCPCH: with transport format channel indicator (TFCI) and without TFCI. It is the UTRAN that determines if the TFCI should be transmitted. It is mandatory for the UE to be able to support the TFCI. Spreading factor (SF) = $256/2^k$ (k = 0,…,6) and data rates are between 32 and 2048 kbps. The SCCPCHs may be transmitted on narrow lobes pointed to the target UE channel (only valid for a Secondary CCPCH carrying the FACH). A few differences are listed out in Table 13.7.

Table 13.7 Comparison of PCCPCH and SCCPCH

PCCPCH	SCCPCH
A fixed rate (32 kbps)	Can support variable rate, based on the value of the associated transport format channel indicator (TFCI)
Transmitted over the whole cell	Only transmitted when there is data to send

3. Synchronization channel (SCH) The synchronization channel is used in allowing UEs to synchronize with the network. It is used for cell search. It consists of two subchannels: primary synchronization channel and secondary synchronization channel. The *primary synchronization channel* (PSCH) is transmitted once per time slot. Contents are the same in each time slot and aligned in time with the broadcast control channel (BCCH). It is unmodulated and the spreading factor is SF = 256. The same code c_p is used for all bursts of the PSCH. The *secondary synchronization channel* (SSCH) consists of repeatedly transmitting a length = 16 sequence of *unmodulated* codes c_s of length 256 chips. The secondary SCH sequence indicates which of the 32 different code the cell's downlink scrambling

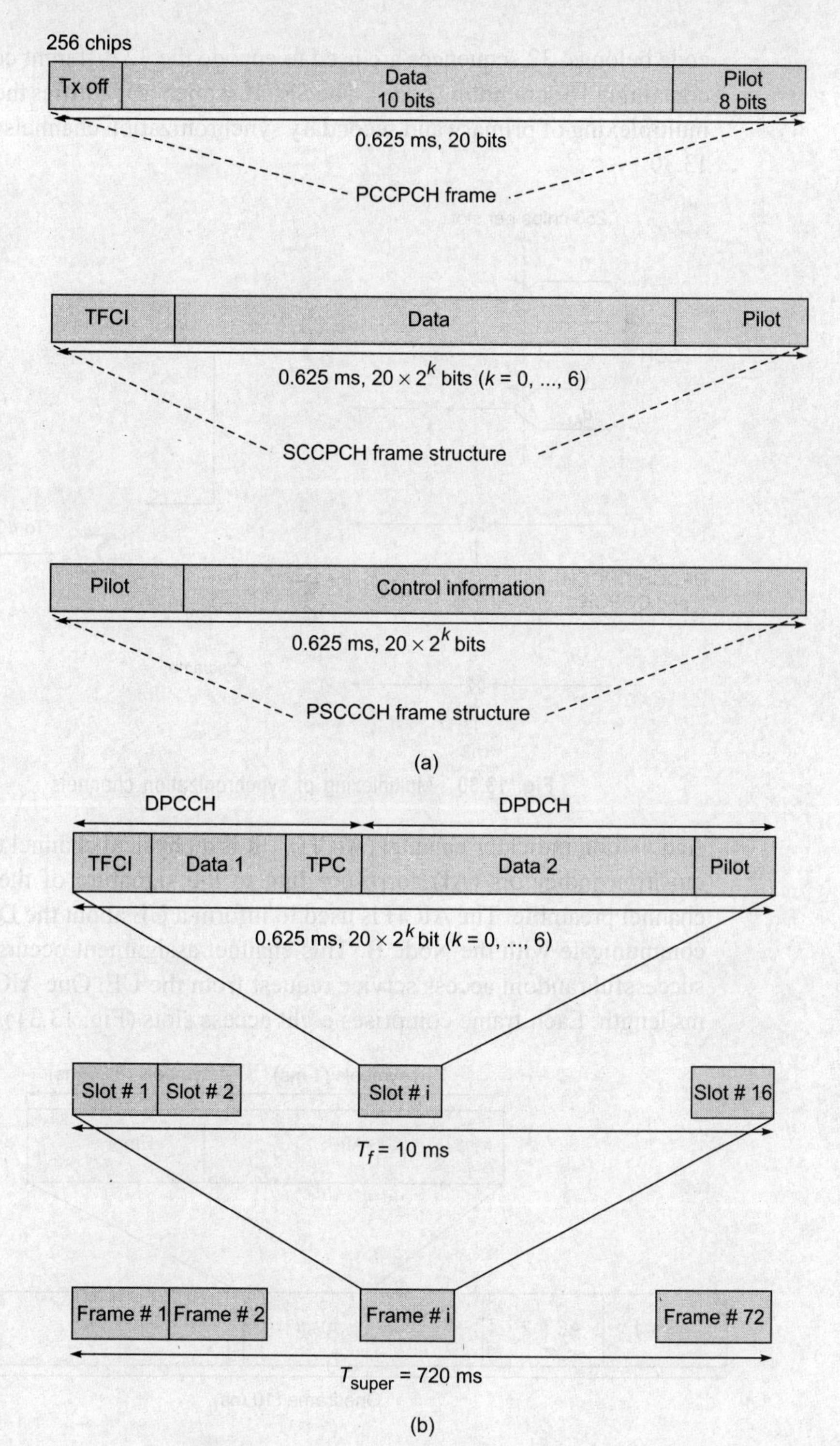

Fig. 13.29 (a) PCCPCH, SCCPCH, and PSCCCH frame structures and (b) DDPCH frame structure (Number of frames and number of slots will remain same and only the frame format for different channels will change.)

code belongs. 32 sequences are used to encode the 32 different code groups each containing 16 scrambling codes. The SSCH sequence identifies the cell group. The multiplexing of primary and secondary synchronization channels is shown in Fig. 13.30.

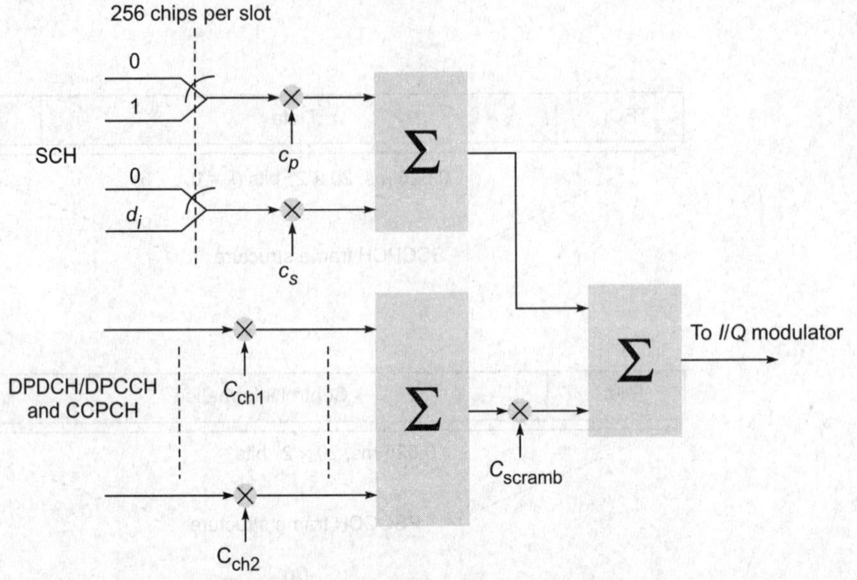

Fig. 13.30 Multiplexing of synchronization channels

4. Acquisition indicator channel (AICH) It is a physical channel used to carry acquisition indicators (AI) corresponding to the signature of the random access channel preamble. The AICH is used to inform a UE about the DCH it can use to communicate with the Node B. This channel assignment occurs as a result of a successful random access service request from the UE. One AICH frame has 10 ms length. Each frame comprises eight access slots (Fig. 13.31).

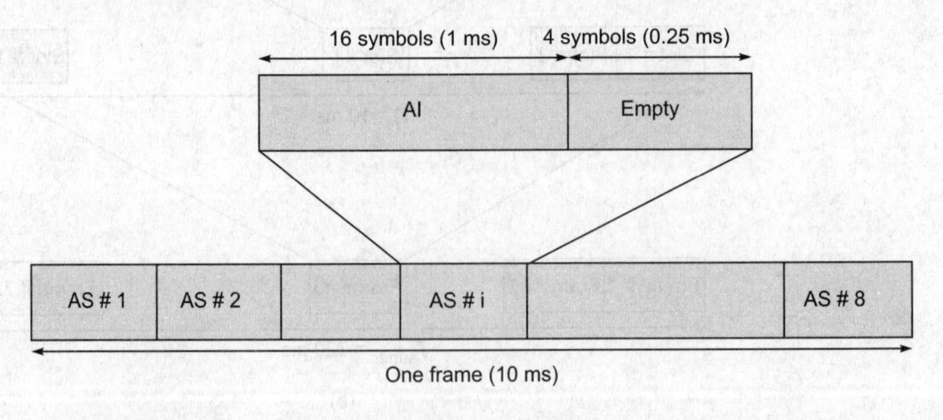

Fig. 13.31 AICH format

5. Paging indication channel (PICH) The PICH provides page indicators (PI) to the user equipment. This channel provides the information to the UE to be able to operate its sleep mode to conserve its battery when listening on the PCH. As the UE needs to know when to monitor the PCH, data is provided on the PICH to assign a UE a paging repetition ratio to enable it to determine how often it needs to 'wake up' and listen to the PCH. One PICH frame has 10 ms length. Each frame comprises eight access slots and 5, 10, or 20 page indicators may be included in an access slot (Fig. 13.32).

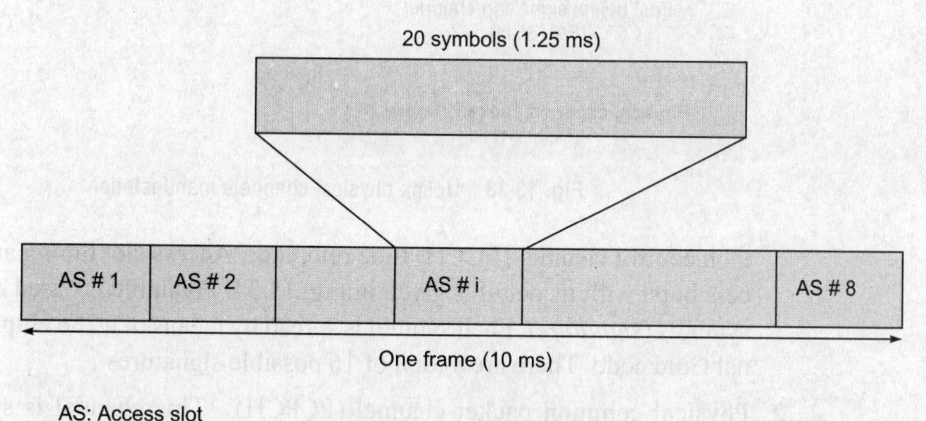

Fig. 13.32 PICH format

6 Physical downlink shared channel (PDSCH) This channel shares control information to UEs within the coverage area of the Node B. It is shared by several UE by means of code multiplexing. It is always associated with another physical channel. Physical shared channel control channel (PSCCCH) contains information shared by several UE [Fig.13.29(a)]. Its control information field includes TPC to be applied by the UE pool. Detailed structure is still under development.
7. Dedicated physical data channel (DPDCH) [Fig. 13.29(b)] This channel is used to transfer user data.
8. Dedicated physical control channel (DPCCH) [Fig. 13.29(b)] This channel carries control information to and from the UE. In both directions the channel carries pilot bits and TFCI. The downlink channel also includes the transmit power control and FBI bits.

Uplink physical channels (Fig. 13.33)–FDD mode On the uplink, dual channel modulation is used so that both data and control are transmitted simultaneously. Here the control elements contain a pilot signal, transport format combination identifier (TFCI), feedback information (FBI), and transmission power control (TPC).

1. Physical random access channel (PRACH) This channel enables the UE to transmit random access bursts in an attempt to access a network/fixed network. Access method is based on the slotted ALOHA principle. Access time slots have 1.25 ms offset. Access time slots boundaries are referred to the broadcast com-

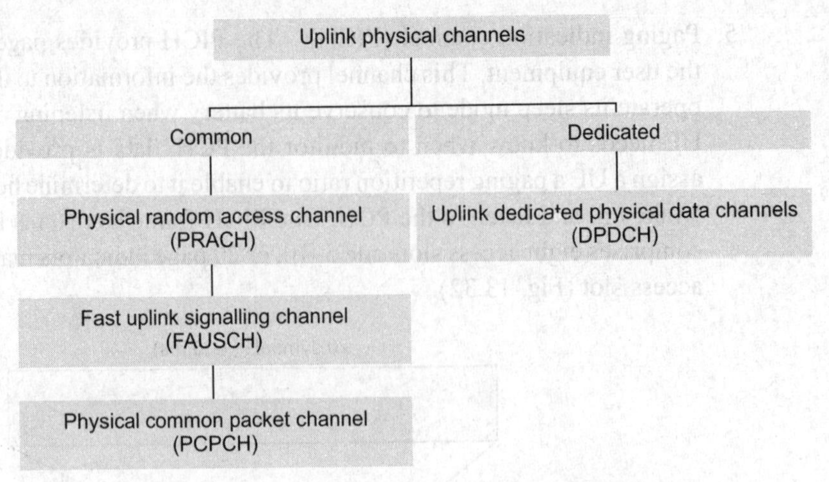

Fig. 13.33 Uplink physical channels manifestation

mon control channel (BCCH) time reference. Access slot timing and random access burst with its detail is given in Fig. 13.34. Preamble is based on 16 complex symbols (*signature)*. Each symbol is spread by means of a 256 chips real orthogonal Gold code. There are a total of 16 possible signatures.

2. Physical common packet channel (PCPCH) This channel is specifically intended to carry packet data. In operation, the UE monitors the system to check if it is busy and if not, it then transmits a brief access burst. This is retransmitted if no acknowledgement is gained with a slight increase in power each time. Once the Node B acknowledges the request, the data is transmitted on the channel.
3. Dedicated physical data channel (DPDCH) This channel is used to transfer user data (Fig. 13.35).
4. Dedicated physical control channel (DPCCH) This channel carries control information to and from the UE. In both directions, the channel carries pilot bits and TFCI. The downlink channel also includes the transmit power control and FBI bits. The TFCI field is optional, UTRAN may request its presence to the user equipment. If it is present, it is represented by a 32-bits word transmitted in each frame. The TFCI value may be negotiated between UTRAN and UE on a frame-to-frame basis (Fig. 13.35).

TDD mode The services provided by layer 1 to layer 2 are called *transport channels* (Fig. 13.37). The part of layer 2 that directly interfaces with layer 1 is the *medium access control* (MAC) sublayer. The upper part of layer 2 that directly interfaces with the MAC sublayer is called *radio link control* (RLC) sublayer. The services provided by the MAC sublayer to the radio link control sublayers are called *logical channels*. Transport channel identification may be based on the physical channel used for data transport. In this case, the physical channel properties are used for identification of the following:

- Carrier frequency and spreading code (FDD)
- Carrier frequency, spreading code and time slot (TDD)

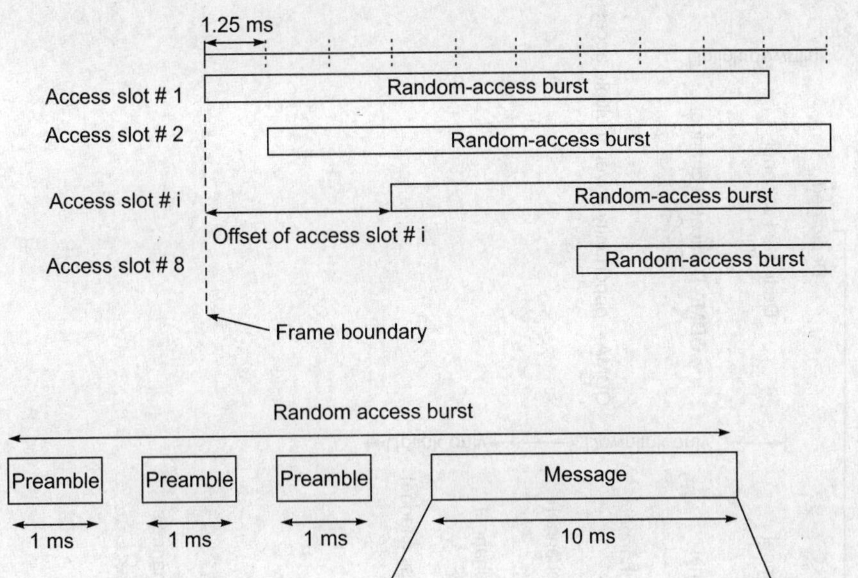

Fig. 13.34 Access slot timing and random access burst structure correlated with message timing

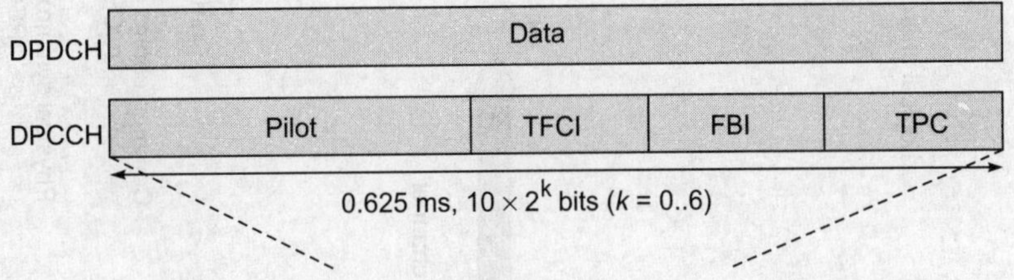

Fig. 13.35 DPDCH frame within the slot time within the superframe

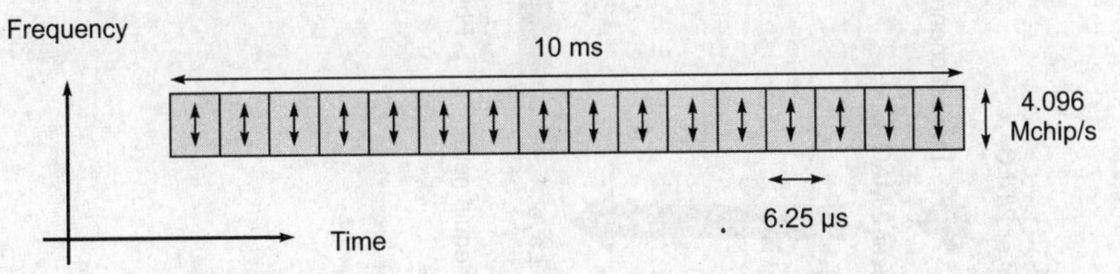

Fig.13.36 TDD frame structure (Each time slot may be used in the Uplink or downlink. At least one TS must be used in uplink and at least one TS must be used in downlink.)

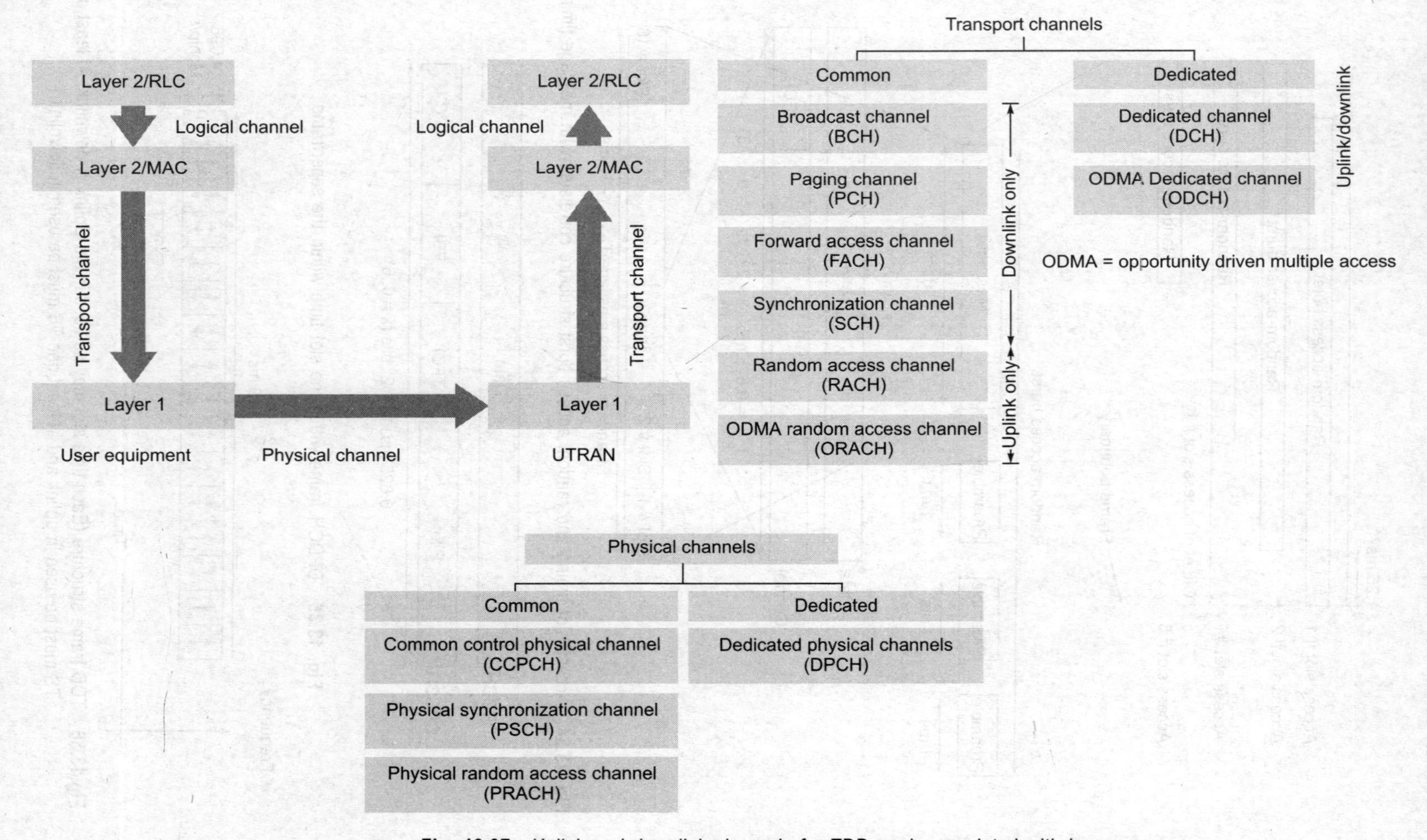

Fig. 13.37 Uplink and downlink channels for TDD mode correlated with layers

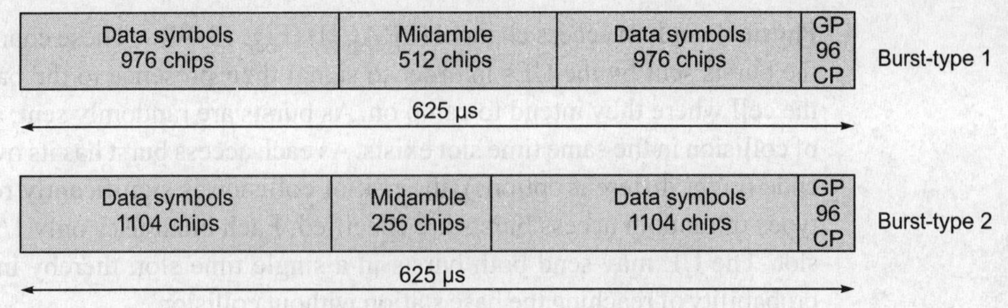

Fig. 13.38 UMTS TDD burst type

This class of channels is called *dedicated channel*. If the in-band identification is there, then the class of channels is called *common channel*.

Physical channels–TDD mode

1. Dedicated physical channel (DPCH) We consider here the DPCH burst-type applications. The burst type one has a longer midamble. Therefore, it allows a more accurate multiuser detection, as is required in the uplink. The burst type two has a shorter midamble, and, therefore, higher user data throughput. It is more suitable for the downlink that will require more capacity in, for example. wireless Internet applications.
2. Common control physical channel (CCPCH) This channel is used to transport the BCH, PCH, and FACH. The burst type is the same as for the dedicated physical channels (DPCHs).
3. Physical synchronization channel (PSCH) It has a similar structure as in the FDD mode. Two code sequences are periodically broadcasted: Primary synchronization code (c_p) and secondary synchronization code (c_s). In each frame, two time slots are allocated for the primary synchronization channel: TS0 and TS7 [Fig. 13.39].

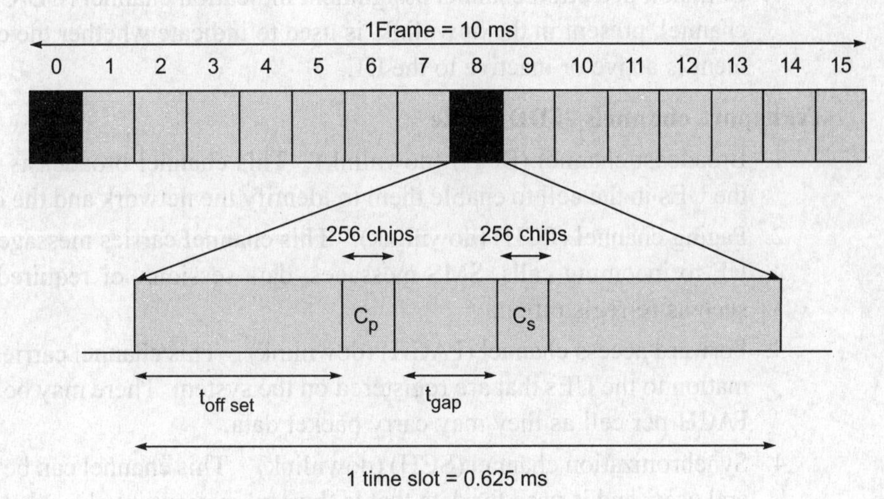

Fig. 13.39 PSCH internal burst structure

4. Physical random access channel (PRACH) (Fig. 13.40) These channels support the bursts sent by the UEs in order to signal their presence to the base station of the cell where they intend to camp on. As bursts are randomly sent, a certain risk of collision in the same time slot exists. As each access burst has its own spreading code (eight different options), the risk of collision is significantly reduced. Two types of random access bursts are specified. Each burst uses only 1/2 of the time slot. The UE may send both bursts in a single time slot, thereby increasing the probability of reaching the base station without collision.

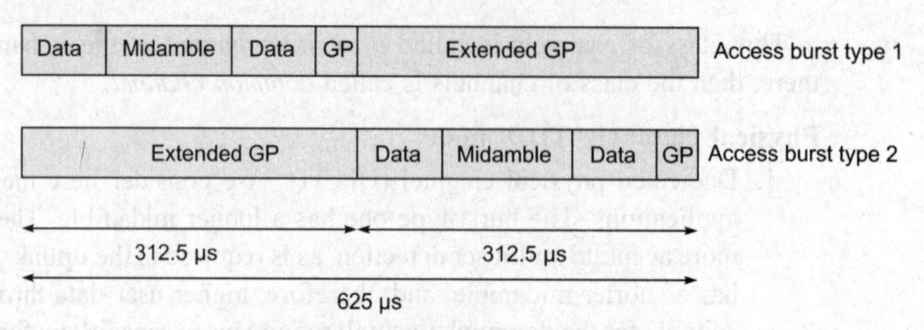

Fig. 13.40 PRACH access bursts

Some other channels are discussed below.

5. Common pilot channel (CPCH) This channel is transmitted by every Node B so that the UEs are able estimate the timing for signal demodulation. Additionally, they can be used as a beacon for the UE to determine the best cell with which to communicate.
6. CPCH status indication channel (CSICH) This channel, which only appears in the downlink, carries the status of the CPCH and may also be used to carry some intermittent or 'bursty' data. It works in a similar fashion to PICH.
7. Collision detection/channel assignment indication channel (CD/CA-ICH) This channel, present in the downlink, is used to indicate whether the channel assignment is active or inactive to the UE.

Transport channels–TDD mode

1. Broadcast channel (BCH) (downlink) This channel broadcasts information to the UEs in the cell to enable them to identify the network and the cell.
2. Paging channel (PCH) (downlink) This channel carries messages that alert the UE to incoming calls, SMS messages, data sessions, or required maintenance, such as re-registration.
3. Forward access channel (FACH) (downlink) This channel carries data or information to the UEs that are registered on the system. There may be more than one FACH per cell as they may carry packet data.
4. Synchronization channel (SCH) (downlink) This channel can be shared by several users and is used for data that is 'bursty' in nature such as that obtained from web browsing, etc.

5. Random access channel (RACH) (uplink) This channel carries requests for service from UEs trying to access the system.
6. Uplink common packet channel (CPCH) (uplink) This channel provides additional capability beyond that of the RACH and for fast power control.
7. Dedicated transport channel (DCH) (uplink and downlink) This is used to transfer data to a particular UE. Each UE has its own DCH in each direction.

Logical channels–TDD mode

1. Broadcast control channel (BCCH) (downlink) This channel broadcasts information to UEs relevant to the cell, such as radio channels of neighbouring cells, etc.
2. Paging control channel (PCCH) (downlink) This channel is associated with the PICH and is used for paging messages and notification information.
3. Common traffic channel (CTCH) (downlink) A unidirectional channel used to transfer dedicated user information to a group of UEs.
4. Dedicated control channel (DCCH) (uplink and downlink) This channel is used to carry dedicated control information in both directions.
5. Common control channel (CCCH) (uplink and downlink) This bi-directional channel is used to transfer control information.
6. Shared channel control Channel (SHCCH) This channel is bi-directional and only found in the TDD form of WCDMA / UMTS, where it is used to transport shared channel control information.
7. Dedicated traffic channel (DTCH) (uplink and downlink) This is a bi-directional channel used to carry user data or traffic.

UMTS/WCDMA Packet Handling, Power Saving, and Handover

This section deals with three independent elements of the system, namely the way packet data is carried, the way power saving is accomplished, and handover, including hard, soft, and softer handover.

Packet data is an increasingly important element within mobile phone applications. The WCDMA is able to carry data in this format in two ways. The first is for short data packets to be appended directly to a random access burst. This method is called common channel packet transmission and it is used for short infrequent packets. It is preferable to transmit short packets in this manner because the link maintenance needed for a dedicated channel would lead to an unacceptable overhead. Additionally, the delay in setting up a packet data channel and transferring the operational mode to this format is avoided.

Larger or more frequent packets have to be transmitted on a dedicated channel. A large single packet is transmitted using a single-packet scheme, where the dedicated channel is released immediately after the packet has been transmitted. In a multi packet scheme the dedicated channel is maintained by transmitting power control and synchronization information between subsequent packets.

Speech coding in UMTS uses a variety of source rates. As a result, a variety of vocoders are employed including the GSM EFR vocoder. When a variety of rates are available, a system known as adaptive multirate (AMR) may be employed, where the rate is chosen according to the system capacity and requirements. This scheme is the same as that used on GSM. The actual vocoder that is chosen is governed by the system.

The speech coding process can be combined with a voice activity detector. This is particularly useful because during normal conversations there are long periods of inactivity. In the same way that discontinuous transmission is applied to GSM, the same is also true for UMTS. It employs the same technique of inserting background noise when there is no speech as when the discontinuous transmission cuts out the transmission, no background noise would otherwise be heard and this can be very disconcerting for the listener.

One of the big issues with mobile phones, in general, is that of battery life. It is one of the key differentiators that people take into account when buying a phone and this gives a measure of its importance. Taking this into consideration while developing the UMTS/WCDMA standard, a discontinuous reception or sleep mode was introduced. This mode allows several non-essential segments of the phone circuitry to power down during periods when paging messages will not be received.

To enable this facility to be introduced into the UMTS UE circuitry, the paging channel is divided into groups or subchannels. The actual number of the paging subchannels to be used by a particular UE is assigned by the network. In this way, the UE only has to listen for part of the time. To achieve this, the paging indicator channel is split into 10 ms frames, each of which comprises 300 bits—288 for paging data and 12 idle bits. At the beginning of each paging channel frame, there is a paging indicator (PI) that identifies the paging group being transmitted. By synchronizing with the paging channels being transmitted, it is able to turn the receiver on only when it needs to monitor the paging channel. As the receiver, with its RF circuitry, will consume power, savings can be made by switching it off.

Within UMTS, handover follows many of the similar concepts to those used for other CDMA systems. There are three basic types of handover: hard, soft, and softer. Hard handover is interfrequency handover, while soft and softer handovers are intrafrequency. All three types are used but under different circumstances.

Hard handover is like that used for the previous generations of systems. Here as the UE moves out of range of one Node B, the call has to be handed over to another frequency channel. In this instance, simultaneous reception of both channels is not possible and there must be a physical break of one channel.

Soft handover is a technique that was not available on the previous generations of mobile phone systems. With CDMA systems, it is possible to have adjacent cell sites on the same frequency and, as a result, it is possible for the UE to receive the signals from two adjacent cells at once, and they are also able to receive the signals from the UE. These base stations form the active set. All active set base stations use the same RF carrier frequency. Each active set base station maintains its own scrambling code. When this occurs and the handover is affected, it is known as soft handover.

Softer handover is the special case of soft handover in which active set base stations are a part of the same physical site. Softer handover allows more efficient combining implementations than soft handover (e.g. to use maximal ratio combining instead of selection combining).

The decisions about handover are generally handled by the RNC. It continually monitors information regarding the signals being received by both the UE and Node B and when a particular link has fallen below a given level and another better radio channel is available, it initiates a handover. As part of this monitoring process, the UE measures the received signal code power (RSCP) and received signal strength indicator (RSSI) and the information is then returned to the Node B and hence to the RNC on the uplink control channel.

If a hard handover is required, then the RNC will instruct the UE to adopt a compressed mode, allowing for short time intervals in which the UE is able to measure the channel quality of other radio channels.

High-Speed Uplink Packet Access (HSUPA)

Work is in progress on the developing of standards for high speed uplink packet access (HSUPA) to improve the data rates on the 3G W-CDMA mobile or cell phone standard. With the cellular telecommunications standards established and work progressing to introduce the equipment for high-speed downlink packet access (HSDPA) (described next), the standards are now starting to be developed to enable the uplink from the mobile handset or UE to the base station (Node B), to be able to handle data at similar speeds. This is known as HSUPA and it will enable new features including full video conferencing to be introduced.

For most applications, including Internet surfing, emails, video downloads, and the like, data flowing in the downlink is far greater than the uplink. However, for applications, such as video conferencing, data flows equally in both directions. It is anticipated that video conferencing will become an increasing requirement and a significant revenue generator for the operators in the near future. To enable high-quality video to be passed, it is, therefore, essential to ensure that the uplink performs as fast as the downlink. Although it is very early days for the standards, work on HSUPA has already started under the auspices of 3GPP, the body that controls the W-CDMA standards.

It is anticipated that many of the same techniques used in HSDPA will be used for HSUPA, but these still need to be formalized. Accordingly, it is expected that adaptive modulation, along with hybrid automatic repeat request (HARQ) will be used. Improvements in the base station similar to those employed on HSDPA are also likely.

Originally, W-CDMA had used only QPSK as the modulation scheme, however, under the new HSUPA system,16-QAM, which can carry a higher data rate, but is less resilient to noise is also used when the link is sufficiently robust. The robustness of the channel and its suitability to use 16-QAM instead of QPSK is determined by analysing information fed back about a variety of parameters. These include details of the channel physical layer conditions, power control, quality of service (QoS), and information specific to HSDPA.

Fast HARQ has also been implemented along with multi code operation and this eliminates the need for a variable spreading factor. By using these approaches, all users, whether near or far from the base station are able to receive the optimum available data rate. It is also likely that within the HSUPA upgrades there will be an additional uplink data channel introduced comparable to that in the downlink.

Many manufacturers are working on implementing HSDPA, with initial equipment deliveries anticipated in 2005. Now, with HSUPA in people's sight, this should be implemented in the following years, making a far faster 3G system than is currently available.

High-Speed Downlink Packet Access (HSDPA)

Improvements and enhancements are being made to the wideband CDMA or UMTS 3G telecommunications system. The HSDPA, the new technology, promises to increase the download data rate five-fold. In addition to this, it also provides a two-fold increase in base station capacity.

The introduction of HSDPA technology has come about as a result of the need to drive down costs as well as increasing the data rates possible. Current trends show the volume of packet switched data rising and overtaking the more traditional circuit switched traffic. By adopting a packet based approach to the delivery of digital content as well as IP based person-to-person digitized voice, a single session can be used for multiple purposes and this can be used to drive revenues upwards. With this approach in mind, the use of HSDPA is a key element in providing the user with a better service as well as increasing revenues as a result of increased capacity and usage for the service providers.

Release 4 of the 3GPP W-CDMA standard provided the efficient IP support to enable provision of services through an all IP core network. Then Release 5 included HSDPA itself with support for the packet based multimedia services. A further enhancement MIMO is then be contained within Release 6. As HSDPA needs to work alongside the original Release 99 systems, the new technology is completely backwards compatible.

One of the keys to the operation of HSDPA is the use of an additional form of modulation. Originally, W-CDMA had used only QPSK as the modulation scheme; however, under the new system,16-QAM, which can carry a higher data rate but is less resilient to noise, is also used when the link is sufficiently robust. The robustness of the channel and its suitability to use 16-QAM instead of QPSK is determined by analysing information fed back about a variety of parameters. These include details of the channel physical layer conditions, power control, quality of service, and information specific to HSDPA.

Fast HARQ has also been implemented along with multicode operation and this eliminates the need for a variable spreading factor. By using these approaches, all users, whether near or far from the base station, are able to receive the optimum available data rate.

Further advances have been made in the area of scheduling. By moving more intelligence into the base station, data traffic scheduling can be achieved in a more dynamic fashion. This enables variations arising from fast fading to be accommodated and the cell is even able to allocate much of the cell capacity for a short period of time to a particular user. In this way, the user is able to receive the data as fast as conditions allow.

A further channel, known as the high-speed downlink shared channel (HS-DSCH), has been introduced. The W-CDMA normally carries data over dedicated transport channels, several of which are multiplexed onto one RF carrier. This approach has been adopted because it provides the optimum performance with continuous user data. Under the new scheme, the 'bursty' nature of the data has been accounted for and more efficient use of the available spectrum has been made.

Using the new HSDPA scheme, it will be possible to achieve peak data rates of 10 Mbps within the five MHz channel bandwidth offered under W-CDMA. The new scheme has a number of benefits. It improves the overall network packet data capacity, improves the spectral efficiency, and will enable networks to achieve a lower delivery cost per bit. The users will see higher data speeds as well as shorter service response times and better availability of services. However, new mobile designs will need to be able to handle the increased data throughput rates. Reports indicate that handsets will need to have at least double the memory currently contained within handsets. Nevertheless, the advantages of HSDPA mean that it will be widely used as networks are upgraded and new phones are introduced.

The QoS for different applications in UMTS also needs separate attention and can be self-studied.

13.8 MOBILE SATELLITE COMMUNICATION

In mobile satellite communication, the satellite and mobile systems are combined as shown in Fig. 13.41. Antenna spot beams can be used for the frequency reuse. The beam projection areas are treated as the cells. But here instead of the normal transmitter and receiver, which are used in GSM or CDMA-based mobile systems, small satellite earth

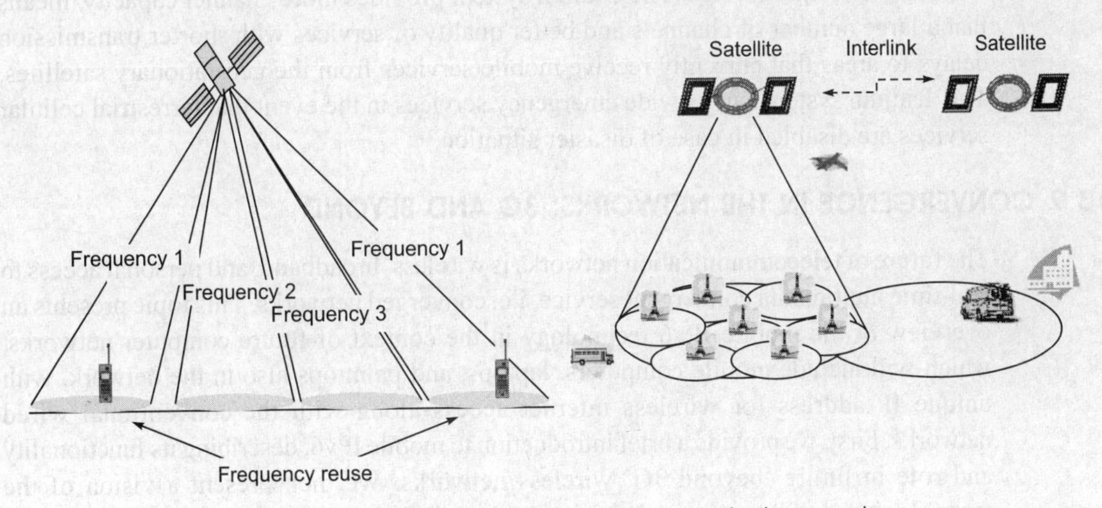

Fig. 13.41 Mobile satellite communication scenario

stations are there to send and receive the signals from the satellite. These signals can be further routed to the MSC, which will then take care of further routing of the signals.

The system offers communication services to mobile users operating within a predefined service area. The users communicate with other mobiles or with fixed users through one of the visible satellites. The users in the fixed network are accessed through large fixed stations called gateways, which carries a large amount of traffic, maybe 100 to 200 E, whereas the mobiles are small portable units capable of supporting less number of channels. Mobile terminals may be mounted on the top of the vehicles like ships, aircrafts, trucks, or maybe carried by individuals.

There may be one or more space segments that may consist of one satellite or a group of interlinked satellites. Depending on the service area and application, the space segments can be utilized. Telemetry and control ground stations, used for monitoring and controlling satellites, constitute a part of the space segment. To simplify the mobile terminals, complexity is shifted to the space segment and hence the satellites used for this system tend to be complex. A 3–3.5 kW geostationary satellite with 5–10 spot beams typically exists for 2G systems. The 5 kW geosatellite with 100–200 spot beams is a typical case of 3G systems. As the geosynchronous satellite is synchronized with the earth in terms of angular velocity, it comprises one or more static footprints of the spot beam. Therefore, fixed stations can operate with a single antenna with minimal tracking and the network topology of the satellites is also simple.

According to the static spot beams, the entries are there in HLR and VLR. Hence, the identification the of mobile unit is not much difficult. The main problem in mobile satellite communication is due to the difficult propagation environment and small mobile terminal size.

The Iridium system cannot become the replacement for the existing terrestrial cellular systems. However, it can be considered as an extension of existing wireless systems to provide mobile services to remote and populated areas, which are not covered by terrestrial cellular services. The Iridium system provides more channel capacity, means that a large number of channels and better quality of services with shorter transmission delays to areas that currently receive mobile services from the geostationary satellites. The Iridium system can provide emergency services in the event that terrestrial cellular services are disabled in case of disaster situation.

13.9 CONVERGENCE IN THE NETWORKS: 3G AND BEYOND

The future of telecommunication networks is wireless, broadband, and personal access to real-time multimedia convergent service, i.e. converged networks. This topic presents an overview of the mobile IPv6 technology in the context of future computer networks, which will include mobile computers, laptops, and palmtops also in the network, with unique IP address for wireless internet access along with the conventional wired networks. First, we provide a brief introduction to mobile IPv6, describing its functionality and role in future 'beyond 3G' wireless networks. We then present a vision of the technologies, architectures and services expected from future wireless networks, also referred to as wireless networks 'beyond 3G'.

The Internet protocol (IP) is becoming the natural language of telecommunication systems. The promise of ubiquity in future telecommunication networks will be achieved by using IP as a convergence layer, together with wireless access technologies and mobility support. This is nowadays more or less a consensual issue. Nevertheless, in order to meet this goal, there has to be extensive work in improving current wireless and IP technologies and in making them cooperatively support QoS criteria in an efficient way. The development of the new Internet protocol, IPv6, began in 1992 with the purpose of eliminating IPv4's shortcomings. The (IPv4 can be studied from any book on computer networks). These were mainly found to lack address space for a booming web-based Internet, header complexity, overhead and extension problems, and the lack of automatic address configuration. The IPv6 introduced all the improvements to overcome these problems, the most important one being the 128-bit wide address format.

In the context of future wireless networks, in order to make IP technology integrate all services and build true convergent networks, it is crucial to have an IP-layer solution to the mobility management problem. The users will demand the same quality of service they can achieve with fixed networks. Mechanisms are being developed at the IP layer to make handovers seamless for users, even when different access technologies are involved (inter technology handovers). One of the most important efforts by the IETF in making the Internet mobility available is mobile IPv6. The use of IPv6 native mechanisms, like address autoconfiguration or extension headers (like the routing header or the destination option header), increased the simplicity of how the mobility protocol is implemented and at the same time its efficiency.

There are five basic elements in an MIPv6 scenario:

- *A mobile node (MN)* This is the device that will travel through the different networks.
- *A home agent (HA)* This is a router with the important mission of keeping track of the MN location and intercepting traffic sent to it.
- *A correspondent node (CN)* This is a regular host with whom the MN is communicating. Associated with this element are the concepts of home network (HN) and foreign network (FN).
- *Home network (HN)* This is the network where the HA is attached. While in this network, the MN is accessible through its Home address and no MIPv6 process is necessary to the communication.
- *Foreign network (FN)* or *visited network* It is a network where the MN travels and because it has to acquire a new IP address, an MIPv6 process is necessary to keep all the connections alive.

Let us consider the simplest scenario as illuctrated, in Fig. 13.42 in which we can describe and explain the mobile IPv6 process.

A mobile node (MN) is always identified by its home address (HA), regardless of its current point of attachment in the network. While the MN is attached to its home network (HN), the communication is made in a standard IPv6 way. However, when the MN moves to a different network, it gets a new address, the *care of address* (CoA), which provides the necessary information about the MN point of attachment, allowing packets

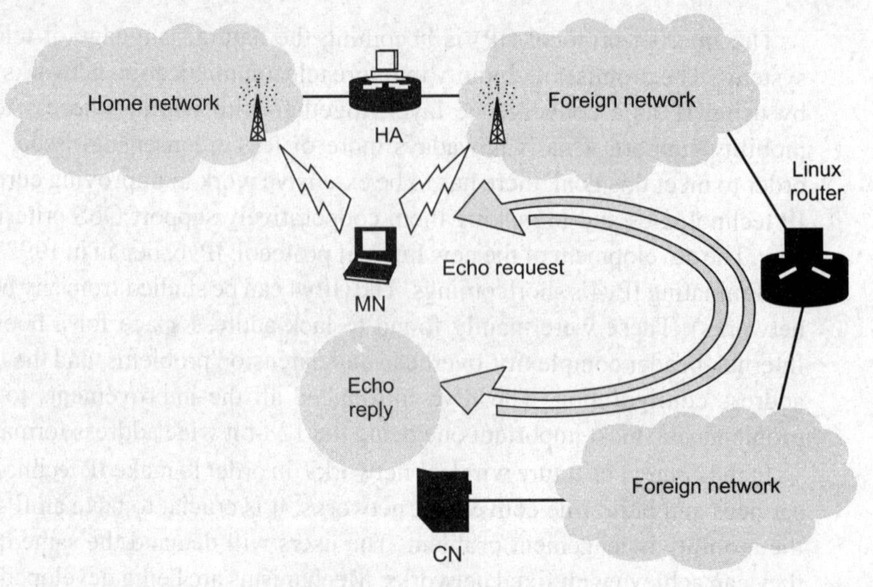

Fig. 13.42 MIPv6 scenario

to be routed transparently to that CoA, and then delivered to the MN. The MN reads the additional information contained on the routing header (RH), translates the CoA into to the home address, and delivers the packet transparently to the upper layers. The main entity that supports this operation, besides the MN, is the home agent (HA), located in the HN, responsible for keeping track of the MN current point of attachment on the network. Furthermore, a packet sent to the MN home address while it in an FN, is intercepted and delivered by the HA to the MN, using IPv6 over IPv6 encapsulation. When the MN receives an encapsulated packet from its HA, it knows that the node that sent this packet, called *correspondent node* (CN), but is not aware of the actual point of attachment of the MN. Then, the MN sends the information about its current association, allowing the CN to send packets directly to the MN, using a RH. The information about the current association (CoA, HA) of the MN, is owned by the HA and also the CN (plus some other information, e. g., association lifetimes), and is saved in a structure, called *binding cache* (BC). The MN requests either to the HA or the CN to add or update associations are called *bindings*. When the CN is aware of the point of attachment of the MN, the communication is made directly, using optimal routing. Now, the packets from the CN are sent to the MN using a RH, while the packets from the MN to the CN are sent using home address. Every time the MN moves to another FN (changing its point of attachment), the process is repeated. When the MN returns to his HN, this process is stopped and normal IPv6 packets are exchanged between the nodes.

The third-generation wireless networks, like UMTS, CDMA2000 and Japan's I-mode, represent a significant step forward in terms of bit rates and multimedia services. Some work is also being done in designing alternative radio technologies that could be able to support even bigger bit rates and wide coverage. Efforts are being made to develop and standardize new aspects of third generation telecommunication networks, which will likely improve current specifications in terms of IP convergence. The most significant aspect is that of the change in core network philosophy to an all-IP network. Future UMTS and beyond third-generation networks intend to use IP to bear all services and even real-time and traditional circuit-switched traffic, like voice, will be carried over IP, using the session initiation protocol. However, 3G systems are still far behind wireless users' expectations. Total service integration with the Internet, real-time multimedia, and broadband access are still a promise, for future beyond 3G networks to deliver. Researchers and vendors are expressing a growing interest in 4G wireless networks that support global roaming across multiple wireless networks. With this feature, users will be able to access different services, have increased coverage, have the convenience of utilizing a single device, receive a common bill, and gain more reliable wireless access, even with the failure or loss of one or more networks. The transparent conjunction of various wireless technologies into a single mobile terminal can further boost the wireless explosion. The vision is to create a unified wireless networking environment, utilizing, in a seamless manner, multiple and different wireless technologies. It is envisaged that the mobile terminal freely roams between wireless networks of the same type and automatically switches to a wireless network of a different type, whenever a need to do so arises.

One of the most challenging problems is how to access several different mobile and wireless networks through a single terminal, equipped with multiple wireless interfaces. A primary question that has to be answered is: Which wireless interfaces could be used for this kind of mobile terminal? Given the wide market acceptance of the GSM/CDMA, GPRS/UMTS/CDMA2000 and IEEE 802.11 (Wireless LAN – WLAN) technologies, as well as their differential strengths and limitations, with respect to coverage area and bandwidth, the aforementioned technologies may be the best candidates. Additionally, supporting QoS in 4G networks is another major challenge, due to varying bit rates, channel characteristics, bandwidth allocation and handover support among heterogeneous wireless networks. The provisioning of adequate end-to-end transport services in an all-IP based communications scenario where (a) the supporting access networks are wireless and varied, (b) core networks may or may not provide QoS support, (c) applications demand differentiated transport services, and (d) terminals are moving between access networks, is a complex, open, and challenging problem.

This problem can be first addressed from the IP QoS (packet switching) point of view. The network elements dealing with IP (e.g., hosts, core routers, IEEE-802.11b associated router, and GGSNs, terminal equipment) need to cooperate and manage their packet queues so that the packets generated by the applications can be transported with some guarantees or differentiation. The two well-known approaches for IP dealing with quality of service are integrated and differentiated.

The IntServ approach aims to provide packets with a maximum transfer delay between a source and a destination. When, for instance, a client process needs to send a sequence of packets (flow) to a server process and these packets require a known maximum delay, some resources must be reserved at every IP network element. The terminal, typically using RSVP, issues a message to the first router indicating the traffic characteristics of the flow.

The DiffServ approach reuses the old ToS field of the IPv4 header, renaming it to DSCP (differentiated services code point). The packets at the entrance of an IP domain have this field marked according to some policy. The scheduler associated to a router output port or to the network interface of a terminal, uses this field to prioritize packets so that each class of packets gets a different share of the output link bandwidth. DiffServ is more easily implemented than IntServ in IP nodes and can be used in independent IP clouds. The wireless access networks addressed, are perceived from IP, as a link layer technology. From the QoS point of view, and besides the maximum bandwidth available, there are no major differences between UMTS, GPRS or WLAN. Despite some technologies that comprise some QoS mechanisms at the transmission layer, like GPRS or UMTS, currently available WLAN technologies, like IEEE- 802.11b, donot provide such support for QoS.

One of the most promising protocols to support interactive services and user mobility in 4G networks is the *session initiation protocol*. The SIP has been developed by the Internet Engineering Task Force (IETF) and is being adopted by other bodies, such as ETSI and 3GPP. The protocol accomplishes establishment and control of sessions among mobile users. To enable multimedia communication, the protocol is quite generic and puts no limitations on types of sessions supported. It can be easily used to set up and maintain telephone calls, distributed virtual reality scenarios, teleconferences, multiparty games, etc. To accomplish global mobility of users, a global e-mail like address is used for identification proposes and users are being registered with location servers as they move, in a fashion similar to the GSM location registers. SIP also provides means with which users can specify and negotiate their expectations from sessions. This is a very important issue, when using it on heterogeneous mobile environments, like the ones mentioned above. SIP has been designed for Internet-wide deployment. With its stateless working mode it is resistant against failures. Multicasting support allows managing sessions with a large number of participants. In order to meet expectations, beyond 3G wireless networks must span multiple technologies, offer real-time broadband services, and converge to all-IP architecture with integrated QoS and Authentication, Authorization and Accounting (AAA) support.

Summary

- There are two types of networks: infrastructure networks based on cellular theory and ad hoc networks without requirement of infrastructure.
- The infrastructure-based networks use wired infrastructure partly along with wireless infrastructure; 100% wireless systems are very difficult to handle.
- The infrastructure-based networks are GSM+GPRS+EDGE, WLL, IS-95 based cell phone system, UMTS, etc.
- Cell-wise base station is required in all these networks to catch the signals from mobile phones. The cells are formed on the basis of geographical situations and population density and cell size dependent bit rates are achieved as radio propagation effects also differ with cell size.
- Handovers are must in such systems. The GSM systems deal with hard handover as adjacent cells use different carriers, while CDMA-based systems use soft handover as

carrier is same for all the cells. The UMTS offers hard, soft, and softer handovers.

- The GSM is a 2G system and with upgradations, it is converted into 2.5G system. The GSM is specially used for voice services while GPRS is specially used for data services. But both can use the same basic infrastructure. The UMTS is a 3G system compatible with existing ones.
- Important components of GSM system are mobile equipment, BSC, BTS, MSC, etc. The MSC is supported by HLR, VLR, EIR, AUC, and OMC.
- The EDGE technology is developed only for data rate enhancement.
- The IS-95 is CDMA-based standard for mobile telephony, which has become popular with the names cdmaOne afterwards and its evolution resulted in CDMA 2000.
- Due to spread spectrum modulation used in CDMA systems, secure communication is possible.
- The WLL is the modified telephone system in which wired local loops are replaced by wireless local loops.
- The WLL can use different existing wireless technologies over cell as it is based on interworking.
- The WLL and GSM differ form each other in terms of cell size and coverage, power, and handovers.
- The CorDECT WLL is especially suitable for communication in rural area.
- The UMTS is a 3G system based on WCDMA.
- Most of the infrastructure of UMTS is similar to that of GSM-GPRS systems.
- The TDD and FDD both are used in UMTS for the first time.
- Satellite systems are to be incorporated in UMTS to create 'anywhere anytime' scenario.
- Universal SIM is required for UMTS.
- Cellular systems, when combined with satellite interlinks, make mobile satellite communication system. Here the earth station is connected with MSC.
- Present technologies can be designed in such a way that interworking can become possible and combinations of the various systems can also be possible, which may result in converged networks.
- Mobile IP is required for mobile Internet for which IPv6 protocol is designed.
- Future trend is towards all-IP networks.

Review Questions

1. What is meant by infrastructure-based networks? How do they differ from ad hoc networks?
2. How does the 2G GSM system differ from the analog mobile and cellular phone systems?
3. What are the upgradations made in the 2G GSM system?
4. List out the 2G GSM services and also the services after its enhancement.
5. How does the GPRS service enhance the GSM services?
6. List out the basic components of the GSM system and additional components due to the GPRS addition.
7. How many identity numbers are required in the GSM system? List out the GSM service providers in your area.
8. Mention the steps for location updation in the GSM network.
9. How does the GPRS differ from computer networks?
10. Why is it necessary to define the GSM and GPRS logical channels?
11. How does the EDGE differ from the 2G GSM?
12. Find out the infrastructure required for IS-95 to CDMA2000-based systems. Draw

the forward and reverse link diagrams for the IS-95 system.

13. What is the type of infrastructure required for a wireless local loop?
14. Why are the separate specifications for a WLL system not required?
15. Explain the interworking in the DECT with other systems like the GSM and CDMA.
16. 'The CorDECT WLL is the best technology for rural areas'. Explain.
17. What will be the access equipment that WLL uses?
18. State the solutions for long-haul transmissions in case of WLL.
19. What is LMDS?
20. List out the basic attributes of WLL systems.
21. Draw the CorDECT WLL architectures.
22. What is WCDMA?
23. What are the features of 3G systems that do not appear in the systems of previous generations?
24. How can you say that UMTS is a 3G GSM? What are the similarities and differences between the 2G GSM and 3G UMTS? List out all of them.
25. Draw the detailed diagram of a mobile satellite communication, showing the interlinks between the satellites.
26. What do you mean by converged networks?
27. Fill in the blanks:
    (a) In a cellular system, one cluster uses f1, f2, f3, f4, f5, f6, f7 frequencies, the frequency reuse factor for that system = ________ .
    (b) Consider total number of duplex channels available is 56, total number of cells in one cluster is 7, and each cell is allocated total ________ number of channels.
    (c) ________ uplink logical channel is used for mobile originating calls.
    (d) In TCH/F 9.6, speech signal is digitized at raw data rate ________ .
    (e) In GSM frame structure one hyperframe = ________ TDMA frame.
    (f) For GSM security ________ key is dynamic.
    (g) Each 20 ms segment of digitized speech signal applied to RPE-LPC speech encoder of GSM consists of ________ samples.
    (h) The additional components for GPRS inclusion in GSM system are ________ and ________ .
    (i) DECT standard use ________ modulation technique.
    (j) The conventional frequency band of DECT standard is ________ .
    (k) DECT supports ________ handoff.
    (l) The transmission bandwidth of IS-95 is ________ .
    (m) The interface between RNC and MSC in UMTS is ________ .
    (n) ________ protocol is used to transfer data packets between SGSN and MS.
    (o) The interface used between SGSN and GGSN is ________ when they are in different network.
    (p) Downloading documents is ________ type of service of UMTS.
    (q) The ________ kHz channel bandwidth is used for EDGE.
    (r) The responsible agency for developing the UMTS is ________ .
    (s) The UMTS equivalent of a GSM BTS is ________ .

28. Match the following in terms of GSM:

A	B
(a) A_3 algorithm	(a) Public number
(b) A_5 algorithm	(b) Authentication algorithm
(c) A_8 algorithm eration of cipher	(c) Used for gen key Kc
(d) MSISDN	(d) Not public number
(e) IMSI	(e) Timely upgrade number
(f) TMSI	(f) Used for encryption

Multiple Choice Questions

1. Which of the following multiple access techniques is used in GSM?
(a) FDMA
(b) FDMA/SDMA
(c) TDMA/FDMA/SDMA
(d) TDMA/SDMA

2. The modulation method for GSM is
(a) GMSK (b) GFSK
(c) MSK (d) FSK

3. The uplink frequency of GSM is
(a) 890–915 MHz
(b) 890–915 kHz
(c) 935–960 MHz
(d) 1800–1900 MHz

4. The channel width for GSM is
(a) 124 kHz (b) 200 kHz
(c) 270 kHz (d) 890 kHz

5. The permanent database of all registered subscribers is maintained in
(a) VLR (b) AUC
(c) EIR (d) HLR

6. Which of the following is a protected database with a secret key also contained in the user's SIM card?
(a) VLR (b) AUC
(c) EIR (d) HLR

7. The EDGE is a
(a) 2 G technology
(b) 3 G technology
(c) 2.5 G technology
(d) 1.5 G technology

8. The GSM-2G uses
(a) NB – TDMA
(b) WB – FDMA
(c) NB – FDMA
(d) WB – TDMA

9. The GMSK modulation scheme used in GSM is a form of
(a) ASK
(b) PSK
(c) (a) and (b)
(d) none of the above

10. FDMA is used in
(a) radio (b) GSM
(c) TV (d) of these

11. How many bit identification numbers known as the electronic serial numbers (ESD) are put in the mobile phone at the time of it manufacture?
(a) 40 (b) 32
(c) 15 (d) 4

12. The frequency reuse
(a) maintains spectrum efficiency
(b) increases spectrum efficiency
(c) decreases spectrum efficiency
(d) does not affect spectrum efficiency

13. 'The SDMA is used only in combination with TDMA, FDMA and CDMA.' This statement is
(a) true
(b) false
(c) true in case of CDMA only
(d) true in case of FDMA only

14. A high-capacity switch that provides handover, cell configuration data, and control of radio frequency (RF) power levels in base transceiver stations is known as
    (a) VLR (b) BSC
    (c) EIR (d) none
15. In G-2 GSM, the rate of voice transmission over TCH/F is
    (a) 11.8 kbps (b) 9.6 kbps
    (c) 13 kbps (d) None.
16. Which is the code used in IS-95 for spreading purpose
    (a) PN (b) Walsh
    (c) PN and Walsh (d) Gold
17. Which of the following is not the UMTS interface?
    (a) Uu (b) Iub
    (c) Iuc (d) Iur
18. Which of the following is not a characteristic of 3G systems?
    (a) Support to only data transmissions
    (b) Worldwide use
    (c) High data rates
    (d) Vehicular mobility
19. The UMTS increases the transmission speed to
    (a) 1 Mbps per mobile user.
    (b) 2 Mbps per mobile user.
    (c) 3 Mbps per mobile user.
    (d) 5 Mbps per mobile user.
20. The most important revolutionary step of GSM towards UMTS is
    (a) CAMEL (b) EDGE
    (c) GPRS (d) GGSN
21. Which of the following enables worldwide access to operator-specific IN applications, such as prepaid, call screening, and supervision?
    (a) 3GPP (b) CN
    (c) Node B (d) CAMEL
22. The Release 99 for UMTS
    (a) migrates the core voice network to VOIP network
    (b) enables IP multimedia service
    (c) adds new radio access network-UTRAN
    (d) defines BTS as node B
23. Which of the following statements is correct?
    (a) The WLL is based on CDMA only.
    (b) The WLL is the direct substitute of wireline local loop.
    (c) The WLL system is designed for rural area only.
    (d) The WLL coverage is limited than GSM.
24. The UMTS is a 3G technology
    (a) being deployed by GSM/GPRS service providers
    (b) that is successor to GSM
    (c) developed for fixed wireless access
    (d) that support both TDD and FDD
25. Which of the following channels are used for downlink in GSM/UMTS?
    (a) BCH, PCH
    (b) DCH, PRACH
    (c) FACH, SCH
    (d) RACH, CCPCH
26. The prevalent versions of UMTS Releases are
    (a) Release 99 and 4
    (b) Release 5 and 6
    (c) Release 9 and 10
    (d) none of the above
27. The GPRS supports
    (a) voice
    (b) data
    (c) voice and data
    (d) video

chapter 14

Wireless Communication Systems and Standards 3: Ad hoc Network, WLAN, and WMAN

Key Topics

- Bluetooth
- Wi-Fi: IEEE802.11 standards
- WiMAX: IEEE802.16 standards
- Zigbee: IEEE 802.15.4
- UWB
- IEEE802.20 and beyond

Chapter Outline

Ad hoc networks are the voice/data networks that are established temporarily without requirement of an infrastructure. The number of users in such systems may be limited. Mostly, the ad hoc networks are established for personal use or for use within the limited domain like an office or a plant. These networks can be established anywhere in the world, maybe on temporary basis and because of this, the carrier frequencies chosen are ISM band frequencies. Bluetooth can make a wireless home network for all personal devices. The WLAN and WMAN are the examples of ad hoc networks, when formed on short ranges using personal laptops. However, the convergence in WLAN and WMAN industry made it necessary to have some infrastructure (cell based) to allow for interoperability and long ranges. Sensor networks are the specially designed networks in which physical changes are sensed by the sensor-based nodes and these changes are noted at a remote computer connected in a wireless way. We can consider the sensor networks as ad hoc networks or can consider it in a special category. Similarly, ultra-wideband is also considered as a special type of network for very high-speed personal area communication. It should be noted that the main focus is on architecture and physical layer along with some key aspects of the systems. The data link layer and other aspects are touched upon as and when required, though not in detail. The reader may self-study them.

14.1 INTRODUCTION

The LANs, without the need for an infrastructure and with a very limited coverage, are being conceived for connecting different small devices in close proximity without expensive wiring and infrastructure. The area of interest can be the personal area about the person who is using the device. This new emerging architecture is known as *wireless personal area network* (WPAN). The concept of a personal area network, hence, refers to a space of small coverage (less than 100 m) around a person where ad hoc communication occurs.

An ad hoc (or 'spontaneous') network is a local area network or some other small network in personal area, especially one with wireless or temporary plug-in connection devices, in which some of the network devices are a part of the network only for the duration of a communications session or in the case of mobile or portable devices are in some close proximity to the rest of the network. The ad hoc wireless networks do not need any infrastructure. In Latin, *ad hoc* literally means 'for this,' further meaning 'for this purpose only,' and thus usually 'temporary'. The term has been applied to future office or home networks in which new devices can be quickly added, using, for example, the proposed Bluetooth technology, in which devices communicate with the computer and perhaps other devices using wireless transmission.

Many WLANs of today need an infrastructure network that provides access to other networks and include MAC. In ad hoc systems, mobile stations may act as a relay station in a multihop transmission environment from distant mobiles to base stations. Mobile stations will have the ability to support base station functionality. The network organization will be based on interference measurements by all mobiles and base stations for automatic and dynamic network organization according to the actual interference and channel assignment situation for channel allocation of new connections and link optimization. These systems will play a complementary role to extend coverage for low power systems and for unlicensed applications. A central challenge in the design of ad hoc networks is the development of dynamic routing protocols that can efficiently find routes between two communication nodes. A mobile ad hoc networking (MANET) working group has been formed within the Internet engineering task force (IETF) to develop a routing framework for IP-based protocols in ad hoc networks. Another challenge is the design of proper MAC protocols for multihop ad hoc networks.

Vendors offer an ad hoc network technology that allows people to come to a conference room and using infrared transmission or radio frequency (RF) wireless signals join their notebook computers with other conferees to a local network with shared data and printing resources. Each user has a unique network address, which is immediately recognized as a part of the network. The technology would also include remote users and hybrid wireless/wire connections. Figure 14.1 shows this technology for the wireless LAN. If only personal laptops are to be interconnected, it is an ad hoc network for personal area. If personal ad hoc network is connected with wired network, the interoperability starts.

The following sections describe some examples of such networks. One observation made is that in most of the ad hoc networks, the range of frequencies is ISM band and hence the network can be established in any part of the world.

Most of the standards described in this chapter are IEEE standards.

14.2 BLUETOOTH

The aim of this section is to provide the reader with knowledge of the Bluetooth technology and its underlying concepts. Bluetooth represents a single-chip, low-cost, radio-based wireless network technology. It is a standard for wireless networking of small peripherals. Bluetooth technology aims at the so-called ad hoc piconets, which are local area

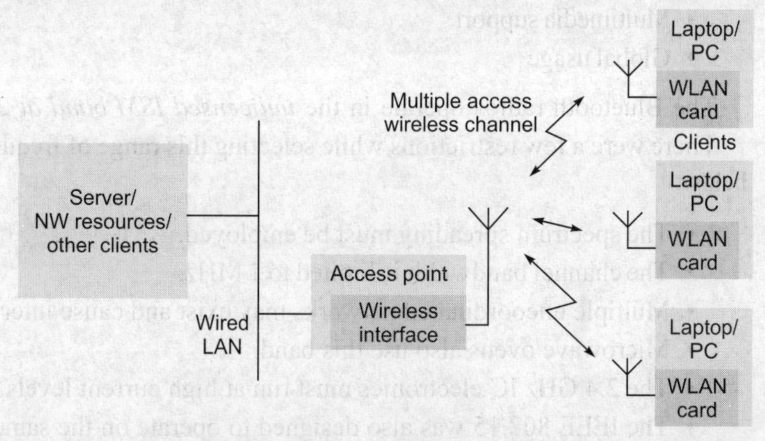

Fig. 14.1 Client access point configuration for WLAN shows the interoperability with other networks/infrastructure

networks with a very limited coverage and without the need for an infrastructure. Bluetooth technology is for personal area communication, maybe a communication between personal Bluetooth supported devices for maximum up to 100 metre in the latest scenario. Bluetooth devices consist of a hardware and protocol stack support as shown in Fig. 14.2.

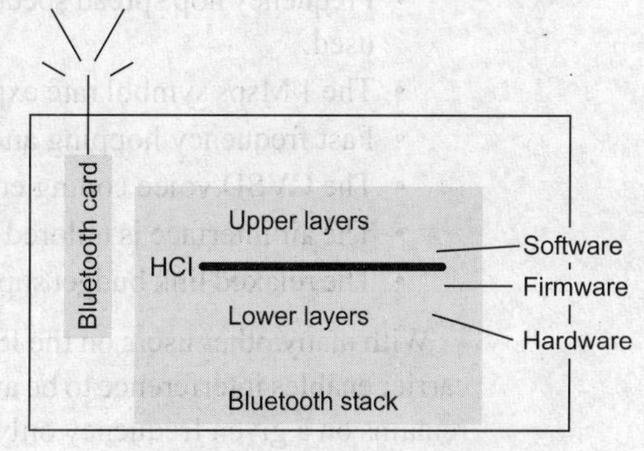

Fig. 14.2 Bluetooth device requirement

This technology is the result of cooperation between the leaders in the telecommunication and computer industries. The founder members or five promoters were IBM, Ericsson, Intel, Nokia, and Toshiba. Their joint name is known as the *Bluetooth special interest group* (SIG). After the release of the Bluetooth specification 1.0B, the SIG has been joined by 3COM, Lucent Technologies, Microsoft, and Motorola. There are over 650 adopters. Many companies and research institutions joined the SIG, whose goal was the development of Bluetooth-enabled mobile phones, laptops, notebooks, headsets, etc. Up to now, Bluetooth is not a standard like IEEE 802.11 or higher and HIPERLAN 1, but it is the de facto standard.

The SIG attempted to make Bluetooth a global standard for wireless connectivity. The design goals were as follows:

- Low cost
- Low energy consumption
- Robust operation
- High aggregate capacity
- Flexible usage

- Multimedia support
- Global usage

The Bluetooth radios operate in the *unlicensed ISM band at 2.4 GHz.*

There were a few restrictions while selecting this range of frequency, which are given below:

- The spectrum spreading must be employed.
- The channel bandwidth is limited to 1 MHz.
- Multiple uncoordinated networks may exist and cause interference.
- Microwave ovens also use this band.
- The 2.4 GHz IC electronics must run at high current levels.
- The IEEE 802.15 was also designed to operate on the same frequency.

The Bluetooth device implementations are found with the following features:

- Frequency hop spread spectrum/TDD with GFSK carrier modulation (BT = 0.5) is used.
- The 1 Msps symbol rate exploits maximum channel bandwidth.
- Fast frequency hopping and short data packets.
- The CVSD voice coding enables operation at high bit error rates.
- The air interface is tailored to minimize current consumption.
- The relaxed link budget supports low-cost single-chip integration.

With many other users on the ISM band from microwave ovens to Wi-Fi, the hopping carrier enables interference to be avoided by Bluetooth devices. A Bluetooth transmission remains on a given frequency only for a short time and if any interference is present, the data will be resent later when the signal has changed, to a different channel, which is likely to be clear of other interfering signals. The standard uses a hopping rate of 1600 hops per second. These are spread over 79 fixed frequencies and they are chosen in a pseudo-random sequence. The fixed frequencies occur at 2400 + n MHz, where the value of n varies from 1 to 79. This gives frequencies of 2401, 2402, …, 2480 MHz. In some countries, the ISM band allocation does not allow the full range of frequencies to be used. In order to enable effective communications to take place in an environment where a number of devices may receive the signal, each device has its own identifier. This is provided by having a 48 bit hard-wired address identity, giving a total of 2.815×10^{14} unique identifiers.

During the development of the Bluetooth standard, it was decided to adopt the use of frequency hopping system rather than a direct sequence spread spectrum approach because it is able to operate over a greater dynamic range. If direct sequence spread spectrum techniques were used, then the other transmitters nearer to the receiver would block the required transmission if it is farther away and weaker. The transmitter power levels for different distances are given in Table 14.1.

In short, *Bluetooth can connect peripheral devices and support their ad hoc networking*. The technology is an open specification for wireless communication of data,

Table 14.1 Bluetooth transmitter power levels

Class	Distance Covered	Power Level
I	Long range up to 100 m	20 dBm
II	Ordinary range up to 10 m	4 dBm
III	Short range up to 10 cm	0 dBm

voice, image, and video clips. The Bluetooth radio technology can be built into both the cellular telephone as well as the laptop and would replace the cumbersome cables to connect a laptop and a cellular telephone. Apart from this, printers, PDAs, desktops, fax machines, keyboards, joysticks, and virtually any other digital device can be a part of the Bluetooth system making all of them wireless.

14.2.1 Bluetooth Network Structure

Some important terms related to the Bluetooth network structure are defined here.

Piconet

It is a topology of devices connected via the Bluetooth technology in an ad hoc fashion as shown in Fig. 14.3. A piconet starts with two connected devices, such as a portable PC (laptop) and a cellular phone, and may grow to eight connected devices. All Bluetooth devices are peer units and have identical implementations. However, when establishing a piconet, one unit will act as a master and the other(s) as slave(s) for the duration of the piconet connection (one master and up to seven slaves). Each piconet is identified by a different frequency hopping sequence and is determined by the Bluetooth device address (BD-ADDR) of the master. All users participating on the same piconet are synchronized to this hopping sequence, phase of which is determined by the Bluetooth clock of the master. Parked and standby devices are not considered as a part of piconet.

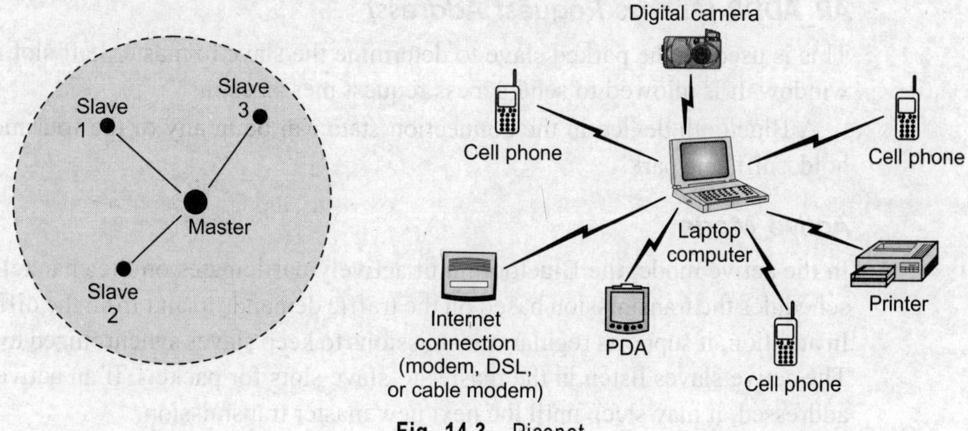

Fig. 14.3 Piconet

Scatternet

Multiple independent and non-synchronized piconets form a scatternet as shown in Fig. 14.4. Normally, ten piconets are allowed in a scatternet.

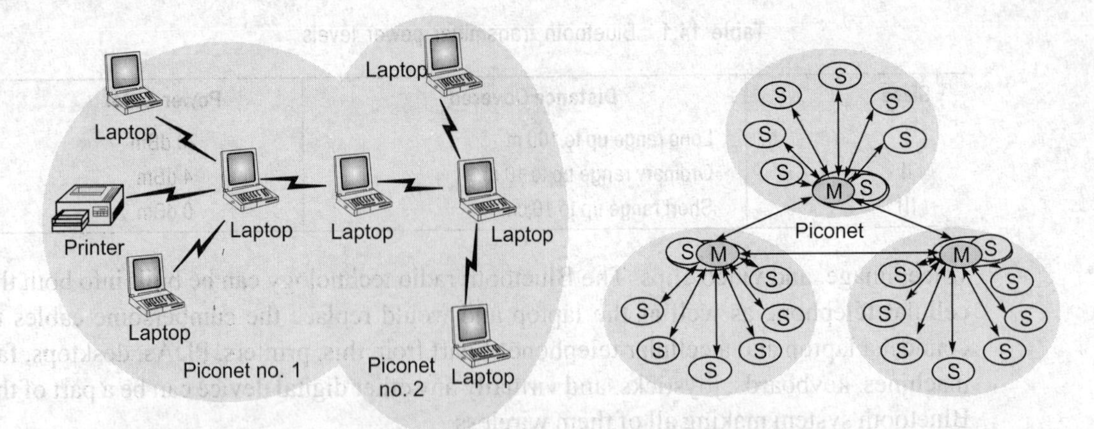

Fig. 14.4 (a) Scatternet concept and (b) master of one piconet as a slave of another piconet

Master Unit

It is the device in a piconet whose clock and hopping sequence are used to synchronize all other devices in the piconet.

Slave Units

All devices in a piconet that are not the master are called slave units.

AM_ADDR: (Active Member Address)

It is a three-bit address to distinguish between active units participating in the piconet.

Parked Units

These are the devices in a piconet that are synchronized but do not have an AM_ADDR. They are identified with PM_ADDR, which is valid only as long as the slave is parked.

AR_ADDR (Access Request Address)

This is used by the parked slave to determine the slave to master half slot in the access window. It is allowed to send access request messages in.

A Bluetooth device in the connection state can be in any of the four modes: active, hold, sniff, and park.

Active Mode

In the active mode, the Bluetooth unit actively participates on the channel. The master schedules the transmission based on the traffic demands to and from the different slaves. In addition, it supports regular transmissions to keep slaves synchronized to the channel. The active slaves listen in the master-to-slave slots for packets. If an active slave is not addressed, it may sleep until the next new master transmission.

Sniff Mode

The devices synchronized to a piconet can enter power-saving modes in which the device activity is lowered. In the sniff mode, a slave device listens to the piconet at reduced rate, thus reducing its duty cycle. The sniff interval is programmable and depends on the application. It has the highest duty cycle (least power efficient) of all three power-saving modes (sniff, hold, and park).

Hold Mode

As mentioned earlier, the devices synchronized to a piconet can enter power-saving modes in which the device activity is lowered. The master unit can put slave units into hold mode, where only an internal timer is running. The slave units can also demand to be put into hold mode. The data transfer restarts instantly when units transit out of the hold mode. It has an intermediate duty cycle (medium power efficient) of the three power saving modes.

Park Mode

In the park mode, a device is still synchronized to the piconet but does not participate in the traffic. The parked devices have given up their MAC (AM_ADDR) address and occasional listen to the traffic of the master to resynchronize and check on the broadcast messages. It has the lowest duty cycle (power efficiency) of all three power saving modes.

To enable a net to be set up, the master transmits an enquiry message every 1.28 seconds to discover whether there are any other devices within the range. If a reply is received, then an invitation to join the net is transmitted to the specific device that has responded. After this, the master allocates each device a member address and then controls all the transmissions. All Bluetooth devices have a clock that runs at twice the hopping speed and this provides synchronization to the whole net. The master transmits in the even numbered time slots whilst the slaves transmit in the odd numbered slots once they have been given permission to transmit.

The Bluetooth system supports both point-to-point and point-to-multipoint connections. There are two ways in which the data is transferred. The first is by using what is termed an *asynchronous connectionless communications link* (ACL). This is used for file and data transfers. A second method is termed a *synchronous connection-oriented communications link* (SCL). This is used for applications such as digital audio. The synchronous connection-oriented (SCO) link is a symmetric point-to-point link between a master and a slave in the piconet (circuit switched type). The master can support up to 3 simultaneous SCO links, while the slave can support 2 or 3 SCO links. The SCO packets are never retransmitted. The ACL link is a point-to-multipoint link between the master and all the slaves.

The ACL enables data to be transferred via Bluetooth at speeds up to the maximum rate of 732.2 kbps. This occurs when it is operating in an asymmetric mode. This is commonly used because for most applications there is far more data transferred in one direction than the other. When a symmetrical mode is needed with data transferred at the same rate in both directions, the data transfer rate falls to 433.9 kbps (Table 14.1). The synchronous links support two bi-directional connections at a rate of 64 kbps. The data rates are adequate for audio and most file transfers. However, the available data rate is

Table 14.2 Data rates in Bluetooth

Packet Type	Symmetric (SCO)	Asymmetric (ACL)	
		Downlink	Uplink
DM1	108.8	108.8	108.8
DH1	172.8	172.8	172.8
DM3	258.1	387.2	54.4
DH3	390.4	585.6	86.4
DM5	286.7	477.8	36.3
DH5	433.9	723.2	57.6

Note: Other packet types are ID, NULL, POLL, FHS, AUX1, HV1, HV2, HV3, and DV.

insufficient for applications like high-rate DVDs, which require 9.8 Mbps and for many other video applications, including games spectacles.

14.2.2 Bluetooth Protocol Stack

The protocol stack associated with Bluetooth is shown in Fig. 14.5. The functions of each layer are mentioned below:

LM or LMP:	Link manager (protocol)
HCI:	Host controller interface
L2CAP:	Logical link control and adaptation protocol
RFCOMM:	Radio frequency comm.
OBEX:	Generalized multi-transport object exchange protocol
BNEP:	Bluetooth network encapsulation protocol
IP:	Internet protocol
TCP/UDP:	Transfer control protocol/user datagram protocol
AT:	Attention sequence
TCS BIN:	Telephony control protocol specification binary
SDP:	Service discovery protocol

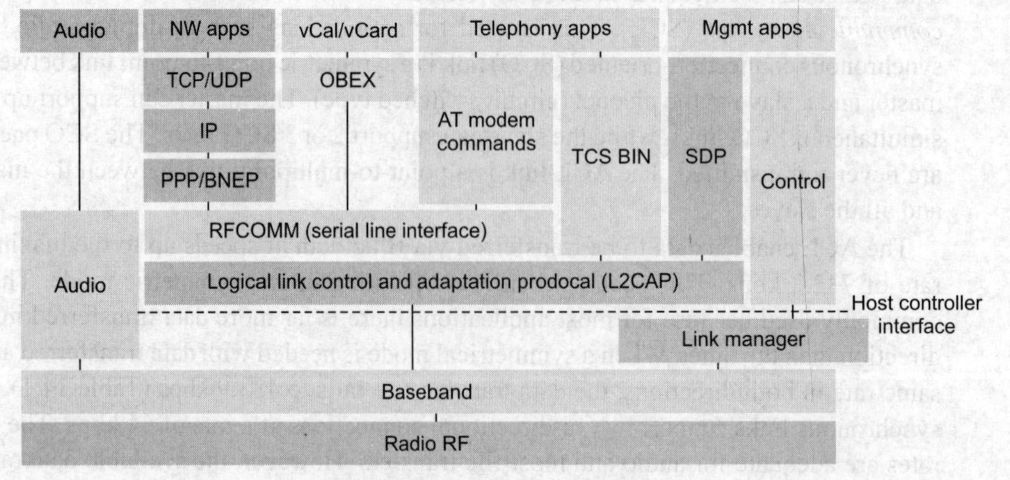

Fig. 14.5 Bluetooth_protocol stack

At the bottom of the stack is the hardware component that implements the Bluetooth *radio*, *baseband*, and *link manager* protocols. Neither an application nor even the host has access to this layer of the stack.

The *HCI* transmits data and commands from the layers above to the Bluetooth module below. Conversely, the HCI receives events from the Bluetooth module and transmits them to the upper layers. The functions of the HCI are implemented in the kernel, which are introduced by various companies on their own. Some companies, like Apple, implements the L2CAP and RFCOMM layers in the kernel. The applications can use objects in the user-level L2CAP and RFCOMM layers to access the corresponding in-kernel objects, although many applications will not need to do so directly.

The *L2CAP layer* (logical link control adaptation protocol) provides the following:

- Multiplexing of data channels
- Segmentation and reassembly of data packets to conform to a device's maximum packet size
- Support for different channel types and channel IDs, such as RFCOMM

These functions are nearer to the functions of the transport layer in the OSI model. The in-kernel L2CAP layer provides the transport for the higher-level protocols and profiles. Using the L2CAP layer's multiplexing feature, it is possible to send and receive data to and from the RFCOMM layer and the SDP layer at the same time. The in-kernel *RFCOMM protocol layer* is a serial port emulation protocol. Its primary mission is to make a data channel appear as a serial port. It also implements the ability to create and destroy RFCOMM channels and to control the speed of the channel as if it were a physical serial port cable.

The *SDP layer* is more of a service than a protocol. It is shown connected to the user level L2CAP layer, because it uses an L2CAP channel to communicate with remote Bluetooth devices to discover their available (location-based) services. Apple provides an SDP API that can be used to discover what services a device support.

Above the user-level RFCOMM layer is the *OBEX protocol layer*. The OBEX protocol is an HTTP-like protocol that supports the transfer of simple objects, like files, between devices. The OBEX protocol uses an RFCOMM channel for transport because of the similarities between IrDA (which defines the OBEX protocol) and serial-port communication. The TCP/UDP and IP protocols support normal Internet services.

14.2.3 Bluetooth MAC Layer Considerations

Just like any other networks, in Bluetooth also, the MAC layer deals with the packet format in which the data transfer takes place. The format is given in Fig. 14.6 and the details of the packet header are given in Fig. 14.7.

The data is organized into packets to be sent across a Bluetooth link. The Bluetooth specification lists out seventeen different formats, which can be used depending upon the requirements. They have options for elements, such as forward error correction data and the like. However, the standard packet consists of a 72-bit access code field, a 54-bit header field, and then the data to be transmitted, which may be between 0 and 2745 bits. This data includes the 16-bit CRC if it is needed.

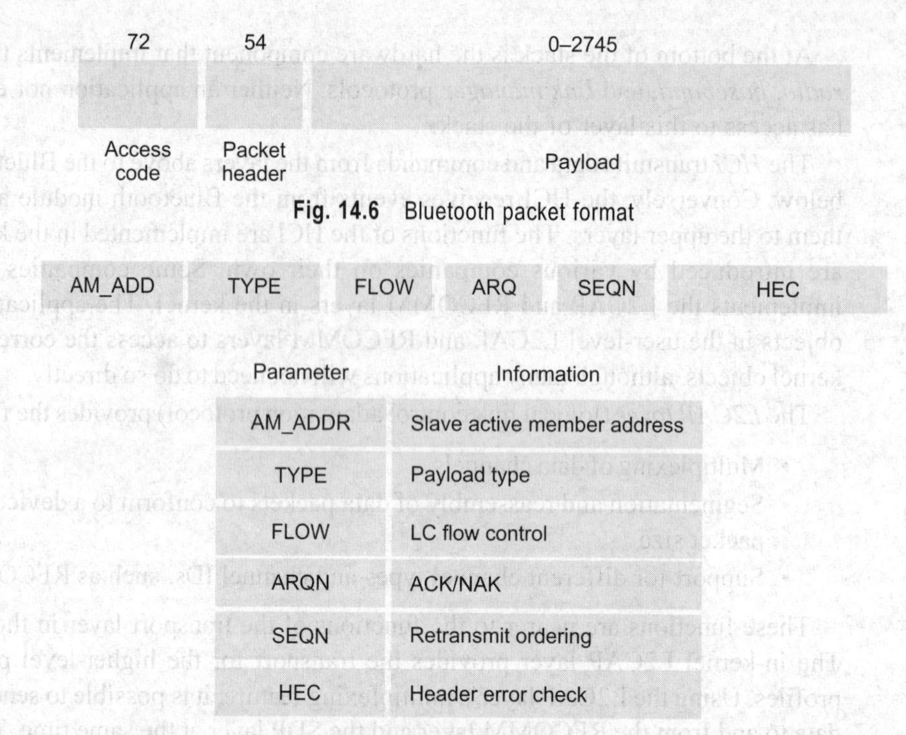

Fig. 14.6 Bluetooth packet format

Fig. 14.7 Packet header

As it is likely that interference will cause errors, error handling is incorporated within the system. For asynchronous links, packet sequence numbers are transmitted. If an error is detected in a packet, then the receiver can request it to be resent. The error coding using a 16-bit CRC is also available. For the synchronous links, packets cannot be resent as there is unlikely to be sufficient bandwidth available to resend data and 'catch up'. However, it is possible to include some forward error control.

As security is becoming an important issue, especially where links to computers are concerned, secure communications are possible over Bluetooth with the devices encrypting the data transmitted. A key up to 128 bits is used and it is claimed that the level of security provided is sufficient for financial transactions. However, in some countries, the length of the key is limited to enable the security agencies to gain access if required.

14.2.4 Modified Version of Bluetooth

Bluetooth is now well established as a wireless technology. It has found a very significant number of applications, particularly in areas such as connecting mobile or cell phones to hands-free headsets. One of the disadvantages of the original version of Bluetooth in some applications was that the data rate was not sufficiently high, especially when compared to other wireless technologies, such as 802.11. Hence, modified version of Bluetooth is introduced. Of all the features included in Bluetooth 2, the most important is the enhanced data rate (EDR) facility. In the new specification, the maximum data rate is able to reach 3 Mbps, a significant increase on what was available in the previous Bluetooth specifications.

One of the main reasons why Bluetooth 2 is able to support a much higher data throughput is that it utilizes a different modulation scheme for the payload data. However, this is implemented in a manner in which compatibility with previous revisions of the Bluetooth standard is still retained.

The Bluetooth data is transmitted as packets made up from a standard format. This consists of four elements: (a) the access code, which is used by the receiving device to recognize the incoming transmission, (b) the header, which describes the packet type and its length, (c) the payload, which is the data that is required to be carried, and finally (d) the interpacket guardband, which is required between the transmissions to ensure that the transmissions from two sources do not collide and to enable the receiver to re-tune. In previous versions of the Bluetooth standard, all three elements of the transmission, i.e. access code, header, and payload, were transmitted using Gaussian frequency shift keying (GFSK), where the carrier is shifted by ±160 kHz indicating a one or a zero and, in this way, one bit is encoded per symbol.

The Bluetooth 2.0 specification uses a variety of forms of modulation. The GFSK is still used for transmitting the access code and header and, in this way, compatibility is maintained. However, other forms of modulation can be used for the payload. There are two additional forms of modulation that have been introduced. One of these is mandatory, while the other is optional. A further small change is the addition of a small guard band between the header and the payload. In addition to this, a short synchronization word is inserted at the beginning of the payload. The first of the new modulation formats, which must be included on any Bluetooth 2 device, gives a two-fold improvement in the data rate and thereby allows a maximum speed of 2 Mbps. This is achieved by using pi/4 differential quaternary phase shift keying (pi/4 DQPSK). This form of modulation is significantly different from the GFSK, which was used on previous Bluetooth standards, in that the new standard uses a form of phase modulation, whereas the previous ones used frequency modulation.

Using quaternary phase shift modulation means that there are four possible phase positions for each symbol. Accordingly, this means that two bits can be encoded per symbol, and this provides for the two-fold data increase over the frequency shift keying used for the previous versions of Bluetooth.

To enable the full three-fold increase in data rate to be achieved, a further form of modulation is used. Eight phase differential phase shift keying (8DPSK) enables eight positions to be defined with 45 degrees between each of them. By using this form of modulation, eight positions are possible and three bits can be encoded per symbol. This enables the data rate of 3 Mbps to be achieved. As the separation between the different phase positions is much smaller than it was with the QPSK used to provide the two-fold increase in speed, the noise immunity has been reduced in favour of the increased speed. Accordingly, this optional form of modulation is used only when a link is sufficiently robust.

The Bluetooth 2 specification defines ten new packet formats for use with the higher data rate modulation schemes, five each of the enhanced data rate schemes. Three of these are for the first, third and fifth slot asynchronous packets used for transferring data.

The remaining two are used for the third and fifth slots extended synchronous connection-oriented (eSCO) packets. These use bandwidth that is normally reserved for voice communications.

The new format for these packets does not incorporate FEC. If this is required, then the system switches back automatically to the standard rate packets. However, many of the links are over a very short range where the signal level is high and the link quality is good. It is necessary for the packet type to be identified so that the receiver can decode them correctly, knowing also the type of modulation being used. An identifier is, therefore, included in the header, which is sent using GFSK. This packet header used for the previous version of Bluetooth used only four bits. This gave sufficient capability for the original system. However, there was insufficient space for the additional information that needed to be sent for Bluetooth 2.

It was not possible to change the header format because backward compatibility would not be possible. Instead, different link modes are defined. When two Bluetooth 2 or EDR devices communicate, the messages are used in a slightly different way, indicating the Bluetooth 2 or EDR modes. In this way, compatibility is retained while still being able to carry the required information.

14.3 WI-FI STANDARDS

The IEEE 802.11 protocols are designed by IEEE and are handled by Wi-Fi forum. The 802.11 tools are now moving into the second stage, where the wireless LAN is being treated as a large wireless communication system. As a system, there is more to consider than simply the communication over the air between a single access point and the associated mobile devices. This has led to innovative changes in the equipment that makes up a wireless LAN. Mainly, two layers are considered: physical and data link. The data link layer consists of two sublayers: logical link control (LLC) and media access control (MAC). The 802.11 protocol uses the same 802.2 LLC and 48-bit addressing as other 802 LANs, allowing for very simple bridging from IEEE wireless to wired networks, but the MAC is unique to WLANs.

The following aspects are to be considered for the study of 802.11 protocols:

- Medium access control, including MAC management and MAC management information base (MIB)
- Security enhancements
- Quality of service enhancements

Medium Access Control

The carrier sense medium access with collision avoidance protocol (CSMA/CA) is utilized, which avoids collision by explicit acknowledgement (ACK). This will result in additional overhead of ACK packets and so slow performance. It also supports *request-to-send/clear-to-send* (RTS/CTS) protocol, which is considered as the solution for 'hidden node' problem. This protocol adds additional overhead by temporarily reserving the medium. So, if used for large size packets, only retransmission would be expensive.

The MAC supports power conservation to extend the battery life of portable devices. There are two power utilization modes—*continuous aware mode*, in which the radio is always on and is continually drawing power, and *power save polling mode*, in which the radio is 'dozing' with access point queue any data for it. The client radio will wake up periodically in time to receive regular beacon signals from the access point. The beacon includes information regarding which stations have traffic waiting for them. The client awakes on the beacon notification and receives its data.

Security Enhancements

Wireless technology does not remove any old security issues but introduces new ones—eavesdropping, man-in-the-middle, attacks, and denial of service. The requirements for Wi-Fi network security can be broken down into two primary components: authentication and privacy.

1. Authentication This is to keep unauthorized users off the network.

User authentication For user authentication, the server is used in which the username and password are sent and checked. The risk is data (username and password) sent before a secure channel is established. It is also prone to passive eavesdropping by attacker. The solution is to establish an encrypted channel before sending the username and password.

Server authentication Here the digital certificate is used. Validation of digital certificate occurs automatically within the client software.

2. Privacy It involves addressing failures of the original wired equivalent privacy (WEP) algorithm with the robust security network (RSN), and transition security network (TSN). Other security techniques are service set identifier (SSID), 802.1X access control, wireless protected access (WPA), and IEEE 802.11i.

WEP It provides same level of security as by wired network. The original security solution is offered by the IEEE 802.11 standard. It uses RC4 encryption with pre-shared keys and 24 bit *initialization vectors* (IVs). The key schedule is generated by concatenating the shared secret key with a random generated 24-bit IV. There is a total of 32 bit ICV (integrity check value). The number of bits in key schedule is equal to the sum of length of the plaintext and ICV. The two shared secret keys are

- 64 bit preshared key—WEP
- 128 bit preshared key—WEP2

These encrypt data only between 802.11 stations. Once the data enters the wired side of the network (between access point), WEP is no longer valid. The WEP has short IV and it is a static key. So, it offers very little security.

SSID It is used to identify an 802.11 network. It can be pre-configured or advertised in beacon broadcast. It is transmitted in clear text and provides very little security.

802.1X access control It is designed as a general-purpose network access control mechanism and not Wi-Fi specific, which authenticates each client connected to AP (for

WLAN) or switch port (for Ethernet). Authentication is done with the RADIUS server, which 'tells' the access point whether access to controlled ports should be allowed or not by following the procedure given below:

- The AP forces the user into an unauthorized state.
- The user sends an extensible authentication protocol (EAP) start message.
- The AP returns an EAP message requesting the user's identity.
- The identity sent by the user is then forwarded to the authentication server by AP.
- The authentication server authenticates user and returns an accept or reject message back to the AP.
- If the accept message is returned, then AP changes the client's state to authorized and normal traffic flows.

The access control scenario is shown in Fig. 14.8.

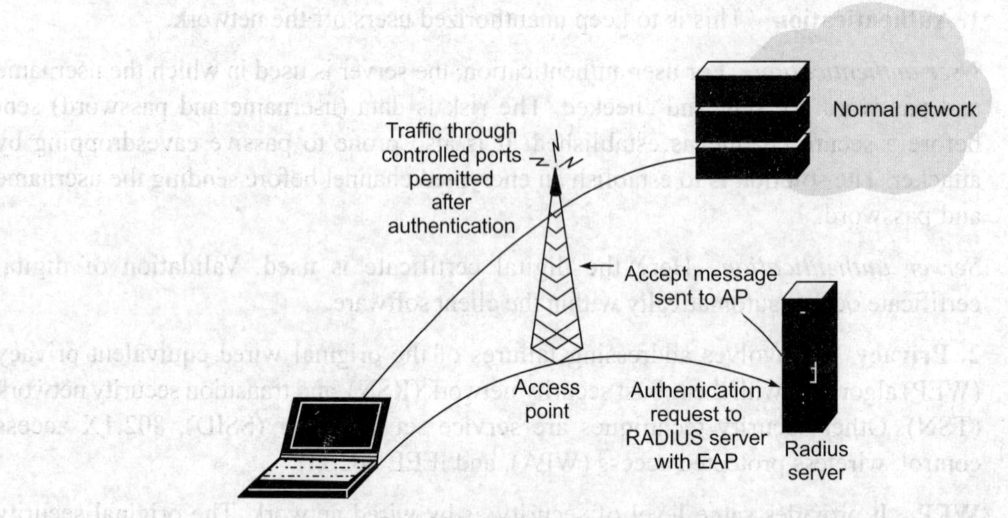

Fig. 14.8 IEEE802.1X access control scenario

Wireless protected access (WPA) It is a specification of standard-based, interoperable security enhancements that strongly increase the level of data protection and access control for existing and future wireless LAN system. The user authentication is done by 802.1x and EAP. Temporal key integrity protocol (TKIP) encryption is used with RC4, dynamic encryption keys (session based), in which 48 bit IV and per packet key mixing function are there. It fixes all issues found from WEP. It also uses message integrity code (MIC) Michael to ensure data integrity. Old hardware should be upgradeable to WPA. The WPA comes in two forms:

1. WPA pre-shared keys (PSK):
   - For small office/home office (SOHO) environments
   - Single master key used for all users
2. WPA enterprise:
   - For large organization

- Most secure method
- Unique keys for each user
- Separate user name and password for each user

IEEE 802.11i It provides standard for WLAN security with authentication. For data encryption, advance encryption standard (AES) protocol is used. It supports secure fast handoff, which allows roaming between APs without requiring the client to fully reauthenticate to every AP. The protocol will require new hardware.

Quality of Service Enhancements

These involve meeting the demands for bandwidth and minimizing network congestion and slowdown. The topic is left for self-study.

The general architecture of Wi-Fi network contains the following elements in general.

Access points (APs) The AP is wireless LAN transceiver or 'base station' that can connect one or many wireless devices simultaneously to the Internet.

Wi-Fi cards They accept the wireless signal and relay information. They can be internal or external. Examples are PCMCIA card for laptop and PCI card for desktop PC.

Safeguards These include firewalls or anti-virus protection software to protect the network from uninvited users and keep information secure.

The basic concept of Wi-Fi is same as of a walkie-talkie. A Wi-Fi hotspot is created by installing an access point to an Internet connection. An access point acts as a base station. When a Wi-Fi enabled device encounters a hotspot, the device can then connect to that network wirelessly. A single access point can support up to 30 users and can function within a range of 100–150 feet indoors and up to 300 feet outdoors. Many access points can be connected to each other via Ethernet cables to create a single large network.

There are three basic topologies for Wi-Fi: AP-based, peer-to-peer, and point-to-multipoint bridge.

AP-based topology (infrastructure mode) The client communicates through the access point. The RF coverage is provided by an AP. The overlapping area consists of two or more RF coverages of APs, which include 10–15% overlap to allow roaming (Fig. 14.9).

Peer-to-peer topology (ad hoc mode) AP is not required in this mode. The client devices within a cell can communicate directly with each other (Fig. 14.10). It is useful for setting up a wireless network quickly and easily.

Point-to-multipoint bridge topology This is used to connect a LAN in one building to LANs in other buildings even if the buildings are miles apart (Fig. 14.11). These conditions receive a clear line of sight between buildings. The line-of-sight range varies based on the type of wireless bridge and antenna used as well as the environmental conditions.

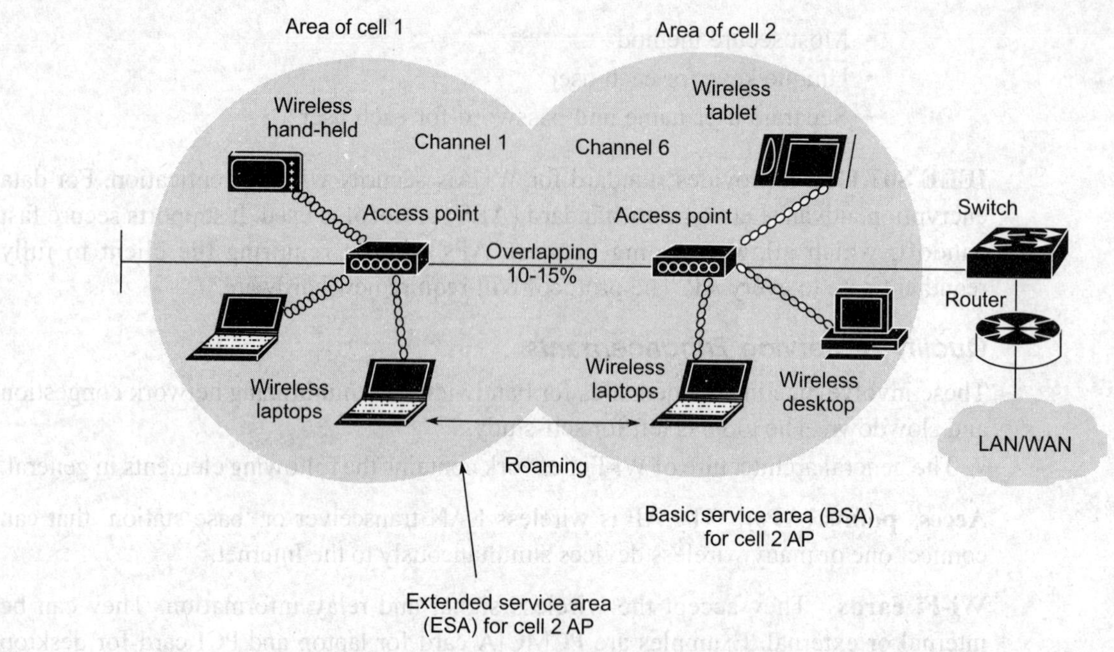

Fig. 14.9 Typical AP-based Wi-Fi infrastructure (An ESA is larger than or equal to a BSA.)

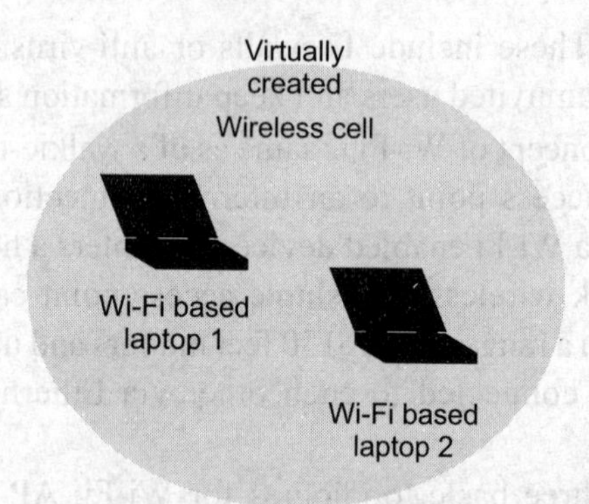

Fig. 14.10 Typical peer-to-peer topology in ad hoc mode

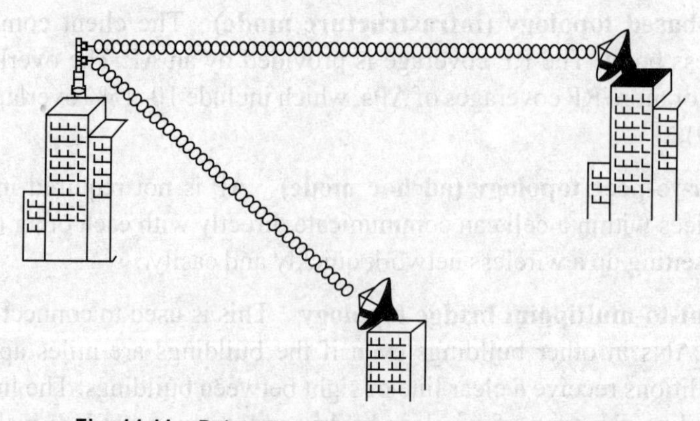

Fig. 14.11 Point-to-multipoint bridge topology for Wi-Fi

The Wi-Fi is able to compete with many wired systems. As a result of the flexibility and performance of the system, many Wi-Fi 'hotspots' have been set up and more are following. These enable people to use their laptops as they wait in hotels, airport lounges, cafes, and many other places using a wireless link rather than needing to use a cable. In view of its convenience, 802.11 Wi-Fi is now incorporated into virtually all new laptops.

Earlier, 802.11 was designed for infrared communication. It has a variety of standards, each with a letter suffix. These cover everything from the wireless standards themselves, to standards for security aspects, quality of service, etc.

Various 802.11 standards are given below:

- 802.11a: Wireless network bearer operating in the 5 GHz ISM band with data rate up to 54 Mbps
- 802.11b: Wireless network bearer operating in the 2.4 GHz ISM band with data rates up to 11 Mbps
- 802.11e: Quality of service and prioritization (QoS extension)
- 802.11f: Handover
- 802.11g: Wireless network bearer operating in 2.4 GHz ISM band with data rates up to 54 Mbps
- 802.11h: Power control
- 802.11i: Authentication and encryption (enhanced security)
- 802.11j: Interworking
- 802.11k: Measurement reporting
- 802.11n: Wireless network bearer operating in the 2.4 and 5 GHz ISM bands with data rates up to 600 Mbps
- 802.11s: Mesh networking

Of these, the standards that are most widely known are the network-bearer standards, 802.11a, 802.11b, and 802.11g. Shortly, the new 802.11n standard will be ratified and it is expected that the products will quickly become available for this. All the 802.11 Wi-Fi standards operate within the ISM frequency bands. These are shared by a variety of other users, but no licence is required for operation within these frequencies. This makes them ideal for a general system for widespread use.

Each of the different standards has different features and they were launched at different times. The first accepted 802.11 WLAN standard was 802.11b. This used frequencies in the 2.4 GHz ISM frequency band, which offered raw, over-the-air data rates of 11 Mbps using a modulation scheme known as *complementary code keying* (CCK) as well as supporting direct sequence spread spectrum (DSSS) from the original 802.11 specification. Almost in parallel with this, a second standard was defined. This was 802.11a, which used a different modulation technique, OFDM, and used the 5 GHz ISM band. Of the two standards, it was the 802.11b variant that caught on. This was primarily because the chips for the lower 2.4 GHz band were easier and cheaper to manufacture.

The 802.11b standard became the main Wi-Fi standard. Looking to increase the speeds, another standard, 802.11g, was introduced and ratified in June 2003. Using the more popular 2.4 GHz band and OFDM, it offered raw data rates of 54 Mbps, the same as 802.11b. In addition to this, it offered backward compatibility to 802.11b.

Then in January 2004, the IEEE announced that it had formed a new committee to develop an even higher speed standard. With much of the work now complete, 802.11n is beginning to establish itself in the same way as 802.11g. The industry came to a substantive agreement about the features for 802.11n in early 2006. This gave many chip manufacturers sufficient information to get their developments under way. As a result, it is anticipated that before long, with ratification of 802.11n expected in 2009, some cards and routers will find their way into the stores.

Although based around the IEEE standard 802.11, the Wi-Fi technology also addresses the European Telecommunications Standards Institute (ETSI) HIPERLAN (high performance radio local area network) standard. Table 14.3 gives the elementary comparison between major Wi-Fi standards.

Table 14.3 Comparison of various significant Wi-Fi standards

	802.11a (1999)	802.11b (1999)	802.11g (2003)	802.11n (2009)
Maximum data rate (Mbps)	54	11	54	~248
Modulation	OFDM (with change in modulation mapping method) (see Table 14.4)	CCK or DSSS	CCK, DSSS, or OFDM	CCK, DSSS, or OFDM
RF band (GHz)	5	2.4	2.4	2.4 or 5
Number of spatial streams	1	1	1	1, 2, 3, or 4
Channel width (MHz)	20	20	20	20 or 40
Typical range (indoors) in metre	~30	~30	Slightly greater than 30 m	>30m
Throughput (Mbps)	23	4.3	19	74

14.3.1 IEEE 802.11a

The 802.11a standard uses basic 802.11 concepts as its base and operates within the 5 GHz ISM band. The modulation is OFDM to enable it to transfer raw data at a maximum rate of 54 Mbps, although a more realistic practical level is in the region of the mid-20 Mbps region. The data rate can be reduced to 48, 36, 24, 18, 12, and 9, then to 6 Mbit/s if required. The 802.11a standard has 12 non-overlapping channels, 8 dedicated to indoors and 4 to point to point.

The OFDM signal used for 802.11 comprises 52 subcarriers. Of these, 48 are used for the data transmission and 4 are used as pilot subcarriers. The separation between the individual subcarriers is 0.3125 MHz. This results from the fact that the 20 MHz bandwidth is divided by 64. Although only 52 subcarriers are used, occupying a total of 16.6 MHz, the remaining space is used as a guard band between the different channels.

A variety of forms of modulation can be used on each of the 802.11a subcarriers. The BPSK, QPSK, 16 QAM, and 64 QAM can be used as the conditions permit (Table 14.4).

Table 14.4 Details of OFDM-based modulation stage in 802.11a

Data Rate (Mbps)	Modulation	Coding Rate	Coded Bits per Subcarrier	Coded Bits per OFDM Symbol	Data Bits per OFDM Symbol
6	BPSK	1/2	1	48	24
9	BPSK	3/4	1	48	36
12	QPSK	1/2	2	96	48
18	QPSK	3/4	2	96	72
24	16-QAM	1/2	4	192	96
36	16-QAM	3/4	4	192	144
48	64-QAM	1/2	6	288	192
54	64-QAM	3/4	6	288	216

For each set data rate, there is a corresponding form of modulation that is used. Within the signal itself, the symbol duration is 4 microseconds and there is a guard interval of 0.8 microseconds.

As with many data transmission systems, the generation of the signal is performed using digital signal processing techniques and a baseband signal is generated. This is then upconverted to the final frequency. Similarly, for signal reception, the incoming 802.11a signal is converted down to baseband and converted to its digital format after which it can be processed digitally.

14.3.2 IEEE 802.11b

The IEEE 802.11b was the first wireless LAN standard to be widely adopted and built in many laptop computers and other forms of equipment. The idea for wireless networking quickly caught on with many Wi-Fi hotspots being set up so that business people could access their e-mails and surf the Internet as required when they were travelling.

The RF signal format used for 802.11b is complementary code keying (CCK). The CCK can be considered as a block code generalization of the lower rate Barker code. This is a slight variation on CDMA, which uses the basic direct sequence spread spectrum as its basis. In view of the fact that the original 802.11 specification used CDMA/DSSS, it was easy to upgrade any existing chipset and other investment to provide the new 802.11b standard. As a result, 802.11b chipsets appeared relatively quickly in the market.

In 1999, CCK was adopted to replace the Barker code. A set of 64 eight-bit code words used to encode data for 5.5 and 11Mbps data rates in the 2.4 GHz band of 802.11b wireless networking. The code words have unique mathematical properties that allow them to be correctly distinguished from one another by a receiver even in the presence of substantial noise and multipath interference. Complementary codes, first introduced by Golay, are sets of finite sequences of equal length, such that the number of pairs of identical elements with any given separation in one sequence is equal to the number of pairs of unlike elements having the same separation in the other sequences.

The CCK works only in conjunction with the DSSS technology, which is specified in the original 802.11 standard. It does not work with FHSS. The CCK applies sophisticated mathematical formulas to the DSSS codes, permitting the codes to represent a greater volume of information per clock cycle. The transmitter can then send multiple bits of information with each DSSS code, enough to make possible the 11Mbps of data rather than the 2Mbps in the original standard.

Although 802.11b cards are specified to operate at a basic rate of 11 Mbps, the system monitors the signal quality. If the signal falls or the interference levels rise, then it is possible for the system to adopt a slower data rate with more error correction that is more resilient. Under these conditions, the system will first fall back to a rate of 5.5 Mbps, then 2, and finally 1 Mbps. This scheme is known as *adaptive rate selection* (ARS).

Although the basic raw data rates for transmitting data seem very good, in reality, the actual data rates achieved over a real-time network are much smaller. Even under reasonably good radio conditions, i.e. good signal and low interference, the maximum data rate that might be expected when the system uses TCP is about 5.9 Mbps. This results from a number of factors. One is the use of CSMA/CA where the system has to wait for clear times on a channel to transmit. Using this technique, when a node wants to make a transmission, it listens for a clear channel and then transmits. It then listens for an acknowledgement and if it does not receive one, it backs off a random amount of time, assuming that another transmission caused interference and then listens for a clear channel and then retransmits the data. Another is associated with the use of TCP and the additional overhead required. If UDP is used rather than TCP, then the data rate can increase to around 7.1 Mbps.

Some 802.11b systems advertise that they support much higher data rates than the basic 802.11b standard specifies. While more recent versions of the 802.11 standard, namely 802.11g, and 802.11n, specify much higher speeds, some proprietary improvements were made to 802.11b. These proprietary improvements offered speeds of 22, 33, or 44 Mbps and were sometimes labelled as '802.11b+'. These schemes were not endorsed by the IEEE and, in any case, they have been superseded by later versions of the 802.11 standard.

The 22 Mbps version of IEEE 802.11b (sometimes called 802.11b+) technology uses packet binary convolutional code (PBCC) CCK. Ordinary 802.11b uses a short block length for its 8QPSK data symbols. The PBCC, on the other hand, uses 64-state symbols. A PBCC symbol can carry more data, but it also requires a more powerful digital signal processor at the access point and the NIC to make that data available.

Another advantage PBCC has over CCK is that its 'convolutional coding' is a method of forward error correcting that enables to reduce the bit rate error without increasing transmission power. In real life, this means that one can get a higher data transmission rate and expand the range, all the while not using any more power than a conventional 802.11b device.

While CCK-OFDM lagged behind PBCC in development, it had the potential to deliver up to 55 Mbps at speeds of 2.4 GHz, whereas PBCC currently would top out at 33Mbps. Figure 14.12 will give an idea about the PBCC. Here the cover sequence is used for keying purpose.

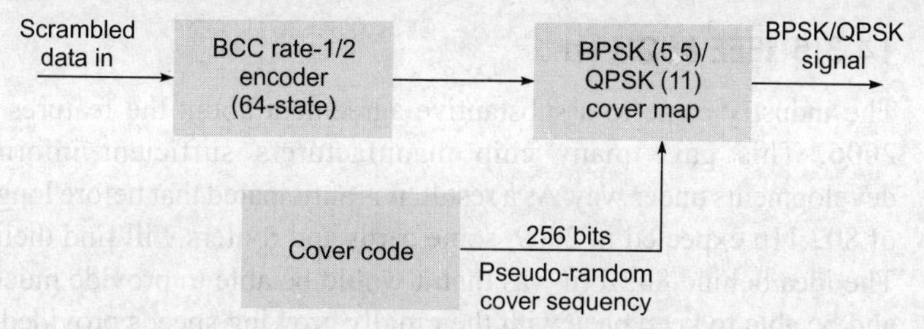

Fig. 14.12 Typical packet binary convolution code generation

14.3.3 IEEE 802.11g

In order to provide the higher speeds of 802.11a while operating on the 2.4 GHz ISM band, a new standard was introduced. This is known as 802.11g. It soon took over from the b standard. Even before the standard was ratified, 802.11g products were available in the market and before long, it became the dominant Wi-Fi technology. The 802.11g standard provided a number of improvements over the 802.11b standard, which was its predecessor.

A variety of modulation schemes can be used by 802.11g. For speeds of 6, 9, 12, 18, 24, 36, 48, and 54 Mbps, OFDM is used, but for 5.5 and 11 Mbps, it uses complementary code keying, and then for one and two Mbps, it uses DBPSK/DQPSK+DSSS.

The maximum range that can be achieved by 802.11g devices is slightly greater than that achieved by 802.11b, but the range at which the full 54 Mbps can be achieved is much shorter than the maximum range of an 802.11 device. Only when the signal levels and interference levels are low, can the maximum specified performance be achieved.

14.3.4 IEEE 802.11e

One of the major shortfalls for the developing applications for Wi-Fi is that it is not possible to allocate a required quality of service for the particular application. Now with IEEE 802.11e, the quality of service (QoS) problem is being addressed.

The issue of quality of service on 802.11 Wi-Fi is of particular importance in some applications and, accordingly, 802.11e is addressing it. For surfing applications, such as Internet web browsing and sending e-mails, delays in receiving responses or sending data does not have a major impact. It results in slow downloads or small delays in e-mails being sent. While it may have a small annoyance to the user, there is no real operational impact on the service being provided. However, for applications like voice or video transmission, e.g., voice over IP (VoIP), there is a far greater impact and this creates a much greater need for 802.11e. Delay, jitter, and missing packets result in the system, which loses the data and the service quality becomes poor. Accordingly, for these time-sensitive applications, it is necessary to be able to prioritize the traffic. This can only be done by allocating a service priority level to the packets being sent, and this is now all being addressed by IEEE standard 802.11e.

14.3.5 IEEE 802.11n

The industry came to a substantive agreement about the features for 802.11n in early 2006. This gave many chip manufacturers sufficient information to get their developments under way. As a result, it is anticipated that before long, with the ratification of 802.11n expected in 2009, some cards and routers will find their way into the stores. The idea behind 802.11n was that it would be able to provide much better performance and be able to keep pace with the rapidly growing speeds provided by technologies like Ethernet.

To achieve this, a number of new features have been incorporated into 802.11n to enable the higher performance. The major innovations are listed below:

- Changes to implementation of OFDM
- Introduction of MIMO
- MIMO power saving
- Antenna technology
- Wider channel bandwidth
- Reduced support for backward compatibility under special circumstances

Although each of these new innovations adds complexity to the system, much of this can be incorporated into the chipsets, enabling a large amount of the cost increase to be absorbed by the large production runs of the chipsets.

One of the major changes to the physical layer of 802.11n is to improve the performance of the implementation of the OFDM modulation. By adapting the way it is set up, the data rate can be increased from 54 Mbps achieved for 802.11a and g to 300 Mbps and more.

Use of MIMO in 802.11n

Multiple input–multiple output (MIMO) is a technique that exploits multipath propagation (refer to Chapter 15). Normally, when a signal is transmitted from A to B, the signal will reach the receiving antenna via multiple paths, causing interference. However, MIMO uses this multipath propagation to increase the data rate by using a technique known as spatial division multiplexing. The data is split into a number of what are termed spatial streams and these are transmitted through separate antennas to corresponding antennas at the receiver. Doubling the number of spatial streams doubles the raw data rate, enabling a far greater utilization of the available bandwidth. The current 802.11n standard allows for up to four spatial streams.

Power Saving

One of the problems with using MIMO is that it increases the power of the hardware circuitry. More transmitters and receivers need to be supported and this entails the use of more current. While it is not possible to eliminate the power increase resulting from the use of MIMO in 802.11n, it is possible to make the most efficient use of it. Data is normally transmitted in a 'bursty' fashion. This means that there are long periods, when the system remains idle or running at a very slow speed. During these periods, when MIMO is not required, the circuitry can be held inactive so that it does not consume power.

Antenna Technology for 802.11n

For 802.11n, the antenna-associated technologies have been significantly improved by the introduction of beam forming and diversity.

The beam forming focuses the radio signals directly along the path for the receiving antenna to improve the range and overall performance. A higher signal level and better signal-to-noise ratio will mean that the full use can be made of the channel.

The diversity uses the multiple antennas available and combines or selects the best subset from a larger number of antennas to obtain the optimum signal conditions. This can be achieved because there are often surplus antennas in an MIMO system. As 802.11n supports any number of antennas between one and four, it is possible that one device may have three antennas while another with which it is communicating will have only two. The supposedly surplus antenna can be used to provide diversity reception or transmission as appropriate.

Increased Bandwidth

An optional mode for the new 802.11 chips is to run using a double-sized channel bandwidth. Previous systems used the 20 MHz bandwidth, but the new ones have the option of using 40 MHz. The main trade-off for this is that there are less channels that can be used for other devices. There is sufficient scope at 2.4 GHz for three 20 MHz channels, but only one 40 MHz channel can be accommodated. Thus, the choice of whether to use 20 or 40 MHz has to be made dynamically by the devices in the net.

Removal of Backward Compatibility

While 802.11n retains the support for backward compatibility, this feature can be removed when all the devices operating in a net are 802.11n devices. This removes an overhead that is not required and enables the maximum efficiency to be maintained. When earlier devices enter the net, then the backward compatibility overhead and features are reintroduced. As with 802.11g, when earlier devices enter a net, the operation of the whole net is considerably slowed. Therefore, operating a net in 802.11n only, offers considerable advantages.

14.3.6 Wi-Fi MAC

The way in which the data is transmitted and controlled has a major impact on the way the QoS is achieved. This is largely determined by the way the MAC layer operates. Within 802.11, there are two options for the MAC layer:

1. The centralized control scheme, referred to as the *point coordination function* (PCF), for delay sensitive service
2. The contention based approach, called *distributed coordination function* (DCF), for best effort delivery service

Of these, few manufacturers of chips and equipment have implemented PCF and the industry seems to have adopted the DCF approach.

The PCF mode supports time-sensitive traffic flows to some degree. Wireless access points periodically send beacon frames to communicate network management and identification that is specific to that WLAN. Between the sending of these frames, PCF splits the time frame into a contention-free period and a contention period (Fig. 14.13). If PCF is enabled on the remote station, it can transmit data during the contention-free polling periods. However, the main reason why this approach has not been widely adopted is that the transmission times are not predictable. Unfortunately, the PCF has limited support and a number of limitations, for example, it does not define classes of traffic.

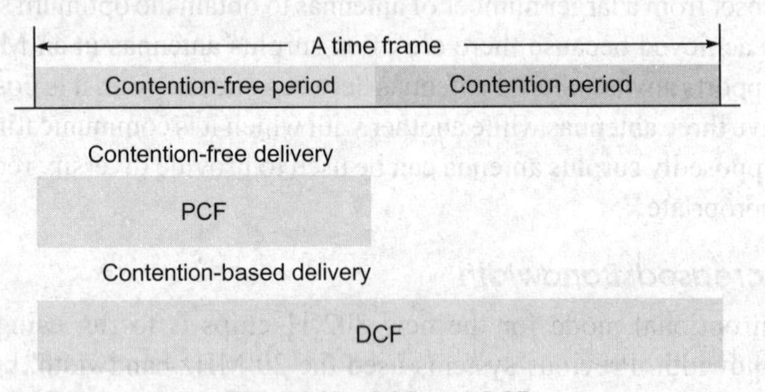

Fig. 14.13 PCF and DCF

The DCF scheme uses carrier sense multiple access with collision avoidance (CSMA/CA). Within this scheme, the MAC layer sends instructions for the receiver to look for other carriers transmitting. If it sees none, then it sends its packet after a given interval and awaits for an acknowledgement. If one is not received, then it knows its packet was not successfully received. It then waits for a given time interval and also checks the channel before retrying to send its data packet.

In more exact terms, the transmitter uses a variety of methods to determine whether the channel is in use, monitoring the activity looking for real signals and also determining whether any signals may be expected. This can be achieved because every packet that is transmitted includes a value indicating the length of time that transmitting station expects to occupy the channel. This is noted by any stations that receive the signal and only when this time has expired may they consider transmitting.

Once the channel appears to be idle, the prospective transmitting station must wait for a period equal to the DCF interframe space (DIFS). If the channel has been active, it must first wait for a time consisting of the DIFS and a random number of back-off slot times. This is to ensure that if two stations are waiting to transmit, then they do not both transmit together and then repeatedly transmit together.

A time, known as a *contention window* (CW), is used for this. This is a random number of back-off slots. If a transmitter intending to transmit senses that the channel becomes active, it must wait until the channel comes free, waiting a random period for the channel to come free, but this time allowing a longer CW.

The DCF has several limitations as discussed below.

- If many stations communicate at the same time, many collisions will occur, which will lower the available bandwidth (just like in Ethernet, which uses CSMA/CD).

  While the system works well in preventing stations transmitting together, the result of using this access system is that if the network usage level is high, then the time that it takes for data to be successfully transferred increases. This results in the system appearing to become slower for the users. There is no notion of high or low priority traffic.
- Once a station 'wins' access to the medium, it may keep the medium for as long as it chooses. If a station has a low bit rate (1 Mbps, for example), then it will take a long time to send its packet and all other stations will suffer from it.
- Generally, there are no QoS guarantees, when real-time data transfer is required.

The problem can be addressed by introducing a QoS identifier into the system. In this way, those applications where a high quality of service is required can tag their transmissions and take priority over the transmissions carrying the data that does not require immediate transmission and response. In this way, the level of delay and jitter on the data, such as that used for VoIP and video, may be reduced. To introduce the QoS identifier, it has been necessary to develop a new MAC layer and this has been undertaken under the IEEE 802.11e standard. In this standard, the traffic is assigned a priority level prior to transmission. This is termed *user priority* (UP) level and there are eight in total. Having done this, the transmitter then prioritizes all the data it has to waiting to be sent, by assigning it one of the four *access categories* (AC).

In order to achieve the required functions, the re-developed MAC layer takes on aspects of both the DCF and PCF from the previous MAC layer alternatives. This is termed the hybrid coordination function (HCF). The modified elements of the DCF are termed the enhanced distributed channel access (EDCA), while the elements of the PCF are termed the HCF controlled channel access (HCCA). Both EDCA and HCCA define traffic classes (TC).

EDCA

The EDCA uses a mechanism, called a *transmit opportunity* (TXOP), which is a bounded time interval during which a station can send as many frames as possible, but the transmission time must not extend beyond the maximum duration of the TXOP. Each priority level is assigned a TXOP, and this mechanism prevents low-speed stations from spending too much time using the media when other clients (including those with traffic in the higher priority queues). The EDCA provides a mechanism whereby traffic can be prioritized, but it remains a contention based system and, therefore, it cannot guarantee a given QoS. In view of this, it is still possible that transmitters with data of a lower importance can still pre-empt data from another transmitter with data of a higher importance.

When using EDCA, a new class of interframe space, called an *arbitration interframe space* (AIFS), has been introduced. This is chosen such that the higher the priority of the message, the shorter the AIFS and associated with this there is also a shorter contention window. The transmitter then gains access to the channel in the normal way, but in view of the shorter AIFS and shorter contention window, this means that the higher the chance of it gaining access to the channel. Although, statistically a higher priority message will usually gain the channel, this will not always be the case.

HCCA

The HCCA is more complex than EDCA—it acts like the PCF in that it uses a contention period (CP) and contention-free period (CFP). The HCCA adopts a different technique, using a polling mechanism. Accordingly it can provide guarantees about the level of service it can provide, and thereby providing a true quality of service level. Although it has interframe space (IFS), what is termed a point coordination IFS, as this is shorter than the DIFS mentioned earlier, it will always gain control of the channel. Once it has taken control, it polls all the stations or transmitters in the network. To do this, it broadcasts as particular frame indicating the start of polling, and it will poll each station in turn to determine the highest priority. It will then enable the transmitter with the highest priority data to transmit, although it will result in longer delays for traffic that has a lower priority.

The HCCA allows for a CFP to be initiated any time during a CP, which is called the *controlled access phase* (CAP). The control station, which is normally the access point (AP), is known as the *hybrid coordinator* (HC). It takes control of the channel. An AP can initiate a CAP whenever it needs to send a frame to a station; this allows the AP to make better decisions about how the wireless medium is used. During the CP, all wireless clients operate in EDCA mode. Another feature of HCCA is that traffic class and *traffic streams* (TS) are defined—allowing the AP to provide a per-session service for QoS on top of the per-station service that EDCA enables. The HCCA is a very powerful but complex coordination function that even allows clients to request specific transmission requirements like jitter and data rate, allowing for very effective implementations of voice over WiFi and video over WiFi.

14.4 WiMAX STANDARDS

The WiMAX is a new broadband wireless data communications technology or mobile Internet based around the IEEE 802.16 standard that will provide high-speed data communications (70 Mbps) over a wide area. The letters of WiMAX stand for *worldwide interoperability for microwave access* and it is a technology for point-to-multipoint (P2MP) wireless networking. The WiMAX technology is expected to meet the needs of a large variety of users—from those who are in developed nations wanting to install a new high-speed wireless data network very cheaply with the minimum cost and time required to those who are in rural areas needing fast access where wired solutions may not be viable because of the distances and costs involved.

The standard for WiMAX is a standard for wireless metropolitan networks (WMANs) that has been developed by working group number 16 of IEEE 802, specializing in broadband wireless access. It is also supported by a wide number of industry companies. The WiMAX forum is a wireless industry consortium with over 100 members, including many industry leaders. It has been set up to support and develop WiMAX technology worldwide, to bring common standards across the globe, to enable WiMAX to become an established worldwide technology.

One of the aims of the forum is to enable a standard to be adopted that will enable full interoperability between the products. Learning from the problems of poor interoperability experienced with previous wireless standards and the impact that this had on take up, the WiMAX forum aims to prevent this from happening. Ultimately, vendors will be able to have products certified under the auspices of the forum and then be able to advertise their products as 'forum certified'.

Although WiMAX technology will support traffic based on transport technologies ranging from Ethernet, Internet protocol (IP), and asynchronous transfer mode (ATM), the Forum will only certify the IP-related elements of the 802.16 products. The focus is on IP operations because this is now becoming the chief protocol to be used.

Although based around the IEEE standard 802.16, the WiMAX technology also addresses the ETSI HIPERMAN (HIgh PERformance Radio Metropolitan Area Network) standard. This will make the standard a truly international standard and one that has the backing of the industry leaders in these fields.

The WiMAX has two important standards/usage models: a fixed usage model IEEE 802.16d-2004 for fixed wireless broadband access (FWBA) and a portable usage model IEEE 802.16e-2005 for mobile wireless broadband access (MWBA). Both are released standards and amendments are available in the form of drafts. The OFDM is made scalable in IEEE 802.16e-2005. A new standard, IEEE 802.16m, is being developed for higher bandwidths. Higher data rate transmissions (@ 100 Mbps) are achieved in IEEE 802.16-2004 WiMAX through LOS communications, which incorporates a stationary transmitter and receiver but NLOS communication is much more complicated. The architecture of WiMAX is given in Fig. 14.14. For LOS, 802.16 protocol is used between two base stations. For fixed wireless access (e.g., Wi-Fi connection), 802.16d protocol is utilized while for mobile applications, 802.16e is more suitable. Normally, point-to-point/point-to-multipoint architecture is a basic one and with so many connections, they may result in a mesh architecture. The basic topologies are given in Fig. 14.15. Also the use of cellular infrastructure and base stations are required to have the continuous broadband access along with vehicular mobiliy. Under this condition, the base stations are facilitated by handover mechanisms.

The IEEE 802.16 protocol architecture has four layers—convergence, MAC, transmission, and physical. The protocol architecture shown in Fig. 14.16 represents mainly two layers equivalent to OSI—physical and data link layers. The WiMAX system can be converged with digital telephony, ATM, IP, etc.

14.4.1 WiMAX Air Interface/Physical Layer

The WiMAX 802.16 standard describes four different RF or air interfaces dependent upon the application envisaged. Of these, the one that is intended for non-line-of-sight applications up to 30 km and for frequencies below 11 GHz is the most widely implemented at the moment. As a result, it is often thought of as the WiMAX air interface.

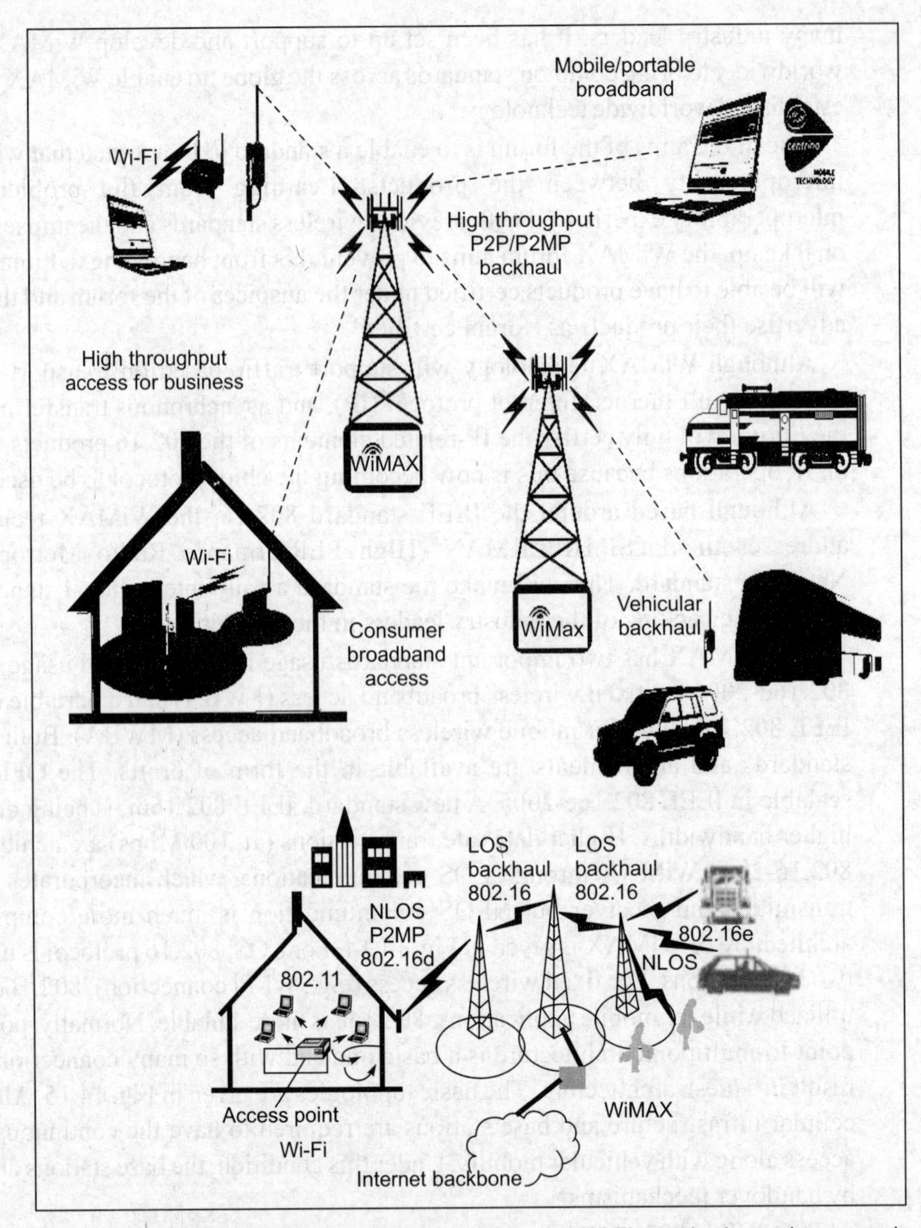

Fig. 14.14 WiMAX scenario (The figure represents that Wi-Fi can also be made communicating with WiMAX devices.)

The basic 802.16 WiMAX standard uses frequencies in the range of 10 to 66 GHz although extensions to the standard allow for the use of other frequencies. The 802.16a and higher standards added support for frequencies in the band 2–11 GHz. At these frequencies, particularly those in the range between 10 and 66 GHz, the transmission path is essentially line of sight and multipath reflections are reduced. This increases the rate at which data can be sent. This is the reason 802.16 is used between base station to base station LOS links in the WiMAX architecture.

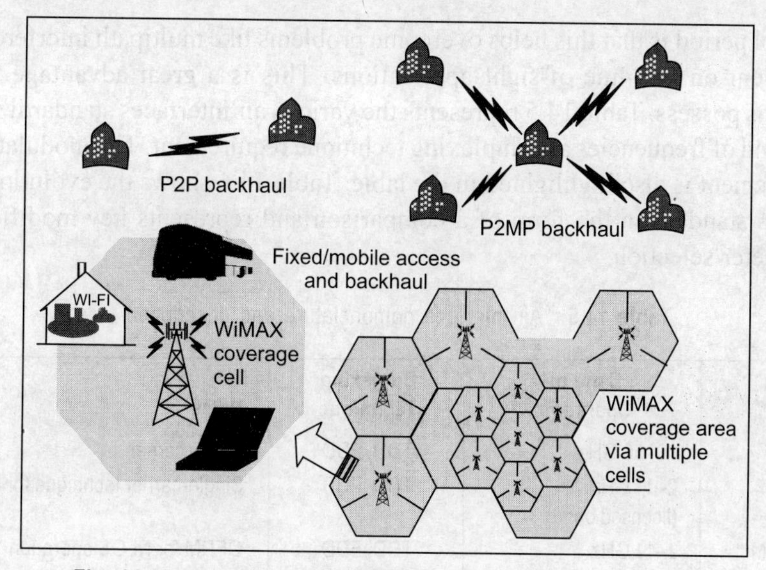

Fig. 14.15 Examples of WiMAX topology with cellular structure

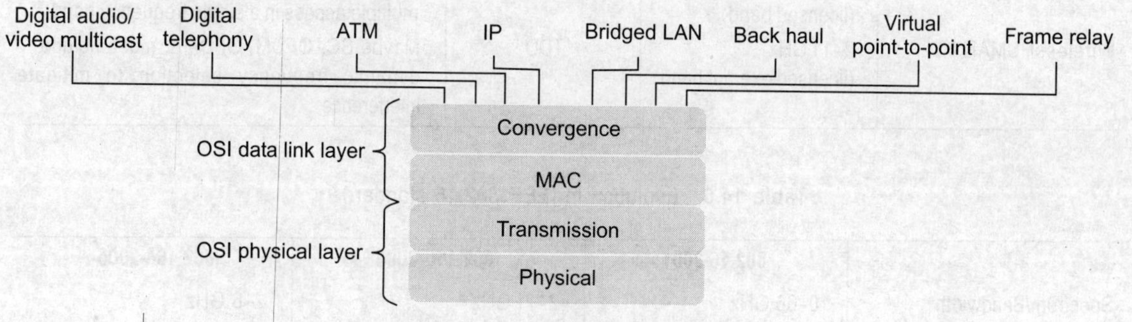

Fig. 14.16 WiMAX protocol architecture

The WiMAX 802.16d RF signal uses OFDM techniques and the signal incorporates 256 carriers in a total signal bandwidth, which may range from 1.25 to 20 MHz. Of the 256 carriers possible, only 200 are actually used. Some are not used as the frequencies that would be occupied by them are used as a guard band, and the centre frequency carrier is not used because it is very susceptible to RF carrier feed-through. The total of 200 carriers used is split between 192 carriers used for data payload and the remaining 8, which are used as pilots. The pilot carriers are always BPSK modulated and the data carriers are BPSK, QPSK, 16QAM, or 64QAM.

The WiMAX signal bandwidth can be set between 1.25 and 20 MHz and regardless of the bandwidth, the WiMAX signal contains the same 200 carriers. Thus, the carrier spacing varies according to the overall bandwidth (adaptive bandwidth concept). This feature is called *scalability*. Due to this, the number of users can be increased from 100 to 1000. As the number of subscribers grows, the spectrum can be reallocated with the process of sectoring. To maintain orthogonality between the individual carriers, the symbol period must be the reciprocal of the carrier spacing. As a result, narrow bandwidth WiMAX systems have a longer symbol period. The advantage of a longer symbol period is that this helps overcome problems like multipath interference, which is prevalent on non-line-of-sight applications. This is a great advantage that WiMAX systems possess. Table 14.5 represents the various air interfaces standardized along with the band of frequencies and duplexing technique requirement. The modulation technique requirement is also highlighted in the table. Table 14.6 shows the evolution of the IEEE 802.16 standard in the form of a comparison and represents key modifications in the parameter selection.

Table 14.5 Air interface nomenclature and description

Designation	Band of Operation	Duplexing Technique	Notes
WirelessMAN SC™	10–66 GHz	TDD, FDD	Single carrier
WirelessMAN SCa™	2–11 GHz (licensed band)	TDD, FDD	Single-carrier technique for NLOS
WirelessMAN OFDM™	2–11 GHz (licensed band)	TDD, FDD	OFDM for NLOS operation
WirelessMAN OFDMA™	2–11 GHz (licensed band)	TDD, FDD	OFDM broken into subgroups to provide multiple access in a single frequency band
WirelessHUMAN™	2–11 GHz (licensed exempt band)	TDD	Maybe SC, OFDM, OFDMA; must include dynamic frequency selection to mitigate interference

Table 14.6 Evolution in IEEE 802.16 standard

	802.16-2001	802.16d-2004	802.16e-2005
Spectrum/Bandwidth	10–66 GHz	2–11 GHz	2–6 GHz
Propagation channel condition	LOS	NLOS	NLOS
Bit rate	Up to 134 Mbps (28 MHz channelization)	Up to 75 Mbps (20 MHz channelization)	Up to 15 Mbps (5 MHz channelization)
Modulation	QPSK, 16QAM (optional in UL)	256 subcarriers, OFDM-BPSK, QPSK, 16QAM, 64QAM	Scalable OFDMA, QPSK, 16QAM, 64QAM, 256QAM (optional)
Mobility	Fixed	Fixed/nomadic	Portable/mobile

Note: Only significant changes are highlighted.

The IEEE 802.16 WiMax standard allows data transmission using multiple broadband frequency ranges. The fact that the technology can work in multiple frequency ranges allows it to avoid interference with other wireless applications, thereby being able to operate satisfactorily in the presence of other transmissions. The frequency bands that are chosen for a particular system affect a number of factors, including the data rate that can be carried and the range that can be achieved. Thus, it is possible to choose the frequency band to be used according to the prevailing conditions and the requirements for the system.

14.4.2 WiMAX Applications in Competition with Wi-Fi

The WiMAX 802.16 standard has a tremendous number of possible applications. The initial plan is to use WiMAX as a point-to-multipoint broadband technology to provide individual users with access to broadband data services. In this way, it would provide an attractive alternative to technologies like DSL. Typically, providers would want to limit the number of subscribers to any one base station, to around 500 to preserve the bandwidth needed by each user. In this application, it is unlikely that ranges will exceed 10 miles.

While it will no doubt be used in this role, with a maximum ranges up to 31 miles, WiMAX may also achieve widespread use and provide a backbone for other services when no wired service exists. In this way, it may be used as a part of the service providers' infrastructure. It is anticipated that services will use directional antennas and, in this way, several services can share the same frequency bands with minimum levels of interference. As a result, the capacity of the system can be huge.

In some instances, it may be thought that there may be competition between WiMAX and Wi-Fi 802.11. However, this is not the intention. Although WiMAX can be used to provide applications like Internet/intranet access in a way similar to Wi-Fi, the range of Wi-Fi is limited up to around 50 metres. Against this, WiMAX can transmit data up to a distance of 50 km. Additionally, WiMAX offers better connecting speeds, although the data rates that can be passed by Wi-Fi are rising.

The WiMAX is not aimed to compete with Wi-Fi but is aimed to address a different market and hence to coexist with it as shown in Fig. 14.17. A general comparison of Wi-Fi and WiMAX is given in Table 14.7. This arises from the fact that WiMAX coverage is measured in square kilometres, while that of Wi-Fi is measured in square metres. With

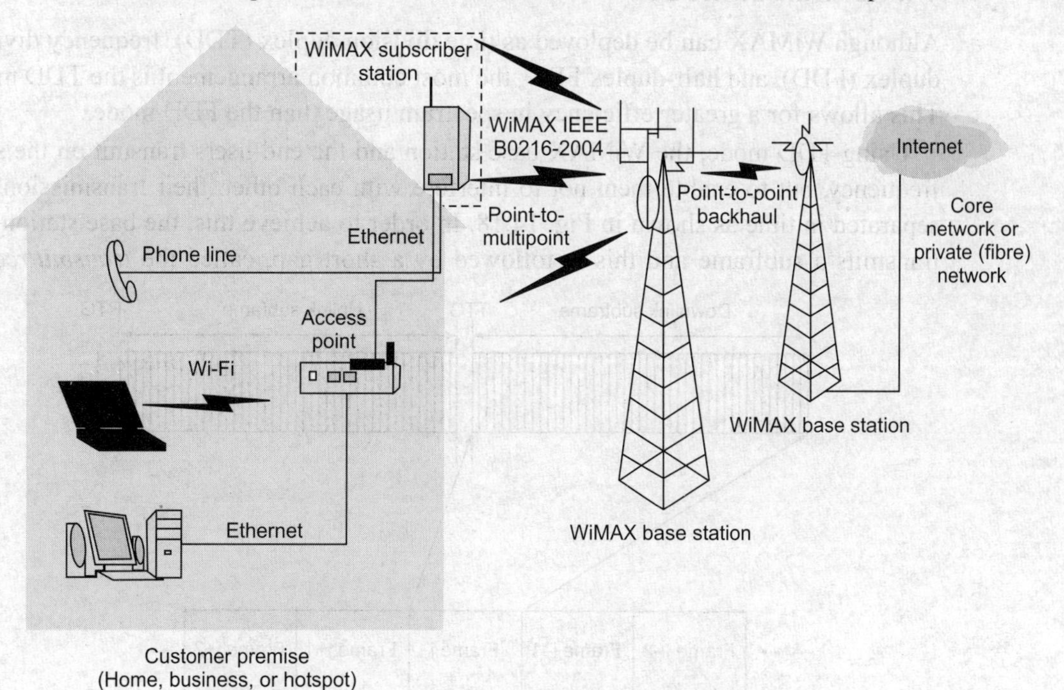

Fig. 14.17 P2MP architecture of WiMAX incorporating Wi-Fi

Table 14.7 General comparison of Wi-Fi and WiMAX

Wi-Fi	WiMAX
1. Wide (20 MHz) frequency channels	1. Channel bandwidth can be chosen by operator (e.g., for sectorization), 1.5 MHz to 20 MHz wide channels, MAC designed for scalability independent of channel bandwidth
2. MAC designed to support 10's of users ~ 2.7 bps/Hz	2. MAC designed to support thousands of users ~5.0 bps/ Hz
3. Contention based MAC (CSMA/CA)=> no guaranteed QoS	3. Grant request MAC, centrally enforced QoS
4. Standard cannot currently guarantee latency for voice, video	4. Designed to support voice and video from ground up
5. Standard does not allow for differentiated levels of service on a per-user basis	5. Supports differentiated service levels, e.g., best effort for residential, T1 for business customers
6. Optimized for ~50+ metres	6. Optimized for up to ~50 km
7. No near–far compensation	7. Designed to handle many users spread over kilometres
8. Optimized for indoor performance; designed to handle indoor multipath delay spread of 0.8 μs	8. Optimized for outdoor NLOS performance; designed to tolerate greater multipath delay spread up to 10 μs
9. No mesh topology support within ratified standards	9. Standard supports mesh network topology and advanced antenna techniques

WiMAX specifications being finalized, chips are now available and the technology is becoming a reality. The WiMAX is likely to become a major wireless technology, but one that complements the others, including 802.11, Bluetooth, Zigbee, and the like, as each one addresses a different market sector.

14.4.3 WiMAX Modes

Although WiMAX can be deployed as time division duplex (TDD), frequency division duplex (FDD), and half-duplex FDD, the most common arrangement is the TDD mode. This allows for a greater efficiency in spectrum usage than the FDD mode.

Using TDD mode, the WiMAX base station and the end-users transmit on the same frequency, but to enable them not to interfere with each other, their transmissions are separated in time as shown in Fig. 14.18. In order to achieve this, the base station first transmits a subframe and this is followed by a short gap, called the *transmit/receive*

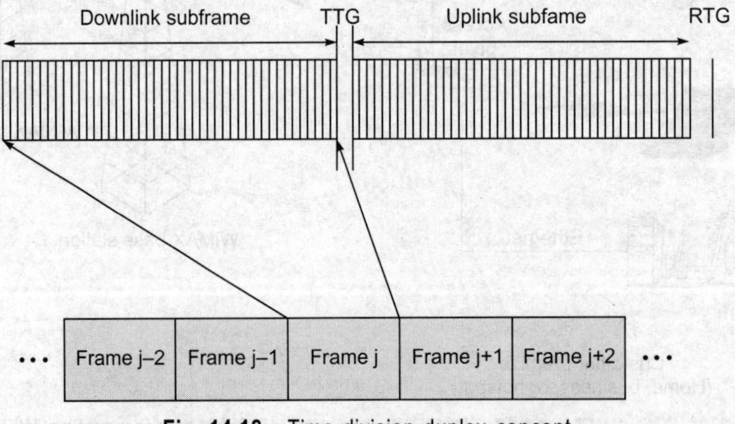

Fig. 14.18 Time division duplex concept

transition gap (TTG). After this gap, the users or remote stations are able to transmit their subframes. The timing of these 'uplink' subframes needs to be accurately controlled and synchronized such that they do not overlap whatever the distance they are from the base station. Once all the uplink subframes have been transmitted, another short gap, known as the *receive/transmit transition gap* (RTG), is left before the base station transmits again.

There are slight differences between the WiMAX subframes transmitted on the uplink and downlink. The downlink subframe begins with a preamble, after which a header is transmitted and this is followed by one or more bursts of data. The modulation within a subframe may change, but it remains the same within an individual burst. Nevertheless, it is possible for the modulation type to change from one burst to the next. The first bursts to be transmitted use the more resilient forms of modulation, such as BPSK and QPSK. Later bursts may use the less resilient forms of modulation, such as 16QAM and 64QAM, which enable more data to be carried.

14.4.4 Different Versions of WiMAX Standard

IEEE 802.16a

This is a version of WiMAX for use in licensed and unlicensed frequency bands between 2 and 11 GHz. It supports mesh deployment in which transceivers can act as relay stations, passing messages on from one station to the next, thereby increasing the range. The use of the lower frequencies allows more flexible implementations of the technology as signals at these frequencies can penetrate walls and other barriers without the levels of attenuation experienced at higher frequencies. As a result, currently these frequencies are mostly being used.

IEEE 802.16b

This increases the spectrum that is specified to include frequencies between 5 and 6 GHz while also providing for quality of service.

IEEE 802.16c

This provides a system profile for operating between 10 and 66 GHz and provides more details for operations within this range. The aim is to enable greater levels of interoperability.

IEEE 802.16d

This provides minor improvements and fixes to 802.16a. This includes the use of 256 carrier OFDM. The profiles for compliance testing are also provided.

IEEE 802.16e

This standard harmonizes the networking between the fixed base stations and the mobile devices, rather than just between the base stations and the fixed recipients. This will enable such activities as handovers, enabling mobile users to receive a high-quality continuous service as their vehicles move.

14.4.5 Quality of Service in WiMAX

The primary purpose of QoS feature is to define transmission ordering and scheduling on the air interface. These features often need to work in conjunction with mechanisms beyond the air interface in order to provide end-to-end QoS.

Requirements for QoS

- A configuration and registration function to pre-configure SS-based QoS service flows and traffic parameters
- A signalling function for dynamically establishing QoS-enabled service flows and traffic parameters
- Utilization of MAC scheduling and QoS traffic parameters for uplink service flows
- Utilization of QoS traffic parameters for downlink service flows

14.5 WIRELESS SENSOR NETWORKS

A wireless ad hoc sensor network consists of a number of sensors spread across a geographical area. Each sensor has wireless communication capability and some level of intelligence for signal processing and networking of the data. Some examples of wireless ad hoc sensor networks are given below:

1. Military sensor networks to detect and gain as much information as possible about enemy movements, explosions, and other phenomena of interest
2. Sensor networks to detect and characterize chemical, biological, radiological, nuclear, and explosive (CBRNE) attacks and material
3. Sensor networks to detect and monitor environmental changes in plains, forests, oceans, etc.
4. Wireless traffic sensor networks to monitor vehicle traffic on highways or in congested parts of a city
5. Wireless surveillance sensor networks for providing security in shopping malls, parking garages, and other facilities
6. Wireless parking lot sensor networks to determine which spots are occupied and which are free

These examples suggest that wireless ad hoc sensor networks offer certain capabilities and enhancements in operational efficiency in civilian applications as well as assist in the national effort to increase alertness to potential terrorist threats.

Two ways to classify wireless ad hoc sensor networks are whether or not the nodes are individually addressable and whether the data in the network is aggregated. The sensor nodes in a parking lot network should be individually addressable so that one can determine the locations of all the free spaces. This application shows that it may be necessary to broadcast a message to all the nodes in the network. If one wants to determine the temperature in a corner of a room, then addressability may not be so important. Any node in the given region can respond. The ability of the sensor network to aggregate the data collected can greatly reduce the number of messages that need to be transmitted across the network.

14.5.1 Goals and Requirements of a Wireless Sensor Network

Goals

The basic goals of a wireless ad hoc sensor network generally depend upon the application, but the following tasks are common to many networks.

To determine the value of some parameter at a given location. In an environmental network, one might want to know the temperature, atmospheric pressure, amount of sunlight, and relative humidity at a number of locations. This example shows that a given sensor node may be connected to different types of sensors, each with a different sampling rate and range of allowed values.

To detect the occurrence of events of interest and estimate parameters of the detected events In the traffic sensor network, one would like to detect a vehicle moving through an intersection and estimate the speed and direction of the vehicle.

To classify a detected object It is a vehicle in a traffic sensor network, a car, a minivan, a light truck, a bus, etc.

To track an object In a military sensor network, an enemy tank is tracked as it moves through the geographic area covered by the network.

In these four tasks, an important requirement of the sensor network is that the required data be disseminated to the proper end-users. In some cases, there are fairly strict time requirements on this communication. For example, the detection of an intruder in a surveillance network should be immediately communicated to the police such that action can be taken.

Network Requirements

The wireless ad hoc sensor network requirements are as follows.

Large number of (mostly stationary) sensors Aside from the deployment of sensors on the ocean surface or the use of mobile, unmanned, robotic sensors in military operations, most nodes in a smart sensor network are stationary. Networks of 10,000 or even 100,000 nodes are envisioned and so scalability is a major issue.

Low energy use Since in many applications the sensor nodes will be placed in a remote area, service of a node may not be possible. In this case, the lifetime of a node may be determined by the battery life, thereby requiring the minimization of energy expenditure.

Network self-organization Given the large number of nodes and their potential placement in hostile locations, it is essential that the network be able to self-organize; manual configuration is not feasible. Moreover, nodes may fail (either from lack of energy or from physical destruction) and new nodes may join the network. Therefore, the network must be able to periodically reconfigure itself so that it can continue to function. Individual nodes may become disconnected from the rest of the network, but a high degree of connectivity must be maintained.

Collaborative signal processing Yet another factor that distinguishes the wireless ad hoc sensor networks from MANETs is that the end goal is detection/estimation of some events of interest and not just communications. To improve the detection/estimation performance, it is often quite useful to fuse data from multiple sensors. This data fusion requires the transmission of data and control messages and so it may put constraints on the network architecture.

Querying ability A user may want to query an individual node or a group of nodes for information collected in the region. Depending on the amount of data fusion performed, it may not be feasible to transmit a large amount of data across the network. Instead, various local sink nodes will collect data from a given area and create summary messages. A query may be directed to the sink node nearest to the desired location.

With the availability of low-cost, short-range radios along with advances in wireless networking, it is expected that wireless ad hoc sensor networks will be more commonly deployed. In these networks, each node may be equipped with a variety of sensors, such as acoustic, seismic, infrared, still/motion video camera, etc. These nodes may be organized in clusters, such that a locally occurring event can be detected by most of, if not all, the nodes in a cluster. Each node may have sufficient processing power to make a decision, and it will be able to broadcast this decision to the other nodes in the cluster. One node may act as the cluster master, and it may also contain a longer range radio using a protocol, such as IEEE 802.11 and Bluetooth.

14.6 IEEE 802.15.4 AND ZIGBEE

Zigbee is a wireless networking standard that is aimed at remote control and sensor applications and is suitable for operation in harsh radio environments and in isolated locations. It builds on IEEE standard 802.15.4, which defines the physical and MAC layers. Above this, Zigbee defines the application and security layer specifications enabling interoperability between the products from different manufacturers. In this way, Zigbee is a superset of the 802.15.4 specification.

With the applications for remote wireless sensing and control growing rapidly, it is estimated that the market size can reach hundreds of millions of dollars in the next a few years. This makes Zigbee a very attractive proposition and one that warrants the introduction of a focused standard.

The Zigbee standard is organized under the auspices of the Zigbee Alliance. This organization has over seventy members, of which some have taken on the status of what they term 'promoter.' These companies are Ember, Honeywell, Invensys, Mitsubishi, Motorola, Philips, and Samsung. Under the umbrella of the Zigbee Alliance, the new standard will be pushed forward, taking on board the requirements of the users, manufacturers, and system developers.

14.6.1 Zigbee Basics

The distances that can be achieved transmitting from one station to the next extend up to about 70 metres, although very much greater distances may be reached by relaying data from one node to the next in a network.

The main applications for 802.15.4 are aimed at control and monitoring applications, where relatively low levels of data throughput are needed, and with the possibility of remote, battery-powered sensors, the low-power consumption is a key requirement. Sensors, lighting controls, security, and many more applications are all candidates for the new technology.

The system is specified to operate in one of the three license-free bands at 2.4 GHz, 915 MHz for North America and 868 MHz for Europe. In this way, the standard is able to operate around the globe, although the exact specifications for each of the bands are slightly different. At 2.4 GHz, there are a total of sixteen different channels available and the maximum data rate is 250 kbps. For 915 MHz, there are ten channels and the standard supports a maximum data rate of 40 kbps, while at 868 MHz, there is only one channel and this can support data transfer at up to 20 kbps.

The modulation techniques also vary according to the band in use. The direct sequence spread spectrum is used in all cases. However, for the 868 and 915 MHz bands, the actual form of modulation is binary phase shift keying. For the 2.4 GHz band, O-QPSK is employed.

The Zigbee systems may operate in heavily congested environments and in areas where levels of extraneous interference are high. The 802.15.4 specification has incorporated a variety of features to ensure exceedingly reliable operation. These include quality assessment, receiver energy detection, and clear channel assessment. The CSMA techniques are used to determine when to transmit and, in this way, unnecessary clashes are avoided.

The data is transferred in packets. These have a maximum size of 128 bytes, allowing for a maximum payload of 104 bytes. Although this may appear low when compared to other systems, the applications in which 802.15.4 and Zigbee are likely to be used should not require very high data rates.

The standard supports 64-bit IEEE addresses as well as 16-bit short addresses. The 64-bit addresses uniquely identify every device in the same way that devices have a unique IP address. Once a network is set up, the short addresses can be used and this enables over 65,000 nodes to be supported.

It also has an optional superframe structure with a method for time synchronization. In addition to this, it is recognized that some messages need to be given a high priority. To achieve this, a guaranteed time slot mechanism has been incorporated into the specification. This enables these high-priority messages to be sent across the network as swiftly as possible.

Above the physical and MAC layers defined by 802.15.4, the Zigbee standard itself defines the upper layers of the system. This includes many aspects, including the messaging, configurations that can be used, security aspects, and application profile layers.

There are three different network topologies that are supported by Zigbee, namely the star, mesh, and cluster tree or hybrid networks. Each has its own advantages and can be used to gain advantage in different situations.

1. The star network is commonly used and has the advantage of simplicity. As the name suggests, it is formed in a star configuration with outlying nodes communicating with a central node.
2. Mesh or peer-to-peer networks enable high degrees of reliability to be obtained. They consist of a variety of nodes placed as needed and nodes within the range being able to communicate with each other form a mesh. The messages may be routed across the network using the different stations as relays. There is usually a choice of routes that can be used and this makes the network very robust. If interference is present on one section of a network, then another can be used instead.
3. Finally, there is what is known as a cluster tree network. This is essentially a combination of star and mesh topologies.

Both 802.15.4 and Zigbee have been optimized to ensure that low power consumption is a key feature. Although nodes with sensors of control mechanisms towards the centre of a network are more likely to have mains power, many towards the extreme may not. The low-power design has enabled battery life to be typically measured in years, enabling the network not to require constant maintenance.

Although there are an increasing number of wireless standards that are appearing, Zigbee has a distinct area upon which it is focused. It is not intended to compete with the standards like 802.11, Bluetooth, etc. Instead, it has been optimized to ensure that it meets its intended requirements, fulfilling the needs for remote control and sensing applications.

14.7 ULTRA-WIDEBAND (UWB) TECHNOLOGY

Just as many wireless technologies seem to be moving into high-volume production and becoming established, a new technology, known as ultra-wideband (UWB), has hit the scene and is making a big impact on the industry. The UWB is used for indoor and short-range outdoor communication. It is a rapidly emerging wireless technology that promises data rates well beyond those possible with the currently deployed technologies, such as 802.11a, b, g, WiMAX, and the like. As such, this new wireless standard is likely to gain a significant market share in the years to come. It is gaining considerable acceptance and being proposed for use in a number of areas. Already Bluetooth, wireless USB, and others are developing solutions and in these areas alone, its use should be colossal.

As the name implies, ultra-wideband is a form of transmission that occupies a very wide bandwidth. Typically, the bandwidth will be of many GHz and it is this aspect that enables it to carry data rates of Gbps. This new technology has much to offer both in the performance and data rates as well as the wide number of application in which it can be used. The UWB can be used in both commercial and military applications as listed below:

Commercial applications:

- High-speed LAN/WAN (>20 Mbps)
- Avoidance radar
- Altimeter (aviation)

- Intrusion detection
- Geolocation
- Tags for intelligent transport systems

Military applications:

- Radar
- Covert communications
- Intrusion detection
- Precision geo-location
- Data links

The rate at which work was undertaken in 1960s increased in the 1970s and 80s as the supporting technologies became available. During this period, it was termed *carrier-free* or *impulse* technology. The term *UWB* was only coined in the later 1980s by the US Department of Defense. However, by this time, many patents had been awarded and a considerable degree of development had been invested in the technology.

The development of UWB technology was primarily intended for military applications and it was classified. As a result, little development took place in the commercial arena. However, in the years following 2000, commercial wireless communications became established. The technologies like 802.11 (Wi-Fi), Bluetooth, and others became established. These paved the way for commercial wireless communications and showed the flexibility offered by wireless communications for a very wide variety of applications from mobile phone peripheral connectivity to mobility and connectivity for laptops. These technologies and others grew rapidly. Accordingly, commercial applications for ultra-wideband (UWB) became very apparent and its commercial exploitation started.

In view of the fact that UWB occupies a wide bandwidth, even though at a low-power level, it has to exist alongside the traditional transmissions without causing any undue interference. Despite this, the UWB transmissions are allowed, provided that they remain within a given power density and frequency profile. This ensures that the allowed transmission levels do not cause any noticeable interference to existing transmissions.

Unlike most other wireless technologies in use today, UWB employs a totally different method of transmission. Rather than using a specified frequency with a carrier, the technique that is used by traditional transmissions, UWB uses what can be termed *time-domain electromagnetics*. In other words, UWB uses pulses that spread out over a wide bandwidth, rather than transmissions that are confined within a given channel.

The fact that UWB transmissions have such a wide bandwidth means that they will cross the boundaries of many of the currently licensed carrier-based transmissions. As such, one of the fears is that UWB transmission may cause interference. However, the very high bandwidth used also allows the power spectral density to be very low, and the power limits on UWB are being strictly limited by the regulatory bodies. In many instances, they are lower than the spurious emissions from electronic apparatus that has been certified. In view of this, it is anticipated that they will cause no noticeable interference to other carrier-based licensed users. Nevertheless, the regulatory bodies are moving forward cautiously, so that the users who already have spectrum allocations are not affected.

Despite the single-named use for the UWB transmissions, there are two very different technologies being developed. One is based around a carrier-free technology, where a series of impulses is transmitted. In view of the very short duration of the pulses, the spectrum of the signal occupies a very wide bandwidth. The alternative technology uses a wideband or multiband OFDM (MBOFDM) signal, which is effectively a 500 MHz wide OFDM signal that is hopped in frequency to enable it to occupy a sufficiently high bandwidth.

Both these systems have their advantages and disadvantages, each one having its supporters and applications for which it is most suited. The impulse-based technology, also called direct sequence ultra-wideband (DS-UWB), in view of some of the techniques used, is being used for a number of high data rate transmissions, such as short-range video transmissions. The MBOFDM, on the other hand, is being adopted for wireless USB, where it performs well.

The FCC in the US has approved the UWB transmissions but with restrictions on the frequencies over which the transmission can spread as well as the power limits. This will enable the UWB transmissions to communicate successfully but without affecting existing 'narrowband' transmissions. To achieve these requirements, the FCC has mandated that UWB transmissions can legally operate in the range from 3.1 GHz to 10.6 GHz, at a limited transmit power of – 41dBm/MHz. Additionally, the transmissions must occupy a bandwidth of at least 500 MHz as well of at least 20% of the centre frequency. To achieve this requirement, a transmission with a centre frequency of 6 GHz, for example, must have a bandwidth of at least 1.2 GHz. Consequently, UWB provides dramatic channel capacity at short range, which limits the interference.

The fact that very low power density levels are transmitted means that the interference to other services will be reduced to the limits that are not noticeable to the traditional transmissions. Additionally, the lowest frequencies for UWB have been set above 3 GHz to ensure that they do not cut across bands currently used for GPS, cellular, and many other services.

14.7.1 UWB Air Interface

The UWB is a revolutionary wireless technology that enables data to be transmitted at speeds well in excess of 100 Mbps. It is often referred to an impulse, baseband, or zero-carrier technology. It operates by sending low-power Gaussian shaped pulses that are coherently received at the receiver. In view of the fact that the system operates using pulses, the transmissions are spread out over a wide bandwidth, typically, many hundreds of megahertz or even several gigahertz. This means that it will overlay the bands and transmissions used by more traditional channel-based transmissions.

Each of the UWB pulses has an extremely short duration. This is typically between 10 and 1000 picoseconds and, as a result, it is shorter than the duration of a single bit of the data to be transmitted. The short pulse duration means that multipath effects can usually be ignored, giving rise to a large degree of resilience in the UWB transmissions when the signal path is within buildings.

In view of the wide bandwidth over which the UWB transmissions are spread, the actual energy density is exceedingly low. In fact, many of the transmissions themselves are less that the unintentional or spurious radiation levels from a typical PC. Typically, a UWB transmitter might transmit less than 75 nW per MHz. When integrated over the total bandwidth of the transmission, it means that transmissions may only be around 0.25 mW. This is very small when compared to 802.11 transmissions, which may be between 25 and 100 mW, or Bluetooth, which may be anywhere between 1 mW and 1 W.

This very low spectral density means that the UWB transmissions do not cause harmful interference to other radio transmissions using the traditional carrier-based techniques and operating in the existing bands. Even in the bands that are likely to be more sensitive to interference, such as the global positioning system (GPS), it is possible to reduce the UWB transmission power density levels even further to ensure that there is no noticeable interference.

There are a number of ways in which the UWB transmissions can be modulated to enable the data to be carried. The strict power density limits placed on any UWB transmissions by the FCC means that the form of modulation applied must be efficient. It must provide the optimum error performance for a given level of energy per bit. The choice of modulation also affects the UWB transmission spectrum and this must be taken into account to ensure that the spectral density limits are not exceeded.

Two of the most popular forms of modulation used for UWB are pulse position modulation (PPM) and BPSK. These provide the best performance, for in terms of modulation efficiency and spectral performance. The PPM encodes the information by modifying the time interval and hence the position of the pulses. The BPSK reverses the phase of the pulse to signify the data to be transmitted. This is a 180 degree reversal. As the pulses consist of an initial upward or downward voltage, this is easy to reverse. Looking at a pulse on an oscilloscope, it would appear that a pulse is either the right way up or upside down.

Ultra-wideband is still in its infancy. Despite this, it is being recognized as a technology with a huge capability and as such it is being adopted in many new areas. Many silicon manufacturers have already developed solutions, which are being demonstrated, and more are being developed. With the growing level of wireless communications, UWB offers significant advantages in many areas. One of the main attractions for WAN/LAN applications is the very high data rates that can be supported. With computer technology requiring ever-increasing amounts of data to be transported, it is likely that standards such as 802.11 and others may not be able to support the data speeds required in some applications. The UWB may well become a major technology of the future in overcoming this problem.

14.8 IEEE 802.20 AND BEYOND

The WiMAX is for fixed broadband wireless access systems, while on December 11, 2002, the IEEE Standards Board approved the establishment of IEEE 802.20, the mobile broadband wireless access working group. It is hoped that such an interface will allow the creation of low-cost, always-on, and truly mobile broadband wireless networks, nicknamed as Mobile-Fi.

The goals of 802.20 and 802.16e, the so-called 'mobile WiMAX', are similar. A draft 802.20 specification was balloted and approved on January 18, 2006. The mission of IEEE 802.20 is to develop the specification for an efficient packet-based air interface that is optimized for the transport of IP-based services. The goal is to enable worldwide deployment of affordable, ubiquitous, always-on, and interoperable multi-vendor mobile broadband wireless access networks that meet the needs of business and residential end-user markets. The IEEE 802.20 standard will be specified according to a layered architecture, which is consistent with other IEEE 802 specifications. The scope of the working group consists of the physical (PHY), medium access control (MAC), and logical link control (LLC) layers.

The specifications of physical and medium access control layers of an air interface are for interoperable mobile broadband wireless access systems, operating in licensed bands below 3.5 GHz, optimized for IP-data transport, with peak data rates per user in excess of 1 Mbps. It supports various vehicular mobility classes up to 250 km/h in a MAN environment and targets spectral efficiencies, sustained user data rates, and the number of active users, which are all significantly higher than achieved by existing mobile systems. Low latency is also a benefit.

The IEEE 802.21 standard is an IEEE emerging standard. The standard supports algorithms enabling seamless handover between the networks of the same type as well as the handover between different network types, also called media independent handover (MIH) or vertical handover. The standard provides information to allow handing over to and from cellular, GSM, GPRS, Wi-Fi, Bluetooth, and 802.11 networks through different handover mechanisms.

The IEEE 802.21 working group started working in March 2004 and produced the first draft of the standard, including the protocol definition in May 2005. The letter ballot process began and subsequent revisions to the draft are in progress as of September 2006, with almost all comments resolved. The cellular networks and 802.11 networks employ handover mechanisms for handover within the same network type (horizontal handover). The mobile IP provides handover mechanisms for handover across subnets of different types of networks but can be slow in the process. Current 802 standards do not support handover between different types of networks. They also do not provide triggers or other services to accelerate mobile IP-based handovers. Moreover, the existing 802 standards provide mechanisms for detecting and selecting network access points but do not allow for detection and selection of network access points in a way that is independent of the network type. The benefits of IEEE 802.21 will be as follows:

- It allows roaming between 802.11 networks and 3G cellular networks.
- It allows users to engage in ad hoc teleconferencing.
- It is applicable to both wired and wireless networks.
- It allows for use by multiple vendors and users.

- It has compatibility and conformity with other IEEE 802 standards.
- It includes definitions for managed objects that are compatible with management standards like SNMP.
- Although security algorithms and security protocols will not be defined in the standard, the authentication, authorization, and network detection and selection will be supported by the protocol.

Summary

- Ad hoc networks are temporarily established.
- In all the ad hoc networks, some hardware and software support is required because most of the networks are data networks.
- Bluetooth supports voice/data services while WLAN and WMAN are especially data networks.
- Bluetooth replaces cables in home.
- Bluetooth is based on piconet and scatternet structures in which communication takes place between the master and slave devices. These devices are identified on the basis of addresses.
- Bluetooth operates on 2.4 GHz.
- The IEEE 802.11a, b, g, and n standards are significant Wi-Fi standards for establishment of WLAN and all are having different physical layers (modulation schemes).
- The IEEE 802.16 is WMAN standard by which the range of communication can be extended compared to Wi-Fi.
- Zigbee is a standard that deals with tiny OS for sensor networks.
- Ultra-wideband can provide very high bit rate of communication for indoor applications.
- The IEEE 802.20 and 21 are upcoming protocols, which will have better features compared to Wi-Fi and WiMAX.

Review Questions

1. How can you say that Bluetooth is an ad hoc network?
2. Describe the concept of piconet and scatternet.
3. Why is it necessary to have a protocol stack along with Bluetooth chip-based hardware?
4. Differentiate the asynchronous connectionless and synchronous connection-oriented modes in Bluetooth. Draw the timing diagrams for both. Justify the suitability of the type of data transferred using these modes.
5. Why is WLAN considered along with ad hoc networks? Is it a pure ad hoc network?
6. What are the major differences between Wi-Fi and WiMAX systems?
7. What do you mean by air interface?
8. Give the data of all the Wi-Fi standards and discuss their key features.
9. What are the upgradations still required in WiMAX? Find out from the drafts of IEEE.
10. How does Zigbee differ from Bluetooth?
11. Find out the different types of applications of sensor networks. Do you think that all the applications will be possible with single standard?
12. How does UWB differ from other types of networks?
13. Fill in the blanks:
    (a) The lowest physical layer topology is identified as ______ in Bluetooth.

(b) The final modulation scheme in Bluetooth is ______.
(c) The number of hops per second used in Bluetooth devices is ______.
(d) The maximum number of devices which can form a piconet is ____.
(e) Security aspects are mainly covered in ________ Wi-Fi standard.
(f) ____ is the original security solution for 802.11.
(g) The EDCA uses a mechanism called ____.
(h) A new modulation scheme ____ is used along with DSSS in IEEE 802.11b.

14. Strike out the improper word:
(a) Bluetooth is a wireless radio **standard/specification** for data and voice communication.
(b) Bluetooth is a **personal/local** area network.
(c) The link set-up and control is managed by **LM/L2CA** protocol.
(d) The physical layer security in Bluetooth is achieved due to **SSM/PIN**.
(e) The IEEE 802.11a standard is based on **CDMA/OFDM** physical layer.
(f) The IEEE 802.16 standard supports **LOS/NLOS** links.
(g) The IEEE 802.16e standard is a **fixed /mobile** wireless broadband access standard.
(h) The LMDS exhibit **LOS/NLOS** links with **GHz/MHz** range frequencies.
(i) The WiMAX is a **LAN/MAN** standard.

15. Define the following devices in terms of Bluetooth:
(a) Master unit (b) Park unit
(c) Hold mode

Multiple Choice Questions

1. In Bluetooth, per channel bandwidth is
(a) 10 MHz (b) 1 MHz
(c) 79 MHz (d) 2.4 GHz

2. The total number of devices in a piconet
(a) is not more than 2
(b) is not more than 7
(c) is not more than 8
(d) has no restriction

3. The RF carrier modulation scheme in Bluetooth is
(a) FFH-SSM (b) GMSK
(c) GFSK (d) FSK

4. The largest distance covered by Bluetooth devices is
(a) 10 m (b) 30 feet
(c) 60 feet (d) 300 feet

5. The hopping rate for Bluetooth is
(a) 1000 hops/second
(b) 1200 hops/second
(c) 1400 hops/second
(d) 1600 hops/second

6. Which of the following layers in the bluetooth protocol stack is used to perform transport layer equivalent tasks?
(a) L2CAP (b) TCS
(c) SDP (d) Link manager

7. Which of the following devices is synchronized but may not have AM_addr in Bluetooth?
(a) Sniff (b) Hold
(c) Park (d) Active

8. The size of Bluetooth packet header is
(a) 68 bits (b) 72 bits
(c) 54 bits (d) 2745 bits

9. In Bluetooth, masters can start communication in

(a) odd number of time slot
(b) any number of time slot
(c) even number of time slot
(d) first number of time slot

10. In Bluetooth devices, polling-based access is used in
(a) ACL links (b) SCO links
(c) voice links (d) active links

11. In which of the following Wi-Fi standards, the MIMO concept is introduced?
(a) 802.11a (b) 802.11n
(c) 802.11e (d) 802.11b

12. In which of the following networks, the physical layer combines spread spectrum and OFDM modulation schemes?
(a) Wi-Fi (b) Zigbee
(c) UWB (d) WiMAX

13. Which of the following modulation schemes is introduced by IEEE 802.16e?
(a) Scalable OFDM (b) MC-CDMA
(c) CCK-DSSS (d) COFDM

14. Which of the following is considered as zero-carrier technology?
(a) Zigbee (b) Wi-Fi
(c) UWB (d) Bluetooth

15. In IEEE 802.16d, the actual number of carriers used to carry data is
(a) 192
(b) 200
(c) 256
(d) none of the above

16. QoS is the main focus in IEEE
(a) 802.11n standard
(b) 802.11b standard
(c) 802.11e standard
(d) 802.11a standard

17. On which of the following layers of 802.11, PCF and DCF are implemented?
(a) Application (b) Link
(c) Physical (d) MAC

18. Which of the following modes is significant in WiMAX systems while communicating between the base station and the mobile receiver?
(a) Half duplex (b) TDM
(c) TDD (d) FDD

19. Which of the following statements is true?
(a) Bluetooth devices can communicate for 10 metres only.
(b) Bluetooth is based on CDMA concept.
(c) The per channel bandwidth in Bluetooth is 1 MHz.
(d) The master device in Bluetooth cannot work as a slave at all.

20. Which of the following statements is/are true?
(a) The Wi-Fi and WiMAX have independent protocols and interworking cannot be established.
(b) The IEEE 802.16d standard supports vehicular mobility.
(c) The IEEE 802.11n standard is designed for power-saving mode only.
(d) All of the above.

21. Zigbee is the protocol for the
(a) wireless sensor network
(b) Wi-Fi
(c) UWB
(d) data networks

22. The UWB physical layer is based on
(a) CDMA
(b) OFDM
(c) CDMA-OFDM
(d) none of the above

chapter 15

MIMO Systems

Key Topics

- Introduction to MIMO
- Space diversity and systems based on space diversity
- MIMO architecture
- How MIMO exploits multipath
- Space-time processing in MIMO
- MIMO antenna considerations
- Smart antenna and comparison with MIMO
- MIMO channel modelling
- MIMO channel capacity
- STBC coding
- MIMO advantages and applications
- MIMO OFDM

Chapter Outline

Multiple input-multiple output (MIMO) is a multiple antenna system. The MIMO technology exploits the use of multiple signals transmitted into the wireless medium and multiple signals received from the wireless medium to improve the wireless channel performance. The MIMO transmission is an extremely spectrum-efficient technology that uses several antennas at both ends of the communication link. The MIMO can also be thought of as a multi-dimensional wireless communications system. Greater spectral efficiency, higher data rates, greater range, an increased number of users, enhanced reliability, or any combination of the preceding factors can be achieved by the MIMO technology. By multiplying spectral efficiency, MIMO opens the door to a variety of new applications and enables more cost-effective implementation for existing applications. The MIMO OFDM is under significant considerations for development of 4G wireless systems.

15.1 INTRODUCTION

There are three basic link performance parameters that completely describe the quality and usefulness of any wireless link: speed (or spectrum), range (or coverage), and reliability (or security). The use of multiple waveforms transmission in parallel constitutes a new type of radio communication—communication using multi-dimensional signals—which is the way to improve all three basic link performance parameters using multiple antenna system. The multiple input-multiple output (MIMO), a multiantenna system, answers the question of how to achieve higher data rates, wider coverage, and increased reliability—all without using additional frequency spectrum. The combination of multiantenna system with multicarrier system gives an excellent performance.

Guglielmo Marconi demonstrated the first non-line-of-sight (NLOS) wireless communication system in 1896 by communicating over a hill. From that day onwards, engineers viewed multipath signals as a serious problem. The first paper describing wireless MIMO's capacity was published 100 years later in 1996 in Global Communications Conference Proceedings. In wireless mobile radio communication, there is an endless quest for increased capacity and improved quality. Hence, MIMO is of greatest importance in today's scenario, especially for 4G. Multiple antenna systems became popular roughly a decade ago. The reason for the great interest is that multiple antennas offer an efficient way to increase the spectral efficiency of mobile radio systems by exploiting the resource 'space'. After their large potential has been widely recognized, they found their way into several standards. As an example, very simple structures can be found in Release 99 of UMTS systems.

The transmission in wireless communication systems is typically organized in packets, with a training sequence (preamble) at the beginning of the packet, to allow for channel estimation and coherent detection at the receiver. When the transmitter is unaware of the channel and the receiver does not give a feedback detailed phase and magnitude information, we speak of 'open-loop' transmission, which is considered in the remainder of this chapter. It is a good match for the wireless MIMO channel that is time-varying, and the rate of feeding back channel information might be low.

There are different modes of operation possible and the preference depends on the SNR, channel conditions, and constraints imposed on the system complexity. Examples of open-loop MIMO techniques include antenna subset selection, maximum ratio combining (MRC), spatial multiplexing (SMX), cyclic delay diversity (CDD), and space-time block coding (STBC). All these methods are described in this chapter.

There exist a multitude of purposes for using multiple antenna systems. Principally, two different categories can be distinguished. The first objective is to improve the link reliability, that is the ergodic error probability or the outage probability is reduced. This can be accomplished by enhancing the instantaneous signal-to-noise ratio (SNR) (beam forming). The second objective is to decrease the variations of the SNR using some methods (diversity). If multiple access or co-channel interference in cellular networks disturbs the transmission, the interferers that are separable in space can be suppressed with multiple antennas, resulting in an improved signal-to-interference and noise ratio (SINR).

The multipath propagation is a well-known phenomenon of all mobile communication environments as described in Chapter 5. The multipath is typically perceived as interference, degrading a receiver's ability to recover the intelligent information. The MIMO, in contrast, takes advantage of multipath propagation to increase throughput, range/coverage, and reliability. This is accomplished by sending and receiving more than one data signal in the same radio channel at the same time.

In a simple smart antenna system, multiple transmitting antennas may carry the same information and the receiving antennas receive multiple signals of the same information without change in the transmission capacity. The strongest signal is considered for the final reception. This is called *spatial diversity*. Most of the diversity techniques follow the same concept as that of smart antenna. The spatial diversity is introduced in Chapter 6 along with other diversity techniques, but described in detail in this chapter. The MIMO differs from smart antenna system as it is an efficient spatial diversity technique with extra capabilities. The MIMO system has two or more transmitting antennas and two or more receiving antennas and MIMO transmits and receives two or more radio signals in a single channel in parallel, where each signal carries unique information and the system delivers two or more times the data rate per channel. This is called *spatial multiplexing*.

In general, MIMO exploits the multipath by spatial diversity as well as spatial multiplexing techniques. The SDMA-based systems are more suitable for the development of multiantenna system like MIMO. The improvement in wireless communication performance, at no cost of extra spectrum (only hardware and complexity are added), is largely responsible for the success of MIMO as a topic for new research and hence it is considered as an independent chapter. The MIMO multiplies speed to deliver high bandwidth application such as streaming multimedia.

15.2 SPACE DIVERSITY AND SYSTEMS BASED ON SPACE DIVERSITY

In space diversity, the signal is transferred over several different propagation paths. In the case of wireless transmission, it can be achieved by antenna diversity using multiple transmitter antennas (transmit diversity) and/or multiple receiving antennas (diversity reception). In the latter case, a diversity combining technique is applied before further signal processing takes place. Four types of wireless systems are listed out below, which are necessary to be categorized to study the space diversity. The space diversity again has two categories. As mentioned in Chapter 6, if the antennas are at far distance, for example, at different cellular base station sites or WLAN access points, the diversity is called *macro-diversity*. If the antennas are at a distance in the order of one wavelength, this is called *micro-diversity*. Generally, the channel reliability is measured through BER performance.

Presently, four different types of systems can be categorized as far as diversity is concerned. (Input and output refers to the number of antennas.)

- Single input-single output (SISO): No diversity
- Single input-multiple outputs (SIMO): Receive diversity
- Multiple inputs-single output (MISO): Transmit diversity
- Multiple Inputs-multiple outputs (MIMO): Transmit receive diversity

The SISO system is very simple and deals with communication between a transmitter and a receiver. In SISO, error probability is critically damaged by fading. Other diversity techniques, such as SIMO, MISO, and MIMO systems, are also represented conceptually in Fig. 15.1.

In SIMO channel, the concept of MRC, as a way to exploit the receive diversity is offered. The error probability achieved by the MRC is to be much smaller than the one corresponding to the SISO channel. To perform MRC, the receiver has to know the fading or, in other words, the receiver has to have access to the channel state information (CSI). Full CSI means the knowledge of the complete channel transfer function. Partial CSI provides limited channel information. This is usually done by sending some known

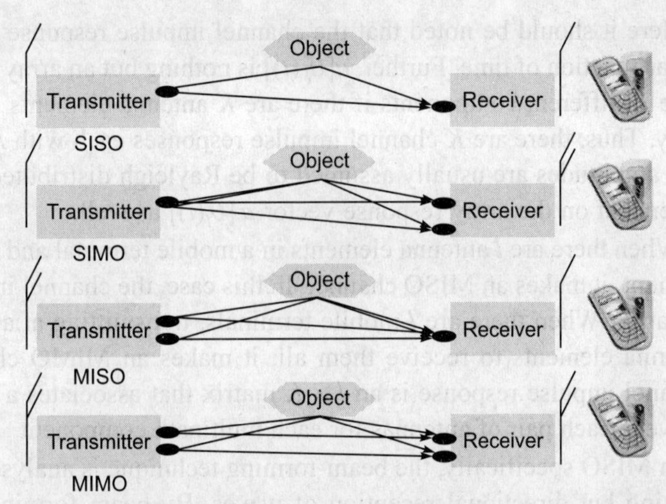

Fig. 15.1 Different types of systems and spatial diversity shown with these systems (Transmitter and receiver both must have digital signal processing unit.)

signal through the channel. A significant amount of research work has been carried out in area of SIMO radio channel models. In these models, a typical cellular environment is considered where it is assumed that the mobile transmitters are relatively simple and with single transmitting antenna and the base station can have a complex receiver with adaptive smart antennas with K antenna elements. Figure 15.2 represents the scenario.

As shown in Fig. 15.2, the multipath environment is such that up to L signals arrive at each base station antenna from different mobile terminals (l) with different amplitudes (α_l) and phases (ϕ_l) at different delays (τ_l) from different directions (θ_l). These are in general time invariant and, as a result, the channel impulse response (CIR) for each antenna is usually represented by

$$h(t) = \sum_{l=1}^{L} [\alpha_l(t) e^{j\phi(t)}]\, \delta[(t - \tau_l(t)] a[\theta_l(t)] \qquad (15.1)$$

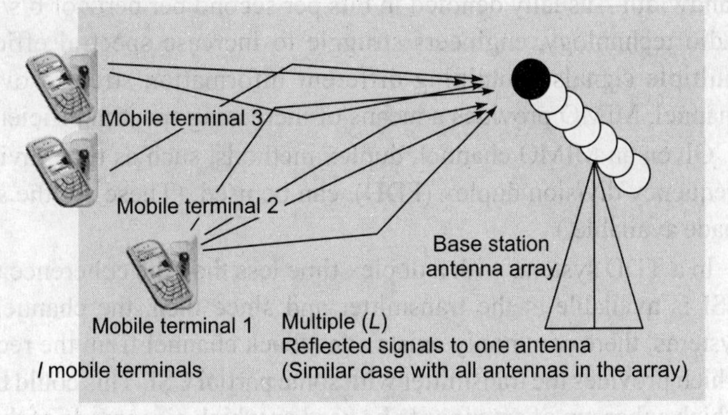

Fig. 15.2 SIMO system example (Mobile terminal has single antenna while base station is complex with multiple antennas.)

Here it should be noted that the channel impulse response is now a vector rather a scalar function of time. Further, $a[\theta_l(t)]$ is nothing but an array response vector and will have K different components if there are K antenna elements of the receiving antenna array. Thus, there are K channel impulse responses each with L multipath components. The amplitudes are usually assumed to be Rayleigh distributed although they are now dependent on the array response vector $a[\theta_l(t)]$ as well.

When there are l antenna elements in a mobile terminal and one base station antenna element, it makes an MISO channel. In this case, the channel impulse response is an $l \times 1$ matrix. When there are l mobile terminals, transmitting at a time and K base station antenna elements to receive them all, it makes an MIMO channel. In this case, the channel impulse response is an $l \times K$ matrix that associates a transmission coefficient between each pair of antennas for each multipath component.

In MISO specifically, the beam-forming technique is analysed. The beam forming is nothing but directional reception of waves. By beam forming, one can increase the average SNR through focusing the energy into desired directions. Transmit beam forming achieves a diversity order of K and an antenna gain of K, same as MRC with K receive antennas. However, note that for transmit beam forming, the transmitter must have the CSI provided through training sequences. This presents us with a bit of a problem, since in order for the transmitter to have the CSI, the receiver must send it to the transmitter, unavoidably reducing the throughput. Using Alamouti's scheme, we can achieve transmit diversity without having to provide the transmitter with the CSI, which is described later.

Bringing together transmit and receive diversity, the MIMO channel is introduced to bring together transmit and receive diversity. Full advantage of both transmit and receive diversity is provided by the MIMO channel in which CSI may not be required. Given multiple antennas, the spatial dimension can be exploited to improve the BER performance of the wireless link depending on the application. Because MIMO transmits multiple signals in parallel across the communications channel (rather than the conventional system's single signal), MIMO has the ability to multiply capacity (which is another word for 'speed' of communication). A common measure of wireless capacity is spectral efficiency—the number of units of information per unit of time per unit of bandwidth—usually denoted in bits per second per hertz or b/s/Hz. Using conventional radio technology, engineers struggle to increase spectral efficiency. By transmitting multiple signals containing different information streams over the same frequency channel, MIMO provides a means of increasing spectral efficiency.

Given an MIMO channel, duplex methods, such as time division duplex (TDD) and frequency division duplex (FDD), can be used. (These are the systems in which CSI is made available.)

In a TDD system, with a duplex time less than the coherence time of the channel, full CSI is available at the transmitter and since then, the channel is reciprocal. In FDD systems, there commonly exists a feedback channel from the receiver to the transmitter, which provides the transmitter with some partial CSI. This could be information regarding which subgroup of antennas to be used or which eigenmode of the channel is strongest. It is also possible to achieve a highly robust wireless link without any CSI at the transmitter by using transmit diversity. This diversity can be achieved through the so-called space-time codes (described later on), like the Alamouti code as in the case of MISO.

The transmitter optimization based on CSI has well-documented advantages. While it is often reasonable to assume that the receiver can acquire accurate CSI, assuming same for the transmitter's end can be very unrealistic for mobile wireless links. In between the perfect CSI and no CSI lies the pragmatic case of partial CSI; the case wherein the transmitter only has a coarse estimate of the actual channel. This coarse estimate can be in the form of a quantization index, denoting the region where the channel vector (as measured at the receiver) falls; or in the form of a noisy channel estimate whose difference from the actual channel is treated as a random vector. The former corresponds to finite-rate feedback, while the latter can also model analog feedback.

Comparison of Channel Capacities of SIMO, MISO, and MIMO Systems

As described in Chapter 2, according to Shannon, the limit on the channel capacity:

$$C = B \log_2(1 + \text{SNR})$$

This is the SISO system.

For the SIMO system, we have M antennas at the receiver end. Suppose the signals received on these antennas have the same amplitude on average. Then they can be added coherently to produce M^2 times increase in the signal power. Hence, the increase in SNR is equivalent to

$$\text{SNR} \approx \frac{M^2 \cdot (\text{Signal power})}{M \cdot \text{Noise}} = M \cdot \text{SNR}$$

So, the channel capacity becomes

$$C = B \log_2(1 + M \cdot \text{SNR})$$

For the MISO system, we have N transmitting antennas. The total transmitted power is divided into N branches. There is only one receiving antenna and the noise level is the same as in the SISO case.

Thus, the overall increase in SNR is approximately

$$\text{SNR} \approx \frac{N^2 \cdot (\text{Signal Power}/N)}{\text{Noise}} = N \cdot \text{SNR}$$

Thus, the channel capacity for this case is

$$C = B \log_2(1 + N \cdot \text{SNR})$$

The MIMO system can be viewed, in effect, as a combination of MISO and SIMO systems. In this case, it is possible to get approximately an MN-fold increase in the SNR yielding a channel capacity equal to

$$C = B \log_2(1 + MN \cdot \text{SNR})$$

By analysing the above equation, it can be concluded that the channel capacity for the MIMO system is higher.

15.3 SMART ANTENNA SYSTEM AND MIMO

Let us first know in detail what is a smart antenna. Smart antennas are designed to help wireless operators cope with variable traffic levels and the network inefficiencies they cause. These systems also allow carriers to change gain settings to expand or contract coverage in highly localized areas, all without climbing a tower or mounting another custom antenna. The wireless carriers can then tailor a cell's coverage to fit its unique traffic distribution. If necessary, the carriers can modify a cell's operation using smart antennas based on the time of day or the day of the week, or to accommodate heavy traffic.

Usually, colocated with the base station, a smart antenna system combines an antenna's array with a digital signal-processing capability to transmit and receive in an adaptive, spatially sensitive manner. In other words, this type of system can automatically change the directionality of its radiation patterns in response to its signal environment. They can increase the performance of a wireless system dramatically. An analogy can be used to explain how smart antenna systems operate. Imagine you are in a room sitting in a chair. Someone in the room is talking to you and while speaking he or she begins moving around the room. Your ears and brain have the ability to track where the user's speech is originating from as they move throughout the room. This is very similar to how smart antenna systems operate; they locate users, track them, and provide optimal RF signals to them as they move through a base station's coverage area.

The terms that are commonly associated with various aspects of smart antenna system technology include phased array, SDMA, spatial processing, digital beam forming, adaptive antenna systems, and others. The smart antenna systems fall into two main categories: switched-beam systems and adaptive-array systems. Generally speaking, each approach directs a main lobe (or radio beam) towards the individual users and attempts to reject interference or noise from outside of that main lobe.

15.3.1 How MIMO Differs from Smart Antenna

The MIMO and 'smart-antenna' systems may look the same on first examination: Both employ multiple antennas, spaced as far apart as practical. But in fact these are fundamentally different. The smart antennas enhance conventional, one-dimensional radio systems. The most common smart-antenna systems use beam forming or transmit diversity (Fig. 15.3) to concentrate the signal energy on the main path and receive the combination (Fig.15.4) to capture the strongest signal at any given moment.

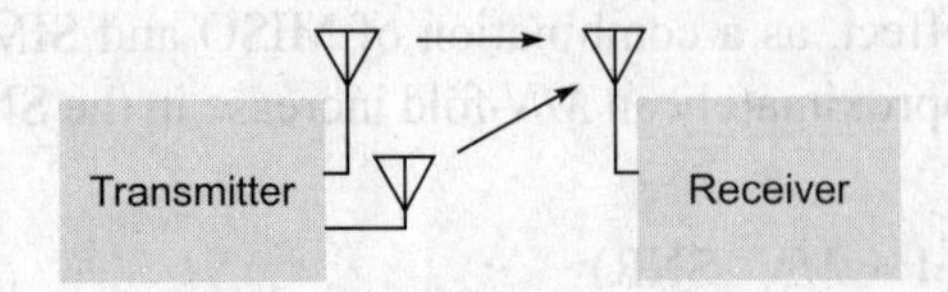

Fig. 15.3 Beam forming (beam steering) employs two transmit antennas to deliver the best multipath signal

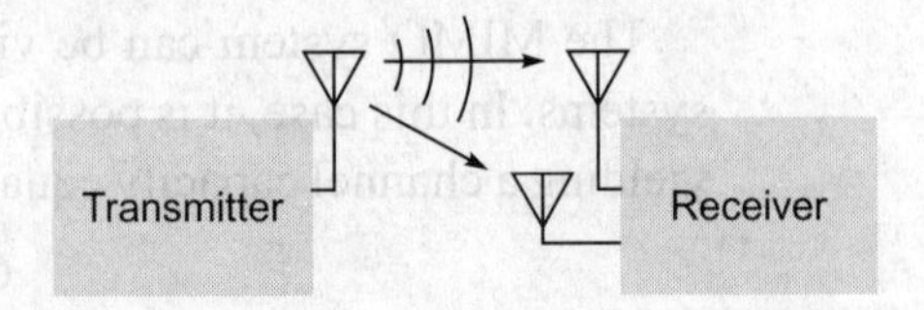

Fig. 15.4 Diversity (receive combining) uses two receive antennas to capture the best multipath signal

Note that beam forming and receive combining are only multipath mitigation techniques and do not multiply data throughput over the wireless channel. Both combined together (Fig. 15.5) demonstrate an ability to improve performance incrementally in point-to-point applications (e.g., outdoor wireless back-haul applications).

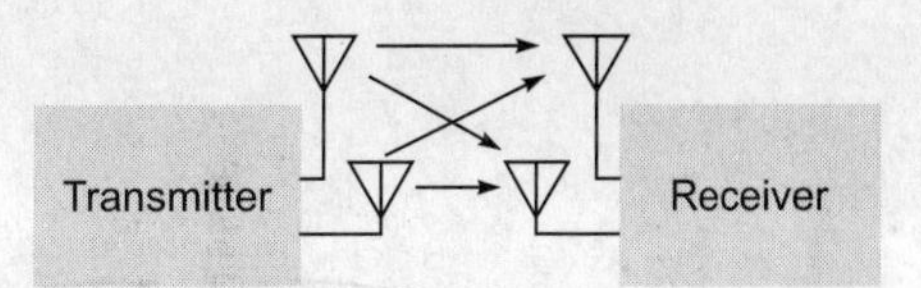

Fig. 15.5 There may be a physical resemblance between radio systems using a combination of beam steering and diversity and MIMO systems

However, while beam forming and receive combining are valuable enhancements to conventional radio systems, MIMO (Fig. 15.6) is a paradigm shift, dramatically changing perceptions of and responses to multipath propagation. The transmitted information by both the antennas is different (shown with the help of separate full and dotted lines in the figure). While receive combining and beam forming increase spectral efficiency by one or two b/s/Hz at a time, MIMO multiplies the b/s/Hz.

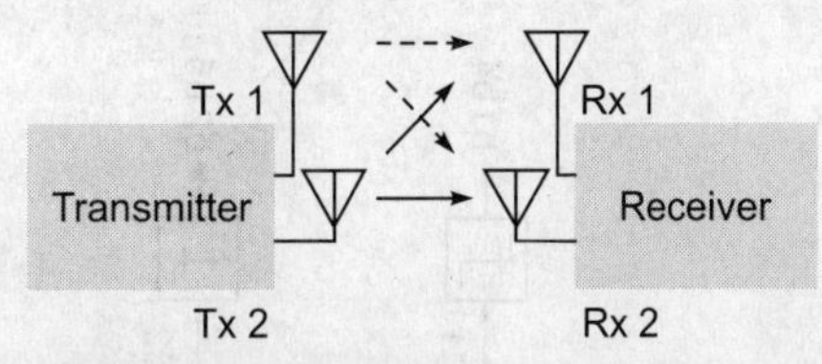

Fig. 15.6 MIMO uses multiple transmitting and receiving antennas to send multiple signals over the same channel, multiplying spectral efficiency

The powerful effect of smart antenna is that in the presence of random fading caused by multipath propagation, the probability of losing the signal vanishes exponentially with the number of decorrelated antenna elements being used. Clearly, in an MIMO link, the benefits of conventional smart antennas are retained since the optimization of the multi-antenna signals is carried out in a larger space, thus providing additional degrees of freedom. In particular, MIMO systems can provide a joint transmit-receive diversity gain, as well as an array gain upon coherent combining of the antenna elements (assuming prior channel estimation). The underlying mathematical nature of MIMO, where data is transmitted over a matrix rather than a vector channel, creates new and enormous opportunities beyond just the added diversity or array gain benefits—the spectrum efficiency. This is shown in Fig. 15.7.

Here *Ai*, *Bi*, and *Ci* represent symbol constellations for the three inputs at the various stages of transmission and reception. A high-rate bit stream (leftmost part of the block diagram) is decomposed into three independent rate bit sequences, which are then transmitted simultaneously using multiple antennas, thus consuming one-third of the nominal spectrum.

As such, MIMO systems can be viewed as an extension of the so-called smart antennas. A strong analogy can be made with CDMA transmission in which multiple users share the same time/frequency channel, which are mixed upon transmission and recovered through their unique codes. Here, however, the advantage of MIMO is that the unique signatures of input streams ('virtual users') are provided by nature in a close-to-orthogonal manner (depending, however, on the fading correlation) without frequency spreading and hence at no cost of spectrum efficiency.

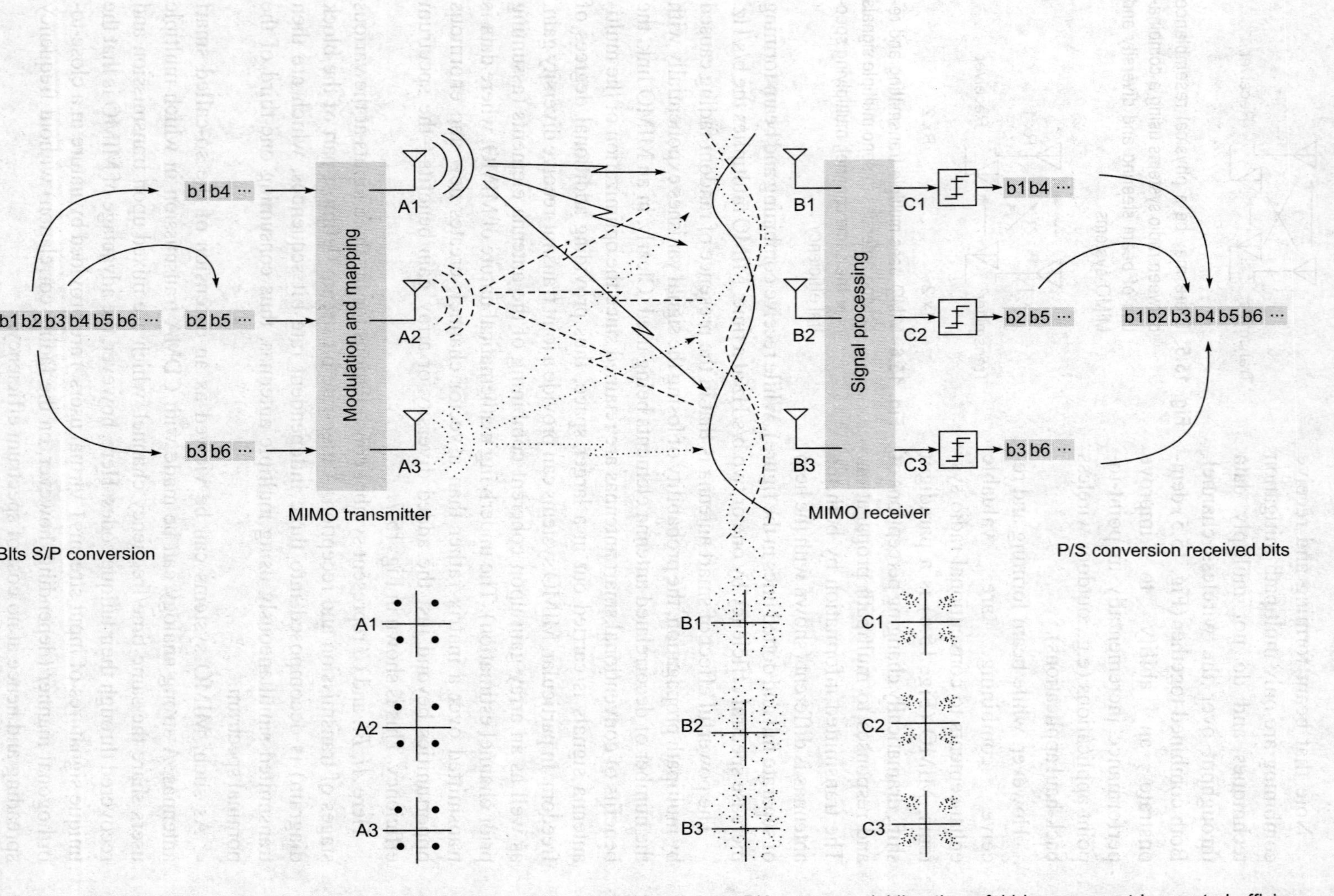

Fig. 15.7 Basic spatial multiplexing (SM) scheme with three TX and three RX antennas yielding three-fold improvement in spectral efficiency

Table 15.1 Comparison between smart antenna and MIMO

Smart Antenna	MIMO Technology
1. In smart-antenna techniques, only transmitter or receiver or both are equipped with more than one antenna. (Typically, it is the transmitter station where the extra cost and space is more easily available and affordable than small phone handset.)	1. In MIMO, both sides of the communication have more than one antenna (the subscriber unit is gradually evolving to become sophisticated wireless Internet access device with maximum features rather than just pocket telephones. This makes multiple antenna a possibility at both sides of the link, even though pushing much of the processing and cost on the transmitter side.)
2. In smart antenna, the data is transmitted over a vector channel.	2. In MIMO, data is transmitted over matrix channel.
3. Some smart antenna system performs better in line of sight (LOS) or close to LOS system. This is especially true when the optimization criterion depends explicitly on angle of arrival departure parameter.	3. Alternatively, MIMO technology can perform well in non-LOS but it really tries to mitigate multipath rather than exploiting it.

Another advantage of MIMO is the ability to jointly code and decode the multiple streams since these are intended for the same use. The MIMO channel relies on the presence of rich multipath, which is needed to make the channel spatially selective.

It was proved during research leading to smart antenna that by deploying an array of antenna either on the transmitter or the receiver end, we can enhance the data rate and can utilize the frequency spectrum more efficiently. Despite this benefit offered by the single arrayed base solution, fourth-generation wireless communication networks could still fall short of market expectations in terms of available capacity for e-commerce and multimedia services. The smart-antenna concept can be extended by multi-element array at both sides of the communication system and can exploit the frequency spectrum more efficiently.

15.4 MIMO-BASED SYSTEM ARCHITECTURE

The MIMO systems can be represented as follows. Given an arbitrary wireless communication system, consider a link for which the transmitting end as well as the receiving end is equipped with multiple antenna elements. Such a set-up is illustrated in Fig. 15.8 with N transmitting and M receiving antennas. A core idea in an MIMO system is space-time signal processing in which time (the natural dimension of digital communication data) is complemented with the spatial dimension inherent in the use of multiple spatially distributed antennas. Single incoming data stream can be converted into parallel streams and can be processed separately. The blocks will be as usual, source coding, channel coding, modulation, and RF up-conversion blocks on the transmitter side and opposite at the receiver side but may be individual for individual antenna element, or some two-dimensional signal-processing methods may be used. Space-time encoder and decoder are explained afterwards.

The digital signal processing is used to separate the multiple streams in MIMO at the receiving end. For example, a 3×3 MIMO system is a 'three-measurement three-unknown problem'. Unless each receive antenna captures a different combination of

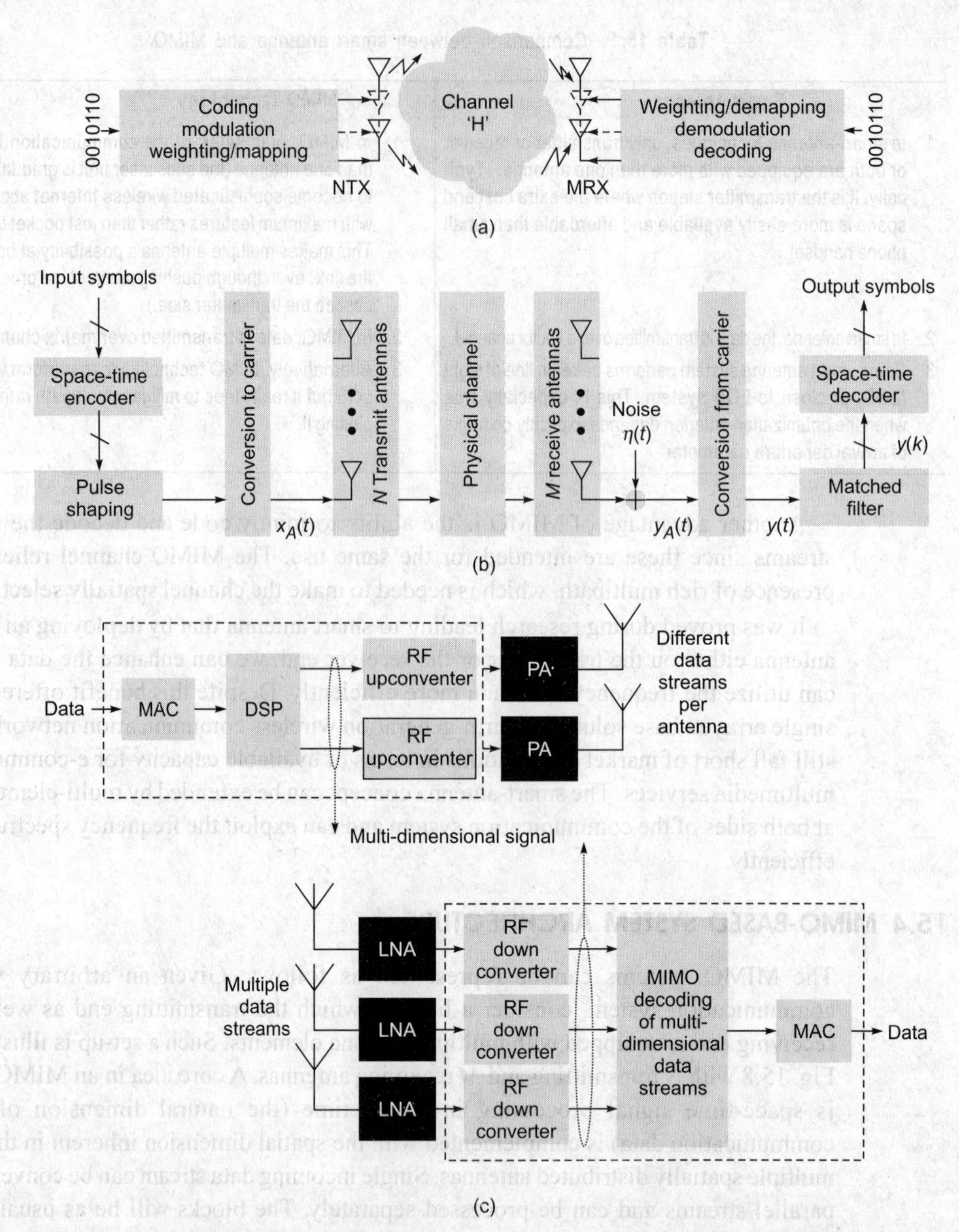

Fig. 15.8 Diagram of a MIMO wireless transmission system: (a) simple concept, (b) detailed diagram showing important blocks including STBC, and (c) diagram showing important blocks at the RF stages

transmitted symbols, this problem cannot be solved because the system of linear equations is dependent, the channel matrix is not full rank, and the antennas are strongly correlated to one another, which are influenced by spacing, polarization, radiation pattern, etc.

15.5 MIMO EXPLOITS MULTIPATH

How is multipath signal treated? There is usually a primary (most direct) path from a transmitter to receiver and some of the transmitted signals take other paths to the receiver, bouncing off objects, the ground, or layers of the atmosphere. The signals traversing less direct paths arrive at the receiver later and are often attenuated as shown in Fig. 15.9. A common strategy for dealing with weaker multipath signals (emerging too late) is to simply ignore them, in which case the energy they contain is wasted. The strongest multipath signals may be too strong to ignore and also can degrade the performance of wireless equipment.

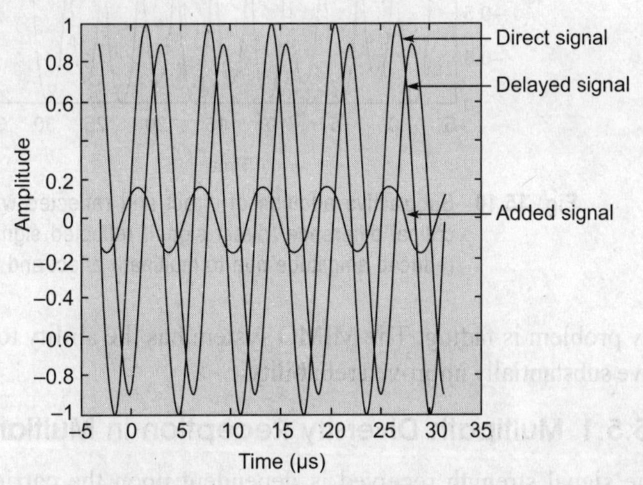

Fig. 15.9 Subtractive addition of direct and reflected waves with out-of-phase delay difference—Direct, delayed (with little attenuation), and added (heavily attenuated) signals are shown on the same axis.

Radio signals can be depicted on a graph in form of sine wave with the vertical axis indicating amplitude and the horizontal axis indicating time. From Figures 15.9 and 15.10, it is clear that when a multipath signal arrives slightly later than the primary signal, its peaks and troughs are not quite aligned with those of the primary signal and the combined signal seen by the receiver is somewhat attenuated and blurred. As shown in the figures, if delay is sufficient to cause the multipath signal's peaks to line up with the primary signal's troughs, the multipath signal will partially or totally cancel out the main signal. The traditional radio systems either do nothing to combat the multipath interference, relying on the primary signal to out-muscle interfering copies, or they employ multipath mitigation techniques. The strongest signal in each moment in time is received or adds different delays to received signals to force the peaks and troughs back into alignment.

The same concept is applied in MIMO and thus MIMO takes advantage of multipath propagation. Rather than combating multipath signals, MIMO puts multipath signals to work carrying more information due to a number of parallel channels. This is accomplished by sending and receiving more than one data signal in the same radio channel at the same time using different antennas. As we discussed, a wireless channel's

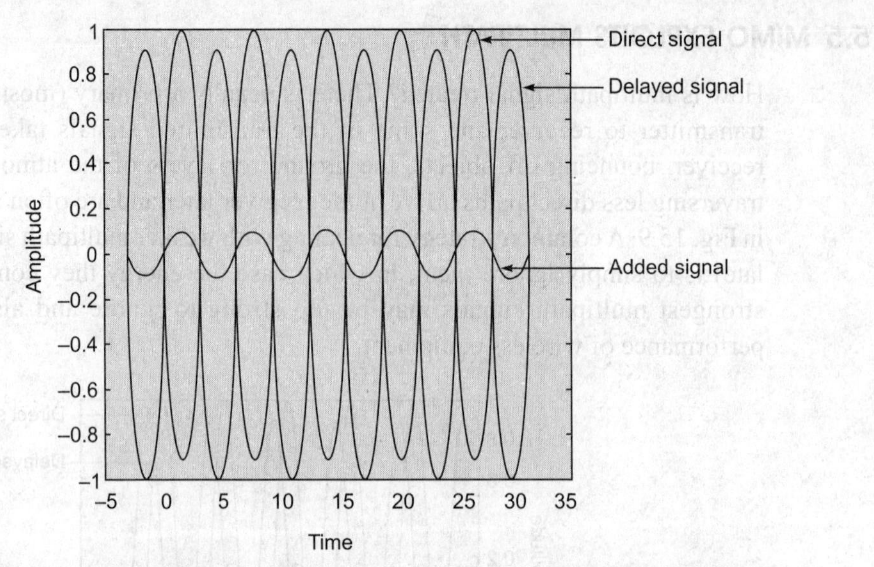

Fig. 15.10 Subtractive addition of direct and reflected waves with very large critical difference (direct signal, reflected signal, and added signal reduced amplitude due to multipath effect and subtractive addition.)

key problem is fading. The MIMO system has the ability to reject fading and ultimately have substantially improved reliability.

15.5.1 Multipath Diversity Reception in Multiantenna Receiver

The signal strength received is dependent upon the carrier frequency as the wireless channel is the dielectric medium in which the phase velocity of different waves will be different and reach the receiver at a different instant of time. A signal transmitted at a particular carrier frequency and at a particular instant of time may be received in a multipath null. The diversity reception reduces the probability of occurrence of communication failures (outages) caused by fades by combining several copies of the same message received over different channels. In general, the efficiency of the diversity techniques reduces if the signal fading is correlated at different branches, i.e. all the received signals must be maximum uncorrelated. One M-branch receiver is shown in Fig. 15.11, which represents the maximum ratio combining. Here phase shifters and attenuators will try to align the signals in terms of phase and amplitude.

The MRC method is already described in Chapter 6. Here it is covered in brief for the continuity of the topic. For $1 \times M$ situations, the maximum ratio combining takes the M received signals and performs a coherent superposition (adder block in Fig. 15.11), corresponding to optimal maximum likelihood detection. Before the coherent superposition stage, each signal branch is

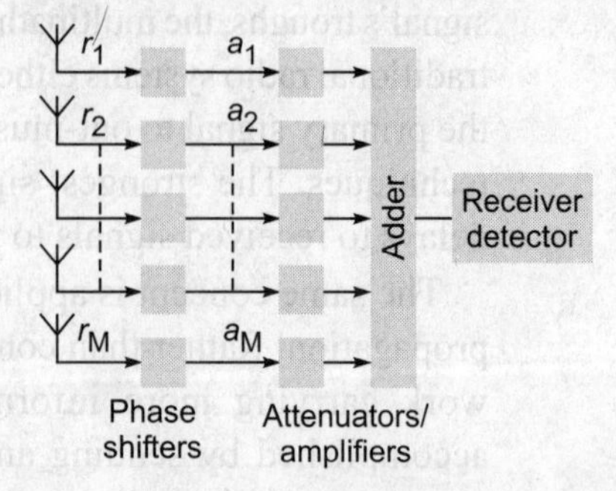

Fig. 15.11 M-branch antenna diversity receiver (can be a maximum ratio combining)

multiplied by a weight factor, which is proportional to the signal amplitude. That is, the branches with a strong signal are further amplified, while the weak signals are attenuated. The rationale behind MRC is that the M signals fade differently and the probability of all being in a deep fade becomes smaller with increasing number of receive antennas. The MRC requires full RF chains with LNA, mixer, and the ADC stage. The combination of the received signals is done in digital baseband, using separate channel estimation for the respective signal paths, resulting in coherent superposition of the wanted signal and non-coherent superposition of the unwanted noise components. For $M = 2$, the gain on an AWGN test channel is 3 dB, going up to about 5 to 7 dB on fading channels.

The idea to boost the strong signal components and attenuate the weak (relatively noisy) components, as performed in MRC diversity, is exactly the same as the type of filtering and signal weighting used in the matched filter receiver. A particularly interesting application of this concept is the Rake receiver for detecting direct-sequence CDMA signals over a dispersive channel. After MRC of independent and identically distributed (IID) Rayleigh-fading signals, the received signal exhibits Nakagami fading with a gamma distribution. This allows relative simple mathematical evaluation of, for instance, outage probabilities.

15.6 SPACE-TIME PROCESSING

The space-time processing (STP) uses the 'spatial dimension' along with the traditional temporal modulation and coding at the transmitter along with advanced decoding at the receiver making parallel transmissions possible and thus launching the concept of STP. A conceptual diagram is given in Fig. 15.12. The STP can improve spectral efficiency as well as coverage area. It makes better use of spectrum and allows support of multiple users and reduces power requirements (and thus PA and battery requirements). It is due to spatial multiplexing and spatial diversity in MIMO.

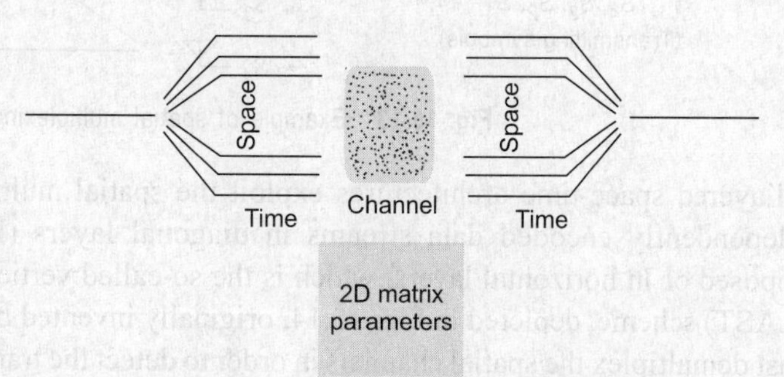

Fig. 15.12 Concept of space-time processing

The space-time processing for MIMO communications is a broad area because of the convergence of information theory, communications, and signal processing research that brought it to fruition. Among its constituent topics, space-time coding has played a prominent role and is well summarized in the following text. Several other topics, however, are also important in order to grasp the bigger picture of space-time processing. These include MIMO wireless channel characterization, modelling, and validation, model-based performance analysis, etc.

Let us first understand the spatial multiplexing and then the difference between spatial multiplexing and spatial diversity in the next few lines.

15.6.1 Spatial Multiplexing

The spatial multiplexing (SM) delivers parallel streams of data by exploiting multipath. It can double (2×2) or quadruple (4×4) capacity and throughput. It gives higher capacity when RF conditions are favourable and the users are closer to the BTS. The concept is already introduced in Fig. 15.7.

In spatial multiplexing, different information is transmitted simultaneously over N transmit antennas increasing the data rate at short distance and providing high spectral efficiency similar to increasing the constellation size, but without suffering from a power penalty. The spatial multiplexing is sometimes also referred to as *direct transmission* or simply MIMO. The receiver has to decouple the N spatial streams to recover the transmitted information. Optimal detection tends to be complex and several suboptimal low-complexity detectors have been devised, ranging from the simple zero forcing to complex soft a posteriori probability detection.

An example of spatial multiplexing is shown in Fig.15.13, which improves the average capacity behaviour by sending as many independent signals as we have antennas for a specific error rate. In the uplink scenario, a base station employs multiple receive antennas and beam forming to separate transmissions from the different mobiles. More recently, transmit beam forming/precoding for multi-user downlink transmission is drawing increasing attention.

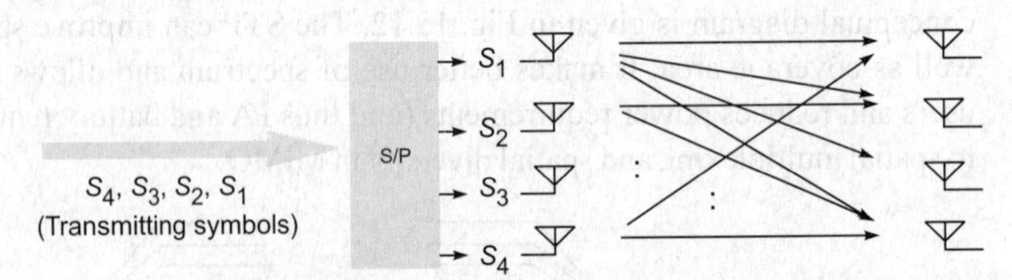

Fig. 15.13 Example of spatial multiplexing

Layered space-time architectures exploit the spatial multiplexing gain by sending independently encoded data streams in diagonal layers (D-BLAST) as originally proposed or in horizontal layers, which is the so-called vertical layered space-time (V-BLAST) scheme, depicted in Fig. 15.14, originally invented by Bell Labs. The receiver must demultiplex the spatial channels in order to detect the transmitted symbols. Various estimation and equalization techniques are zero-forcing (ZF), which use simple matrix inversion but give poor results when the channel matrix is ill conditioned, minimum mean square error (MMSE), more robust in that sense but provides limited enhancement if knowledge of the noise/interference is not used, and maximum likelihood (ML), which is optimal in the sense that it compares all possible combinations of symbols but can be too complex, especially for high-order modulation.

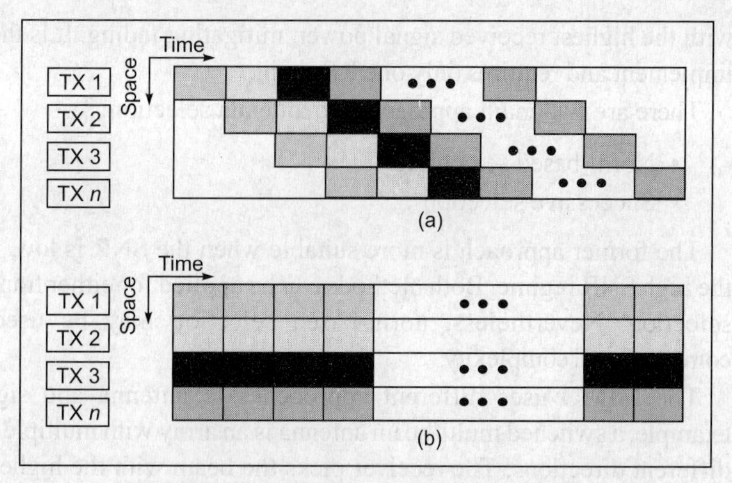

Fig. 15.14 (a) D-BLAST and (b) V-BLAST

15.6.2 Spatial Multiplexing vs Spatial Diversity

It has already been described. In contrast to spatial multiplexing, spatial diversity is used to increase the diversity order of an MIMO link to mitigate fading by coding a signal across space and time so that a receiver can receive the replicas of the signal and combine those received signals constructively to achieve a diversity gain.

The trade-off exists between the spatial multiplexing and the spatial diversity. Pure spatial multiplexing allows for full independent use of the antennas. However, it gives limited diversity benefit and is rarely the best transmission scheme for a given BER target. The possibility of a linear capacity growth with a number of antennas is good, especially knowing that increasing power (SNR), coding the symbols within a block can result in additional coding and diversity gain, which can help improve the performance and robustness, even though the data rate is kept at the same level. It is also possible to sacrifice some data rate for more diversity.

In general, it can be shown that the throughput grows linearly with a number of eigenmodes of the channel and that is the fundamental for MIMO. The throughput can be improved by more spatially multiplexed channels optimally selected.

15.7 ANTENNA CONSIDERATIONS FOR MIMO

Multiple-antenna systems can improve the capacity and reliability of radio communication. However, the multiple RF chains associated with multiple antennas are costly in terms of size, power, and hardware. The antenna selection is a low-cost, low-complexity alternative to capture many of the advantages of MIMO systems. It can reduce hardware complexity and cost, achieve full diversity and, in the case of transmit antenna selection, achieve capacity. The MIMO signalling can improve wireless communication in two different ways as we showed previously: diversity methods and spatial multiplexing, which are again dependent upon the antenna selection. Thus, it is called antenna selection diversity. The selection diversity approach selects the antenna with the highest received signal power, mitigating fading. It is the simplest technique to implement and requires only one RF chain.

There are two main approaches to antenna selection:

- Norm-based selection
- Successive selection

The former approach is more suitable when the SNR is low, whereas the latter suits the high SNR regime. Both methods can be applied for either transmit or receive antenna selection. Nevertheless, norm-based selection may be used because of its low computational complexity.

The MIMO uses different approaches to antenna and signal optimization. For example, a switched multi-beam antenna is an array with multiple fixed beams pointing in different directions. The receiver picks the beam with the highest signal-to-noise ratio. With an adaptive array, the signals received by each antenna are weighted and combined to improve output signal performance. If the signals are combined to maximize output SNR, the technique is maximal ratio combining as described previously. If the signals are combined to maximize the signal-to-interference plus noise ratio, the technique is referred to as MMSE combining.

Assuming equal power transmission from antennas, the capacity as a function of the channel matrix is the ideal antenna selection technique. One of the algorithms is incremental algorithm. An outline of the incremental selection algorithm (for high SNR) is as follows. Start by selecting the row vector with highest norm. At each selection step, project each remaining row vector on the orthogonal complement of the span of the previously chosen vectors and choose the one whose projection has the largest magnitude. Continue until exactly desired antennas are selected. Successive selection is a greedy algorithm for maximizing capacity.

The antenna selection has certain inherent limitations. One of the most important limitations arises whenever the system bandwidth is larger than the coherence bandwidth of the channel (i.e. when the channel is frequency selective). The different response of the channel at different frequencies implies that at each band, a different antenna selection is optimal. So, whenever the channel is highly frequency selective with many uncorrelated frequency bands, antenna selection may not be feasible or useful.

15.7.1 Antenna Separation in Micro-diversity System

In antenna micro-diversity, the signal from antennas mounted at separate locations are combined. Typically, these antennas are located on the vehicle or at the same base station tower and their spacing is a few wavelengths. The received signal amplitude is correlated, depending on the antenna separation d relative to the wavelength. The received multipath signal becomes practically uncorrelated if the antennas at the mobile are spaced by more than, say, half a wavelength.

In the analysis of this correlation, it was assumed that the mobile antenna is mounted at low height and close to all kinds of reflecting and scattering objects. The base station antenna, however, is mostly located well above such obstacles. Hence, at the base station, all multipath waves arrive from approximately the same direction. If the antenna is mc, ed over a certain small distance d, the phase shift is almost identical for all arriving waves. This is in sharp contrast to the situation at the mobile, where motion over half a wavelength leads to almost uncorrelated signal phases. To ensure effective antenna diversity at the base station, antennas must be separated much farther than the fraction of the wavelength required for diversity at the mobile.

15.8 MIMO CHANNEL MODELLING

It is common to model a wireless channel as a sum of two components—an LOS component and an NLOS component.

LOS Component Model

The Rician factor is the ratio between the power of the LOS component and the mean power of the NLOS component. For MIMO systems, however, higher the Rician factor K, the more dominant NLOS becomes. Since NLOS is time invariant, it allows high antenna correlation and low spatial degree of freedom. Hence, a lower MIMO capacity for the same SNR. The Rician distribution factor is a function of season, antenna heights, antenna beamwidth, and distance K. So, in a fixed wireless network (macrocell), MIMO improves the quality of service in areas that are far away from the base station or are physically limited to using low antennas. In a metropolitan city (microcell), antenna height is low, factored to be smaller than that in a macrocell area. In an indoor environment, many simulations and measurements have shown that typically the multipath scattering is rich enough that the LOS component rarely dominates. This plays in favour of in-building MIMO deployments (e.g., WLAN).

Correlation Model for NLOS Component

In the absence of an LOS component, the channel matrix is modelled with Gaussian random variables (i.e. Rayleigh fading). The antenna elements can be correlated, often due to insufficient antenna spacing and existence of few dominant scatterers. Such antenna correlation is considered the leading cause of rank deficiency in the channel matrix to obtain the diversity benefits. Note that in order for the antenna correlation to be low, one desires a large antenna spacing at the base station; on the other hand, phase-array beam forming will only perform well if the antennas are closely spaced in order to prevent spatial aliasing. Thus, at deployment, one must make a choice between the two.

In the Rician channel case, the channel matrix can be represented as a sum of the line-of-sight (LOS) and non-line-of-sight (NLOS) components:

$$\boldsymbol{H} = \boldsymbol{H}_{\text{LOS}} + \boldsymbol{H}_{\text{NLOS}} \tag{15.2}$$

where $\boldsymbol{H}_{\text{LOS}} \triangleq \text{E}\{\boldsymbol{H}\}$ and $\boldsymbol{H}_{\text{NLOS}} \triangleq \boldsymbol{H} - \boldsymbol{H}_{\text{LOS}}$

A commonly used, although quite specific, model for correlated Rician fading channels assumes independent transmit and receive correlation matrices. According to this model,

$$\boldsymbol{H}_{\text{NLOS}} = (\boldsymbol{R}_T)^{1/2}\ \boldsymbol{H}_w(\boldsymbol{R}_R)^{1/2} \tag{15.3}$$

where $\boldsymbol{R}_R$ is the $M \times M$ correlation matrix of the receive antennas, $\boldsymbol{R}_T$ is the $N \times N$ correlation matrix of the transmit antennas, and H_w is a complex $N \times M$ matrix whose elements are zero-mean independent and identically distributed (IID) complex Gaussian random variables.

Secondly, it is common to represent the input/output relations of a narrowband, single-user MIMO link by the complex baseband vector notation:

$$y = \boldsymbol{H}\boldsymbol{x} + \boldsymbol{\eta} \tag{15.4}$$

where $\boldsymbol{x}$ is the $(1 \times N)$ transmit vector, $\boldsymbol{y}$ is the $(M \times 1)$ receive vector, $\boldsymbol{H}$ is the $(M \times N)$ channel matrix, and $\boldsymbol{\eta}$ is the $(M \times 1)$ additive white Gaussian noise (AWGN) vector at a given instant of time. It is assumed that the channel matrix is random, the receiver has perfect channel knowledge, and the channel is memoryless. General entropy of the channel matrix is denoted by $\{h_{ij}\}$. This represents the complex gain of the channel between the jth transmitter and the ith receiver. For an MIMO system, the channel matrix is written as

$$\begin{pmatrix} h_{11} & \cdots & h_{1N} \\ h_{21} & & h_{2N} \\ \vdots & \ddots & \vdots \\ h_{M1} & \cdots & h_{MN} \end{pmatrix} \tag{15.5}$$

where $\mathrm{h}_{ij} = \alpha + j\beta$ (complex representation)

In a rich scattering environment with no LOS, the channel gains $|h_{ij}|$ are usually Rayleigh distributed, where α and β are independent and normally distributed random variables. According to Fig. 15.8(b), the M-element receive array samples this field and generates the $M \times 1$ signal vector $\boldsymbol{y}_A(t)$ at the array terminals. Noise in the system is typically generated in the physical propagation channel (interference) and the receiver front-end electronics (thermal noise). To simplify the discussion, we will lump all additive noise into a single contribution represented by the $M \times 1$ vector $\boldsymbol{\eta}(t)$, which is injected at the receive antenna terminals. The resulting signal plus noise vector $\boldsymbol{y}_A(t)$ is then

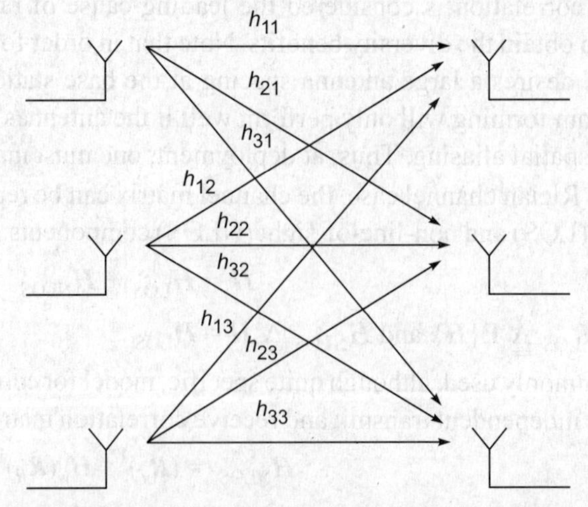

Fig. 15.15 Model for deriving the (3 × 3) channel matrix

downconverted to produce the $M \times 1$ baseband output vector $y(t)$. Finally, $y(t)$ is passed through a matched filter whose output is sampled once per symbol to produce $y(k)$, after which the space-time decoder produces estimates of the originally transmitted symbols. (The variables y's are arbitrarily selected to represent the differences between the stages.)

This topic focuses on the characteristics of the channel. In some cases, we wish to focus on the physical channel impulse response and use it to generate a channel matrix relating the signals $\mathbf{x}_A(t)$ and $\mathbf{y}_A(t)$. A common assumption will be that all scattering in the propagation channel is in the far field and that a discrete number of propagation 'paths' connect transmit and receive arrays. Under this assumption, the physical channel response for L paths may be written as from Eq. (15.1):

$$h(t, \tau, \theta_R, \phi_R, \theta_T, \phi_T) = \sum_{\ell=1}^{L} A_\ell \, \delta(\tau - \tau_l)\delta(\theta_T - \theta_{T,l})$$

$$\times \, \delta(\phi_T - \theta_{T,l})\delta(\theta_R - \theta_{R,l})\delta(\theta_R - \delta_{R,l}) \tag{15.6}$$

where A_l is the gain of the lth path with angle of departure (AOD) (θ_T, l, Φ_T, l), angle of arrival (AOA) (θ_R, l, Φ_R, l), and time of arrival (TOA) τ_l. The term $\delta(\cdot)$ represents the Dirac delta function. The time variation of the channel is included by making the parameters of the multipaths $(L, A_l, \tau_l, \theta_T, l, \ldots)$ time dependent.

To use this response to relate $\mathbf{x}_A(t)$ to $\mathbf{y}_A(t)$, it is easier to proceed in the frequency domain. We take the Fourier transform of the relevant signals to obtain $x_A(\omega)$, $y_A(\omega)$, and $\eta(\omega)$, and take the Fourier transform of $h(\cdot)$ with respect to the delay variable τ to obtain $\bar{h}\,(t, \omega, \theta_R, \Phi_R, \theta_T, \Phi_T)$ where ω is the radian frequency. Assuming single-polarization array elements, the frequency domain radiation patterns of the nth transmit and mth receive array elements are $e_{T,n}(\omega, \theta_T, \Phi_T)$ and $e_{R,m}(\omega, \theta_R, \Phi_R)$, respectively. We must convolve these patterns with $\bar{h}$ in the angular coordinates to obtain

$$y_A(\omega) = H(\omega)\, x_A(\omega) + \eta(\omega) \tag{15.7}$$

where

$$H_{mn}(\omega) = \sum_{l=1}^{L} B_l \, e_{R,m}(\omega, \theta_{R,l}, \phi_{R,l})\, \ell_{T,n}(\omega, \theta_{T,l}, \phi_{T,l}) \tag{15.8}$$

and $B_l = A_l \exp\{-j\omega\tau\, l\}$ is the complex gain of the lth path. $H(\omega)$ is referred to as the *channel transfer matrix* or simply *channel matrix* in frequency domain. Although this representation is usually inconvenient for the closed-form analysis of space-time code behaviour, it explicitly uses the antenna properties in constructing the channel matrix and, therefore, facilitates examination of a variety of antenna configurations for a single physical propagation channel. Also, because it is based on the physical channel impulse response, it is appropriate for wideband or frequency selective channels for which the elements of $H(\omega)$ vary significantly over the bandwidth of interest.

While the physical channel model in Eq. (15.4) is useful for certain cases, in most signal processing analyses, the channel is taken to relate the input to the pulse shaping block $x^{(k)}$ to the output of the matched filter $y^{(k)}$. This model is typically used for cases where the frequency domain channel transfer function remains approximately constant over the bandwidth of the transmitted waveform, often referred to as the *frequency non-selective* or *flat fading* scenario. In this case, the frequency domain transfer functions can be treated as complex constants that simply scale the complex input symbols. We can, therefore, write the input/output relationship as

$$y^{(k)} = H^{(k)} x^{(k)} + \eta^{(k)} \tag{15.9}$$

where $\eta^{(k)}$ denotes the noise that has passed through the receiver and has been sampled at the matched-filter output. The term $H^{(k)}$ represents the channel matrix for the kth transmitted symbol, with the superscript explicitly indicating that the channel can change over time. We emphasize here that $H^{(k)}$ is based on the value of $H(\omega)$ evaluated at the carrier frequency but includes the additional effects of the transmit and receive electronics. This model forms the basis of the random matrix models and is very convenient for the closed-form analysis. The main drawbacks of this approach are the modelling inaccuracy, the difficulty of specifying $H^{(k)}$ for all systems of practical interest, and the fact that it does not lend itself to the description of frequency selective channels.

The discussion thus far has ignored certain aspects of the MIMO system that may be important in some applications. For example, realistic microwave components will experience complicated interactions due to coupling and noise—the factors that may be treated with advanced network models. The effects of non-ideal matched filters and sampling may also be of interest, which can be analysed with the appropriate level of modelling details.

The sensitivity of MIMO algorithms with respect to the channel matrix properties makes the channel modelling particularly critical to assess the relative performance of the various MIMO architectures. Key modelling parameters, for which results from measurements of MIMO can be exploited, include path loss, shadowing, Doppler spread, and the Rician factor distribution.

15.9 MIMO CHANNEL MEASUREMENT

The most direct way to gain an understanding of the MIMO wireless channel is to experimentally measure the $M \times N$ channel matrix H. These measurements include the effects of the RF subsystems and the antennas and, therefore, the results are dependent on the array configurations used. A variety of such measurements have been reported and the results obtained include channel capacity, signal correlation structure (in space, frequency, and time), channel matrix rank, path loss, delay spread, and a variety of other quantities. True array systems, where all antennas operate simultaneously, most closely model a real-world MIMO communication and can accommodate channels that vary in time. However, the implementation of such a system comes at significant cost and complexity because of the requirement of multiple parallel transmit and receive electronic subsystems.

Other measurement systems are based on either *switched array* or *virtual array* architectures. The switched array designs use a single transmitter and single receiver, sequentially connecting all array elements to the electronics using high-speed switches. The switching times for such systems are necessarily low (2 μs to 100 ms) to allow the measurement over all antenna pairs to be completed before the channel changes. The virtual array architectures use precision displacement (or rotation) of a single-antenna element rather than a set of fixed antennas connected via switches. Although this method has the advantage of eliminating mutual coupling, a complete channel matrix measurement often takes several seconds or minutes. Therefore, such a technique is appropriate for fixed indoor measurement campaigns when the channel remains constant for relatively long periods.

Conceptually, the simplest method for measuring the physical channel response is to use two steerable (manually or electronically) high-gain antennas. For each fixed position of the transmit beam, the receive beam is rotated through 360°, thus mapping out the physical channel response as a function of azimuth angle at transmit and receive antenna locations. Typically, broad probing bandwidths are used to allow resolution of the multipath plane waves in time as well as angle. The resolution of this system is proportional to the antenna aperture (for directional estimation) and the bandwidth (for delay estimation). Unfortunately, because of the long time required to rotate the antennas, the use of such a measurement arrangement is limited to channels that are highly stationary.

To avoid the difficulties with the steered-antenna systems, it is more common and convenient to use the same measurement architecture as is used to directly measure the channel transfer matrix. However, attempting to extract more detailed information about the propagation environment requires a much higher level of postprocessing. Assuming far-field scattering and if the radiation patterns for the antennas are known, theoretically, the information regarding phases, delays, etc., can be obtained by applying an optimal ML estimator. However, since many practical scenarios will have tens to hundreds of multipath components, this method quickly becomes computationally intractable.

15.10 MIMO CHANNEL CAPACITY

The propagation environment plays a dominant role in determining the capacity of an MIMO channel. However, robust MIMO performance also depends on proper implementation of the antenna system. It is important to emphasize that the transfer matrix H depends not only on the propagation environment but also on the array configurations. The question becomes which array topology is best in terms of maximizing capacity or minimizing symbol error rates. This is difficult to answer definitively, since the optimal array shape depends on the site-specific propagation characteristics. One rule of thumb is to place antennas as far apart as possible to reduce the correlation between the received signals. As the array size increases, i.e., antenna is placed closer, the simulated capacity continues to grow while the measured capacity per antenna decreases because of higher correlation between adjacent elements.

Outage Capacity

Another measure of channel capacity that is frequently used is outage capacity. Capacity is treated as a random variable, which depends on the channel instantaneous response and remains constant during the transmission of a finite-length coded block of information. So, it is the short-term behaviour of the MIMO channel achieved by coding within fading interval. If the channel capacity falls below the outage capacity, there is no possibility that the transmitted block of information can be decoded with no errors, whichever coding scheme is employed.

15.11 CYCLIC DELAY DIVERSITY (CDD)

In CDD, delayed replicas of the message are transmitted over different antennas. In a way, it is like deliberately creating a single-antenna intersymbol interference channel from an otherwise perfectly frequency flat multiple antenna channel. The upside is that all transmitted symbols benefit from the full spatial diversity. From the transmit matrix given for $N=2$, constellation symbol $s1$ is transmitted from antenna 1 at time instant 1, and from antenna 2 at time instant 2, the constellation symbol $s2$ is transmitted from antenna 1 at time instant 2, and from antenna 2 at time instant 3, and so on. This can be extended to any number of transmit antennas, corresponding to a rate $1/N$ repetition code. The downside is that detection at the receiver is still rather complex, applying conventional methods known from equalization of ISI channels, such as optimal Viterbi equalization. Of course, any of the SMX detectors will do the job as well. So, we have suffered a rate loss, giving up some spectral efficiency to collect some more spatial diversity but have not got simpler detection in return. Does this pay off? Can we not get both—the benefit of transmit diversity and simple detection at the receiver? Another space-time technique comes to the rescue, commonly referred to as space-time coding, which is discussed next.

15.12 SPACE-TIME CODING

A space-time code (STC) is a method employed to improve the reliability of data transmission in wireless communication systems using multiple transmit antennas. The STCs rely on transmitting multiple, redundant copies of a data stream to the receiver in the hope that at least some of them may survive the physical path between the transmission and the reception in a good enough state to allow reliable decoding.

A pioneering work in the area of space-time coding for MIMO wireless channels has been done by Tarokh, et al., in which two code design criteria have been proposed for flat fading channels with coherent receivers and high-performance space-time trellis codes have been designed. However, these codes suffer from rather high decoding complexity. In the same year, Alamouti proposed his celebrated space-time block coding (STBC) scheme for two transmit and multiple receive antennas. The maximum likelihood (ML) decoder for Alamouti's code has very low complexity.

The space-time codes may be split into two types: space-time trellis codes and space-time block codes.

Space-time trellis codes (STTC) distribute trellis code over multiple antennas and multiple time slots providing both coding and diversity gains. This scheme transmits multiple, redundant copies of a trellis (or convolutional) code distributed over time and a number of antennas ('space'). These multiple 'diverse' copies of the data are used by the receiver to attempt to reconstruct the actual transmitted data. For an STC to be used, there must necessarily be multiple transmit antennas, but only a single receive antennas is required; nevertheless, multiple receive antennas are often used since performance of the system is improved by doing so.

Space-time block codes (STBC) act on a block of data at a time and provide only diversity gain, but they are much less complex in implementation than STTCs.

The STTCs rely on Viterbi decoder at the receiver whereas STBCs require only linear processing.

The space-time codes can also be subdivided according to whether the receiver knows channel impairments as discussed below.

In *coherent STC*, the receiver knows the channel impairments through training or some other form of estimation. These codes have been studied more widely because they are less complex than their non-coherent counterparts.

In *non-coherent STC*, the receiver does not know the channel impairments but knows the statistics of the channel.

In *differential STC*, neither the channel nor statistics of the channel are available to the receiver. These STCs are usually based on space-time block codes and transmit one block code from a set in response to a change in the input signal. The differences among the blocks in the set are designed to allow the receiver to extract the data with good reliability.

From Fig.15.16, for each input symbol from the information source, the space-time encoder chooses the constellation point and it simultaneously transmits it from different antennas at different time slots giving coding and diversity gains. For STTC, both coding and diversity gains can be obtained, but for STBC, only diversity and some coding gain depending upon the code rate can be obtained.

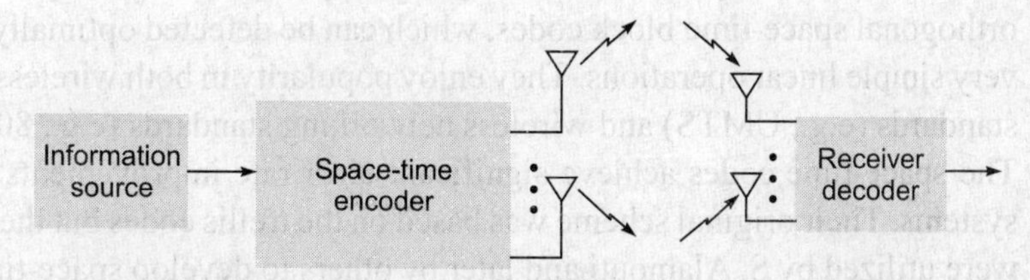

Fig. 15.16 How to incorporate space-time coding in MIMO

In the model, if there are N transmitting and M receiving antennas, then the channel is made up of $N \times M$ sub-channels and in each channel, fading effects are there. At any time slot, N signals are transmitted simultaneously, one from each transmit antenna. The transmitted signals in the sub-channels undergo independent fading and the fading coefficients are assumed to be fixed during a fixed time slot and are independent from one slot to another.

The transmitted code vector is $\vec{s} = [s_1, s_2,..., s_{Nt}]$ and any channel can be modelled as multiplication by complex number h, which is called the *fade coefficient*. Moreover, the receiver antenna will also pick up some noise, which is modelled by a complex number η. So, the received code vector can be expressed as

$$r = hs + \eta \tag{15.10}$$

If there are N transmitting antennas and a single receiving antenna, then the received signal can be written as

$$r = h_1 s_1 + h_2 s_2 + \cdots + h_N s_N + \eta \tag{15.11}$$

If there are M receiving antennas, then we have fade coefficients

$$h_{ij} \; (0 < i <= M \text{ and } 0 < j <= N)$$

and for each corresponding path between transmitter j and receiver i, we have noise at each receiver. Then we have equation of ith receiver signal as

$$r_i = h_{i1} s_1 + h_{i2} s_2 + \cdots + h_{iN} s_N + \eta_i \tag{15.12}$$

If we consider the channel as a whole by expressing M equations as a single matrix equation, then we have

$$\begin{pmatrix} r_1 \\ r_2 \\ \vdots \\ r_M \end{pmatrix} = \begin{pmatrix} h_{11} & h_{12} & \cdots & h_{1N} \\ h_{21} & h_{22} & \cdots & h_{2N} \\ \vdots & \vdots & \ddots & \vdots \\ h_{M1} & h_{M2} & \cdots & h_{MN} \end{pmatrix} \begin{pmatrix} s_1 \\ s_2 \\ \vdots \\ s_N \end{pmatrix} + \begin{pmatrix} \eta_1 \\ \eta_2 \\ \vdots \\ \eta_M \end{pmatrix} \tag{15.13}$$

In vector form, the equation can be written as

$$\vec{r} = H\vec{s} + \vec{\eta} \tag{15.14}$$

15.12.1 Space-Time Block Code (STBC)

As with CDD, space-time block codes can be viewed as repetition codes over space and time, simultaneously transmitting the same data over different antennas. Similar to MRC, a fading channel can be made more AWGN-like using this technique, providing increased robustness and range extension. A quite attractive variant of space-time codes are the orthogonal space-time block codes, which can be detected optimally at the receiver with very simple linear operations. They enjoy popularity in both wireless cellular networking standards (e.g., UMTS) and wireless networking standards (e.g., 802.11n and 802.16e). The space-time codes achieve significant error rate improvements over single-antenna systems. Their original scheme was based on the trellis codes but the simpler block codes were utilized by S. Alamouti and later by others to develop space-time block codes.

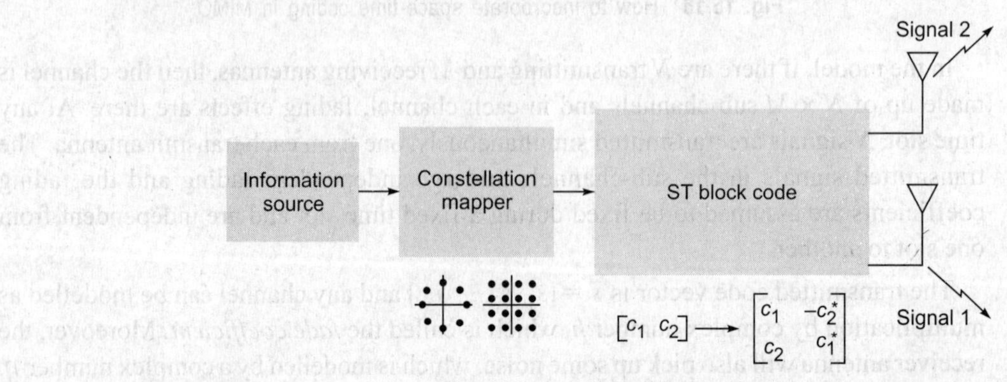

Fig. 15.17 STBC coding in detail

The space time block code in itself is not a code but is a technique to transmit multiple copies of a data stream across a number of antennas and to exploit the various received versions of the data to improve the reliability of data transfer. The fact that transmitted data must traverse a potentially difficult environment with scattering, reflection, refraction, and so on as well as be corrupted by thermal noise in the receiver means that some of the received copies of the data will be 'better' than others. This redundancy results in a higher chance of being able to use one or more of the received copies of the data to correctly decode the received signal. In fact, space-time coding combines all the copies of the received signal in an optimal way to extract as much information from each of them as possible. The STBC can be represented by a matrix showing the symbols transmitted through a given antenna at a given time slot. This matrix can be shown as in Fig.15.18.

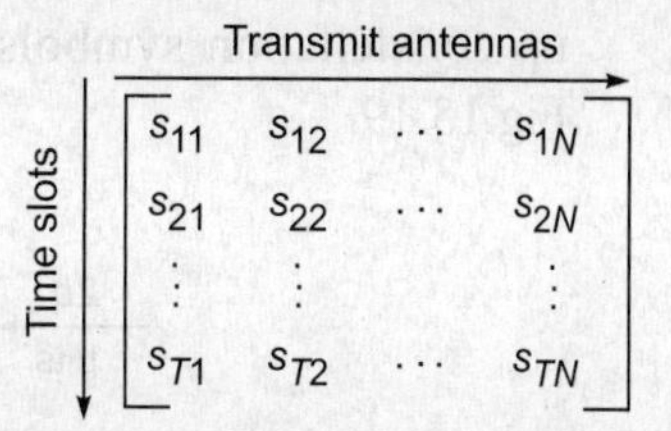

Fig. 15.18 STBC matrix showing the symbols to be transmitted by given antenna in a given time slot

Here s_{ij} is the modulated symbol to be transmitted in time slot i from antenna j. There are to be T time slots and N transmit antennas as well as M receive antennas. This block is usually considered to be of length T. The code rate of an STBC measures how many symbols per time slot it transmits on average over the course of one block. If a block encodes k symbols, the code rate is

$$r = \frac{k}{T} \tag{15.15}$$

Alamouti's Code

The simplest form of space-time block codes was invented by Alamouti in 1998. He proposed this technique for two transmitter antennas and one receiver antenna. Alamouti's code uses a complex orthogonal design, in which the transmission matrix is square and satisfies the conditions for complex orthogonality in both space and time dimensions. This is a very special STBC. It is the only orthogonal STBC that achieves rate-1, that is it is the only STBC that can achieve its full diversity gain without needing to sacrifice its data rate. This code is designed for two transmit antenna system and the code matrix is given as

$$C_2 = \begin{bmatrix} s_1 & s_2 \\ -s_2^* & s_1^* \end{bmatrix} \tag{15.16}$$

where * denotes complex conjugate

Here, two time slots are required to transmit two symbols. Thus, $k = 2$ and $T = 2$. Hence, the code rate for Alamouti's code is 1.

Encoder An encoder consists of symbol calculation and then transmitting the symbols over the transmitter antennas over different time slots. Consider a digital modulation scheme with 2^b constellation elements, where b is the number of bits per symbol, e.g., QPSK, BPSK, and 16-QAM. At time t_1, say, $2b$ bits arrive at the encoder and they pick up constellation symbols s_1 and s_2. This is shown in the form of a block diagram in Fig.15.19.

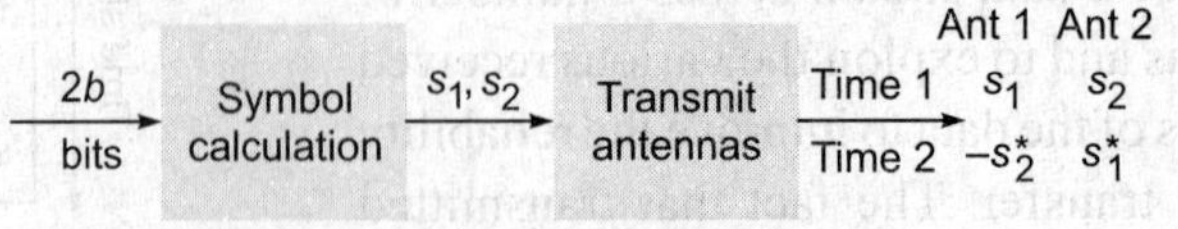

Fig. 15.19 An example of STBC encoder model for Alamouti's code

Decoder The STBC receiver linearly processes the received symbols and uses maximum likelihood decoding. The received signal is the linear superposition of the transmitted elements corrupted by AWGN and channel fading. The receiver model for Alamouti's code can be shown as in Fig. 15.20.

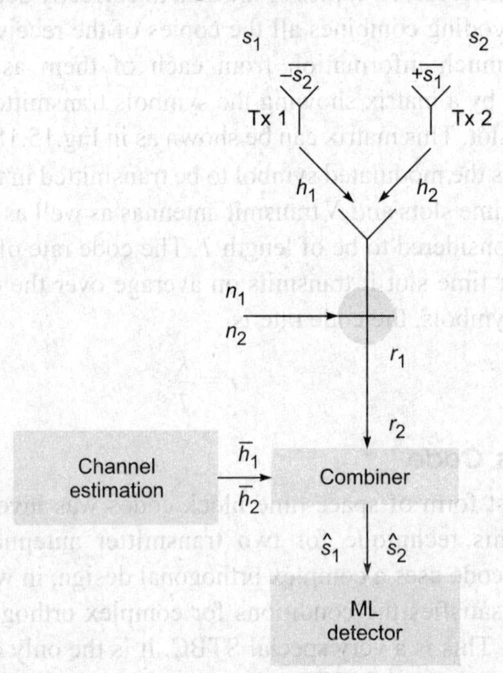

Fig. 15.20 An example of STBC receiver model for Alamouti's code ('−' represents conjugate)

As shown in Fig. 15.20, The two transmitters Tx 1 and Tx 2 transmit the signals simultaneously. Also, it is assumed that the fade coefficients of the channel are constant throughout one time slot. Independent noise samples are added to the transmitted signal in each time slot. Thus, the received signals r_1 and r_2 can be written as

$$r_1 = h_1 s_1 + h_2 s_2 + n_1 \tag{15.17}$$

$$r_2 = -h_1 \bar{s}_2 + h_2 \bar{s}_1 + n_1 \tag{15.18}$$

In the combiner aided by the channel estimator, which provides perfect estimation of the diversity channels in this example, simple signal processing is performed in order to separate the signals s_1 and s_2. Specifically, the maximum likelihood detector minimizes the decision metric

$$|r_1 - h_1 s_1 - h_2 s_2|^2 + |r_1 + h_1 \bar{s}_2 - h_2 \bar{s}_1|^2 \tag{15.19}$$

for all received code words over all possible values of s_1 and s_2.

Expanding the above metric and deleting the terms independent of code words, we get

$$-\left[r_1 \bar{h}_1 \bar{s}_1 + \bar{r}_1 h_1 s_1 + r_1 \bar{h}_2 \bar{s}_2 + \bar{r}_1 h_2 s_2 - r_2 \bar{h}_1 \bar{s}_2 - \bar{r}_2 h_1 \bar{s}_2 + r_2 \bar{h}_2 s_1 + \bar{r}_2 h_2 \bar{s}_1\right] + (|s_1|^2 + |s_2|^2)(|h_1|^2 + |h_2|^2) \tag{15.20}$$

which can be further decomposed into two parts

$$-\left[r_1 \bar{h}_1 \bar{s}_1 + \bar{r}_1 h_1 s_1 + r_2 \bar{h}_2 s_1 + \bar{r}_2 h_2 \bar{s}_1\right] + |s_1|^2 (|h_1|^2 + |h_2|^2) \tag{15.21}$$

which is a function of s_1 only, and the other part

$$-\left[r_1 \bar{h}_2 \bar{s}_2 + \bar{r}_1 h_2 s_2 - r_2 \bar{h}_1 s_2 - \bar{r}_2 h_1 \bar{s}_2\right] + |s_2|^2 (|h_1|^2 + |h_2|^2) \tag{15.22}$$

which is a function of s_2 only.

Minimizing the two parts separately, we have

$$|(r_1 \bar{h}_1 + \bar{r}_2 h_2) - s_1|^2 + (-1 + |h_1|^2 + |h_2|^2)|s_1|^2 \tag{15.23}$$

for detecting s_1, and for s_2, we have

$$|(r_1 \bar{h}_2 - \bar{r}_2 h_1) - s_2|^2 + (-1 + |h_1|^2 + |h_2|^2)|s_2|^2 \tag{15.24}$$

Using the above two values, the optimum soft values of code words s_1 and s_2 are:

$$(r_1 \bar{h}_1 + \bar{r}_2 h_2) = \hat{s}_1 \tag{15.25a}$$

$$(r_1 \bar{h}_2 + \bar{r}_2 h_1) = \hat{s}_2 \tag{15.25b}$$

Thus, we have separated signals s_1 and s_2 by simple multiplications and additions. Due to orthogonality of the code, s_1 is cancelled out in the expression of s_2 and vice versa. The combined signal is then passed through maximum likelihood detector. The most likely transmitted symbol is determined by the maximum likelihood detector based on the Euclidean distances between the combined signal and all possible transmitted symbols. The simplified decision rule for choosing s_i symbol is

$$\text{dist}(\hat{s}, s_i) \le \text{dist}(\hat{s}, s_j), \forall i \ne j \tag{15.26}$$

where $\text{dist}(a, b)$ is the Euclidean distance between the signals a and b and the index j spans all possible transmitted signals. Thus, the detected symbol is the one having minimum Euclidean distance from the combined signal.

Advantages of Alamouti's scheme are given below:

- No feedback is required from receiver to transmitter. No CSI is required.
- There is no bandwidth expansion because it is a rate-1 code.
- There is low complexity of decoders due to orthogonality of the code.

Some major drawbacks of Alamouti STBC scheme are given below:

- It does not provide a code gain.
- Rate-1 code cannot be constructed for complex signal constellations with more than two transmitter antennas.

- The simple decoding rule is valid only for flat fading channel where channel gain is constant over two consecutive symbols.

But these disadvantages are overruled by the simplicity and efficiency of STBC for MIMO systems.

Higher-Order STBC Schemes

The Alamouti scheme can be applied for a two-transmitter antenna system. Tarokh developed the schemes for STBC for three- or four-antenna systems. For three-antenna systems, the code matrix for rate-1/2 and rate-3/4 codes can be shown as

$$C_{3,1/2} = \begin{bmatrix} s_1 & s_2 & s_3 \\ -s_2 & s_1 & s_4 \\ -s_3 & s_4 & s_1 \\ -s_4 & -s_3 & s_2 \\ s_1^* & s_2^* & s_3^* \\ -s_2^* & s_1^* & s_4^* \\ -s_3^* & s_4^* & s_1^* \\ -s_4^* & -s_3^* & s_2^* \end{bmatrix} \tag{15.27}$$

and

$$C_{3,3/4} = \begin{bmatrix} s_1 & s_2 & \frac{s_3}{\sqrt{2}} \\ -s_2^* & s_1^* & \frac{s_3}{\sqrt{2}} \\ \frac{s_3^*}{\sqrt{2}} & \frac{s_3^*}{\sqrt{2}} & \frac{(-s_1 - s_1^* + s_2 - s_2^*)}{2} \\ \frac{s_3^*}{\sqrt{2}} & -\frac{s_3^*}{\sqrt{2}} & \frac{(s_2 + s_2^* + s_1 - s_1^*)}{2} \end{bmatrix} \tag{15.28}$$

For the first matrix, four symbols are transmitted in a block within eight time slots. Thus, it is a rate-1/2 code. Similarly, for the second matrix, three symbols are transmitted in four time slots in a block; hence it is a rate 3/4 code. For four-antenna systems, the codes are

$$C_{4,1/2} = \begin{bmatrix} s_1 & s_2 & s_3 & s_4 \\ -s_2 & s_1 & s_4 & s_3 \\ -s_3 & s_4 & s_1 & -s_2 \\ -s_4 & -s_3 & s_2 & s_1 \\ s_1^* & s_2^* & s_3^* & s_4^* \\ -s_2^* & s_1^* & s_4^* & s_3^* \\ -s_3^* & s_4^* & s_1^* & -s_2^* \\ -s_4^* & -s_3^* & s_2^* & s_1^* \end{bmatrix} \tag{15.29}$$

and
$$C_{4,3/4} = \begin{bmatrix} s_1 & s_2 & \dfrac{s_3}{\sqrt{2}} & +\dfrac{s_3}{\sqrt{2}} \\ -s_2^* & s_1^* & \dfrac{s_3}{\sqrt{2}} & -\dfrac{s_3}{\sqrt{2}} \\ \dfrac{s_3^*}{\sqrt{2}} & \dfrac{s_3^*}{\sqrt{2}} & \dfrac{(-s_1 - s_1^* + s_2 - s_2^*)}{2} & \dfrac{(-s_2 - s_2^* + s_1 - s_1^*)}{2} \\ \dfrac{s_3^*}{\sqrt{2}} & -\dfrac{s_3^*}{\sqrt{2}} & \dfrac{(s_2 + s_2^* + s_1 - s_1^*)}{2} & -\dfrac{(s_1 + s_1^* + s_2 - s_2^*)}{2} \end{bmatrix} \tag{15.30}$$

For both the cases of three- and four-antenna systems, rate-3/4 code had the disadvantage of uneven power distribution among the symbols. Hence, an improved version of the code matrix is given as

$$C_{4,3/4} = \begin{bmatrix} s_1 & s_2 & s_3 & 0 \\ -s_2^* & s_1^* & 0 & s_3 \\ -s_3^* & 0 & s_1^* & -s_2 \\ 0 & -s_3^* & s_2^* & s_1 \end{bmatrix} \tag{15.31}$$

which has equal power from all antennas in all time slots.

Quasi-orthogonal STBC (Q-STBC)

These codes exhibit partial orthogonality and provide a part of diversity gain. An example of such a code proposed by Hamid Jafarkhani is

$$C_{4,1} = \begin{bmatrix} s_1 & s_2 & s_3 & s_4 \\ -s_2^* & s_1^* & -s_4^* & s_3^* \\ -s_3^* & -s_4^* & s_1^* & s_2^* \\ s_4 & -s_3 & -s_2 & s_1 \end{bmatrix} \tag{15.32}$$

Here, the orthogonality criterion only holds for columns (1 and 2), (1 and 3), (2 and 4), and (3 and 4). However, the code is full rate and still only requires linear processing at the receiver, although decoding is slightly more complex than for orthogonal STBCs. The performance results give that Q-STBC outperforms a fully orthogonal four-antenna STBC over a good range of SNRs. However, at higher SNRs, particularly over 22 dB, for this case, orthogonal STBC yields a better BER.

STBC Concatenated with Trellis-coded Modulation

The space-time block coding is a simple technique to achieve diversity; however, there is no significant coding gain. An outer channel code is required to yield coding gain. Trellis-coded modulation (TCM) is a bandwidth-efficient technique that combines coding and modulation, without reducing the data rate. Concatenating STBC with TCM provides coding gain with a reasonable increase in complexity. The STBC-TCM concatenation system is shown in Fig.15.21.

First, the TCM encoder encodes the source data. Next, the encoded data is interleaved and then mapped according to the desired signal constellation. Finally, the space-time

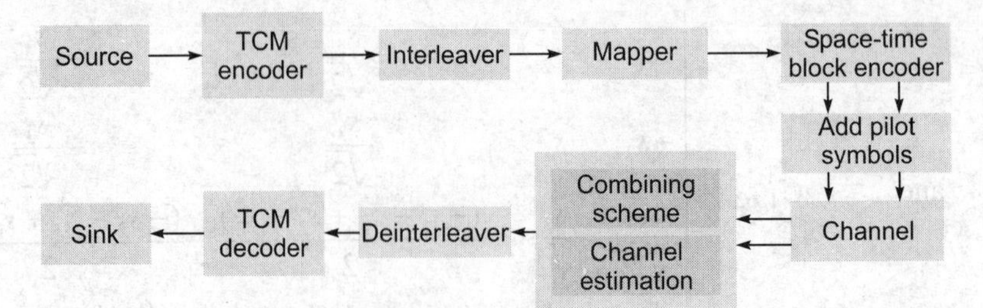

Fig. 15.21 Typical STBC-TCM concatenated system

encoder encodes the data. At each time interval, the symbols are modulated and transmitted simultaneously over different transmit antennas. At the receiver, the received data is combined according to the combining techniques described for STBC. The soft output of the combiner is sent directly to the deinterleaver. Finally, a TCM decoder, such as the Viterbi algorithm, decodes the data.

A New STBC-Based Cellular System Structure

Due to the effectiveness of STBC for MIMO systems, it is used in a variety of applications, especially for mobile environment. Alamouti STBC has been adopted in wireless standards such as WCDMA and CDMA2000. Some of the applications of STBC are elaborated here.

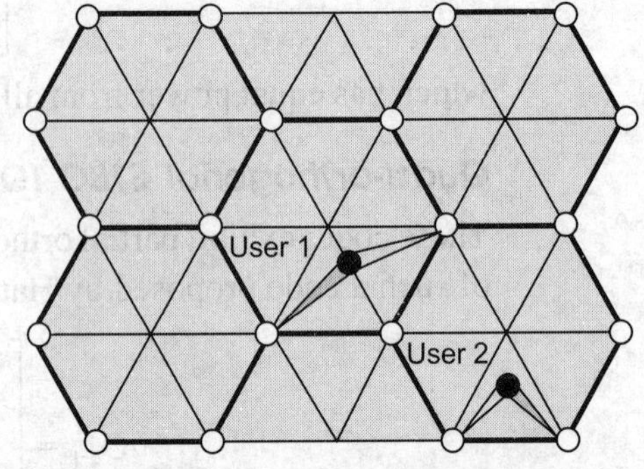

Fig. 15.22 STBC-based CDMA SDMA cellular structure

As shown in Fig.15.22, each cell of the mobile system is divided into three or six sectors depending upon the number of users. The cells are edge excited by three base stations located at the corners. User 1 corresponds to a system with three sectors, whereas User 2 corresponds to a six-sector system. Each base station covers two sectors for both the divisions. We use an STBC to achieve transmit diversity. The difference here is that we use two antenna arrays at two different base stations instead of two antennas at the same base station. One can transmit different STBC code words from one base station if an STBC with more than two dimensions is employed. In this case, the use of proper interleaving is important to guarantee system performance. By separating the base station antennas, our system suffers far less from correlation between transmit channels.

STBC Decoding

The decoding of ST block codes above requires knowledge of the channel at the receiver. Both STTC and STBC were first designed assuming a narrowband wireless system, i.e. a flat fading channel. However, when used over frequency-selective channels, a channel equalizer has to be used at the receiver along with the space-time decoder. Using classical equalization methods with space-time coded signals is a difficult problem. For example, for STTC designed for two TX antennas and a receiver with one RX antenna, we need to design an equalizer that will equalize two independent channels (one for each TX antenna) from one receive signal. For the case of the STBC, the non-linear and non-causal nature of the code makes the use of classical equalization methods (such as the MMSE linear equalizer) is a challenging problem.

The STBC finds a variety of applications in many MIMO communication systems. Some of the systems where STBC can be applied are listed below:

- Wavelet OFDM
- MC-CDMA systems
- Space-time multilevel codes
- Interference cancellation for many MIMO systems

The STBC is currently used mostly in mobile communications. Nokia has launched its mobile phones enabled with STBC reception. With the development in the STBC technology, there is a parallel development in the design of RF components. The antenna design has been confined to a very small area. This increases the mobility of the receivers with reduction in cost.

15.13 ADVANTAGES AND APPLICATIONS OF MIMO IN PRESENT CONTEXT

The major benefits of MIMO are summarized below.

Low Power Requirements and/or Cost Reduction

Optimizing transmission toward the wanted user (transmit beam-forming gain) achieves low power consumption and amplifier costs.

Increased Range/Coverage

The array or beam-forming gain is the average increase in signal power at the receiver due to a coherent combination of the signals received at all antenna elements.

Improved Link Quality/Reliability

Diversity gain is obtained by receiving independent replicas of the signal through independently fading signal components. Based on the fact that it is highly probable that at least one or more of these signal components will not be in a deep fade, the availability of multiple independent dimensions reduces the effective fluctuations of the signal. The maximum spatial diversity order of a non-frequency-selective fading MIMO channel is equal to the product of the numbers of receive and transmit antennas.

Increased Spectral Efficiency

Increased data rates and, therefore, increased spectral efficiency can be achieved by exploiting the spatial multiplexing gain. The spatial multiplexing creates the possibility to simultaneously transmit multiple data streams, exploiting the multiple independent dimensions, the so-called spatial signatures or MIMO channel eigenmodes.

15.14 MIMO APPLICATIONS IN 3G WIRELESS SYSTEMS AND BEYOND

Several techniques, seen as complementary to MIMO in improving throughput, performance, and spectrum efficiency, are drawing interest, especially as enhancements to present 3G mobile systems, e.g., HSDPA. The gains in throughput that MIMO offers are for ideal conditions and are known to be sensitive to channel conditions. In particular, the conditions in urban channels that give rise to uncorrelated fading amongst antenna elements are known to be suitable for MIMO. The MIMO-OFDM technology delivers significant performance improvements for wireless LANs, enabling them to serve existing applications more cost-effectively as well as making new, more demanding applications possible. The three parameters were interrelated according to strict rules. The speed can be increased only by sacrificing range and reliability. The range can be extended at the expense of speed and reliability and the reliability can be improved by reducing the speed and range. The MIMO-OFDM has redefined the trade-offs, clearly demonstrating that it can boost all three parameters simultaneously. While MIMO will ultimately benefit every major wireless industry, including mobile telephone, the wireless LAN industry is leading the way in exploiting MIMO innovations. That is why MIMO-OFDM is the foundation of all proposals for the IEEE 802.11n standard MIMO-OFDM.

The areas where MIMO techniques add significant value to wireless systems include Wi-Fi, WiMAX, cellular, RFID, mobile satellite TV, and satellite radio.

Wi-Fi

Small devices with mainly indoor coverage generally favour adaptive arrays. The main benefits include range increase, interference mitigation (particularly with unlicensed band operation), and uniform coverage. Achieving higher data rates makes MIMO an attractive solution.

WiMAX

Multi-beam antennas (perhaps in combination with adaptive arrays) are favoured at base stations, while adaptive arrays are favoured at the client. Here, the base station antenna provides greater range and allows a capacity increase through spatial reuse, i.e. separate beams. Adaptive arrays on the client side can provide gains to compensate for building penetration losses, permitting use of WiMax for in-building clients and eliminating outside the building installation. Many of these systems use time-division duplexing, allowing for equal gains in both directions with implementation on the client alone.

Cellular

Like WiMAX, multi-beam and adaptive arrays are useful at base stations, with adaptive arrays useful at the handsets. Increased coverage and higher system capacity along with the higher data rates are the biggest gains.

RFID

Smart antennas, either multi-beam or adaptive arrays, can be used on readers to increase range for which a response from an RFID can be received.

Mobile Satellite TV

A multi-beam antenna permits a low-profile antenna that can track the received signal even while the vehicle is in motion and yet can also be incorporated into the roof of a vehicle, unseen from inside or outside the vehicle. The adaptive array technology can further improve performance by mitigating multipath and adapting to implementation/environmental variations.

Satellite Radio

The adaptive arrays can provide more uniform coverage, multipath mitigation, and robustness on the satellite radio receiver. In addition, adaptive arrays can allow for low-profile antennas that can be hidden on the vehicle with higher gain than existing antennas (even though each antenna element may have lower gain than the existing antenna). Furthermore, the adaptive arrays can improve indoor reception and eliminate the need for manual pointing of the indoor unit.

Still some of the open issues and remaining hurdles on the way to a full-scale commercialization of MIMO systems are antenna issues, reciever complexity, system integration and signalling , and CSI at transmitter although MIMO is going to be the 4G technology.

15.15 MIMO-OFDM

Figure 15.23 illustrates the block diagram of an STBC-OFDM system.

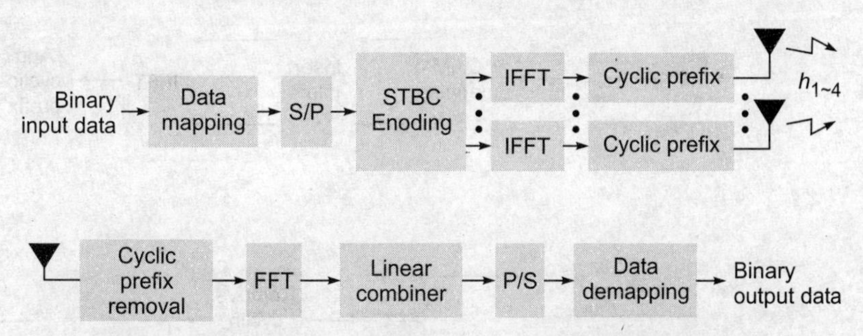

Fig. 15.23 Block diagram of a typical STBC-OFDM system

Binary input data are first mapped into modulation symbols. An $N \times M$ data matrix is formed through serial-to-parallel conversion of N successive modulation symbols, where N and M represent the FFT size and the number of OFDM symbols in each slot, respectively. The OFDM symbols are then space-time encoded for each row using two successive symbols for Alamouti's scheme and four successive symbols for the quasi-orthogonal scheme (or eight symbols for Tarokh's scheme). A preamble is inserted at the beginning of each slot for channel estimation. The size of the preamble should be as small as possible to avoid the serious reduction of transmission efficiency. To this end, we assign one OFDM symbol for preamble and encode the preamble in the frequency domain.

The performance degradation of the four-transmit antenna systems is rather obvious for the designed OFDM system in case where the delay spread is large or the mobile speed is relatively high, which may sometimes limit their use. However, the performance of four-transmit antenna systems may be improved by using the combination of space-time and space-frequency coding schemes.

If a broadband wireless connection is desired, the symbol rate must be increased further which at some point will lead to a frequency-selective channel. Then, there are two ways to go—either we employ pre- or post-equalization of the channel or we divide the channel into many narrowband flat fading subchannels, a technique utilized by OFDM and transmits our data on these substreams, without the need for channel equalization. Hence, it is always possible to convert a frequency-selective channel to many flat fading channels using OFDM and apply the developed flat fading MIMO signalling techniques to each of these subchannels. Figure 15.24 shows MIMO-OFDM transmitter and receiver.

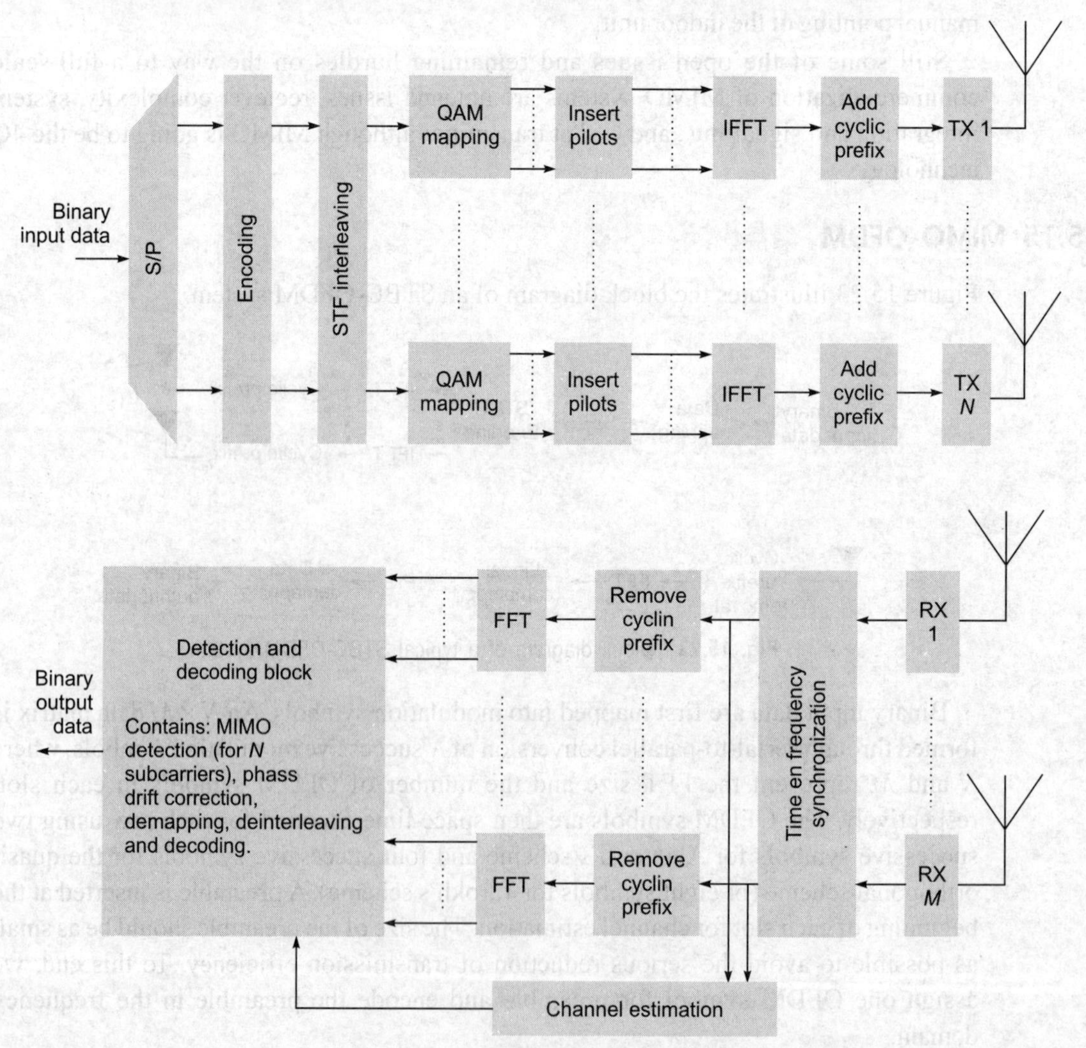

Fig. 15.24 MIMO-OFDM transmitter and receiver

The use of OFDM modulation within MIMO-structured systems creates a strong system that has the ability to successfully reject fading as well as ISI and fulfils the need for high throughput. The increased capacity of MIMO channels can be translated into increased throughput provided that proper coding is used prior to transmission. The coding procedure essentially pads the transmitted data with some protection bits that help the receiver decide whether errors occurred during transmission. The LDPC codes are highly efficient (with respect to encoding and decoding complexity) capacity-approaching codes. The upcoming IEEE 802.11n wireless local area networks (WLANs) standard uses MIMO architecture along with OFDM and LDPC coding. The standard has the ability to provide a stunning throughput of up to 600 Mbps as opposed to 54 Mbps provided by the older, non-MIMO IEEE 802.11a standard.

Summary

- There are four types of wireless communication systems based on space diversity: SISO, SIMO, MISO, and MIMO.
- The MIMO is a multiantenna system both at the transmitter and the receiver side.
- The MIMO deals with multidimensional signal. Data is de-multiplexed and sent on multiple transmitters, maybe at the same carrier frequency (except in OFDM with multiple subcarriers).
- Due to multipath, each receive antenna captures a different combination of transmitted stream. The DSP is used to separate the multiple stream.
- The MIMO increases speed, range, and reliability.
- The MIMO is normally open loop type system and hence sometimes channel state information is required.
- The open loop MIMO technique includes MRC, SMX, CDD, and STBC.
- The SNR can be increased by beam forming and diversity can be achieved by decreasing variation of SNR.
- There is difference between spatial multiplexing and spatial diversity.
- The MIMO exploits multipath. The MIMO capacity increases with the amount of multipath.
- The MIMO channel representation is a two-dimensional matrix and exhibits space-time processing.
- The MIMO differs from smart antenna.
- Antenna selection can capture many advantages.
- Alamouti's STBC code achieves diversity without CSI at the transmitter.
- The MIMO-OFDM has become a popular system for 4G.

Review Questions

1. How are the various types of communication systems classified on the basis of the number of antennas used? Describe their mathematical representations for channel impulse responses.
2. How does MIMO exploit multipath?
3. Give the concept of spatial multiplexing.
4. What do you mean by space diversity?
5. How can we say that 'MIMO increases speed, range, and reliability'?
6. How does a smart antenna differ from MIMO technology?

7. Define the following terms:
   (i) Beam forming
   (ii) Cyclic delay diversity
   (iii) Maximal ratio combining
8. What is the difference between micro-diversity and macro-diversity? Which one will be preferable for MIMO and why?
9. What are the antenna selection criteria for MIMO?
10. Explain the STBC coding. Can we expect the same performance of conventional block codes and space-time block codes?
11. How do STTC and STBC differ?
12. Using the concept of parallel transmission in OFDM (discussed in Chapter 9), draw a complete diagram of MIMO-OFDM.
13. Draw a STBC coder and decoder diagram.
14. Justify the statement: 'MIMO-OFDM is the solution of wireless transmission'.
15. Explain in brief the different methods to measure the channel matrix in MIMO.

Problem

1. Compare the channel capacities of the SISO, SIMO, MISO, and MIMO systems when the single-channel SNR is 12 dB and 1 MHz signal is to be transmitted. For MIMO, the number of transmitting antennas is 4 and the number of receiving antennas is 4.

Multiple Choice Questions

1. The MIMO is based on
   (a) space diversity
   (b) spatial multiplexing
   (c) both (a) and (b)
   (d) OFDM
2. The MIMO channel capacity can be increased by
   (a) space diversity
   (b) STBC coding
   (c) increasing the number of antennas
   (d) having more data
3. The MIMO is incorporated in IEEE
   (a) 802.11a (b) 802.11b
   (c) 802.11g (d) 802.11n
4. The MIMO represents
   (a) multi-dimensional signal
   (b) single-dimensional signal
   (c) two-dimensional signal
   (d) none of the above
5. Which of the following statements is true?
   (a) The MIMO is same as smart antenna system.
   (b) The MIMO improves channel capacity but degrades the performance.
   (c) Channel matrix of MIMO depends upon the channel impulse responses of individual paths.
   (d) The MIMO is always a closed loop system.

chapter 16

Simulation of Communication Systems and Software-Defined Radio

Key Topics

- Simulation and its need
- Simulation methodology
- Multidisciplinary aspects of simulation
- Modelling of a system
- Examples of deterministic and Monte Carlo simulation
- General steps of simulation
- Miscellaneous topics in simulation
- Software-defined radio
- Need for software radio
- General structure of an SDR transceiver
- Third-generation SDR system
- Present trends in SDR development
- The future of SDR

Chapter Outline

In this chapter, two independent parts are covered. First part is simulation of the whole communication system by using a software simulator (block diagram approach) or by developing the codes using associated mathematics with the help of C, C++, or MATLAB tools. Proper modelling is important in this case. Estimation of system performance can be done easily with the simulations. Accuracy of the results is dependent upon modelling and parameter selection. Second part is partly related to the first part as far as modelling and code development is concerned but discusses a different approach of system implementation. Software radio uses digital signal processors and/or FPGA in place of hardware. Software-based waveform generation with hard RF front end is the special feature. Multiple antenna systems, like MIMO, are also possible using SDR approach. Software-defined radio will bring a revolution in the field of wireless communication.

16.1 SIMULATION AND ITS NEED

There are basically two ways for designing or the development of communication (or any other) systems, which are discussed below.

1. Go on implementing the hardware stages and test each of them in sequence. This is a tedious and expensive job. Dry soldering, loose contacts, and other unpredictable problems may lead to unexpected results.
2. Develop a system equivalent scenario on computer using a system model, block diagram, and associated mathematics. This approach, called *simulation*, is easy

and less time consuming, as it is software based. However, one cannot expect 100% accurate results with this approach because it is an estimated equivalent design and response. Accuracy of the results will depend upon the accuracy of modelling and associated mathematics. Practical problems of actual hardware implementation may lead to slightly varied performance.

There are rapid advances in two related technologies—communications and computers. Over the past few decades, communication systems have increased in complexity to the point where system design and performance analysis can no longer be conducted without a significant level of computer support. Many of the communication systems of fifty years ago were either power or noise limited. A significant degrading effect in many of those systems was thermal noise, which was modelled using the additive Gaussian noise channel.

Modern communication systems are required to operate at high data rates with constrained power and bandwidth. These conflicting requirements lead to complex modulation techniques, such as spread spectrum modulation and OFDM, along with an increased level of signal processing at the receiver. Many modern communication systems, such as wireless cellular system, operate in environments that are interference and bandwidth limited. In addition, the desire for wideband channels (to accommodate maximum possible users) and their miniature components along with transmission frequencies into the GHz range, where propagation characteristics are more complicated and multipath-induced fading is a common problem. In order to combat these effects, complex receiver structures, such as those using complicated synchronization structures, symbol estimators, and rake processors, are often used. Even, because of digital communication, coding techniques and error-detection correction schemes are also used in the systems.

The process of making the hardware and testing the above-mentioned complicated systems is totally time consuming because of circuit designing, PCB designing, collecting components, soldering of components, etc., and rather impossible to test without known results. Many of these systems are not analytically tractable using non-computer based techniques and simulation is often necessary for the design and analysis of these systems.

The same advances in technology that made modern communication systems possible, such as microprocessors and DSP techniques (along with the growth of VLSI designs), provided us with high-speed digital computers. The modern work stations and personal computers (PCs)/laptops have computational capabilities greatly exceeding the mainframe computers used just a few years ago. In addition, they are now inexpensive and, therefore, available at the desktop of design engineers. As a result, simulation-based design and analysis techniques are practical tools widely used throughout the communications industry as well as for research and development activities. Powerful software packages targeted to communication systems are now easily available with lots of features. Through the use of simulation, one can study the operating characteristics of the systems that are more complex and more real world can be created than the traditional analysis.

The growth in the computer technology and Internet has also been accompanied by a rapid growth in simulation theory. As a result, the tools and methodologies required for the successful application of simulation to design and analysis problems are more accessible and better understood than was the case few decades ago. A large number of technical papers and several books that illustrate the application of these tools to design and analyse communication systems are now available.

Advantages of Simulation

An important motivation for the use of simulation is that it is a valuable tool for gaining insight into the system behaviour. A properly developed simulation is much like a laboratory implementation of a system. Measurements can easily be made at various points in the system under study. Parametric studies are easily conducted, since parameter values, such as filter bandwidths and signal-to-noise ratios, can be changed and the effects of these changes on the system performance can quickly be observed. Time-domain waveforms, signal spectra, eye diagrams, signal constellations, histograms, and many other graphical displays can easily be generated and, if desired, a comparison can be made between these graphical products and the equivalent displays generated by system hardware. System design testing is performed economically to generate predictions about the results.

In addition, an understanding of simulation techniques supports the research programmes of many graduate and postgraduate students working in the communications area. Finally, students going into the communications industry upon graduation have there required skills needed by industry. The latest communication systems are mainly wireless and optical fibre-based systems. The conventional wired line based systems are not totally obsolete; even in the Internet and also in ISDN, wired line support is needed. The physical layer of all the above systems can be simulated.

There are two basic approaches by which the communication systems can be simulated.

1. By writing codes (and/or functions) in C, C++, VC++, MATLAB, etc., for each and every mathematical relationships available as given below:

   MATLAB program for AM generation

```
Fs = 800; % Sampling rate is 800 samples per second
Fc = 3000; % Carrier frequency in Hz
t = [0:.1*Fs]'/Fs;   % Sampling times for 1 second
x = sin (20*pi*t); % Representation of the signal
y = amod (x,Fc,Fs,'amdsb-tc'); % Modulate x to produce y
figure;
subplot (2,1,1), plot (t,x); % Plot x on top
subplot (2,1,2), plot(t,y); % Plot y below
r = fft (y);
figure
plot (abs(r));
```

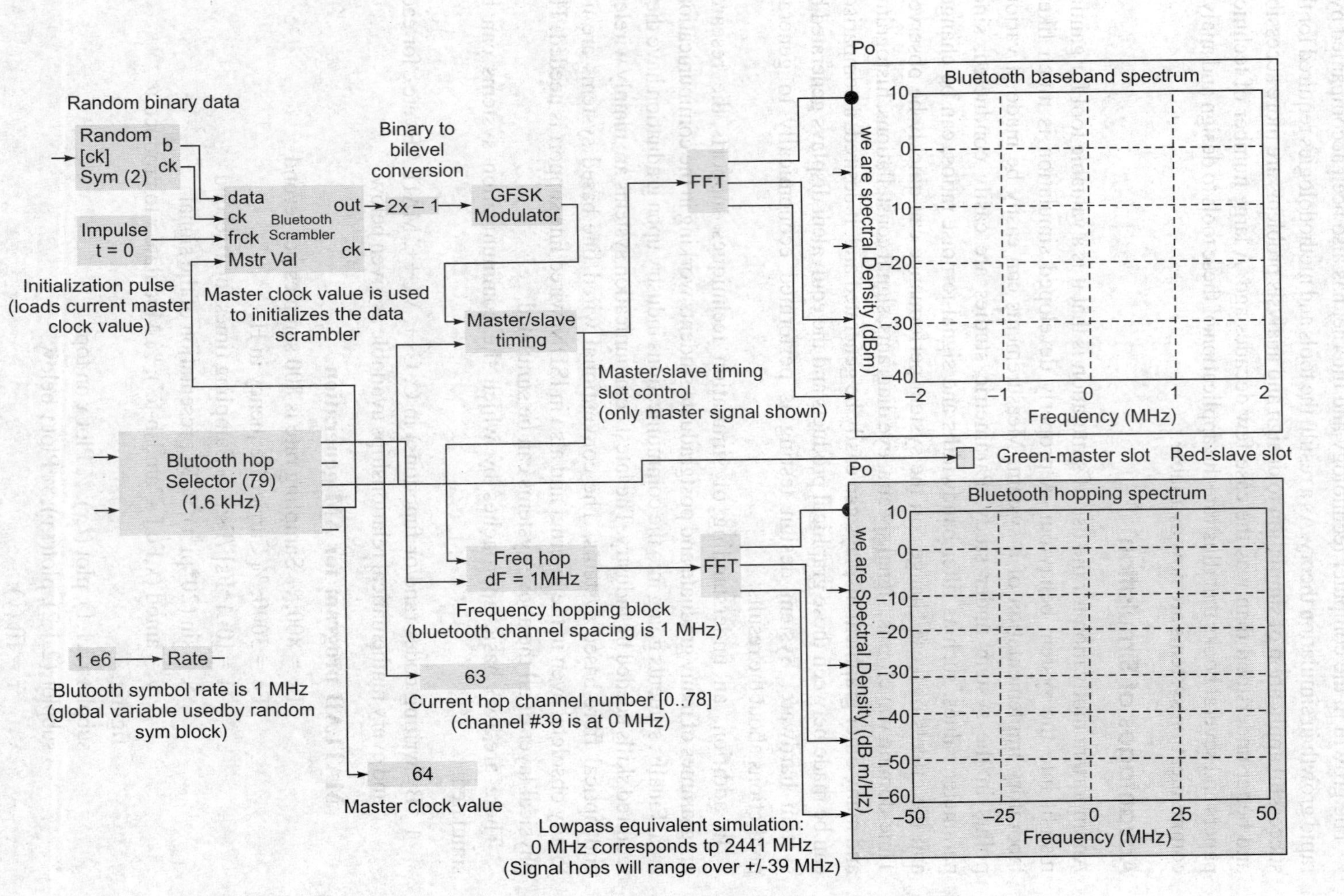

Fig. 16.1 A simulated block diagram in COMMSIM7

2. By block diagram approach, using SIMULINK, COMMSIM, labview software, etc.

For field strength measurements and analysis in mobile communication, various software systems are available. Figures 16.1 to 16.4 show some examples of simulation.

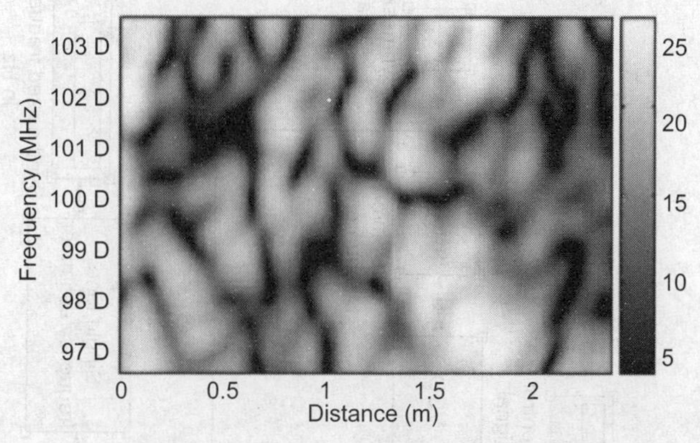

Fig. 16.2 MATLAB example of field strength measurement output, where at various distances and various frequencies different SNR can be observed

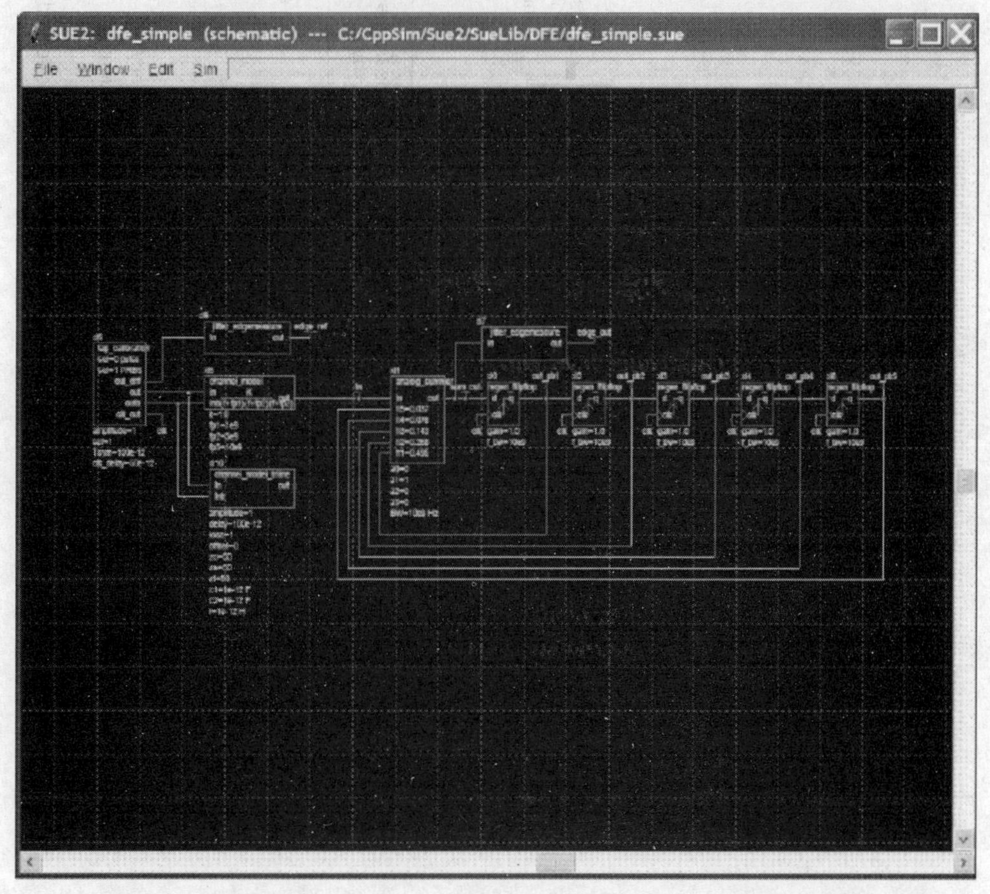

Fig. 16.3 A CppSim example, in which decision feedback equalizer is simulated

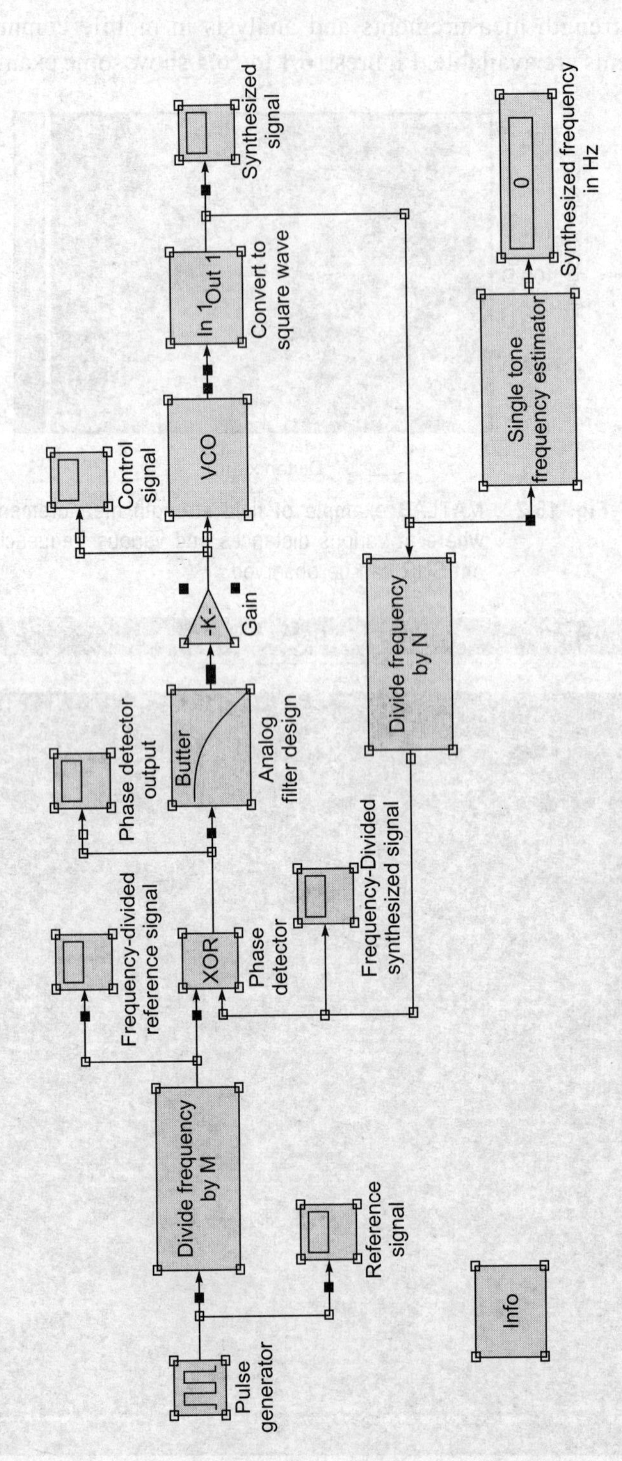

Fig. 16.4 An example of SIMULINK: phased-locked loop (PLL) based frequency synthesis

16.2 SIMULATION METHODOLOGY

In general, there are different types of communication systems based on wirelines and wireless. A few of them are listed below and all are using different protocols.

- Copper media-based traditional telephone system and modems
- Traditional AM and FM broadcasting systems
- Satellite systems
- Data communication and Internet system
- Optical fibre-based various applications, such as B-ISDN LAN
- WLL and mobile telephony systems
- In general, baseband or broadband systems with many different modulation schemes

There is no single approach of system modelling and simulation. At the same time, without deep knowledge, mathematics, or basic theory, even simulation is not an easy task, though there are readily available functions or block simulators, because without the knowledge, one cannot decide the appropriate input-output relationships or parameters. It is necessary to understand for which conditions the blocks or functions are designed.

Category-wise, three types of communication systems exist:

1. Analytically tractable (linear)
2. Analytically complex (non-linear)
3. Analytically intractable

All can be simulated in a different manner. Their exact simulations may not be possible but maybe with approximations and assumptions, simplified models can be derived and simulated for tentative results to find out the behaviour of the system. One must try to identify the system before the simulation category under which it falls. Identification of the system category is very important for deciding the approach of simulation.

16.2.1 Analytically Tractable System

The features of an analytically tractable system are as follows:

- This is a simple basic communication system studied in the first course on communication theory.
- Data is a sequence of discrete symbols and memoryless source, i.e. *k*th generated symbol is independent of others.
- The modulator simply maps the source symbols onto other waveforms representing the source symbols (like AM, FM, PCM, and PSK).
- The transmitter is simply a linear amplifier in this case, according to the desired energy.
- Noise is assumed to be AWGN.
- The receiver observes and estimates the symbols of the original data signal such that the probability of error is minimized.

- We refer this system analytically tractable because with the knowledge of basic communication theory, analysis of the system is carried out with ease (traditional analysis techniques) and the receiver is linear.
- Perfect symbol synchronization is assumed, so that we have exact knowledge of the beginning and ending time of the data symbols.
- The parameters that can be analysed are probability of error, BER, etc.

Figure 16.5 shows an analytically tractable system.

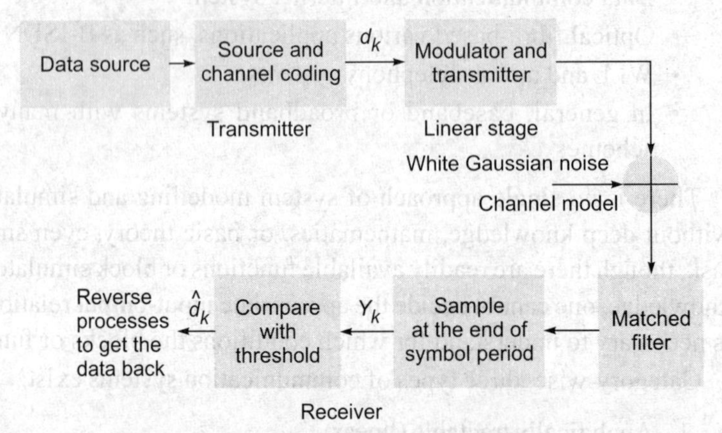

Fig. 16.5 An example of analytically tractable system, which is a simple communication system with a channel, without any non-linear parts like high-power amplifier (It may also be a baseband system.)

16.2.2 Analytically Complex System

Its features are discussed below:

- It is somewhat complex.
- It is same as analytically tractable system except the non-linear high-power amplifier and the filter in the transmitter.
- The non-linear amplifiers exhibit much higher power efficiency than the linear amplifiers and hence are preferred for use in an environment where power is limited, e.g., space applications and mobile cellular systems, where battery power must be conserved. A linear amplifier preserves the spectrum, while a non-linear amplifier will generate harmonic as well as inter-modulation distortion results in much wider spectrum in the output. The BPF is required with the centre frequency equal to the desired carrier frequency. The filter leads to time dispersion of the data signals and leads to ISI. As a result of ISI, the probability of error of the *i*th symbol is dependent upon one or more previous symbols upon which the decision is being made.
- Case study: One must compute $2k$ different error probabilities for a sequence length k and $2k$ different probable sequences and average the k results. The system has important property, the system is linear from the point at which the noise is injected to the point at which the statistic threshold level Y_k appears. Since the

channel is AWGN, each of the $2k$ error probabilities is a Gaussian Q-function. It is a straight forward but tedious procedure to calculate the argument of each Q-function and therefore simulation is often used. If $Y_k = S_k + I_k + N_k$ (signal plus noise plus interfering signal) It is difficult to find out I_k.. The fact that the noise passes only through the linear portion of the system, the mean of Y_k can quickly be determined using a noise free simulation. Variance of Y_k can be determined analytically, as a result, PDF of Y_k is known and error probability is easily determined.

- Semi-analytic simulation and this way the system is somewhat approximated.

Figure 16.6 shows an analytically complex system.

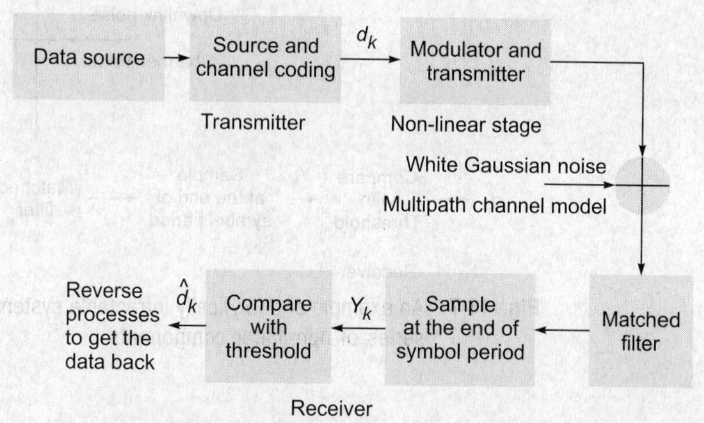

Fig. 16.6 An example of analytically complex (or tedious) system, which is a communication system with a channel with high power amplifier and filter (Here for simplicity, certain assumptions are needed.)

16.2.3 Analytically Intractable System

The following features make the analytically intractable system the most complex system:

- It is a model of a two-hop satellite communication system.
- It has two noise sources rather than one.
- It has a wireless cellular radio link operating in a high interference and multipath environment.
- It is difficult to decide Y_k, though not impossible. Most complex systems.

Figure 16.7 shows an analytically intractable system.

16.3 MULTIDISCIPLINARY ASPECTS OF SIMULATION

While dealing with the communication systems, we must have the knowledge of other related fields, like digital signal processing methods, numerical methods for computation, etc. A communication system has different blocks, which are given with some input-output relationships and associated transfer functions. While simulating the system, various mathematical equations and different techniques are also used (see Fig. 16.8).

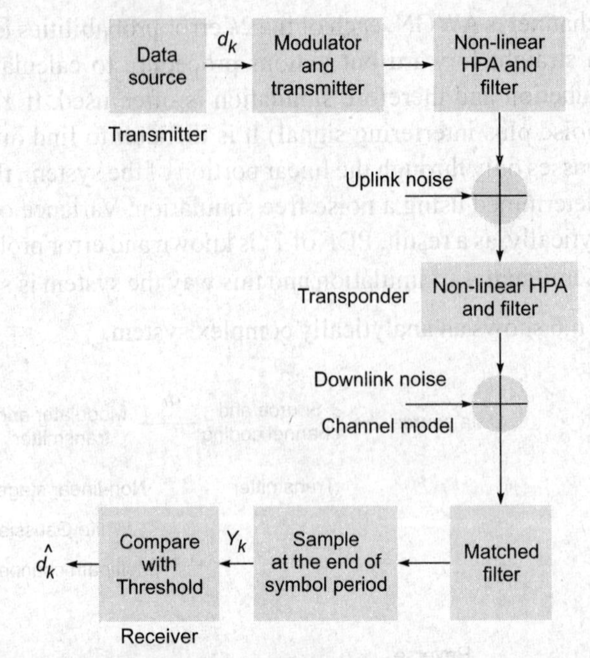

Fig. 16.7 An example of analytically intractable system with a series of non-linear components

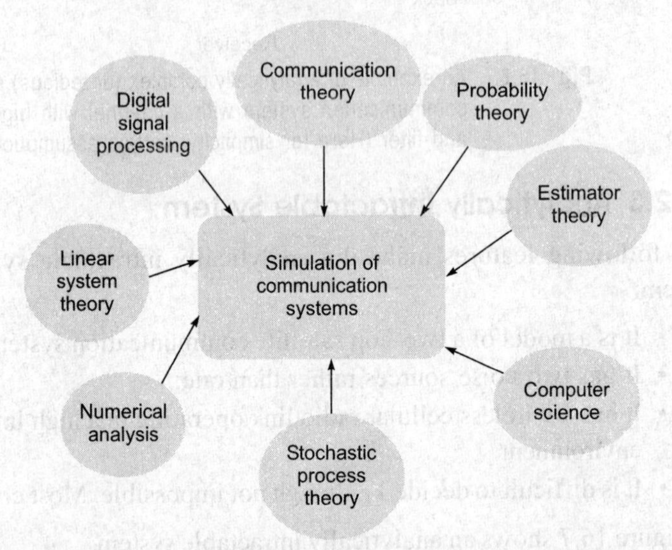

Fig. 16.8 Multidisciplinary aspects of simulation as mentioned in Fig.1.1

16.4 MODELLING OF THE SYSTEM

- Models describe the input-output relationship or behaviour of the physical systems or devices.
- Models are typically expressed in mathematical form also for code development.

- The art of modelling is to develop the behavioural model because the model captures the input-output behaviour of the device under specific conditions.
- Models must be sufficiently detailed to maintain the essential features of the system.
- Trade-offs between accuracy, complexity, and computational requirements are, therefore, usually observed because accuracy is totally dependent upon modelling.
- It is useful to consider analytical models and simulation models in the work simultaneously.

Figure 16.9 shows how simulation models are developed.

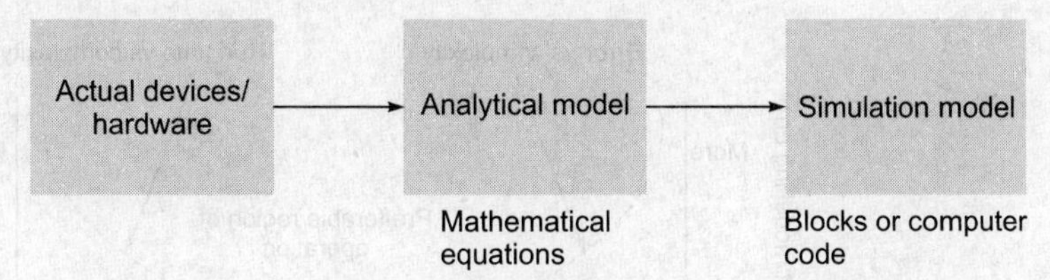

Fig. 16.9 Developing the simulation models for the systems

16.4.1 Modelling Considerations

While modelling a system, the following points should be considered:

- The first and most important step in modelling is to identify those attributes and operational characteristics of the physical device that are to be represented in the model.
- The identification of these essential features often requires considerable engineering judgement and a thorough understanding of the application for which the model is to be developed.
- The accuracy required of any mathematical analysis or any computer simulation based on the model is limited by the accuracy of the system model.

Some of the basic concepts of the computer science will be useful in simulation as follows:

- The word length and the format of words used to represent the samples of signals will impact simulation accuracy, but this would not be of much importance in floating point processors.
- The choice of language is important in the development of commercial simulators.
- Available memory and its organization will impact the manner in which data and instructions are passed from one part of the simulation to another.
- Graphic requirements and capabilities will determine how waveforms are displayed and will impact the transportability of the simulation code from one computer platform to another.

In commercially available simulators, the designers specify the computer system requirements.

16.4.2 Effects of Model Complexity and Trade-off

The trade-off exists in the following manner:

- The desirable attribute of a simulation is fast execution of the simulation code—simple models will be executed faster than more complex models.
- Simple models may not fully characterize the important attributes of a device. Therefore, the simulation may yield inaccurate results, which may lead to more modelling errors.
- Perfect models may be complex and can increase the accuracy, but with a cost of increased run time (see Fig. 16.10).

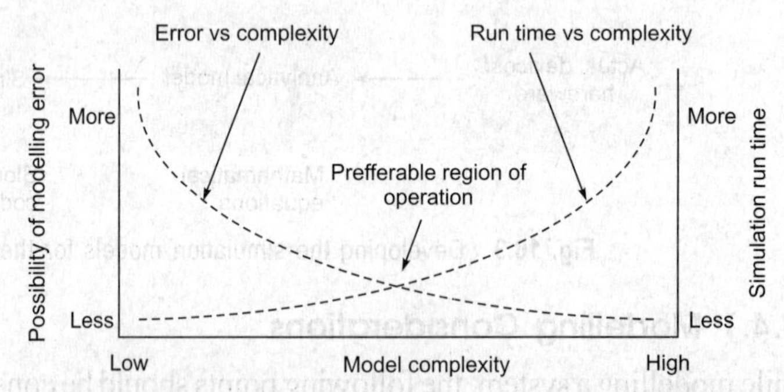

Fig. 16.10 Modelling errors reduce when simulation run time is long and vice versa and hence modelling complexity is a compromise between error and run time

16.5 DETERMINISTIC SIMULATION: AN EXAMPLE

The behaviour is known for PLL as shown in Fig. 16.11. For all those circuits whose behaviour is known, the mathematical part is established. Also their results are known in advance. Hence, while the simulation is being done, the results can be readily compared and modifications can be applied to bring the results nearer to the expected ones. As we know, initially PLL tries to lock the phase. So, frequency deviation is obvious but as the locking is achieved, it remains at the constant frequency (Fig. 16.11).

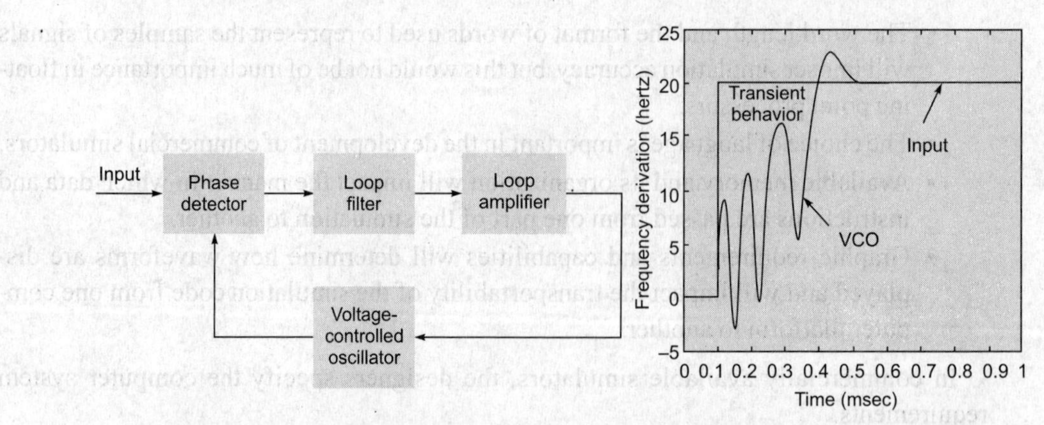

Fig. 16.11 Phase lock loop, which is a known circuit with known acquisition behaviour

16.6 STOCHASTIC SIMULATION: AN EXAMPLE

In general, stochastic processes are random processes, which are analysed on the basis of probability theory. Hence, estimations for particular parameters are made on the basis of probability functions. One typical example is given in Fig. 16.12. The QPSK-based transmitter and receiver are used for the simulation. Here the Monte Carlo simulation method is used.

The Monte Carlo simulation describes a simulation in which a parameter of a system, such as BER, is estimated using the Monte Carlo techniques. It is the process of estimating the value of a parameter by performing an underlying stochastic or random

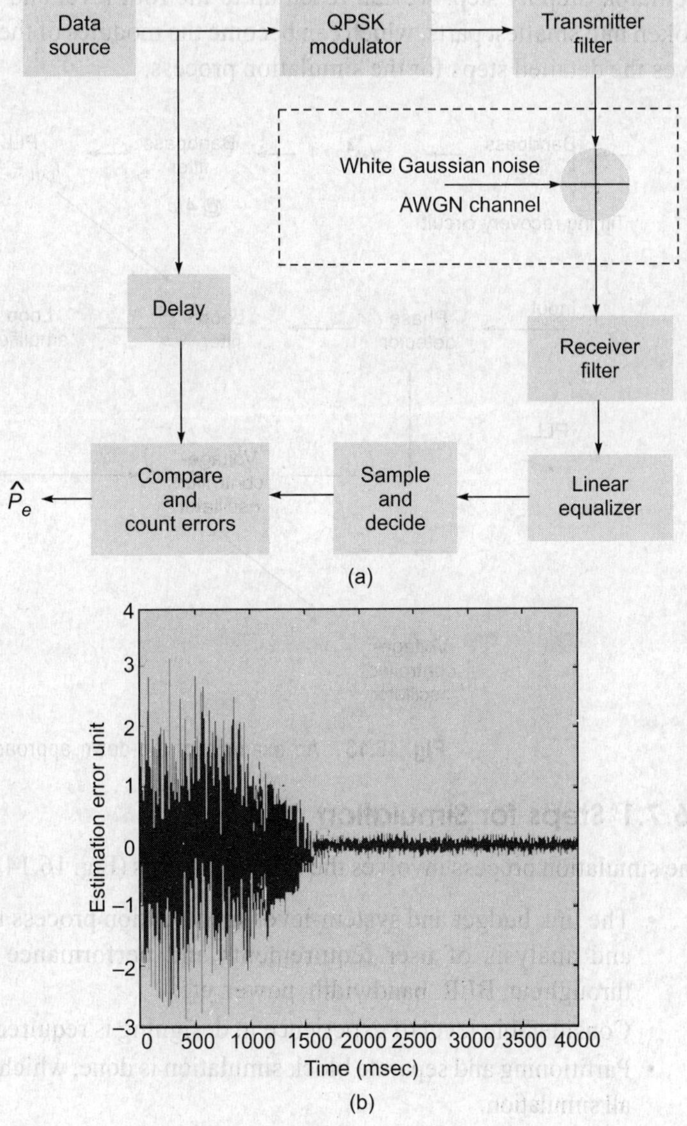

Fig. 16.12 (a) Stochastic simulation model using QPSK transceiver and (b) Monte Carlo technique for error estimation

experiment. Here, in the example, the transmitted and received data patterns are compared. If codes are developed for the Monte Carlo simulations, 'FOR' loop is a must in most of the cases. The estimated parameters may be E_b/N_o, probability of error, etc.

16.7 GENERAL STEPS OF SIMULATION

Simulation development is a 'top-down' approach, where hardware implementation proceeds from the 'bottom up'. In Fig. 16.13, a typical example of simulation is given, where the first level block diagram includes one of the blocks, PLL. In the second level, we can expand the PLL block diagram, in which one of the blocks is a voltage-controlled oscillator. Step by step, we can reach up to the root level and the whole system can be broken into smallest parts, which can become the modules of the simulation. Figure 16.14 gives the detailed steps for the simulation process.

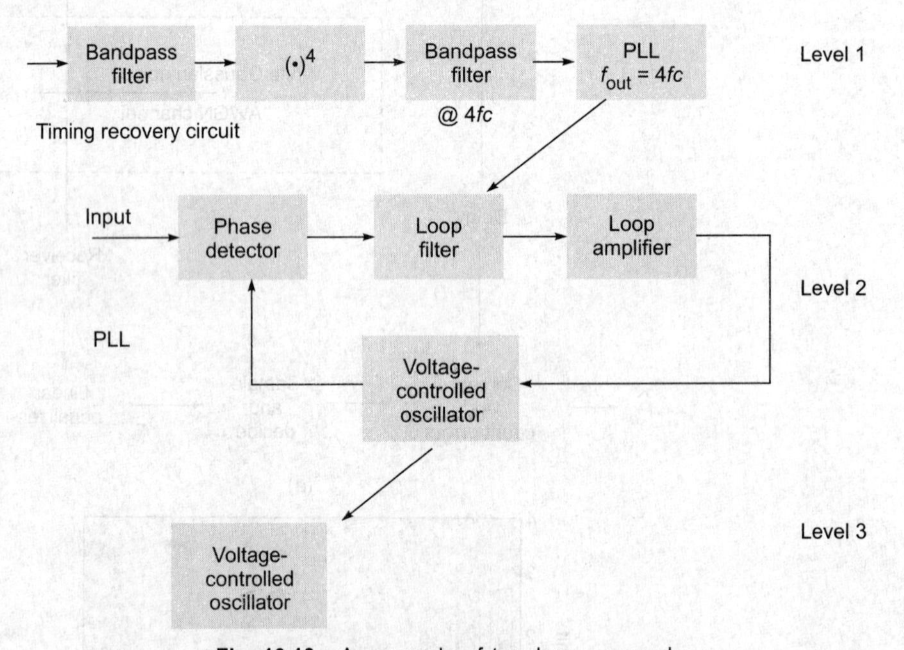

Fig. 16.13 An example of top-down approach

16.7.1 Steps for Simulation

The simulation process involves the following steps (Fig. 16.14):

- The link budget and system-level specification process begins with the statement and analysis of user requirements and performance expectations, including throughput, BER, bandwidth, power, etc.
- Considerable level of experience in designing is required. It may be a paper work.
- Partitioning and separate block simulation is done, which is then followed by overall simulation.

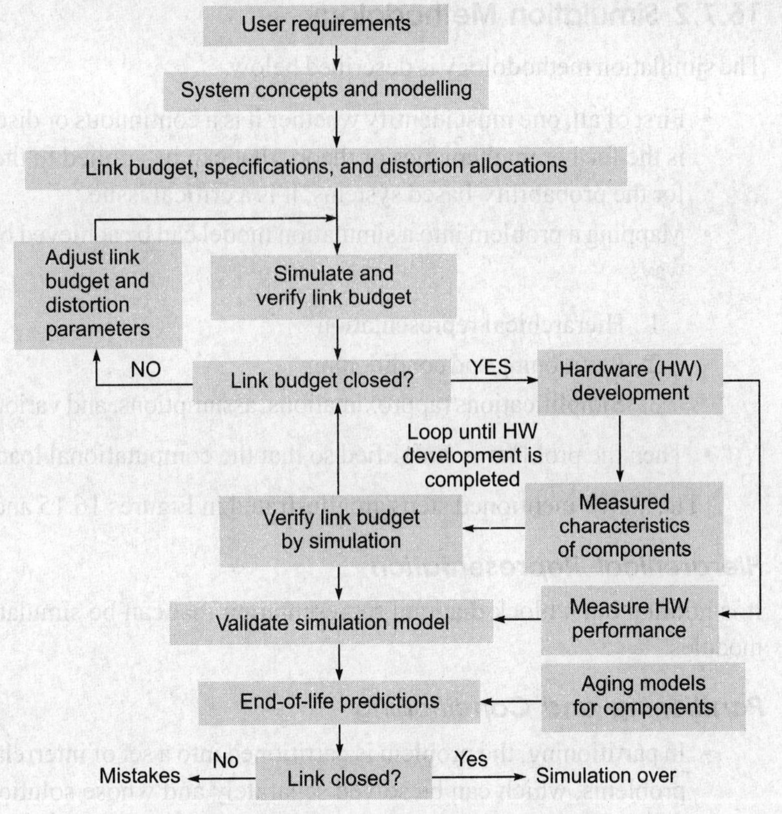

Fig. 16.14 Flow chart for simulation

- Simulate and run the simulations. If the link budget closes, then follow the next step.
- Implementation is started with testing of key components. Completion of the hardware prototype will decide the validation or acreditation of the simulation model. (Otherwise the steps are repeated from the beginning modifying the parameters.)
- End-of-life predictions are done using the aging models.

The simulation has both qualitative and quantitative aspects of simulation and hence it is a combination of both an art and a science and both are interrelated. Some steps are theory based and quantitative in nature, while some may be heuristic in nature, which cannot be quantified. Hence, the 'methodology of simulation' is necessary. The overall approach or methodology used to solve a design or performance-estimation problem depends on the nature of the specific problem. Hence, it is difficult to present the methodology as an independent set of rules or algorithms and some generic aspects that can be applied to wide variety of simulation problems.

16.7.2 Simulation Methodology

The simulation methodology is described below.

- First of all, one must identify whether it is a continuous or discrete model and what is the further mathematics or theory that can be applied to the system. Especially, for the probability-based systems, it is a critical issue.
- Mapping a problem into a simulation model can be achieved by the following three ways:

  1. Hierarchical representation
  2. Partitioning and conditioning
  3. Simplifications (approximations, assumptions, and various simplifications)

- Then the problem is simplified so that the computational load may be reduced.

The above-mentioned steps are illustrated in Figures 16.15 and 16.16.

Hierarchical Representation

It is nothing but a block diagram representation that can be simulated independently as modules.

Partitioning and Conditioning

- In partitioning, the problem is partitioned into a set of interrelated but independent problems, which can be solved separately and whose solutions can be combined later. It is a useful method for reducing the complexity and the computational burden.
- In conditioning, we simply fix the condition or state of a portion of the system and simulate the rest of the system under various values of the conditioned variables or states. The conditioned part of the system is simulated separately and the results obtained are averaged with respect to the distribution of the conditioning variable.

Simplifications and Approximations

- One can omit some of the blocks and add it afterwards, e.g., simulate modulator/demodulator without a channel and add the channel afterwards.
- First use the normal conditions and thereafter find out worst-case conditions using simulations.
- Non-linear systems can be approximated for its linear portion.
- Assume a time-invariant system (stationary).
- The simplifications and approximations vary with the systems and leads to completion of basic-level simulation but maybe far from an actual system. Then, one by one add the actual situations and step by step make it complex.

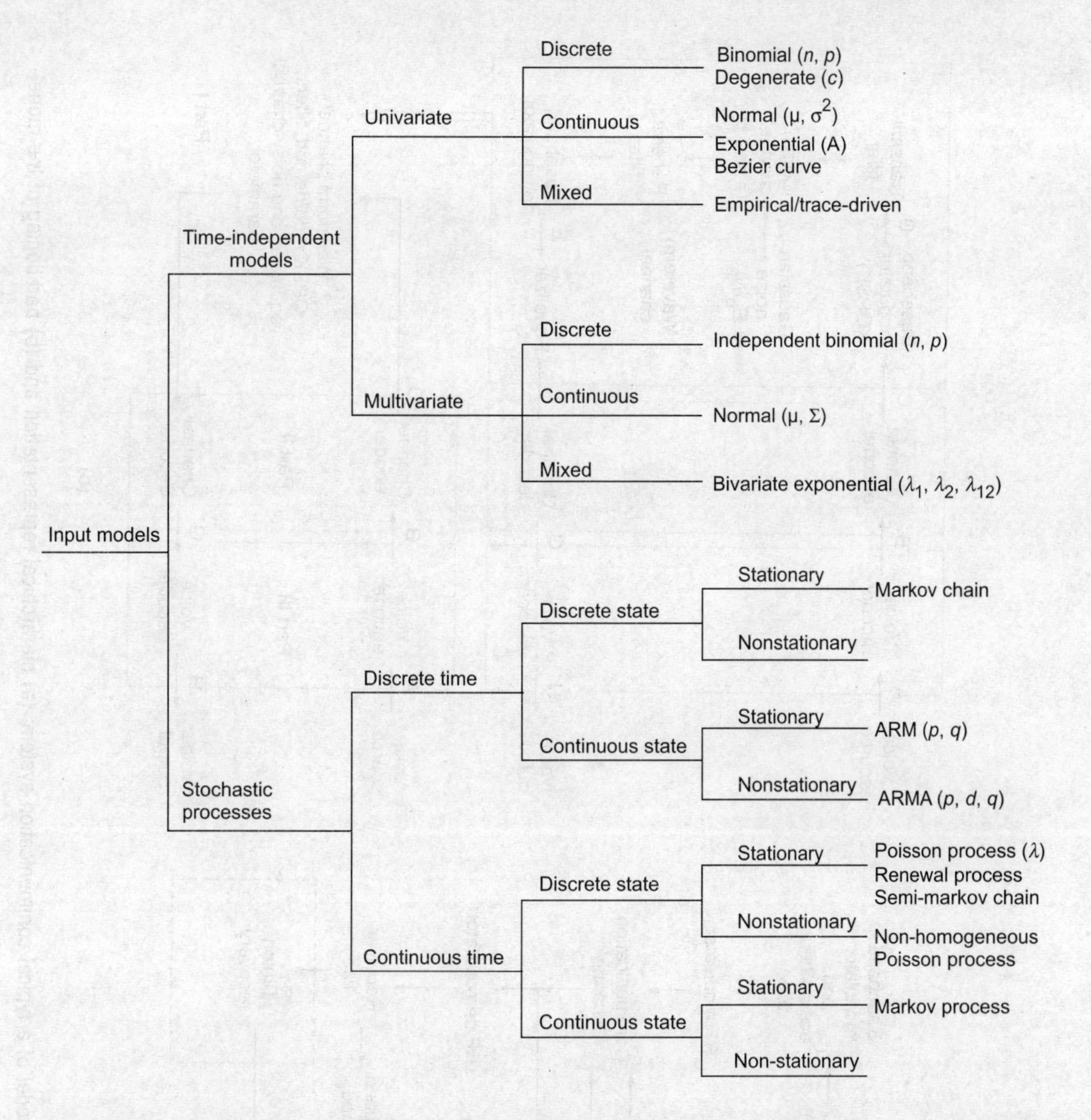

Fig. 16.15 Identification of type of system and associated theory or mathematics if the system is based on probability

16.8 SOME MISCELLANEOUS CONSIDERATIONS FOR SIMULATION

Time-Domain and Frequency-Domain Simulation

- It is a common practice to use time-domain samples to represent the input and output signals. The frequency domain models, such as filters simulated using FFT, will require internal buffering of the time-domain input samples, taking the transform of the input vector stored in the buffer followed by frequency domain processing, inverse FFT, and buffering at the output.
- Buffering is required since the transform is a block processing operation based on a set of samples rather than on a single sample.

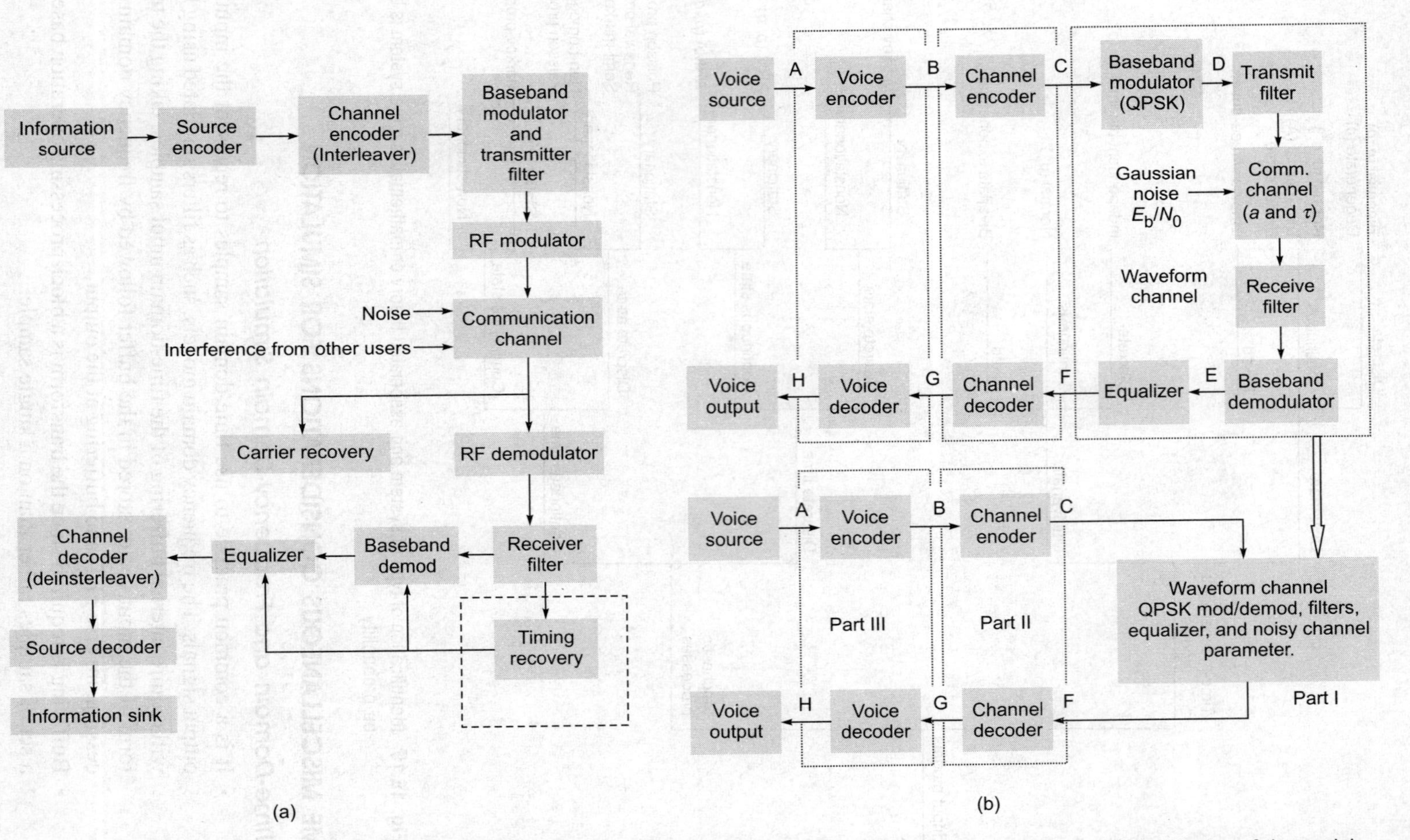

Fig. 16.16 System level model of a typical communication system: (a) hierarchical representation and (b) partitioning of the model

Design Optimization

Design optimization means finding the optimum value of critical parameters, such as the bandwidth of a receiver filter, the operating point of an amplifier, and the number of quantization levels. To do this, models have to parameterized properly and the key design parameters should be made visible externally, enabling to design the parameters iteratively during the simulations.

Validation Techniques

- Animation or graphs should meet the expected requirements; graphical comparison may be done.
- Comparison to other models should be done.
- Degenerate tests should be conducted.
- Event validity should be done, i.e. block diagram or flow diagram must be corrected.
- Extreme condition tests should be performed.
- Use fixed known values rather than variables and run the program. Observe the behaviour.
- Validity of the internal stages—multistage validation.
- Parameter variability, sensitivity analysis.
- Predictive validation there should be correct approximation.
- Conceptual model validation and model verification.
- Operational validity.

Figure 16.17 shows the validation technique.

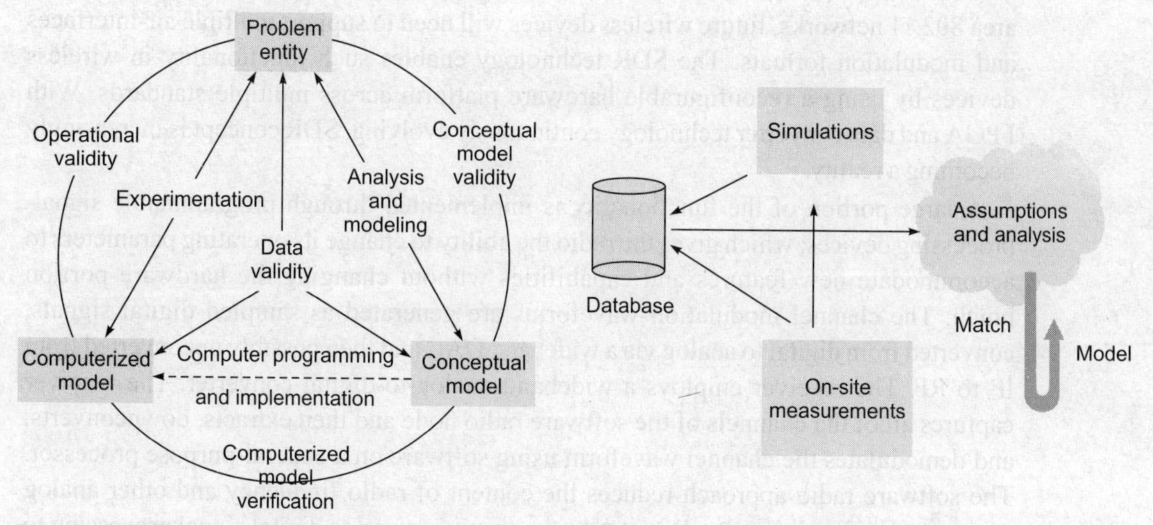

Fig. 16.17 Validation of the simulation (Simulation results and on-site measurements should match. It is possible only if the assumptions and modelling for simulation are correct.)

Emulators vs Simulators

Emulators are hardware systems with software support. Some of the functions will be performed by hardware and their results will be displayed by the software, depending upon the feasibility of the tasks.

16.9 SOFTWARE-DEFINED RADIO (SDR)

A radio means any kind of device that intentionally transmits or receives signals in the radio frequency (RF) part of the electromagnetic spectrum in a wireless way. This, of course, includes our everyday AM and FM radios, such as those in our homes and cars. TVs are also radios that happen to turn the signals they receive into moving pictures and sound. Cell phones, cordless phones, garage door openers, car door openers, wireless Internet cards (Wi-Fi/802.11), shortwave, satellite, pagers, GPS, and radar—all are radios.

We all know that without the role of digital signal processing, most of the above-mentioned communication systems are not possible in the present scenario. When the digital signal processor itself is introduced in the wireless communication transmitter and receiver, a new methodology can be introduced by designing the transceivers with the help of programming the DSP (processors) for the required modulated waveform generation, which can be communicated by the supporting hardware for RF conversion and antenna.

There are many ways to describe a software radio, but one definition that seems to encompass the essence of the software-defined radio (SDR) is that it has a generic hardware platform on which software runs to provide functions, including modulation and demodulation, filtering (including bandwidth changes), and other functions, such as frequency selection and, if required, frequency hopping.

With the proliferation of wireless standards, including wide area 3G, 2.5G, and local area 802.11 networks, future wireless devices will need to support multiple air-interfaces and modulation formats. The SDR technology enables such functionality in wireless devices by using a reconfigurable hardware platform across multiple standards. With FPGA and data converter technology continuously evolving, SDR concept is increasingly becoming a reality.

A large portion of the functionality is implemented through programmable signal-processing devices, which gives the radio the ability to change its operating parameters to accommodate new features and capabilities without changing the hardware portion much. The channel modulation waveforms are generated as sampled digital signals, converted from digital to analog via a wideband DAC and then possibly upconverted from IF to RF. The receiver employs a wideband analog-to-digital converter. The receiver captures all of the channels of the software radio node and then extracts, downconverts, and demodulates the channel waveform using software on a general-purpose processor. The software radio approach reduces the content of radio frequency and other analog components of traditional radios and the emphasis is given to digital signal processing to enhance the overall receiver features and flexibility.

A conceptual transmitter of SDR would be very simple. A computer program would generate a stream of numbers in binary form. These would be sent to a digital-to-analog converter connected to RF converter and to a radio antenna, and the receiver will follow the opposite tasks.

This change in the design paradigm for new radios has occurred so rapidly that it is necessary to provide the literature to students and engineers to train them. Conventional radio engineering books give more emphasis on analog component level design with little emphasis on the increasingly important role of digital signal processing in performing the main functions of the radio transceiver. Hence, the literature on software radio must be developed.

Three excellent starting points for further information on SD radio are as follows:

- Software-defined radio forum
- GNU software radio project
- Amateur radio relay league
- Altera SDR design

Given the constraints of today's technology, there is still some RF hardware involved, but the idea is to get the software as close to the antenna as is feasible. The antennas should also be controlled by the software. Ultimately, we are turning the hardware problems into software problems. The ideal scheme would be to attach an analog-to-digital converter to an antenna. A computer would read the converter and then software would transform the stream of data from the converter to any other form.

Software radios are emerging in commercial as well as military infrastructure because of numerous advantages, such as ease of design, ease of manufacture, multimode operation, use of advance digital signal processing techniques, fewer discrete components, and flexibility to incorporate additional functionality.

16.10 NEED FOR SOFTWARE RADIO

The most significant asset of SDR is versatility. Wireless systems employ protocols that vary from one service to another. Even in the same type of service, for example, in a wireless fax, the protocol often differs from country to country. A single SDR set with an all-inclusive software repertoire can be used in any mode, anywhere in the world. One can make portable SDRs. Interoperability is another issue. As so much of the radio is contained within the software, it will be possible to integrate other associated software functions into the system more easily. Thus, the aim is to be able to integrate other third-party software applications into the basic radio, thereby increasing the functionality.

There are a few new things that software radios can do, which have not been possible before. These are discussed below.

- The SDR can be reconfigured 'on-the-fly', that is depending on what you need, your universal communication device would reconfigure itself appropriately for your environment. It can be a cordless phone, cell phone, wireless Internet gadget, or GPS receiver. In other words, the device exhibits mulifunctionality, compactness, and power efficiency.
- The SDR can be quickly and easily upgraded with enhanced features according to the requirements. In fact, the upgrade can be delivered over the air. The SDR can

talk and listen to multiple channels at the same time. Today, there are many places where public safety people from one organization cannot talk to those from another. The locals cannot talk to the emergency crew from the next town because they have different kinds of radios. Software radio solves this problem as it has global mobility.

- We can build new kinds of radios that have never before existed. Smart radios or cognitive radios can look at the utilization of the RF spectrum in their immediate neighbourhood and configure themselves for best performance.

Current digital technologies are too slow to receive typical radio signals, which range from 10 kHz to 2 GHz. An ideal software radio would have to collect and process samples at twice the maximum frequency at which it is to operate. Real software radios solve this problem by using a mixer and a reference oscillator to heterodyne the radio signal to a lower frequency. The mixer changes the frequency of the signal. The phase information becomes more difficult to detect in it. Many digital encoding systems depend on the phase encoding. The classic solution is to mix and digitize two channels, using a reference oscillator that produces two signals of same frequency. However, one of the frequency outputs lags the other by 90 degrees of a cycle. Thus, the two sets of samples provide the needed phase information.

Another related problem is that the information about the bit timing is lost when the frequency changes. The phase information helps recover that as well. The sampling works best if it is at a simple multiple of the protocol's symbol rate. Since the distant transmitter and the receiver are linked only by the radio, the sampling speed should somehow adapt to the distant radio's symbol rate. The phase information may, therefore, be used to adjust the effective sampling rate as well. Any signals above the sampling frequency would 'interfere' with the sampling, causing spurious signals to appear in the data stream at a frequency, that is the difference between the signal and the sampling frequency. For this reason, a low-pass analog electronic filter must precede the digital conversion step.

A good software radio must operate at any symbol rate within a wide range of rates in order to be compatible with many protocols and this adaptive control is crucial. It can be implemented either with a hardware linkage to the converter or in software.

16.11 GENERAL STRUCTURE OF THE TRANSCEIVER FOR SDR

There are mainly three stages in SDR—a baseband section, an IF section, and an RF section and an antenna. A typical voice SDR transmitter, such as that used in mobile two-way radio or cellular telephone communication, consists of the following components:

- Microphone
- Audio amplifier
- Analog-to-digital converter (ADC), which converts the voice audio to ASCII (or digital) data*
- Modulator which, impresses the ASCII intelligence onto a radio-frequency (RF) carrier *

- Series of amplifiers, which boosts the RF carrier to the power level necessary for the transmission
- Transmitting antenna*

(The parts of the system shown with * represent computer-controlled circuits whose parameters are determined by the programming software.)

A typical receiver designed to intercept the above-mentioned voice SDR signal would employ the following components, essentially reversing the transmitter's action:

- Receiving antenna*
- Super-heterodyne system, which boosts incoming RF signal strength and converts it to a constant frequency*
- Demodulator, which separates the ASCII intelligence from the RF carrier *
- Digital-to-analog converter (DAC), which generates a voice waveform from the ASCII data*
- Audio amplifier
- Speaker, earphone, or headset

(Again, the items followed by asterisks represent programmable circuits.)

In order to meet the objectives set for a protected signal and flexibility for the radio platform, we have defined a general structure for the transmitter and receiver. Each application or waveform (existing or under development) can use these general structures.

The general structure of an SDR transmitter is represented in Fig.16.18. If the transmitter contains an adaptive antenna array, it can independently transmit each waveform to its desired direction (maybe in a multiantenna system); otherwise a single-antenna system is the normal requirement. The waveforms are generated digitally using simulation methods, mathematical expressions, and digital signal processor programming. Beam forming is required in multiple-antenna system using software methods; otherwise a direct D-to-A converter converts the digitally generated signal into the analog form for transmitting it through hardware-based RF upconverter and single antenna. For a multiantenna system, such hardware is required for each antenna.

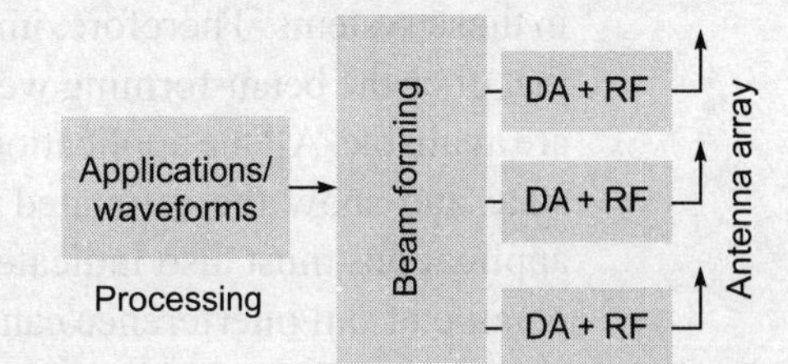

Fig. 16.18 Simplified block diagram of SDR transmitter using multiple antennas

The possible structures of the receiver in multiple-antenna and single-antenna cases are shown in Fig. 16.19. The general structure at the application level is shown in Fig. 16.19(c). The Interference cancellers (ICs) used in the single-antenna case can typically mitigate interference of signals, which have a narrower bandwidth than the desired signal. The adaptive antennas can typically mitigate both wideband and narrowband interference and thus offer enhanced performance.

Importantly, the receiver consists of a matched filter for each application. The matched filters provide fast synchronization, which has always been a necessary requirement in communication systems. The matched filters are realized in frequency domain not as a correlator or a bank of correlators in the time domain, which is the case with typical implementations. The frequency domain realization was selected—first, due to its flexibility because it allows an easy use of signals even with different lengths and second because it is most promising because it offers the lowest computational burden of all matched filter implementations for general waveforms. The reason for the rareness of frequency domain realizations is that so far, a sufficiently fast technology has not been available. For our purposes, on the receiver side, the idea is to get a wideband ADC as close to the antenna as is convenient, get the samples into something we can program, and then process them in software.

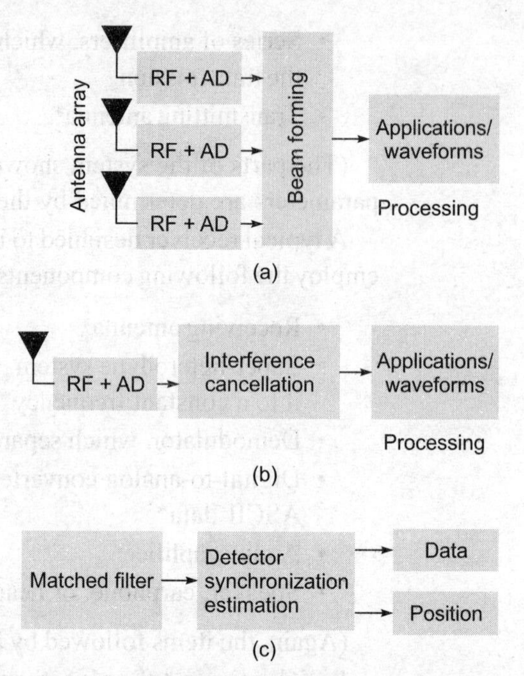

Fig. 16.19 Simplified block diagram of an SDR receiver: (a) receiver structure with multiple antennas, (b) receiver structure with single antenna, and (c) receiver structure at application/waveform level

Fully adaptive versions cannot be implemented at a reasonably low cost with the current technology because of the limitations set by a rather high bandwidth used in these systems. Therefore, implementation can be defined as 'semi-adaptive', meaning that once the beam-forming weights are computed, they are kept fixed until new weights are available. All the applications using the SDR platform have to inform users about their state and show the estimated SNR so that the channel quality can be surveyed. The applications must also indicate if the channel is jammed or not. This knowledge can be obtained from interference cancellers or from the direction-of-arrival (DOA) estimation circuit.

16.12 THIRD-GENERATION (3G) SDR SYSTEM ARCHITECTURE

The SDR is a good example of multirate signal processing. Figure 16.20 illustrates a typical hardware partitioning of an SDR-based 3G base station that can be reconfigured to support multiple standards. In order to reconfigure the entire system, an ideal SDR base station would perform all signal processing tasks in the digital domain. The same diagram is applicable to a mobile SDR also. However, current-generation wideband data converters cannot support the processing bandwidth and the dynamic range required across different wireless standards. As a result, the analog-to-digital converter and the digital-to-analog converter are usually operated at the intermediate frequency (IF) and separate wideband analog front ends are used for subsequent signal processing to the RF stages, as shown in Fig.16.20.

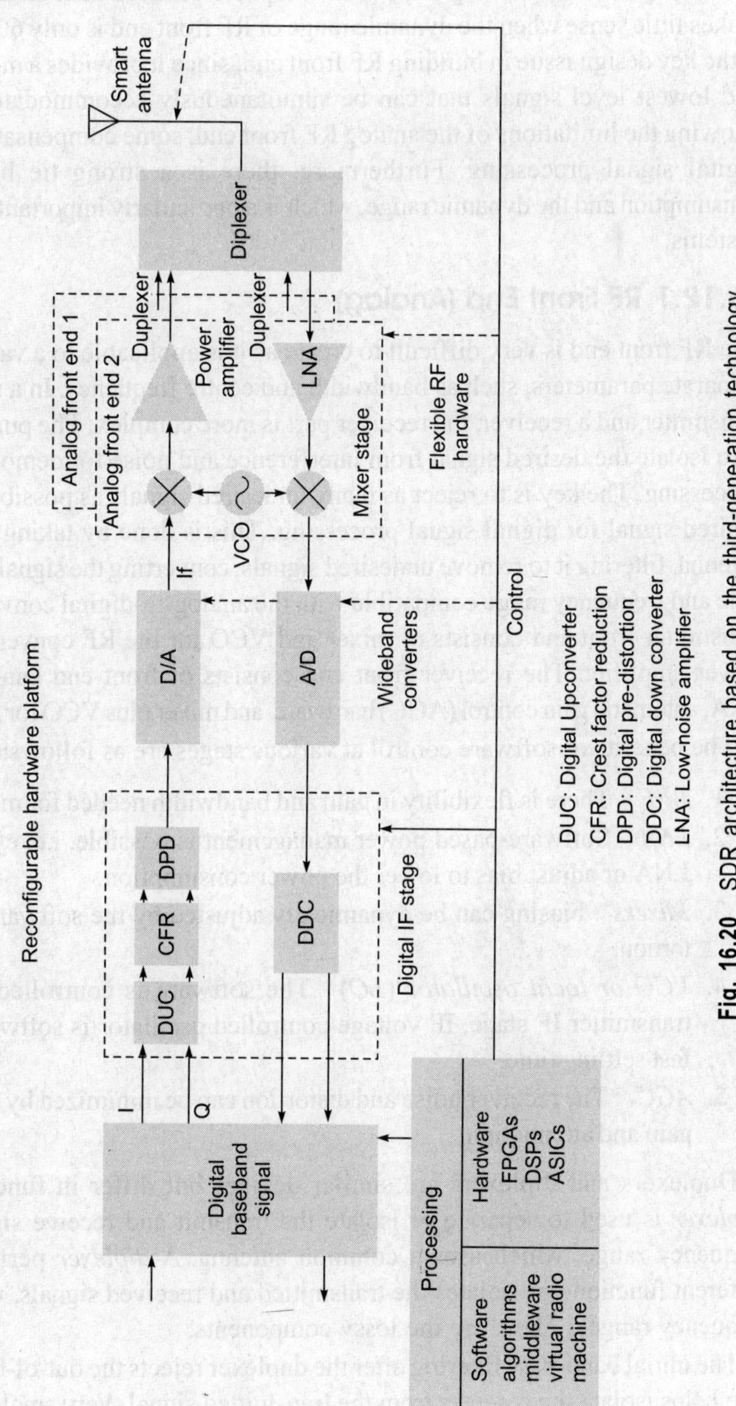

Fig. 16.20 SDR architecture based on the third-generation technology

A good communication system design must have the right balance in its components. For example, having an expensive ADC capable of more than 85 dB of dynamic range makes little sense when the dynamic range of RF front end is only 60 dB. Dynamic range is the key design issue in building RF front end, since it provides a measure of the highest and lowest level signals that can be simultaneously accommodated by the radio. By knowing the limitations of the analog RF front end, some compensation can be made via digital signal processing. Furthermore, there is a strong tie between the battery consumption and the dynamic range, which is a particularly important trade-off for mobile systems.

16.12.1 RF Front End (Analog)

The RF front end is very difficult to create as it is applicable to a variety of signals with disparate parameters, such as bandwidth and centre frequency. In a transceiver having a transmitter and a receiver, the receiver part is more complex. The purpose of the receiver is to isolate the desired signal from interference and noise for demodulation and further processing. The key is to reject as many undesired signals as possible and condition the desired signal for digital signal processing. This is done by taking the signal from the antenna, filtering it to remove undesired signals, converting the signal to IF with an amplitude and frequency range compatible with the analog-to-digital conversion process. The transmitter front end consists of mixer and VCO for the RF conversion along with the power amplifier. The receiver front end consists of front end band-pass filter (BPF), LNA, automatic gain control (AGC) hardware, and mixer plus VCO for RF-to-IF conversion.

The benefits of software control at various stages are as follows:

1. *BPF* There is flexibility in gain and bandwidth needed for multimode operation.
2. *LNA* Software-based power management is possible, i.e. cycle ON or OFF the LNA or adjust bias to lower the power consumption.
3. *Mixers* biasing can be dynamically adjusted by the software to reduce the distortion.
4. *VCO or local oscillator* (*LO*) The software is controlled for tuning. In the transmitter IF stage, IF voltage-controlled oscillator is software controlled for a fast settling time.
5. *AGC* The receiver noise and distortion can be minimized by software-controlled gain and attenuation.

Duplexers and diplexers are similar devices but differ in functions. Normally, a *duplexer* is used to separate or isolate the transmit and receive signals in a common frequency range, which uses a common antenna. A *diplexer* performs a similar but different function—it isolates the transmitted and received signals, which lie in distinct frequency ranges. These are the lossy components.

The initial band-pass filtering after the duplexer rejects the out-of-band interference. It also helps isolate the receiver from the transmitted signal. Very small amount of noise is injected with low loss and much selectivity is provided without limiting the bandwidth needed to support multiple modes of the software radio. Low-noise amplifier boosts the signal power level into a range compatible with other components in the circuit without adding excessive noise into the signal. This must be traded for power consumption and dynamic range. Image reject filter is located before the mixer, which reduces noise and protects the mixer from interference, including any signals located at the image frequency, which after IF conversion may lie in the same band as the desired signal. The mixer is used to downconvert the signal and can be a major source of intermoduation distortion as it is a non-linear device. Harmonics due to the mixer can be removed by the IF pass band. The mixer is driven by a local oscillator or VCO, whose frequency determines the channel selection. The LO should have a good tuning range and good phase stability to minimize the phase noise. The AGC is primarily required to ensure that the signal has a voltage level that is compatible with the ADC input range. It should be fast enough to account for changing signal levels (due to fast channel fading) and is placed to ensure that minimal noise is injected into the system and also to ensure that the signal is not clipped off by the ADC.

16.12.2 Digital IF Processing

Digital IF extends the scope of digital signal processing beyond the baseband domain out to the antenna—to the RF domain. This increases the flexibility of the system while reducing manufacturing costs. Moreover, digital frequency conversion provides greater flexibility and higher performance (in terms of attenuation and selectivity) than traditional analog techniques. Latest FPGAs, with high-performance embedded digital signal processing block, embedded soft processors, memory architecture, and high-speed interfaces, provide a highly flexible and integrated platform to implement computationally intensive digital IF functions, including digital upconverter and downconverter (DUC and DDC), while reducing the risk involved in introducing new techniques, such as DPD, CER, and smart antennas.

Digital Upconverter (DUC)

The data formatting, often required between the baseband processing elements and the upconverter, can be seamlessly added at the front end of the upconverter as shown in Fig. 16.21. This technique provides a fully customizable front end to the upconverter and allows for channelization of high-bandwidth input data, which is found in many 3G systems. Custom logic or an embedded processor can be used to control the interface between the upconverter and the baseband processing element.

In digital upconversion, the input data is baseband filtered and interpolated before it is quadrature modulated with a tunable carrier frequency. To implement the interpolating baseband finite impulse response (FIR) filter, the compiler with the optimal fixed or

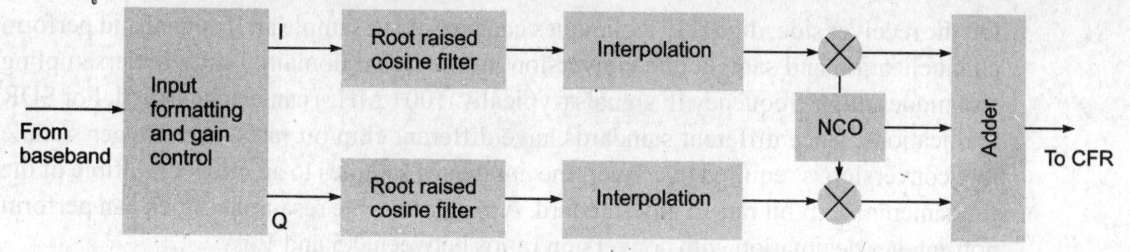

Fig. 16.21 Digital upconverter

adaptive filter architectures can be built for a particular standard, optimizing speed-area trade-offs. The numerically controlled oscillator (NCO) compiler IP core can generate a wide range of architectures for oscillators with good dynamic range and very high performance. Depending on the number of frequency assignments to be supported, the right number of digital upconverters can be easily instantiated in a programmable logic device.

Crest Factor Reduction (CFR)

The 3G CDMA-based systems and multicarrier systems, such as OFDM, exhibit signals with high crest factors (peak-to-average ratios). Large peaks (Fig. 9.33) increase the amount of inter-modulation distortion resulting in an increase in the error rate. The average signal power must be kept low in order to prevent the transmitter amplifier limiting. The variability of a signal is normally measured by its crest factor (CF), which corresponds to a measure of the peak-to-average envelope power of the modulated RF carrier. Minimizing the CF allows a higher average power to be transmitted for a fixed peak power, improving the overall signal-to-noise ratio at the receiver.

Such signals drastically reduce the efficiency of PAs used in the base stations.

Digital Predistortion (DPD)

The 3G standards and their high-speed mobile data versions employ the non-constant envelope modulation techniques, such as QPSK and QAM. This places stringent linearity requirements on the power amplifiers. Class A amplifiers with appropriate backoff can achieve the level of linear fidelity but at the cost of power efficiency. Class AB, B, and C amplifiers provide much better power efficiency but have a non-linear response, which leads to a broadening of the transmitted signal spectrum. One remedy to this is to predistort the signal before power amplification, such that when the signal undergoes the amplifier distortion, the resulting signal approximates the original signal. The predistortion may be performed either in digital or in analog domain. Basic idea is to compensate for the non-linear gain of the amplifier with the predistortion function.

The DPD linearization techniques, including both the look-up table and polynomial approaches, can be implemented efficiently. The multipliers in the DSP blocks can have speeds up to 380 MHz and can be effectively time-shared to implement complex multiplications. When used in SDR base stations, an FPGA can be reconfigured to implement the appropriate DPD algorithm, which efficiently linearizes the PA used for a specific standard.

Digital Downconverter (DDC)

On the receiver side, digital IF techniques can be used to sample an IF signal and perform channelization and sample rate conversion in the digital domain. Using undersampling techniques, high-frequency IF signals (typically, 100+ MHz) can be quantified. For SDR applications, since different standards have different chip/bit rates, non-integer sample rate conversion is required to convert the number of samples to an integer multiple of the fundamental chip/bit rate of any standard. A programmable resampler block can perform non-integer decimation with conversion ratios between 0.5 and 1.

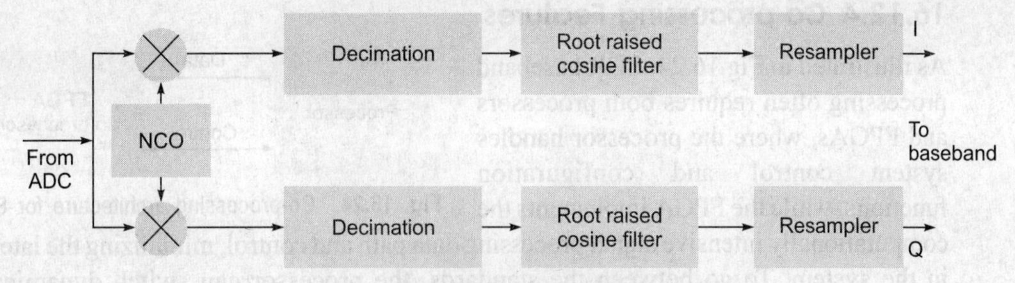

Fig. 16.22 Digital downconverter

16.12.3 Baseband Processing

Wireless standards are continuously evolving to support higher data rates through the introduction of advanced baseband processing techniques, such as adaptive modulation and coding, space-time coding, beam forming, and MIMO techniques.

The baseband components also need to be flexible enough to enable SDR functionality, which is required to support migration between enhanced versions of the same standard as well as the capability to support a completely different standard. Figure 16.23 illustrates an example scenario where FPGAs can be easily reconfigured to support the baseband transmit functions for either WCDMA/HSDPA or 802.16a standards through available mega-core functions and reference designs, such as the turbo encoder, Reed–Solomon encoder, and IFFT.

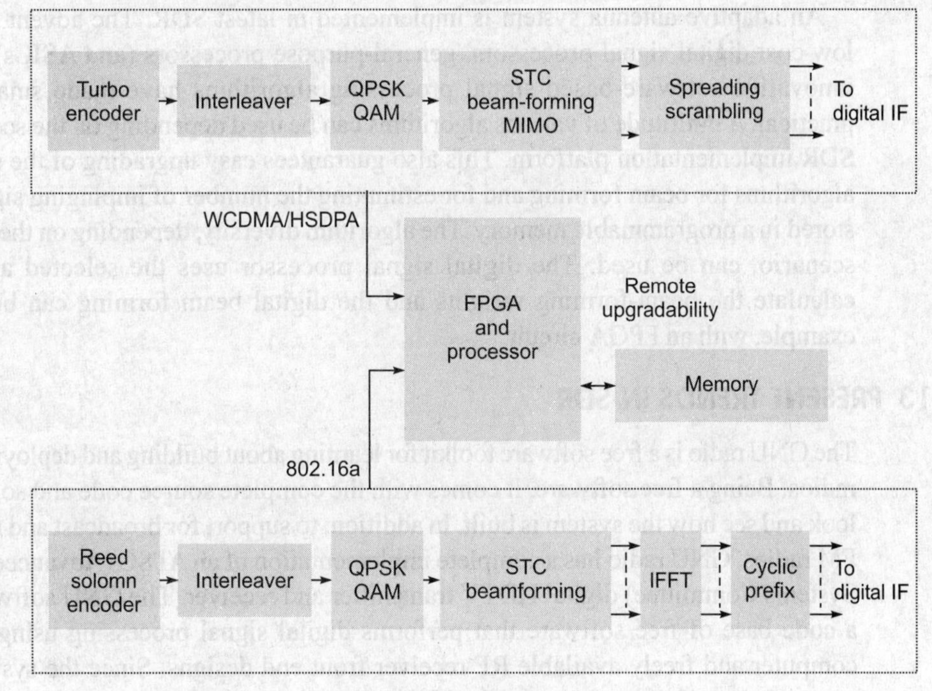

Fig. 16.23 An example of SDR baseband data path reconfiguration

16.12.4 Co-processing Features

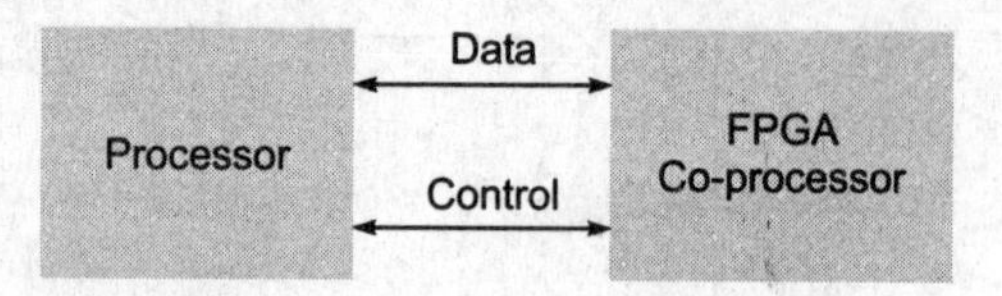

Fig. 16.24 Co-processing architecture for SDR

As illustrated in Fig.16.24, SDR baseband processing often requires both processors and FPGAs, where the processor handles system control and configuration functions, while the FPGA implements the computationally intensive signal processing data path and control, minimizing the latency in the system. To go between the standards, the processor can switch dynamically between major sections of software, while the FPGA can be completely reconfigured, as necessary, to implement the data path for the particular standard.

16.12.5 Adaptive Antenna over SDR Platform

The antenna is perhaps one of the most important links in the overall design. Much of the link gain can be achieved or lost in the selection of the antenna. For a software radio, which is designed to support multiple modes and multiple bands, the antenna choice becomes an even more crucial issue. Most antennas support bandwidths of about 10–12% of the carrier frequency. Thus, multiband radios, which cover the cellular range of 900 MHz and 2 GHz, are difficult to support with a single antenna. The form factor of antenna is very important. Exploitation of diversity or smart antennas can be performed at the handset or at the base station. As mentioned previously, the diversity can be achieved by using multiple receiver chains and spatially separated antennas, antennas with different polarization characteristics or different gain patterns. For mobile terminals, the type of antennas may be dipole, monopole, loop, and patch antennas.

An adaptive antenna system is implemented in latest SDR. The advent of powerful low-cost digital signal processors, general-purpose processors (and ASICs), as well as innovative software-based signal processing algorithms have made smart antennas practical. A multitude of various algorithms can be used depending on the scenario by an SDR implementation platform. This also guarantees easy upgrading of the system. The algorithms for beam forming and for estimating the number of impinging signals can be stored in a programmable memory. The algorithm diversity, depending on the operational scenario, can be used. The digital signal processor uses the selected algorithm to calculate the beam-forming weights and the digital beam forming can be made, for example, with an FPGA circuit.

16.13 PRESENT TRENDS IN SDR

The GNU radio is a free software toolkit for learning about building and deploying software radios. Being a free software, it comes with the complete source code and so anyone can look and see how the system is built. In addition, to support for broadcast and narrowband FM radios, GNU radio has a complete implementation of an ATSC (Advanced Television Systems Committee) digital HDTV transmitter and receiver. The GNU software radio is a code base of free software that performs digital signal processing using a personal computer and freely available RF receiver front end designs. Since the system utilizes open source software and a standard PC under Linux, the GNU software radio receiver is an ideal platform for learning and experimenting with SDR concepts.

The NCASSR (National Center For Advanced Secure Systems Research) SDR project at NCSA (National Center For Supercomputing Applications) has extended the GNU software radio receiver design to a 900 MHz narrowband software-defined radio transceiver, that is a radio capable of transmitting and receiving. The project team developed the SDR transceiver to facilitate further research and development of SDR, including new front-end hardware, algorithms, protocols, security, and operational visualization.

An ideal SDR can operate in any radio frequency band, limited only by national regulatory agencies and the characteristics of its RF 'front-end' analog hardware and antenna. The researchers' current approach to the front-end analog hardware is to support a common IF and design RF hardware to fit between the antenna and the common IF. Dr. Jennifer Bernhard of the University of Illinois at the Urbana-Champaign Electrical and Computer Engineering Department, is leading an effort to design an antenna for the extensible sensor platform that would cover a wideband from 50 MHz to 2.4 GHz, which would be wider than any antenna commercially available today.

One area of SDR research already underway at NCSA is the SDR operational visualization software that serves as both an educational tool for introducing concepts of radio communications to novice users and an architectural overview of our SDR system. This real-time visualization provides a qualitative, high-level block diagram of the operation of the SDR as it receives a still image. The visualization software taps into an active SDR, extracts the signal after each stage of reception, and superimposes an image of the signal onto the block diagram.

Another new application is a reconfigurable communication protocol stack. Reminiscent of the Unix system V-streams architecture, the transmit and receive data paths are composed of processing blocks that may be dynamically swapped in and out of the flow at run time. Currently, the communication stack elements are discrete blocks that handle the network transport protocol, security processing, end-user applications, and radio hardware management.

An example of the flexibility and usefulness of this architecture is a simple security application. The security processing block in the receiver data stream uses the DES algorithm to decrypt the received data. An authenticated command to change to the AES algorithm is received by the radio. The radio management software swaps out the security block implementing DES and replaces it with the security block implementing AES. The data processing continues seamlessly.

To provide security for SDR, the NCSA SDR project team is developing a voice-authentication application that identifies and authorizes SDR users, allowing specific radio capabilities to be unlocked depending on the user. For example, imagine a software-defined radio for emergency response with this voice-authentication application—an ordinary citizen could be allowed to use the radio to contact authorities to report incidents, while an identifier response commander could use the radio to communicate on a number of private bands reserved for the responding teams, such as police, fire, medical, etc.

16.14 FUTURE OF SDR

Traditionally, different countries and different kinds of units have used non-compatible radio systems. This has led to the situation where there are multiple radio systems operating in the same geographical area without ability to move messages efficiently from one system to another. The use of SDR-based multi-mode multi-band radio (MMMB radio) will decrease the amount of radio equipment needed and enable efficient communication between different players at the operational area. The digital battle space of the future is based on a flexible communication platform that connects separate systems together.

The waveforms will be implemented to the SDR platform with software communications architectures. In addition, software-defined radio platform is going to increase processing capacity to implement time and frequency-domain interference cancellation algorithms. Adaptive antenna-based systems are also under development to enhance system performance. The software radio will offer resources to adaptively control the radiation pattern of an antenna array using beam forming.

A software radio will be able to handle dynamic allocation of processing resources between different applications. If the processing resources run out, the radio must have means to determine the priority of various connections. The RF parts of a software-defined radio must be able to handle wide bandwidths varying from HF frequencies to microwaves. All these will be possible with SDR. It is presumed that a software radio has a very bright future.

16.15 COGNITIVE RADIO (CR)

Cognitive radio (CR) is one of the new long-term developments taking place in radio receiver technology. After SDR, which is slowly becoming more of a reality, CR will be the next step ahead. The only difference between SDR and CR is that CR needs to scan a wide range of frequency spectra before deciding which band to use, instead of a predefined one, as an SDR terminal does.

The idea for cognitive radio is developed from the need to utilize the radio spectrum more efficiently and to be able to maintain the most efficient form of communication for the prevailing conditions. By using the levels of processing that are available today, it is possible to develop a radio that is able to look at the spectrum, detect which frequencies are clear, and then implement the best form of communication for the required conditions. In this way, cognitive radio is able to select the frequency band, type of modulation, and power levels most suited to the requirements and prevailing conditions. In view of this, cognitive radio is effectively an environment-friendly technology, adapting to the prevailing conditions.

Work is underway to determine the best methods of developing a radio system that would be able to fulfil the requirements of a CR system. Although the level of processing required may not be fully understood yet, it is clear that a significant level of processing will be needed. The radio will need to determine the occupancy of the available spectrum and then decide the best power level, mode of transmission, and other necessary characteristics. Additionally, the radio will need to be able to judge the level of interference it may cause to other users. In addition to the level of processing required for the cognitive radio, the RF sections will need to be particularly flexible. Not only that, they need to swap frequency bands, possibly moving between the portions of the radio spectrum that are widely different in frequency, but they may also need to change between the transmission modes that can occupy different bandwidths.

To achieve the required level of performance, a very flexible front end is needed. Traditional front-end technology cannot handle these requirements because they are generally band limited, both for the form of modulation used and the frequency band in which they operate. Even the so called wideband receivers have limitations and generally operate by switching front ends as required. Accordingly, the required level of performance can only be achieved by converting to and from the signal as close to the antenna as possible. In this way, no analog signal processing will be needed, all the processing being handled by the digital signal processing.

The conversion to and from the digital format is handled by DACs and ADCs. To achieve the performance required for a cognitive radio, not only do the DACs and ADCs require enormous dynamic range and need to be able to operate over a very wide range, extending up to many GHz, but in the case of the transmitter, they must also be able to handle significant levels of power. Currently, these requirements are beyond the limits of the technology available. Thus, the full vision for cognitive radio cannot yet be met. Nevertheless, in the future, the required DAC and ADC technology will undoubtedly become available, thereby making cognitive radio a reality.

With the growth in radio and wireless communications, ideas like cognitive radio will become more important and the way has been opened from this viewpoint to assist the development of cognitive radio technology in many countries.

Summary

- Simulation is nothing but an equivalent scenario of the system created on the PC.
- Simulation is done for estimation of system performance under certain parametric condition.
- Simulation is advantageous for the system design because hardware-based design for the complex system is a tedious job.
- The results of simulation are totally dependent upon the parameter selection and modelling or assumptions. Hence, 100% results may not be achieved.
- Different types of communication systems require different approaches. Common rules or methods are not there; hence, the system identification is a very important issue.
- In most of the wireless digital communication systems, Monte Carlo simulations are required for BER estimation, channel estimation, etc., as the channel being time varying makes the whole receiver scenario probability based.
- Validation techniques are required for checking the correctness of the simulation.
- Software-defined radio (SDR) needs the concepts of simulation partially, especially for waveform generation.
- The SDR is a different approach for radio system design, in which channel modulation waveforms are generated by software and the RF communication takes place by using hardware and antenna.
- Digital signal processor and FPGA can be mainly used in SDR because they provide flexibility and reconfigurable platform.
- The ADC and DAC are very important parts of SDR.
- Digital upconverter and digital downconverter are the special devices used in SDR for RF up/downconversions directly without using ADC/ DAC.
- Running many protocols in one radio device will be possible by SDR-based devices.
- Multi-antenna based systems are also possible with SDR.

Review Questions

1. What do you mean by simulation?
2. 'Simulation may not give correct results. Justify this statement.
3. 'Sometimes simulation-based system design may not work practically.' Comment.
4. What are the various approaches by which the simulation can be done?
5. List out the advantages and the drawbacks/limitations of the simulation.
6. How does the processor speed sometimes affect the simulation results?
7. What is the importance of modelling in simulation?
8. What do you mean by validation of a model? List out the techniques for validation.
9. Assume any particular communication system and give your own steps for simulating that scenario.
10. What will be the role of ADC/DAC conversion time in the designing of software-defined radio system?
11. Draw the block diagram of a single antenna-based SDR system transmitter and receiver. List out the functions of each of the blocks in the diagram.
12. Will it be possible to have very high-frequency communication with SDR devices and under what conditions?
13. What is the role of digital signal processor in software-defined radio system?
14. What is the role of FPGA part in software-defined radio system? How is it possible to have flexibility and reconfigurability with FPGA?

Multiple Choice Questions

1. A QPSK modem without power amplifier forms
   (a) analytically complex system
   (b) analytically tractable system
   (c) analytically intractable system
   (d) none of the above
2. The last stage of simulation is
   (a) accreditation (b) modelling
   (c) partitioning (d) analysis
3. Which of the following statements is true?
   (a) Simulation is a bottom-to-top approach.
   (b) Simulation gives 100% results.
   (c) Simulation results are dependent upon the parameter selection.
   (d) Simulation is always based on the approximations.
4. Which of the following parts in SDR is hardware?
   (a) Modulator
   (b) ADC/DAC
   (c) Beam-forming controller
   (d) Waveform generator,
5. The flexibility and reconfigurability in SDR is due to
   (a) DDC and DUC
   (b) DSP and FPGA
   (c) ADC/DAC
   (d) all of the above
6. Which of the following statements is incorrect?
   (a) An SDR can work with systems with different protocols
   (b) An SDR is some as cognitive radio.
   (c) An SDR performs analog as well as digital operations.
   (d) An SDR can use MIMO.

Appendix A

Linear Systems Theory

A linear system is given by

$$y(t) = \int_{-\infty}^{\infty} x(\tau)h(t,\tau)d\tau \tag{A.1}$$

where $x(t)$ is the input and $h(t, \tau)$ is the system weighting function.

A linear system model (Fig. A.1) is the model of a linear systems with input-output relationship represented by a transfer function shown in the box. It is tried to correlate this model with the general diagram of a communication system in which the transmit-receive relationship is decided by the channel transfer function (or Impulse response). The whole theory with mathematical equations is described with the help of this model.

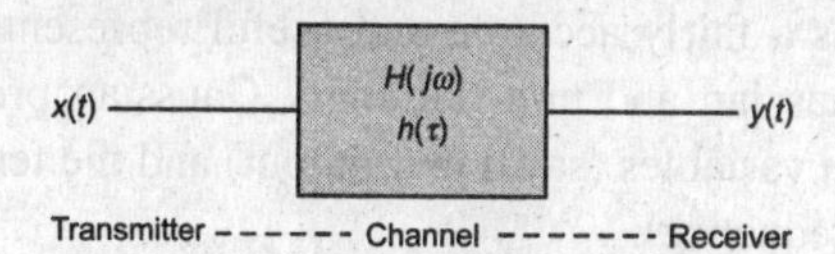

Fig. A.1 Linear system model

If the system is time invariant, then Eq. (A.1) becomes

$$y(t) = \int_{-\infty}^{\infty} h(\tau)\, x(t-\tau)\, d\tau \tag{A.2}$$

This type of integral is called *convolution integral.*

Differential equations involving random processes are called *stochastic differential*. A linear stochastic differential equation as a model of an RP (random process) with initial conditions has the general form

$$\bar{x}(t) = F(t)x(t) + G(t)w(t) + C(t)u(t) \tag{A.3}$$

$$y(t) = H(t)x(t) + v(t) + D(t)u(t) \tag{A.4}$$

where

$\bar{x}(t)$ = the next state estimation

$x(t)$ = $n \times 1$, state vector (input)

$y(t)$ = $l \times 1$, measurement vector (received output of the channel)

$u(t) = r \times 1$, deterministic input vector (known control input—not present in our system)
$F(t) = n \times n$, time-varying dynamic coefficient matrix
$C(t) = n \times r$, time-varying control input coupling matrix (not present in our system)
$H(t) = l \times n$, time-varying measurement sensitivity matrix (i.e. transfer function—here channel)
$D(t) = l \times r$, time-varying output coupling matrix (for further control input—not in our system)
$G(t) = n \times r$, time-varying process noise coupling matrix (related to input—omitted in our system)
$w(t) = r \times 1$, zero mean uncorrelated 'plant noise' process (to input—omitted in our system)
$v(t) = l \times 1$, zero mean uncorrelated 'measurement noise' process (here channel noise or interference)

The expected values (mean) are as follows:

$$E[w(t)] = 0, \text{ zero mean Gaussian noise at input}$$
$$E[v(t)] = 0, \text{ zero mean Gaussian noise at output}$$
$$E[w(t_1)w^T(t_2)] = Q(t_1)\delta(t_2 - t_1)$$
$$E[v(t_1)v^T(t_2)] = R(t_1)\delta(t_2 - t_1)$$
$$E[w(t_1)v^T(t_2)] = M(t_1)\delta(t_2 - t_1)$$

The simplified suitable equations for the system are

$$\bar{x}(t) = F(t)x(t) \tag{A.5}$$
$$y(t) = H(t)x(t) + v(t) \tag{A.6}$$

The symbols Q, R, and M represent $r \times r$, $l \times l$, and $r \times l$ matrices, respectively, and δ represents the Dirac delta 'function' (a measure).

The values over time of variable $x(t)$ in the differential equation model define vector-valued Markov processes. This model is a fairly accurate and useful representation for many real-world processes, including stationary Gaussian and non-stationary Gaussian processes, depending on the statistical properties of the random variables (say, input/output) and the temporal properties of the deterministic variables (say, training sequence).

Various filters can be derived on the basis of difference equations mentioned in Equations (A.5) and (A.6).

Discrete model of a random sequence or frame-based data without initial conditions or control input can be given in the form

$$x_k = \Phi_{k-1} x_{k-1} \tag{A.7}$$
$$y_k = H_k x_k + v_k \tag{A.8}$$

Here

$x_k = n \times 1$, state vector (input)
$y_k = l \times 1$, measurement vector (received output of the channel)
$\Phi_{k-1} = n \times n$, time-varying dynamic coefficient matrix
$H_k = l \times n$, time-varying measurement sensitivity matrix (i.e. transfer function—here channel)
$E[v_k] = 0$, zero mean Gaussian noise at output
$E[w_{k1} w^T_{k2}] = Q_{k1}\,\Delta(t_2 - t_1)$
$E[v_{k1} v^T_{k2}] = R_{k1}\Delta(t_2 - t_1)$
$E[w_{k1} v^T_{k2}] = M_{k1}\,\Delta(t_2 - t_1)$

Appendix B

Algebra for the Linear System

Many times, it is convenient to use vector and matrix notation for the representation of the discrete signals and operations performed on them. This representation simplifies many of the mathematical expressions and their solutions. Such representation is also useful while studying mathematics of MIMO systems.

A vector is an array of real-valued or complex-valued numbers or functions. Vectors can be denoted by lower-case bold letters and, in all cases, these vectors will be assumed to be column vectors. For example,

$$\mathbf{x} = \begin{bmatrix} x_1 \\ x_2 \\ \vdots \\ x_N \end{bmatrix} \tag{B.1}$$

is a column vector containing N scalars. If the elements are real, then it is a real, vector otherwise a complex vector of N dimensions. The transpose of a vector, $\mathbf{x}^T$ is a row vector and Hermitian transpose $\mathbf{x}^H$ is the complex conjugate of the transpose of $\mathbf{x}$. The vector elements can be represented with time index values in case of finite length sequence $x(n)$ as

$$\mathbf{x} = \begin{bmatrix} x(0) \\ x(1) \\ \vdots \\ x(N-1) \end{bmatrix} \quad \text{or} \quad \mathbf{x}(n) = \begin{bmatrix} x\ (n) \\ x\ (n-1) \\ \vdots \\ x\ (n-N+1) \end{bmatrix} \tag{B.2}$$

The convolution between $h(n)$ and $x(n)$, where $h(n)$ can be represented in the same vector form as $x(n)$, results in $y(n)$. Then $y(n)$ may be written as the inner product

$$y(n) = \mathbf{h}^{\mathrm{T}}\mathbf{x}(n) \tag{B.3}$$

A set of linear equations can be better represented in matrix form. An $n \times m$ matrix is an array of numbers (real or complex) or functions having n rows and m columns. For example,

$$\mathbf{A} = \{a_{ij}\} = \begin{bmatrix} a_{11} & a_{12} & a_{13} & \cdots & a_{1m} \\ a_{21} & a_{22} & a_{23} & \cdots & a_{2m} \\ a_{31} & a_{32} & a_{33} & \cdots & a_{3m} \\ \vdots & \vdots & \vdots & \vdots & \vdots \\ a_{n1} & a_{n2} & a_{n3} & \cdots & a_{nm} \end{bmatrix} \tag{B.4}$$

The rank of **A** is defined to be the number of linearly independent columns in **A**. Special matrix forms are square matrix, diagonal matrix, and identity matrix (where the diagonal elements are 1's). Other important terms, such as Toeplitz matrix, eigenvalues, and eigenvectors, can be studied from appropriate references.

Appendix C

Probability Theory

The following concepts are important to study the randomly fading channel behavioural mathematics and signal detection process mathematics.

Discrete Probability Distributions

Definition C.1

Suppose we have an experiment whose outcome depends on the chance. We represent the outcome of the experiment by a letter, such as X, called a *random variable*. The sample space of the experiment is the set of all possible outcomes. If the sample space is either finite or countably infinite, the random variable is said to be *discrete*. We generally denote a sample space by the capital Greek letter Ω. There are two additional definitions. These are subsidiary to the definition of sample space and serve to make precise some of the common terminology used in conjunction with sample spaces. First, we define the elements of a sample space to be outcomes. Second, each subset of a sample space is defined to be an event. Normally, we shall denote outcomes by lower-case letters and events by capital letters.

If a sample space has an infinite number of points, then the way a distribution function is defined depends upon whether or not the sample space is countable. A sample space is countably infinite if the elements can be counted, i.e. can be put in one-to-one correspondence with the positive integers, and uncountably infinite otherwise.

Definition C.2

Let X be a random variable, which denotes the value of the outcome of a certain experiment, assuming that this experiment has only finitely many possible outcomes. Let Ω be the sample space of the experiment (i.e. the set of all possible values of X or equivalently, the set of all possible outcomes of the experiment). A distribution function for X is a real-valued function m whose domain is Ω and which satisfies the following conditions:

1. $m(\omega) \geq 0 \quad$ for all $\omega \in \Omega$
2. $\sum_{\omega \in \Omega} m(\omega) = 1$

For any subset E of Ω, we define the probability of E to be the number $P(E)$ given by

$$P(E) = \sum_{\omega \in E} m(\omega) \tag{C.1}$$

Note that an immediate consequence of the above definitions is that for every $\omega \in \Omega$,

$$P(\{\omega\}) = m(\omega) \tag{C.2}$$

That is, the probability of the elementary event $\{\omega\}$, consisting of a single outcome ω, is equal to the value $m(\omega)$ assigned to the outcome ω by the distribution function.

It is important to consider ways in which probability distributions are determined in practice. One way is by symmetry. In general, considerations of symmetry often suggest the uniform distribution function. Care must be taken here. We should not always assume that just because we do not know any reason to suggest that one outcome is more likely than another, it is appropriate to assign equal probabilities. Statistical estimates for probabilities are fine if the experiment under consideration can be repeated a number of times under similar circumstances.

Definition C.3

The uniform distribution on a sample space Ω containing n elements is the function m defined by

$$m(\omega) = 1/n \quad \text{for every } \omega \in \Omega \tag{C.3}$$

Definition C.4

If $P(E) = p$, the odds in favour of the event E occurring are $r : s$ (r to s), where $r/s = p/(1-p)$. If r and s are given, then p can be found by using the equation $p = r/(r + s)$.

Set Theory

In many cases, events can be described in terms of other events through the use of the standard constructions of set theory. We will briefly review the definitions of these constructions. The Venn diagrams shown in Fig. C.1 illustrate these constructions.

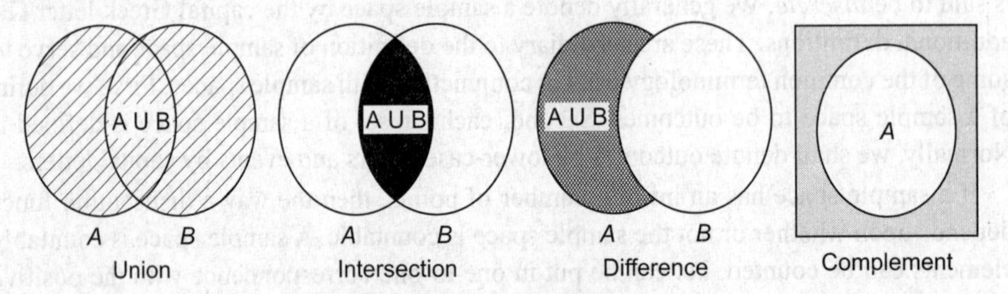

Fig. C.1 Venn diagrams

Let A and B be two sets.

Then the union of A and B is the set

$$A \cup B = \{x \mid x \in A \text{ or } x \in B\} \tag{C.4}$$

The intersection of A and B is the set

$$A \cap B = \{x \mid x \in A \text{ and } x \in B\} \tag{C.5}$$

The difference of A and B is the set

$$A - B = \{x \mid x \in A \text{ and } x \notin B\} \tag{C.6}$$

The set A is a subset of B, written as $A \subset B$, if every element of A is also an element of B.

Finally, the complement of A is the set

$$A = \{x \mid x \in \Omega \text{ and } x \notin A\} \tag{C.7}$$

For any two events A and B,

$$P(A) = P(A \cap B) + P(A \cap \overline{B}) \tag{C.8}$$

Properties

Theorem C.1 The probabilities assigned to events by a distribution function on a sample space Ω satisfy the following properties:

1. $P(E) \geq 0$ for every $E \subset \Omega$
2. $P(\Omega) = 1$
3. If $E \subset F \subset \Omega$, then $P(E) \leq P(F)$
4. If A and B are disjoint subsets of Ω, then $P(A \cup B) = P(A) + P(B)$
5. $P(\overline{A}) = 1 - P(A)$ for every $A \subset \Omega$

Theorem C.2 If $A_1, \ldots, A_n$ are pairwise disjoint subsets of Ω (i.e. no two of the A_i's have an element in common), then

$$P(A_1 \cup \ldots \cup A_n) = \sum_{i=1}^{n} P(A_i) \tag{C.9}$$

Theorem C.3 Let $A_1, \ldots, A_n$ be pairwise disjoint events with $\Omega = A_1 \cup \ldots \cup A_n$ and E be any event. Then

$$P(E) = \sum_{i=1}^{n} P(E \cap A_i) \tag{C.10}$$

Theorem C.4 If A and B are subsets of Ω, then

$$P(A \cup B) = P(A) + P(B) - P(A \cap B) \tag{C.11}$$

Density Functions of Continuous Random Variables

Definition C.5

Let X be a continuous real-valued random variable. A density function for X is a real-valued function f that satisfies the following condition:

$$P(a \leq X \leq b) = \int_a^b f(x)\, dx \qquad \text{for all } a, b \in R \text{ (set of real values)} \tag{C.12}$$

We note that it is not the case that all continuous real-valued random variables possess density functions. We will only consider continuous random variables for which the density functions exist. The experiments in which the coordinates are chosen at random can be described by constant density functions. Such density functions are called *uniform* or *equiprobable*.

Cumulative Distribution Functions of Continuous Random Variables

The density functions are useful when considering continuous random variables. There is another kind of function, closely related to these density functions, which is also of great importance. These functions are called *cumulative distribution functions*.

Definition C.6

Let X be a continuous real-valued random variable. Then the cumulative distribution function of X is defined by the equation

$$F_X(x) = P(X \leq x) \tag{C.13}$$

If X is a continuous real-valued random variable that possesses a density function, then it also has a cumulative distribution function. Theorem C.5 shows that the two functions are related in a very nice way.

Theorem C.5 Let X be a continuous real-valued random variable with density function $f(x)$. Then the function defined by

$$F(x) = \int_{-\infty}^{x} f(t)\, dt \tag{C.14}$$

is the cumulative distribution function of X. Furthermore, we have

$$\frac{d}{dx} F(x) = f(x) \tag{C.15}$$

In many experiments, the density function of the relevant random variable is easy to write down. However, it is quite often the case that the cumulative distribution function is easier to obtain than the density function (Fig. C.2). (Of course, once we have the cumulative distribution function, the density function can easily be obtained by differentiation, as the above theorem shows.)

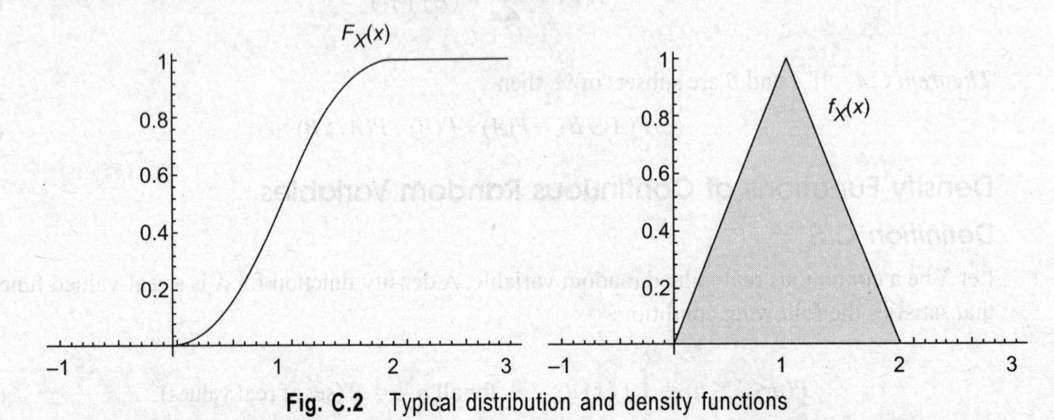

Fig. C.2 Typical distribution and density functions

A fundamental question in practice is: How shall we choose the probability density function in describing any given experiment? The answer depends to a great extent on the amount and kind of information available to us about the experiment. In some cases, we can see that the outcomes are equally likely. In some cases, we can see that the experiment resembles another experiment already described by a known density. In some cases, we can run the experiment a large number of times and make a reasonable guess at the density on the basis of the observed distribution of outcomes. In general, the problem of choosing the right density function for a given experiment is a central problem for the experimenter and is not always easy to solve.

Definition C.7

Let $X_1, X_2, \ldots, X_n$ be random variables associated with an experiment. Suppose the sample space (i.e. the set of possible outcomes) of X_i is the set R_i. Then the joint random variable $\overline{X} = (X_1, X_2, \ldots, X_n)$ is defined to be the random variable whose outcomes consist of ordered n-tuples of outcomes, with the ith coordinate lying in the set R_i. The sample space Ω of X is the Cartesian product of the R_i's

$$\Omega = R_1 \times R_2 \times \ldots \times R_n \tag{C.16}$$

The joint distribution function of $\overline{X}$ is the function that gives the probability of each of the outcomes of $\overline{X}$.

Definition C.8

The random variables $X_1, X_2, \ldots, X_n$ are mutually independent if

$$P(X_1 = r_1, X_2 = r_2, \ldots, X_n = r_n) \\ = P(X_1 = r_1)P(X_2 = r_2) \ldots P(X_n = r_n) \tag{C.17}$$

for any choice of $r_1, r_2, \ldots, r_n$. Thus, if $X_1, X_2, \ldots, X_n$ are mutually independent, then the joint distribution function of the random variable

$$\overline{X} = (X_1, X_2, \ldots, X_n) \tag{C.18}$$

is just the product of the individual distribution functions. When two random variables are mutually independent, we can say more briefly that they are independent.

Continuous Probability Densities Simulation

It is sometimes desirable to estimate quantities whose exact values are difficult or impossible to calculate exactly. In some of these cases, a procedure involving chance, called a *Monte Carlo procedure*, can be used to provide such an estimate. The Monte Carlo method is a technique that involves using random numbers and probability to solve problems. The term *Monte Carlo method* was coined by S. Ulam, von Neumann, and Nicholas Metropolis in reference to games of chance, a popular attraction in Monte Carlo, Monaco (Metropolis, N. and S. Ulam, 'The Monte Carlo Method', *J. Amer. Stat. Assoc.*, 44, 335–41, 1949).

The *Monte Carlo simulation* is a method for iteratively evaluating a deterministic model using sets of random numbers as inputs. This method is often used when the model is complex, non-linear, or involves more than just a couple of uncertain parameters. A simulation can typically involve over 10,000 evaluations of the model, a task which in the past was only practical using supercomputers. By using *random inputs*, we are essentially turning the deterministic model into a stochastic model. The Monte Carlo method is just one of many methods for analysing *uncertainty propagation*, where the goal is to determine how the random variation, lack of knowledge, or error affects the sensitivity, performance, or reliability of the system that is being modelled. The Monte Carlo simulation is categorized as a sampling method because the inputs are randomly generated from probability distributions to simulate the process of sampling from an actual population. So, we try to choose a distribution for the inputs that most closely matches data we already have, or best represents our current state of knowledge. The data generated from the simulation can be represented as probability distributions (or histograms) or converted to error bars, reliability predictions, tolerance zones, and confidence intervals.

The steps in the Monte Carlo simulation correspond to the uncertainty propagation. All we need to do is follow the five simple steps listed below:

Step 1 Create a parametric model, $y = f(x_1, x_2, \ldots, x_p)$.

Step 2 Generate a set of random inputs, $x_{i1}, x_{i2}, \ldots, x_{ip}$.

Step 3 Evaluate the model and store the results as y_i.

Step 4 Repeat Steps 2 and 3 for $i = 1$ to n.

Step 5 Analyse the results using histograms, summary statistics, confidence intervals, etc.

Readers may refer to appropriate references for further reading on the following topics:

1. **Random processes:**
   - Bayes formula
   - Bernoulli processes
   - Markov processes
   - Poisson processes
2. **Distributions:**
   - Binomial distribution
   - Geometric distribution
   - Negative binomial distribution
   - Poisson distribution
3. **Densities:**
   - Continuous uniform density
   - Exponential and gamma densities
   - Maxwell and Rayleigh densities

Normal Density

The normal density is the most important density function for communication systems.

A very important theorem in probability theory, called the central limit theorem, states that under very general conditions, if we sum a large number of mutually independent random variables, then the distribution of the sum can be closely approximated by a certain specific continuous density, called the normal density. The normal density function with parameters μ and σ is defined as follows:

$$f_x(x) = \frac{1}{\sqrt{2\pi\sigma}} e^{-(x-\mu)^2/2\sigma^2} \tag{C.19}$$

The parameter μ represents the 'centre' of the density (it is the average or expected value of the density). The parameter σ is a measure of the 'spread' of the density and thus it is assumed to be positive (σ is the standard deviation of the density). We note that it is not at all obvious that the above function is a density, i.e. that its integral over the real line equals 1. The cumulative distribution function is given by the formula

$$F_x(x) = \int_{-\infty}^{x} \frac{1}{\sqrt{2\pi\sigma}} e^{-(u-\mu)^2/2\sigma^2} du \tag{C.20}$$

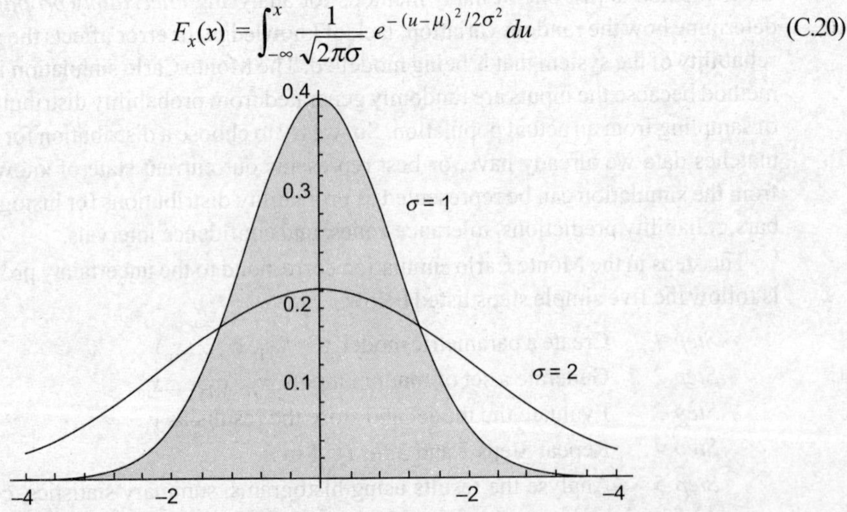

Fig. C.3 Normal density for the cases (a) μ = 0 and σ = 1 and (b) μ = 0 and σ = 2

Definition C.9

Let X be a numerically valued discrete random variable with sample space Ω and distribution function $m(x)$. The expected value $E(X)$ is defined by

$$E(X) = \sum_{x \in \Omega} x m(x) \tag{C.21}$$

provided this sum converges absolutely. We often refer the expected value as the *mean* and denote $E(X)$ by μ for short. If the above sum does not converge absolutely, then we say that X does not have an expected value.

It is easy to prove by mathematical induction that the expected value of the sum of any finite number of random variables is the sum of the expected values of the individual random variables.

Definition C.10

Let X be a numerically valued random variable with expected value $\mu = E(X)$. Then the variance of X, denoted by $V(X)$, is

$$V(X) = E[(X-\mu)^2] \tag{C.22}$$

The standard deviation of X, denoted by $D(X)$, is

$$D(X) = \text{sqrt}\,[V(X)]$$

We often write σ for $D(X)$ and σ^2 for $V(X)$.

Appendix D

DSP Fundamentals Applied to OFDM Processing

The orthogonal frequency division multiplexing (OFDM) system is based on the fundamentals of digital signal processing. The subcarrier setting in frequency domain is considered as discrete one. Hence, it is necessary to deal with discrete time signal and discrete time systems (the reason behind it will be clear with the mathematics in which the whole system is represented). Since a discrete time signal is an indexed sequence of real or complex numbers, a discrete time signal is a function of an integer valued index n, denoted by $x(n)$. A discrete time system is a mathematical operator or mapping that transforms one signal (the input) into another signal (the output) by means of a fixed set of rules or functions

$$y(n) = T[x(n)] \tag{D.1}$$

where T is the transformation. The relation between input and output can be represented by difference equation.

The OFDM is based on time and frequency domain transformations. The discrete time Fourier transform (DTFT) of a signal $x(n)$ is the complex-valued function with continuous (frequency) variabl ω defined by

$$X(e^{j\omega}) = \sum_{n=-\infty}^{\infty} x(n) e^{-jn\omega} \tag{D.2}$$

The DTFT is, in general, complex-valued function of ω Therefore, it is normally represented in polar form in terms of its magnitude ad phase $X(e^{j\omega}) = |X(e^{j\omega})| e^{-j\phi_x\omega}$. The DTFT of the unit sample (impulse) response of a linear shift-invariant stable system (in our case, a channel) is

$$H(e^{j\omega}) = \sum_{n=-\infty}^{\infty} h(n) e^{-jn\omega} \tag{D.3}$$

This is called frequency response of the filter. It defines how a complex exponential is changed in amplitude and phase by the system. The channel can be represented by a filter. The DTFT is invertible transformation and the reverse transformation is achieved by inverse DTFT (IDTFT).

$$x(n) = \frac{1}{2\pi} \int_{-\pi}^{\pi} X(e^{j\omega}) e^{jn\omega} d\omega \tag{D.4}$$

The DTFT of a convolution of two signals is equal to the product of their transforms:

$$y(n) = x(n) * y(n) \tag{D.5}$$

or

$$Y(e^{j\omega}) = X(e^{j\omega})H(e^{j\omega}) \tag{D.6}$$

Since the DTFT is a function of a continuous variable ω it is not directly amenable to digital computation. For finite length sequences, there is another representation, called discrete Fourier transform (DFT), which is a function of an integer variable k. For a finite length sequence $x(n)$ of length N, which is equal to zero outside the interval [0, N-1], the N-point DFT is

$$X(k) = \sum_{n=0}^{N-1} x(n)e^{-j2\pi kn/N} \tag{D.7}$$

The IDFT is

$$x(n) = \frac{1}{N}\sum_{k=0}^{N-1} X(k)e^{-j2\pi kn/N} \tag{D.8}$$

The fast algorithm (due to special structures) of DFT is called fast Fourier transform (FFT) and its inverse is IFFT. The signal transmission over the channel exhibits the convolution process. The product of the DTFT of two signals corresponds, in time domain, to linear convolution of the two signals. For the DFT, however, if $H(k)$ and $X(k)$ are the N-point DFTs of $h(n)$ and $x(n)$, respectively, and if $Y(k) = X(k)H(k)$, then

$$y(n) = \sum_{k=0}^{N} x((k))_N h((n-k))_N \tag{D.9}$$

which is N-point circular convolution of $x(n)$ and $h(n)$. In general, circular convolution of two sequences is not the same as linear convolution. However, there is a special and important case in which the two are the same. Specifically, if $x(n)$ is a finite length sequence of length N_1 and $h(n)$ is a finite length sequence of length N_2, then the linear convolution of $x(n)$ and $h(n)$ is of length $L = N_1 + N_2 - 1$. In this case, the N-point circular convolution of $x(n)$ and $h(n)$ will be equal to the linear convolution provided $N \geq L$.

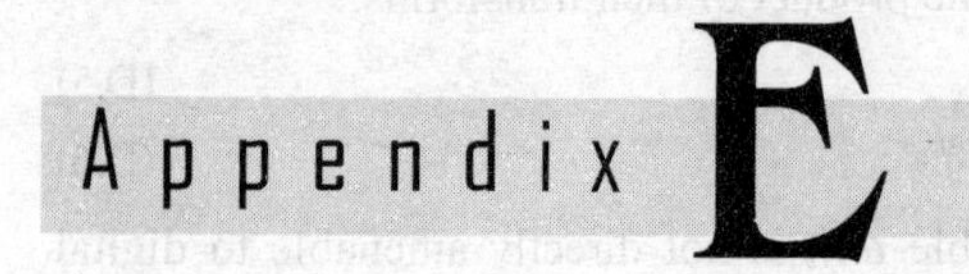

Model Question Papers

Model Paper 1

Time: 3 hr *Marks: 100*

Answer all questions.

Part A ($10 \times 2 = 20$ marks)

1. What do you mean by 2G and 3G?
2. State the band-pass sampling theorem.
3. Explain why compressors are used in PCM.
4. A TDM signal with a bit time of 0.5 μs is to be transmitted using a channel with raised cosine roll-off factor of 0.5. Calculate the required bandwidth.
5. What is the difference between Rayleigh and Rician channels?
6. How is the security provided by spread spectrum system?
7. What will be the advantages of multicarrier transmission?
8. State the condition for a set of basis functions to be orthonormal.
9. Write the expression for bit error rate for coherent BFSK.
10. State the standard interfaces of UMTS.

Part B ($5 \times 16 = 80$ marks)

1. (a) What is meant by ideal Nyquist channel? What are its merits and limitations?

or

What is the condition for zero ISI?

or

Explain the expression for Nyquist criterian for distortionless base-band transmission for zero ISI. [8]

(b) What do you mean by zero forcing equalizer? Give the other equalizers in comparison with zero forcing equalizer. [8]

or

(a) Explain what is meant by likelihood function. [4]

(b) What is meant by moment? [4]

(c) Explain the various schemes for quantization in brief. [8]

2. (a) Draw the diagram of a 1/2 rate convolutional encoder with generator polynomials $g^{(1)}(D) = 1 + D^2$ and $g^{(2)}(D) = 1 + D + D^2$. Compute the encoder output for the input sequence 1010. [8]

or

Draw the diagram of the rate ½ convolutional encoder with constraint length of 4. What is the generator polynomial of the encoder? Find the encoded sequence you have drawn, corresponding to the message sequence.

(b) Draw the block diagram of MSK transmitter-receiver and explain the function of each block. [8]

3. (a) State and explain the properties of maximal length sequences. [6]

(b) Draw the block diagram of DS-spread spectrum system transmitter and receiver and explain the function performed by each block in brief. Describe anti-jam property of DSSS system. [10]

or

(a) Enumerate the different hand-off strategies when a mobile moves into a different cell while a conversation is in progress. [8]

(b) Explain how the coverage and capacity in cellular systems can be improved. [8]

4. (a) Explain in detail the block diagram of OFDM system and state the importance of each stage. [8]

(b) Explain in detail the spectral setting and frequency to time-domain conversion at the transmitter end. [8]

or

(a) Justify that OFDM is a spectrally efficient technique compared to FDM. [4]

(b) Give the general architecture of software-defined radio transmitter and receiver. [6]

(c) How is MIMO suitable with OFDM? [6]

5. (a) Describe GSM phase 2+ protocol HSCSD highlighting the key modifications in the initial phase GSM. [4]

(b) Explain the protocol architecture of DECT system and explain the significance of C-plane (signalling) and U-plane (application process). [6]

(c) Show that the GPRS architecture is based on the GSM architecture. Show all standard interfaces clearly along with key components of the architecture. [6]

or

(a) What is LMDS? Draw CorDECT WLL architecture. [6]

(b) State the role of PCF and DCF in Wi-Fi. [6]

(c) State the advantages of software radio. [4]

Model Paper 2

Time: 3 hr *Marks: 100*

Answer all questions.

Part A (10 × 2 = 20 marks)

1. What are the two limitations of delta modulation?
2. What do you mean by synchronous and asynchronous transmissions?
3. How is the transfer function of the matched filter related to the spectrum of the input signal?

4. What is meant by processing gain of DS spread spectrum system?
5. How can the space diversity be exploited in MIMO?
6. Why is the IFFT/FFT stage required in OFDM scheme?
7. What do you mean by aliasing? What should be the pass band for antialiasing and smoothing filters used with pulse modulation/demodulation systems?
8. Why validation is required in simulations?
9. What is the role of BTS in GSM system?
10. State the equations for CDMA capacity calculation.

Part B (5 × 16 = 80 marks)

1. (a) Sketch the time response and frequency response of signal with raised cosine pulse. Correlate them with mathematics. Why is the raised cosine filter applied? [8]
   (b) Explain the evolution of the wireless communication systems from 1G to 4G. [8]
2. (a) Explain with the block diagram the ADPCM method. [8]
   (b) What do you understand by channel coding? [2]
   (c) Discuss any method for bit synchronization. [6]

or

   (a) Explain in detail the method of sub-band coding. [6]
   (b) Draw the block diagram of the 8QAM transmitter and corresponding receiver and explain their operations. [6]
   (c) Find the 3 dB bandwidth for a Gaussian low-pass filter used to produce BT = 0.25 GMSK. The channel data rate is 200 kbps. What is the 90% power bandwidth in the RF channel? *Hint:* For 0.25 GMSK, the occupied RF bandwidth is 0.57 times the bit rate. [4]
3. (a) Explain the differences between (i) slow fading and fast fading, (ii) small-scale fading and large-scale fading, and (iii) frequency-selective fading and flat fading. [6]
   (b) Explain the modelling of Rayleigh fading channel. [5]
   (c) How is it suitable to use FIR filter for simulation of the channel? [5]

or

   (a) Draw the block diagram of an adaptive filter used in an equalizer and explain it. [6]
   (b) Discuss the ways in which fast hopping scheme and slow frequency hopping spread spectrum schemes can be used to mitigate multipath effect. [4]
   (c) Describe the various diversity techniques. [6]
4. (a) Describe the various handover strategies for cellular systems. [4]
   (c) How can you define spectrum efficiency and power efficiency? [4]
   (b) Write a short note on digital audio broadcasting. [8]

or

   (a) Explain with relevant information the principles of CDMA digital cellular standard IS-95. [8]
   (b) Why is it necessary to send pilot carriers? In OFDM system, state the methods of sending pilots. What will be the drawbacks of sending pilots? [8]
5. (a) Explain in detail the Bluetooth protocol stack. [8]
   (b) Draw and explain the methods of extending the range of WLL service. [8]

or

   (a) For OFDM signalling with $Nc = 64$, find out the bandwidth of this OFDM signal if the input data rate is 10 Mbps and each carrier uses 16PSK modulation. Also find out the PSD for the signal. [6]

(b) Describe in brief the modelling methods for the simulation of communication systems. [6]

(c) Explain the MIMO channel modelling. [4]

Model Paper 3

Time: 3 hrs *Marks: 100*

Answer all questions.

Part A (10 × 2 = 20 marks)

1. What is meant by pseudo-ternary signalling?
2. What information is found from the eye pattern?
3. State the difference between coherent and non-coherent binary modulation schemes.
4. Define trunking efficiency and grade of service.
5. Define small-scale propagation model.
6. A transmitter produces 50 W of power. Express the transmitter power in units of dBm and dBW.
7. How can we remove ISI and ICI in OFDM?
8. If a mobile is moving at 60 km/hr and the frequency is 900 MHz, what will be the Doppler spread?
9. Give the different modes in Bluetooth.
10. Prove that for a particular coverage area, if a cell size reduces, there will be an increase in the cellular system capacity. Consider the bandwidth as 25 MHz and each channel of 200 kHz bandwidth.

Part B (5 × 16 = 80 marks)

1. (a) Give the various definitions of bandwidth. [6]

(b) Explain the concept of autocorrelation and cross correlation and their importance in communication systems. [6]

(c) What do you mean by wide sense stationary processes? [4]

2. (a) The antenna for a television station is located at the top of a 1600-feet tower for transmission. Compute the LOS coverage for the TV station if the receiving antenna (in the fringe area) is 25 feet above the ground. [4]

(b) Explain in detail low bit rate coders and their importance in telecommunication systems. [8]

(c) Define the terms shadowing and signal outages. [4]

or

(a) Define the following terms related to multipath:

(i) Power delay profile, (ii) first arrival delay, (iii) mean access delay, (iv) RMS delay, and (v) maximum access delay. [5]

(b) Explain the difference between hard and soft decision channel decoding. [5]

(c) Explain the principles of turbo coding. [6]

3. (a) Differentiate between control and voice channels. [2]

(b) Describe the frequency reuse techniques in allocating channel groups of base stations within a system. [6]

(c) Compare FDMA, TDMA, and CDMA schemes, including in terms of their capacities. [8]

or

(a) What do you mean by common channel signalling? [2]

(b) Derive the equation for co-channel interference in cellular system. Find out the co-channel interference at the desired base station (cell site) if $N = 7$ cell pattern. [6]

(c) Describe in brief the random access schemes for packet radio systems. [8]

4. (a) Explain the frequency hopping spread spectrum system with the help of transmitter and receiver diagrams and derive the equation for finding out the processing gain of the system. [8]

(b) Explain the importance of adding cyclic prefix or guard interval in the OFDM symbols. [6]

(c) Why is windowing applied to OFDM transmission bandwidth? [2]

or

(a) Write a note on mobile satellite communication. [8]

(b) Explain the key features of a rake receiver. [8]

5. (a) Explain the mobile WiMAX (802.16e) architecture. [4]

(b) Explain the basic concept of UWB. [4]

(c) Discuss in detail the WAP. [8]

or

(a) Differentiate between the following by mentioning two important points of difference:

(i) Wi-Fi and WiMAX

(ii) UMTS and WiMAX

(iii) UMTS and GSM

(iv) Bluetooth and Wi-Fi [8]

(b) What are the features of a smart-antenna system? [4]

(c) What is the motivation behind the evolution of space-time processing? [4]

Model Paper 4

Time: 3 hrs *Marks: 100*

Answer all questions.

Part A (10 × 2 = 20 marks)

1. Define coherence bandwidth.
2. What do you mean by central limit theorem?
3. What is offset QPSK? Give its advantage over QPSK modulation.
4. What do you mean by wideband systems?
5. What is the purpose of VLR in a GSM system?
6. Explain the near–far problem in a CDMA system.
7. For $i = 2$ and $j = 1$, find out the group size N and draw the cellular structure depicting the co-channel cells.
8. What is the period of the maximum length sequence generated using a three-bit shift register?
9. What do you mean by micro-diversity and macro-diversity? Differentiate between them with the help of examples of systems.
10. Why is DAB called a single-frequency network?

Part B (5 × 16 = 80 marks)

1. (a) Explain the speech signal properties that can be exploited in the source coding. Do you find the same properties in images as well as videos? [6]

(b) Draw and explain the general block diagram of digital communication link? Will you have the same link for wireless communication? Where will the differences be observed? [6]

(c) What is the difference between the pulse shaping and the windowing applied to a signal? [4]

2. (a) A mobile phone keypad has the digits 0 to 9 and the * and # keys. Assume that the probability of sending * or # is 0.005 and that of sending 0 to 9 is 0.099 each. If the keys are pressed at a rate of 1.5 keys/sec, compute the data rate for this source. [6]

(b) Explain the use of an interleaver in the turbo codes. [4]

(c) What do you mean by low-density parity check (LDPC) codes? [6]

or

(a) What will be the advantages of code puncturing? [4]

(b) Explain any one differential modulation scheme in brief. [6]

(c) Explain different types of CELP coders. How do they differ? [6]

3. (a) Discuss the suitability of digital modulation scheme in frequency-selective mobile channels. [4]

(b) Compare in brief Rayleigh, Rician, and Nakagami channel fading models. [6]

(c) Explain the different pilot-based channel estimators for OFDM scheme. [6]

or

(a) Define the spectral efficiency of cellular system. [4]

(b) What are the constraints of designing CDMA multi-user system on cellular infrastructure? [6]

(c) What is the difference between the time hopping and the chirping in SSM? [6]

4. (a) How does IEEE 802.16e Mobile WiMAX protocol improve upon IEEE 802.16d fixed WiMAX? [6]

(b) Explain space diversity exploitation in MIMO. [4]

(c) What are the different types of system models in which the significant difference is observed during communication system simulation? [6]

or

(a) Explain the components of the GSM system along with their functions. [8]

(b) Explain major aspects of DVB-T and DVB-H and differentiate between the two standards. [8]

5. Write short notes on any four of the following systems:
(i) Wi-Fi (ii) DTH (iii) GPRS (iv) IS-95 (V) Bluetooth

Appendix F

Answers to Model Question Papers

Model Paper 1

Part A

1. The 2G and 3G are the generations by which the significant difference in communication systems can be identified. Partially, analog and digital are classified as 2G systems, where audio and images were communicated. The bit rate was very low around 10 to 50 kbps. Fully digital systems with audio, image, and video are classified as 3G with tremendous rise in the bit rate, of the order of 2 to 10 Mbps.
2. The band-pass sampling theorem states that f_s is the sampling frequency and the band-pass signal bandwidth is W_{signal}, then
$$f_s \geq 2W_{singal}$$
It means that the sampling frequency must be equal to or greater than twice the input analog band-pass signal bandwidth.
3. The compressors are used in PCM to compress the higher amplitude analog signal (crossing the threshold), which are amplified less than the lower amplitude signals before the transmission.
4. The TDM signal bit time $T_b = 0.5$ µs. We have,
$$W = 2B_o - B_o(1 - a)$$
Now,
$$B_o = \frac{1}{2T_b} = 1 \text{ MHz and } \alpha = 0.5$$
So, the bandwidth, $W = 1.5$ MHz
5. In Rician channel, the significant line-of-sight (LOS) component is there along with the multipath fading NLOS components, which is not there in Rayleigh channel.
6. The spread-spectrum modulation uses unique user code for the spreading and despreading purpose. If the unintended user tries to detect the signal without knowing the code, it will not be correlated and the despreading will not be possible. Thus, the security is provided by the spread-spectrum scheme.
7. The multicarrier transmission can make the parallel transmissions by splitting the incoming bit steam, because of which the higher data rates are possible. Due to splitting of bits, the effective duration between the consecutive bits increases, which in turn reduces the effect of ISI.
8. The $s_1(t)$ and $s_2(t)$ form an orthonormal set if $\int_0^T s_1^*(t)\, s_2(t) = 0$.

9. The equation for coherent BFSK is

$$P_e = \frac{1}{2}\, erfc\sqrt{\frac{E_b}{2N_0}} \quad \text{(in terms of standard notations)}$$

10. The core network of UMTS is same as that of GPRS. The air interface is totally different. The air interfaces in UMTS are listed below:

(i) Uu: UE to Node B (UTRA, the UMTS WCDMA air interface)

(ii) Iu: RNC to GSM phase 2+ CN interface (MSC/VLR or SGSN)

Iu-CS for circuit-switched data

Iu-PS for packet-switched data

(iii) Iub: RNC to Node B interface

(iv) Iur: RNC to RNC interface, not comparable to any interface in GSM

The Iu, Iub, and Iur interfaces are based on ATM transmission principles.

Part B

1. (a) The ideal Nyquist channel is a channel that exhibits ideal channel filtering (maybe with a little loss of harmonics) resulting in zero ISI while passing the pulses through it.

The channel uses an equivalent transfer function $H_e(f)$ such that the impulse response satisfies the condition

$$h_e(kT_s + \tau) = \begin{cases} C, k = 0 \\ 0, k \neq 0 \end{cases}$$

where k is an integer, T_s is the symbol (sample) clocking period, τ is the offset in the receiver sampling clock times compared with the clock times of the input symbols, and C is a non-zero constant. Now, if we choose a sin x/x function for $h_e(t)$, then the impulse response satisfies the Nyquist first criterion for zero ISI. Consequently, if the transmit and receive filters are designed so that the overall transfer function is

$$H_e(f) = \frac{1}{f_s}\Pi\left(\frac{f}{f_s}\right) \text{ where } f_s = 1/T_s$$

then there will be no ISI. Furthermore, the absolute bandwidth of this transfer function is $W = f_s/2$.

Merits and Limitations

- No ISI means higher data rate is supported.
- The overall amplitude and transfer characteristic $H_e(f)$ has to be flat over $-W < f < W$ and zero elsewhere. This is physically unrealizable, i.e. the impulse response would be non-causal and of infinite duration. $H_e(f)$ is difficult to approximate because of the steep skirts in the filter transfer function at $f = \pm W$.
- The synchronization of the clock in the decoding sampling circuit has to be almost perfect, since the sin x/x pulse decays only as $1/x$ and is zero in adjacent time slots only when t is exactly at correct sampling time. Thus, the inaccurate sync will cause ISI.
- Because of these difficulties, instead of sin x/x pulse, we are forced to consider other pulse shapes that have a slightly wider bandwidth. The idea is to find pulse shapes that go through zero at adjacent sampling points and yet have an envelope that decays much faster than $1/x$,

so that the clock jitter in the sampling times does not cause appreciable ISI. One solution for the equivalent transfer function, which has many desirable features, is the raise cosine roll-off Nyquist filter.

(b) The zero-forcing equalizer applies the inverse of the channel to the received signal to restore the signal before the channel. It is not useful for practical applications (there are some exceptions) and is just a textbook study case, which is useful to explain more realistic equalizer concepts. In a zero-forcing equalizer, the coefficients c_n are chosen to force the samples of the combined channel and equalizer impulse response to zero.

$$e(t) = \sum_{n=-N}^{N} C_n \delta(t - nT_s) \Leftrightarrow E(f) = \sum_{n=-N}^{N} C_n e^{-j2\pi f n T_s}$$

$$g(kT) = \sum_{n=-N}^{N} C_n x\,[(kT - nT_s)], \qquad k = 0, \pm 1, \pm 2, \ldots, \pm N$$

For no ISI, let $g(kT) = \begin{cases} 1 & k = 0 \\ 0 & k = \pm 1, \pm 2, \cdots, \pm N \end{cases}$

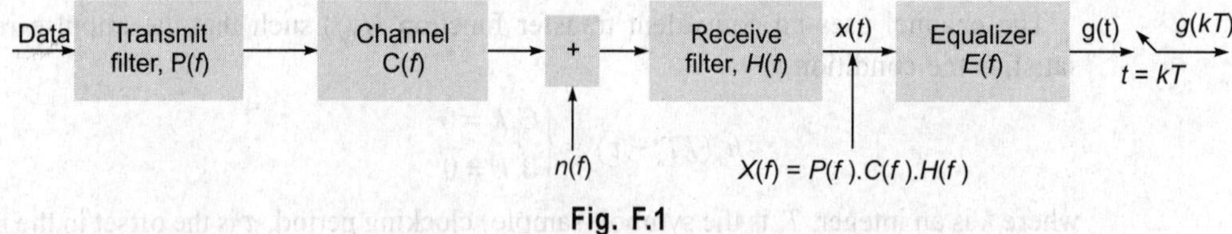

Fig. F.1

Comparison with Other Equalizers

- Zero-forcing equalizers ignore the additive noise and may significantly amplify noise for channels with spectral nulls.
- Minimum mean square error (MMSE) equalizers minimize the mean-square error between the output of the equalizer and the transmitted symbol. They require knowledge of some autocorrelation and cross correlation functions, which in practice can be estimated by transmitting a known signal over the channel.
- Adaptive equalizers are needed for the channels that are time varying.
- Blind equalizers are needed when no preamble is allowed.
- Decision-feedback equalizers (DFEs) use tentative symbol decisions to eliminate ISI.
- Ultimately, the optimum equalizer is a maximum-likelihood sequence estimator.

or

(a) In statistics, the likelihood function is a function of the parameters of a statistical model that plays a key role in statistical inference. If probability allows us to predict unknown outcomes based on the known parameters, the likelihood function allows us to estimate unknown parameters based on known outcomes. In a sense, the likelihood works backwards from probability. For a given parameter Y, we use the conditional probability $P(X/Y)$ to reason about outcome X, and for given outcome X, we use likelihood function $L(Y/X)$ to reason about parameter Y. In statistics, a likelihood function is a conditional probability function.

(b) The concept of moment in mathematics (probability theory) is evolved from the concept of moment in physics. The nth moment of a real-valued function $f(x)$ of a real variable about a value c is $\mu_n = \int_{-\infty}^{\infty} (x-c)^n f(x)\, dx$.

It is possible to define moment for random variables in more general form rather than real values. The moments are defined as ensemble averages of some specific functions used for $f(x)$. The nth moment of the random variable x taken about point $x = x_0$ is given by

$$\overline{(x-x_0)^n} = \int_{-\infty}^{\infty} (x-x_0)^n f(x)\, dx$$

The mean μ is the first moment taken about the origin, i.e. $x_0 = 0$. So,

$$\mu \cong \overline{x} = \int_{-\infty}^{\infty} xf(x) dx$$

The variance σ^2 is the second moment taken about the mean. Thus,

$$\sigma^2 = \overline{(x-\overline{x})^2} = \int_{-\infty}^{\infty} (x-\overline{x})^2 f(x)\, dx$$

The standard deviation is the square root of the variance.

(c) Refer to Section 3.3, Chapter 3.

2. (a) The diagram of a 1/2 rate convolutional encoder is shown in Fig. F.2.

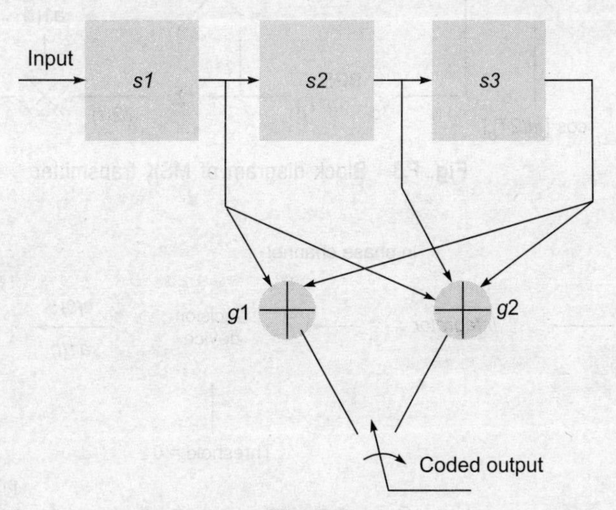

Fig. F.2

Input 1010,

Generator polynomials:

$$g^{(1)}(D) = 1 + D^2 \qquad g^{(2)}(D) = 1 + D + D^2$$

g \ u	1	0	1	0		
1	1	0	1	0		
0		0	0	0	0	
1			1	0	1	0
	1	0	0	0	1	0

g \ u	1	0	1	0		
1	1	0	1	0		
1		1	0	1	0	
1			1	0	1	0
	1	1	0	1	1	0

1 0 0 0 1 0

1 1 0 1 1 0

$v =$ 11 01 00 01 11 00

(b) The block diagrams of MSK transmitter and receiver are shown in Figures F.3 and F.4.

At the transmitter end, multiplying a carrier signal with cos $(\pi t/2T_b)$ produces two phase coherent signals at $f_c + 1/4T_b$ and $f_c - 1/4T_b$. These two FSK signals are separated using two narrow band-pass filters and appropriately combined to form the in-phase and quadrature carrier components $\phi 1(t)$ and $\phi 2(t)$, respectively. These carriers are multiplied with the odd and even bit streams $a1(t)$ and $a2(t)$, and summed up as shown above to produce the MSK signal.

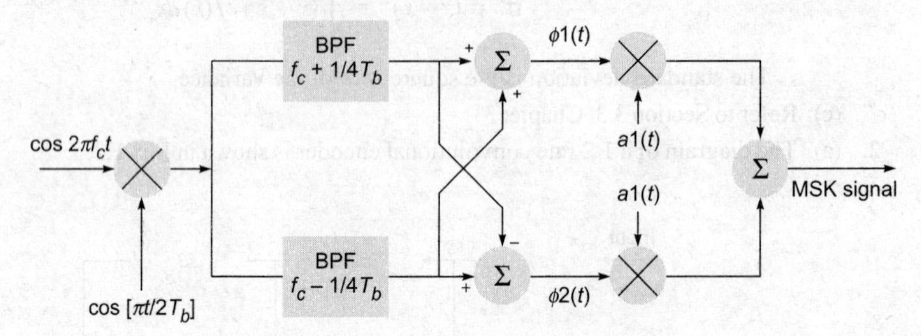

Fig. F.3 Block diagram of MSK transmitter

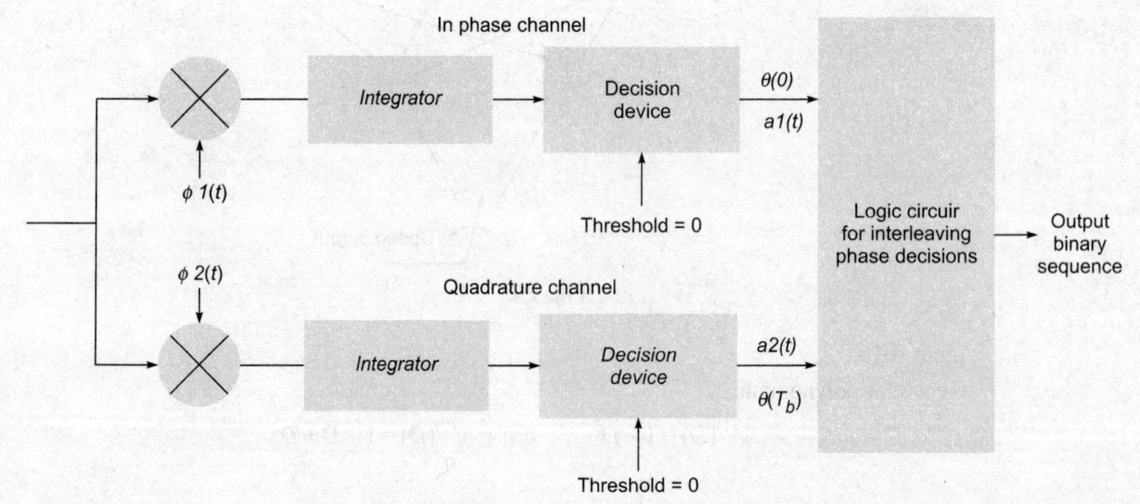

Fig. F.4 Block diagram of MSK receiver

The received signal (in the absence of noise and interference) is multiplied by the respective in phase and quadrature carriers $\phi 1(t)$ and $\phi 2(t)$. The output of the multipliers is integrated over two bit periods and given to a decision circuit at the end of each two-bit period. Based on the level of the signal at the output of the integrator, the threshold detector decides whether the signal is 0 or 1. The output data streams corresponds to $a1(t)$ and $a2(t)$, which are offset combined to obtain the demodulated signal.

3. (a) Refer to Section 8.4.1, Chapter 8.

(b) Refer to Sections 8.5 and 8.10.2, Chapter 8.

or

(a) Refer to Section 11.5.5, Chapter 11.

(b) Cell splitting, frequency reuse, and cell sectorization improve the coverage and cell capacity. With three-directional antennas based sectorization, the capacity triples and so on. Refer to Sections 11.2 and 11.9, Chapter 11.

4. (a) Refer to Section 9.3, Chapter 9.

(b) Refer to Section 9.1.1, especially Figures 9.16, 9.23, and 9.25 and the associated discussion.

or

(a) The OFDM is a spectrally efficient technique compared to FDM. (Refer to Section 9.1.2, Chapter 9.)

(b) For the general architecture of software-defined radio transmitter and receiver, refer to Section 16.11, Chapter 16.

(c) The MIMO is suitable with OFDM due to the following reasons.

The multiple input-multiple output orthogonal frequency division multiplexing (MIMO-OFDM) technology uses multiple antennas to transmit and receive radio signals. The MIMO-OFDM will allow service providers to deploy a broadband wireless access (BWA) system that has non-line-of-sight (NLOS) functionality. Specifically, MIMO-OFDM takes advantage of the multipath properties of environments using base station antennas that do not have LOS.

In NLOS environment, radio signals bounce off buildings, trees, and other objects as they travel between the two antennas. This bouncing effect produces multiple 'echoes' or 'images' of the signal. As a result, the original signal and the individual echoes each arrive at the receiver antenna at slightly different times causing the echoes to interfere with one another, thus degrading the signal quality.

The MIMO system uses multiple antennas to simultaneously transmit data, in small pieces to the receiver, which can process the data flows and put them back together. This process, called spatial multiplexing, proportionally boosts the data-transmission speed by a factor equal to the number of transmitting antennas. In addition, since all data is transmitted both in the same frequency band and with separate spatial signatures, this technique utilizes spectrum very efficiently. Multiple OFDM frames can be transmitted in parallel as shown in Section 15.15 in Chapter 15. The OFDM itself is a method combating the channel and along with MIMO, the spectral efficiency as well as capacity will increase manifold.

5. (a) Refer to Section 13.2.5, Chapter 13.

(b) The DECT system protocol architecture is shown in Fig. F.5.

The DECT system protocol architecture mapped in terms of OSI management plane is there over all the layers.

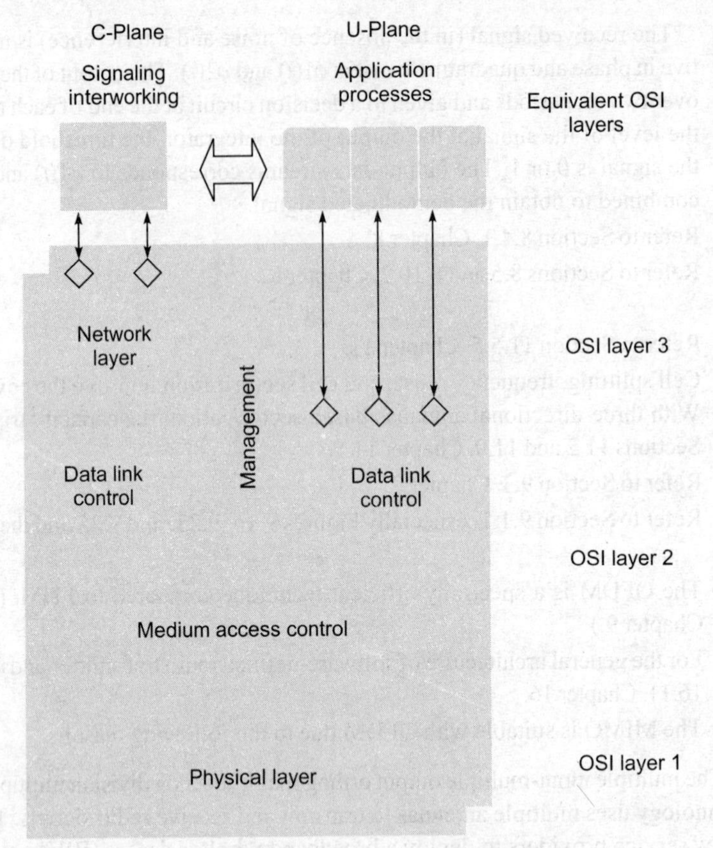

Fig. F.5 DECT system protocol architecture

There are several services in C (Control) and U (User) planes DECT layers.

Physical Layer Functions

- Modulation/demodulation
- Controlling of radio transmission
- Generation of the physical channel structure with a guaranteed throughput
- Channel assignment on request of the MAC layer
- Detection of incoming signals
- Sender/receiver synchronization
- Collecting status information for the management plane

MAC Layer Functions

- Maintaining basic services and activating/deactivating physical channels
- Multiplexing of logical channels, e.g., C: signalling, I: user data, P: paging, and Q: broadcast
- Segmentation/reassembly
- Error control/error correction

Data Link Control Layer Functions

- Creating and keeping up reliable connections between the mobile terminal and base station

There are two DLC protocols for the control plane (C-plane):

1. Connectionless broadcast service: paging functionality
2. Point-to-point protocol

There are several services specified for the user plane (U-plane):

1. Null-service: offers unmodified MAC services
2. Frame relay: simple packet transmission
3. Frame switching: time-bound packet transmission
4. Error correcting transmission: uses FEC for delay-critical and time-bound services
5. Bandwidth adaptive transmission
6. Services for future enhancements of the standard

Network Layer

- Similar to ISDN (Q.931) and GSM (04.08)
- Offers services to request, check, reserve, control, and release resources at the base station and mobile terminal
- Resources
  (i) Necessary for a wireless connection
  (ii) Necessary for the connection of the DECT system to the fixed network
- Main tasks
  (i) Call control: setup, release, negotiation, and control
  (ii) Call independent services: call forwarding, accounting, and call redirecting
  (iii) Mobility management: identity management, authentication, and management of the location register

(c) GPRS architecture is based on the GSM architecture. (Refer to Section 13.3 and Fig. 13.7, Chapter 13.)

or

(a) For LMDS, refer to Section 13.6, Chapter 13.
For CorDECT WLL architecture, refer to Section 13.6.2, Chapter 13.

(b) Refer to Section 14.3.5, Chapter 14.

(c) Refer to Section 16.10, Chapter 16.

Model Paper 2

Part A

1. The two limitations of DM scheme are slope overload noise and granular noise. The slope overload noise occurs when the step size is too small for the accumulator output to follow quick changes in the input waveform. The granular noise occurs for any step size but is smaller for small step size. So, we would like to have the step size as small as possible to minimize the granular noise. The granular noise is similar to the granular noise in PCM.
2. Whenever an electronic device transmits digital (and sometimes analog) data to another electronic device, there must be a certain rhythm established between the two devices. For digital transmission, all information is converted into binary codes of 0 and 1 for transmission. It is necessary to notify the other end of the precise information, such as where the data starts and how long the interval of data is. Synchronous and asynchronous transmissions are two different methods of transmission synchronization. The synchronous transmissions are synchronized by an external clock, while the asynchronous transmissions are synchronized by special signals along the trans-

mission medium. In synchronous signals, the signals that play the role of signs are added to the top of the data to be transmitted (preamble). In the asynchronous transmission, the first position of the character is identified. Two bits are added to each character, a start bit for the beginning and an end bit for the end, making the transmission speed slower.

3. The given transmitter pulse shape $g(t)$ of duration T, matched filter is given by

$$h_m(t) = kg^*(T-t) \text{ for all } k$$

where * represents the conjugate

The duration and shape of the impulse response of the optimal filter is determined by pulse shape $g(t)$. Here $h_m(t)$ is the scaled, time reversed, and shifted version of $g(t)$. If we consider their transforms, the spectrum relations can be established.

4. The processing gain for DS spread spectrum system is defined as

$$PG = B_s/B_m$$

where B_s = Bandwidth of the SSM or PN signal
B_m = Bandwidth of the message signal

It gives the amount of spreading of the signal.

5. The MIMO is an efficient spatial diversity technique with extra capabilities. The MIMO system has two or more transmitting antennas and two or more receiving antennas and MIMO transmits and receives two or more radio signals in a single channel in parallel, where each signal carries unique information and transmits diversity as well as receive diversity. In the latter case, a diversity combining technique is applied before further signal processing takes place. The diversity reception reduces the probability of occurrence of communication failures (outages) caused by fades by combining several copies of the same message received over different channels.

6. By IFFT stage in the transmitter, the frequency domain setting of subcarriers along with the assigned data (which are treated as discrete samples in frequency domain) is converted into time domain. The IFFT stage acts as an adder. Different subcarriers act as different harmonics of OFDM signals, which are orthogonal. Such harmonics along with the assigned data is combined by IFFT stage. The FFT stage does the reverse function at the receiver end.

7. The sampling frequency less than twice the bandwidth of the input analog signal creates an overlap between the frequency spectrum of the samples and the input analog signal. This is nothing but the effect of undersampling. The low-pass output filter used to reconstruct the original input signal is not smart enough to detect this overlap. So, it creates a new signal that did not originate from the source. This creation of a false signal while sampling is called aliasing. In other words, the phenomenon of a high frequency in the spectrum of the original signal seemingly taking on the identity of a lower frequency in the spectrum of the sampled signal is called aliasing. The pass band for anti-aliasing and smoothing filter used with pulse modulation-demodulation systems is 2W.

8. Refer to Section 16.8, Chapter 16.
9. Refer to Section 13.2.1, Chapter 13.
10. Refer to Section 10.2.3, Chapter 10.

Part B

1. **(a) Time Response**

The raised cosine time response is given in terms of impulse response (Fig. F.6).

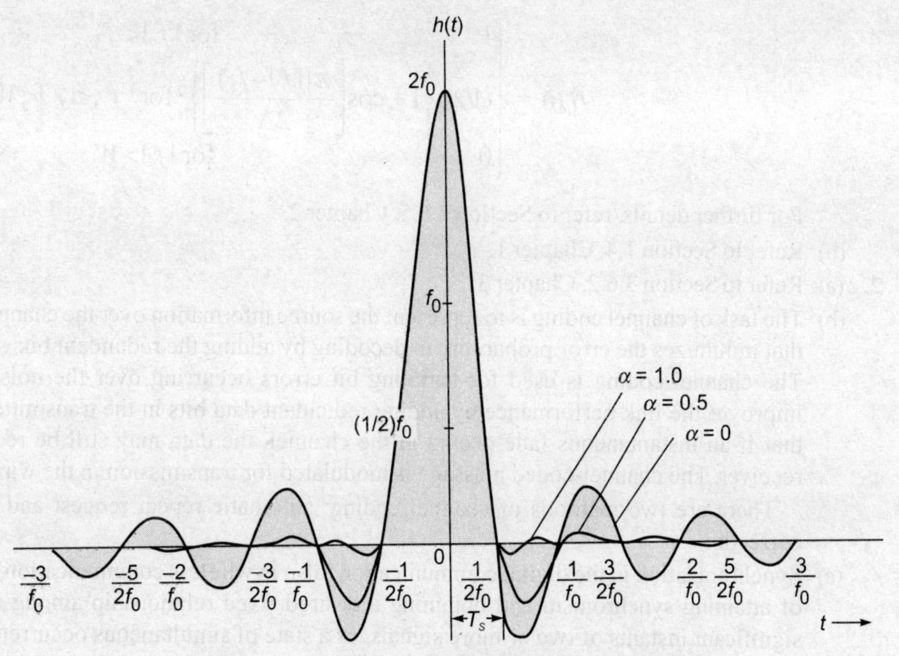

Fig. F.6 Time response for different roll-off factors

Let W is the absolute bandwidth and

$$\Delta = W - f_0 \text{ and } f_1 = f_0 - \Delta$$

Here, f_0 is the 6 dB bandwidth of the filter. The roll-off factor is defined to be

$$\alpha = \Delta / f_0$$

$$h(t) = F^{-1}[H(f)] = 2f_0\left(\frac{\sin 2\pi f_0 t}{2\pi f_0 t}\right)\left[\frac{\cos 2\pi \Delta t}{1-(4\Delta t)^2}\right]$$

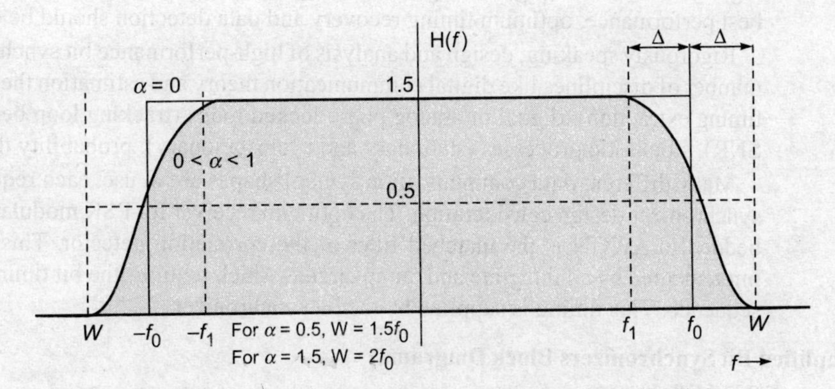

Fig. F.7 Raised cosine roll-off Nyquist filter characteristics

Frequency Response

The transfer function of raised cosine roll-off Nyquist filter (Fig. F.7) is

$$H_e(f) = \begin{cases} 1 & \text{for } |f| < f_1 \\ (1/2)\left\{1+\cos\left[\dfrac{\pi(|f|-f_1)}{2\Delta}\right]\right\} & \text{for } f_1 < |f| < W \\ 0 & \text{for } |f| > W \end{cases}$$

For further details, refer to Section 2.6.5, Chapter 2.

(b) Refer to Section 1.4, Chapter 1.

2. (a) Refer to Section 3.6.2, Chapter 3.

(b) The task of channel coding is to represent the source information over the channel in a manner that minimizes the error probability in decoding by adding the redundant bits systematically. The channel coding is used for handling bit errors occurring over the noisy channels. It improves the link performance by adding redundant data bits in the transmitted message so that if an instantaneous fade occurs in the channel, the data may still be recovered at the receiver. The channel-coded message is modulated for transmission in the wireless channel.

There are two methods of channel coding: automatic repeat request and forward error correction.

(c) Synchronization in the digital communication (also in wireless communication) is the process of attaining synchronism and obtaining a desired fixed relationship among corresponding significant instants of two or more signals, or a state of simultaneous occurrences of significant instants among two or more signals.

Bit synchronization is the synchronization in which the decision instant is brought into alignment with the received bit, i.e. basic signalling element. Decision instant in the reception of a digital signal is the instant at which a decision is made by a receiving device as to the probable value of a signal condition. Many times, the element of a message header is used to synchronize all the bits and characters that follow.

In digital communication systems, optimum estimation and detection algorithms require precise knowledge of the bit transition time be known to the receiver before bit-by-bit detection takes place. Bit synchronizer is a device/circuit that recovers or reconstructs the underlying digital signals timing, thereby allowing high-quality data bit detections to be made. For the best performance, optimum timing recovery and data detection should be done.

Rigorously speaking, design and analysis of high-performance bit synchronizers involve a number of disciplines like digital communication theory and estimation theory (detection and timing extraction), digital or analog phase locked loops (tracking loop behaviour with poor SNR), stochastic processes (stationary and cyclostationary), probability theory, etc.

Many different data communication symbol shapes are in use, each requiring separate bit synchronizer design consideration. The optimum receiver for PSK modulated waveform embedded in AWGN is the matched filter or the correlation detector. This device is usually implemented by an integrate and dump circuit, which requires the bit timing of the incoming sequence. This timing is supplied by the bit synchronizer.

Simplified Bit Synchronizers Block Diagrams

(Figures F.8 and F.9)

1. **Indirect method**

This employs non-linear processing of baseband signal to extract data rate. Adjustment must be made in the delay line to properly centre the sampling operation in the data eye. Sampling point is, in general, not self-correcting. Improper selection of the two band-pass filters and non-linearity can result in poor performance due to high self-noise. The results from noise processes are being cyclostationary rather than stationary.

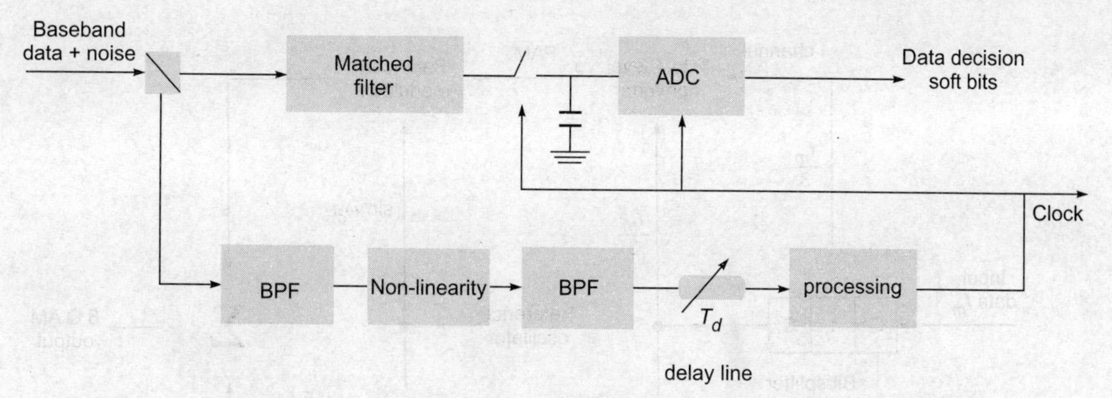

Fig. F.8

2. **Direct method**

This is a self-synchronizing system that acquires bit synchronization timing directly from the incoming modulated data sequences. The optimal sampling point in the data eye pattern is tracked directly. A control loop provides precise selection of the VCO phase for proper sampling. No bandpass filter is required. Considerably, less hardware than indirect approach and self-correcting circuit are the key features.

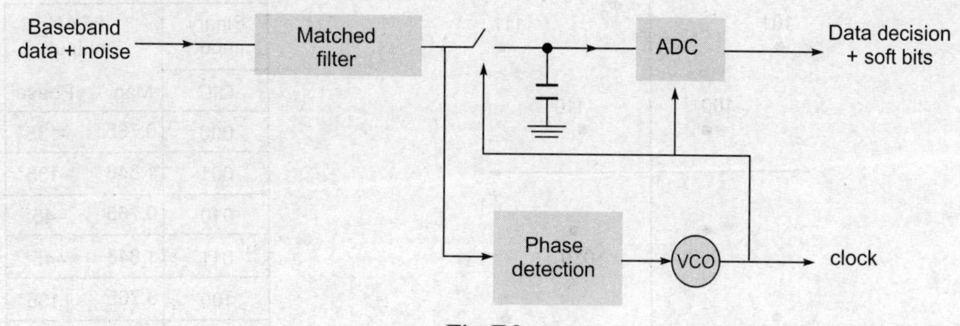

Fig. F.9

One more configuration is to use square law device → narrowband filter → comparator with a threshold. This synchronizer is used for filtered polar NRZ waveform.

or

(a) Refer to Section 3.10.1, Chapter 3.

(b)

8QAM Transmitter

The block diagram of 8QAM transmitter is shown in Fig. F.10.

The output of the product/balanced modulators is

$$I = 0.541 \sin(\omega_c t), \text{ and } Q = 0.541 \cos(\omega_c t)$$

The output of the linear summer is

$$0.765 \sin(\omega_c t + \pi/4)$$

The split-bit sequences are further decomposed into four PAM levels as shown in Fig. F.10. The levels will be then modulated by two orthogonal sinusoidal signals of a carrier frequency and then linearly summed up to form the 8QAM signal.

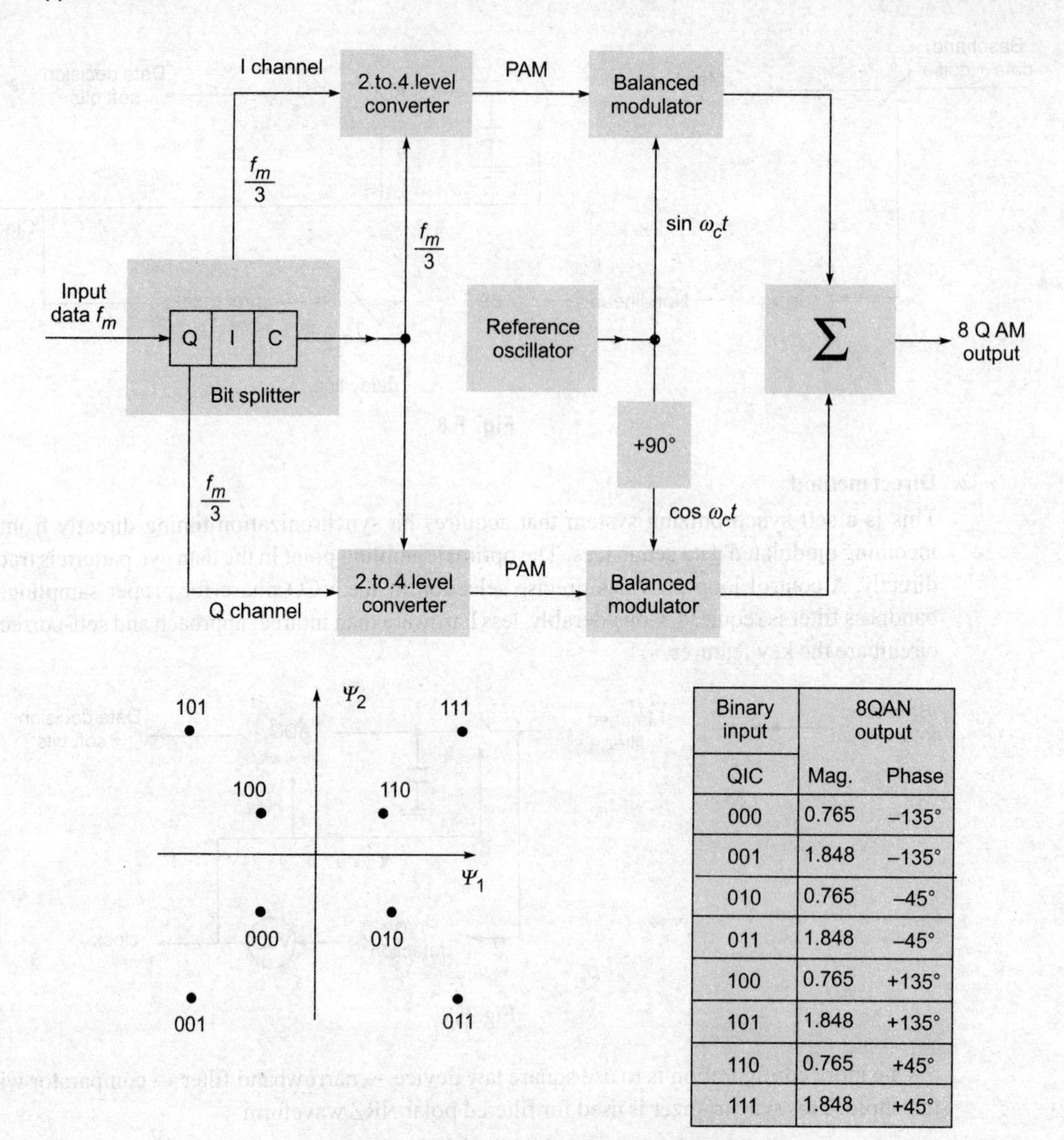

Binary input	8QAN output	
QIC	Mag.	Phase
000	0.765	–135°
001	1.848	–135°
010	0.765	–45°
011	1.848	–45°
100	0.765	+135°
101	1.848	+135°
110	0.765	+45°
111	1.848	+45°

Fig. F.10 Block diagram of 8QAM transmitter

I/Q	C	Output
0	0	–0.541
0	1	–1.307
1	0	+0.541
1	1	+1.307

Fig. F.11 Truth table for the *I* and *Q* channels 2 to 4 level converter

8QAM Receiver

The block diagram of 8QAM receiver is shown in Fig. F.12.

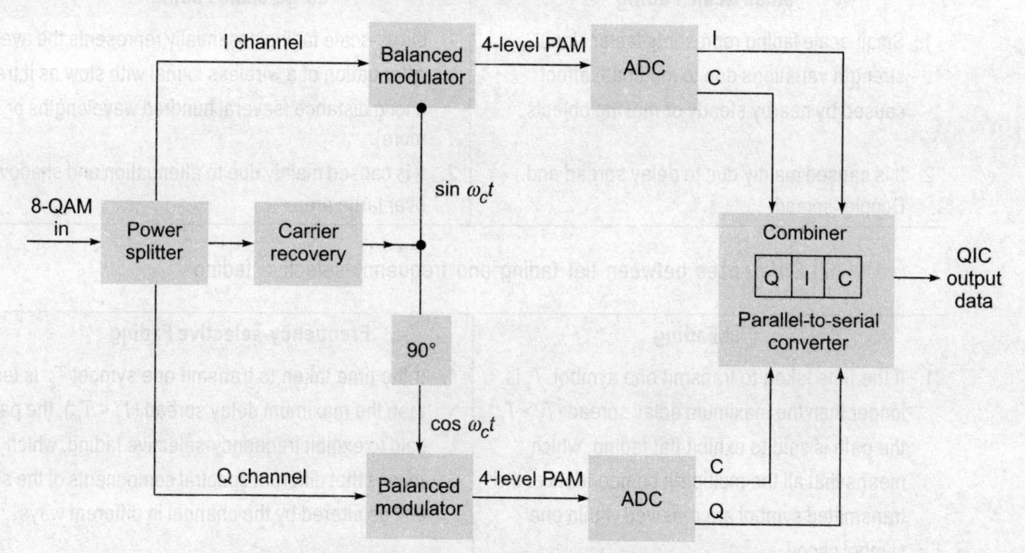

Fig. F.12 Block diagram of 8QAM transmitter

The 8-QAM receiver is almost the same as the 8-PSK. The differences are the PAM levels at the output of the product detectors (balanced modulators) and the binary signals at the output of the analog-to-digital converters (ADCs). In the 8QAM, we have two transmit amplitudes and four phases. The four demodulated PAM levels in 8-QAM are different from those of the 8-PSK. With 8-QAM, the binary output signals from I-channel are I and C bits, while the binary output signals from Q-channel are Q and C.

(c) Bit rate = 200 kbps => $T_b = 1/R_b = 1/(200 \times 10^3) = 5\ \mu s$

Now, $BT = 0.25 \Rightarrow B = 0.25/T_b = 0.25/(5 \times 10^{-6}) = 50$ kHz

Thus, the 3 dB bandwidth is 50 kHz. To determine 90% power bandwidth, we can take $0.57R_b$ as the desired value. The occupied RF spectrum for 90% power bandwidth is given by

$$\text{RF BW} = 0.57R_b = 0.57 \times 200 \times 10^3 = 114 \text{ kHz}$$

3. (a)

(i) Differences between slow fading and fast fading

Slow Fading	Fast Fading
1. There are slow changes in signal strength. The user bandwidth is greater than the Doppler spread.	1. The signal strength variation is fast. The user bandwidth is smaller than the Doppler spread.
2. If $T_s < T_0$ (or $W > f_d$), the channel exhibits slow fading, meaning that the channel conditions are stable and predictable during the time that the symbol is transmitted.	2. If the symbol period is longer than the coherence time ($T_s > T_0$), the channel exhibits fast fading because the fading conditions are coming and going faster than the symbols are being transmitted.

(ii) Differences between small-scale fading and large-scale fading

Small-scale Fading	Large-scale Fading
1. Small-scale fading represents faster signal strength variations due to multipath effect caused by nearby steady or moving objects.	1. Large-scale fading essentially represents the average attenuation of a wireless signal with slow as it travels a long distance (several hundred wavelengths or more).
2. It is caused mainly due to delay spread and Doppler spread.	2. It is caused mainly due to attenuation and shadowing over large area.

(iii) Differences between flat fading and frequency-selective fading

Flat Fading	Frequency-selective Fading
1. If the time taken to transmit one symbol T_s is longer than the maximum delay spread ($T_s > T_m$), the path is said to exhibit flat fading, which means that all the multipath components of the transmitted symbol are received within one symbol period.	1. If the time taken to transmit one symbol T_s is less than the maximum delay spread ($T_s < T_m$), the path is said to exhibit frequency-selective fading, which means that different spectral components of the signal will be altered by the channel in different ways.
2. $W < B_c$, i.e. the user bandwidth is less than the coherence bandwidth.	2. $W > B_c$, i.e. the user bandwidth is grater than the coherence bandwidth..

Note Reader may refer to Sections 5.9 and 5.10, Chapter 5, for more points of difference.

(b) Refer to Section 5.10, Chapter 5, and Section 6.5, Chapter 6.

(c) Refer to Section 6.2, Chapter 6.

or

(a) Refer to Sections 6.12.2 and 6.12.3, Chapter 6.

(b) In a frequency hopping spread spectrum system, the performance is improved in presence of multipath. The effect of multipath is diminished, provided that the carrier frequency of the transmitted signal hops fast enough relative to the differential time delay between the desired signal from the direct path and the undesired signals from the indirect paths. With this condition, all of the multipath energy will fall in frequency slots that are orthogonal to the currently occupied slot by the desired signal. This way, the degradation due to multipath is reduced.

(c) Refer to Section 6.10, Chapter 6.

4. (a) Refer to Section 11.5.5, Chapter 11.

(b) **Spectrum Efficiency**

The spectral efficiency of a digital signal is given by the number of bits per second of the data that can be supported by each hertz of bandwidth. Thus,

$$\text{Spectral efficiency, } \eta = R_b/W \text{ (bits/second)/Hz}$$

where R_b is the data rate and W is the bandwidth.

The maximum possible spectral efficiency is limited by the channel noise if the error is to be small. As per Shannon's formula, the maximum spectrum efficiency is given by

$$\eta_{\max} = C/W = \log_2(1+\text{SNR})$$

Shanon's theory does not state how to achieve a system with maximum spectrum efficiency. However, practically, this spectral efficiency can be approached using the channel coding and multilevel signalling methods.

Power Efficiency

Power efficiency describes the ability of a modulation technique to preserve the fidelity of the digital information at low power levels. In order to increase noise immunity over the channel, it is necessary to increase the signal power. However, the amount of power for certain level of fidelity (acceptable BER performance, say, upto 10^{-5}) depends on the particular modulation scheme.

Power efficiency, P = Signal energy per bit/Noise power spectral density = E_b/N_o

(c) Refer to Section 12.2, Chapter 12.

or

(a) The US digital cellular system based on direct sequence spread spectrum CDMA was standardized as an Interim Standard 95 (IS-95) by the US Telecommunications Industry Association (TIA). The IS-95 system has increased the capacity and offers many advantages over FDMA and TDMA. It was designed to be compatible with the US analog cellular system (AMPS) frequency band. Mobiles and base stations can be operated on dual mode for both applications.

Further refer to Sections 13.5 and 13.5.1, Chapter 13.

(b) The pilot carriers are used for the offset control purpose. They are usually transmitted with more power so that they can be identified with surety. Also the pilots are used to send the training sequences, which are extracted and used at the receiving end for the channel-estimation purpose. The channel estimation is very important to eliminate phase errors due to the channel.

In OFDM system, the methods of sending pilot are block type, comb type, scattered pattern, etc. (*Refer to Fig. 9.31 and Section 9.9, Chapter 9.*) Superimposed pilots type is one latest method.

Drawbacks of Sending Pilots

The information transmission efficiency reduces when the training sequences are embedded to data symbols in time domain, which are called pilot symbols, and the spectral efficiency reduces when additional pilot subcarriers are used only for sending the training sequences.

5. (a) Refer to Section 14.2.2, Chapter 14.

(b) Refer to Section 13.6, Chapter 13.

or

(a) Bit rate = 10 Mbps = 10^7 bps

So, Bit duration, T_b = 0.1 μsec

In 16PSK scheme, 4 bits per symbol are read and assigned to the carrier. So, the effective pulse duration on each subcarrier is 0.4 μsec or the symbol rate is 2.5 MHz if one symbol is assigned to one carrier In this case, the OFDM frame consist of 256 bits and so the frame duration T_s is 25.6 μsec.

Then, the total bandwidth will be

$$(N_c+1)/T_s = 65/25.6 = 2.53 \text{ MHz}$$

(If more than one symbol is assigned to one carrier, then this calculation will differ.)

Alternative approach:

For $N_c > 10$, the reasonable approximation for bandwidth is

$$W = (N_c+1)/T_s = R_s = 2.5 \text{ MHz}$$

Here OFDM frame duration is

$$(N_c+1)/D_s = 65/2.5 = 26 \ \mu\text{sec}$$

which matches the previous calculations.

PSD calculation: $$f_n = \frac{1}{T}\left(n - \frac{N_c - 1}{2}\right) = \frac{1}{25.6}\left(n - \frac{64-1}{2}\right) = 0.039n - 1.23$$

$$\text{PSD} = A_c^2 \overline{|w_n|^2} T_s \sum_{n=0}^{N_c-1} \left| \frac{\sin[\pi(f - f_n) T_s]}{[\pi(f - f_n) T_s]} \right|^2$$

where A_c is the carrier amplitude (can be assumed unity) and w_n are the elements applied to N_c carriers, with the mean value zero for binary signal.

Simplifying the above equation, we get

$$\text{PSD} = 25.6 \overline{|w_n|^2} \sum_{n=0}^{N_c-1} \left| \frac{\sin[80.424(f - (0.039n - 1.23))]}{[80.424(f - (0.039n - 1.23)]} \right|^2$$

(b) Refer to Section 16.4, Chapter 16.

(c) Refer to Section 15.8, Chapter 15.

Model Paper 3

Part A

1. It is a bipolar format in which the positive and negative pulses are used alternately for the transmission of 1's (with the alternation taking place every occurrence of 1) and no pulses for the transmission of 0's.
2. The eye pattern provides the following information:
   - The timing error allowed on the sampler at the receiver is given by the width inside the eye, called the eye opening. Of course, the preferred time for sampling is at the point when the vertical opening of the eye is largest.
   - The sensitivity to timing error is given by the slope of the open eye (evaluated at or near the zero crossing point).
   - The noise margin of the system is given by the height of the eye opening.
3. The coherent detection is performed by gross correlation of the received signal with each one of the replicas and then making division based on the comparisons with pre-selected thresholds. Here the phase of the carrier wave is very important.

   In non-coherent detection, knowledge of the carrier wave's phase is not required, The complexity of the receiver is thereby reduced, but at the cost of an inferior error performance in comparison with coherent schemes.
4. The trunking efficiency is a measure of the number of users. The grade of service (GOS) is a measure of the abilities of a system to provide a user the access to the trunked system at the busy hour. The GOS is also defined as the blocking probability for initiating the call in busy hour.

5. The small-scale propagation models are the models that characterize the rapid fluctuations in the signal strength over the short travel distance. It is due to the multipath environment. Small scale fading mainly resulted due to the combined effect of the delay spread and the Doppler spread.

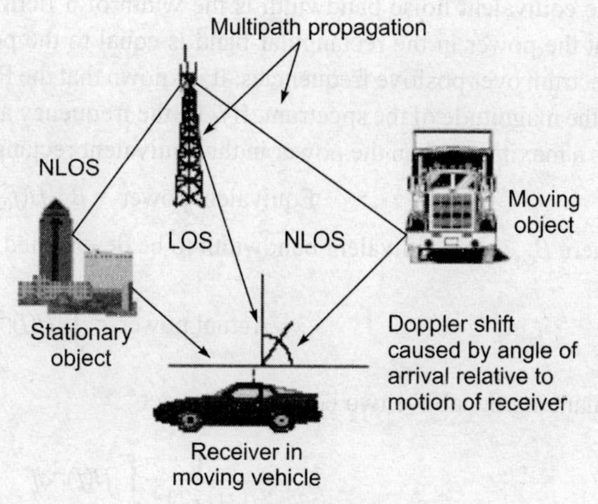

Fig. F.13

6. Given, transmitter power = 50 W

$$P_t(\text{dBm}) = 10\log(P_t\,\text{mW}/1\,\text{mW})$$

$$= 10\log\left(\frac{50\times 10^3}{1\times 1}\right) = 47.0\ \text{dBm}$$

$$P_t(\text{dBW}) = 10\log[P_t\,\text{W}/1\,\text{W}] = 10\log 50 = 17.0\ \text{dBW}$$

7. Due to the multicarrier transmission in parallel, after symbol mapping, each parallel line will have the effective pulse duration larger than the actual input bit duration, i.e. sufficient gap is provided between two consecutive pulses on each line. This minimizes the merging of the consecutive pulses eliminating ISI. For ICI removal, multiple subcarriers are set in the spectrum such that they are orthogonal to each other and at the receiver end, the control is provided to remove offset of the carrier to maintain the same orthogonality.

8. $D1 = -fv/c$ and $D2 = +fv/c$

So, the total Doppler spread = $D2 - D1 = 100$ Hz

9. Different modes in Bluetooth are active mode, sniff mode, hold mode, and parked mode. For details, refer to Section 14.2.1, Chapter 14.

10. Given, bandwidth = 25 MHz

So, the number of channels possible is 125. If the cluster size $N = 7$, then the number of channels available to each cell = $125/7 \approx 17$. Now, if the cluster size $N = 4$, then the available channels will be ≈ 31.

Part B

1. (a) There are six engineering definitions and one legal definition of the bandwidth. The various definitions are (if f_2 and f_1 are two extreme frequencies of a band-pass signal and for baseband signal, f_1 is zero/DC) given below:

- The absolute bandwidth is $f_2 - f_1$, where the spectrum is zero outside the interval $f_1 < f < f_2$ along the positive frequency axis.

- The 3-dB bandwidth or half-power bandwidth is $f_2 - f_1$, where the frequencies are inside the bands $f_1 < f < f_2$, the magnitude spectra, say, $|H(f)|$, fall no lower than 0.707 times the maximum value of $|H(f)|$, and the maximum value occurs at a frequency inside the band.
- The equivalent noise bandwidth is the width of a fictitious rectangular spectrum such that the power in the rectangular band is equal to the power associated with the actual spectrum over positive frequencies. It is known that the PSD is proportional to the square of the magnitude of the spectrum. If f_0 be the frequency at which the magnitude spectrum has a maximum, then the power in the equivalent rectangular band is proportional to

$$\text{Equivalent power} = B_{\text{eq}}|H(f_0)|^2$$

where B_{eq} is the equivalent bandwidth to be determined.

$$\text{Actual power} = \int_0^{\infty} |H(f)|^2 df$$

Balancing the above two equations, we get

$$B_{\text{eq}} = \frac{1}{|H(f_0)|^2} \int_0^{\infty} |H(f)|^2 df$$

- The null-to-null bandwidth or zero crossing bandwidth is $f_2 - f_1$, where f_2 is the first null in the envelope of the magnitude spectrum above f_0 and, for band-pass systems, f_1 is the first null in the envelope below f_0. For baseband systems, f_1 is usually zero.
- The bounded spectrum bandwidth is $f_2 - f_1$ such that outside the band $f_1 < f < f_2$, the PSD (proportional to $|H(f)|^2$) must be down by at least a certain amount, say, 50 dB or so, below the maximum value of the PSD.
- The power bandwidth is $f_2 - f_1$, where $f_1 < f < f_2$ defines the frequency band in which 99% of the total power resides. This is similar to FCC (Federal Communications Commission) definition of the oacupied bandwidth, which states that the power above f_2 is ½ % and below the lower edge is ½ %, leavine 99% of the total power within the occupied band (FCC Rules and Regulations, Sec. 2.202).
- *Legal definition* The FCC bandwidth is an authorized bandwidth parameter assigned by the FCC. When the FCC bandwidth parameter is substituted into FCC formula, the minimum attenuation is given dor the power level allowed in a 4 kHz band at the edges of the band with respect to the total average signal power.

(b) The autocorrelation of the real physical waveform is defined as

$$R_a(\tau) = \langle (s(t)s(t+\tau) \rangle = \lim_{T\to\infty} \frac{1}{T} \int_{-T/2}^{T/2} (s(t)s(t+\tau)\, dt$$

Furthermore, it can be shown that the PSD and the autocorrelation function are Fourier transform pairs (Wiener–Khintchine theorem).

The cross correlation can be defined for two different real waveforms $s_1(t)$ and $s_2(t)$ as

$$R_c(\tau) = \langle (s_1(t)s_2(t+\tau) \rangle = \lim_{T\to\infty} \frac{1}{T} \int_{-T/2}^{T/2} (s_1(t)s_2(t+\tau)dt$$

For further details, refer to Section 2.5.6, Chapter 2.

(c) The autocorrelation function of a real process $p(t)$ considering two different times t_1 and t_2 is

$$R_p(t_1, t_2) = \int_{-\infty}^{\infty}\int_{-\infty}^{\infty} p(t_1)\,p(t_2)\,f_p\,[p(t_1), p(t_2)]\,dp(t_1)\,dp(t_2)]$$

If the process is stationary to the second order, the autocorrelation function is a function only of the time difference $t = t_2 - t_1$.

In this case,

$$R_p(\tau) = \overline{(s_1(t)s_2(t+\tau))} \quad \text{(ensemble average)}$$

if $p(t)$ is second-order stationary.

[The time average of the autocorrelation function is identical to the ensemble average autocorrelation function when the process is ergodic. In ergodic processes, the time average is equal to the ensemble average (expectations) for the first and second moments.]

Now, a random process is said to be wide sense stationary if

$$\overline{p(t)} \text{ constant} \quad \text{and} \quad R_p(t_1, t_2) = R_p(\tau), \text{ where } \tau = t_2 - t_1$$

A process stationary to second or greater order is wide sense stationary. However, converse is not necessarily true. An exception occurs for the Gaussian random process, in which wide sense stationary does not imply because the process is strict sense stationary, since the $N \to \infty$-dimensional Gaussian PDF is completely specified by the mean, variance, and covariance of $p(t_1), p(t_2), \ldots, p(t_N)$.

2. (a) The distance from the TV transmission tower to the radio horizon is (refer to **Fig. 5.3**)

$$d_1 = \sqrt{2h_T}$$

We have $d^2 + r^2 = (r + h_T)^2$, where r is the radius of the earth equal to 3960 statute miles. However, at LOS radio frequencies, the effective radius of the earth is (4/3) 3960 miles. So,

$$d_1 = \sqrt{2 \times 1600} = 56.57 \text{ miles}$$

The distance from the receiving antenna to the radio horizon is

$$d_2 = \sqrt{2 \times 25} = 7.07 \text{ miles}$$

Then the total radius for the LOS coverage contour, which is a circle around the transmission tower, is

56.57 miles + 7.07 miles = 63.64 miles

(b) Refer to Section 3.9.1, Chapter 3.

Low bit rate coders are required in the wireless telecommunication systems especially dealing with voice communication, such as GSM and UMTS.

(c) For the definition of shadowing, refer to Section 5.11, Chapter 5.

Signal outages occur due to channel fades. For the definition, refer to Section 5.12, Chapter 5.

or

(a)

(i) *Power delay profile* The power delay profile provides an indication of the dispersion or distribution of transmitted power over various paths of the multipath structure. The mobile radio channel continuously exhibits multipath and so the power delay profile can be thought of a density function.

(ii) *First arrival delay* This is the delay of the first arriving path, which is measured at the receiver. This delay usually approximates the minimum possible path delay from the transmitter to the receiver. This is served as a reference and all delay measurements are taken with respect to it. Any delay measured after this delay is called excess delay.

(iii) *Mean excess delay* This is the average delay measured with respect to the first arrival delay.

(iv) *RMS delay* This is the term normally used to measure the delay spread. It is the standard deviation about the mean excess delay.

(v) *Maximum access delay* It is measured with respect to a certain power level. The maximum excess delay spread can be specified as the excess delay for which the power level reduces to -30 dB of its peak value.

(b) Refer to Section 4.1.6, Chapter 4.

(c) Refer to Section 4.7, Chapter 4.

3. (a) Control channel is a channel used for the transmission of digital control information from a land station to a mobile station or vice versa. It is of two types: 1) forward control channel and reverse control channel.

Voice channel is a channel used to transmit the voice to a mobile unit to a landline unit and vice versa. It is also classified as forward and reverse voice channels.

(b) Cellular radio system relies on an intelligent allocation and reuse of channels throughout the coverage region. The design process of selecting and allocating channel groups for all of the cellular base stations within a system is called frequency reuse or frequency planning.

For the detailed answer, refer to Section 11.2.2, Chapter 11.

(c) Refer to Section 10.2, Chapter 10.

or

(a) There are two ways of signaling, in-channel and common-channel signalling. In common channel signalling, there are separate channels for sending data as well as control information. The SS7 is the signalling scheme used in GSM system with separate common control channel

(b) The relation between D (the distance between the co-channel cell sites) and R (the cell radius or coverage radius of the base station transmitter of one cell) for hexagonal cells sharing N frequencies is given by

$$D = \text{sqrt}(3N)R$$

The CCI-caused C/I ratio or S/I ratio at the base station is

$$\frac{s}{I} = \frac{s}{\sum_{i=1}^{i_{\text{int}}} I_i} = \frac{(D/R)^n}{6} = \frac{\left(\sqrt{3N}\right)^n}{6}$$

For derivation, refer to Section 11.7, Chapter 11.

Substituting $N = 7$, S/I becomes 73.5. Converting into dB, it is 18.7 dB.

(c) Refer to Section 10.3, Chapter 10.

4. (a) Refer to Section 8.11, Chapter 8.

(b) Refer to Section 9.3.3, Chapter 9.

(c) Refer to Section 9.7, Chapter 9.

or

(a) Refer to Section 13.8, Chapter 13.

(b) In a multipath environment, the direct wave is not necessarily the best signal. So, by synthesizing the delayed signal, the receiver performance can be improved. In the rake receiver, the delay elements try to simulate the multipath effect, i.e. the time-delayed versions of the signal. This means that the cross correlator resolves the signals from the input code, equal to the number of delay elements.

Normally, the receivers add the received versions cumulatively and a receiver has no way to know which signal has taken direct path or which signal has reflected versions. The artificially created time delays in PN codes at the rake receiver for the correlation purpose will make the receiver intelligent enough to resolve the delayed components of the signal that will make the whole received signal.

For further details, refer to Section 8.8, Chapter 8.

5. (a) The IEEE 802.16e is a standard for Mobile WiMAX. It was released in 2005. This standard also supports NLOS communication with vehicular mobility. It supports around 15 Mbps data rate with various modulation schemes, such as scalable OFDMA (with 1.25 to 20 MHz bandwidth support), QPSK, 16QAM, 64QAM, and 256QAM (optional).

The WiMAX IEEE 802.16e supports diversity, spatial multiplexing, beam forming, virtual/collaborative/multi-user MIMO, precoding, and space-time coding (STC). Its deployment is mainly foreseen in licensed bands. The frequency-selective scheduling and subchannelization with multiple permutation options provides connection quality. The WiMAX, in general, can potentially be deployed in a variety of spectrum bands, such as 2.3 GHz, 2.5 GHz, 3.5 GHz, and 5.8 GHz. The IEEE 802.16e is just a modified version of IEEE 802.16d standard. It supports handoff and roaming.

(b) Refer to Section 14.7, Chapter 14.

(c) The wireless application protocol (WAP) is an open international standard developed to provide Internet contents on wireless devices, like mobile phones. It was preceded with mobile data service and web browsing capabilities in 1970's as a revolution in cellular telephony in Japan. The WAP Forum was established in 1977 by Motorola, Ericson, Nokia, and Unwired Planet, where the main aim was to improve Internet-based wireless communication. The WML is the language used to create pages to be displayed in a WAP browser.

The Forum was supported to have solutions that must be efficient and secure with scalability and reliability features and integrity of the user data. The architecture of WAP (protocol suite) includes the application environment (WAE), session protocol (WSP), transaction protocol (WTP), transport layer security (WTLS), and datagram protocol (WDP). The WAP protocol is the standard based on the Internet standards HTML, XML, and TCP/IP. The protocol suite allows the interoperability of WAP equipment and software with many different network technologies, such as GSM and IS-95 networks. The different bearer services are the basis for data transmission and WAP will integrate further services, such as SMS of GSM and packet-switched data as GPRS.

The WAP has the following features:

- It is an application communication protocol.
- It is used to access services and information.
- It is inherited from the Internet standards.
- It is suitable for handheld devices like as mobile phones.
- It is a protocol designed for micro-browser to fit into a small wireless terminal. The WAP uses a micro-browser, which is a small piece of software that makes minimal demands on hardware, memory, and CPU.
- It enables the creating of web applications for mobile devices.
- It uses the mark-up language WML, which is defined as an XML 1.0 application.

The important services provided by WAP are as follows:

- E-mail by mobile phones
- Tracking on stock market prices
- Sports results
- News headlines
- Music downloads

or

(a)

(i)

Wi-Fi	WiMAX
1. Local area network	1. Metropolitan area network

For other points, refer to Table 14.7, Chapter 14.

(ii)

UMTS	WiMAX
1. Universal telecommunication system for voice and data communication	1. Metropolitan area network for data exchange and Internet access
2. Based on WCDMA mainly	2. Based on OFDM technology mainly

(iii)

UMTS	GSM
1. 3G system	1. 2G system
2. Supports both voice and data at high bit rate	2. Mainly voice communication and limited data service
3. WCDMA modulation scheme	3. GMSK modulation scheme

(iv)

Bluetooth	Wi-Fi
1. A complete personal communication system, covering the area, maximum up to 100 m	1. A personal communication system that can be connected to hub to access external data, covering an area of 30 m indoor and 100 m outdoor.
2. Main topologies based on piconet and scatternet; point (ad hoc), or point-to-multipoint (bridge) topologies possible	2. AP based (infrastructure based), point-to- no infrastructure required; completely ad hoc
3. Uses FHSS with GFSK carrier modulation scheme	3. Uses CDMA and OFDM modulation schemes

(b) The antennas are in fact not smart—antenna systems are smart. Generally, co-located with a base station, a smart-antenna system combines an antenna array with a digital signal processing facilities and capability to transmit and receive in an adaptive, spatially sensitive manner, i.e. such a system can automatically change the directionality of its radiation patterns in response to its signal environment. A simple antenna works for a simple RF environment. The smart antenna solutions are required as the number of users, interference, and propagation complexity grow. The dual purpose of a smart-antenna system is to augment the signal quality of the radio-based system through more focused transmission of radio signals while enhancing capacity (performance characteristics) through increased frequency reuse. A smart-antenna system has the following main features:

1. It can increase the number of subscribers that can be serviced by a single cell.
2. It helps the wireless operators to cope up with the traffic levels and traffic inefficiencies.

There are two main categories of smart-antenna systems as per the transmit strategy

- Switched beam—a finite number of fixed, predefined patterns or combining strategies (sectors- directional beams) Switched beam systems combine the outputs of multiple antennas in such a way as to form finely sectorized (directional) beams with more spatial selectivity than can be achieved with conventional, single-element approaches.
- Adaptive array—an infinite number of patterns (scenario-based) that are adjusted in real time using new signal processing algorithm in adaptive manner The adaptive system takes advantage of its ability to effectively locate and track various types of signals to dynamically minimize interference and maximize intended signal reception.

Other features of smart-antenna systems are discussed below.

Better Signal Gain, hence Better Range/Coverage

Inputs from multiple antennas are combined to optimize the available power required to establish the given level of coverage. Focusing the energy sent out into the cell increases base station range and coverage. Lower power requirements also enable a greater battery life and smaller/lighter handset size.

Interference Rejection, hence Increased Capacity

Antenna pattern can be generated toward co-channel interference sources, improving the signal-to-interference ratio of the received signals. Precise control of signal nulls quality and mitigation of interference combine to reduce the frequency reuse distance (or cluster size), improving the capacity. Certain adaptive technologies (such as SDMA) support the reuse of frequencies within the same cell.

Spatial Diversity, hence Multipath Rejection

Composite information from the array is used to minimize fading and other undesirable effects of multipath propagation. This can reduce the effective delay spread of the channel, allowing higher bit rates to be supported without the use of an equalizer

Power Efficiency, hence Reduced Expense

A smart-antenna system combines the inputs to multiple elements to optimize the available processing gain in the downlink (toward the user). Lower amplifier costs, power consumption, and higher reliability will result.

(c) Refer to Section 15.6, Chapter 15.

Model Paper 4

Part A

1. The coherence bandwidth is a defined relation derived from the rms delay spread. It is a statistical measure of the range of frequencies over which the channel can be considered flat.
2. The central limit theorem states that given a distribution with a mean μ and variance σ^2, the sampling distribution of the mean approaches a normal distribution with a (same) mean μ and a variance σ^2/N as N, the sample size increases, i.e. the distribution of an average tends to be normal, even when the distribution from which the average is computed is decidedly non-normal. (Normal distributions are a family of distributions that have the shape like Gaussian. These distributions are symmetric with scores more concentrated in the middle than in the tails.)
3. It is the modulation technique that is used to prevent the reservation of the side lobes and spectral widening. It supports more efficient amplification.
4. If the transmission bandwidth of a single channel is much larger than the coherence bandwidth of the channel, the frequency-selective fading occurs, i.e. multipath fading will affect the different frequency components differently. Such a channel is called wideband channel and the system with frequency-selective channel response is called wideband system. The CDMA and OFDM are the examples of wideband systems.
5. The VLR is a visitor location register that stores the IMSI and customer information for each roaming subscriber.
6. The near–far problem is a problem that arises from varying received powers of the users due to different distances between the transmitter and receivers in that domain.
7. Given, $i = 2$ and $j = 1$

   So, $$N = i^2 + ij + j^2 = 2^2 + (2 \times 1) + 1^2$$
   $$= 4 + 2 + 1 = 7$$

   The cellular structure depicting the co-channel cells is shown in Fig. F.14.

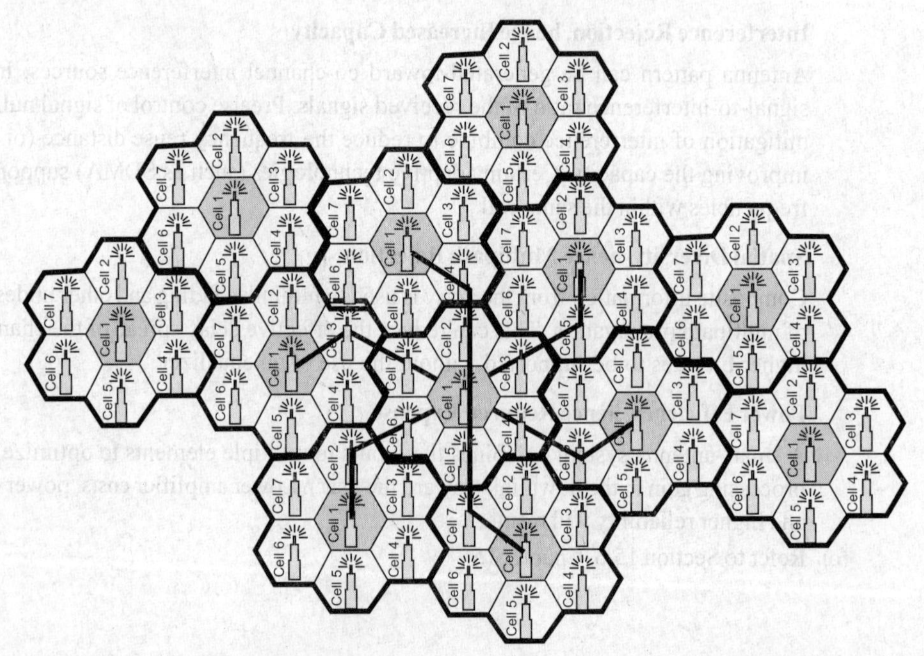

Fig. F.14 Cellular structure depicting the co-channel cells

8. The period is $2^3 - 1 = 8 - 1 = 7$ chips

9. Refer to Section 6.10, Chapter 6 and Section 15.7, Chapter 15.

10. A single-frequency network in conventional manner is a broadcast network where several transmitters simultaneously send the same signal over the same frequency at the same time, e.g., analog AM and FM. A simplified SFN is achieved by a low-power co-channel repeater. In DAB, because the OFDM scheme is utilized, multiple subcarriers make this type of transmission possible with cancellation of interference. The OFDM uses a number of slow, low-bandwidth modulators instead of fast wideband one. Thus, in SFN, 'human-made multipath' is created. All subcarriers will be combined and transmitted as single OFDM channel.

Part B

1. (a) Refer to Section 3.1.1, Chapter 3.

(b) Refer to Section 2.7, Chapter 2. General digital link may be considered over wired as well as wireless channels. The wired channel is more reliable requiring less complex receivers. There are some additional stages, such as RF stage and channel-estimation stage, which are not there in wired link. Such many points of difference can be identified.

(c) Refer to Sections 2.6.5 and 2.6.6, Chapter 2.

2. (a)

$$H(X) = \sum_i P\{x_i\} I\{x_i\};$$

and

$$I(x_i) = \log_2\left(\frac{1}{P\{x_i\}}\right)$$

So, the entropy is

$$H(X) = \frac{1}{\log_{10}(1.5)}\left[10(0.099)\log_{10}\left(\frac{1}{0.099}\right) + 2(0.005)\log_{10}\left(\frac{1}{0.005}\right)\right]$$

or

$$H(X) = 0.994 + 0.023$$
$$= 5.68\,(1.017) = 5.776 \text{ bits/key}$$

Now, $T = 1/(1.5 \text{ keys/sec})$

Then $R = H(X)/T = 5.776 \times 1.5 = 8.66 \text{ bits/sec}$

(b) The permutation of the payload data is carried out by the interleaver. Refer to Chapter 4.

(c) the LDPC code is an error-correcting code. It was the first code to approach theoretically maximum Shannon limit and impractical to implement when developed in 1963. Turbo codes discovered in 1993 became the coding scheme of choice and used for applications like deep-space satellite communications. In last few years, LDPC has become popular.

The LDPC codes are defined by a sparse parity check matrix. This sparse matrix is often randomly generated, subject to sparsity constraints. Refer to http://en.wikipedia.org/wiki/Low-density-parity-check_code for the example with calculations.

or

(a) For the special case of $k = 1$, the codes of rates 1/2, 1/3, 1/4, 1/5, and 1/7 are sometime called mother codes. We can combine these single-bit input codes to produce punctured codes, which gives us code rates other than $1/n$, e.g., in Fig. F.15, by using two rate 1/2 coders together and then just not transmitting one of the output bits, we can convert this rate 1/2 implementation into 2/3 rate codes. On the receive side, the dummy bits that do not affect the decoding metric are inserted in the appropriate places before decoding.

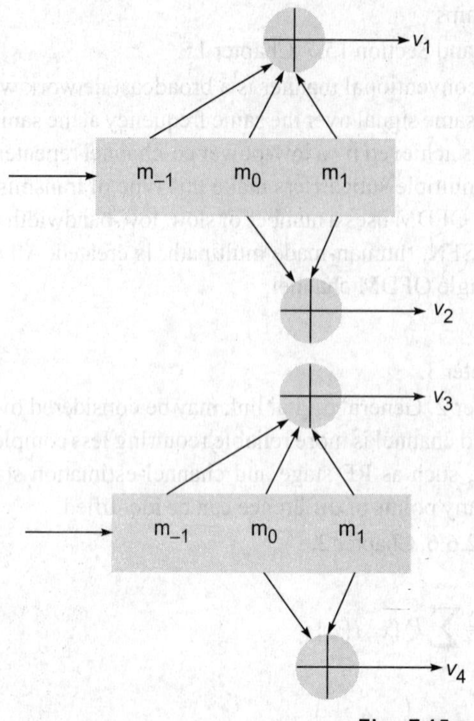

Fig. F.15

This technique allows us to produce codes of many different rates using just one simple hardware. The direct construction of rate 2/3 coder is possible. However, the advantage of the punctured code is that the rates can be changed dynamically (through software), depending on the channel condition. This is possible with additional cost of implementation complexity compared to fixed rate coders.

(b) Refer to Chapter 7.

(c) Refer to Chapter 3.

3. (a) The frequency-selective fading due to multipath time delay spread causes ISI, which results in an irreducible BER floor for mobile systems. Even with the time- varying Doppler spread due to motion creates an irreducible BER floor due to random spectral spreading. These factors impose bounds on the data rate and BER that can be transmitted reliably over frequency-selective channel. Many researchers have studied the effects of various modulation schemes through simulations.

Comparison on the Basis of Symbol

Irreducible BER floor in frequency-selective channel is mainly caused by the errors due to ISI, which interferes with the signal component at the receiver sampling instants. The errors due to frequency-selective channel tend to be bursty. Normally, the BER performance of BPSK is best among all modulation schemes compared. This is because symbol offset interference does not exist in BPSK. (This is also called cross rail interference due to the fact that the eye diagram has multiple rails.) Both OQPSK and MSK have $T/2$ timing offset between two bit sequences and hence the cross rail ISI is more severe. Also their performances are similar to QPSK.

Comparison on the Basis of Bit

By comparing on a bit rather than symbol basis, it becomes easier to compare different modulations. It is proved that four-level modulations (QPSK, OQPSK, and MSK) are more resistant to delay spread than BPSK for constant information throughput. The 8-ary keying has been found to be less resistant than the 4-ary keying. So, the 4-ary keying has been chosen for many 2G and 3G systems.

(b) Refer to Chapter 6.

(c) A few pilot-based estimators for OFDM are described below.

LS

Without using any knowledge of the statistics of the channels, the LS estimators are calculated with very low complexity, but they suffer from a high mean square error. Suppose X is pilot signal matrix, $(\bullet)^H$ is the conjugate transpose operation, H is specified channel condition matrix (from previous OFDM block estimation), and Y is received signal matrix then for block type pilot structure, then LS estimator minimizes the parameter $(Y-XH)^H(Y-XH)$.

For the block method, it is shown that the LS estimator of H is given by

$$H_{\text{LS}} = [(X_k/Y_k)]^T \qquad \text{where } k = 0, 1, \ldots, N_c - 1$$

For comb-type pilot-based estimation for each transmitted symbol, the N_p pilot signals are uniformly inserted into $x(k)$, i.e. a total of N subcarriers are divided into N_p groups, each with $L = N/N_p$ adjacent subcarriers.

$$H_{p\text{LS}} = [(Xp/Yp)]^T \qquad \text{where } k = 0, 1, \ldots, Np-1$$

MMSE

The MMSE estimator employs the second-order statistics of the channel conditions to minimize the mean square error and yields much better performance than LS estimators, especially under low-SNR scenarios. A major drawback of the MMSE estimator is its high computational complexity, especially if the matrix inversions are needed each time data in X changes. It can be derived that

$$H_{\text{MMSE}} = R_{\text{HH}}[R_{\text{HH}} + \sigma_n^2(XX^H)^{-1}]^{-1}H_{\text{LS}}$$

where H_{MMSE} is the estimated channel by MMSE algorithm, H_{LS} is the LS-based estimate, R_{HH} is the auto-covariance matrix of H (channel conditions), and σ_n^2 denotes noise variance.

There are other methods of modified MMSE with low-rank approximation, singular value decomposition (SVD), etc. The MMSE methods are applicable for slow time-varying channel only. Also, the knowledge of the channel transfer function for the previous OFDM data blocks is needed.

or

(a) In a line-of-sight (LOS) microwave, satellite and many other cable systems, the spectral efficiency is defined in terms of bits/second/hertz. We use this term for modulated systems also. For cellular non-line-of-sight (NLOS) applications, the basic spectral efficiency concept of modulated systems must be extended to the spectral efficiency of a complete geographic coverage or service area in terms of b/s/Hz/m^2 or erlang/Hz/ m^2.
For further details, refer to Section 11.11, Chapter 11.

(b) Refer to Chapters 8, 10, and 11. (Soft handover, co-user interference, near–far problem, etc. are the major constraints.)

(c) Refer to Chapter 8.

4. (a) The IEEE 802.16e-2005 standard improves upon the IEEE 802.16-2004 standard by employing the following features:

- It adds support for mobility (soft and hard handover between base stations). This is seen as one of the most important aspects of 802.16e-2005 and is the very basis of 'Mobile WiMAX'.
- The 802.16e version also supports SOFDMA. The OFDM and S-OFDMA work slightly differently. The S-OFDMA is akin to mobile version only, while the OFDM is in both the cases. In OFDM, when we increase the bandwidth, we increase the channel bandwidth of each tone as the number of tones remains constant. But in S-OFDMA, we instead increase the number of FFT points by increasing the channel bandwidth, keeping the bandwidth of the tones constant in the mobile environment. The FFT size and the number of carriers are equal in both fixed and mobile WiMAX based on OFDM (256), but they are different in S-OFDMA. The fixed bandwidth is a compromised solution in mobile environment
- It scales the fast Fourier transform (FFT) to the channel bandwidth in order to keep the carrier spacing constant across different channel bandwidths (typically, 1.25 MHz, 5 MHz, 10 MHz, or 20 MHz). The constant carrier spacing results in higher spectrum efficiency in wide channels and a cost reduction in narrow channels, also known as scalable OFDMA (S-OFDMA). Other bands that are not multiples of 1.25 MHz are defined in the standard, but because the allowed FFT subcarrier numbers are only 128, 512, 1024, and 2048, other frequency bands will not have exactly the same carrier spacing, which might not be optimal for implementations.
- It improves non-line-of-sight propagation coverage by utilizing advanced antenna diversity schemes and hybrid automatic repeat request (HARQ).
- It improves the capacity and coverage by introducing adaptive antenna systems (AAS) and MIMO technology (spatial multiplexing, beam forming, virtual/collaborative/multi-user MIMO).
- It increases the system gain by use of denser subchannelization, thereby improving indoor penetration and frequency-selective scheduling.
- It introduces high-performance coding techniques, such as turbo coding, low-density parity check (LDPC), and space-time coding (STC), enhancing security and NLOS performance.
- It introduces downlink subchannelization, allowing administrators to trade coverage for the capacity or vice versa.
- Enhanced fast Fourier transform algorithm can tolerate larger delay spreads, increasing resistance to multipath interference.
- It adds an extra QoS class (enhanced real-time polling service) more appropriate for VoIP applications.

(b) Refer to Section 15.2, Chapter 15.

or

(a) Refer to Section 13.2.1, Chapter 13.

(b) Refer to Section 12.5, Chapter 12.

5. Refer to (i) Section 14.3, (ii) Section 12.6, (iii) Section 13.3, (iv) Section 13.5, and (v) Section 14.2.

Appendix G

Answers to Multiple Choice Questions

Chapter 1

1. (d) 2. (a) 3. (d) 4. (d) 5. (c) 6. (a) 7. (b)

Chapter 2

1. (b) 2. (c) 3. (a) 4 (d) 5. (b) 6. (b) 7. (c) 8. (a)
9. (d) 10. (c) 11. (a) 12. (b) 13. (d) 14. (b) 15. (c) 16. (d)
17. (c) 18. (a) 19. (a) 20. (a) 21. (c) 22. (a) 23. (b) 24. (b)
25. (b) 26. (d) 27. (c) 28. (a) 29. (c) 30. (c)

Chapter 3

1. (b) 2. (b) 3. (d) 4. (b) 5. (b) 6. (d) 7. (c) 8. (c)
9. (c) 10. (b) 11. (d) 12. (c) 13. (d) 14. (a) 15. (b) 16. (c)
17. (c) 18. (c) 19. (d) 20. (d) 21. (d) 22. (a) 23. (d) 24. (b)

Chapter 4

1. (b) 2. (b) 3. (d) 4. (b) 5. (c) 6. (c) 7. (a) 8. (b)
9. (b) 10. (a) 11. (b)

Chapter 5

1. (b) 2. (c) 3. (c) 4. (c) 5. (b) 6. (b) 7. (a) 8. (c)
9. (d) 10. (d) 11. (a) 12. (b) 13. (a) 14. (b) 15. (b) 16. (b)
17. (a) 18. (c) 19. (c) 20. (d) 21. (d) 22. (a) 23. (d) 24. (a)
25. (c) 26. (b) 27. (c) 28. (d)

Chapter 6

1. (d) 2. (c) 3. (a) 4. (c) 5. (c) 6. (c) 7. b 8. (a)
9. (c) 10. (c) 11. (c)

Chapter 7

1. (a) 2. (a) 3. (a) 4. (b) 5. (d) 6. (d) 7. d 8. (b)
9. (a) 10. (c) 11. (d)

Chapter 8

1. (a) 2. (d) 3. (b) 4. (b) 5. (d) 6. (b) 7. (b) 8. (d)
9. (c) 10. (b) 11. (d)

Chapter 9

1. (c) 2. (a) 3. (c) 4. (a) 5. (d) 6. (d) 7. (d) 8. (b)

Chapter 10

1. (b) 2. (b) 3. (b) 4. (d) 5. (d) 6. (d) 7. (d) 8. (c)
9. (c) 10. (b) 11. (d) 12. (d) 13. (b) 14. (c) 15. (a) 16. (b)

Chapter 11

1. (d) 2. (d) 3. (c) 4. (d) 5. (c) 6. (b) 7. (a) 8. (c)
9. (c) 10. (b) 11. (d) 12. (c) 13. (b) 14. (b) 15. (b) 16. (d)

Chapter 12

1. (b) 2. (b) 3. (d) 4. (c) 5. (d) 6. (c) 7. (d) 8 (a)
9. (c) 10. (b) 11. (c) 12. (c) 13. (d) 14. (a) 15. (d) 16. (d)

Chapter 13

1. (c) 2. (a) 3. (a) 4. (b) 5. (d) 6. (b) 7. (c) 8. (c)
9. (d) 10. (d) 11. (b) 12. (b) 13. (b) 14. (b) 15. (b) 16. (c)
17 (c) 18. (a) 19. (b) 20. (c) 21. (d) 22. (c) 23. (d) 24. (c)
25. (a) 26. (a) 27. (b)

Chapter 14

1. (b) 2. (c) 3. (c) 4. (d) 5. (d) 6. (a) 7. (c) 8. (c)
9. (c) 10. (a) 11. (b) 12. (c) 13. (a) 14. (c) 15. (a) 16. (c)
17. (d) 18. (c) 19. (c) 20. (d) 21. (a) 22. (c)

Chapter 15

1. (c) 2. (c) 3. (d) 4. (a) 5. (c)

Chapter 16

1. (b) 2. (a) 3. (c) 4. (b) 5. (d) 6. (b)

Further Readings

Books

Bahai, A.R.S. and B.R. Saltzberg, *Multi-carrier Digital Communications: Theory and Applications of OFDM*, 2nd ed., Springer, 2004

Bhattacharya, Amitabha, *Digital Communication*, Tata McGraw-Hill, 2005

Bingham, J.A.C., *The Theory and Practice of Modem Design*, John Wiley & Sons, New York, 1988

Boithias, L., *Radio Wave Propagation*, McGraw-Hill Inc., New York, 1987

Boucher, J.R., *Voice Teletraffic Systems Engineering*, Artech House, Inc., 1988

Boucher, N., *Cellular Radio Handbook*, Quantum Publishing, Inc., 1991

Calhoun, G., *Digital Cellular Radio*, Artech House, Inc., 1988.

Carlson, A. Bruce, *Communication Systems*, McGraw Hill International Editions, 1986

Cattermole, K.W., *Principles of Pulse Code Modulation*, Elsevier, New York, 1969

Cooper, G.R. and C.D. McGillem, *Modern Communications and Spread Spectrum*, McGraw Hill, New York, 1986

Couch, L.W., *Digital and Analog Communication Systems*, 4th ed., Macmillan, New York, 1993

Deller, J.R, J.G. Proakis, and J.H.L. Hansen, *Discrete-Time Processing of Speech Signals*, Macmillan, New York, 1993

Dixon, R.C., *Spread Spectrum Systems with Commercial Applications*, 3rd ed., John Wiley & Sons, New York, 1994

Dixon, R.C., *Spread Spectrum Systems*, 2nd ed., John Wiley & Sons, New York, 1984

Feher, K., *Applications of Digital Wireless Technologies to Global Wireless Communications*; Prentice Hall, 1997

Forney, G.D., Jr., *Concatenated Coding*, MIT Press, Cambridge, MA 1966

Garg, V.K. and J.E. Wilkes, *Principles and Applications of GSM*, Pearson Education Asia, LPE /Prantice Hall, NJ, 1999

Garg, V.K., *IS-95 CDMA and cdma2000: Cellular/PCS Systems Implementation*, Pearson Education Asia, LPE/Prantice Hall, NJ, 2000

Garg, V.K., *Wireless Network Evolution: 2G to 3G*, Second Indian Reprint, Pearson Education Asia, LPE, 2004

Gershman, A.B. and N.D. Sidiropoulos (Eds), *Space Time Processing for MIMO Communications*, John Wiley & Sons, 2005

Glisic, S.G., *Advanced Wireless Networks: 4G Technologies*, John Wiley & Sons, 2006

Gravano, Salvatore, *Introduction to Error Control Codes*, Oxford University Press, 2001 (reprint 2007)

Griffiths, J., *Radio Wave Propagation and Antennas*, Prentice Hall International, 1987

Guizani, Mohsen, *Wireless Communications Systems and Networks*, First Indian Reprint, Springer, 2006

Haykin, S., *Adaptive filter Theory*, Prentice Hall, Englewood Cliffs, NJ, 1986

Haykin, S., *Communication Systems*, John Wiley and Sons, New York, 1994

Hess, W., *Pitch Determination of Speech Signals*, Springer-Verlag, Berlin 1983

Jayant, N.S. and P. Noll, *Digital Coding of Waveforms*, Prentice-Hall, NJ, 1984

Kennedy, George and Bernard Davis, *Electronic Communication Systems*, 4th ed., Tata McGraw-Hill, 1999

Kim, K.I. (Ed.), *Handbook of CDMA System Design, Engineering, and Optimization*, Pearson Education Asia, LPE/Prentice Hall, NJ, 2000

Krauss, J.D., *Antennas*, McGraw-Hill, New York, 1950

Lathi, B.P., *Modern Digital and Analog Communication Systems*, 3rd ed., Oxford University Press, 1998

Lee, E.A. and D.G. Messerschmitt, *Digital Communication*, Kluwer Academic Publishers, 1994

Lee, J.S., and L.E. Miller, *CDMA Systems Engineering Handbook*, Artech House, 1998

Lee, William C.Y., *Mobile Cellular Telecommunications*, 2nd ed., McGraw Hill, New York, 1995

Lee, William C.Y., *Mobile Communications Engineering*, McGraw-Hill, New York, 1985

Lee, William C.Y., *Wireless and Cellular Telecommunications*, 3rd ed., McGraw-Hill, New York, 2005

Liberti, J.C. Jr. and T.S. Rappaport, *Smart Antennas for Wireless Communications: IS-95 and Third Generation Applications*, Pearson Education Asia, LPE/Prentice Hall, NJ, 1999

Lin, S. and D.J. Costello, *Error Control Coding: Fundamentals and Applications*, Prentice Hall, NJ, 1983

Liu, H., and G. Li, *OFDM Based Broadband Wireless Networks: Design and Optimization*, John Wiley and Sons, Inc., 2005

McEliece, R.J., *The Theory of Information and Coding*, 2nd ed., Cambridge University Press, Cambridge, UK, 2002

Metropolis, N. and S. Ulam, 'The Monte Carlo Method', *J. Amer. Stat. Assoc.*, 44, 335–41, 1949

Molisch, A., *Wideband Wireless Digital Communication,* Pearson Education Asia, LPE, 2002

Oestges, Claude and Bruno Clerckx, *MIMO Wireless Communications: From Real World Propagation to Space Time Code Design*, Academic, 2007

Owens, F.J., *Signal Processing of Speech*, McGraw-Hill, New York, 1993

Pahlavan, K. and A.H. Levesque, *Wireless Information Networks*, John Wiley & Sons, Inc., 1995

Pahlavan, Kaveh and Prashant Krishnamurthy, *Principles of Wireless Networks: A Unified Approach*, Pearson Education Asia, LPE, 2001

Pandya, Raj, *Mobile and Personal Communication Services and Systems*, Wiley-IEEE Press, 2000

Papulis, A., *Probability, Random Variables, and Stochastic Processes*, 3rd ed., McGraw-Hill, New York, 1991

Proakis, J.G. and Demitris K. Manolakis, *Digital Signal Processing: Principles, Algorithms and Applications*, 3rd ed., Prantice-Hall of India, 2004

Proakis, J.G. and M. Salehi, *Communication Systems Engineering*, Prentice Hall, 1994

Proakis, J.G., *Digital Communications*, 3rd ed., McGraw-Hill, New York, 1995

Rao, K.R., Z.S. Bojkovic, and D.A. Milovanovic, *Multimedia Communication Systems*, Prentice-Hall of India, New Delhi, 2005

Rappaport, Theodore S., 2002, *Wireless Communications: Principles and Practice*, 2nd ed., Prentice Hall/Pearson Education Asia

Reed, Jeffrey H., *Software Radio: A Modern Approach to Radio Engineering,* 1st Indian Reprint, Pearson Education Asia, LPE, 2002

Rhee, M.Y., *Error Correction Coding Theory*, McGraw-Hill, New York, 1989

Rogers, Gary S. and John Edwards, *An Introduction to Wireless Technology*, Second Impression, Pearson Education Asia, LPE, 2008

Roth, R.M., *Introduction to Coding Theory*, Cambridge University Press, Cambridge, UK, 2006

Schulze, H. and C. Luders, *Theory and Applications of OFDM and CDMA: Wideband Wireless Communications*, John Wiley & Sons, 2005

Sharma, Sanjay, *Wireless Communications*, Katson Books, 2006

Sklar, Bernad, *Digital Communications: Fundamentals and Applications*, 2nd ed., Pearson Education (LPE)/Prentice Hall, NJ, 2001

Stallings, William,, *Wireless Communications and Networking*, 5th Indian Reprint, Pearson Education Asia, LPE, 2004

Stark, H. and J.W. Woods, *Probability, Random Variables, and Estimation Theory for Engineers,* Prentice Hall, Englewood Cliffs, NJ, 1986

Steele, R., *Mobile Radio Communications*, IEEE Press, 1994

Stutzman, W.L. and G.A. Thiele, *Antenna Theory and Design*, John Wiley & Sons, New York, 1981

Tafazolli, R. (Ed.), *Technologies for the Wireless Future*, Vol. 2, Wireless World Research Forum, John Wiley & Sons, 2006

Tanenbaum, A.S., *Computer Networks*, Prentice Hall, Inc. 1981

Taub, Herbert and Donald L. Schilling, *Principles of Communication Systems*, 2nd ed., Tata McGraw-Hill, 1991

Toh, C.K., *Ad Hoc Mobile Wireless Networks: Protocols and Systems*, Second Impression, Pearson Education Asia, 2009

Tranter, W.H., K. Shanmugan, T.S. Rappaport, and and K. Kosbar, *Computer Aided Design and Analysis of Communication Systems with Wireless Applications*, Pearson Education Asia (LPE)/ Prentice Hall, NJ, 2002

Tsui, James, *Digital Techniques for Wideband Receivers*, 2nd ed., Prentice-Hall of India, New Delhi, 2005

van Nee, R. and R. Prasad, *OFDM for Wireless Multimedia Communications*, Artech House, Boston, USA, 2000

Vankka, J., *Digital Synthesizers and Transmitters for Software Radio*, Springer, 2005

Viterbi, A.J. and J.K. Omura, *Principles of Digital Communication and Coding*, McGraw-Hill, New York, 1979

Viterbi, A.J., 'Error Bounds for Convolutional Codes and An Asymptotically Optimum Decoding Algorithm', *IEEE Transactions on Information Theory*, Vol. IT-13, pp. 260–69, April 1967

Volker, K., *Wireless Communications over MIMO Channels*, John Wiley & Sons, 2006

Widrow, B. and S.D. Stearns, *Adaptive Signal Processing*, Prentice Hall, 1985

Yacoub, M.D., *Foundations of Mobile Radio Engineering*, CRC Press, 1993

Ziemer, R.E. and R.L. Peterson, *Digital Communications*, Prentice Hall, NJ, 1990

Ziemer, R.E., and R.L. Peterson, *Introduction to Digital Communications*, Macmillan Publishing, 1992

Papers

Akaiwa, Y. and Y. Nagata, 'Highly Efficient Digital Mobile Communications with a Linear Modulation Method', *IEEE Journal on Selected Areas in Communications*, Vol. SAC-5, No. 5, pp. 890–95, June 1987

Alexiou, Angeliki and Martin Haardt, 'Smart Antenna Technologies for Future Wireless System: Trends and Challenges', Wireless Research Forum, *IEEE Communication Magazine*, pp. 90–97, September 2004.

Amoroso, F., 'The Bandwidth of Digital Data Signals', *IEEE Communications Magazine*, pp. 13–24, November 1980

Anderson, J.B., Thoedore S. Rappaport, and S. Yoshida, S., 'Propagation Measurements and Models for Wireless Communications Channels', *IEEE Communications Magazine*, November 1994

Ashitey, D., A. Sheikh, and K.M.S. Murthy, 'Intelligent Personal Communication System', 43rd ed., *IEEE Vehicular Technology Conference*, pp. 696–99, 1993

Atal, B.S., 'High Quality Speech at Low Bit Rates: Multi-pulse and Stochastically Excited Linear Predictive Coders', *Proceedings of ICASSP*, pp. 1681–84, 1986

Bayless, J.W., 'Voice Signals: Bit-by-Bit', *IEEE Spectrum*, pp.28–34, October 1973

Belfiori, C.A. and J.H. Park, 'Decision Feedback Equalization', *Proceedings of IEEE*, Vol. 67, pp. 1143–56, August 1979

Benedetto, S., D. Divsalar, DMontorsi, and F. Pollara, 'Serial Concatenation of Interleaved Codes', *IEEE Transactions on Information Theory,* pp. 909–26, May 1998

Bernhardt, R.C., 'Macroscopic Diversity in Frequency Reuse Systems', *IEEE Journal on Selected Areas in Communications*, Vol. SAC-5, pp. 862–78, June 1987

Berrou, C. and A. Glavieux, 'Near Optimum Error Correcting Coding and Decoding: Turbo Codes', *IEEE Transaction on Communications*, Vol. 44, pp. 126–27, 1996

Berrou, C., A. Glavieux, and P. Thitimajshimha, 'Near Shannon Limit Error Correcting Coding and Decoding: Turbo Codes', *IEEE International Communication Conference (ICC)*, Geneva, pp. 1064–70, May 1993

Bingham, John A.C., 'Multicarrier Modulation for Data Transmission: An Idea Whose Time Has Come', *IEEE Communication Magazine*, Vol. 28, No. 5, pp. 5–14, May 1990

Bose, R.C. and D.K. Ray-Chaudhari, 'On a Class of Error Correcting Binary Group Codes', *Information and Control*, Vol. 3, pp. 68–70, March 1960

Brady, D.M., 'An Adaptive Coherent Diversity Receiver for Data Transmission through Dispersive Media', *Proceedings of IEEE International Conference on Communications*, pp. 21–35, June 1970

Braley, R.C., I.C. Gifford, and R.F. Heile, 'Wireless Personal Area Networks: An Overview of the IEEE P802.15 Working Group', *ACM Mobile Computing and Communications Review*, Vol. 4, No. 1, pp. 20–27, February 2000

Buckley, S., '3G Wireless: Mobility Scales New Heights', *Telecommunications Magazine*, November 2000

Cellular Telephony Industry Association, *Cellular Digital Packet Data System Specification,* Release 1.0, July 1993

Chang, R.W. and R.A. Gibby, 'A Theoretical Study of Performance of an Orthogonal Multiplexing Data Transmission Scheme', *IEEE Transactions on Communication Technology*, Vol. 16, pp. 529–40, August 1968

Chang, R.W., 'Orthogonal Frequency Division Multiplexing', US Patent 3488445, January 1970

Chang, R.W., 'Synthesis of Band Limited Orthogonal Signals for Multichannel Data Transmission', *Bell System Technical Journal*, Vol. 45, pp. 1775–96, December 1966

Chiariglione, L., 'The Development of an Integrated Audiovisual Coding Standard: MPEG', *Proceedings of IEEE,* Vol. 83, No. 2, pp. 151–57, February 1995

Chuanng, J., 'The Effects of Time Delay Spread on Portable Communications Channels with Digital Modulation', *IEEE Journal on Selected Areas in Communications*, Vol. SAC-5, No. 5, pp. 879–89, June 1987

Clarke, R.H., 'A Statistical Theory of Mobile Radio Reception', *Bell Systems Technical Journal*, Vol. 47, pp. 957–1000, 1968

Coleri, Sinem, Mustafa Ergan, Ąnuj Puri, and Ahmed Bahai, 'Channel Estimation Techniques based on Pilot Arrangement in OFDM Systems', *IEEE Transaction on Broadcasting*, Vol. 48, No. 3, pp. 223–29, September 2002

Correia, L. and R. Prasad, 'An Overview of Wireless Broadband Communications', *IEEE Communications Magazine*, pp. 28–33, January 1997

Coulson, A.J., A.G. Willianson, and R.G. Vaughan, 'A Statistical Basis for Log-normal Shadowing Effects in Multipath Fading Channels', *IEEE Transactions on Communications*, Vol. 46, pp. 494–502, April 1998

Cox, D.C. W. Arnold, and P.T. Porter, 'Universal Digital Portable Communications: A System Perspective', *IEEE Journal on Selected Areas of Communications*, Vol. SAC-5, No. 5, p. 764, 1987

Cox, D.C., 'Portable Digital Radio Communication—An Approach to Tetherless Access', *IEEE Communication Magazine*, pp. 30–40, July 1989

Crochiere, R.E., et al., 'Digital Coding of Speech in Sub-bands', *Bell Systems Technical Journal*, Vol. 55, No. 8, pp. 1069–85, October 1976

Dalal, U.D. and Y.P. Kosta, 'Reinvestigation of Pilot Transmission Issues For OFDM with An Efficient Solution', *International Journal of Engineering Research and Industrial Applications*, Vol. 1, No.VII, pp. 283–98, 2008

Dalal, U.D., Y.P. Kosta, and K.S. Dasgupta, 'A Study of Various Issues of Pilot Based Channel Estimation Methods for OFDM System over Time Varying Channel', *Proceedings of International Conference TIES2007-Technical Collaborator IEEE*, Chennai, Vol. 1, pp. 6–11, November 2007

Dalal, U.D., Y.P. Kosta, and K.S. Dasgupta, 'Study of Issues Related to Adaptive Channel Estimation in OFDM System-MWBA Aspects', *Proceedings of IEEE International Conference, WCSN-07*, IIIT Allahabad, pp. 149–54, December 2007

deBuda, R., 'Coherent Demodulation of Frequency Shift Keying with Low Deviation Ratio', *IEEE Transactions on Communications*, Vol. COM-20, pp. 466–70, June 1972

Dechaux, C. and R. Scheller, 'What are GSM and DECT?', *Electrical Communication*, pp. 118–27, 2nd quarter, 1993

Deygout, J., 'Multiple Knife-edge Diffraction of Microwaves', *IEEE Transactions on Antennas and Propagation*, Vol. AP-14, No. 4, pp. 480–89, 1966

Dinis, Manuel and Jose Fernandes, 'Provision of Sufficient Transmission Capacity for Broadband Mobile Multimedia: A Step Towards 4G', *IEEE Communications Magazine*, August 2001

Durgin, G.D. and Theodore S. Rappaport, 'Theory of Multipath Shape Factors for Small Scale Fading Wireless Channels', *IEEE Transactions on Antennas and Propagation*, Vol. 48, No. 5, pp. 682–93, May 2000

Eng, T. and L.B. Milstein, 'Capacities of Hybrid FDMA/CDMA Systems in Multipath Fading', *IEEE MILCOM Conference Records*, pp. 753–57, 1993

Ertel, R., P. Cardieri, K.W. Sowerby, Theodore S. Rappaport, and J.H. Reed, 'Overview of Spatial Channel Models for Antenna Array Communication Systems', *Special Issue IEEE Personal Communications*, Vol. 5, No. 1, pp. 10–22, February 1998

ETSI, 'Radio Broadcasting Systems: Digital Audio Broadcasting (DAB) to Mobile, Portable and Fixed Receivers', *European Telecommunications Standard 300 401*, 2nd ed., European Telecommunication Standard Institute, Valbonne, France, 1998

Fano, R.M., 'A Heuristic Discussion on Probabilistic Coding', *IEEE Transactions on Information Theory*, Vol. IT-9, pp. 64–74, April 1963

Feher, K., 'Modems for Emerging Digital Cellular Mobile Radio Systems', *IEEE Transactions on Vehicular Technology*, Vol. 40, No. 2, pp. 355–65, May 1991

Fernando, W.A.C., R.M.A.P. Rajtheva, and K.M. Ahmed, 'Performance of Coded OFDM with Higher Modulation Schemes', *Communication Technology Proceedings*, Beijing, China, Vol. 2, October 1998

Feuerstein, M.J., K.L. Blackard, Theodore S. Rappaport, S.Y. Seidal, and H.H. Xia, 'Path Loss, Delay Spread and Outage Models as Functions of Antenna Height for Microcellular System Design', *IEEE Transactions on Vehicular Technology*, Vol. 43, No. 3, pp. 487–98, August 1994

Flanagan, J.L., 'Speech Coding', *IEEE Transactions on Communications*, Vol. COM-27, No. 4, pp. 710–35, April 1979

Forney, G.D., 'The Viterbi Algorithm', *Proceedings of the IEEE*, Vol. 61, No. 3, pp. 268–78, March 1973

Foschini, G.J. and M.J. Gans, 'On Limits of Wireless Communications in a Fading Environment when Using Multiple Antennas', *Wireless Personal Communications*, Vol. 6, No. 3, pp. 311–35, January 1998

Foschini, G.J., 'Layered Space time Architecture for Wireless Communications in a Fading Environment When Using Multi-element Antennas', *Bell Labs Technical Journal,* Vol. 1, Issue 2, pp. 41–59, 1996

Frederiksen, Flemming and Ramjee Prasad, 'An Overview of OFDM and Related Techniques Towards Development of Future Wireless Multimedia Communications', *IEEE Radio and Wireless Conference*, pp.19–22, August 2002

Fulghum, T. and K. Molnar, 'The Jakes Fading Model Incorporating Angular Spread for a Disk of Scatterers', *IEEE Vehicular Technology Conference*, Ottawa, Canada, pp. 489–93, May 1998

Gans, M.J., 'A Power Spectral Theory of Propagation in the Mobile Radio Environment', *IEEE Transactions on Vehicular Technology*, Vol. VT-21, pp. 27–38, February 1972

Garber F.D. and M.B. Pursley, 'Optimal Phases of Maximal Length Sequences for Asynchronous Spread-Spectrum Multiplexing', *Electronic Letter*, Vol. 16, pp. 756–57, September 1980

Geraniotis, E. and B. Ghaffari, 'Performance of Binary and Quaternary Direct Sequence Spread-spectrum Multiple-Access Systems with Random Signature Sequences', *IEEE Trans. Comm.*, COM-39, No. 5, pp. 713–24, May 1991

Geraniotis, E. and M.B., Purseley, 'Error Probabilities for Direct Sequence Spread Spectrum Multiple Access Communications–Part II: Approximations', *IEEE Transaction Communication*, Vol. COM-30, pp. 985–95, May 1982

Gesbert, D., M. Kountouris, R.W. Heath, Jr., C.B. Chae, and T. Salzer, 'From Single User to Multiuser Communications: Shifting the MIMO Paradigm', *IEEE Signal Processing Magazine*, Vol. 24, No. 5, pp. 36–46, October 2007

Gesbert, David, Mansoor Shafi, Da-shan Shiu, and Peter J. Smith, 'From Theory to Practice: An Overview of MIMO Space-Time Coded Wireless Systems', *IEEE Journal on Selected Areas in Communication,* Vol. 21, No. 3, p. 281, April 2003

Gilhousen, et al., 'On the Capacity of Cellular CDMA System', *IEEE Transactions on Vehicular Technology*, Vol. 40, No. 2, pp. 303–11, May 1991

Gitlin, R.D. and S.B. Weinstein, 'Fractionally Spaced Equalization: An Improved Digital Transversal Filter', *Bell Systems Technical Journal*, Vol. 60, pp. 275–96, February 1981

Golay, M.J.E., 'Notes on Digital Coding', *Proceedings of IRE*, Vol. 37, p. 657, June 1949

Gold R., 'Optimal Binary Sequences for Spread Spectrum Multiplexing', *IEEE Trans. Information Theory*, Vol. IT-13, pp. 619–21, 1967

Goodman, D.J., R.A. Valenzula, K.T. Gayliard, and B. Ramamurthy, 'Packet Reservation Multiple Access for Local Wireless Communication', *IEEE Transactions on Communications*, Vol. 37, No. 8, pp. 885–90, August 1989

Gray, R.M., 'Vector Quantization', *IEEE ASSP Magazine*, pp. 4–29, April 1984

Greenstein, L.J. and R.D. Gitlin, 'A Microcell/Macrocell Cellular Architecture for Low and High Mobility Wireless Users', *IEEE Vehicular Technology Transactions*, pp. 885–91, August 1993

Gudmundson, B., J. Skold, and J.K. Ugland, 'A Comparison of CDMA and TDMA Systems', *Proceedings of the 42nd IEEE Vehicular Technology Conference*, Vol. 2, pp.732–35, 1992

Hamming, R.W., 'Error Detecting and Error Correcting Codes', *Bell System Technical Journal*, April 1950

Hashemi, H., 'The Indoor Radio Propagation Channel', *Proceedings of the IEEE*, Vol. 81, No. 7, pp. 943–68, July 1993

Hata, Masaharu, 'Empirical Formula for Propagation Loss in Land Mobile Radio Services', *IEEE Transactions on Vehicular Technology*, Vol. VT-29, No. 3, pp. 317–25, August 1980

Hirosaki, B., 'An Orthogonally Multiplexed QAM System Using the Discrete Fourier Transform', *IEEE Pans. Comm.,* Vol. COM-29, pp. 982–89, July 1981

Hoges, M.R.L., 'The GSM Radio Interface', *British Telecom Technological Journal*, Vol. 8, No.1, pp. 31–43, January 1990

Hotler, Bengt, 'On the capacity of the MIMO channel: A Tutorial Introduction', Norwegian University of Science and Technology, Department of Telecommunications

Hsieh, M. and C. Wei, 'Channel Estimation for OFDM Systems Based on Comb-type Pilot Arrangement in Frequency Selective Fading Channels', *IEEE Transactions on Consumer Electronics*, Vol. 44, No.1, February 1998

Hutter, A., R. Hasholzner, and J. Hammerschmidt, 'Channel Estimation for Mobile OFDM System', *IEEE Vehicular Technology Conference*, pp. 305–09, 1999

IEEE Communications Magazine, 'Special Issue on Satellite Communication Systems and Services for Travelers', November 1991

International Telecommunication Union (ITU), Radio Regulations 1982 ed., rev. 1985, 1986, 1988, vol. 2, appendix 30A

International Telecommunication Union, 'Digital Multiprogramme Television Systems for Use by Satellites Operating in the 11/12 GHz Frequency Range', *Recommendation ITU-R BO.1516*, 2001

International Telecommunication Union, 'ITU World Telecommunications Report', *ITU Documents*, Radiocommunications Sector, Geneva

Ishizuka, M. and K. Hirade, 'Optimum Gaussian Filter and Deviated Frequency Locking Scheme for Coherent Detection of MSK', *IEEE Transactions on Communications*, Vol. COM-28, No. 6, pp. 850–57, June 1980

Jain, Mehul and M. Mani Roja, 'Comparison of OFDM with CDMA System in Wireless Telecommunication for Multipath Delay Spread', *The First IEEE and IFIP International Conference in Central Asia*, September 26-29, 2005

Jakes, W.C., 'New Techniques for Mobile Radio', *Bell Laboratory Rec.*, pp. 326–30, December 1970

Jayant, N.S., 'Coding Speech at Low Bit Rates', *IEEE Spectrum*, pp. 58–63, August 1986

Jayant, N.S., 'High Quality Coding of Telephone Speech and Wideband Audio', *IEEE Communications Magazine*, pp. 10–19, January 1990

Kleinrock, L. and Tobagi, F.A., 'Packet Switching in Radio Channels, Part 1: Carrier Sense Multiple Access Models and Their Throughput-Delay Characteristics', *IEEE Transactions on Communications,* Vol. 23, No. 5, pp.1400–16, 1975

Kucar, A.D., 'Mobile Radio–An Overview', *IEEE Communications Magazine*, pp.72–85, November 1991

Lee, William C.Y. and S.Y. Yeh, 'Polarization Diversity System for Mobile Radio', *IEEE Transactions on Communications*, Vol. 20, pp. 912–22, October 1972

Lee, William C.Y., 'Elements of Cellular Mobile Radio Systems', *IEEE Transactions on Vehicular Technology*, Vol.VT-35, No. 2, pp. 48–56, May 1986

Lee, William C.Y., 'Overview of Cellular CDMA', *IEEE Transactions on Vehicular Technology*, Vol. 40, No. 2, May 1991

Lee, William C.Y., 'Smaller Cells for Greater Performance', *IEEE Communications Magazine*, pp. 19–23, November 1991

Lee, William C.Y., 'Spectrum Efficiency in Cellular', *IEEE Transactions on Vehicular Technology,* Vol. 38, No. 2, pp. 69–75, May 1989

Li, Y., 'Pilot-Symbol-Aided Channel Estimation for OFDM in Wireless Systems', *IEEE Transactions on Vehicular Technology*, Vol. 49, No. 4, pp. 1207–15, July 2000

Liberti, J.C. and Theodore S., Rappaport, 'Analytical Results for Capacity Improvements in CDMA', *IEEE Transactions on Vehicular Technology*, Vol. 43, No. 3, pp. 680–90, August 1994

Linde, Y., A. Buzo, and R.M. Gray, 'An Algorithm for Vector Quantizer Design', *IEEE Transactions on Communications*, pp. 702–10, January 1980

Linnartz, J.P.M.G., Louis C. Yun, and M. Couture, 'Multi-user and Self-interference Effects in a QPSK DS-CDMA Downlink with a Rician Dispersive Channel', *Wireless Personal Communications*, Vol. 1, No. 2, pp. 95–102, 1995

Lo, N.K.W., D.D. Falconer, and A.U.H. Sheikh, 'Adaptive Equalization and Diversity Combining for a Mobile Radio Channel', *IEEE Globecom*, San Diego, December 1990

Lucky, R.W., 'Automatic Equalization for Digital Communication', *Bell System Technical Journal*, Vol. 44, pp. 547–88, 1965

MacDonald, V.H., 'The Cellular Concept', *The Bell Systems Technical Journal,* Vol. 58, No.1, pp. 15–43, January 1979

Makhoul, J., 'Linear Prediction: A Tutorial Review', *Proceedings of IEEE*, Vol. 63, pp. 561–80, April 1975

Mallat, S., 'A Theory for Multiresolutional Signal Decomposition: the Wavelet Representation', *IEEE Trans. Pattern Analysis and Machine Intelligence*, No. 7, pp. 674–93, 1989

Millington, G., R. Hewitt, and F.S. Immirzi, 'Double Knife-edge Diffraction in Field Strength Predictions', *Proceedings of IEE*, 109C, pp. 419–29, 1962

Modarressi, A.R. and R.A. Skoog, 'An Overview of Signal System No. 7', *Proceedings of the IEEE*, Vol. 80, No. 4, pp. 590–606, April 1992

Monsen, P., 'MMSE Equalization of Interference on Fading Diversity Channels', *IEEE Transactions on Communications*, Vol. COM-32, pp. 5–12, January 1984

Mosier, R.R. and R.G. Clabaugh, 'Kineplex, A Bandwidth Efficient Binary Transmission System', *AIEE Trans.,* Vol. 76, pp. 723–28, January 1958

Mulder, R.J., 'DECT–A Universal Cordless Access System', *Philips Telecommunications Review*, Vol. 49, No. 3, pp. 68–73, September 1991

Murota, K. and K. Hirade, 'GMSK Modulation for Digital Mobile Radio Telephony', *IEEE Transactions on Communications*, Vol. COM-29, No.7, pp. 1044–50, July 1981

Nasrabadi, N.M. and R.A. King, 'Image Coding Using Vector Quantization: A Review', *IEEE Trans. on Communications*, Vol. COM-36, pp. 957–71, August 1988

Noble, D., 'The History of Land Mobile Radio Communications', *IEEE Vehicular Technology Transactions*, pp. 1406–16, May 1962

Nyguib, Ayman F., Vahid Tarokh, Nambi Seshadri, and A.R. Calder, 'Space Time Coding and Signal Processing for High Data rate Wireless Communication', AT & T Labs Research, *IEEE Trans. Commun.*, vol. 47, pp. 199–207, February 1999

Nyquist, H., 'Certain Topics in Telegraph Transmission Theory', *Transactions of the AIEE*, Vol. 47, pp. 617–44, February 1928

Ochsner, H., 'DECT–Digital European Cordless Telecommunications', *IEEE Vehicular Technology 39th Conference*, pp. 718–21, 1989

Oeting, J., 'Cellular Mobile Radio–An Emerging Technology', *IEEE Communications Magazine*, pp.10–15, November 1983

Owen, F.C. and C. Pudney, 'DECT: Integrated Services for Cordless Telecommunications', *IEEE Conference Publications*, No. 315, pp. 152–56, 1991

Padovani, R., 'Reverse Link Performance of IS-95 Based Cellular Systems', *IEEE Personal Communications*, pp. 28–34, 3rd quarter, 1994

Padyett, J., E. Gunther, and T. Hattari, 'Overview of Wireless Personal Communications', *IEEE Communications Magazine*, January 1995

Papke, L. and P. Robertson, 'Improved Decoding with the SOVA in a Parallel Concatenated (Turbo Code) Scheme', *IEEE Int. Conference on Communications ICC '96*, pp. 102–06, 1996

Parsons, J.D., 'Diversity Techniques for Mobile Radio Reception', *IEEE Transactions on Vehicular Technology*, Vol.VT-25, No. 3, pp. 75–84, August 1976

Pasupathy, S., 'Minimum Shift Keying: A Spectrally Efficient Modulation', *IEEE Communications Magazine*, pp.14–22, July 1979

Patel, J.N. and U.D. Dalal, 'A Comparative Performance Analysis of OFDM Using MATLAB Simulation with M-PSK and M-QAM Mapping', *Proceedings of IEEE International conference ICCIMA 2007*, IEEE Computer Society Publication, Vol. 4, pp. 406–10, December 2007

Pickholtz, R.L., L.B. Milstein, and D. Schilling, 'Spread Spectrum for Mobile Communications', *IEEE Transactions on Vehicular Technology*, Vol. 40, No. 2, pp. 313–22, May 1991

Pritchard, W.L. and M. Ogata, 'Satellite Direct Broadcast', *Proceedings of IEEE*, Vol. 78, No. 7, pp. 1116–40, July 1990

Proakis, J.G., 'Adaptive Equalization for TDMA Digital Mobile Radio', *IEEE Transactions on Vehicular Technology*, Vol. 40, No. 2, pp. 333–41, May 1991

Pursley, M.B., 'Performance Evaluation for Phase Coded Spread-Spectrum Multiple Access Communication–Part I: System Analysis', *IEEE Transaction on Communication*, COM-25, pp. 795–99, August 1977

Qureshi, S.U.H., 'Adaptive Equalization', *Proceedings of IEEE*, Vol. 37, No. 9, pp. 1340–87, Sept. 1985

Raith, K. and J. Uddenfeldt, 'Capacity of Digital Cellular TDMA Systems', *IEEE Transactions on Vehicular Technology*, Vol. 40, No. 2, pp. 323–31, May 1991

Rappaport, Theodore S., 'The Wireless Revolution', *IEEE Communications Magazine*, pp. 52–71, November 1991

Reudik, D.O., 'Properties o Mobile Radio Propagation Above 400 MHz', *IEEE Transactions on Vehicular Technology*, Vol. 23, No. 2 , pp. 1–20, November 1974

Rice, S.O., 'Mathematical Analysis of Random Noise', *Bell Systems Technical Journal*, Vol. 23, pp. 282–332, July 1944 and Vol. 24, pp. 46–156, January 1945

Rice, S.O., 'Statistical Properties of a Sine Wave Plus Random Noise', *Bell System Technical Journal*, Vol. 27, pp. 109–57, January 1948

Robert, M., 'Bluetooth: A Short Tutorial', *Wireless Personal Communications: Bluetooth Tutorial and Other Technologies*, Tranter, W.H. et al. (Eds.), Kluwer Academic Publishers, pp. 249–70, 2001

Salmasi, A. and K.S. Gilhousen, 'On the System Design Aspects of Code Division Multiple Access (CDMA) Applied to Digital Cellular and Personal Communications Networks', *IEEE Vehicular Technology Conference*, pp. 57–62, 1991

Saltzberg, Burton R., 'Comparison of Single Carrier and Multitone Digital Modulation for ADSL Application', *IEEE Communication Magazine*, Vol. 36, Issue 11, pp. 114–21, November 1998

Sarwate D.V., M.B. Pursley, and T.U. Basar, 'Partial Correlation Effects in Direct Sequence Spread-Spectrum Multiple Access Communication Systems', *IEEE Transaction on Communication,* Vol. COM-32, No. 5, pp. 567–73, May 1984

Sarwate, D.V. and M.B. Pursley, 'Crosscorrelation Properties of Pseudorandom and Related Sequences', *Proceedings of the IEEE*, Vol. 68, No. 5, pp. 593–619, May 1980

Schaubach, K.R., N.J. Davis, and Theodore S. Rappaport, 'A Ray Tracing Method for Predicting Path Loss and Delay Spread in Microcellular Environments', *42nd IEEE Vehicular Technology Conference*, Denver, pp. 932–35, May 1992

Schiibinger, M. and S.R. Meier, 'DSP-Based Signal Processing for OFDM Transmission', *Acoustics, Speech, and Signal Processing Proceedings (ICASSP '01), IEEE*, Vol. 2, pp.1249–52, May 2001

Schroeder, M.R. and B.S. Atal, 'Code-Excited Linear Prediction (CELP): High Quality Speech at Very Low Bit Rates', *Proceedings of ICASSP*, pp. 937–40, 1985

Schroeder, M.R., 'Linear Predictive Coding of Speech: Review and Current Directions', *IEEE Communications Magazine*, Vol. 23, No. 8, pp. 54–61, August 1985

Shannon, C.E., 'A Mathematical Theory of Communications', *Bell Systems Technical Journal*, Vol. 27, pp. 379–423 and 623–56, 1948

Shen, Yushi and Ed Martinez, 'Channel Estimation in OFDM Systems', *Freescale Semiconductor Application Note*, 2006

Sklar, B., 'Defining, Designing and Evaluating Digital Communication Systems', *IEEE Communications Magazine*, pp. 92–101, November 1993

Smith, B., 'Instantaneous Companding of Quantized Signals', *Bell System Technical Journal,* Vol. 36, pp. 653–709, May 1957

Smith, J.I., 'A Computer Generated Multipath Fading Simulation for Mobile Radio', *IEEE Transactions on Vehicular Technology*, Vol. VT-24, No. 3, pp. 39–40, August 1975

Spencer, Q.H., C.B. Peel, A.L. Swindlehurst, and M. Haardt, 'An Introduction to the Multi-user MIMO Downlink', *IEEE Communications Magazine*, Vol. 42, Issue 10, pp. 60–67, October 2004

Steele, R., 'Speech Codecs for Personal Communications', *IEEE Communications Magazine*, pp.76–83, November 1993

Stefanov, A. and T. Duman, 'Turbo-Coded Modulation for Systems with Transmit and Receive Antenna Diversity over Block Fading Channels: System Model, Decoding approaches and practical Considerations', *IEEE Journal on Selected Areas in Communications*, Vol. 19, No. 5, pp. 958–68, 2001

Stein, S., 'Fading Channel Issues in System Engineering', *IEEE Journal on Selected Areas in Communications*, Vol. SAC-5, No. 2, February 1987

Sundberg, C., 'Continuous Phase Modulation', *IEEE Communications Magazine*, Vol. 24, No. 4, pp. 25–38, April 1986

Sung, C.W., and W.S. Wong, 'User Speed Estimation and Dynamic Channel Allocation in Hierarchical Cellular System', *Proceedings of IEEE 1994 Vehicular Technology Conference*, Stockholm, Sweden, pp. 91–95, 1994

Suzuki, H., 'A Statistical Model for Urban Radio Propagation', *IEEE Transactions on Communications*, Vol. 25, pp. 673–80, July 1977

Tekiney, S. and B. Jabbari, 'Handover and Channel Assignment in Mobile Cellular Networks', *IEEE Communications Magazine*, pp. 42–46, November 1991

Tiedemann, E.G., 'CDMA2000-1X: New Capabilities for CDMA Networks', *IEEE Vehicular Technology Society News Letter*, Vol. 48, No. 4, November 2001

Tobagi, F.A. and L. Kleinrock, 'Packet Switching in Radio Channels, Part-II: The Hidden Terminal Problem in Carrier Sense Multiple Access and the Busy Tone Solution', *IEEE Transactions on Communication*, Vol. 23, No. 5, pp. 1417–33, 1975

Tribolet, J.M. and R.E. Crochiere, 'Frequency Domain Coding of Speech', *IEEE Transactions on Acoustics, Speech and Signal Processing*, Vol. ASSP-27, pp. 512–30, October 1979

Tuch, B., 'Development of WaveLAN and ISM Band Wireless LAN', *AT & T Technical Journal*, pp. 27–37, July/August 1993

Tufuessan, F. and T. Maseng, 'Pilot Assisted Channel Estimation for OFDM in Mobile Cellular Systems', *Proceedings of IEEE 47th Vehicular Technology Conference*, Phoenix, USA, pp. 1639–43, May 1997

Ungerboeck, G., 'Trellis Coded Modulation with Redundant Signal Sets, Part 1: Introduction', *IEEE Communication Magazine*, Vol. 25, No. 2, pp. 5–21, February 1987

Valenti, M.C. and J. Sun, 'The UMTS Turbo Code and an Efficient Decoder Implementation Suitable for Software Defined Radios', *International Journal on Wireless Information Networks 8*, pp. 203–16, October 2001

van de Beek, J.J., O. Edfors, M. Sandell, S.K. Wilson, and P.O. Borjesson, 'On Channel Estimation in OFDM Systems', *Proc. IEEE 45th Vehicular Technology Conference*, pp. 815–19, Chicago, IL, July 1995

Van Nielen, M.J.J., 'UMTS: A Third Generation Mobile System', *IEEE Third International Symposium on Personal Indoor and Mobile Radio Communications*, pp. 17–21, 1992

van Zelst, Albert, R. van Nee, and G. Awater, 'Turbo-BLAST and its Performance', *IEEE Vehicular Technology Conference*, Rhodes, Greece, Spring 2001

Wang, Zhengdao, 'OFDM or Single Carrier Block Transmission', *IEEE Transaction on Communication*, Vol. 52, No. 3, pp. 480–394, March 2004

Weinstein, S.B. and P.M. Ebert, 'Data Transmission by Frequency Division Multiplexing Using the Discrete Fourier Transform', *IEEE Transactions on Communication Technology*, Vol. COM-19, pp. 628–34, October 1971

Widrow, B., 'Adaptive Filter, 1: Fundamentals', *Tech. Rep. 6764-6*, Stanford Electronics Laboratory, Stanford University, Stanford, CA, December 1966

Winters, J.H., 'Smart Antennas for Wireless Systems', *IEEE Personal Communications*, Vol.1, pp. 23–27, February 1998

Xia, H., 'Radio Propagation Characteristics for Line-of-Sight Microcellular and Personal Communications', *IEEE Transactions on Antennas and Propagation*, Vol. 41, No. 10, pp. 1439–47, October 1993

Xiong, F., 'Modem Techniques in Satellite Communications', *IEEE Communications Magazine*, pp. 84–97, August 1994

Zheng, L. and D.N.C. Tse, 'Diversity and Multiplexing: A Fundamental Tradeoff in Multiple Antenna Channels', *IEEE Transactions on Information Theory*, Vol. 49, No. 5, pp. 1073–96, May 2003

Zou, William Y. and Yiyan Wu, 'COFDM: An Overview', *IEEE Transaction on Broadcasting,* Vol. 41, No. 1, pp. 1–8, March 1995

Important Websites/Links

cictr.ee.psu.edu
www.wirelesstoday.com
www.mwjournal.com
www.amwireless.com
www.gpsworld.com
www.dectweb.com
www.webproforum.com
www.intel.com
www.worlddab.org/ (For DAB details)
www.sbca.com (Satellite Broadcasting and Communications Association)
www.directv.com
www.kvh.com (For mobile satellite TV)
www.dvb.org (For DVB details)
www.wikipedia.org (General website for many details including wireless communications and systems)
www.wirelesscommunication.nl (JPL's Reference Website)
www.wirelessvalley.com (Wireless Products)
www.cnp-wireless.com (Cellular Networking Perspectives)

Organizations Links For Standards/Specifications

IEEE: http://standards.ieee.org, www.ieee.org
IEE: www.iee.org.uk IET www.theiet.org
ITU (Formerly CCITT): www.itu.int and ATM Forum
ETSI: www.etsi.org
IMT-2000: www.imt-2000.com
IETF: www.ietf.org
Wireless Forums: www.wirelessforums.org
Wi-Fi Forum: www.wifi-forum.com
WiMAX Forum: www.wimaxforum.org
3GPP: www.3gpp.org
Wireless World Research Forum: www.wireless-world-research.org
ARRL: www.arrl.org
CDMA Development Group: www.cdg.org
CTIA: www.ctia.org
Inarte: www.narte.org
PCIA: www.pcia.com
IEC: www.iec.org
ATSC: www.atsc.org

DVB: www.dvb.org
TD-SCDMA: www.tdscdma-forum.org
WAP Forum: www.wapforum.org

Wireless Related Companies

Lucent Technologies, Bellcore, BT Labs, GTE Labs, NORTEL, Motorola, Ericsson, Siemens AG, NTT, Alcatel, Qualcomm, Nokia, AT & T, Bell, Atlantic/Nynex Mobile, GTE Mobilnet Florida, Canadian Wireless Communications, Communications Entry Points, ALTERA, Agilent

Reputed Journals and Magazines

IEEE Publications, IEE, SPIE, Elsevier, Transactions on Vehicular Technology, IEEE JSAC, IEEE Communications Magazine, IEEE Personal Communications Magazine, IEEE Network Magazine, Wiley Interscience, Bell Labs Technical Journals, Oxford (IEICE) Transactions on Communications, IBM Research and Technical Journal, IETE Research, Technical Review and Tutorials (Indian), IE (Indian)

Useful Websites

Rutgers University WINLAB, Virginia Tech MPRG, Georgia Tech Wireless Systems Laboratory, CITO Research, CITR Broadband Wireless Major Project, Ohio State Site, Stanford University Smart Antennas Research Group

Reputed Conferences

ICC, GLOBECOM, INFOCOM, MOBICOM, ACM conferences

Index